Lecture Notes in Computer Scie

Edited by G. Goos, J. Hartmanis and J. van L

Springer
Berlin
Heidelberg
New York
Barcelona
Hong Kong
London
Milan
Paris
Singapore
Tokyo

Samson Abramsky (Ed.)

Typed Lambda Calculi and Applications

5th International Conference, TLCA 2001
Kraków, Poland, May 2-5, 2001
Proceedings

Springer

Series Editors

Gerhard Goos, Karlsruhe University, Germany
Juris Hartmanis, Cornell University, NY, USA
Jan van Leeuwen, Utrecht University, The Netherlands

Volume Editor

Samson Abramsky
Oxford University Computing Laboratory
Wolfson Building, Parks Road, Oxford OX1 3QD
E-mail: Samson.Abramsky@comlab.ox.ac.uk

Cataloging-in-Publication Data applied for

Die Deutsche Bibliothek - CIP-Einheitsaufnahme

Typed lambda calculi and applications : 5th international conference ; proceedings / TLCA 2001, Kraków, Poland, May 2 - 5, 2001. Samson Abramsky (ed.). - Berlin ; Heidelberg ; New York ; Barcelona ; Hong Kong ; London ; Milan ; Paris ; Singapore ; Tokyo : Springer, 2001
(Lecture notes in computer science ; Vol. 2044)
ISBN 3-540-41960-8

CR Subject Classification (1998): F.4.1, F.3, D.1.1, D.3

ISSN 0302-9743
ISBN 3-540-41960-8 Springer-Verlag Berlin Heidelberg New York

Springer-Verlag Berlin Heidelberg New York
a member of BertelsmannSpringer Science+Business Media GmbH

http://www.springer.de

Typesetting: Camera-ready by author, data conversion by PTP Berlin, Stefan Sossna
Printed on acid-free paper SPIN 10782565 06/3142 5 4 3 2 1 0

Preface

This volume contains the proceedings of the Fifth International Conference on Typed Lambda Calculi and Applications, held in Kraków, Poland on May 2–5, 2001. It contains the abstracts of the four invited lectures, plus 28 contributed papers. These were selected from a total of 55 submissions. The standard was high, and selection was difficult.

The conference programme also featured an evening lecture by Roger Hindley, on "The early days of combinators and lambda".

I would like to express my gratitude to the members of the Program Committee and the Organizing Committee for all their dedication and hard work. I would also like to thank the many referees who assisted in the selection process. Finally, the support of Jagiellonian University, Warsaw University, and the U.S. Office of Naval Research is gratefully acknowledged.

The study of typed lambda calculi continues to expand and develop, and touches on many of the key foundational issues in computer science. This volume bears witness to its continuing vitality.

February 2001 Samson Abramsky

Program Committee

S. Abramsky (Oxford) (Chair)
P.-L. Curien (Paris)
P. Dybjer (Gothenburg)
T. Ehrhard (Marseille)
M. Hasegawa (Kyoto)
F. Honsell (Udine)
D. Leivant (Bloomington)
S. Ronchi della Rocca (Turin)
H. Schwichtenberg (Munich)
P. Scott (Ottawa)
J. Tiuryn (Warsaw)

Organizing Committee

M. Zaionc (Kraków)
P. Urzyczyn (Warsaw)
J. Wielgut-Walczak (Kraków)

Referees

Peter Aczel
Klaus Aehlig
Fabio Allesi
Thorsten Altenkirch
Andrea Asperti
Patrick Baillot
Franco Barbanera
Gilles Barthe
Nick Benton
Stefano Berardi
Ulrich Berger
Gavin Bierman
Rick Blute
Viviana Bono
Wilfried Buchholz
Juliusz Chroboczek
Alberto Ciaffaglione
Robin Cockett
Loic Colson
Adriana Compagnoni
Mario Coppo
Thierry Coquand
Ferrucio Damiani
Vincent Danos
Ewen Denney
Mariangola Dezani
Roberto Di Cosmo
Pietro Di Gianantonio
Roy Dyckhoff
Andrzej Filinski
Bernd Finkbeiner
Gianluca Franco
Herman Geuvers
Paola Giannini
Jeremy Gibbons
Bruno Guillaume
Esfandiar Haghverdi
Peter Hancock
Russ Harmer
Ryu Hasegawa
Michael Hedberg
Peter Moller Heergard
Hugo Herbelin
Roger Hindley
Martin Hofmann
Doug Howe
Radha Jagadeesan
Patrik Jansson
Felix Joachimski
Jan Johannsen
Jean-Baptiste Joinet
Thierry Joly
Delia Kesner
Yoshiki Kinoshita
Josva Kleist
Jean-Louis Krivine
Francois Lamarche
Olivier Laurent
Marina Lenisa
Ugo de'Liguoro
Luigi Liquori
John Longley
Zhaohui Luo
Harry Mairson
Jean-Yves Marion
Simone Martini
Ralph Matthes
Paul-Andre Mellies
Marino Miculan
Larry Moss
Koji Nakazawa
Susumu Nishimura
Vincent Padovani
Luca Paolini
Michel Parigot
C. Paulin-Mohring
Dusko Pavlovic
Adolfo Piperno
Jaco van der Pol
Jeff Polakow
Randy Pollack
Laurent Regnier
Eike Ritter
Kristoffer Rose

Jiri Rosicky
Luca Roversi
Martin Ruckert
Don Sanella
Ivan Scagnetto
Alan Schmitt
Robert Seely
Jonathan Seldin
Harold Simmons
Thomas Streicher
Eijiro Sumii
Makoto Takeyama
Robert Tennent
Wolfgang Thomas
Christian Urban
Pawel Urzyczyn
Tarmo Uustalu
Betti Venneri
Roel de Vrijer
Stan Wainer
Benjamin Werner

Table of Contents

Many Happy Returns

Olivier Danvy

BRICS*
Department of Computer Science
University of Aarhus
Ny Munkegade, Building 540, DK-8000 Aarhus C, Denmark
E-mail: danvy@brics.dk
Home page: http://www.brics.dk/~danvy

Abstract. Continuations occur in many areas of computer science: logic, proof theory, formal semantics, programming-language design and implementation, and programming. Like the wheel, continuations have been discovered and rediscovered many times, independently. In programming languages, they represent of "the rest of a computation" as a function, and proved particularly convenient to formalize control structures (sequence, gotos, exceptions, coroutines, backtracking, resumptions, etc.) and to reason about them. In the lambda-calculus, terms can be transformed into "continuation-passing style" (CPS), and the corresponding transformation over types can be interpreted as a double-negation translation via the Curry-Howard isomorphism. In the computational lambda-calculus, they can simulate monads. In programming, they provide functional accumulators.

Yet continuations are remarkably elusive. They can be explained in five minutes, but grasping them seems to require a lifetime. Consequently one often reacts to them to an extreme, either loving them ("to a man with a hammer, the world looks like a nail") or hating them ("too many lambdas").

In this talk, we will first review basic results about continuations, starting with Plotkin's Indifference and Simulation theorems (evaluating a CPS-transformed program yields the same result independently of the evaluation order). Thus equipped, we will identify where continuations arose and how they contributed to solving various problems in computer science. We will conclude with the state of the art today, and present a number of examples, including an illustration of how applying the continuation of a procedure several times makes this procedure return several times—hence the title of the talk.

* Basic Research in Computer Science (www.brics.dk), funded by the Danish National Research Foundation.

S. Abramsky (Ed.): TLCA 2001, LNCS 2044, p. 1, 2001.

From Bounded Arithmetic to Memory Management: Use of Type Theory to Capture Complexity Classes and Space Behaviour

Martin Hofmann

Laboratory for the Foundations of Computer Science
Division of Informatics, University of Edinburgh

Bounded arithmetic [3] is a subsystem of Peano arithmetic defining exactly the polynomial time functions. As Gödel's system T corresponds to Peano arithmetic Cook and Urquhart's system PV_ω [4] corresponds to bounded arithmetic. It is a type system with the property that all definable functions are polynomial time computable.

PV_ω as a programming language for polynomial time is, however, unsatisfactory in several ways. Firstly, it requires to maintain explicit size bounds on intermediate results and secondly, many obviously polynomial time algorithms do not fit into the type system. The attempt to alleviate these restrictions has lead to a sequence of new type systems capturing various complexity classes (PTIME, PSPACE, EXPTIME, LINSPACE) without explicit reference to bounds. Among them are Cook-Bellantoni's [2] and Bellantoni-Niggl-Schwichtenberg's systems of safe recursion [1], tiered systems by Leivant and Marion [12,11], subsystems of Girard's linear logic [6,5], and various systems by myself [9,7,8].

The most recent work [10] has shown that one of these systems can be adapted to allow for explicit memory management including in-place update while still maintaining a functional semantics.

The talk will give a bird's eye overview of the above-mentioned calculi and then discuss in some more detail the recent applications to memory management. This will include recent yet unpublished results about the expressive power of higher-order linear functions and general recursion in the context of [10]. These results suggests that the expressive power equals $\bigcup_c \mathrm{DTIME}(2^{n^c})$.

References

1. S. Bellantoni, K.-H. Niggl, and H. Schwichtenberg. Ramification, Modality, and Linearity in Higher Type Recursion. *Annals of Pure and Applied Logic*, 2000. to appear.
2. Stephen Bellantoni and Stephen Cook. New recursion-theoretic characterization of the polytime functions. *Computational Complexity*, 2:97–110, 1992.
3. Samuel R. Buss. *Bounded Arithmetic*. Bibliopolis, 1986.
4. S. Cook and A. Urquhart. Functional interpretations of feasibly constructive arithmetic. *Annals of Pure and Applied Logic*, 63:103–200, 1993.
5. J.-Y. Girard. Light Linear Logic. *Information and Computation*, 143, 1998.
6. J.-Y. Girard, A. Scedrov, and P. Scott. Bounded linear logic. *Theoretical Computer Science*, 97(1):1–66, 1992.

S. Abramsky (Ed.): TLCA 2001, LNCS 2044, pp. 2–3, 2001.

7. Martin Hofmann. Linear types and non size-increasing polynomial time computation. To appear in Theoretical Computer Science. See www.dcs.ed.ac.uk/home/papers/icc.ps.gz for a draft. An extended abstract has appeared under the same title in Proc. Symp. Logic in Comp. Sci. (LICS) 1999, Trento, 2000.
8. Martin Hofmann. Programming languages capturing complexity classes. *SIGACT News Logic Column*, 9, 2000. 12 pp.
9. Martin Hofmann. Safe recursion with higher types and BCK-algebra. *Annals of Pure and Applied Logic*, 104:113–166, 2000.
10. Martin Hofmann. A type system for bounded space and functional in-place update. *Nordic Journal of Computing*, 2001. To appear, see www.dcs.ed.ac.uk/home/mxh/papers/nordic.ps.gz for a draft. An extended abstract has appeared in *Programming Languages and Systems*, G. Smolka, ed., Springer LNCS, 2000.
11. D. Leivant and J.-Y. Marion. Predicative Functional Recurrence and Poly-Space. In *Springer LNCS 1214: Proc. CAAP*, 1997.
12. Daniel Leivant. Stratified Functional Programs and Computational Complexity. In *Proc. 20th IEEE Symp. on Principles of Programming Languages*, 1993.

Definability of Total Objects in PCF and Related Calculi

Dag Normann

Department of Mathematics
University of Oslo

We let PCF be Plotkin's [8] calculus based on Scott's [10,11] LCF, and we consider the standard case with base types for the natural numbers and for the Booleans. We consider the standard interpretation using algebraic domains. Plotkin [8] showed that a finite object in general will not be definable, and isolated two nondeterministic constants PAR and $\exists_\omega$ such that each computable object is definable in $PCF + PAR + \exists_\omega$.

The first result to be discussed is

Theorem 1. *If Φ is computable and hereditarily total, then there is a PCF definable $\Psi \sqsubseteq \Phi$ that is also total.*

For details, see [4,5]

Escardó [1,2] extended PCF to $R - PCF$, adding base types for the reals and the unit interval I, using continuous domains for the interpretation.
We investigate the hereditarily total objects and obtain

Theorem 2. *The hereditarily total objects in the semantics for $R-PCF$ posess a natural equivalence relation, and the typed structure of equivalence classes can be characterized in the category of limit spaces.*

For details, see [6]

PAR is definable in $R-PCF$, but $\exists_\omega$ is not. It is an open problem if Theorem 1 can be generalized to $R - PCF$.
We will discuss a partial solution of the problem in

Theorem 3. *$\exists_\omega$ is not uniformly $R-PCF$-definable from any hereditarily total object.*

Uniformly definable will mean that the object is definable by one term from each element of the equivalence class.
For details, see [7]

The final result to be discussed is joint with Christian Rørdam [9].
We will compare PCF with Kleene's classical approach from 1959, and see that when we restrict ourselves to μ-recursion in higher types of continuous functionals, the differences are only cosmetical. Niggl [3] devised a calculus $\mathcal{M}^\omega$ that essentially is

$$(PCF \text{ - Fixpoints}) + PAR + \mu\text{-operator}.$$

Theorem 4. *$\mathcal{M}^\omega$ is strictly weaker than $PCF + PAR$.*

S. Abramsky (Ed.): TLCA 2001, LNCS 2044, pp. 4–5, 2001.

References

1. Escardó, M. H., *PCF extended with real numbers: a domain-theoretic approach to higher-order exact number computation*, Thesis, University of London, Imperial College of Science, Technology and medicine (1996).
2. Escardó, M. H., *PCF extended with real numbers*, Theoretical Computer Science 162 (1) pp. 79 - 115 (1996).
3. Niggl, K.-H., *$\mathcal{M}^\omega$ considered as a programming language*, Annals of Pure and Applied Logic 99, pp. 73-92 (1999)
4. Normann, D., *Computability over the partial continuous functionals*, Journal of Symbolic Logic 65, pp. 1133 - 1142, (2000)
5. Normann, D.,*The Cook-Berger Problem. A Guide to the solution*,In Spreen, D. (ed.): Electronic Notes in Theoretical Computer Science. 2000-10; 35 : 9
6. Normann, D., *The continuous functionals of finite types over the reals*, To appear in Keimel, Zhang, Liu and Chen (eds.) *Domains and Processes* Proceedings of the 1st International Symposium on Domain Theory, Luwer Academic Publishers
7. Normann, D., *Exact real number computations relative to hereditarily total functionals*, To appear in Theoretical Computer Science.
8. Plotkin, G.,*LCF considered as a programming language*, Theoretical Computer Science 5 (1977) pp. 223 - 255.
9. Rørdam, C., *A comparison of the simply typed lambda calculi $\mathcal{M}^\omega$ and $\mathcal{L}_{PA}$*, Cand. Scient. Thesis, Oslo (2000)
10. Scott, D. S., *A theory of computable functionals of higher type*, Unpublished notes, University of Oxford, Oxford (1969).
11. Scott, D. S., *A type-theoretical alternative to ISWIM, CUCH, OWHY*, Theoretical Computer Science 121 pp. 411 - 440 (1993).

Categorical Semantics of Control

Peter Selinger

Department of Computer Science
Stanford University

In this talk, I will describe the categorical semantics of Parigot's $\lambda\mu$-calculus [7]. The $\lambda\mu$-calculus is a proof-term calculus for classical logic, and at the same time a functional programming language with control operators. It is equal in power to Felleisen's $\mathcal{C}$ operator [2,1], except that it allows both a call-by-name and call-by-value semantics. The connection between classical logic and continuation-like control operators was first observed by Griffin [4].

The categorical semantics of the $\lambda\mu$-calculus has been studied by various authors in the last few years [6,5,10]. Here, we give a semantics in terms of *control categories*, which combine a cartesian-closed structure with a premonoidal structure in the sense of Power and Robinson [8]. The call-by-name $\lambda\mu$-calculus (with disjunctions) is an internal language for control categories, in much the same way the simply-typed lambda calculus is an internal language for cartesian-closed categories. Moreover, the call-by-value $\lambda\mu$-calculus is an internal language for the dual class of co-control categories. As a corollary, one obtains a syntactic duality result in the style of Filinski [3]: there exist syntactic translations between call-by-name and call-by-value which are mutually inverse and which preserve the operational semantics.

References

1. P. De Groote. On the relation between the $\lambda\mu$-calculus and the syntactic theory of sequential control. Springer LNCS 822, 1994.
2. M. Felleisen. *The calculi of λ_v-conversion: A syntactic theory of control and state in imperative higher order programming languages.* PhD thesis, Indiana University, 1986.
3. A. Filinski. Declarative continuations and categorical duality. Master's thesis, DIKU, Computer Science Department, University of Copenhagen, Aug. 1989. DIKU Report 89/11.
4. T. G. Griffin. A formulae-as-types notion of control. In *POPL '90: Proceedings of the 17th Annual ACM SIGPLAN-SIGACT Symposium on Principles of Programming Languages*, 1990.
5. M. Hofmann and T. Streicher. Continuation models are universal for $\lambda\mu$-calculus. In *Proceedings of the Twelfth Annual IEEE Symposium on Logic in Computer Science*, pages 387–397, 1997.
6. C.-H. L. Ong. A semantic view of classical proofs: Type-theoretic, categorical, and denotational characterizations. In *Proceedings of the Eleventh Annual IEEE Symposium on Logic in Computer Science*, pages 230–241, 1996.
7. M. Parigot. $\lambda\mu$-calculus: An algorithmic interpretation of classical natural deduction. In *Proceedings of the International Conference on Logic Programming and Automated Reasoning, St. Petersburg*, Springer LNCS 624, pages 190–201, 1992.

S. Abramsky (Ed.): TLCA 2001, LNCS 2044, pp. 6–7, 2001.

8. J. Power and E. Robinson. Premonoidal categories and notions of computation. *Math. Struct. in Computer Science*, 7(5):445–452, 1997.
9. P. Selinger. Control categories and duality: on the categorical semantics of the lambda-mu calculus. *Math. Struct. in Computer Science*, 11(2), 2001. To appear.
10. H. Thielecke. *Categorical Structure of Continuation Passing Style*. PhD thesis, University of Edinburgh, 1997.

Representations of First Order Function Types as Terminal Coalgebras

Thorsten Altenkirch

School of Computer Science and Information Technology
University of Nottingham, UK
txa@cs.nott.ac.uk

Abstract. We show that function types which have only initial algebras for regular functors in the domains, i.e. first order function types, can be represented by terminal coalgebras for certain nested functors. The representation exploits properties of ω^{op}-limits and local ω-colimits.

1 Introduction

The work presented here is inspired by discussions the author had some years ago with Healfdene Goguen in Edinburgh on the question *Can function types be represented inductively?* or maybe more appropriately: *Can function types be represented algebraically?*.

In programming and type theory the universe of types can be divided as follows:

- function types (cartesian closure)
- algebraic types
  - inductive types (initial algebras)
  - coinductive types (terminal coalgebras)

In programming the difference between inductive and coinductive types is often obliterated because one is mainly interested in the collection of partial objects of a certain type. Inspired by Occam's razor it would be interesting if we could explain one class of types by another. Here we try to reduce function types to algebraic types.

The first simple observation is that function spaces can be eliminated using products if the domain is finite. Here we show that function spaces $A \to B$ can be eliminated using coinductive types if the domain A is defined inductively. It is interesting to note that ordinary coinductive types are sufficient only for functions over linear inductive types (i.e. where the signature functor has the form $T(X) = A_1 \times X + A_0$) but in general we need to construct functors defined by terminal coalgebras in categories of endofunctors. Those correspond to nested or nested datatypes which have been the subject of recent work [BM98,AR99, Bla00].

S. Abramsky (Ed.): TLCA 2001, LNCS 2044, pp. 8–21, 2001.

1.1 Examples

We give some examples which are instances of our general construction , proposition 8. We use the usual syntax for products and coproducts and $\mu X.F(X)$ to denote initial algebras and $\nu X.F(X)$ for terminal coalgebras. See section 2 for the details. The isomorphisms stated below exist under the condition that the ambient category has the properties introduced later, see section 3. A category which satisfies these properties is the category of sets whose cardinality is less or equal $\aleph_1$. Note that this category is not cartesian closed.

Natural Numbers. Natural numbers are given by $\mathrm{Nat} = \mu X.1 + X$, i.e. they are the initial algebra of the functor[1] $T(X) = 1 + X$.We have that

$$(\mu X.1 + X) \to B \simeq \nu Y.B \times Y,$$

where $\nu Y.B \times Y$ is the terminal coalgebra of the functor $T'(X) = B \times X$ — this is the type of streams or infinite lists over B.

Using the previous isomorphism we obtain a representation of countable ordinals using only algebraic types. The type $\mathrm{Ord} = \mu X.1 + X + (\mathrm{Nat} \to X)$ can be represented as

$$\mathrm{Ord} \simeq \mu X.1 + X + \nu Y.X \times Y$$

Note, however, that there is no representation for functions over Ord and hence there seems to be no representation of the next number class using only coinductive types.

Lists. We assume as given a type A for which we already know how to construct function spaces. Then lists over A are given by $\mathrm{List}(A) = \mu X.1 + A \times X$ and we have

$$(\mu X.1 + A \times X) \to B \simeq \nu Y.B \times (A \to Y)$$

The right hand side defines B-labelled, A-branching non-wellfounded trees. Combining this with the first case we obtain a representation for functions over lists of natural numbers:

$$\begin{aligned}\mathrm{List}(\mathrm{Nat}) \to B &\simeq \nu Y.B \times \mathrm{Nat} \to Y \\ &\simeq \nu Y.B \times (\nu Z.Y \times Z)\end{aligned}$$

[1] We only give the effect on objects since the morphism part of the functor can be derived from the fact that all the operations we use are functorial in their arguments, i.e. T can be extended on morphisms by $T(f) = 1 + f$.

Binary Trees. By considering functions over binary trees $\text{BTree} = \mu X.1+X\times X$ we leave the realm of linear inductive types. The type of functions over trees is given by:

$$(\mu X.1 + X \times X) \to B \simeq (\nu F.\Lambda X.X \times F(F(X)))(B)$$

Here the right hand side is read as the terminal coalgebra over the endofunctor $H(F) = \Lambda X.X\times F(F(X))$ on the category of endofunctors. There seem to be two ways to extend H to morphisms, i.e. given a natural transformation $\alpha \in F \to G$

$$H_1(\alpha)_A = \alpha_A \times (G(\alpha_A) \circ \alpha_{FA})$$
$$H_2(\alpha)_A = \alpha_A \times (\alpha_{GA} \circ F(\alpha_A))$$

However, it is easy to see that the naturality of α implies $H_1(\alpha) = H_2(\alpha)$.

The type $\nu F.\Lambda X.X \times F(F(X))$ has a straightforward representation in a functional programming language like Haskell which allows nested datatypes. A variation of this type, namely $\mu F.\Lambda X.1 + X \times F(F(X))$, is used in [BM98] as an example for nested datatypes under the name *Bushes*. We can represent $\nu F.\Lambda X.X \times F(F(X))$ as

```
data BTfun x  = Case x (BTfun (BTfun x))
```

Here we consider only total elements of a type which entails that we have to differentiate between inductive and coinductive interpretations of recursively defined type. We interpret `BTfun` coinductively, which is sensible since the inductive interpretation is empty.

We assume that binary trees `BT` are defined as

```
data BT = L | Sp BT BT
```

This time we interpret the type inductively!

The two parts of an isomorphism which we call `lamBT` and `appBT` can be programmed in Haskell [2] :

```
lamBT :: (BT → a) → BTfun a
lamBT f = Case (f L) (lamBT (λ t → lamBT (λ u → f (Sp t u))))

appBT :: BTfun a → BT → a
appBT (Case a f) t = case t of L → a
                                Sp tl tr → appBT (appBT f tl) tr
```

Since we use polymorphic recursion it is essential to give the types of the two functions explicitly.

[2] We take the liberty of writing λ for \ and → for `->`.

Finite Trees. As a last example we shall consider functions over finitely branching trees which are interesting because they are defined using *interleaving* inductive types, i.e.

$$\begin{aligned}\mathrm{FTree} &= \mu X.\mathrm{List}(X)\\ &= \mu X.\mu Y.1 + X \times Y\end{aligned}$$

The function type over finite trees can be represented as follows:

$$(\mu X.\mu Y.1 + X \times Y) \to B \simeq (\nu F.\nu G.\Lambda Z.Z \times F(G(Z)))(B)$$

1.2 Related Work

After having completed most of the technical work presented in this paper it was brought to our attention that Ralf Hinze had already discovered what amounts to essentially the same translation in the context of generic functional programming [Hin00b,Hin00a]. His motivation was of a more practical nature: he was looking for a generic representation of memo functions. One of the anonymous referees pointed out that this construction was anticipated in a note by Geraint Jones [Jon98].

The present paper can be viewed as providing a more detailed categorical underpinning of Hinze's construction. In some regards however, our approaches differ fundamentally:

- We adopt a categorical perspective in which functions are *total*, in that exponentiation is right adjoint to products, where Hinze deals with partial functions (and hence a monoidally closed structure).
- As a consequence of this we differentiate between inductive and coinductive types. It also means that we cannot use fixpoint induction (as suggested by Hinze) but have to rely on using ω-limits and colimits explicitly.
- We show the existence of the exponentials whereas Hinze only shows that they are isomorphic to already existing ones.
- Hinze's programs require 2nd order impredicative polymorphism whereas our construction takes place in a predicative framework (compare also section 6).

Acknowledgments. I would like to thank Healfdene Goguen for inspiring this line of research, Thomas Streicher for valuable help on categorical questions, Peter Hancock for interesting email discussions and for pointing out Ralf Hinze's work to me, and Roland Backhouse for discussions on the use of fusion in this context. Dirk Pattinson provided valuable feedback on a draft version. I would also like to thank the anonymous referees whose comments I tried to incorporate to the best of my abilities.

2 Preliminaries

We work with respect to some ambient category $\mathbb{C}$ whose properties we shall make precise below. We assume that $\mathbb{C}$ is bicartesian, i.e. has all finite products (written $\mathbf{1}, A \times B$) and finite coproducts (written $\mathbf{0}, A+B$). We write $!_A \in A \to \mathbf{1}$ and $?_A \in \mathbf{0} \to A$ for the universal maps. Notationally, we use $\to$ for homsets and $\Rightarrow$ for exponentials. We do not assume that $\mathbb{C}$ is cartesian closed.

We assume the existence of ω^{op}-limits and ω-colimits. Here ω stands for the posetal category $(\omega, \leq)$ and by ω^{op}-completeness we mean that limits of all functors $F \in \omega^{\mathrm{op}} \to \mathbb{C}$ exist, i.e.

$$A \to \lim(F) \simeq \Delta(A) \dot{\to} F$$

where $\Delta(A) = \Lambda X.A$ is the constant functor. We write $\pi_F \in \Delta(\lim(F)) \to F$ for the projection and for the back direction of the isomorphism: $\mathrm{prod}_F(\alpha) \in A \to \lim(F)$ given $\alpha \in \Delta(A) \dot{\to} F$.

Dually, by ω-cocompleteness we mean that all colimits of functors $F \in \omega \to \mathbb{C}$ exist, i.e.

$$\mathrm{colim}(F) \to A \simeq F \dot{\to} \Delta(A)$$

We write $\mathrm{inj}_F \in F \dot{\to} \Delta(\mathrm{colim}(F))$ for the injection and for the inverse $\mathrm{case}(\alpha) \in \mathrm{colim}(F) \to A$ given $\alpha \in F \dot{\to} \Delta(A)$.

A functor $T \in \mathbb{C} \to \mathbb{C}$ is called ω^{op}-continuous (ω-cocontinuous) if it preserves all ω^{op}-limits (ω-colimits) up to isomorphism. We write

$$\Psi_T \in T(\lim(F)) \simeq \lim(T \circ F)$$
$$\Psi^T \in T(\mathrm{colim}(F)) \simeq \mathrm{colim}(T \circ F)$$

It is easy to see that coproducts preserve colimits and products preserve limits and hence the appropriate operations on functors are (co-)continuous. We will later identify the precise circumstances under which products preserve colimits.

Given an endofunctor $T \in \mathbb{C} \to \mathbb{C}$ the category of T-algebras has as objects $(A \in \mathbb{C}, f \in T(A) \to A)$ and morphisms $h \in (A, f) \to (B, g)$ are given by $h \in A \to B$ s.t. $g \circ T(h) = h \circ f$. We denote the initial T-algebra by $(\mu T, \mathrm{in}_T \in T(\mu T) \to \mu T)$. Given any T-algebra (A, f) the unique morphism (often called a *catamorphism*) is written as $\mathrm{fold}_T(f) \in \mu T \to A$. Dually, the category of T-coalgebras has as objects $(A \in \mathbb{C}, f \in A \to T(A))$ and morphisms $h \in (A, f) \to (B, g)$ are given by $h \in A \to B$ s.t. $g \circ h = T(h) \circ f$. The terminal T-coalgebra is written as $(\nu T, \mathrm{out}_T \in \nu T \to T(\nu T))$ and given a coalgebra (A, f) the unique morphism (often called *anamorphism*) is written $\mathrm{unfold}_T(f) \in A \to \nu T$.

For completeness we review some material from [Ada74,PS78] Given an endofunctor $T \in \mathbb{C} \to \mathbb{C}$ and $i \in \omega$ we write $T^i \in \mathbb{C} \to \mathbb{C}$ for the ith iteration of T.

We define $\mathrm{Chain}_T \in \omega \to \mathbb{C}$ and $\mathrm{Chain}^T \in \omega^{\mathrm{op}} \to \mathbb{C}$:

$$\begin{aligned}\mathrm{Chain}_T(i) &= T^i(\mathbf{1})\\ \mathrm{Chain}_T(i \le j) &= T^i(!_{\mathrm{Chain}_T(j-i)})\\ \mathrm{Chain}^T(i) &= T^i(\mathbf{0})\\ \mathrm{Chain}^T(i \ge j) &= T^j(?_{\mathrm{Chain}^T(i-j)})\end{aligned}$$

Proposition 1 (Adamek,Plotkin-Smyth). *Given a ω^{op}-complete and ω-cocomplete category $\mathbb{C}$ and an an endofunctor $T \in \mathbb{C} \to \mathbb{C}$ we have that:*

1. *If T is ω-cocontinuous then the initial algebra exists and*
$$mu(T) \simeq \mathrm{colim}(\mathrm{Chain}^T).$$
2. *If T is ω-continuous then the terminal coalgebra exists and*
$$\nu(T) \simeq \lim(\mathrm{Chain}_T)$$

3 Locality

Since we do not assume that our ambient category is closed we have to be more precise w.r.t coproducts, colimits and initial algebras. We require that all those concepts exist locally. Given an object Γ which corresponds to a type context the local category wrt. Γ has the same objects as $\mathbb{C}$ and as morphisms $f \in \Gamma \times A \to B$. The local identity is just the projection $\pi_2 \in \Gamma \times A \to A$ and composition of $f \in \Gamma \times A \to B$ and $g \in \Gamma \times B \to C$ is given by $g \circ (1, f) \in \Gamma \times A \to C$. We say that X are local , if X exists in all local categories and coincide with global X.

A local functor is given by a function on objects and a natural transformation:

$$\mathrm{st}^T \in (\Gamma \times A \to B) \dot{\to} (\Gamma \times T(A) \to T(B))$$

natural in Γ which preserves local identity and composition:

$$\begin{aligned}\mathrm{st}^T(\pi_2) &= \pi_2\\ \mathrm{st}^T(g \circ (1, f)) &= \mathrm{st}^T(g) \circ (1, \mathrm{st}^T(f))\\ &\quad \text{where } f \in \Gamma \times B \to C \text{ and } g \in \Gamma \times A \to B.\end{aligned}$$

Alternatively this can be formalized via a natural transformation

$$\theta^T_{\Gamma,A} \in \Gamma \times T(A) \dot{\to} T(\Gamma \times A)$$

subject to the appropriate conditions but this can easily be seen to be equivalent.

Traditionally, local functors are called strong [CS92]. We diverge from this use because we want to apply the idea of locality also to other concepts like colimits and coalgebras and here the word strong is already used to signal that the uniqueness condition holds.

Proposition 2.

1. *Products are local.*
2. *ω-limits are local.*
3. *Terminal coalgebras of local functors are local.*

Proof. (Sketch): 3. Given a local T-coalgebra $f \in \Gamma \times A \to T(A)$ the local unfold is given by

$$\begin{aligned}\text{unfold}^*_T(f) &\in \Gamma \times A \to \nu T \\ \text{unfold}^*_T(f) &= \text{unfold}_T(\theta^T_{\Gamma,A} \circ (1, f))\end{aligned}$$

However, the same does not hold for coproducts, colimits or initial algebras. E.g. coproducts are not local in $\text{CPO}_\perp$. This asymmetry is caused by the fact that the notion of local morphisms is not self dual.

Local coproducts are given by the following families of isomorphisms:

$$\begin{aligned}\Gamma \times \mathbf{0} \to &X \simeq \mathbf{1} \\ \Gamma \times (A + B) \to &X \simeq (\Gamma \times A \to X) \times (\Gamma \times B \to X)\end{aligned}$$

natural in Γ. Given a functor $F \in \omega^{\text{op}} \to \mathbb{C}$ (not necessary local) the ω-colimit is local if the following family of isomorphisms exist:

$$\Gamma \times \text{colim}(F) \to C \simeq \Delta(\Gamma) \times F \dot{\to} \Delta(C)$$

We say that a functor is locally cocontinuous if it preserves local colimits and again it is easy to see that local coproducts preserve local colimits. In the special case of local ω-colimits we also have

Proposition 3. *Products preserve local ω-colimits:*

$$\text{colim}(F) \times \text{colim}(G) \simeq \text{colim}(F \times G)$$

Proof. (Sketch) For simplicity we only consider the case of $\Gamma = \mathbf{1}$. Using locality we show that

$$\text{colim}(F) \times \text{colim}(G) \to A \simeq (\Lambda(i,j) \in \omega \times \omega . F(i) \times G(j)) \dot{\to} \Delta(A)$$

Using the fact that either $i \leq j$ or $j \leq i$ we can show that the right hand side is isomorphic to

$$(\Lambda i \in \omega . F(i) \times G(i)) \dot{\to} \Delta(A)$$

Assuming that T is a local endofunctor, a local T-algebra with respect to Γ is given by $f \in \Gamma \times T(A) \to A$ and given another local algebra $g : \Gamma \times T(B) \to B$ then a morphism $h \in f \to g$ is given by $h \in \Gamma \times A \to B$ s.t. $h \circ (1, f) = g \circ (1, \text{st}(h))$. Saying that an initial algebra $(\mu T, \text{in}_T \in T(\mu T) \to \mu T)$ is local means that $\text{in}_T \circ \pi_2 \in \Gamma \times T(\mu T) \to \mu T$ is an initial local T-algebra.

Definition 1. *We call a category $\mathbb{C}$ locally ω-bicomplete if the following conditions hold:*

1. $\mathbb{C}$ has all finite products.
2. $\mathbb{C}$ is ω^{op}-complete, i.e. it has all ω^{op}-limits.
3. $\mathbb{C}$ has all local finite coproducts.
4. $\mathbb{C}$ is locally ω-cocomplete, i.e. it has all local ω-colimits.

We assume that the ambient category $\mathbb{C}$ is locally ω-complete. We note that the initial algebra representation theorem can be localized:

Proposition 4. *Given a cocontinuous local endofunctor T: Then in the presence of local ω-colimits the representation of proposition 1 gives rise to a local initial algebra.*

Finally, we remark that the reason that the question of locality has so far got only very little attention in programming language theory is because here the ambient category is usually assumed to be cartesian closed and we have:

Proposition 5. *Assuming that our ambient category $\mathbb{C}$ is cartesian closed we have*

1. *Coproducts are local.*
2. *ω-colimits are local.*
3. *Initial algebras of local functors are local.*

Proof. (Sketch): 3. Let

$$\overline{\mathrm{app}}_{\Gamma,A} \in \Gamma \times (\Gamma \Rightarrow A) \to A$$
$$\overline{\lambda}_{\Gamma,A}(f) \in B \to (\Gamma \Rightarrow A) \qquad \text{given } f \in \Gamma \times B \to A.$$

be twisted versions of the usual morphisms. Now, given $f \in \Gamma \times T(A) \to A$ we define

$$\begin{aligned}\mathrm{fold}^*_T(f) &= \overline{\mathrm{app}}(\mathrm{fold}_T(\overline{\lambda}(f \circ \mathrm{st}(\overline{\mathrm{app}}))))\\ &\in \Gamma \times \mu(T) \to A\end{aligned}$$

4 The μ-ν Property

We shall now establish the main technical lemma of this paper which relates function spaces whose domains are initial algebras to terminal coalgebras. We say that an object $A \in \mathbb{C}$ is exponentiable if for all $B \in \mathbb{C}$: $A \Rightarrow B$ exists and there is an isomorphism

$$\Gamma \times A \to B \simeq \Gamma \to A \Rightarrow B$$

which is natural in Γ. We define $\mathbb{C}^*$ as the full subcategory of exponentiable objects.

Given a functor $F \in \omega^{\mathrm{op}} \to \mathbb{C}^*$ and an object $C \in \mathbb{C}$ we define

$$F \Rightarrow C \in \omega \to \mathbb{C}$$
$$(F \Rightarrow C)(i) = F(i) \Rightarrow C$$

Note that $F \Rightarrow C \neq F \Rightarrow \Delta(C)$.

Lemma 1.

$$\Gamma \times \mathrm{colim}(F) \to C \simeq \Gamma \to \lim(F \Rightarrow C)$$

natural in Γ.

Proof. Straightforward unfolding of definitions.

Note that local colimits are essential here. We also know that limits in functor categories can be calculated pointwise:

Lemma 2. *Let* $F \in \omega \to (\mathbb{C} \Rightarrow \mathbb{C})$ *then we have*

$$\lim(F)(C) \simeq \lim(\Lambda i.F(i, C))$$

natural in C.

Lemma 3. *Given an* ω*-cocontinuous local functor* $F \in \mathbb{C}^* \to \mathbb{C}^*$ *which preserves exponentiability and an* ω*-continuous functor* $G \in (\mathbb{C} \Rightarrow \mathbb{C}) \to (\mathbb{C} \Rightarrow \mathbb{C})$ *s.t. for all exponentiable objects* A

$$F(A) \Rightarrow B \simeq G(\Lambda X.A \Rightarrow X)(B) \tag{H}$$

which is natural in B *then we have:*

1. *For all* $i \in \omega$*:*

$$\mathrm{Chain}^F(i) \Rightarrow C \simeq \mathrm{Chain}_G(i)(C)$$

   natural in C.

2.

$$\Gamma \times \mu(F) \to C \simeq \Gamma \to (\nu G)(C)$$

   natural in Γ, C.

Proof.

1. By induction over i:
   0

$$\begin{aligned} \mathrm{Chain}^F(0) \Rightarrow C &= \mathbf{0} \Rightarrow C \\ &\simeq \mathbf{1} && \text{Since } \mathbf{0} \text{ is local.} \\ &= \mathrm{Chain}_G(0)(C) && \text{Since } \mathrm{Chain}_G(0) = \Delta(\mathbf{1}). \end{aligned}$$

$i+1$

$$\begin{aligned} \mathrm{Chain}^F(i+1) \Rightarrow C &= F(\mathrm{Chain}^F(i)) \Rightarrow C \\ &\simeq G(\Lambda X.\mathrm{Chain}^F(i) \Rightarrow X)(C) && \text{(H)} \\ &\simeq G(\mathrm{Chain}_G(i))(C) && \text{ind.hyp.} \\ &\simeq \mathrm{Chain}_G(i+1)(C) \end{aligned}$$

2.

$$\begin{aligned} \Gamma \times \mu(F) \to C &\simeq \Gamma \times \mathrm{colim}(\mathrm{Chain}^F) \to C && \text{by prop. 4.} \\ &\simeq \Gamma \to \lim(\mathrm{Chain}^F \Rightarrow C) && \text{by lemma 1.} \\ &\simeq \Gamma \to \lim(\mathrm{Chain}_G)(C) && \text{by 1.} \\ &\simeq \Gamma \to \nu(G)(C) && \text{by prop. 1.} \end{aligned}$$

5 The Representation Theorem

We will now establish that function spaces whose domain is an inductive regular type can be isomorphically represented by coinductive nested types.

The set of inductive regular functors of arity n: $\mathcal{IND}_n \subseteq \mathbb{C}^n \to \mathbb{C}$ is inductively defined by the following rules:

$$\frac{0 \leq i < n}{\Lambda \boldsymbol{X}.X_i \in \mathcal{IND}_n} \qquad \frac{}{\Lambda \boldsymbol{X}.\mathbf{0}, \Lambda \boldsymbol{X}.\mathbf{1} \in \mathcal{IND}_n}$$

$$\frac{F, G \in \mathcal{IND}_n}{\Lambda \boldsymbol{X}.F(\boldsymbol{X}) + G(\boldsymbol{X}), \Lambda \boldsymbol{X}.F(\boldsymbol{X}) \times G(\boldsymbol{X}) \in \mathcal{IND}_n}$$

$$\frac{F \in \mathcal{IND}_{n+1}}{\Lambda \boldsymbol{X}.\mu Y.F(\boldsymbol{X}, Y) \in \mathcal{IND}_n}$$

An inductive regular type is just an inductive regular functor of arity 0.

Proposition 6. *All inductive regular functors are local and locally ω-cocontinuous.*

Proof. (Sketch): By induction over the structure of $\mathcal{IND}$. Locality simply follows from the fact that we use local coproducts and colimits and that projections and products are local anyway.

ω-cocontinuity follows from the fact that local coproducts preserve colimits and local initial algebras preserve local colimits since they correspond to local colimits (proposition 4). The case of products is covered by proposition 3.

We now define the set of coinductive nested functors of arity n: $\mathcal{COIND}_n \subseteq (\mathbb{C} \Rightarrow \mathbb{C})^n \to \mathbb{C} \Rightarrow \mathbb{C}$ inductively:

$$\frac{0 \leq i < n}{\Lambda \boldsymbol{F}.F_i \in \mathcal{COIND}_n} \qquad \frac{}{\Lambda \boldsymbol{F}, X.\mathbf{1}, \Lambda \boldsymbol{F}, \Lambda X.X \in \mathcal{COIND}_n}$$

$$\frac{G, H \in \mathcal{COIND}_n}{\Lambda \boldsymbol{F}, X.G(\boldsymbol{F}, X) \times H(\boldsymbol{F}, X), \Lambda \boldsymbol{F}.G(\boldsymbol{F}) \circ G(\boldsymbol{F}) \in \mathcal{COIND}_0}$$

$$\frac{G \in \mathcal{COIND}_{n+1}}{\Lambda \boldsymbol{F}.\nu H.G(\boldsymbol{F}, H) \in \mathcal{COIND}_n}$$

A coinductive nested type is a coinductive nested functor of arity 0 applied to any type (i.e. definable object).

Proposition 7. *All coinductive nested functors are* ω^{op}*-continuous.*

Proof. (Sketch): Follows from the fact that products and limits preserve limits.

We now assign to every inductive regular functor $F \in \mathcal{IND}_n$ a coinductive nested type $\hat{F} \in \mathcal{COIND}_n$ which represents the function space in the sense made precise below.

$$\begin{array}{ll}
F(\boldsymbol{X}) = X_i & \hat{F}(\boldsymbol{H}) = H_i \\
F(\boldsymbol{X}) = \mathbf{0} & \hat{F}(\boldsymbol{H}) = \Lambda X.\mathbf{1} \\
F(\boldsymbol{X}) = F_1(\boldsymbol{X}) + F_2(\boldsymbol{X}) & \hat{F}(\boldsymbol{H}) = \Lambda X.\hat{F}_1(\boldsymbol{H}) \times \hat{F}_2(\boldsymbol{H}) \\
F(\boldsymbol{X}) = \mathbf{1} & \hat{F}(\boldsymbol{H}) = \Lambda X.X \\
F(\boldsymbol{X}) = F_1(\boldsymbol{X}) \times F_2(\boldsymbol{X}) & \hat{F}(\boldsymbol{H}) = \hat{F}_1(\boldsymbol{H}) \circ \hat{F}_2(\boldsymbol{H}) \\
F(\boldsymbol{X}) = \mu Y.F'(\boldsymbol{X}, Y) & \hat{F}(\boldsymbol{H}) = \nu G.\hat{F}'(\boldsymbol{H}, G)
\end{array}$$

Proposition 8. *Given* $F \in \mathcal{IND}_n$ *and* $\boldsymbol{A} \in \mathbb{C}^*$ *define* $H_i = \Lambda X.A_i \Rightarrow X$. *We have that*

$$\Gamma \times F(\boldsymbol{A}) \to B \simeq \Gamma \to \hat{F}(\boldsymbol{H}, B)$$

which is natural in B

Proof. By induction over the structure of F:

$F((X)) = X_i$

$$\begin{aligned}
\Gamma \times F(\boldsymbol{A}) \to B &= \Gamma \times A_i \to B \\
&\simeq \Gamma \to A_i \Rightarrow B \\
&= \Gamma \to \hat{F}(\boldsymbol{H}, B)
\end{aligned}$$

$F(\boldsymbol{X}) = \mathbf{0}$

$$\begin{aligned} \Gamma \times F(\boldsymbol{A}) \to B &= \Gamma \times 0 \to B \\ &\simeq \Gamma \to \mathbf{1} && \text{strong initial object} \\ &= \Gamma \to \hat{F}(\boldsymbol{H}, B) \end{aligned}$$

$F(\boldsymbol{X}) = F_1(\boldsymbol{X}) + F_2(\boldsymbol{X})$

$$\begin{aligned} \Gamma \times F(\boldsymbol{A}) \to B &= \Gamma \times F_1(\boldsymbol{A}) + F_2(\boldsymbol{A}) \to B \\ &\simeq (\Gamma \times F_1(\boldsymbol{A}) \to B) \times (\Gamma \times F_2(\boldsymbol{A}) \to B) && \text{strong coproducts} \\ &\simeq (\Gamma \to \hat{F}_1(\boldsymbol{H}, B)) \times (\Gamma \to \hat{F}_2(\boldsymbol{H}, B)) && \text{ind.hyp.} \\ &\simeq \Gamma \to \hat{F}_1(\boldsymbol{H}, B) \times \hat{F}_2(\boldsymbol{H}, B) \\ &= \Gamma \to \hat{F}(\boldsymbol{H}, B) \end{aligned}$$

$F(\boldsymbol{X}) = \mathbf{1}$

$$\begin{aligned} \Gamma \times F(\boldsymbol{A}) \to B &= \Gamma \times \mathbf{1} \to B \\ &\simeq \Gamma \to B \\ &= \Gamma \to \hat{F}(\boldsymbol{H}, B) \end{aligned}$$

$F(\boldsymbol{X}) = F_1(\boldsymbol{X}) \times F_2(\boldsymbol{X})$

$$\begin{aligned} \Gamma \times F(\boldsymbol{A}) \to B &= \Gamma \times F_1(\boldsymbol{A}) \times F_2(\boldsymbol{A}) \to B \\ &\simeq \Gamma \times F_1(\boldsymbol{A}) \to F_2(\boldsymbol{A}) \Rightarrow B \\ &\simeq \Gamma \to F_1(\boldsymbol{A}) \Rightarrow F_2(\boldsymbol{A}) \Rightarrow B \\ &\simeq \Gamma \to \hat{F}_1(\boldsymbol{H}, F_2(\boldsymbol{A})) \Rightarrow B) && \text{ind.hyp}(F_1) \\ &\simeq \Gamma \to \hat{F}_1(\boldsymbol{H}, \hat{F}_2(\boldsymbol{H}, B)) && \text{ind.hyp}(F_2) \\ &= \Gamma \to (\hat{F}_1(\boldsymbol{H}) \circ \hat{F}_2(\boldsymbol{H}))(B) \\ &= \Gamma \to \hat{F}(\boldsymbol{H}, B) \end{aligned}$$

$F(\boldsymbol{X}) = \mu Y.F'(\boldsymbol{X}, Y)$

$$\begin{aligned} \Gamma \times F(\boldsymbol{A}) \to B &= \Gamma \times \mu Y.F'(\boldsymbol{A}, Y) \to B \\ &\simeq \Gamma \to (\nu G.\hat{F}'(\boldsymbol{H}, G))(B) && (*) \\ &= \Gamma \to \hat{F}(\boldsymbol{H}, B) \end{aligned}$$

To justify (*) we apply lemma 3,2. to $\Lambda Y.F'(\boldsymbol{A}, Y)$ and $\Lambda G.\hat{F}'(\boldsymbol{H}, G)$. Preservation of exponentials and (H) follows from the ind.hyp.

Corollary 1. *Every function space $A \Rightarrow B$ where $A \in \mathcal{IND}_0$ is an inductive regular type can be represented as a coinductive nested type $\hat{A}(B)$.*

6 Using Fusion?

Roland Backhouse remarked that the central lemma 3 could be proven using the fusion theorem of [BBvGvdW96], pp.76:

Proposition 9 (Fusion). *Given a left adjoint functor* $F \in \mathbb{C} \to \mathbb{D}$ *and functors* $G \in \mathbb{C} \to \mathbb{C}$ *and* $H \in \mathbb{D} \to \mathbb{D}$ *s.t.*

$$F \circ G \simeq H \circ F$$

then

$$F(\mu(G)) \simeq \mu(H)$$

Using

$$F \in \mathbb{C} \Rightarrow (\mathbb{C} \Rightarrow \mathbb{C})^{\mathrm{op}}$$
$$F(X) = \Lambda Y.X \Rightarrow Y$$

we may obtain lemma 3 as a corollary (w.o. requiring that the functors involved are continuous or cocontinuous) if we can show that F has a right adjoint. This right adjoint can be written as

$$F^{\#} \in (\mathbb{C} \Rightarrow \mathbb{C})^{\mathrm{op}} \to \mathbb{C}$$
$$F^{\#}(G) = G \dot{\to} \lambda X.X$$

This requires that there is an internal representation of $G \dot{\to} \lambda X.X$ which depends on impredicative quantification as present in the Calculus of Constructions.

There is a very close connection between the construction sketched above and the Haskell programs (section 1). The programs seem not to use impredicative quantification explicitly because this is hidden by polymorphic recursion. However, if we attempt to present e.g. `appBT` using categorical combinators (e.g. `fold`) there seems to be no way to avoid impredicative polymorphism (which also has the consequence that this cannot be encoded in the current Haskell type system).

This also raises the question whether explaining polymorphic recursion which arises naturally when using nested types does in some natural cases require impredicative polymorphism. The specific case considered here shows that impredicativity can be avoided by using ω-completeness properties. It may be the case that similar explanations can be found for all sensible applications of polymorphic recursion.

7 Further Work

There is a certain asymmetry in our construction: we construct function types of regular (inductive) types using nested (coinductive) types. It seems natural to ask what happens if we look at nested inductive types in the domain. It seems reasonable to look at functors definable in a simply typed language where type constructors like $\times$ or μ are just basic constants. The construction presented here can be generalized to this case (which we may call higher dimensional nested types). We plan to present details of this in a forthcoming paper.

In the current form our result is not applicable to categories of constructive functions like ω-Set. However, it seems likely that our result still holds when moving to an appropriate internal notion of limits and colimits.

The categorical features used here, e.g. initial and terminal algebras but no function types can be syntactically encoded in a calculus which for obvious reasons does not deserve the name λ-calculus. We believe that this calculus deserves further investigation because it represents the algorithms which can be defined using only algebraic types. It would be interesting to determine the precise proof-theoretic strength of this calculus which almost certainly exceeds that of first order arithmetic.

References

[Ada74] J. Adamek. Free algebras and automata realizations in the language of categories. *Comment. Math. Univ. Carolinae*, 15:589–602, 1974.

[AR99] T. Altenkirch and B. Reus. Monadic presentations of lambda terms using generalized inductive types. In *Computer Science Logic*, 1999.

[BBvGvdW96] R. Backhouse, R. Bijsterveld, R. van Geldrop, and J. van der Woude. Category theory as coherently constructive lattice theory. available from `http://www.cs.nott.ac.uk/~rcb/papers/papers.html`, December 1996. Working Document.

[Bla00] P. Blampied. *Structured recursion for non-uniform data-types.* PhD thesis, School of Computer Science and IT at the University of Nottingham, UK, 2000.

[BM98] R. Bird and L. Meertens. Nested datatypes. In J. Jeuring, editor, *Mathematics of Program Construction*, number 1422 in LNCS, pages 52 – 67. Springer Verlag, 1998.

[CS92] J. R. B. Cockett and D. Spencer. Strong categorical datatypes I. In R. A. G. Seely, editor, *Proceedings Intl. Summer Category Theory Meeting, Montréal, Québec, 23–30 June 1991*, volume 13 of *Canadian Mathematical Society Conf. Proceedings.* American Mathematical Society, 1992.

[Hin00a] R. Hinze. Generalizing generalized tries. *Journal of Functional Programming*, 2000.

[Hin00b] R. Hinze. Memo functions, polytypically! In Johan Jeuring, editor, *Proceedings of the Second Workshop on Generic Programming, WGP 2000*, 2000.

[Jon98] G. Jones. Tabulation for type hackers. Available from `ftp://ftp.comlab.ox.ac.uk/`, 1998.

[PS78] G. D. Plotkin and M. B. Smyth. The category-theoretic solution of recursive domain equations. *SIAM Journal on Computing*, 11, 1978.

A Finitary Subsystem of the Polymorphic λ-Calculus

Thorsten Altenkirch[1] and Thierry Coquand[2]

[1] School of Computer Science and Information Technology
University of Nottingham, UK
txa@cs.nott.ac.uk

[2] Department of Computing Science
Chalmers University of Technology, Sweden coquand@cs.chalmers.se

Abstract. We give a finitary normalisation proof for the restriction of system F where we quantify only over first-order type. As an application, the functions representable in this fragment are exactly the ones provably total in Peano Arithmetic. This is inspired by the reduction of Π^1_1-comprehension to inductive definitions presented in [Buch2] and this complements a result of [Leiv]. The argument uses a finitary model of a fragment of the system AF_2 considered in [Kriv,Leiv].

1 The Polymorphic λ-Calculus

We let D be the set of all untyped, maybe open, λ-terms, with β-conversion as equality. We let c_n be the lambda term $\lambda x \lambda f\ f^n\ x$. We consider the following types

$$T \ ::= \ \alpha \mid T \to T \mid (\Pi\alpha)T$$

where in the quantification, T has to be built using only α and $\to$. We use T, U, V to denote over types.

We use the notation $T_1 \to T_2 \to T_3$ for $T_1 \to (T_2 \to T_3)$ and similarly $T_1 \to T_2 \to \ldots \to T_n$ for $T_1 \to (T_2 \to (\ldots \to T_n))$.

Let us give some examples to illustrate the restriction on quantification. We can have $T = (\Pi\alpha)[\alpha \to \alpha]$ or $(\Pi\alpha)[\alpha \to (\alpha \to \alpha) \to \alpha]$ or even $(\Pi\alpha)[((\alpha \to \alpha) \to \alpha) \to \alpha]$ but a type such as $(\Pi\alpha)[[(\Pi\beta)[\alpha \to \beta]] \to \alpha]$ is not allowed.

We have the following typing rules

$$\frac{}{\Gamma \vdash x : T}\ x : T \in \Gamma$$

$$\frac{\Gamma, x : T \vdash t : U}{\Gamma \vdash \lambda x\ t : T \to U} \qquad \frac{\Gamma \vdash u : V \to T \quad \Gamma \vdash v : V}{\Gamma \vdash u\ v : T}$$

$$\frac{\Gamma \vdash t : (\Pi\alpha)T}{\Gamma \vdash t : T(U)} \qquad \frac{\Gamma \vdash t : T}{\Gamma \vdash t : (\Pi\alpha)T}$$

where Γ is a type context, i.e. an assignment of types to a finite set of variables, and in the last rule, α does not appear free in any type of Γ. We write $T(U)$ for

S. Abramsky (Ed.): TLCA 2001, LNCS 2044, pp. 22–28, 2001.

a substitution where the variable which is substituted for is obvious form the context.

We let N be the type $(\Pi\alpha)[\alpha \to (\alpha \to \alpha) \to \alpha]$. We have $\vdash c_n : N$ for each n. The goal of this note is to provide a finitary proof of the following result.

Theorem 1. *If* $\vdash t : N \to N$ *then for each* n *there exists* m *such that* $t\ c_n\ x\ f = f^m\ x$ *for* x, f *variables.*

This result can be seen as a special case of the normalisation property. We concentrate on this simplified case to illustrate the principle of our argument. From a proof theoretical view point this special case is as hard as the normalisation property.

This result follows from [Leiv] if in the formation of $(\Pi\alpha)T$ we restrict T to be of rank ≤ 2. We extend this to cover types such as

$$(\Pi\alpha)[((\alpha \to \alpha) \to \alpha) \to \alpha]$$

One non finitary proof of this result is the following. Each type is interpreted by a subset of D. We define $[\![T]\!]_\rho \subset D$ where ρ is a function assigning subsets of D to type variables.

$$[\![T \to U]\!]_\rho = \{v \in D \mid (\forall t \in [\![T]\!]_\rho)\ v\ t \in [\![U]\!]_\rho\}$$

$$[\![\alpha]\!]_\rho = \rho(\alpha)$$

and

$$[\![(\Pi\alpha)T]\!] = \bigcap_{X \subseteq D} [\![T]\!]_{\rho,\alpha=X}$$

We prove then, by induction on derivations

Lemma 1. *If* $x_1 : T_1, \ldots, x_n : T_n \vdash t : T$ *and* $u_i \in [\![T_i]\!]$ *then* $t(u_1, \ldots, u_n) \in [\![T]\!]$

Corollary 1. *If* $\vdash t : N$ *then* $t \in [\![N]\!]$

Lemma 2. *If* $u \in [\![N]\!]$ *then there exists* m *such that* $u\ x\ f = f^m\ x$ *for* x, f *variables.*

Proof. Consider the subset $S = \{t \mid (\exists m)\ t = f^m\ x\}$. We have $x \in S$ and $f\ t \in S$ if $t \in S$. Hence the result.

We can now prove the theorem. If $\vdash t : N \to N$ we have then $\vdash t\ c_n : N$ because $\vdash c_n : N$. But this implies, by the two lemmas that there exist m such that $t\ c_n\ x\ f = f^m\ x$. Let us write $m = \phi(n)$. We say then that the function ϕ is *represented by* the term t.

2 Second-Order Functional Arithmetic

The proof above is not finitary, because of the use of intersection over all subsets, which requires, a priori, Π^1_1-comprehension.

In some cases however, we can replace Π^1_1-comprehension by arithmetical comprehension. For instance, if T is $\alpha \to (\alpha \to \alpha) \to \alpha$ then we have

$$\bigcap_{X \subseteq D} [\![T]\!]_{\alpha=X} = \{t \in D \mid (\exists n)(\forall u, v \in D)\ t\ u\ v = v^n\ u\ \}$$

Indeed, if t belongs to all $[\![T]\!]_{\alpha=X}$ then we can take x, y variables not free in t and take X to be the set of all terms of the form $y^n\ x$. We have then $t\ x\ y \in X$ and hence $t\ x\ y = y^n\ x$ for some n. Since x, y does not occur free in t this implies $t\ u\ v = v^n\ u$ for all $u, v \in D$. Conversely if we have $t\ u\ v = v^n\ u$ for all $u, v \in D$ then it is direct to check that we have $t \in [\![T]\!]_{\alpha=X}$ for all $X \subseteq D$.

More generally if T is of rank ≤ 2 then we can directly define $[\![(\Pi\alpha)T]\!]$ by using arithmetical comprehension only. But this does not seem to extend simply to the general case.

To analyze the proof in general, we first translate it in the language of second-order logic over D. We introduce the following logic AF_2: we have two sorts, subsets and terms. We use $X, Y, \ldots$ for variables over subsets and $x, y, \ldots$ for variables over terms. The terms are elements of D. The formulae are

$$A \ ::= \ t \in X \mid A \to A \mid (\forall x)A \mid (\forall X)A$$

In forming $(\forall X)A$, the formula A should be a first-order formula having at most X as a subset variable.

A *model* of AF_2 consists in an implicative algebra $(H, \to, \wedge, 1)$[1] and a valuation function $[\![A]\!]_\nu \in H$ where ν assigns a function $D \to H$ to each predicate variable. We write as usual $(\nu, X = f)$ for the *update* of ν. The valuation function should be such that $[\![(\forall x)A]\!]_\nu$ is the greatest lower bound of all $[\![A(d)]\!]_\nu$ for $d \in D$ and $[\![(\forall X)A]\!]_\nu$ is the greatest lower bound of all $[\![A]\!]_{(\nu,X=f)}$ for $f \in D \to H$. Notice, that we don't require H to be complete. Furthermore, we should have

$$[\![A_1 \to A_2]\!]_\nu = [\![A_1]\!]_\nu \to [\![A_2]\!]_\nu$$

and

$$[\![t \in X]\!]_\nu = \nu(X)(t).$$

To each type T we can associate a formula $C_T(x)$ with one term variable x, by taking $C_\alpha = x \in X_\alpha$, $C_{T \to U} = (\forall y)[C_T(y) \to C_U(x\ y)]$ and $C_{(\Pi\alpha)T} = (\forall X_\alpha)C_T(x)$. To each context $\Gamma = x_1 : T_1, \ldots, x_n : T_n$ we associate the set of formulae $C_\Gamma = C_{T_1}(x_1), \ldots, C_{T_n}(x_n)$ and we have

Lemma 3. *If* $\Gamma \vdash t : T$ *then* $\wedge_{A \in C_\Gamma} [\![A]\!] \leq [\![C_T(t)]\!]$ *in any model of* AF_2. *In particular if* $\vdash t : T$ *we have* $[\![C_T(t)]\!] = 1$.

Next we are going to build a model of AF_2 in a finitary way.

[1] That is $(H, \wedge, 1)$ is a meet semilattice and we have $x \wedge y \leq z$ iff $x \leq y \to z$.

3 A Finitary Model

We consider now only first-order formulae

$$A ::= t \in X \mid A \to A \mid (\forall x)A$$

We define a first-order logic AF_1 on these formulae. We let $L, M, \ldots$ denote finite sets of formulae, and we write L, M for $L \cup M$ and L, A for $L \cup \{A\}$. We have the rules

$$\frac{}{L \vdash A}\ (A \in L)$$

$$\frac{L \vdash A_1 \quad L, A_2 \vdash A}{L \vdash A}\ (A_1 \to A_2 \in L) \qquad \frac{L, A_1 \vdash A_2}{L \vdash A_1 \to A_2}$$

$$\frac{L, A_1(t) \vdash A}{L \vdash A}\ ((\forall x)A_1 \in L) \qquad \frac{L \vdash A}{L \vdash (\forall x)A}$$

In the last rule, x should not occur free in L.

We write $L \leq M$ iff $L \vdash A$ for all A in M. It can be proved in a finitary way that this defines a poset, which we call S_0. We use this poset to give a finitary Kripke model of AF_2.

The subsets are interpreted as functions $D \to \mathrm{Down}(S_0)$ where $\mathrm{Down}(S_0)$ is the set of downward closed subsets of S_0. If A is a first-order formula we let $[A]$ be $\{L \in S_0 \mid L \vdash A\}$, and if L is a finite set of formulae $A_1, \ldots, A_n$ we let $[L]$ be $[A_1] \cap \ldots \cap [A_n]$. If X is a variable we let $F_X : D \to \mathrm{Down}(S_0)$ be the function $F_X(t) = [X\ t]$. An assignment ν associates functions $D \to \mathrm{Down}(S_0)$ to subset variables. To any first order formula A we assign $[\![A]\!]_\nu \in \mathrm{Down}(S_0)$:

$$[\![t \in X]\!]_\nu = \nu(X)(t) \qquad [\![(\forall x)A]\!]_\nu = \cap_{u \in D}[\![A(u)]\!]_\nu$$

$$[\![A_1 \to A_2]\!]_\nu = \{L \in S_0 \mid (\forall M \in [\![A_1]\!]_\nu) L, M \in [\![A_2]\!]_\nu\}$$

We let $[\![A_1, \ldots, A_n]\!]$ to be $[\![A_1]\!] \cap \ldots \cap [\![A_n]\!]$.

Lemma 4. *If $L \vdash A$ in AF_1 we have $[\![L]\!]_\nu \subseteq [\![A]\!]_\nu$.*

Proof. Since AF_1 is intuitionistic, its derivations are valid in any Kripke model.

Lemma 5. *If A is a first-order formula then $[\![A]\!]_\nu = [A]$ if $\nu(X) = F_X$ for X free in A. Also, $[\![L]\!]_\nu = [L]$ if $\nu(X) = F_X$ for X free in L.*

Proof. By induction on A. This follows from the equalities

$$[A_1 \to A_2] = \{L \in S_0 \mid (\forall M \in [A_1]) L, M \in [A_2]\}$$

and $[(\forall x)A] = \cap_{u \in D}[A(u)]$.

Lemma 6. *If A is a first-order formula with at most X as a subset variable then*

$$\bigcap_{F\in D\to \mathrm{Down}(S_0)} [\![A]\!]_{X=F} \in \mathrm{Down}(S_0)$$

can be finitary described as the set of all $L \in S_0$ such that $L \vdash A(Z)$ for Z not free in L.

Proof. If $L \vdash A(Z)$ for Z not free in L then we have $[\![L]\!]_\nu \subseteq [\![A(Z)]\!]_\nu$ for any interpretation by lemma 4. If we take $\nu(Z) = F$ and $\nu(Y) = F_Y$ for $Y \neq Z$ we get $[\![L]\!] = [L]$ and hence $[L] \subseteq [\![A(Z)]\!]_{Z=F}$ so that $L \in [\![A]\!]_{X=F}$ for all F.

Conversely, if $L \in [\![A]\!]_{X=F}$ for all F, then in particular $L \in [\![A]\!]_{X=F_Z}$ and so $L \in [A(Z)]$ that is $L \vdash A(Z)$ for Z not free in L, since $[\![A]\!]_{X=F_Z} = [\![A(Z)]\!]_{Z=F_Z} = [A(Z)]$ by lemma 5.

By finitary, we mean here that the functions we consider in $D \to \mathrm{Down}(S_0)$ if looked as relations on $D \times S_0$, are only formed by using arithmetical comprehension.

Using lemma 6 we can build in a finitary way a model of AF_2 by taking $H = \mathrm{Down}(S_0)$ and

$$1 = S_0, \quad X \wedge Y = X \cap Y, \quad X \to Y = \{L \in S_0 \mid M \in X \to L, M \in Y\}$$

By lemma 3 we have that, if $\vdash u : U$ then $1 = [\![C_U(u)]\!]$. In particular, if $\vdash t : N \to N$ then $1 = [\![C_N(t\ c_n)]\!]$, and so $1 = F(t\ c_n\ x\ f)$ if we have $1 = F(x)$ and $F(u) \subseteq F(f\ u)$ for all u. In particular we can take

$$F(u) = \bigcup_{m\in N} \{L \in S_0 \mid u = f^m\ x\}$$

and we have $1 = F(t\ c_n\ x\ f)$ which implies $t\ c_n\ x\ f = f^m\ x$ for some m.

4 An Application

We work now in SAS_0: second order arithmetic with arithmetical comprehension. It is known that this system is conservative over Peano Arithmetic [Troe]. It is possible to represent D and the poset S_0 in SAS_0. The argument above however cannot be formalised as it is in SAS_0 because of the lemma 4 which requires the definition of semantics of formulae.

We consider a fixed derivation of a typing judgement of the form $\vdash t : N \to N$. In this derivation occurs only a finite set of quantified types $T_1, \ldots, T_n$ and we consider the set SF of subformulae of $C_{T_1}(t_1), \ldots, C_{T_n}(t_n)$. We let then S_1 be the subposet of S_0 which consists only of finite sets of such subformulae.

Given any poset defined in SAS_0 we can define $[\![A]\!]_\nu$ for $A \in SF$ in SAS_0, see [Troe], p. 37.

Lemma 7. *If $M \vdash A$ with $A \in SF$, $M \subseteq SF$ then $[\![M]\!]_\nu \subseteq [\![A]\!]_\nu$ and this is provable in* SAS_0.

We consider then the model $H = \text{Down}(S_1)$ and

$$1 = S_1, \quad X \wedge Y = X \cap Y, \quad X \to Y = \{L \in S_1 \mid M \in X \to L, M \in Y\}$$

for all $X, Y \subseteq D$.

Corollary 2. *If A is a first-order formula in SF with at most X as a subset variable then*

$$\bigcap_{F \in D \to H} [\![A]\!]_{X=F} \in H$$

can be finitary described as the set of all $L \in S_1$ such that $L \vdash A(Z)$ for Z not free in L, and this is provable in SAS_0.

Theorem 2. *If ϕ is represented by a term then ϕ is provably total in Peano Arithmetic.*

Proof. Suppose that ϕ is represented by a term t. We have a derivation of $\vdash t : N \to N$. The previous results allow us to transform this derivation to a proof in SAS_0 that $[\![C_N]\!](t\ c_n)$ holds for all n.

It follows from [Glad] that we can represent all functions provably total in Peano Arithmetic, using only the quantified type $N = (\Pi\alpha)[\alpha \to (\alpha \to \alpha) \to \alpha]$. Indeed, [Glad] shows that we can program the predecessor function, and indeed all primitive recursive functions, using only iteration. If follows from this that all functions of Gödel's system T can be programmed using only iteration. A more direct way of seeing this is that in Gödel's system T, it is possible to encode pairing of integers using the type $N \to N \to N$, and it is standard how to reduce primitive recursion to iteration and pairing.

Theorem 3. *The set of representable functions is exactly the set of functions provably total in Peano Arithmetic.*

5 Discussion and Further Work

In an email to the second author, W. Buchholz suggests a simplification of the construction here, which *avoids the detour through the logic* AF_2. He also suggests to show a more general result, i.e. that every term typable in the fragment described here β-reduces to a normal form. Although we agree that his proof is very elegant, we believe that our presentation explains better how the standard infinitary construction can be turned into a finitary one in this special case. We hope to expand on the connections between the two approaches in further work.

We are also interested to extend the construction presented here to full Π^1_1 comprehension. This would amount to showing a normalisation result for the fragment of System F where all Π-types are closed using only iterated inductive definitions. For instance, the introduction of a type such as

$$(\Pi\alpha)[\alpha \to ((N \to \alpha) \to \alpha) \to \alpha]$$

will correspond to the use of a generalised inductive definition and the normalisation will require ID_1. We hope that this work sheds some light on the question at which point a predicative[2] normalisation proof of System F breaks down.

Acknowledgments. We would like to thank W. Buchholz for his comments on this paper and for suggesting an alternative construction. We would also like to point out that this paper has been inspired by [Buch1,Buch2] and to thank the anonymous referees for helpful comments on the paper.

References

[Buch1] W. Buchholz. The $\omega_{\mu+1}$-rule. In *Iterated Inductive Definitions and Subsystems of Analysis: Recent Proof-Theoretical Studies*, volume 897 of *Lecture Notes in Mathematics*, pages 188–233. 1981.

[Buch2] W. Buchholz and K. Schütte. *Proof theory of impredicative subsystems of analysis.* Studies in Proof Theory. Monographs, 2. Bibliopolis, Naples, 1988.

[Glad] M.D. Gladstone. A reduction of the recursion scheme. J. Symbolic Logic 32 1967 505–508.

[Kriv] J.L. Krivine. *Lambda-calcul. Types et modèles.* Masson, Paris, 1990.

[Leiv] D. Leivant. Peano's Lambda Calculus: The Functional Abstraction Implicit in Arithmetic to be published in the Church Memorial Volume.

[Troe] A. Troelstra. *Metamathematical Investigations of Intuitionistic Arithmetic and Analysis.* Lecture Notes in Mathematics 344, 1973.

[2] In the sense of Martin-Löf's Type Theory.

Sequentiality and the π-Calculus

Martin Berger[1], Kohei Honda[1], and Nobuko Yoshida[2]

[1] Queen Mary, University of London, U.K
[2] University of Leicester, U.K.

Abstract. We present a type discipline for the π-calculus which precisely captures the notion of sequential functional computation as a specific class of name passing interactive behaviour. The typed calculus allows direct interpretation of both call-by-name and call-by-value sequential functions. The precision of the representation is demonstrated by way of a fully abstract encoding of PCF. The result shows how a typed π-calculus can be used as a descriptive tool for a significant class of programming languages without losing the latter's semantic properties. Close correspondence with games semantics and process-theoretic reasoning techniques are together used to establish full abstraction.

1 Introduction

This paper studies a type discipline for the π-calculus which precisely captures the notion of sequential functional computation. The precision of the representation is demonstrated by way of a fully abstract encoding of PCF. Preceding studies have shown that while operational encodings of diverse programming language constructs into the π-calculus are possible, they are rarely fully abstract [28,32]: we necessarily lose information by such a translation. The translation of a source term M will generally result in a process containing more behaviour than M. Type disciplines for the π-calculus with significant properties such as linearity and deadlock-freedom have been studied before [9,16,21,22,29,30,37], but, to our knowledge, no previous typing system for the π-calculus has enabled a fully abstract translation of functional sequentiality. The present work shows that a relatively simple typing system suffices for this purpose. Despite its simplicity, the calculus is general enough to give clean interpretations of both call-by-name and call-by-value sequentiality, offering a basic articulation of functional sequentiality without relying on particular evaluation strategies. The core idea of the typing system is that *affineness* and *stateless replication* ensure deterministic computation. Sequentiality is guaranteed by controlling the number of threads through restricting the shape of processes. While the idea itself is simple, the result would offer a technical underpinning for the potential use of typed π-calculi as meta-languages for programming language study: having fully abstract descriptions in this setting means ensuring the results obtained in the meta-language to be transferable, in principle, to object languages. In a later exposition we wish to report how the proposed typed syntax can be a powerful tool for language analysis when coupled with process-theoretic reasoning techniques.

S. Abramsky (Ed.): TLCA 2001, LNCS 2044, pp. 29–45, 2001.

From the viewpoint of the semantic study of sequentiality [6,10,27], our work positions sequentiality as a sub-class of the general universe of name passing interactive behaviour. This characterisation allows us to delineate sequentiality against the background of a broad computational universe which, among others, includes concurrency and non-determinism, offering a uniform basis on which various semantic findings can be integrated and extensions considered. A significant point in this context is the close connection between the presented calculus and game semantics [3,20,23]: the structure of interaction of typed processes (with respect to typed environments) precisely conforms to the intensional structures of games introduced in [23] and studied in e.g. [2,11,20,25,26]. It is notable that the type discipline itself does not mention basic notions in game semantics such as visibility, well-bracketing and innocence (although it does use a syntactic form of IO-alternation): yet they are derivable as operational properties of typed processes. We use this correspondence combined with process-theoretic reasoning techniques to establish full abstraction. While we expect a direct behavioural proof would be possible, the correspondence, in addition to facilitating the proof, offers deeper understanding of the present type discipline and game semantics.

We briefly give comparisons with related work. Hyland and Ong [24] presented a π-calculus encoding of innocent strategies of their games and show operational correspondence with a π-calculus encoding of PCF. Fiore and Honda [11] propose another π-calculus encoding for call-by-value games [20]. Our work, while being built on these preceding studies, is novel in that it puts forward a general type discipline where typability ensures functional sequentiality. In comparison with game semantics, our approach differs as it is based on a syntactic calculus representing a general notion of concurrent, communicating processes. In spite of the difference, our results do confirm some of the significant findings in game semantics, such as the equal status owned by call-by-name and call-by-value evaluation. From a different viewpoint, our work shows an effective way to apply game semantics to the study of basic typing systems for the π-calculus, in particular for the proof of full abstraction of encodings. Concerning the use of the π-calculus as the target language for translations, [28] was the first to point out the difficulty of fully abstract embeddings of functional sequentiality and [32] showed that the same problems arise even with the higher-order π-calculus. While some preceding work studies the significance of replication and linearity of channels [9,16,22,29,31,34,37], none offers a fully abstract interpretation of functional sequentiality.

In the remainder, Section 2 and 3 introduce the typed calculus. Section 4 analyses operational structures of typed terms. Based on them Section 5 establishes full abstraction. The technical details, including proofs omitted from the main sections of the paper, can be found in the full version [4].

Acknowledgements. We thank Makoto Hasegawa and Vasco Vasconcelos for their comments. The work of the first two authors is partially supported by EPSRC grant GR/N/37633.

2 Processes

2.1 Syntax

We use a variant of the π-calculus as our base syntax. As in typed λ-calculi, we start from the leanest untyped syntax. The following gives the reduction rule of the asynchronous version of the π-calculus, introduced in [8,18]:

$$x(\boldsymbol{y}).P \mid \overline{x}\langle \boldsymbol{v}\rangle \longrightarrow P\{\boldsymbol{v}/\boldsymbol{y}\} \tag{1}$$

Here $\boldsymbol{y}$ denotes a potentially empty vector $y_1 ... y_n$, $\mid$ denotes parallel composition, $x(\boldsymbol{y}).P$ is input, and $\overline{x}\langle \boldsymbol{v}\rangle$ is asynchronous output. Operationally, this reduction represents the consumption of an asynchronous message by a receptor. The idea extends to a receptor $!\,x(\boldsymbol{y}).P$ with recursion or replication:

$$!\,x(\boldsymbol{y}).P \mid \overline{x}\langle \boldsymbol{v}\rangle \longrightarrow !\,x(\boldsymbol{y}).P \mid P\{\boldsymbol{v}/\boldsymbol{y}\}, \tag{2}$$

where the replicated process remains in the configuration after reduction. Types for processes prescribe usage of names [29,36]. To be able to do this with precision, it is important to control dynamic sharing of names. For this purpose it is essential to distinguish *free name passing* and *bound (private) name passing*: the latter allows tight control of sharing and can control name usage in more stringent ways. In the present study, using bound name passing alone is sufficient. Further, to have tractable inference rules, it is vital to specify bound names associated with the concerned output. Thus, instead of $(\boldsymbol{\nu}\,\boldsymbol{y})(\overline{x}\langle \boldsymbol{y}\rangle|P)$, we write $\overline{x}(\boldsymbol{y})\,P$, and replace (1) by the following reduction rule.

$$x(\boldsymbol{y}).P \,|\, \overline{x}(\boldsymbol{y})\,Q \longrightarrow (\boldsymbol{\nu}\,\boldsymbol{y})(P \mid Q) \tag{3}$$

Here "$\overline{x}(\boldsymbol{y})\,Q$" indicates that $\overline{x}(\boldsymbol{y})$ is an asynchronous output exporting $\boldsymbol{y}$ which are local to Q. The rule corresponding to (2) is given accordingly. To ensure asynchrony of outputs, we add the following rule to the standard closure rules for $\mid$ and $(\boldsymbol{\nu}\,x)$.

$$P \longrightarrow P' \;\;\Rightarrow\;\; \overline{x}(\boldsymbol{y})\,P \longrightarrow \overline{x}(\boldsymbol{y})\,P' \tag{4}$$

Further, the following structural rules are added to allow inference of interaction under an output prefix.

$$\overline{x}(\boldsymbol{z})\,(P|Q) \equiv (\overline{x}(\boldsymbol{z})\,P)|Q \qquad \text{if } \mathsf{fn}(Q) \cap \{\boldsymbol{z}\} = \emptyset, \tag{5}$$

$$\overline{x}(\boldsymbol{z})\,(\boldsymbol{\nu}\,y)P \equiv (\boldsymbol{\nu}\,y)\overline{x}(\boldsymbol{z})\,P \qquad \text{if } y \notin \{x, \boldsymbol{z}\}. \tag{6}$$

By these rules we maintain the dynamics based on the original asynchronous calculus (up to the equation $\overline{x}(\boldsymbol{z})\,P \equiv (\boldsymbol{\nu}\,\boldsymbol{z})(\overline{x}\langle \boldsymbol{z}\rangle|P)$), while enabling output actions to be typed with the same ease as input actions. Name-passing calculi using only bound name passing, called πI-calculi, have been studied in [7,33].

Another useful construct for typing is *branching*. Branching is similar to the "case" construct in typed λ-calculi and can represent both base values such as booleans or integers and conditionals. While binary branching has some merit, we use indexed branching because it simplifies the description of base value passing. The branching variant of the reduction (3) becomes:

$$x[\&_{i\in I}(\boldsymbol{y}_i).P_i] \mid \overline{x}\mathtt{in}_j(\boldsymbol{y}_j)\,Q \longrightarrow (\boldsymbol{\nu}\,\boldsymbol{y}_j)(P_j \mid Q) \tag{7}$$

where we assume $j \in I$, with $I\,(\neq \emptyset)$ denoting a finite or countably infinite indexing set. Accordingly we define the rule for replicated branching. Branching constructs of this kind have been studied in tyco [35] and other calculi [12,15, 17] (the corresponding type structure already appeared in Linear Logic [1,13]).

Augmenting the original asynchronous syntax with bound output and branching, we now arrive at the following grammar.

$P ::= x(\boldsymbol{y}).P$	input	$\mid P \mid Q$	parallel
$\mid\ \overline{x}(\boldsymbol{y})\,P$	output	$\mid (\boldsymbol{\nu}\,x)P$	hiding
$\mid\ x[\&_{i\in I}(\boldsymbol{y}_i).P_i]$	branching input	$\mid \mathbf{0}$	inaction
$\mid\ \overline{x}\mathtt{in}_i(\boldsymbol{z})\,P$	selection	$\mid\ !P$	replication

In $!P$ we require P to be either a unary or branching input. The bound/free names/variables are defined as usual and we assume the variable convention for bound names. The structural rules are standard except for the omission of $!P \equiv\ !P|P$ and the incorporation of (5) as well as (6) together with the corresponding rules for branching output. The reduction rules are as explained above, which also include variants of (3) and (7) for replicated branching inputs.

2.2 Examples

Henceforth we omit trailing zeros and null arguments and write $x[\&_i P_i]$ for $x[\&_i().P_i]$.

(i) $[\![\mathtt{n}]\!]_u \stackrel{\text{def}}{=}\ !\,u(a).\overline{a}\mathtt{in}_n$. Each time $[\![\mathtt{n}]\!]_u$ is invoked, it replies by telling its number, $\mathtt{n}$. Here a natural number becomes a *stateless server.*

(ii) $[\![\mathtt{succ}]\!]_u \stackrel{\text{def}}{=}\ !\,u(ya).\overline{y}(b)\,b[\&_n\,\overline{a}\mathtt{in}_{n+1}]$. $[\![\mathtt{succ}]\!]_x$ describes the behaviour of a successor function, which queries for its argument, a natural number as in (i) above, and returns its increment. This is another stateless server but this time it asks its client for an input.

(iii) $!u(xa).\overline{x}(zb)\,([\![1]\!]_z \mid b[\&_i \overline{a}\mathtt{in}_i])$. This represents a type-2 functional $\lambda x.x1$: $(\mathsf{Nat}\Rightarrow\mathsf{Nat})\Rightarrow\mathsf{Nat}$. When the process is invoked, it queries for its argument (which is a function itself), that function then asks back for its own argument, to which $[\![1]\!]_z$ replies. Finally the process receives, at b, an answer to its own question, based on which it answers to the initial question.

3 Typing

3.1 Action Modes

Functional computation is *deterministic*. There are two basic ways to realise this in interacting processes. One is to have (at most) one input and (at most) one output at a given channel (such a channel is called *affine*). Another is to have a unique stateless replicated input with zero or more dual outputs. These ideas have been studied in the past [13,15,16,21,22,31,34,37]. To capture them in typing, we use the following *action modes*, denoted $p, q, \ldots$:

$!_1$	Affine input	$?_1$	Affine output
$!_\omega$	Replicated input	$?_\omega$	Output to replicated input

We also use $\bot$ to denote the presence of both input and output at an affine channel. In the table above, the mode on the left and that on the right in the same row are *dual* to each other, denoted $\overline{p}$ (for example, $\overline{!_1} = ?_1$).

3.2 Channel Types

Channel types indicate possible usage of channels. We use sorting [29] augmented with branching [1,13,15,17,35] and action modes. The grammar follows.

$$\alpha ::= \langle\tau, \overline{\tau}\rangle \qquad \tau_{\mathtt{I}} ::= (\boldsymbol{\tau})^{!_1} \mid (\boldsymbol{\tau})^{!_\omega} \mid [\&_{i\in I}\, \boldsymbol{\tau}_i]^{!_1} \mid [\&_{i\in I}\, \boldsymbol{\tau}_i]^{!_\omega}$$
$$\tau ::= \tau_{\mathtt{I}} \mid \tau_{\mathtt{O}} \qquad \tau_{\mathtt{O}} ::= (\boldsymbol{\tau})^{?_1} \mid (\boldsymbol{\tau})^{?_\omega} \mid [\oplus_{i\in I}\, \boldsymbol{\tau}_i]^{?_1} \mid [\oplus_{i\in I}\, \boldsymbol{\tau}_i]^{?_\omega}$$

In the first line $\overline{\tau}$ denotes the *dual* of τ, which is the result of dualising all action modes and exchanging $\oplus$ and $\&$. A type of form $\langle\tau, \overline{\tau}\rangle$ is called *pair type*, which we regard as a set. $[\&_{i\in I}\ldots]$ corresponds to branching and $[\oplus_{i\in I}\ldots]$ corresponds to selection. As an example of types, let $\mathsf{Nat}^\bullet \stackrel{\text{def}}{=} [\oplus_{i\in\mathrm{N}}]^{?_1}$ and $\mathsf{Nat}^\circ \stackrel{\text{def}}{=} (\mathsf{Nat}^\bullet)^{!_\omega}$. Then in $!a(x).\overline{x}\mathtt{in}_n$, x is used as $\mathsf{Nat}^\bullet$ while a is used as Nat°.

A further idea in functional computation is asking a question and receiving a unique answer [3,23]. A type is *sequential* when for each subexpression:

(i) In $(\boldsymbol{\tau})^{!_\omega}$, if $\boldsymbol{\tau} \neq \varepsilon$ then there is a unique τ_i of mode $?_1$, while each τ_j $(i \neq j)$ is of mode $?_\omega$. Dually for $(\boldsymbol{\tau})^{?_\omega}$. The same applies to $[\&_{i\in I}\, \boldsymbol{\tau}_i]^{!_\omega}$ and $[\oplus_{i\in I}\, \boldsymbol{\tau}_i]^{?_\omega}$.

(ii) In $(\boldsymbol{\tau})^{!_1}$, each τ_i is of mode $?_\omega$, dually for $(\boldsymbol{\tau})^{?_1}$. The same applies to $[\&_{i\in I}\, \boldsymbol{\tau}_i]^{!_1}$ and $[\oplus_{i\in I}\, \boldsymbol{\tau}_i]^{?_1}$.

As an example, $(\overline{\mathsf{Nat}^\circ}\mathsf{Nat}^\bullet)^{!_\omega}$ is a sequential type for $[\![\mathsf{succ}]\!]_u$ in §2.2 (ii).

3.3 Action Types and IO-Modes

The sequents we use have the form $\Gamma \vdash_\phi P \triangleright A$. Γ is a *base*, i.e. a finite map from names to channel types, P is a process with type annotations on binding names, A is an *action type*, and ϕ is an *IO-mode*. Intuitively, an action type witnesses the real usage of channels in P with respect to their modes specified in Γ (thus controlling determinacy); an IO-mode ensures P contains at most one active thread (thus controlling sequentiality). Below in (i) we use a symmetric partial operator $\odot$ on action modes generated from $!_1 \odot ?_1 = \bot$, $?_\omega \odot ?_\omega = ?_\omega$ and $!_\omega \odot ?_\omega = !_\omega$. Thus, for example, $!_\omega \odot !_\omega$ is undefined. This partial algebra ensures that only one-one (resp. one-many) connection is possible at an affine (resp. replicated) channel.

(i) An *action type* assigns action modes to names. Each assignment is written px. $\mathsf{fn}(A)$ denotes the set of names in A. A partial operator $A \odot B$ is defined iff $p \odot q$ is defined whenever $px \in A$ and $qx \in B$; then we set $A \odot B = (A \backslash B) \cup (B \backslash A) \cup \{(p \odot q)x \mid px \in A, qx \in B\}$. We write $A \asymp B$ when $A \odot B$ is defined. The set of modes used in A is $\mathsf{md}(A)$.

(ii) An *IO-mode* is one of $\{\mathtt{I}, \mathtt{O}\}$. We set $\mathtt{I} \odot \mathtt{I} = \mathtt{I}$ and $\mathtt{I} \odot \mathtt{O} = \mathtt{O} \odot \mathtt{I} = \mathtt{O}$. Note $\mathtt{O} \odot \mathtt{O}$ is not defined. When $\phi_1 \odot \phi_2$ is defined we write $\phi_1 \asymp \phi_2$.

In IO-modes, $\mathtt{O}$ indicates a unique active output (consider it as a thread): thus $\mathtt{O} \not\asymp \mathtt{O}$ shows that we do not want more than one thread in a process.

3.4 Typing Rules

•• ••••
$$\frac{\Gamma \text{ Sequential}}{\Gamma \vdash_{\mathtt{I}} \mathbf{0} \triangleright \emptyset}$$

•• •••
$$\frac{\Gamma \vdash_{\phi_i} P_i \triangleright A_i \quad (i=1,2) \qquad A_1 \asymp A_2 \quad \phi_1 \asymp \phi_2}{\Gamma \vdash_{\phi_1 \cdot \phi_2} P_1 | P_2 \triangleright A_1 \odot A_2}$$

•• •••
$$\frac{\Gamma \cdot x : \alpha \vdash_\phi P \triangleright A \otimes px \qquad p \in \{\bot, !_\omega\}}{\Gamma \vdash_\phi (\boldsymbol{\nu}\, x : \alpha) P \triangleright A}$$

•••$^{!_1}$• $(C/\boldsymbol{y} = ?A)$
$$\frac{\Gamma \vdash x : (\boldsymbol{\tau})^{!_1} \qquad \Gamma \cdot \boldsymbol{y} : \boldsymbol{\tau} \vdash_{\mathtt{O}} P \triangleright C^{-x}}{\Gamma \vdash_{\mathtt{I}} x(\boldsymbol{y} : \boldsymbol{\tau}).P \triangleright A \otimes !_1 x}$$

•• ••$^{?_1}$• $(C/\boldsymbol{y} = A \asymp ?_1 x)$
$$\frac{\Gamma \vdash x : (\boldsymbol{\tau})^{?_1} \qquad \Gamma \cdot \boldsymbol{y} : \boldsymbol{\tau} \vdash_{\mathtt{I}} P \triangleright C}{\Gamma \vdash_{\mathtt{O}} \overline{x}(\boldsymbol{y} : \boldsymbol{\tau}) P \triangleright A \odot ?_1 x}$$

•• ••••⊥•
$$\frac{\Gamma \vdash x : !_1, ?_1 \qquad \Gamma \vdash_\phi P \triangleright A^{-x}}{\Gamma \vdash_\phi P \triangleright A \otimes \bot x}$$

•••$^{!_\omega}$• $(C/\boldsymbol{y} = ?_\omega A)$
$$\frac{\Gamma \vdash x : (\boldsymbol{\tau})^{!_\omega} \qquad \Gamma \cdot \boldsymbol{y} : \boldsymbol{\tau} \vdash_{\mathtt{O}} P \triangleright C^{-x}}{\Gamma \vdash_{\mathtt{I}} !x(\boldsymbol{y} : \boldsymbol{\tau}).P \triangleright A \otimes !_\omega x}$$

•• ••$^{?_\omega}$• $(C/\boldsymbol{y} = A \asymp ?_\omega x)$
$$\frac{\Gamma \vdash x : (\boldsymbol{\tau})^{?_\omega} \qquad \Gamma \cdot \boldsymbol{y} : \boldsymbol{\tau} \vdash_{\mathtt{I}} P \triangleright C}{\Gamma \vdash_{\mathtt{O}} \overline{x}(\boldsymbol{y} : \boldsymbol{\tau}) P \triangleright A \odot ?_\omega x}$$

•• ••••$?_\omega$•
$$\frac{\Gamma \vdash x : ?_\omega \qquad \Gamma \vdash_\phi P \triangleright A^{-x}}{\Gamma \vdash_\phi P \triangleright A \otimes ?_\omega x}$$

Fig. 1. Sequential Typing System

The typing rules are given in Figure 1. The rules for branching/selection are defined similarly and left to Appendix A. The following notation is used:

$?_\omega A$	A s.t. $\mathsf{md}(A) = \{?_\omega\}$	A^{-x}	A s.t. $x \notin \mathsf{fn}(A)$
$?A$	A s.t. $\mathsf{md}(A) = \{?_\omega, ?_1\}$	$A \otimes B$	$A \cup B$ s.t. $\mathsf{fn}(A) \cap \mathsf{fn}(B) = \emptyset$
$A/\boldsymbol{x}$	$A \backslash \{p\boldsymbol{x}\}$ s.t. $\{\boldsymbol{x}\} \subseteq \mathsf{fn}(A)$	$\Gamma \cdot \Delta$	$\Gamma \cup \Delta$ s.t. $\mathsf{fn}(\Gamma) \cap \mathsf{fn}(\Delta) = \emptyset$

$\Gamma \vdash x : \tau$ denotes $x : \tau$ or $x : \langle \tau, \overline{\tau} \rangle$ in Γ, while $\Gamma \vdash x : p$ indicates $\Gamma \vdash x : \tau$ such that the mode of τ is p. Typed processes are often called *sequential processes.* The sequent $\Gamma \vdash_\phi P \triangleright A$ is often abbreviated to $\Gamma \vdash_\phi P$.

We briefly illustrate each typing rule. In (Zero), we start in $\mathtt{I}$-mode since there is no active output. In (Par), "$\asymp$" controls composability, ensuring that at most one thread is active in a given term. In (Res), we do not allow $?_1$, $?_\omega$ or $!_1$-channel to be restricted since these actions expect their dual actions exist in the environment (cf. [16,19,22]). ($\mathsf{In}^{!_1}$) ensures that x occurs precisely once (by C^{-x}) and no free input is suppressed under prefix (by $C/\boldsymbol{y} = ?A$). ($\mathsf{Out}^{?_1}$) also ensures an output at x occurs precisely once, but does not suppress the body by prefix since output is asynchronous (essentially the rule composes the output prefix and the body in parallel). (Weak-$\bot$) allows assigning the same type after a pair of dual affine channels disappears following an interaction. This is essential for subject reduction. ($\mathsf{In}^{!_\omega}$) is the same as ($\mathsf{In}^{!_1}$) except no free $?_1$-channels are suppressed (note that if a $?_1$-channel is under replication then it can be used more than once). ($\mathsf{Out}^{?_\omega}$) and (Weak-$?_\omega$) say $?_\omega$-channels occur zero or more times, and it does not suppress any actions. Finally, in ($\mathsf{Out}^{?_1}$) and ($\mathsf{Out}^{?_\omega}$), the premise must have $\mathtt{I}$-mode for otherwise we would end up with more than one thread. Note that, for input, we require the premise to be $\mathtt{O}$-mode. This together ensures single-threadedness to be invariant under reduction, as we discuss later.

3.5 Examples

The following examples indicate how the present type discipline imposes strong constraints on term structure.

(i) Given $\Gamma = a : ()^{?_1} \cdot b : \langle ()^{!_1}, ()^{?_1} \rangle \cdot c : ()^{?_1}$, we build sequential processes one by one, starting from inaction. (1) $\Gamma \vdash_{\mathtt{I}} \mathbf{0} \triangleright \emptyset$, (2) $\Gamma \vdash_{\mathtt{O}} \overline{a} \triangleright ?_1 a$, and (3) $\Gamma \vdash_{\mathtt{I}} b.\overline{a} \triangleright ?_1 a \otimes !_1 b$. Then we have:

$$\Gamma \vdash_{\mathtt{O}} \overline{b} \mid b.\overline{a} \triangleright ?_1 a \otimes \bot b \quad \text{with} \quad ?_1 b \odot !_1 b = \bot b \quad \text{and} \quad \mathtt{O} \odot \mathtt{I} = \mathtt{O}$$

where "$\bot b$" means name b is no longer composable. Note for any ϕ, $\Gamma \not\vdash_\phi b.\overline{a} \mid b.\overline{c}$ since b is affine.

(ii) Given $\Gamma = a : ()^{?_\omega} \cdot b : \langle ()^{!_\omega}, ()^{?_\omega} \rangle$, we have:
 – $\Gamma \vdash_{\mathtt{O}} \overline{a} \mid !b.\overline{a} \triangleright ?_\omega a \otimes !_\omega b$ with $?_\omega a \odot ?_\omega a = ?_\omega a$ and $\mathtt{O} \odot \mathtt{I} = \mathtt{O}$; and
 – $\Gamma \vdash_{\mathtt{O}} !b.\overline{a} \mid \overline{b} \triangleright ?_\omega a \otimes !_\omega b$ with $?_\omega b \odot !_\omega b = !_\omega b$.

 However, for any ϕ, $\Gamma \not\vdash_\phi \overline{a} \mid !b.\overline{a} \mid \overline{b}$ since $\mathtt{O} \odot \mathtt{O}$ is undefined. This example shows control by modes is essential even if $?_\omega$-mode channel does not appear in parallel; we can check after one step interaction between $!b.\overline{a}$ and $\overline{b}$, two messages to a will appear in parallel.

(iii) For $[\![\mathtt{n}]\!]_u$ in Example 2.2 (i), we have $u : \mathsf{Nat}^\circ \vdash_{\mathtt{I}} [\![\mathtt{n}]\!]_u$ (see § 3.2 for Nat°).

(iv) For $[\![\mathsf{succ}]\!]_u$ in Example 2.2 (ii), we can derive $u : (\overline{\mathsf{Nat}^\circ} \mathsf{Nat}^\bullet)^{!_\omega} \vdash_{\mathtt{I}} [\![\mathsf{succ}]\!]_u$.

(v) For the process in Example 2.2 (iii), let $\tau \stackrel{\text{def}}{=} ((\mathsf{Nat}^\circ\overline{\mathsf{Nat}^\bullet})^{?_\omega}\mathsf{Nat}^\bullet)^{!_\omega}$. Then we have $u:\tau \vdash_{\mathtt{I}} !u(xa).\overline{x}(zb)\,([\![1]\!]_z \mid b[\&_{i\in\mathbb{N}}\overline{a}\mathtt{in}_i]) \triangleright !_\omega u$.

(vi) A *copy-cat* $[x \to y]^\tau \stackrel{\text{def}}{=} !x(a).\overline{y}(b)b.\overline{a}$ copies all behaviour starting at one channel to those starting at another. Let $\tau = (()^{?_1})^{!_\omega}$ and $\Gamma = x:\tau\cdot y:\overline{\tau}$. Then (1) $\Gamma\cdot a:()^{?_1}\cdot b:()^{!_1} \vdash_{\mathtt{I}} b.\overline{a} \triangleright ?_1 a \otimes !_1 b$, (2) $\Gamma\cdot a:()^{?_1} \vdash_{\mathtt{O}} \overline{y}(b)b.\overline{a} \triangleright ?_1 a \otimes ?_\omega y$, with $(?_1 a \otimes !_1 b)/b = ?_1 a$, and (3) $\Gamma \vdash_{\mathtt{I}} [x \to y]^\tau \triangleright !_\omega x \otimes ?_\omega y$.

Taking for example $(\boldsymbol{\nu}\,x)(P|[x \to y]^\tau)$ with $P \stackrel{\text{def}}{=} \overline{x}(a)a.\overline{c}$, we can check that all actions of P are copied from x to y (this does not include c which is emitted by P).

(vii) Let $\Delta = x:\langle\tau,\overline{\tau}\rangle\cdot y:\langle\tau,\overline{\tau}\rangle\cdot z:\langle\tau,\overline{\tau}\rangle$ and $\tau = (()^{?_1})^{!_\omega}$. Then we have:
- connection of two links: $\Gamma \vdash_{\mathtt{I}} [x \to y]^\tau \mid [y \to z]^\tau \triangleright !_\omega x \otimes !_\omega y \otimes ?_\omega z$ with $!_\omega y \otimes ?_\omega y = !_\omega y$.
- links to a shared resource at z: $\Gamma \vdash_{\mathtt{I}} [x \to z]^\tau \mid [y \to z]^\tau \triangleright !_\omega x \otimes !_\omega y \otimes ?_\omega z$ with $?_\omega z \otimes ?_\omega z = ?_\omega z$.

However, for any ϕ and environent, $[x \to z]^\tau \mid [x \to y]^\tau$ which represents non-deterministic forwarding is untypable since $!_\omega x \odot !_\omega x$ is undefined.

(viii) Let $\rho \stackrel{\text{def}}{=} ([\oplus_{i\in\mathbb{N}}]^{?_1})^{!_\omega}$ and $\Omega^\rho_z \stackrel{\text{def}}{=} (\boldsymbol{\nu}\,xy)([x \to y]^\rho|[y \to x]^\rho|\overline{x}(a)\,a[\&_{i\in\mathbb{N}}\overline{z}\mathtt{in}_i])$. Then $u:\rho \vdash_{\mathtt{I}} !u(z).\Omega^\rho_z \triangleright !_\omega u$. Unlike $[\![\mathtt{n}]\!]_u$, it returns nothing when asked, representing the undefined.

3.6 Basic Syntactic Properties

The type discipline satisfies the following standard properties. In (i) and (ii) below, the partial order $\leq$ on bases is generated from set inclusion and the rule $\Gamma \leq \Delta \;\Rightarrow\; \Gamma\cdot x:\tau \leq \Delta\cdot x:\langle\tau,\overline{\tau}\rangle$. The order on action types is simply set inclusion. In (iii) we let $\twoheadrightarrow \stackrel{\text{def}}{=} \equiv \cup(\longrightarrow)^*$.

Proposition 1. (i) (weakening) *If $\Delta \leq \Gamma$ and $\Delta \vdash_\phi P$ then $\Gamma \vdash_\phi P$.*

(ii) (minimal type) *A typable process has a minimum base and action type. Further, if $\Gamma \vdash_\phi P$ and $\Delta \vdash_\psi P$ then $\phi = \psi$.*

(iii) (subject reduction) *If $\Gamma \vdash_\phi P$ and $P \twoheadrightarrow Q$ then $\Gamma \vdash_\phi Q$.*

We say an occurrence (subterm) in a process is an *active input* (resp. *active output*) if it is an input-prefixed (resp. output-prefixed) term which neither occurs under an input prefix nor has its subject bound by an output prefix.

Proposition 2. (i) *Let $\Gamma \vdash_\phi P \triangleright A \otimes px$ such that $p \in \{!_\omega, !_1\}$. Then there is an active input with free subject x in P.*

(ii) *Let $\Gamma \vdash_\phi P$. (1) If $\phi = \mathtt{I}$ there is no active output in P; (2) If $\phi = \mathtt{O}$ there is a unique active output in P; and (3) In both cases, two input processes never share the same name for their subjects, either bound or free.*

Corollary 1. (determinacy) *If $\Gamma \vdash_\phi P$ and $P \longrightarrow Q_i$ $(i = 1, 2)$ then $Q_1 \equiv Q_2$ and $\phi = \mathtt{O}$.*

3.7 Contextual Equality

Corollary 1 suggests non-deterministic state change (which plays a basic role in e.g. bisimilarity and testing/failure equivalence) may safely be ignored in typed equality, so that a Morris-like contextual equivalence suffices as a basic equality over processes. Let us say x is *active* when it is the free subject of an active input/output, e.g. x in $(\boldsymbol{\nu}\, \boldsymbol{w})(\overline{x}(\boldsymbol{y})P \mid R)$ assuming $x \notin \boldsymbol{w}$. We first define:

$$\Gamma \vdash_\phi P \Downarrow_x \;\stackrel{\text{def}}{\Leftrightarrow}\; \Gamma \vdash_\phi P \twoheadrightarrow P' \text{ with } x \text{ active in } P' \text{ and } \Gamma \vdash_\phi P \rhd A \otimes ?_1 x.$$

Choosing only affine output as observables induces a strictly coarser (pre-)congruence than if we had also included non-affine output ($?_\omega$-actions are not considered since, intuitively, they do not affect the environment). We can now define a typed equality. Below, a relation over sequential processes is *typed* if it relates only processes with identical base, action type and IO-mode. A relation $\cong \supseteq \equiv$ is a *typed congruence* when it is a typed equivalence closed under typed contexts and, moreover, it satisfies: if $\Gamma \geq \Delta$ and $\Delta \vdash_\phi P \cong Q$ then $\Gamma \vdash_\phi P \cong Q$.

Definition 1. $\cong_{\ldots}$ is the maximum typed congruence on sequential processes such that: if $\Gamma \vdash_\phi P \cong_{\ldots} Q$ and $\Gamma \vdash_\phi P \Downarrow_x$ then $\Gamma \vdash_\phi Q \Downarrow_x$.

4 Analysis of Sequential Interactive Behaviour

4.1 Preamble

The purpose of the rest of the paper is to demonstrate that our typed processes precisely characterise the notion of functional sequentiality. By functional sequentiality we mean the class of computational dynamics that is exhibited by, for example, call-by-name and call-by-value PCF. Concretely we show, via an interpretation $u : \alpha^\circ \vdash_{\mathrm{I}} [\![M_i : \alpha]\!]_u$ that, for a PCF term $\vdash M_i : \alpha$ $(i = 1, 2)$, we have $M_1 \cong M_2$ iff $u : \alpha^\circ \vdash_{\mathrm{I}} [\![M_1 : \alpha]\!]_u \cong_{\ldots} [\![M_2 : \alpha]\!]_u$. Here $\cong$ is the standard contextual equality on PCF-terms [14]. To this end we first introduce typed transitions to give a tractable account of processes interacting in typed contexts (the latter, like the former, must be input-output alternating). We then show that these transitions satisfy central properties of the intensional structures of games introduced in [23], namely visibility, bracketing and innocence. In particular, by innocence, any sequential process is representable by the corresponding innocent function up to redundant τ-actions. Further, the typed behaviour of a composite process $P|Q$ is completely determined by that of P and Q. Finally we show, *à la* game semantics, that any difference between typed processes in $\cong_{\ldots}$ can be detected by sequential "tester" processes whose graphs as innocent functions are finite. But finite processes in (the interpretation of) PCF types are in turn representable by PCF-terms up to $\cong$, leading to the completeness of the interpretation. Since soundness is easy by operational correspondence, this establishes full abstraction. In the following we illustrate key steps of reasoning to reach finite definability.

Note on terminology. In this section, correspondence with typed transition and intensional structures of games is a central topic. Since there is some difference in terminology between process calculi and game semantics, we list the correspondence for reference.

O's Question (OQ)	[	$!_\omega$	P's Answer (PA)	]	$?_1$
P's Question (PQ)	(	$?_\omega$	O's Answer (OA)	)	$!_1$

Note that "O" is usually used to indicate "Opponent" in game semantics, which corresponds to *input* in our (process-algebraic) terminology. To avoid confusion, we shall consistently use "input" and "output" rather than "Opponent" and "Player".

4.2 Typed Transitions

Let $P \stackrel{\text{def}}{=} !x(yz).\overline{y}(c)c.\overline{z}$ and $Q \stackrel{\text{def}}{=} \overline{x}(yz)(!y(c).\overline{c}|z.\overline{w})$. Then $P|Q$ is well-typed, and we have:

$$\begin{aligned} P\,|\,Q &\longrightarrow (\boldsymbol{\nu}\, yz)((P|\overline{y}(c)c.\overline{z}) \mid (!y(c).\overline{c}|z.\overline{w})) \\ &\longrightarrow (\boldsymbol{\nu}\, yzc)((P|c.\overline{z}) \mid (\overline{c}|z.\overline{w}|!y(c).\overline{c})) \\ &\longrightarrow (\boldsymbol{\nu}\, yzc)((P|\overline{z}) \mid (z.\overline{w}|!y(c).\overline{c})) \\ &\longrightarrow (\boldsymbol{\nu}\, yzc)(P \mid (\overline{w}|!y(c).\overline{c})). \end{aligned}$$

This example suggests that input and output alternate in typed interaction. Indeed this is the only way sequential processes interact: if P does an output and Q does an input, then the derivatives of P and Q should now be in ɪ-mode and ᴏ-mode, respectively. If they interact again, input and output are reversed. Typed transitions are built on this idea.

First we generate *untyped transitions* $P \stackrel{l}{\longrightarrow} Q$, with labels τ, $x(\boldsymbol{y})$, $\overline{x}(\boldsymbol{y})$, $x\mathtt{in}_i(\boldsymbol{y})$ and $\overline{x}\mathtt{in}_i(\boldsymbol{y})$ by the following rules.

$$(\text{In}) \quad x(\boldsymbol{y}).P \stackrel{x(\boldsymbol{y})}{\longrightarrow} P \qquad\qquad (\text{Out}) \quad \overline{x}(\boldsymbol{z})P \stackrel{\overline{x}(\boldsymbol{z})}{\longrightarrow} P$$

$$(\text{Bra}) \quad x[\&_{i\in I}(\boldsymbol{y}_i).P_i] \stackrel{x\mathtt{in}_i(\boldsymbol{y}_i)}{\longrightarrow} P_i \qquad\qquad (\text{Sel}) \quad \overline{x}\,\mathtt{in}_i(\boldsymbol{z})P \stackrel{\overline{x}\mathtt{in}_i(\boldsymbol{z})}{\longrightarrow} P$$

The rules for replicated input are defined similarly. The contextual rules are standard except for closure under asynchronous output (we omit the corresponding rule for branching).

$$(\text{Out-}\xi) \quad P \stackrel{l}{\longrightarrow} P' \text{ with } \mathsf{fn}(l) \cap \{\boldsymbol{y}\} = \emptyset \quad \Rightarrow \quad \overline{x}(\boldsymbol{y})\, P \stackrel{l}{\longrightarrow} \overline{x}(\boldsymbol{y})\, P'$$

To turn this into typed transitions, we first restrict the transitions of a process of mode ᴏ to only τ-actions and outputs since (as discussed at the outset) the interacting party should always be in ɪ-mode. Secondly, if a process has $\bot x$ (resp. $!_\omega x$) in its action type, then both input and output at x (resp. output at x) are excluded since, again, such actions can never be observed in a typed context. It is easy to check that sequential processes are closed under the restricted transition relation. The resulting typed transitions are written:

$$\Gamma \vdash_\phi P \stackrel{l}{\longrightarrow} \Gamma \cdot \boldsymbol{y}{:}\boldsymbol{\tau} \vdash_\psi Q$$

where $\boldsymbol{y}:\boldsymbol{\tau}$ assigns names introduced in l as prescribed by Γ. Typed τ-transitions coincide with untyped τ-transitions, hence typing of transitions restricts only observability of actions, not computation. Basic properties of transitions follow.

Proposition 3. (i) (IO-alternation) *Let* $\Gamma \vdash_\phi P \stackrel{l_1 l_2}{\Longrightarrow} \Delta \vdash_\psi Q$. *Then* (1) $\phi = \psi$, *and* (2) l_1 *is input iff* l_2 *is output and vice versa.*

(ii) (determinacy) *If* $\Gamma \vdash_\phi P \stackrel{l}{\longrightarrow} \Delta \vdash_\psi Q_i$ $(i = 1,2)$ *then* $Q_1 \equiv_\alpha Q_2$.

(iii) (unique output) *If* $\Gamma \vdash_{\mathtt{O}} P \stackrel{l_i}{\longrightarrow} P_i$ $(i = 1,2)$ *then* $l_1 \equiv_\alpha l_2$.

As an example of typed transitions, let $\tau \stackrel{\text{def}}{=} (\overline{\mathsf{Nat}^\circ \mathsf{Nat}^\bullet})^{!_\omega}$. Then, using the notation in Examples 2.2 (ii), we have:

$$\begin{array}{rl} x:\tau \vdash_{\mathtt{I}} [\![\mathtt{succ}]\!]_u \stackrel{u(ya)}{\longrightarrow} & u:\tau, y:\overline{\mathsf{Nat}^\circ}, a:\mathsf{Nat}^\bullet \vdash_{\mathtt{O}} \overline{y}(b)\, b[\&_{i\in\mathrm{N}}\, \overline{a}\mathtt{in}_{i+1}] \mid [\![\mathtt{succ}]\!]_u \\ \stackrel{\overline{y}(b)}{\longrightarrow} & u:\tau, y:\overline{\mathsf{Nat}^\circ}, a:\mathsf{Nat}^\bullet, b:\overline{\mathsf{Nat}^\bullet} \vdash_{\mathtt{I}} b[\&_{i\in\mathrm{N}}\, \overline{a}\mathtt{in}_{i+1}] \mid [\![\mathtt{succ}]\!]_u \\ \stackrel{b\mathtt{in}_j}{\longrightarrow} & u:\tau, y:\overline{\mathsf{Nat}^\circ}, a:\mathsf{Nat}^\bullet, b:\overline{\mathsf{Nat}^\bullet} \vdash_{\mathtt{O}} \overline{a}\mathtt{in}_{j+1} \mid [\![\mathtt{succ}]\!]_u \\ \stackrel{\overline{a}\mathtt{in}_{j+1}}{\longrightarrow} & u:\tau, y:\overline{\mathsf{Nat}^\circ}, a:\mathsf{Nat}^\bullet, b:\overline{\mathsf{Nat}^\bullet} \vdash_{\mathtt{I}} \mathbf{0} \mid [\![\mathtt{succ}]\!]_u \end{array}$$

4.3 Visibility and Well-Bracketing

Let us write $\Gamma \vdash_\phi P \stackrel{l_1..l_n}{\Longrightarrow} \Delta \vdash_\psi Q$ if $\Gamma \vdash_\phi P \stackrel{\tau^*}{\longrightarrow}\stackrel{l_1}{\longrightarrow}\stackrel{\tau^*}{\longrightarrow} \ldots \stackrel{\tau^*}{\longrightarrow}\stackrel{l_n}{\longrightarrow}\stackrel{\tau^*}{\longrightarrow} \Delta \vdash_\psi Q$ with $l_i \neq \tau$ $(0 \leq i \leq n)$. For $i \lneq j$, we write $l_i \curvearrowright l_j$ (read: l_i binds l_j) when the subject of l_j is bound by l_i (e.g. $x(y) \curvearrowright \overline{y}\mathtt{in}_n$). Clearly, in typable processes, input only binds output and vice versa. $\curvearrowright$ corresponds to justification of moves in games. Now we define the notion of *views* as follows. $\ulcorner l_1 ... l_n \urcorner^{\mathtt{O}}$ is defined first, with $s, t, \ldots$ ranging over sequences of labels.

$$\begin{array}{lll} \ulcorner \epsilon \urcorner^{\mathtt{O}} & = \emptyset & \\ \ulcorner s \cdot l_n \urcorner^{\mathtt{O}} & = \{n\} \cup \ulcorner s \urcorner^{\mathtt{O}} & l_n \text{ output} \\ \ulcorner s \cdot l_n \urcorner^{\mathtt{O}} & = \{n\} & l_n \text{ input}, \forall i. i \not\curvearrowright n \\ \ulcorner s_1 \cdot l_i \cdot s_2 \cdot l_n \urcorner^{\mathtt{O}} & = \{i, n\} \cup \ulcorner s_1 \urcorner^{\mathtt{O}} & l_n \text{ input}, i \curvearrowright n \end{array}$$

Input view, denoted $\ulcorner s \urcorner^{\mathtt{I}}$, is defined dually by exchanging Os and Is as well as input and output. We often confuse $\ulcorner s \urcorner^{\mathtt{O}}$ and $\ulcorner s \urcorner^{\mathtt{I}}$ with the corresponding sequences. We now define:

Definition 2. (visibility) Let $\Gamma \vdash_\phi P \stackrel{s}{\Longrightarrow} \Delta \vdash_\psi Q$. Then $s = l_1 \cdots l_n$ is *input-visible* if whenever l_{i+1} is input such that $l_j \curvearrowright l_i$, we have $j \in \ulcorner l_1 \cdots l_i \urcorner^{\mathtt{I}}$. Dually we define output-visibility. We say $\Gamma \vdash_\phi P$ is *visible* if whenever $\Gamma \vdash_\phi P \stackrel{s}{\Longrightarrow}$ and s is input-visible then it is output-visible.

The first key result follows.

Proposition 4. $\Gamma \vdash_\phi P$ *is visible.*

The proof proceeds by first establishing that it suffices to consider only *well-knit* traces where the only free input (if any) is an initial one. We then use induction on the typing rules to show that well-knit traces are visible. The only non-trivial cases are input prefixes and parallel composition. For input prefixes we use Proposition 3.2 (i). For parallel composition, we use *composite transitions* of $\Gamma \vdash_\phi P|Q$ which record the transitions of P and Q contributing to the those of $P|Q$ as a whole. Such transitions can be written in a matrix with four rows. For example, a composite transition of a sequential process (omitting types) $!x(c).\overline{y}(e).e[\&_{i\in\mathbb{N}}\overline{c}\mathtt{in}_{i+1}] \mid !y(e).\overline{z}(e').e'[\&_{i\in\mathbb{N}}\overline{e}\mathtt{in}_i]$ is given as follows, writing P and Q for the first and second components of parallel composition:

P-visible :	$x(c)$					$\overline{c}\mathtt{in}_3$
P-τ :		$\overline{y}(e)$			$e\mathtt{in}_2$	
Q-τ :		$y(e)$			$\overline{e}\mathtt{in}_2$	
Q-visible :			$\overline{z}(e')$	$e'\mathtt{in}_2$		

If such a sequence is well-knit and input-visible in its observable part (i.e. the first and fourth rows), then it satisfies the *switching condition* [3,23], i.e. the action of P (resp. Q) moving from one row to another is always an output. To establish this we use IO-modes of derivatives and input-visibility. Then output visibility is immediate using standard game semantics technique [20,23,26].

Next, *well-bracketing* [3,23] says that later questions are always answered first, i.e. nesting of bracketing is always properly matched. Below, following the table in §4.1, we call actions of mode $!_\omega$ and $?_\omega$ *questions* while actions of mode $!_1$ and $?_1$ are *answers*.

Definition 3. Let $\Gamma \vdash_\phi P \overset{s}{\Longrightarrow} \Delta \vdash_\psi Q$ be input-visible. Then s is *well-bracketing* if, whenever $s' = s_0 \cdot l_i \cdot s_1 \cdot l_j$ for a prefix s' of s is such that (1) l_i is a question and (2) l_j is an answer free in $s_1 \cdot l_j$, we have $l_i \curvearrowright l_j$.

Now we say $\Gamma \vdash_\phi P$ is *well-bracketing* if whenever $\Gamma \vdash_\phi P \overset{sl}{\Longrightarrow}$, s is well-bracketing and l is output, then sl is well-bracketing. Then we have:

Proposition 5. *$\Gamma \vdash_\phi P$ is well-bracketing.*

The proof uses induction on typing rules, noting that it suffices to consider well-knit sequences. The non-trivial cases are input by $!_\omega$ and parallel composition. The former holds because a $!_\omega$-prefix does not suppress a free output with action mode $?_1$, while the latter follows from the switching condition [20,23,26].

Definition 4. (legal trace) Let $\Gamma \vdash_\phi P \overset{s}{\Longrightarrow}$. Then s is *legal* if it is both input-visible and well-bracketing.

4.4 Innocence

Innocence [23] says that a process does the same action whenever it is in the same "context", i.e. in the same output-view. To establish innocence of traces of typed processes we begin with the following lemma, proved by analysis of possible redexes relying on the shape of the syntax imposed by the type discipline.

Lemma 1. (permutation) *Let* $\Gamma \vdash_{\mathrm{I}} P \overset{l_1 l_2 l_3 l_4}{\Longrightarrow} \Delta \vdash_{\mathrm{I}} Q$ *such that* $l_1 \not\curvearrowleft l_4$ *and* $l_2 \not\curvearrowleft l_3$. *Then* $\Gamma \vdash_{\mathrm{I}} P \overset{l_3 l_4 l_1 l_2}{\Longrightarrow} \Delta \vdash_{\mathrm{I}} Q$.

By the above lemma and visibility, we can transform any transition of form $\Gamma \vdash_\phi P \overset{sl}{\Longrightarrow}$, with l output, to $\Gamma \vdash_\phi P \overset{tl}{\Longrightarrow}$ where $t = \ulcorner s \urcorner^{\mathsf{O}}$. Since an output is always unique (cf. Proposition 2 (ii)), we can now conclude:

Proposition 6. (innocence) *Let* $\Gamma \vdash_\psi P \overset{s_i l_i}{\Longrightarrow}$ $(i = 1, 2)$ *such that:* (1) *both sequences are legal;* (2) *both* l_1 *and* l_2 *are output; and* (3) $\ulcorner s_1 \urcorner^{\mathsf{O}} \equiv_\alpha \ulcorner s_2 \urcorner^{\mathsf{O}}$. *Then we have* $\ulcorner s_1 \urcorner^{\mathsf{O}} \cdot l_1 \equiv_\alpha \ulcorner s_2 \urcorner^{\mathsf{O}} \cdot l_2$.

Note that *contingency completeness* in [23] corresponds to the property that any legal trace ending in an output has a legal extension ending with an input, which is immediate by Proposition 3.2 (i) and typability of transitions. Therefore, up to redundant τ-actions, a sequential process is precisely characterised by the function mapping a set of output views to next actions. This is the *innocent function representation* of a sequential process.

It is now easy to see that well-knit legal traces of $\Gamma \vdash_\phi P_1|P_2$ are uniquely determined by those of $\Gamma \vdash_{\psi_i} P_i$ $(i = 1, 2)$ in the same way that innocent strategies are composed in the appropriate category of games [23].

4.5 Factoring Observables

An important property of $\cong_{\dots}$ is that any violation of $\cong_{\dots}$ can be detected by a tester process which is finite in the sense that the cardinality of the graph of its induced innocent function is finite. In particular, for our full abstraction result, we need finite processes which are type-wise translatable to (the interpretation of) PCF terms. To this end, we first show that the congruence $\cong_{\dots}$ can be obtained by only closing terms under $|$, given an appropriate base (*Context Lemma*, cf. [27]). Then we use the following result to unfold replication.

Proposition 7. (open replication) *Assume* $\Gamma \vdash_\psi P_1 \mid P_2 \mid R$ *where* R *is a replication with subject* x. *Then* $\Gamma \vdash_\psi P_1 \mid P_2 \mid R \cong_{\dots} (P_1 \mid R) \mid (\boldsymbol{\nu}\, x)(P_2 \mid R)$.

The proof of Proposition 7 uses a bisimulation induced by the typed transition (which stays within $\cong$). We can then establish the following proposition where $\overline{\Gamma}$ denotes the result of dualising each type occurring in Γ.

Proposition 8. (finite testability) *Assume* $\Gamma \vdash_{\mathrm{I}} P_i \triangleright ?_\omega y_1 \otimes \cdots ?_\omega y_n \otimes !_\omega z$ $(i = 1, 2)$ *such that* $\mathsf{fn}(\Gamma) = \{\boldsymbol{y}, z\}$. *Then* $\Gamma \vdash_{\mathrm{I}} P_1 \not\cong_{\dots} P_2$ *iff there exist finite* $\overline{\Gamma} \vdash_{\mathrm{I}} R_j \triangleright !_\omega y_j$ $(1 \leq j \leq n)$ *and a finite* $\overline{\Gamma} \cdot x : \mathsf{Nat}^\bullet \vdash_{\mathsf{O}} S \triangleright ?_\omega z \otimes ?_1 x$ *such that* $(\Pi_j R_j | P_1 | S) \Downarrow_x$ *and* $(\Pi_j R_j | P_2 | S) \not\Downarrow_x$, *or its symmetric case.*

Towards the proof, we first take, by the Context Lemma mentioned above, a tester of form $\overline{\Gamma} \cdot x : \mathsf{Nat}^\bullet \vdash T'$ which, when composed with P_i, gives different observables. We then make, using Proposition 7, all shared replicated processes private to their "clients". This gives processes R'_i and S' which have the same

$$\textbf{(Type)}\quad \bullet\bullet\bullet^{\bullet} \stackrel{\text{def}}{=} [\oplus_{i\bullet\mathbb{N}}]^{?_1} \qquad [\alpha_1..\alpha_{n\bullet 1}\bullet\bullet\bullet]^{\bullet} \stackrel{\text{def}}{=} (\overline{\alpha_1^{\bullet}}..\overline{\alpha_{n\bullet 1}^{\bullet}}\bullet\bullet\bullet^{\bullet})^{!_\omega}$$

$$\textbf{(Base)}\quad \emptyset^{\bullet} \stackrel{\text{def}}{=} \emptyset \qquad (E\cdot x{:}\alpha)^{\bullet} \stackrel{\text{def}}{=} E^{\bullet}\cdot x{:}\overline{\alpha^{\bullet}}$$

(Terms) Below we set $\beta = [\alpha_1..\alpha_{n\bullet 1}\bullet\bullet\bullet]$.

$$[\![x:\alpha]\!]_u \stackrel{\text{def}}{=} [u \to x]^{\alpha^\circ}$$

$$[\![\lambda x_0:\alpha_0.M:\alpha_0\Rightarrow\beta]\!]_u \stackrel{\text{def}}{=} !\,u(x_0x_1..x_{n\bullet 1}z).(\boldsymbol{\nu}\,u^{\bullet})([\![M]\!]_{u'} \mid \bullet\bullet\bullet\langle u^{\bullet}x_1...x_{n\bullet 1}z\rangle^{\beta})$$

$$[\![MN:\beta]\!]_u \stackrel{\text{def}}{=} !\,u(x_1..x_{n\bullet 1}z).(\boldsymbol{\nu}\,u^{\bullet}x_0)([\![M:\alpha\Rightarrow\beta]\!]_{u'} \mid [\![N:\alpha]\!]_{x_0} \mid \bullet\bullet\bullet\langle u^{\bullet}x_0...x_{n\bullet 1}z\rangle^{\alpha\bullet\,\beta})$$

$$[\![\mathtt{n}:\bullet\bullet\bullet]\!]_u \stackrel{\text{def}}{=} !\,u(z).\overline{z}\mathtt{in}_n$$

$$[\![\mathtt{succ}(M):\bullet\bullet\bullet]\!]_u \stackrel{\text{def}}{=} !\,u(z).(\boldsymbol{\nu}\,x)([\![M]\!]_x \mid \overline{x}(y)y[\&_{n\bullet\mathbb{N}}\,\overline{z}\mathtt{in}_{n+1}])$$

$$[\![\mathtt{pred}(M):\bullet\bullet\bullet]\!]_u \stackrel{\text{def}}{=} !\,u(z).(\boldsymbol{\nu}\,x)([\![M]\!]_x \mid \overline{x}(y)y[\&_{n\bullet\mathbb{N}}\,\overline{z}\mathtt{in}_{n\bullet 1}])$$

$$[\![\mathtt{ifzero}\ M\ \mathtt{then}\ N\ \mathtt{else}\ L:\beta]\!]_u$$
$$\stackrel{\text{def}}{=} !\,u(x_1..x_{n\bullet 1}z).(\boldsymbol{\nu}\,m)([\![M]\!]_m \mid \overline{m}(z^{\bullet})z^{\bullet}[\&_i(\boldsymbol{\nu}\,u^{\bullet})(P_i \mid \bullet\bullet\bullet\langle u^{\bullet}x_1..x_{n\bullet 1}z\rangle^{\beta})])$$
$$\text{where } P_0 \stackrel{\text{def}}{=} [\![N]\!]_{u'} \text{ else } P_i \stackrel{\text{def}}{=} [\![L]\!]_{u'}.$$

$$[\![\mu x:\alpha.M:\alpha]\!]_u \stackrel{\text{def}}{=} (\boldsymbol{\nu}\,m)([u\to m]^{\alpha^\circ} \mid [\![M:\alpha]\!]_m \mid [x\to m]^{\alpha^\circ})$$

$$\bullet\bullet\bullet\langle x\boldsymbol{y}z\rangle^{[\alpha\bullet\bullet\bullet]} \stackrel{\text{def}}{=} \overline{x}(\boldsymbol{y}^{\bullet}z^{\bullet})(\Pi_i[y_i^{\bullet}\to y_i]^{\alpha_i^\circ} \mid [z^{\bullet}\to z]^{\overline{\bullet\bullet\bullet}})$$

Fig. 2. Encoding of PCF

types as R_i and S above. Finally, the shapes of types allow to consider processes R_i, P_i and S (to be precise by turning S to $u(x).S$) as strategies in games. We can now appeal to finite testability in games, cf. [23], from which, by retranslating finite innocent strategies to finite processes, we conclude that finite testers suffice. Alternatively we can directly reason at the level of the π-calculus and its typed transitions, showing that any behaviour characterised by a finite innocent function (which is enough for testability) is realisable by (typable) syntactic processes [5,38].

5 Full Abstraction

5.1 Interpretation

We consider PCF with a single base type, Nat, without loss of generality. Let $\alpha ::= \mathsf{Nat} \mid \alpha\Rightarrow\beta$. We write $[\alpha_1..\alpha_n\mathsf{Nat}]$ $(n \le 0)$ for $\alpha_1 \Rightarrow (...(\alpha_n \Rightarrow \mathsf{Nat})..)$. Now the syntax of PCF terms are given by:

$$M ::= x \mid \lambda x:\alpha.M \mid MN \mid n \mid \mathtt{succ}(M) \mid \mathtt{pred}(M)$$
$$\mid\ \mathtt{ifzero}\ M\ \mathtt{then}\ N\ \mathtt{else}\ L \mid \mu x:\alpha.M$$

We omit operational semantics and the typing rules [14]. The mappings from PCF types and terms to π-types and terms, which are due to Hyland and Ong

[24], are given in Figure 2. Copy-cat processes are given by $[x \to x']^{[\&_i(\boldsymbol{\tau}_i)]^{!_1}} \stackrel{\mathrm{def}}{=} x[\&_i(\boldsymbol{y}_i).\overline{x'}\mathtt{in}_i(\boldsymbol{y'}_i)\Pi_{ij}[y'_{ij} \to y_{ij}]^{\overline{\tau_{ij}}}]$ and for replicated types: $[x \to x']^{[\&_i(\boldsymbol{\tau}_i)]^{!_\omega}} \stackrel{\mathrm{def}}{=} !x[\&_i(\boldsymbol{y}_i).\overline{x'}\mathtt{in}_i(\boldsymbol{y'}_i)\Pi_{ij}[y'_{ij} \to y_{ij}]^{\overline{\tau_{ij}}}]$. Copy-cats for unary types are special cases where the indexing sets are singletons. The interpretation of $[\alpha_1..\alpha_n\mathsf{Nat}]$ says a process, when asked for its value, asks back questions at types $\alpha_1,..,\alpha_n$, receives the results to these questions, and finally returns a natural number as the answer to the initial question.

5.2 Soundness

This is by the standard computational adequacy [27], which is proved by both-way operational correspondence, cf. [28]. Below let $\bot^{\bullet\bullet\bullet\circ} \stackrel{\mathrm{def}}{=} !\, u(z).\Omega_z^{\bullet\bullet\bullet\circ}$ where Ω_u^τ is given in Example 3.5 (v).

Theorem 1. (computational adequacy) $M\!:\!\mathsf{Nat} \Downarrow$ *iff* $[\![M : \mathsf{Nat}]\!]_u \not\cong_{\ldots} \bot^{\bullet\bullet\bullet\circ}$.

Corollary 2. (soundness) $E \vdash M \cong N : \alpha$ *if* $E^\circ \cdot u\!:\!\alpha^\circ \vdash_{\mathtt{I}} [\![M\!:\!\alpha]\!]_u \cong_{\ldots} [\![N\!:\!\alpha]\!]_u$

5.3 Completeness

Assume P is typed under (the interpretation of) a PCF-type and, moreover, it is finite, i.e. is representable by an innocent function. By [3,23] or by a direct syntactic transformation, P can be mapped into a so-called *finite canonical-form*, which in turn is easily transformed to a standard PCF term without changing meaning in its interpretation up to $\cong_{\ldots}$. Thus we obtain:

Theorem 2. (finite definability) *Let* $E^\circ\!\cdot\! u\!:\!\alpha^\circ \vdash_{\mathtt{I}} P \rhd !_\omega u$ *be finite. Then* $E^\circ\!\cdot\! u\!:\!\alpha^\circ \vdash [\![M\!:\!\alpha]\!]_u \cong_{\ldots} P$ *for some* M.

This result indicates that, in essence, *only sequential functional behaviour inhabits each type.* Now suppose $\vdash M_1 \cong M_2 : \alpha$ but $u : \mathsf{Nat}^\circ \vdash_{\mathtt{I}} [\![M_1]\!]_u \not\cong_{\ldots} [\![M_2]\!]_u$. Then the latter's difference is detectable by finite processes (Proposition 8). By Theorem 2 we can consider these finite testers as interpretations of PCF-terms so that we know, for example, $[\![C[M_1] : \mathsf{Nat}]\!] \Downarrow$ and $[\![C[M_2] : \mathsf{Nat}]\!] \Uparrow$. But this means, by Theorem 1, $C[M_1] : \mathsf{Nat} \Downarrow$ and $C[M_2] : \mathsf{Nat} \Uparrow$, contradicting our assumption. We have now reached the main result of the paper.

Theorem 3. (full abstraction) $E^\circ \cdot u\!:\!\alpha^\circ \vdash [\![M_1 : \alpha]\!]_u \cong_{\ldots} [\![M_2 : \alpha]\!]_u$ *if and only if* $E \vdash M_1 \cong M_2 : \alpha$.

By replacing $\cong_{\ldots}$ and $\cong$ with the corresponding precongruences, we similarly obtain inequational full abstraction. It is also notable that a fully abstract interpretation of call-by-value sequentiality is easily gotten by simply changing the interpretation of types. The following comes from [20]. (1) $\mathsf{Nat}^\star \stackrel{\mathrm{def}}{=} [\oplus_{i\in\mathbb{N}}]^{?_1}$ and $(A \Rightarrow B)^\star \stackrel{\mathrm{def}}{=} ((A \Rightarrow B)^\diamond)^{?_1}$; (2) $(\mathsf{Nat} \Rightarrow B)^\diamond \stackrel{\mathrm{def}}{=} [\&_{i\in\mathbb{N}} B^\star]^{!_\omega}$ and, when $A \neq \mathsf{Nat}$, $(A \Rightarrow B)^\diamond \stackrel{\mathrm{def}}{=} (\overline{A^\diamond} B^\star)^{!_\omega}$. For example, $\mathsf{Nat} \Rightarrow \mathsf{Nat}$ is interpreted as

$([\&_{i\in\mathbb{N}}[\oplus_{i\in\mathbb{N}}]^{?_1}]^{!_\omega})^{?_1}$, where the function first signals itself, receives a natural number, then returns the result. Again the only inhabitants of $A^\diamond$ are easily the encodings of call-by-value PCF terms, from which we obtain full abstraction. The result also extends to recursive types [11]. Further, another change in interpretation of types allows us to fully abstractly capture the semantics of call-by-name PCF with observability at higher-order types. These results may suggest the power and flexibility of the present framework for the semantic analysis of sequentiality.

References

1. Abramsky, S., Computational interpretation of linear logic. *TCS*, Vol. 111, 1993.
2. Abramsky, S., Honda, K. and McCusker, G., A Fully Abstract Game Semantics for General References. *LICS*, 334-344, IEEE, 1998.
3. Abramsky, S., Jagadeesan, R. and Malacaria, P., Full Abstraction for PCF. *Info. & Comp.*, Vol. 163, 2000.
4. Berger, M. Honda, K. and N. Yoshida. *Sequentiality and the π-Calculus.* To appear as a QMW DCS Technical Report, 2001.
5. Berger, M. Honda, K. and N. Yoshida. *Genericity in the π-Calculus.* To appear as a QMW DCS Technical Report, 2001.
6. Berry, G. and Curien, P. L., Sequential algorithms on concrete data structures *TCS*, 20(3), 265-321, North-Holland, 1982.
7. Boreale, M. and Sangiorgi, D., Some congruence properties for π-calculus bisimilarities, *TCS*, 198, 159–176, 1998.
8. Boudol, G., Asynchrony and the pi-calculus, INRIA Research Report 1702, 1992.
9. Boudol, G., The pi-calculus in direct style, *POPL'97*, 228–241, ACM, 1997.
10. Curien, P. L., Sequentiality and full abstraction. Proc. of *Application of Categories in Computer Science*, LNM 177, 86–94, Cambridge Press, 1995.
11. Fiore, M. and Honda, K., Recursive Types in Games: axiomatics and process representation, *LICS'98*, 345-356, IEEE, 1998.
12. Gay, S. and Hole, M., Types and Subtypes for Client-Server Interactions, *ESOP'99*, LNCS 1576, 74–90, Springer, 1999.
13. Girard, J.-Y., Linear Logic, *TCS*, Vol. 50, 1–102, 1987.
14. Gunter, C., *Semantics of Programming Languages: Structures and Techniques*, MIT Press, 1992.
15. Honda, K., Types for Dyadic Interaction. *CONCUR'93*, LNCS 715, 509-523, 1993.
16. Honda, K., Composing Processes, *POPL'96*, 344-357, ACM, 1996.
17. Honda, K., Kubo, M. and Vasconcelos, V., Language Primitives and Type Discipline for Structured Communication-Based Programming. *ESOP'98*, LNCS 1381, 122–138. Springer-Verlag, 1998.
18. Honda, K. and Tokoro, M., An Object Calculus for Asynchronous Communication. *ECOOP'91*, LNCS 512, 133–147, Springer-Verlag 1991.
19. Honda, K. Vasconcelos, V., and Yoshida, N. Secure Information Flow as Typed Process Behaviour, *ESOP '99*, LNCS 1782, 180–199, Springer-Verlag, 2000.
20. Honda, K. and Yoshida, N. Game-theoretic analysis of call-by-value computation. *TCS* Vol. 221 (1999), 393–456, North-Holland, 1999.
21. Kobayashi, N., A partially deadlock-free typed process calculus, *ACM TOPLAS*, Vol. 20, No. 2, 436–482, 1998.

22. Kobayashi, N., Pierce, B., and Turner, D., Linear Types and π-calculus, *POPL'96*, 358–371, ACM Press, 1996.
23. Hyland, M. and Ong, L., On Full Abstraction for PCF: I, II and III. 130 pages, 1994. To appear in *Info. & Comp.*
24. Hyland, M. and Ong, L., Pi-calculus, dialogue games and PCF, *FPCA'95*, ACM, 1995.
25. Laird, J., Full abstraction for functional languages with control, *LICS'97*, IEEE, 1997.
26. McCusker, G., Games and Full Abstraction for FPC. *LICS'96*, IEEE, 1996.
27. Milner, R., Fully abstract models of typed lambda calculi. *TCS*, 4:1–22, 1977.
28. Milner, R., Functions as Processes. *MSCS*, 2(2), 119–146, CUP, 1992.
29. Milner, R., Polyadic π-Calculus: a tutorial. *Proceedings of the International Summer School on Logic Algebra of Specification*, Marktoberdorf, 1992.
30. Pierce, B.C. and Sangiorgi. D, Typing and subtyping for mobile processes. *LICS'93*, 187–215, IEEE, 1993.
31. Quaglia, P. and Walker, D., On Synchronous and Asynchronous Mobile Processes, *FoSSaCS 00*, LNCS 1784, 283–296, Springer, 2000.
32. Sangiorgi, D., *Expressing Mobility in Process Algebras: First Order and Higher Order Paradigms.* Ph.D. Thesis, University of Edinburgh, 1992.
33. Sangiorgi, D. π-calculus, internal mobility, and agent-passing calculi. *TCS*, 167(2):235–271, North-Holland, 1996.
34. Sangiorgi, D., The name discipline of uniform receptiveness, *ICALP'97*, LNCS 1256, 303–313, Springer, 1997.
35. Vasconcelos, V., Typed concurrent objects. *ECOOP'94*, LNCS 821, 100–117. Springer, 1994.
36. Vasconcelos, V. and Honda, K., Principal Typing Scheme for Polyadic π-Calculus. *CONCUR'93*, LNCS 715, 524–538, Springer-Verlag, 1993.
37. Yoshida, N., Graph Types for Monadic Mobile Processes, *FST/TCS'16*, LNCS 1180, 371–387, Springer-Verlag, December, 1996.
38. Yoshida, N., Berger, M. and Honda, K., *Strong Normalisation in the π-Calculus*, To appear as a MCS Technical Report, University of Leicester, 2001.

A Typing Rules for Branching

$$\bullet\bullet\ \bullet\bullet^{!_1}\bullet\quad (C_i/\boldsymbol{y}_i = ?A) \qquad\qquad \bullet\bullet\bullet\bullet^{?_1}\bullet\quad (C_i/\boldsymbol{y}_i = A \asymp ?_1 x)$$

$$\frac{\Gamma \vdash x : [\&_{i\cdot I}\boldsymbol{\tau}_i]^{!_1} \qquad \Gamma \cdot \boldsymbol{y}_i : \boldsymbol{\tau}_i \vdash_{\mathtt{O}} P_i \triangleright C_i^{-x}}{\Gamma \vdash_{\mathtt{I}} x[\&_{i\cdot I}(\boldsymbol{y}_i : \boldsymbol{\tau}_i).P_i] \triangleright A \otimes !_1 x} \qquad \frac{\Gamma \vdash x : [\oplus_{i\cdot I}\,\boldsymbol{\tau}_i]^{?_1} \qquad \Gamma \cdot \boldsymbol{y}_i : \boldsymbol{\tau}_i \vdash_{\mathtt{I}} P \triangleright C}{\Gamma \vdash_{\mathtt{O}} \overline{x}\mathtt{in}(\boldsymbol{y}_i : \boldsymbol{\tau}_i)P \triangleright A \odot ?_1 x}$$

(Bra$^{!_\omega}$) and (Sel$^{?_\omega}$) are similarly defined.

Logical Properties of Name Restriction

Luca Cardelli
Microsoft Research

Andrew D. Gordon
Microsoft Research

Abstract. We extend the modal logic of ambients described in [7] to the full ambient calculus, including name restriction. We introduce logical operators that can be used to make assertions about restricted names, and we study their properties.

1 Introduction

The π-calculus notion of name restriction [12], initially intended to represent hidden communication channels, has been used also to represent hidden encryption keys [2] and as the basis for definitions of secrecy [2, 4]. In the context of the ambient calculus [6], name restriction can be used to represent hidden locations and (by extrapolating [4] and [5]) secret locations. In general, we would like to have process calculi where we can represent protocols for creating shared encryption keys and secret locations; name restriction seems crucial to all this.

In π-calculus notation, $(\nu n)P$ is a restriction of the name n in the process P, meaning that n is not currently known outside the scope of P. The prefix (νn) is more a bookkeeping device than a barrier. It is quite possible for P to communicate n to some external process; then the restriction (νn) must be formally pushed outwards to encompass the new scope of n and maintain the scoping invariant; this procedure is called *name extrusion*. Processes are considered equivalent up to extrusion; that is, extrusion is not regarded as a computational step. Conversely, when a name is forgotten in part of a process, the scope of (νn) may be restricted; this is called name *intrusion*. Manipulation of (νn) prefixes includes, in particular, renaming and swapping of prefixes, so that there is no obvious way of talking about "the first restricted name" or any particular restricted name of a process.

The ambient calculus can be regarded essentially as an extension of the π-calculus with dynamic location structures. In [7] we present a modal logic for describing properties of ambient calculus processes, with particular emphasis on expressing the structure and evolution of hierarchies of locations. Much of that logic can be applied directly to the π-calculus. However, in [7] we left out name restriction; we now intend to fill that gap in a way that can be applied both to the π-calculus, where names are channels, and to the ambient calculus, where names are locations. In both cases, we need to investigate the logical properties of name restriction.

In our existing logic we can describe detailed properties of processes. If we now consider restriction, what does it mean to describe properties of restricted names? We would like to be able to say, for example, "a shared key is established between locations a and b", or "a secret location is created that only a and b can access". In a protocol that establishes such shared secrets, the secrets are typically represented by restricted names. The problem is that there is no obvious way to talk about such restricted names in the specification of the protocol. We might be tempted to use ordinary existential quantification, and say "there exists a name shared between locations a and b". But this is not good enough, because we want that name to be fresh and unknown to other locations or potential attackers.

Therefore, we want a new form of quantification that can be read as "in the process there exists a restricted name which we shall call x, and such that $\mathcal{A}$", where x is a variable that ranges over names, and $\mathcal{A}$ is some property that may involve x. Let us indicate this quantifier as $(\nu x)\mathcal{A}$; this formula is meant to correspond somehow to a process of the form $(\nu n)P$ where x denotes n. However, since (νn) can float, the matching of (νx) to any particular (νn) is not obvious.

This means that the logical rules of our tentative $(\nu x)\mathcal{A}$ quantifier are going to be fairly complex, or at least unfamiliar. We have approached this complexity by splitting $(\nu x)\mathcal{A}$ into two op-

S. Abramsky (Ed.): TLCA 2001, LNCS 2044, pp. 46–60, 2001.

erators; one for quantifying over fresh names, and one for mentioning restricted names. The first operator is the Gabbay-Pitts quantifier, $\text{И}x.\mathcal{A}$, adapted to our context: it quantifies over all names that do not occur free either in the formula $\mathcal{A}$ or in the described process. The second is a binary operator (not a quantifier) called *revelation*, $n®\mathcal{A}$, which means that it is possible to reveal a restricted name as the given name n, and then assert $\mathcal{A}$. (Revelation fails to hold if it would lead to a name clash in the process.)

We investigate the properties of $n®\mathcal{A}$ and $\text{И}x.\mathcal{A}$ separately. We combine them to define $(\nu x)\mathcal{A}$ as $\text{И}x.x®\mathcal{A}$, and then we study the derived properties of $(\nu x)\mathcal{A}$.

2 Summary of the Ambient Logic

In this section, we provide a quick summary of the ambient calculus. Although this summary is technically self-contained, we assume some knowledge of [6]: see that paper for discussion and motivation. We also summarize the ambient logic studied in [7]. Again, this is self-contained, but knowledge of that paper will help. Two new operators, *revelation* and its adjunct *hiding*, are introduced here, and are discussed in the following sections.

2.1 The Calculus

The syntax of the ambient calculus is defined in the following table:

Processes

$P,Q,R ::=$	processes	$M ::=$	capabilities
$(\nu n)P$	restriction	n	name
$\mathbf{0}$	void	$in\ M$	can enter into M
$P \mid Q$	composition	$out\ M$	can exit out of M
$!P$	replication	$open\ M$	can open M
$M[P]$	ambient	ε	null
$M.P$	capability action	$M.M'$	path
$(n).P$	input action		
$\langle M\rangle$	output action		

The set of free names of a process P, written $fn(P)$ is defined as usual, where the only binders are restriction and the input action, so that $fn((\nu n)P) = fn((n).P) = fn(P) - \{n\}$.

We write $P\{n\leftarrow M\}$ for the substitution of the capability M for each free occurrence of the name n in the process P. Similarly for $M\{n\leftarrow M'\}$. We identify processes up to renaming of bound names; that is, we assume, for $m \notin fn(P)$, that $(\nu n)P = (\nu m)P\{n\leftarrow m\}$ and $(n).P = (m).P\{n\leftarrow m\}$.

We use some syntactic conventions. We use parentheses for precedence. The process $\mathbf{0}$ is often omitted in the contexts $n[\mathbf{0}]$ and $M.\mathbf{0}$, yielding $n[]$ and M. Composition has the weakest binding power, so that the expression $(\nu n)P \mid Q$ is read $((\nu n)P) \mid Q$, the expression $!P \mid Q$ is read $(!P) \mid Q$, the expression $M.P \mid Q$ is read $(M.P) \mid Q$, and the expression $(n).P \mid Q$ is read $((n).P) \mid Q$.

Structural congruence is a relation between processes used as an aid in the definition of reduction. With respect to [6], the structural rules for replication have been refined.

The reduction relation describes the dynamic behavior of ambients. In particular, the rules (Red In), (Red Out) and (Red Open) represent mobility, while (Red Comm) represents local communication (see [6] for an extended discussion). For example, the process $a[p[out\ a.\ in\ b.\ \langle m\rangle]] \mid b[open\ p.\ (n).\ n[]]$ represents a packet p that travels out of host a and into host b, where it is opened, and its contents m are read and used to create a new ambient. The process reduces in four steps (illustrating each of the four reduction rules) to the residual process $a[] \mid b[m[]]$.

Structural Congruence

$P \equiv P$	(Struct Refl)	$(\nu n)(\nu m)P \equiv (\nu m)(\nu n)P$	(Struct Res Res)
$P \equiv Q \Rightarrow Q \equiv P$	(Struct Symm)	$(\nu n)\mathbf{0} \equiv \mathbf{0}$	(Struct Res Zero)
$P \equiv Q, Q \equiv R \Rightarrow P \equiv R$	(Struct Trans)	$(\nu n)(P \mid Q) \equiv P \mid (\nu n)Q$ if $n \notin fn(P)$	(Struct Res Par)
		$(\nu n)(m[P]) \equiv m[(\nu n)P]$ if $n \neq m$	(Struct Res Amb)
$P \equiv Q \Rightarrow (\nu n)P \equiv (\nu n)Q$	(Struct Res)		
$P \equiv Q \Rightarrow P \mid R \equiv Q \mid R$	(Struct Par)	$P \mid \mathbf{0} \equiv P$	(Struct Par Zero)
$P \equiv Q \Rightarrow\ !P \equiv\ !Q$	(Struct Repl)	$P \mid Q \equiv Q \mid P$	(Struct Par Comm)
$P \equiv Q \Rightarrow n[P] \equiv n[Q]$	(Struct Amb)	$(P \mid Q) \mid R \equiv P \mid (Q \mid R)$	(Struct Par Assoc)
$P \equiv Q \Rightarrow M.P \equiv M.Q$	(Struct Action)		
$P \equiv Q \Rightarrow (n).P \equiv (n).Q$	(Struct Input)	$!\mathbf{0} \equiv \mathbf{0}$	(Struct Repl Zero)
		$!(P \mid Q) \equiv\ !P \mid\ !Q$	(Struct Repl Par)
$\varepsilon.P \equiv P$	(Struct ε)	$!P \equiv P \mid\ !P$	(Struct Repl Copy)
$(M.M').P \equiv M.M'.P$	(Struct .)	$!P \equiv\ !!P$	(Struct Repl Repl)

Reduction

$n[in\ m.\ P \mid Q] \mid m[R] \longrightarrow m[n[P \mid Q] \mid R]$	(Red In)
$m[n[out\ m.\ P \mid Q] \mid R] \longrightarrow n[P \mid Q] \mid m[R]$	(Red Out)
$open\ n.\ P \mid n[Q] \longrightarrow P \mid Q$	(Red Open)
$(n).P \mid \langle M \rangle \longrightarrow P\{n \leftarrow M\}$	(Red Comm)
$P \longrightarrow Q \Rightarrow (\nu n)P \longrightarrow (\nu n)Q$	(Red Res)
$P \longrightarrow Q \Rightarrow P \mid R \longrightarrow Q \mid R$	(Red Par)
$P \longrightarrow Q \Rightarrow n[P] \longrightarrow n[Q]$	(Red Amb)
$P' \equiv P, P \longrightarrow Q, Q \equiv Q' \Rightarrow P' \longrightarrow Q'$	(Red $\equiv$)
$\longrightarrow^*$	reflexive and transitive closure of $\longrightarrow$

2.2 The Logic

The syntax of logical formulas is summarized below. This is a modal predicate logic with classical negation. As usual, many standard connectives are interdefinable; we take $\mathbf{T}$, $\neg$, $\vee$, $\Diamond$, $\forall$ as primitive, and $\mathbf{F}$, $\Rightarrow$, $\wedge$, $\Box$, $\exists$ as derived.

Logical Formulas

η	a name n or a variable x		
$\mathcal{A}, \mathcal{B}, \mathcal{C} ::=$	formulas	$\eta[\mathcal{A}]$	location
$\mathbf{T}$	true	$\mathcal{A}@\eta$	location adjunct
$\neg\mathcal{A}$	negation	$\eta®\mathcal{A}$	revelation
$\mathcal{A} \vee \mathcal{B}$	disjunction	$\mathcal{A}\oslash\eta$	revelation adjunct
$\mathbf{0}$	void	$\Diamond\mathcal{A}$	sometime modality
$\mathcal{A} \mid \mathcal{B}$	composition	$✧\mathcal{A}$	somewhere modality
$\mathcal{A} \triangleright \mathcal{B}$	composition adjunct	$\forall x.\mathcal{A}$	universal quantification

The meaning of the formulas will be given shortly in terms of a satisfaction relation. Informally, the first three formulas (true, negation, disjunction) give propositional logic. The next five (void, composition and its adjunct, location and its adjunct) describe tree-like structures of locations. Revelation and its adjunct are new to this paper, and are discussed in detail later. The two spatial and temporal modalities make assertions about states that may happen "further away" in space or time respectively. Quantified variables range only over names: these variables may appear in the location and revelation constructs, and their adjuncts.

The collections of free names, $fn(\mathcal{A})$, and free variables, $fv(\mathcal{A})$, of a formula $\mathcal{A}$ are defined along standard lines, keeping in mind that there are no name-binding constructs and just one variable-binding construct ($\forall x.\mathcal{A}$).

A formula $\mathcal{A}$ is closed if $fv(\mathcal{A}) = \emptyset$. Substitution $\mathcal{A}\{\eta\leftarrow\mu\}$ of a name or variable μ for another name or variable η in a formula $\mathcal{A}$, is defined in the usual way. We identify formulas up to renaming of bound variables, that is, we assume the identity $\forall x.\mathcal{A} = \forall y.\mathcal{A}\{x\leftarrow y\}$, where $y \notin fv(\mathcal{A})$. We often write $\eta[]$ for $\eta[\mathbf{0}]$, $\mathcal{A}^{\mathbf{F}}$ for $\mathcal{A}\triangleright\mathbf{F}$, and $\mathcal{A}^{\neg}$ for $\neg\mathcal{A}$.

2.3 Satisfaction

The satisfaction relation $P \vDash \mathcal{A}$ means that the process P satisfies the closed formula $\mathcal{A}$. The definition of satisfaction is based heavily on the structural congruence relation. The satisfaction relation is defined inductively in the following tables, where Π is the sort of processes, Φ is the sort of formulas, ϑ is the sort of variables, and Λ is the sort of names. We use similar syntax for logical connectives at the meta-level and object-level, but this is unambiguous.

The meaning of the temporal modality is given by reductions in the operational semantics of the ambient calculus. For the spatial modality, we need the following definitions. The relation $P{\downarrow}P'$ indicates that P contains P' within exactly one level of nesting. Then, $P{\downarrow}^*P'$ is the reflexive and transitive closure of the previous relation, indicating that P contains P' at some nesting level. Note that P' constitutes the entire contents of an enclosed ambient.

$$P{\downarrow}P' \text{ iff } \exists n, P''.\ P \equiv n[P'] \mid P''$$

${\downarrow}^*$ is the reflexive and transitive closure of ${\downarrow}$

Satisfaction

$\forall P\in\Pi.$	$P \vDash \mathbf{T}$	
$\forall P\in\Pi, \mathcal{A}\in\Phi.$	$P \vDash \neg\mathcal{A}$	$\triangleq \neg P \vDash \mathcal{A}$
$\forall P\in\Pi, \mathcal{A},\mathcal{B}\in\Phi.$	$P \vDash \mathcal{A}\vee\mathcal{B}$	$\triangleq P \vDash \mathcal{A} \vee P \vDash \mathcal{B}$
$\forall P\in\Pi.$	$P \vDash \mathbf{0}$	$\triangleq P \equiv \mathbf{0}$
$\forall P\in\Pi, \mathcal{A},\mathcal{B}\in\Phi.$	$P \vDash \mathcal{A} \mid \mathcal{B}$	$\triangleq \exists P',P''\in\Pi.\ P \equiv P' \mid P'' \wedge P' \vDash \mathcal{A} \wedge P'' \vDash \mathcal{B}$
$\forall P\in\Pi, \mathcal{A},\mathcal{B}\in\Phi.$	$P \vDash \mathcal{A}\triangleright\mathcal{B}$	$\triangleq \forall P'\in\Pi.\ P' \vDash \mathcal{A} \Rightarrow P \mid P' \vDash \mathcal{B}$
$\forall P\in\Pi, n\in\Lambda, \mathcal{A}\in\Phi.$	$P \vDash n[\mathcal{A}]$	$\triangleq \exists P'\in\Pi.\ P \equiv n[P'] \wedge P' \vDash \mathcal{A}$
$\forall P\in\Pi, \mathcal{A}\in\Phi.$	$P \vDash \mathcal{A}@n$	$\triangleq n[P] \vDash \mathcal{A}$
$\forall P\in\Pi, n\in\Lambda, \mathcal{A}\in\Phi.$	$P \vDash n®\mathcal{A}$	$\triangleq \exists P'\in\Pi.\ P \equiv (\nu n)P' \wedge P' \vDash \mathcal{A}$
$\forall P\in\Pi, \mathcal{A}\in\Phi.$	$P \vDash \mathcal{A}⊘n$	$\triangleq (\nu n)P \vDash \mathcal{A}$
$\forall P\in\Pi, \mathcal{A}\in\Phi.$	$P \vDash \Diamond\mathcal{A}$	$\triangleq \exists P'\in\Pi.\ P \rightarrow^* P' \wedge P' \vDash \mathcal{A}$
$\forall P\in\Pi, \mathcal{A}\in\Phi.$	$P \vDash ✧\mathcal{A}$	$\triangleq \exists P'\in\Pi.\ P{\downarrow}^*P' \wedge P' \vDash \mathcal{A}$
$\forall P\in\Pi, x\in\vartheta, \mathcal{A}\in\Phi.$	$P \vDash \forall x.\mathcal{A}$	$\triangleq \forall m\in\Lambda.\ P \vDash \mathcal{A}\{x\leftarrow m\}$

Again, all these logical connectives are described and discussed in [7], except for revelation and its adjunct, which are the subject of Section 3.

Remark: Given our policy of identifying formulas up to the renaming of bound variables, we need to check that satisfaction is well defined with respect to the equation $\forall x.\mathcal{A} = \forall y.\mathcal{A}\{x\leftarrow y\}$, where $y \notin fv(\mathcal{A})$. We need to show for all processes P, formulas $\mathcal{A}$, and variables x and y such that $y \notin fv(\mathcal{A})$ that $P \vDash \forall x.\mathcal{A}$ if and only if $P \vDash \forall y.\mathcal{A}\{x\leftarrow y\}$. By definition, $P \vDash \forall y.\mathcal{A}\{x\leftarrow y\}$ if and only if $\forall m\in\Lambda.\ P \vDash \mathcal{A}\{x\leftarrow y\}\{y\leftarrow m\}$. Since $y \notin fv(\mathcal{A})$, we have $\mathcal{A}\{x\leftarrow y\}\{y\leftarrow m\} = \mathcal{A}\{x\leftarrow m\}$. Therefore, $P \vDash \forall y.\mathcal{A}\{x\leftarrow y\}$ if and only if $\forall m\in\Lambda.\ P \vDash \mathcal{A}\{x\leftarrow m\}$ This is the definition of satisfaction for $\forall x.\mathcal{A}$. So it follows that $y \notin fv(\mathcal{A})$ implies that $P \vDash \forall x.\mathcal{A}$ if and only if $P \vDash \forall y.\mathcal{A}\{x\leftarrow y\}$. □

Fundamental Lemmas

The following lemmas are crucial in what follows.

2-1 Lemma (Satisfaction is up to ≡)

$(P \vDash \mathcal{A} \wedge P \equiv P') \Rightarrow P' \vDash \mathcal{A}$ □

2-2 Lemmas (Inversion)

(1) $P \equiv Q \;\Rightarrow\; fn(P) = fn(Q)$

(2) $(\nu n)P \equiv \mathbf{0} \;\Rightarrow\; P \equiv \mathbf{0}$

(3) $(\nu n)P \equiv m[Q] \;\Rightarrow\; \exists R \in \Pi.\ P \equiv m[R] \wedge Q \equiv (\nu n)R$ (for $n \neq m$)

(4) $(\nu n)P \equiv Q' \mid Q'' \;\Rightarrow\; \exists R', R'' \in \Pi.\ P \equiv R' \mid R'' \wedge Q' \equiv (\nu n)R' \wedge Q'' \equiv (\nu n)R''$ □

Remark. It is not true that $(\nu n)P \equiv (\nu n)Q$ implies $P \equiv Q$. Take $P = n[]$ and $Q = (\nu n)n[]$; then $(\nu n)n[] \equiv (\nu n)(\nu n)n[]$ but $n[] \not\equiv (\nu n)n[]$. □

2-3 Lemma (Fresh renaming preserves ⊨)

For all closed formulas $\mathcal{A}$, processes P, and names m, m',
if $m' \notin fn(P) \cup fn(\mathcal{A})$ then $P \vDash \mathcal{A} \Leftrightarrow P\{m \leftarrow m'\} \vDash \mathcal{A}\{m \leftarrow m'\}$. □

The proof of Lemma 2-1 is an induction on the structure of $\mathcal{A}$. See [8] for Lemmas 2-2. The proof of Lemma 2-3 is by induction on the number of symbols in the closed formula $\mathcal{A}$, which is unchanged by substituting a name for a variable or another name; this proof is an extension of the analogous one from [7] with cases for revelation and hiding. It is common for semantic properties of the π-calculus and its descendants to be preserved by fresh renaming; an early example is a fresh renaming lemma for strong bisimulation in the original article on the π-calculus [13, 9].

2.4 Validity

Valid Formulas, Sequents, and Rules

A closed formula is valid when it is satisfied by all processes. A general formula is valid when it is valid under any closed instantiation of its free variables with names.

More precisely, if $fv(\mathcal{A}) = \{x_1, ..., x_k\}$ are the free variables of $\mathcal{A}$ and $\varphi \in \vartheta \rightarrow \Lambda$ is a substitution of names for variables such that $dom(\varphi) \supseteq fv(\mathcal{A})$, then we write $\mathcal{A}_\varphi$ for $\mathcal{A}\{x_1 \leftarrow \varphi(x_1), ..., x_k \leftarrow \varphi(x_k)\}$, and we define:

Valid Formulas

$\boldsymbol{vld}(\mathcal{A})_\varphi \triangleq \forall P \in \Pi.\ P \vDash \mathcal{A}_\varphi$ for $\varphi \in \vartheta \rightarrow \Lambda$ with $dom(\varphi) \supseteq fv(\mathcal{A})$

$\boldsymbol{vld}(\mathcal{A}) \triangleq \forall \varphi \in fv(\mathcal{A}) \rightarrow \Lambda.\ \boldsymbol{vld}(\mathcal{A})_\varphi$

We use validity for interpreting logical inference rules, as described in the following tables. We use a linearized notation for inference rules, where the usual horizontal bar separating antecedents from consequents is written '$\}$' in-line, and ';' is used to separate antecedents.

Sequents are interpreted as follows. A simple sequent $\mathcal{A} \vdash \mathcal{B}$ is interpreted as the validity of the formula $\mathcal{A} \Rightarrow \mathcal{B}$. Sequents with conditions about disjointness of variables, or disjointness of variables from names, are reduced to simple sequents, as described below. Equality of names $\eta = \mu$ is definable in the logic as $\eta[\mathbf{T}]@\mu$ [7].

Sequents

$\mathcal{A} \vdash \mathcal{B} \triangleq \boldsymbol{vld}(\mathcal{A} \Rightarrow \mathcal{B})$

$\mathcal{A} \vdash \mathcal{B}\ (\eta_1 \neq \mu_1, ..., \eta_n \neq \mu_n) \triangleq (\eta_1 \neq \mu_1 \wedge ... \wedge \eta_n \neq \mu_n \wedge \mathcal{A}) \vdash \mathcal{B}$

$\mathcal{A} \dashv\vdash \mathcal{B}\ (\Xi) \triangleq (\mathcal{A} \vdash \mathcal{B}\ (\Xi)) \wedge (\mathcal{B} \vdash \mathcal{A}\ (\Xi))$ where $\Xi = \eta_1 \neq \mu_1, ..., \eta_n \neq \mu_n$

For example: $\mathcal{A} \vdash \mathcal{B}$ means $\forall \varphi \in fv(\mathcal{A} \Rightarrow \mathcal{B}) \rightarrow \Lambda.\ \forall P \in \Pi.\ P \vDash \mathcal{A}_\varphi \Rightarrow P \vDash \mathcal{B}_\varphi$.

Logical rules are interpreted as follows, where $\mathcal{S}$ are sequents (any of the three forms above, including sequents with side conditions and double sequents):

Rules

$\mathcal{S}_1; ...; \mathcal{S}_n \nmid \mathcal{S}_0 \triangleq (\mathcal{S}_1 \wedge ... \wedge \mathcal{S}_n) \Rightarrow \mathcal{S}_0$
$\mathcal{S}_1 \{\nmid \mathcal{S}_2 \triangleq \mathcal{S}_1 \nmid \mathcal{S}_2 \wedge \mathcal{S}_2 \nmid \mathcal{S}_1$

The definition of validity for formulas with free variables allows us to handle quantification over names. We obtain the validity of the following standard rules for the universal quantifier, and for the definable existential quantifier:

Quantification

(∀ L)	$\mathcal{A}\{x\leftarrow\eta\} \vdash \mathcal{B} \nmid \forall x.\mathcal{A} \vdash \mathcal{B}$	where η is a name or a variable
(∀ R)	$\mathcal{A} \vdash \mathcal{B} \nmid \mathcal{A} \vdash \forall x.\mathcal{B}$	where $x \notin fv(\mathcal{A})$
(∃ L)	$\mathcal{A} \vdash \mathcal{B} \nmid \exists x.\mathcal{A} \vdash \mathcal{B}$	where $x \notin fv(\mathcal{B})$
(∃ R)	$\mathcal{A} \vdash \mathcal{B}\{x\leftarrow\eta\} \nmid \mathcal{A} \vdash \exists x.\mathcal{B}$	where η is a name or a variable

***Remark:* (∀ R).** The distinction between variables and names in formulas, and the use of variables (as opposed to names) in quantification is crucial for (∀ R). The version of (∀ R) with names instead of variables:

$\mathcal{A} \vdash \mathcal{B} \nmid \mathcal{A} \vdash \forall n.\mathcal{B}$ where $n \notin fn(\mathcal{A})$,

is not sound. Consider the valid sequent $m[\mathbf{T}] \vdash \neg n[\mathbf{T}]$. If quantification binders were names, then the rule (∀ R) could be used to produce $m[\mathbf{T}] \vdash \forall n.\neg n[\mathbf{T}]$, which is not valid. Since quantification binders are variables, one can only deduce $m[\mathbf{T}] \vdash \forall x.\neg n[\mathbf{T}]$. □

***Remark:* (∀ L).** The use of substitutions that admit variables, in addition to names, in (∀ L), is crucial. Otherwise, if (∀ L) is formulated as $\mathcal{A}\{x\leftarrow m\} \vdash \mathcal{B} \nmid \forall x.\mathcal{A} \vdash \mathcal{B}$, there does not seem to be any way to derive, for example:

$\mathcal{A} \vdash \mathcal{B} \nmid \forall x.\mathcal{A} \vdash \forall x.\mathcal{B}$

which is obtained by starting from $\mathcal{A}\{x\leftarrow x\} \vdash \mathcal{B}$ and applying (∀ L) and then (∀ R). □

A number of proof principles can be derived from the definition of validity:

Instantiation Principle. Let $\mathcal{S}$ be a one-directional sequent, then, for any x,n:

(Inst) $\mathcal{S} \nmid \mathcal{S}\{x\leftarrow n\}$

Substitution Principle. Let $\mathcal{B}\{-\}$ be a formula with a set of formula holes, indicated by –, and let $\mathcal{B}\{\mathcal{A}\}$ denote the formula obtained by filling those holes with $\mathcal{A}$, after renaming the bound variables of $\mathcal{B}$ so they do not capture free variables of $\mathcal{A}$.

(Subst) $\mathcal{A}' \dashv\vdash \mathcal{A} \nmid \mathcal{B}\{\mathcal{A}'\} \dashv\vdash \mathcal{B}\{\mathcal{A}\}$

Case Analysis Principle. A case analysis principle is useful for proofs involving equality and inequality; inequalities often occur as side-conditions of primitive and derived rules. A predicate $\mathcal{A}$ is *classical* iff $\forall \varphi \in fv(\mathcal{A}) \rightarrow \Lambda.\ \{P \mid P \vDash \mathcal{A}_\varphi\} \in \{\Pi, \emptyset\}$. Note that $\mathbf{T}$, $\mathbf{F}$, and $\eta=\mu$ are classical predicates; so is $\mathcal{A}^{\mathbf{F}}$, for any $\mathcal{A}$ (meaning that $\mathcal{A}$ is unsatisfiable), and so is the conjunction, disjunction, and negation of classical predicates. Let $\mathcal{S}\{-\}$ be a one-directional sequent with a set of formula holes, and $\mathcal{A}$ be a classical predicate. Then:

(Case Analysis) $\mathcal{S}\{\mathbf{T}\}; \mathcal{S}\{\mathbf{F}\} \nmid \mathcal{S}\{\mathcal{A}\}$

3 Revelation

We now study the logical connectives $\eta \circledR \mathcal{A}$ (*revelation*), and $\mathcal{A} \oslash \eta$ (*revelation adjunct* or *hiding*). These connectives make assertions about restricted names that occur in processes.

3.1 Satisfaction

The formula $\eta \circledR \mathcal{A}$ is used to *reveal* a restricted name; it is read "reveal η then $\mathcal{A}$", where η is either a name (n) or the occurrence of a variable (x) that denotes a name. A process P satisfies the formula $n \circledR \mathcal{A}$ if it is possible to pull a restricted name occurring in P to the top and rename it n, and then strip off the restriction to leave a residual process that satisfies $\mathcal{A}$.

We cannot rename a top-level restricted name of P to n if n is already free in P. Therefore, a revelation formula provides a way of testing for the free names of the underlying process P, as we discuss below.

The inverse (technically, the adjunct) of revelation is called *hiding*: $\mathcal{A} \oslash \eta$, which is read "hide η then $\mathcal{A}$". A process P satisfies the formula $\mathcal{A} \oslash n$ if $(\nu n)P$ satisfies $\mathcal{A}$, that is, if it is possible to hide n in P and then satisfy $\mathcal{A}$. The satisfaction relation $P \models \mathcal{A}$ for revelation and hiding is repeated below:

Satisfaction for Revelation and Hiding

$$P \models n \circledR \mathcal{A} \triangleq \exists P' \in \Pi.\ P \equiv (\nu n)P' \wedge P' \models \mathcal{A}$$
$$P \models \mathcal{A} \oslash n \triangleq (\nu n)P \models \mathcal{A}$$

Here are some simple examples:

$(\nu n)n[] \models n \circledR \mathbf{T}$	because $n[] \models \mathbf{T}$
$\mathbf{0} \models n \circledR \mathbf{T}$	because $\mathbf{0} \equiv (\nu n)\mathbf{0}$ and $\mathbf{0} \models \mathbf{T}$
$\neg\, n[] \models n \circledR \mathbf{T}$	because there is no process $(\nu n)P' \equiv n[]$
$(\nu m)m[] \models n \circledR n[]$	because $(\nu m)m[] \equiv (\nu n)n[]$ and $n[] \models n[]$
$m[] \models (n \circledR n[]) \oslash m$	because $(\nu m)m[] \models n \circledR n[]$

Revelation gives us a way to talk about the free (or "known") names of a process. This can be embodied in a derived operator $©n$, satisfied by a process P iff $n \in fn(P)$. Several other derived connectives can be imagined:

Examples of Derived Connectives

$©\eta$	$\triangleq \neg \eta \circledR \mathbf{T}$	contains η free (it is not possible to reveal η)
closed	$\triangleq \neg \exists x. ©x$	no free names
separate	$\triangleq \neg \exists x. (©x \mid ©x)$	no shared free names
atmostfree η	$\triangleq$ *closed* $\oslash \eta$	contains at most η free

For clarity, we expand the definitions of these derived connectives:

Expanded Definitions

$\forall P \in \Pi.\ P \models ©n$	iff $\neg \exists P' \in \Pi.\ P \equiv (\nu n)P'$	iff $n \in fn(P)$
$\forall P \in \Pi.\ P \models$ *closed*	iff $\forall n \in \Lambda.\ \exists P' \in \Pi.\ P \equiv (\nu n)P'$	iff $fn(P) = \emptyset$
$\forall P \in \Pi.\ P \models$ *separate*	iff $\neg \exists n \in \Lambda.\ \exists P', P'' \in \Pi.\ P \equiv P' \mid P'' \wedge n \in fn(P') \wedge n \in fn(P'')$	
$\forall P \in \Pi.\ P \models$ *atmostfree* n	iff $\forall m \in \Lambda.\ \exists P' \in \Pi.\ (\nu n)P \equiv (\nu m)P'$	iff $fn(P) \subseteq \{n\}$

Examples:

$n[] \models ©n$	because $\neg \exists P' \in \Pi.\ n[] \equiv (\nu n)P'$
$(\nu m)m[] \models$ *closed*	because $\forall n \in \Lambda.\ (\nu m)m[] \equiv (\nu n)(\nu m)m[]$
$n[] \mid m[] \mid (\nu p)(p[] \mid p[]) \models$ *separate*	

3.2 Rules

Before giving our set of primitive rules of revelation and hiding, we discuss the most interesting properties of ® and ⊘ that are derived in this section. In order to emphasize some symmetries, we use here a combination of primitive and derived rules.

First, the cancellation and swapping properties of double restriction, $(\nu n)(\nu n)P \equiv (\nu n)P$ and $(\nu n)(\nu m)P \equiv (\nu m)(\nu n)P$, are inherited by both ® and ⊘:

$$n®n®\mathcal{A} \dashv\vdash n®\mathcal{A} \qquad n®m®\mathcal{A} \vdash m®n®\mathcal{A}$$
$$\mathcal{A}⊘n⊘n \dashv\vdash \mathcal{A}⊘n \qquad \mathcal{A}⊘m⊘n \vdash \mathcal{A}⊘n⊘m$$

Next, consider the combinations:

$$n®(\mathcal{A}⊘n) \qquad (n®\mathcal{A})⊘n$$

We see easily that $P \vDash n®(\mathcal{A}⊘n)$ means that $P \vDash \mathcal{A}$ and that $n \notin fn(P)$, where $n \notin fn(P)$ can be written also as $P \vDash n®\mathbf{T}$. Instead, $P \vDash (n®\mathcal{A})⊘n$ means that, although P may not satisfy $\mathcal{A}$, if we hide n in P we obtain something where we can reveal n and satisfy $\mathcal{A}$. For example, $(\nu m)n[m[]] \nvDash m®m[n[]]$, but $(\nu m)n[m[]] \vDash (n®m®m[n[]])⊘n$, because $(\nu n)(\nu m)n[m[]] \equiv (\nu n)(\nu m)m[n[]] \vDash n®m®m[n[]]$. In other words, $P \vDash (n®\mathcal{A})⊘n$ means that we can satisfy $\mathcal{A}$ by hiding the name n of P, and revealing a possibly different restricted name of P as n. We obtain the properties:

$$n®(\mathcal{A}⊘n) \dashv\vdash \mathcal{A} \wedge n®\mathbf{T}$$
$$n®(\mathcal{A}⊘n) \vdash \mathcal{A} \qquad \mathcal{A} \vdash (n®\mathcal{A})⊘n$$
$$n®(\mathcal{A}⊘n) \vdash \mathcal{A}⊘n \qquad \mathcal{A}⊘n \vdash (n®\mathcal{A})⊘n$$
$$n®(\mathcal{A}⊘n) \vdash n®\mathcal{A} \qquad n®\mathcal{A} \vdash (n®\mathcal{A})⊘n$$

The interactions of ® and ⊘ with | are the most interesting, and the most complex. There are basically three distribution rules: distribution of ® over | in both directions (with a constraint), unrestricted distribution of ⊘ over | in one direction, and distribution of $n®((-)⊘n)$ over | in both directions.

$$n®(\mathcal{A} \mid n®\mathcal{B}) \dashv\vdash n®\mathcal{A} \mid n®\mathcal{B}$$
$$(\mathcal{A} \mid \mathcal{B})⊘n \vdash \mathcal{A}⊘n \mid \mathcal{B}⊘n$$
$$n®((\mathcal{A} \mid \mathcal{B})⊘n) \dashv\vdash n®(\mathcal{A}⊘n) \mid n®(\mathcal{B}⊘n)$$

The first rule embodies the scope extrusion rule, $(\nu n)(P \mid Q) \equiv ((\nu n)P) \mid Q$ if $n \notin fn(Q)$. This can be seen more clearly if we note that $n \notin fn(Q)$ is equivalent to $Q \equiv (\nu n)Q$; then the extrusion rule can be written as $(\nu n)(P \mid (\nu n)Q) \equiv ((\nu n)P) \mid ((\nu n)Q)$ with no side condition.

The second rule implies that if $(\nu n)(P \mid Q) \vDash \mathcal{A} \mid \mathcal{B}$ then it is possible to distribute the restriction so that $(\nu n)P \vDash \mathcal{A}$ and $(\nu n)Q \vDash \mathcal{B}$; this is a consequence of Lemma 2-2(4).

The last rule looks mysterious, but has a simple interpretation. According to one of the equivalences above, it can be rewritten as $(\mathcal{A} \mid \mathcal{B}) \wedge n®\mathbf{T} \dashv\vdash (\mathcal{A} \wedge n®\mathbf{T}) \mid (\mathcal{B} \wedge n®\mathbf{T})$; that is, the name n does not occur in a parallel composition iff it does not occur in either component. The right-to-left direction is actually a derivable rule.

A similar set of rules holds for distribution of ® and ⊘ over $n[-]$:

$$n®m[\mathcal{A}] \dashv\vdash m[n®\mathcal{A}] \qquad (n \neq m)$$
$$m[\mathcal{A}]⊘n \dashv\vdash m[\mathcal{A}⊘n] \qquad (n \neq m)$$
$$n[\mathcal{A}]⊘n \dashv\vdash \mathbf{F}$$
$$n®(m[\mathcal{A}]⊘n) \dashv\vdash m[n®(\mathcal{A}⊘n)] \qquad (n \neq m)$$

The distribution of $n®-$ over $m[-]$ (first rule) holds in both directions as long as $n \neq m$.

The distribution of $-⊘n$ over $m[-]$ (second and third rules) comes in two cases, depending on whether $n{=}m$. In each case, the right-to-left direction is derivable. From $n[\mathbf{T}]⊘n \vdash \mathbf{F}$ we can derive $n®\mathbf{T} \vdash \neg n[\mathbf{T}]$, which means that if a name n does not occur free in a process, the process cannot be a location named n.

The distribution of $n\circledR((-)\oslash n)$ over $m[-]$ (fourth rule) is derivable in both directions, from the first two rules. Again, this rule can be rewritten as $m[\mathcal{A}] \wedge n\circledR\mathbf{T} \dashv\vdash m[\mathcal{A} \wedge n\circledR\mathbf{T}]$ $(n \neq m)$; that is, the name n does not occur in a location iff it is distinct from the name of the location and it does not occur inside the location.

Finally, $\circledR$ and $\oslash$ commute in one direction:

$$m\circledR(\mathcal{A}\oslash n) \vdash (m\circledR\mathcal{A})\oslash n$$

We now take the following set of rules as primitive (i.e., we verify their validity in the model). The first group handles double revelation, distribution of $\circledR$ over $\vee$, congruence of $\circledR$ with $\vdash$, the adjunction rule connecting $\circledR$ and $\oslash$, and the rather curious but very useful fact that $\neg$ commutes with $\oslash$. The other three groups deal with the interactions of $\circledR$ and $\oslash$ with $\mathbf{0}$, $\mid$, and $n[-]$.

3-1 Proposition (Validity: Revelation Rules)

$(\circledR)$	$\} \; x\circledR x\circledR\mathcal{A} \dashv\vdash x\circledR\mathcal{A}$	$(\circledR \mid)$	$\} \; x\circledR(\mathcal{A} \mid x\circledR\mathcal{B}) \dashv\vdash x\circledR\mathcal{A} \mid x\circledR\mathcal{B}$
$(\circledR\,\circledR)$	$\} \; x\circledR y\circledR\mathcal{A} \vdash y\circledR x\circledR\mathcal{A}$	$(\oslash \mid)$	$\} \; (\mathcal{A} \mid \mathcal{B})\oslash x \vdash \mathcal{A}\oslash x \mid \mathcal{B}\oslash x$
$(\circledR \vee)$	$\} \; x\circledR(\mathcal{A}\vee\mathcal{B}) \vdash x\circledR\mathcal{A} \vee x\circledR\mathcal{A}$	$(\circledR \oslash \mid)$	$\} \; x\circledR((\mathcal{A} \mid \mathcal{B})\oslash x) \vdash x\circledR(\mathcal{A}\oslash x) \mid x\circledR(\mathcal{B}\oslash x)$
$(\circledR \vdash)$	$\mathcal{A} \vdash \mathcal{B} \; \} \; x\circledR\mathcal{A} \vdash x\circledR\mathcal{B}$		
$(\circledR \oslash)$	$\eta\circledR\mathcal{A} \vdash \mathcal{B} \; \{\} \; \mathcal{A} \vdash \mathcal{B}\oslash\eta$	$(\circledR\, n[])$	$\} \; x\circledR y[\mathcal{A}] \dashv\vdash y[x\circledR\mathcal{A}] \quad (x \neq y)$
$(\oslash \neg)$	$\} \; (\neg\mathcal{A})\oslash x \dashv\vdash \neg(\mathcal{A}\oslash x)$	$(\oslash\, n[])$	$\} \; y[\mathcal{A}]\oslash x \vdash y[\mathcal{A}\oslash x] \quad (x \neq y)$
$(\oslash \triangleright \mathbf{F})$	$\} \; \mathcal{A}^{\mathbf{F}}\oslash x \dashv\vdash \mathcal{A}^{\mathbf{F}}$	$(\oslash\, n[])$	$\} \; x[\mathcal{A}]\oslash x \vdash \mathbf{F}$
$(\circledR\, \mathbf{0})$	$\} \; x\circledR\mathbf{0} \dashv\vdash \mathbf{0}$		
$(\oslash\, \mathbf{0})$	$\} \; \mathbf{0}\oslash x \vdash \mathbf{0}$ □		

From the rules that we have validated in Proposition 3-1, we can derive a large collection of facts by logical deduction, including the following:

3-2 Logical Corollaries (Case Analysis)

Let Cl be a classical predicate (typically, Cl is a side condition of the form $x \neq y$).

(CA $\mid \wedge$**)**	$\} \; (Cl \wedge \mathcal{A}) \mid (Cl \wedge \mathcal{B}) \dashv\vdash Cl \wedge (\mathcal{A} \mid \mathcal{B})$	
(CA $\mid \Rightarrow$**)**	$\} \; (Cl \Rightarrow \mathcal{A}) \mid (Cl \Rightarrow \mathcal{B}) \vdash Cl \Rightarrow (\mathcal{A} \mid \mathcal{B})$	
(CA $n[] \wedge$**)**	$\} \; z[Cl \wedge \mathcal{A}] \dashv\vdash Cl \wedge z[\mathcal{A}]$	where z may occur in Cl
(CA $\circledR \wedge$**)**	$\} \; z\circledR(Cl \wedge \mathcal{A}) \dashv\vdash Cl \wedge z\circledR\mathcal{A}$	where z may occur in Cl □

3-3 Logical Corollaries (Revelation)

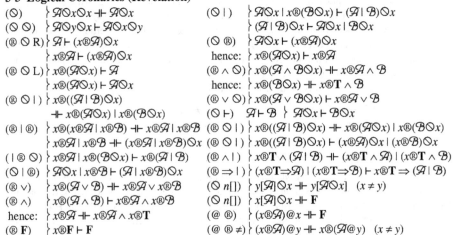

$(\oslash)$	$\} \; \mathcal{A}\oslash x\oslash x \dashv\vdash \mathcal{A}\oslash x$	$(\oslash \mid)$	$\} \; \mathcal{A}\oslash x \mid x\circledR(\mathcal{B}\oslash x) \vdash (\mathcal{A} \mid \mathcal{B})\oslash x$
$(\oslash\,\oslash)$	$\} \; \mathcal{A}\oslash y\oslash x \vdash \mathcal{A}\oslash x\oslash y$		$\} \; (\mathcal{A} \mid \mathcal{B})\oslash x \vdash \mathcal{A}\oslash x \mid \mathcal{B}\oslash x$
$(\circledR \oslash \mathrm{R})$	$\} \; \mathcal{A} \vdash (x\circledR\mathcal{A})\oslash x$	$(\oslash \circledR)$	$\} \; \mathcal{A}\oslash x \vdash (x\circledR\mathcal{A})\oslash x$
	$\} \; x\circledR\mathcal{A} \vdash (x\circledR\mathcal{A})\oslash x$	hence:	$\} \; x\circledR(\mathcal{A}\oslash x) \vdash x\circledR\mathcal{A}$
$(\circledR \oslash \mathrm{L})$	$\} \; x\circledR(\mathcal{A}\oslash x) \vdash \mathcal{A}$	$(\circledR \wedge \oslash)$	$\} \; x\circledR(\mathcal{A} \wedge \mathcal{B}\oslash x) \dashv\vdash x\circledR\mathcal{A} \wedge \mathcal{B}$
	$\} \; x\circledR(\mathcal{A}\oslash x) \vdash \mathcal{A}\oslash x$	hence:	$\} \; x\circledR(\mathcal{B}\oslash x) \dashv\vdash x\circledR\mathbf{T} \wedge \mathcal{B}$
$(\circledR \oslash \mid)$	$\} \; x\circledR((\mathcal{A} \mid \mathcal{B})\oslash x)$	$(\circledR \vee \oslash)$	$\} \; x\circledR(\mathcal{A} \vee \mathcal{B}\oslash x) \vdash x\circledR\mathcal{A} \vee \mathcal{B}$
	$\dashv\vdash x\circledR(\mathcal{A}\oslash x) \mid x\circledR(\mathcal{B}\oslash x)$	$(\oslash \vdash)$	$\mathcal{A} \vdash \mathcal{B} \; \} \; \mathcal{A}\oslash x \vdash \mathcal{B}\oslash x$
$(\circledR \mid \circledR)$	$\} \; x\circledR(x\circledR\mathcal{A} \mid x\circledR\mathcal{B}) \dashv\vdash x\circledR\mathcal{A} \mid x\circledR\mathcal{B}$	$(\circledR \oslash \mid)$	$\} \; x\circledR((\mathcal{A} \mid \mathcal{B})\oslash x) \dashv\vdash x\circledR(\mathcal{A}\oslash x) \mid x\circledR(\mathcal{B}\oslash x)$
	$\} \; x\circledR\mathcal{A} \mid x\circledR\mathcal{B} \dashv\vdash (x\circledR\mathcal{A} \mid x\circledR\mathcal{B})\oslash x$	$(\circledR \oslash \mid)$	$\} \; x\circledR((\mathcal{A} \mid \mathcal{B})\oslash x) \vdash (x\circledR\mathcal{A})\oslash x \mid (x\circledR\mathcal{B})\oslash x$
$(\mid \circledR \oslash)$	$\} \; x\circledR\mathcal{A} \mid x\circledR(\mathcal{B}\oslash x) \vdash x\circledR(\mathcal{A} \mid \mathcal{B})$	$(\circledR \wedge \mid)$	$\} \; x\circledR\mathbf{T} \wedge (\mathcal{A} \mid \mathcal{B}) \dashv\vdash (x\circledR\mathbf{T} \wedge \mathcal{A}) \mid (x\circledR\mathbf{T} \wedge \mathcal{B})$
$(\oslash \mid \circledR)$	$\} \; \mathcal{A}\oslash x \mid x\circledR\mathcal{B} \vdash (\mathcal{A} \mid x\circledR\mathcal{B})\oslash x$	$(\circledR \Rightarrow \mid)$	$\} \; (x\circledR\mathbf{T} \Rightarrow \mathcal{A}) \mid (x\circledR\mathbf{T} \Rightarrow \mathcal{B}) \vdash x\circledR\mathbf{T} \Rightarrow (\mathcal{A} \mid \mathcal{B})$
$(\circledR \vee)$	$\} \; x\circledR(\mathcal{A} \vee \mathcal{B}) \dashv\vdash x\circledR\mathcal{A} \vee x\circledR\mathcal{B}$	$(\oslash\, n[])$	$\} \; y[\mathcal{A}]\oslash x \dashv\vdash y[\mathcal{A}\oslash x] \quad (x \neq y)$
$(\circledR \wedge)$	$\} \; x\circledR(\mathcal{A} \wedge \mathcal{B}) \vdash x\circledR\mathcal{A} \wedge x\circledR\mathcal{B}$	$(\oslash\, n[])$	$\} \; x[\mathcal{A}]\oslash x \dashv\vdash \mathbf{F}$
hence:	$\} \; x\circledR\mathcal{A} \dashv\vdash x\circledR\mathcal{A} \wedge x\circledR\mathbf{T}$	$(@\, \circledR)$	$\} \; (x\circledR\mathcal{A})@x \dashv\vdash \mathbf{F}$
$(\circledR\, \mathbf{F})$	$\} \; x\circledR\mathbf{F} \vdash \mathbf{F}$	$(@\, \circledR \neq)$	$\} \; (x\circledR\mathcal{A})@y \dashv\vdash x\circledR(\mathcal{A}@y) \quad (x \neq y)$

($\oslash$ T)	$\}$ $\mathbf{T} \dashv\vdash \mathbf{T}\oslash x$	(® $\oslash$ $n[]$)	$\}$ $x®(y[\mathcal{A}]\oslash x) \dashv\vdash y[x®(\mathcal{A}\oslash x)]$ $(x \neq y)$
($\oslash$ F)	$\}$ $\mathbf{F}\oslash x \dashv\vdash \mathbf{F}$	(® $\wedge$ $n[]$)	$\}$ $x®\mathbf{T} \wedge y[\mathcal{A}] \dashv\vdash y[x®\mathbf{T} \wedge \mathcal{A}]$ $(x \neq y)$
($\oslash$ $\vee$)	$\}$ $(\mathcal{A} \vee \mathcal{B})\oslash x \dashv\vdash \mathcal{A}\oslash x \vee \mathcal{B}\oslash x$	($\oslash$ ® $\neq$)	$\}$ $x®(\mathcal{A}\oslash y) \vdash (x®\mathcal{A})\oslash y$
($\oslash$ $\wedge$)	$\}$ $(\mathcal{A} \wedge \mathcal{B})\oslash x \dashv\vdash \mathcal{A}\oslash x \wedge \mathcal{B}\oslash x$	(® $\exists$)	$\}$ $\exists x.y®\mathcal{A} \dashv\vdash y®\exists x.\mathcal{A}$ where $x \neq y$
($\oslash$ 0)	$\}$ $\mathbf{0} \vdash \mathbf{0}\oslash x$	(® $\forall$)	$\}$ $y®\forall x.\mathcal{A} \vdash \forall x.y®\mathcal{A}$ where $x \neq y$ □
(® $\neg$0)	$\}$ $x®\neg\mathbf{0} \vdash \neg\mathbf{0}$		

Remark. The derived rule (® $\neg$**0**) says that if we reveal a restricted name and find non-**0**, then the original process is also non-**0**. That is, non-**0**-ness cannot be hidden by restriction. Consider, for example, the process $P = (\nu n)n[]$. Under many standard behavioral equivalences $\approx$ we have $P \approx \mathbf{0}$ [11]. However, we have $P \vDash n®\neg\mathbf{0}$, and hence by (® $\neg$**0**), we have that $P \vDash \neg\mathbf{0}$. This example shows quite clearly that our logic is finer than standard behavioral equivalences, and that it can inspect the structure of restricted processes. □

4 Fresh-Name Quantifier

In this section we define a formula, $\text{И}x.\mathcal{A}$, with the meaning "for fresh x, $\mathcal{A}$ holds". Here, "fresh" means, informally, distinct from any name that might clash with an existing name.

The set of free (i.e., non-fresh) names that occur in a process or formula is always finite; hence sets of fresh names are always cofinite. (A cofinite set is the complement of a finite set with respect to an infinite universe, which, in our case, is the countable universe of names Λ.) If there is a suitable fresh x, then there are infinitely many of them, since a fresh name can be replaced by any other fresh name. Therefore, "freshness" can be expressed formally as the existence of a cofinite set of interchangeable names [10]. We use $Fin(S)$ for the collection of finite subsets of a set S.

4.1 The Gabbay-Pitts Property

We would like to obtain the following property for $\text{И}x.\mathcal{A}$:

$$P \vDash \text{И}x.\mathcal{A} \quad \Leftrightarrow \quad \exists m \in \Lambda.\ m \notin fn(P,\mathcal{A}) \wedge P \vDash \mathcal{A}\{x \leftarrow m\}$$

That is, $P \vDash \text{И}x.\mathcal{A}$ iff there exists a fresh name m such that $P \vDash \mathcal{A}\{x \leftarrow m\}$.

This definition is given by existential quantification over fresh names. Remarkably, there is an equivalent definition based on universal quantification. The equivalence of these two definitions is based on a deep property of the logic (Lemma 2-3), and will be used to great effect later. We state the equivalence as follows: there exists a fresh name m such that $P \vDash \mathcal{A}\{x \leftarrow m\}$, if and only if for all fresh names m we have $P \vDash \mathcal{A}\{x \leftarrow m\}$:

4-1 Proposition (Gabbay-Pitts Property)

$\forall P \in \Pi, \mathcal{A} \in \Phi, N \in Fin(\Lambda).$

$N \supseteq fn(P,\mathcal{A}) \wedge fv(\mathcal{A}) \subseteq \{x\} \Rightarrow$

$(\exists m \in \Lambda.\ m \notin N \wedge P \vDash \mathcal{A}\{x \leftarrow m\}) \quad \Leftrightarrow \quad (\forall m \in \Lambda.\ m \notin N \Rightarrow P \vDash \mathcal{A}\{x \leftarrow m\})$

Proof

Assume $N \supseteq fn(P,\mathcal{A})$ and $fv(\mathcal{A}) \subseteq \{x\}$.

Case $\Leftarrow$) Assume $\forall m \in \Lambda.\ m \notin N \Rightarrow P \vDash \mathcal{A}\{x \leftarrow m\}$. Since N is finite and Λ is infinite, there is a $p \in \Lambda$ such that $p \notin N$. Then, by assumption, $P \vDash \mathcal{A}\{x \leftarrow p\}$. We have shown $(\exists p \in \Lambda.\ p \notin N \wedge P \vDash \mathcal{A}\{x \leftarrow p\})$.

Case $\Rightarrow$) Assume $\exists m \in \Lambda.\ m \notin N \wedge P \vDash \mathcal{A}\{x \leftarrow m\}$; in particular, $m \notin fn(P,\mathcal{A})$. Take any $p \in \Lambda$ and assume $p \notin N$. If $p=m$ we have by assumption that $P \vDash \mathcal{A}\{x \leftarrow p\}$. Otherwise, if $p \neq m$ then $p \notin N \cup \{m\}$; since $fn(P,\mathcal{A}\{x \leftarrow m\}) \subseteq N \cup \{m\}$, we have that $p \notin fn(P,\mathcal{A}\{x \leftarrow m\})$. By applying Lemma 2-3 to the assumption $P \vDash \mathcal{A}\{x \leftarrow m\}$ we obtain $P\{m \leftarrow p\} \vDash \mathcal{A}\{x \leftarrow m\}\{m \leftarrow p\}$; that is, $P \vDash \mathcal{A}\{x \leftarrow p\}$. In both cases, we have shown that $(\forall p \in \Lambda.\ p \notin N \Rightarrow P \vDash \mathcal{A}\{x \leftarrow p\})$. □

4.2 A Gabbay-Pitts Logical Rule

We now want to formulate a Gabbay-Pitts property similar to Proposition 4-1, but expressible within the logic. We are going to use extensively the idiom $x\#N \wedge x®\mathbf{T}$, for a quantified variable x. The first part of this conjunction says that the name x is fresh with respect to a given set of names N that usually includes the set of free names of a formula of interest. The second part says that x is fresh in the "underlying process", because $P \models n®\mathbf{T}$ iff $n \notin fn(P)$. For a suitable choice of N, the whole conjunction can be understood as saying that x is "completely fresh", both at the formula and process level, in a given situation.

Notation

- For $N \in Fin(\Lambda \cup \vartheta)$ we define the formula $\eta\#N \triangleq \bigwedge_{\mu \in N}(\eta \neq \mu)$.
  For any P and closed $m\#N$, we have $P \models m\#N$ iff $m \notin N$.
- Let $fnv(\mathcal{A}) \triangleq fn(\mathcal{A}) \cup fv(\mathcal{A})$, so that $fnv(\mathcal{A}) \in Fin(\Lambda \cup \vartheta)$

With this understanding, the following proposition states the single rule (schema) that we add to our logic in order to capture "freshness", and establishes its soundness. Note that this rule holds for open formulas.

4-2 Proposition (Validity: Gabbay-Pitts)

(GP) $\succ \exists x.\ x\#N \wedge x®\mathbf{T} \wedge \mathcal{A} \dashv\vdash \forall x.\ (x\#N \wedge x®\mathbf{T}) \Rightarrow \mathcal{A}$
where $N \in Fin(\Lambda \cup \vartheta)$ and $N \supseteq fnv(\mathcal{A})\text{-}\{x\}$ and $x \notin N$

Proof

Assume $N \supseteq fnv(\mathcal{A})\text{-}\{x\}$ and $x \notin N$. We need to show that the sequent is valid, that is that $\forall \varphi \in (fv(\mathcal{A})\text{-}\{x\}) \rightarrow \Lambda,\ P \in \Pi.\ P \models (\exists x.\ x\#N \wedge x®\mathbf{T} \wedge \mathcal{A})_\varphi \Leftrightarrow P \models (\forall x.\ x\#N \wedge x®\mathbf{T} \Rightarrow \mathcal{A})_\varphi$.

(1) $\succ \exists x.\ x\#N \wedge x®\mathbf{T} \wedge \mathcal{A} \vdash \forall x.\ x\#N \wedge x®\mathbf{T} \Rightarrow \mathcal{A}$

Take any $\varphi \in (fv(\mathcal{A})\text{-}\{x\}) \rightarrow \Lambda$ and $P \in \Pi$, and assume $P \models (\exists x.\ x\#N \wedge x®\mathbf{T} \wedge \mathcal{A})_\varphi$. That is, assume $\exists m \in \Lambda.\ m \notin N_\varphi \cup fn(P) \wedge P \models \mathcal{A}_\varphi\{x \leftarrow m\}$, where $N_\varphi \cup fn(P) \supseteq fn(P, \mathcal{A}_\varphi)$ and $fv(\mathcal{A}_\varphi) \subseteq \{x\}$. By Proposition 4-1, we obtain $\forall m \in \Lambda.\ m \notin N_\varphi \cup fn(P) \Rightarrow P \models \mathcal{A}_\varphi\{x \leftarrow m\}$, that is $P \models (\forall x.\ x\#N \wedge x®\mathbf{T} \Rightarrow \mathcal{A})_\varphi$.

(2) $\succ \forall x.\ x\#N \wedge x®\mathbf{T} \Rightarrow \mathcal{A} \vdash \exists x.\ x\#N \wedge x®\mathbf{T} \wedge \mathcal{A}$

Take any $\varphi \in (fv(\mathcal{A})\text{-}\{x\}) \rightarrow \Lambda$ and $P \in \Pi$ and assume $P \models (\forall x.\ x\#N \wedge x®\mathbf{T} \Rightarrow \mathcal{A})_\varphi$; that is assume $(\forall m \in \Lambda.\ m \notin N_\varphi \cup fn(P) \Rightarrow P \models \mathcal{A}_\varphi\{x \leftarrow m\})$, where $N_\varphi \cup fn(P) \supseteq fn(P, \mathcal{A}_\varphi)$ and $fv(\mathcal{A}_\varphi) \subseteq \{x\}$. By Proposition 4-1, we obtain $\exists m \in \Lambda.\ m \notin N_\varphi \cup fn(P) \wedge P \models \mathcal{A}_\varphi\{x \leftarrow m\}$, that is $P \models (\exists x.\ x\#N \wedge x®\mathbf{T} \wedge \mathcal{A})_\varphi$. □

Remark. (GP) gives us a way to prove that $\forall x.\mathcal{A} \vdash \exists x.\mathcal{A}$. This depends on the fact that the set of names is non-empty, and is obviously not derivable from the normal quantifier rules. Take $N = fnv(\mathcal{A})\text{-}\{x\}$. Starting from $\mathcal{A} \vdash \mathcal{A}$, by right weakening and quantifier introduction we obtain $\forall x.\ \mathcal{A} \vdash \forall x.\ x\#N \wedge x®\mathbf{T} \Rightarrow \mathcal{A}$. Again starting from $\mathcal{A} \vdash \mathcal{A}$, by left weakening and quantifier introduction we obtain $\exists x.\ x\#N \wedge x®\mathbf{T} \wedge \mathcal{A} \vdash \exists x.\ \mathcal{A}$. By (GP) we have $\forall x.\ x\#N \wedge x®\mathbf{T} \Rightarrow \mathcal{A} \vdash \exists x.\ x\#N \wedge x®\mathbf{T} \wedge \mathcal{A}$. Hence, by transitivity we obtain $\forall x.\ \mathcal{A} \vdash \exists x.\ \mathcal{A}$. □

4.3 Fresh-Name Quantifier

Without extending the syntax of our logic, we can define quantification over fresh names, $\text{И}x.\mathcal{A}$, as follows:

4-3 Definition (Fresh-Name Quantifier)

$\text{И}x.\mathcal{A} \triangleq \exists x.\ x\#(fnv(\mathcal{A})\text{-}\{x\}) \wedge x®\mathbf{T} \wedge \mathcal{A}$ □

Hence $fn(\text{И}x.\mathcal{A}) = fn(\mathcal{A})$ and $fv(\text{И}x.\mathcal{A}) = fv(\mathcal{A})\text{-}\{x\}$.

Note that the right-hand side of this definition depends on the set of free names and variables of $\mathcal{A}$. Therefore, this is not a definition within the logic, but rather a meta-theoretical definition (or abbreviation) that should always be understood in its expanded form. Any general theorem or derived rule involving $И x.\mathcal{A}$ will in fact be a schematic theorem or rule with respect to the free names and variables of $\mathcal{A}$, in the same way that (GP) is a rule schema.

By (GP) (Proposition 4-2) we have:

$$И x.\mathcal{A} \dashv\vdash \forall x.\ x\#fnv(\mathcal{A})\text{-}\{x\} \wedge x®\mathbf{T} \Rightarrow \mathcal{A}$$

In terms of satisfaction, we obtain:

4-4 Lemma ($P \vDash И x.\mathcal{A}$)

$P \vDash И x.\mathcal{A}$

iff $\exists m \in \Lambda.\ m \notin fn(P,\mathcal{A}) \wedge P \vDash \mathcal{A}\{x \leftarrow m\}$

iff $\forall m \in \Lambda.\ m \notin fn(P,\mathcal{A}) \Rightarrow P \vDash \mathcal{A}\{x \leftarrow m\}$ □

Therefore, $И x.\mathcal{A}$ can be understood as saying either that there is a fresh name x such that $\mathcal{A}$ holds, or that for any fresh name x we have that $\mathcal{A}$ holds. These formulations are equivalent because of the cofinite nature of sets of fresh names. If there is a suitably fresh x such that $\mathcal{A}$ holds, then any other fresh name will work equally well, so all fresh names will work. Conversely, if for all suitably fresh names $\mathcal{A}$ holds, since any set of fresh names is (cofinite and hence) non-empty, there exists a fresh name for which $\mathcal{A}$ holds.

Remark. The meaning of $И x.\mathcal{A}$ when $\mathcal{A}$ has free variables other than x is subtle. When we write $И x.\ ...n...$ we intend x to be fresh w.r.t. any existing name, and in particular n; similarly, when we write $И x.\ ...y...$ we intend x to be fresh with respect to any name denoted by y. Consider $И y.\ И x.\ y{=}x$; this formula should not be valid. In fact, it is contradictory because, by definition, it means, $\exists y.\ y®\mathbf{T} \wedge \exists x.\ x{\neq}y \wedge x®\mathbf{T} \wedge y{=}x$. Similarly, $\forall y.\ И x.\ y{=}x$ and $\exists y.\ И x.\ y{=}x$ are contradictory. (Instead, $И x.\ \exists y.\ x{=}y$ is valid.) □

The following rules are now derivable entirely within the logic:

4-5 Logical Corollaries (Fresh-Name Quantifier)

($И\exists$)	$\} \ И x.\mathcal{A} \dashv\vdash \exists x.\ x\#N \wedge x®\mathbf{T} \wedge \mathcal{A}$	where $N \supseteq fnv(\mathcal{A})\text{-}\{x\}$ and $x \notin N$
($И\forall$)	$\} \ \forall x.\ x\#N \wedge x®\mathbf{T} \Rightarrow \mathcal{A} \dashv\vdash И x.\mathcal{A}$	where $N \supseteq fnv(\mathcal{A})\text{-}\{x\}$ and $x \notin N$
($И\neg$)	$\} \ \neg И x.\mathcal{A} \dashv\vdash И x.\neg\mathcal{A}$	
($И \mid$)	$\} \ И x.(\mathcal{A} \mid \mathcal{B}) \dashv\vdash (И x.\mathcal{A}) \mid (И x.\mathcal{B})$	
($И\vdash$)	$\mathcal{A} \vdash \mathcal{B} \ \} \ И x.\mathcal{A} \vdash И x.\mathcal{B}$	
($И fv$)	$\} \ И x.\mathcal{A} \dashv\vdash \mathcal{A}$	where $x \notin fv(\mathcal{A})$
($И n[]$)	$\} \ И x.y[\mathcal{A}] \dashv\vdash y[И x.\mathcal{A}]$	where $x \neq y$
($И$R)	$\mathcal{A} \wedge x\#N \wedge x®\mathbf{T} \vdash \mathcal{B} \ \} \ \mathcal{A} \vdash И x.\mathcal{B}$	where $N \supseteq fnv(\mathcal{B})\text{-}\{x\}$ and $x \notin N \cup fv(\mathcal{A})$
($И$L)	$\mathcal{A} \wedge x\#N \wedge x®\mathbf{T} \vdash \mathcal{B} \ \} \ И x.\mathcal{A} \vdash \mathcal{B}$	where $N \supseteq fnv(\mathcal{A})\text{-}\{x\}$ and $x \notin N \cup fv(\mathcal{B})$
($И$E)	$\mathcal{A} \vdash И x.\mathcal{B};\ \mathcal{B} \wedge x\#N \wedge x®\mathbf{T} \vdash \mathcal{C} \ \} \ \mathcal{A} \vdash \mathcal{C}$	where $N \supseteq fnv(\mathcal{B})\text{-}\{x\}$ and $x \notin N \cup fv(\mathcal{C})$ □

Remark. Of particular interest (and difficulty) is the distribution of $И$ over $\mid$, rule ($И \mid$):

$$\} \ И x.(\mathcal{A} \mid \mathcal{B}) \dashv\vdash (И x.\mathcal{A}) \mid (И x.\mathcal{B})$$

Distribution over $\mid$ holds in one direction for universal quantification, in the other direction for existential quantification, and in both directions for fresh-name quantification. This rule can be understood informally as follows (this is a sketch of the formal derivation). In the left-to-right direction we use the existential interpretation of $И$. Take any P; if $P \vDash И x.(\mathcal{A} \mid \mathcal{B})$ then there are a fresh name x and processes P', P'' such that $P \equiv P' \mid P''$ and $P' \vDash \mathcal{A}$ and $P'' \vDash \mathcal{B}$. Hence, there is a fresh name x such that $P' \vDash \mathcal{A}$ and again a fresh name x such that $P'' \vDash \mathcal{B}$; that is, $P' \vDash И x.\mathcal{A}$ and $P'' \vDash И x.\mathcal{B}$. Therefore, $P \equiv P' \mid P'' \vDash (И x.\mathcal{A}) \mid (И x.\mathcal{B})$. In the right-to-left direction we use the universal interpretation of $И$. Take any P; if $P \vDash (И x.\mathcal{A}) \mid (И x.\mathcal{B})$ then there are processes P', P'' such that $P \equiv P' \mid P''$ and $P' \vDash И x.\mathcal{A}$ and $P'' \vDash И x.\mathcal{B}$. This means that for all names x' fresh in P' and $\mathcal{A}$, we have $P' \vDash \mathcal{A}\{x \leftarrow x'\}$ and for all names x'' fresh in P'' and $\mathcal{B}$, we have $P'' \vDash \mathcal{B}\{x \leftarrow x''\}$.

Now, for all names y that are fresh in $P', \mathcal{A}, P'', \mathcal{B}$; we have that $P' \vDash \mathcal{A}\{x \leftarrow y\}$ and $P'' \vDash \mathcal{B}\{x \leftarrow y\}$. That is, $P \equiv P' \mid P'' \vDash \text{И}y.(\mathcal{A}\{x \leftarrow y\} \mid \mathcal{B}\{x \leftarrow y\}) = \text{И}x.(\mathcal{A} \mid \mathcal{B})$. □

5 Hidden-Name Quantifier

As discussed in the introduction, a *hidden-name quantifier* should be a construct of the logic that allows us to talk about restricted names in processes. We would like to define a formula $(\nu x)\mathcal{A}$ to mean, informally, that "for hidden name x" (hidden in the underlying process), $\mathcal{A}$ holds. The intention is that there should be some correspondence between the binder (νx) in the formula, and a binder (νn) in a process that satisfies the formula. We take:

5-1 Definition (Hidden-Name Quantifier)

$(\nu x)\mathcal{A} \triangleq \text{И}x.x®\mathcal{A}$ □

Hence $fn((\nu x)\mathcal{A}) = fn(\mathcal{A})$ and $fv((\nu x)\mathcal{A}) = fv(\mathcal{A})\text{-}\{x\}$. Moreover, by definition of И:

$$(\nu x)\mathcal{A} \;=\; \exists x.\ x\#fnv(\mathcal{A})\text{-}\{x\} \wedge x®\mathbf{T} \wedge x®\mathcal{A}$$

and, because of Logical Corollary 3-3(® ∧), we can simplify this to:

$$(\nu x)\mathcal{A} \;\dashv\vdash\; \exists x.\ x\#fnv(\mathcal{A})\text{-}\{x\} \wedge x®\mathcal{A}$$

In terms of satisfaction, we obtain:

5-2 Lemma ($P \vDash (\nu x)\mathcal{A}$)

$P \vDash (\nu x)\mathcal{A}$ iff

$\exists m \in \Lambda.\ m \notin fn(P,\mathcal{A}) \wedge \exists P' \in \Pi.\ P \equiv (\nu m)P' \wedge P' \vDash \mathcal{A}\{x \leftarrow m\}$ □

We have in fact experimented with several plausible definitions for the hidden-name quantifier, before converging on the one above. We have found that the following property, (νx-proper), distinguishes the definition above from other definitions of $(\nu x)\mathcal{A}$ that turned out to be unsatisfactory or flawed:

5-3 Proposition (νx-proper)

For all $n \in \Lambda$, $x \in \vartheta$, $P \in \Pi$, and closed $\mathcal{A} \in \Phi$:

$n \notin fn(P) \wedge P \vDash (\nu x)(\mathcal{A}\{n \leftarrow x\}) \quad \Leftrightarrow \quad \exists P' \in \Pi.\ P \equiv (\nu n)P' \wedge P' \vDash \mathcal{A}$ □

Corollary: $P' \vDash \mathcal{A} \;\Rightarrow\; (\nu n)P' \vDash (\nu x)(\mathcal{A}\{n \leftarrow x\})$. □

This property can be written in logical form as $n®\mathbf{T} \wedge (\nu x)(\mathcal{A}\{n \leftarrow x\}) \dashv\vdash n®\mathcal{A}$, for all n.

Remark. It is natural to first consider the simpler property:

$$(\nu n)P' \vDash (\nu x)(\mathcal{A}\{n \leftarrow x\}) \quad \Leftrightarrow \quad P' \vDash \mathcal{A} \qquad (\nu x\text{-}1)$$

The ⇐ direction is equivalent to (νx-proper⇐). However, the ⇒ direction is inconsistent with the fundamental Lemma 2-1. Start with $n[] \vDash n[]$. By (νx-1⇐) we obtain $(\nu n)n[] \vDash (\nu x)x[]$. Since $(\nu n)n[] \equiv (\nu n)(\nu n)n[]$, by Lemma 2-1 we obtain that $(\nu n)(\nu n)n[] \vDash (\nu x)x[]$. Then, by ($\nu x$-1⇒) we obtain $(\nu n)n[] \vDash n[]$, that is $(\nu n)n[] \equiv n[]$, which is contradictory by Lemma 2-2(1). The problem is that we cannot expect a (νx) in the formula to match any (νn) in the process, but only an appropriate one. Hence the refined statement of (νx-proper). □

The following rules for $(\nu x)\mathcal{A}$ can be derived by the rules for revelation and fresh-name.

5-4 Logical Corollaries (Hidden-Name Quantifier)

(ν ∀)	$\succ \forall x.\ (x\#N \wedge x®\mathbf{T}) \Rightarrow x®\mathcal{A} \dashv\vdash (\nu x)\mathcal{A}$	where $N \supseteq fnv(\mathcal{A})\text{-}\{x\}$ and $x \notin N$
(ν ∃)	$\succ (\nu x)\mathcal{A} \dashv\vdash \exists x.\ x\#N \wedge x®\mathbf{T} \wedge x®\mathcal{A}$ $\dashv\vdash \exists x.\ x\#N \wedge x®\mathcal{A}$	where $N \supseteq fnv(\mathcal{A})\text{-}\{x\}$ and $x \notin N$
(ν R)	$\mathcal{A} \wedge x\#N \wedge x®\mathbf{T} \vdash x®\mathcal{B} \succ \mathcal{A} \vdash (\nu x)\mathcal{B}$	where $N \supseteq fnv(\mathcal{B})\text{-}\{x\}$ and $x \notin N \cup fv(\mathcal{A})$
(ν L)	$x®\mathcal{A} \wedge x\#N \vdash \mathcal{B} \succ (\nu x)\mathcal{A} \vdash \mathcal{B}$	where $N \supseteq fnv(\mathcal{A})\text{-}\{x\}$ and $x \notin N \cup fv(\mathcal{B})$

(ν E)	$\mathcal{A} \vdash (\nu x)\mathcal{B};\ x®\mathcal{B} \wedge x\#N \vdash C\ \}\ \mathcal{A} \vdash C$	where $N \supseteq fnv(\mathcal{B})\text{-}\{x\}$ and $x \notin N \cup fv(C)$
($\nu \vdash$)	$\mathcal{A} \vdash \mathcal{B}\ \}\ (\nu x)\mathcal{A} \vdash (\nu x)\mathcal{B}$	
(ν *fv*)	$\}\ (\nu x)(\mathcal{A}⊘x) \dashv\vdash \mathcal{A}$	where $x \notin fv(\mathcal{A})$
(ν *fv*)	$\}\ \mathcal{A} \vdash (\nu x)\mathcal{A}$	where $x \notin fv(\mathcal{A})$
(ν ®)	$\}\ (\nu x)(x®\mathcal{A}) \dashv\vdash (\nu x)\mathcal{A}$	
(ν ⊘)	$\}\ (\nu x)(\mathcal{A}⊘x) \vdash (\nu x)\mathcal{A}$	
(ν **0**)	$\}\ (\nu x)\mathbf{0} \dashv\vdash \mathbf{0}$	
(ν $n[]$)	$\}\ (\nu x)y[\mathcal{A}] \dashv\vdash y[(\nu x)\mathcal{A}]$	where $x \neq y$
(ν |)	$\}\ (\nu x)(\mathcal{A} \mid x®\mathcal{B}) \dashv\vdash ((\nu x)\mathcal{A}) \mid ((\nu x)\mathcal{B})$	
(ν ⊘ |)	$\}\ (\nu x)(\mathcal{A}⊘x) \mid (\nu x)(\mathcal{B}⊘x) \dashv\vdash (\nu x)((\mathcal{A} \mid \mathcal{B})⊘x)$ □	

Remark. We obtain $\forall x.\ x®\mathcal{A} \vdash (\nu x)\mathcal{A} \vdash \exists x.\ x®\mathcal{A}$. However, there are no interesting rules for $\neg(\nu x)\mathcal{A}$. □

Remark. This fails:

$\}\ (\nu x)\mathcal{A} \vdash \mathcal{A}$ where $x \notin fv(\mathcal{A})$

because $(\nu n)(n[] \mid n[]) \models Иx.x®(\neg\mathbf{0} \mid \neg\mathbf{0})$ but $(\nu n)(n[] \mid n[]) \not\models \neg\mathbf{0} \mid \neg\mathbf{0}$. This is ®'s fault, not И's: $n®\mathcal{A} \vdash \mathcal{A}$ fails with the same counterexample. □

Example

As an example of a specification containing a hidden-name quantifier, consider a situation where a secret is shared by two locations *n* and *m*, but is not known outside those locations.

We can state this as follows (recall that $©\eta \triangleq \neg\eta®\mathbf{T}$ and that $P \models ©n$ iff $n \in fn(P)$):

$$(\nu x)\ (n[©x] \mid m[©x])$$

It reads: for a fresh *x*, the name *x* is known at *n* and *m*, and is restricted anywhere else.

Expanding the definitions, we obtain:

$$P \models (\nu x)\ (n[©x] \mid m[©x])$$
$$\Leftrightarrow \exists r \in \Lambda.\ r \notin fn(P) \cup \{n,m\} \wedge \exists R',R'' \in \Pi.\ P \equiv (\nu r)(n[R'] \mid m[R'']) \wedge r \in fn(R') \wedge r \in fn(R'')$$

The last line reads: *P* satisfies the specification iff there exists a name *r* that is fresh (not conflicting with *n* and *m* or public to *P*), such that *r* is known to the processes *R'* and *R''* located at *n* and *m*, and is restricted inside *P*.

Here is a simple example of an implementation of this specification:

$$P = (\nu p)\ (n[p[]] \mid m[p[]])$$

6 Related Work and Conclusions

We have introduced a logic for describing concurrent processes with restricted names. Most previous logics for concurrency have strived to describe properties that are invariant under some coarse process equivalence, such as bisimulation. Because of our original motivation in describing location structures in detail, the properties described by our logic are much finer, and are invariant only up to structural congruence (see also [14] for a recent characterization). Because of this, our logic is closely related to intuitionistic linear logic and to bunched logics: see [7] for a comparison. Our logic is unusual also because it handles variables ranging over a countable universe of names; these variables can be the subject of universal, existential, fresh-name, and hidden-name quantification.

Our logic is built directly out of a process model, so logical soundness is easy to check. Logical completeness is a much more difficult question. We do not expect the full logic to be complete with respect to our model (even for finite behaviors). Silvano Dal Zilio is investigating some small, complete fragments of the logic. So far, we have mostly tried to discover as many true logical facts as possible (a measure of which is, for example, to be able to embed other logics into ours [7]), and to minimize the collection of basic rules. We have concentrated in particular on commutation and distribution properties of operators that can be useful in formal proofs.

In the present paper, fresh-name quantification is modeled after Gabbay and Pitts [10], adapted to our context; it provides logical rules for reasoning abstractly about freshness. Hidden-name quantification is obtained by combining fresh-name quantification with a revelation operator (not a quantifier) for revealing restricted process names. Most novel axioms have to do with revelation; they often reflect and resemble well-known properties of π-calculus restriction. Technically, we have added to our previous ambient logic just the revelation operator (and its adjunct) and an axiom schema expressing the Gabbay-Pitts property. In particular, fresh-name quantification, hidden-name quantification, and their properties, are derived.

Recently, we have become aware of related work by Luís Caires (both [3] and more recent unpublished work). Our aims are quite similar, but we are currently using different formal techniques; we are in the process of comparing results.

References

[1] Martín Abadi, **Secrecy by Typing in Security Protocols.** JACM 46, 5 (September 1999), 749-786.

[2] Martín Abadi and Andrew D. Gordon, **A Calculus for Cryptographic Protocols: the Spi Calculus**. Information and Computation 148(1999):1-70.

[3] Luís Caires and Luís Monteiro, **Verifiable and Executable Logic Specifications of Concurrent Objects in $\mathcal{L}_\pi$.** In Programming Languages and Systems, Proceedings of ESOP'98, Chris Hankin (Ed.), LNCS, Springer, 1998. pp. 42-56.

[4] Luca Cardelli, Giorgio Ghelli, and Andrew D. Gordon, **Secrecy and Group Creation**. Catuscia Palamidessi, editor. Proceedings of CONCUR 2000. LNCS 1877, Springer 2000, pp. 365-379.

[5] Luca Cardelli, Giorgio Ghelli, and Andrew D. Gordon, **Ambient Groups and Mobility Types.** Proceedings of IFIP TCS2000. J. van Leeuwen, O. Watanabe, M. Hagiya, P.D. Mosses, T. Ito (Eds.). Theoretical Computer Science; Exploring New Frontiers in Theoretical Informatics. LNCS 1872, Springer, 2000. pp. 333-347.

[6] Luca Cardelli and Andrew D. Gordon, **Mobile Ambients.** Foundations of Software Science and Computational Structures, Maurice Nivat (Ed.), LNCS 1378, Springer, 1998. pp. 140-155.

[7] Luca Cardelli and Andrew D. Gordon, **Anytime, Anywhere. Modal Logics for Mobile Ambients.** Proceedings of the 27th ACM Symposium on Principles of Programming Languages, 2000. pp 365-377.

[8] Silvano Dal Zilio, **Spatial Congruence for Ambients is Decidable**. Technical Report MSR-TR-2000-41, Microsoft Research, May 2000.

[9] Furio Honsell, Marino Miculan and Ivan Scangnetto, **π-Calculus in (Co)Inductive Type Theory**. TCS 2000.

[10] Murdoch J. Gabbay and Andrew M. Pitts, **A New Approach to Abstract Syntax Involving Binders**. In Proceedings 14th Annual IEEE Symposium on Logic in Computer Science, Trento, Italy, July 1999. IEEE Computer Society Press, 1999. pp 214-224.

[11] Andrew D. Gordon and Luca Cardelli, **Equational Properties of Mobile Ambients.** Wolfgang Thomas, Editor. Foundations of Software Science and Computational Structures, Second International Conference, FOSSACS'99. LNCS 1578. Springer, 1999. pp 212-226.

[12] Robin Milner, **Communicating and Mobile Systems: the π-Calculus**. Cambridge University Press, 1999.

[13] Robin Milner, Joachim Parrow and David Walker, **A Calculus of Mobile Processes, Parts 1-2**. Information and Computation, 100(1), 1-77. 1992

[14] Davide Sangiorgi, **Extensionality and Intensionality in the Ambient Logics**. Proceedings of the 28th ACM Symposium on Principles of Programming Languages, 2001. pp 4-13.

Subtyping Recursive Games

Juliusz Chroboczek

Université de Paris VII
Paris, France

Abstract. Using methods drawn from Game Semantics, we build a sound and computationally adequate model of a simple calculus that includes both subtyping and recursive types. Our model solves recursive type equations up to equality, and is shown to validate a subtyping rule for recursive types proposed by Amadio and Cardelli.

Introduction

Subtyping is an ordering relation over types that is an essential feature of a wide range of programming languages. While at first order subtyping corresponds to inclusion of the carriers, there is no simple set-theoretic interpretation of subtyping at higher order.

Many programming languages also include recursive types — types that are defined implicitly, as fixpoints of maps over types. The interaction of recursive types with subtyping has been studied before [4], and shown to present a number of interesting challenges. While both are important features of many programming languages, there are only few interpretations that satisfactorily model both.

Game Semantics is a framework for modelling programming languages that combines the elegant mathematical structure of Denotational Semantics with explicitly operational notions. Due to the blend of the two, Game Semantics has been successful at modelling a wide range of programming language features. In a previous work [7], we have shown how the simple feature of adding explicit error elements to Game Semantics allows us to model subtyping; in this paper, we extend this model to include recursive types, and show that it validates the subtyping rule for recursive types proposed by Amadio and Cardelli [4].

There are two main new results in this paper. First, we show how a minor modification of the operational semantics of the untyped model presented in an earlier makes the model computationally adequate (Sections 1.1 and 2.2), thus solving the main open problem in our previous paper [7]. Second, we show how the space of games, used for modelling types, can be equipped with a metric that allows us to construct recursive types; the metric is shown to interact with the order structure related to subtyping so as to validate the desired subtyping rules (Sections 4 and 5).

S. Abramsky (Ed.): TLCA 2001, LNCS 2044, pp. 61–75, 2001.

1 A λ-Calculus with Errors

We consider an untyped λ-calculus with ground values, defined by the following syntax:

$$\begin{aligned} M, N, N' ::= {} & x \mid \lambda x.M \mid (M\ N) \\ \mid {} & (M, N) \mid \pi_l(M) \mid \pi_r(M) \\ \mid {} & \mathbf{tt} \mid \mathbf{ff} \mid \mathbf{top} \mid \mathbf{if}\ M\ \mathbf{then}\ N\ \mathbf{else}\ N'\ \mathbf{fi}. \end{aligned}$$

The only unusual feature of this calculus is the presence of a ground value **top** that will be used for representing the result of badly-typed terms.

Our calculus may be equipped with an operational semantics *e.g.* by defining a *one step reduction* relation $\rightsquigarrow$ on terms. For our calculus, a common choice — the *call-by-name* semantics — consists of the rules

$$((\lambda x.M)\ N) \rightsquigarrow M[x\backslash N]$$

$$\pi_l((M, N)) \rightsquigarrow M \qquad \pi_r((M, N)) \rightsquigarrow N$$

$$\mathbf{if\ tt\ then}\ N\ \mathbf{else}\ N'\ \mathbf{fi} \rightsquigarrow N \qquad \mathbf{if\ ff\ then}\ N\ \mathbf{else}\ N'\ \mathbf{fi} \rightsquigarrow N'$$

$$\frac{M \rightsquigarrow M'}{E[M] \rightsquigarrow E[M']}$$

where the set of *evaluation contexts* $E[\cdot]$ is defined by

$$E[\cdot] ::= ([\cdot]\ N) \mid \pi_l([\cdot]) \mid \pi_r([\cdot]) \mid \mathbf{if}\ [\cdot]\ \mathbf{then}\ N\ \mathbf{else}\ N'\ \mathbf{fi}.$$

In general, we will be interested in computations that take more than one step. The *reduction relation* $\rightsquigarrow^*$ is the transitive reflexive closure of $\rightsquigarrow$.

We say that a term M *reduces to value* V, written $M \downarrow V$, if $M \rightsquigarrow^* V$ where V is a value. We write $M \downarrow$ when there exists a value V such that $M \downarrow V$, and $M \uparrow$ otherwise.

1.1 Errors in the Calculus

The relation $\downarrow$ is not total; a number of terms do not reduce to values. This is expected, as we have done nothing whatsoever to prevent the formation of meaningless terms.

Let δ be the term $\lambda x.(x\ x)$. The term $(\delta\ \delta)$ does not reduce to a value; $(\delta\ \delta)$ leads to an infinite sequence of one-step reductions:

$$(\delta\ \delta) \rightsquigarrow (\delta\ \delta) \rightsquigarrow (\delta\ \delta) \rightsquigarrow \cdots$$

A very different example of a term that fails to reduce to a value is

$$M = \mathbf{if}\ \lambda x.x\ \mathbf{then\ tt\ else\ ff\ fi}$$

In this case, small-step semantics shows that the reduction remains "stuck" at a non-value term: there is no M' such that $M \rightsquigarrow M'$. In our view, this situation corresponds to a runtime error — an exceptional situation detected during the reduction of a term. As our calculus contains no constructs that allow us to handle ("trap") such errors, we shall use the term *untrappable error*.

We will use the term **top** to represent untrappable errors. Intuitively, a term should reduce to **top** whenever its reduction gets "stuck" with no applicable rule; unfortunately, such a simple extension does not quite work. Indeed, consider the "identity on the Booleans" $\mathbf{I_{Bool}} = \lambda x.\mathbf{if}\ x\ \mathbf{then\ tt\ else\ ff\ fi}$; this term behaves as the identity when applied to a Boolean, but returns an error when applied to a function or a pair. Let now **Y** be a fixpoint combinator, and consider the "looping Boolean" $(\mathbf{Y}\ \mathbf{I_{Bool}})$; intuitively, we would expect this term to loop when invoked in a Boolean context, but return an error when *e.g.* applied. The reduction relation $\downarrow$, augmented as suggested above, causes it to loop (this problem is expressed technically by the failure of computational adequacy of our model w.r.t. $\downarrow$ [7, Section 2.3]).

We therefore define a different reduction relation, written $\Downarrow$, which is explicitly decorated with the locus of the computation. To do so, we introduce a notion of *initial component*, a finite sequence over $\{1, l, r\}$ (the empty sequence is written ϵ). Walking the syntax tree of a type, computation happening on the right-hand-side of an arrow is marked by 1; computation happening on the left-hand-side (resp. right-hand-side) of a product is marked by l (resp. r). A family of reduction relations, indexed by initial components, is defined in Fig. 1.

As usual, we write $M \Uparrow_c$ when there is no V such that $M \Downarrow_c V$. We write ϵ for the empty component, and $|c|$ for the length of component c.

To clarify this definition, note that the form of a value resulting from reduction at initial component c is determined by c. More precisely, if M is a closed term and c an initial component such that $M \Downarrow_c V$, then one of the following is true:

- $V = \mathbf{top}$; or
- $c = \epsilon$ and $V = \mathbf{ff}$ or $V = \mathbf{tt}$; or
- c is of the form $1 \cdot c'$ and V is of the form $\lambda x.M'$; or
- c is of the form $l \cdot c'$ or $r \cdot c'$, and V is of the form (N, P).

Conversely,

- if $(M, N) \Downarrow_c P$, then either c is of the form $l \cdot c'$ or $r \cdot c'$, or $P = \mathbf{top}$;
- if $\lambda x.M \Downarrow_c N$, then either c is of the form $1 \cdot c'$ or $N = \mathbf{top}$;
- $\mathbf{tt} \Downarrow_c N$ or $\mathbf{ff} \Downarrow_c N$, then either $c = \epsilon$ or $N = \mathbf{top}$.

There is also a simple relationship between $\Downarrow$ and the simple reduction relation $\downarrow$; it shows that the extension that we introduce only concerns erroneous reductions. Roughly speaking, the relations $\Downarrow$ and $\downarrow$ coincide, except in the case in which $\Downarrow$ yields an untrappable error and $\downarrow$ diverges. More precisely, $M \downarrow V$ implies that for some initial component c, $M \Downarrow_c V$. Conversely, $M \uparrow$ implies that for all c, either $M \Uparrow_c$ or $M \Downarrow_c \mathbf{top}$. Finally, if for some c, $M \Uparrow_c$, then $M \uparrow$.

1.2 Errors and Denotational Semantics

It is not immediately obvious how to model errors in Denotational Semantics. Consider for example the domain of Booleans. One choice would be to add an

$$\mathbf{tt} \Downarrow_\epsilon \mathbf{tt} \qquad \mathbf{ff} \Downarrow_\epsilon \mathbf{ff} \qquad \mathbf{tt} \Downarrow_c \mathbf{top} \quad (c \neq \epsilon) \qquad \mathbf{ff} \Downarrow_c \mathbf{top} \quad (c \neq \epsilon)$$

$$\lambda x.M \Downarrow_{1\cdot c} \lambda x.M \qquad \lambda x.M \Downarrow_c \mathbf{top} \quad (c \neq 1 \cdot c^{\bullet})$$

$$(M,N) \Downarrow_{l\cdot c} (M,N) \qquad (M,N) \Downarrow_{r\cdot c} (M,N) \qquad (M,N) \Downarrow_c \mathbf{top} \quad (c = \epsilon \text{ or } c = 1 \cdot c^{\bullet})$$

$$\frac{M \Downarrow_{1\cdot c} \lambda x.M^{\bullet} \qquad M^{\bullet}[x \backslash N] \Downarrow_c P}{(M\ N) \Downarrow_c P} \qquad \frac{M \Downarrow_{1\cdot c} \mathbf{top}}{(M\ N) \Downarrow_c \mathbf{top}}$$

$$\frac{M \Downarrow_{l\cdot c} (N,P) \qquad N \Downarrow_c N^{\bullet}}{\pi_l(M) \Downarrow_c N^{\bullet}} \qquad \frac{M \Downarrow_{r\cdot c} (N,P) \qquad P \Downarrow_c P^{\bullet}}{\pi_r(M) \Downarrow_c P^{\bullet}}$$

$$\frac{M \Downarrow_{l\cdot c} \mathbf{top}}{\pi_l(M) \Downarrow_c \mathbf{top}} \qquad \frac{M \Downarrow_{r\cdot c} \mathbf{top}}{\pi_r(M) \Downarrow_c \mathbf{top}}$$

$$\frac{M \Downarrow_\epsilon \mathbf{tt} \qquad N \Downarrow_\epsilon N^{\bullet}}{\mathbf{if}\ M\ \mathbf{then}\ N\ \mathbf{else}\ P\ \mathbf{fi} \Downarrow_\epsilon N^{\bullet}} \qquad \frac{M \Downarrow_\epsilon \mathbf{ff} \qquad P \Downarrow_\epsilon P^{\bullet}}{\mathbf{if}\ M\ \mathbf{then}\ N\ \mathbf{else}\ P\ \mathbf{fi} \Downarrow_\epsilon P^{\bullet}}$$

$$\frac{M \Downarrow_\epsilon \mathbf{top}}{\mathbf{if}\ M\ \mathbf{then}\ N\ \mathbf{else}\ P\ \mathbf{fi} \Downarrow_\epsilon \mathbf{top}} \qquad \mathbf{if}\ M\ \mathbf{then}\ N\ \mathbf{else}\ P\ \mathbf{fi} \Downarrow_c \mathbf{top} \quad (c \neq \epsilon)$$

Fig. 1. Big-step semantics with errors and initial components

error value **error** "on the side" (Fig. 2(a)); another one would be to add a value **top** as a top element (Fig. 2(b)).

It is our view that errors on the side model (trappable) exceptions, while errors at top model untrappable errors. Consider, indeed, the addition to our calculus of a term **ignore-errors** that would satisfy

$$\mathbf{ignore\text{-}errors}\ \mathbf{tt} \Downarrow_\epsilon \mathbf{tt}$$
$$\mathbf{ignore\text{-}errors}\ \mathbf{ff} \Downarrow_\epsilon \mathbf{tt}$$
$$\mathbf{ignore\text{-}errors}\ \mathbf{top} \Downarrow_\epsilon \mathbf{ff}$$

Denotationally, such a term would have to map **tt** to **tt** while mapping **top** to **ff**, which would be a non-monotone semantics. On the other hand, modelling an analogous term using **error** instead of **top** would cause no problem at all.

Errors "on the side," or exceptions, have been studied before [6]; in this paper, we adopt the domain in Fig. 2(b).

The addition of a top value to Scott domains was a common feature of early Denotational Semantics. However, this value does not seem to be used

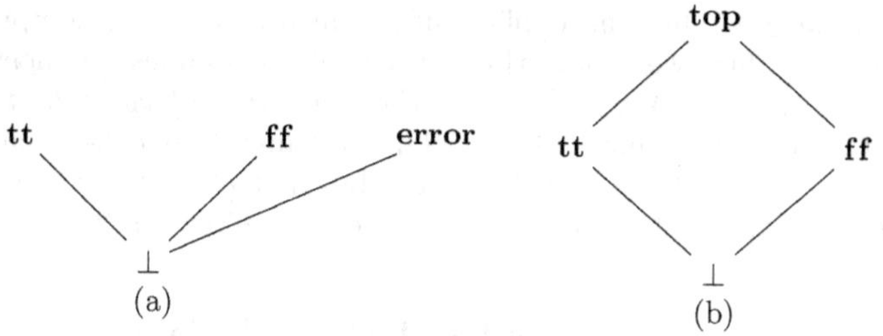

Fig. 2. Two domains of Booleans with errors

for modelling anything, but is just added to domains in order to turn them into complete lattices.

1.3 Observational Preorder

In order to complete the definition of the semantics of our calculus, we need to introduce a notion of equivalence of terms. This is usually done by defining a set of *observations*, which is then used to define a congruent preorder on terms known as the *observational preorder*. We choose our set of observations to consist of the observations "reduction to **top** at ϵ," "reduction to **tt** at ϵ" and "reduction to **ff** at ϵ," ordered analogously to Fig. 2(b).

Definition 1. *(Observational preorder)*

$$M \lesssim N \textit{ iff } \forall C[\cdot] \begin{cases} C[M] \Downarrow_\epsilon \mathbf{top} \Rightarrow C[N] \Downarrow_\epsilon \mathbf{top}; \\ C[M] \Downarrow_\epsilon \mathbf{tt} \Rightarrow C[N] \Downarrow_\epsilon \mathbf{tt} \textit{ or } C[N] \Downarrow_\epsilon \mathbf{top}; \\ C[M] \Downarrow_\epsilon \mathbf{ff} \Rightarrow C[N] \Downarrow_\epsilon \mathbf{ff} \textit{ or } C[N] \Downarrow_\epsilon \mathbf{top}. \end{cases}$$

We say that two terms M and N are observationally equivalent, *and write $M \cong N$, when $M \lesssim N$ and $N \lesssim M$.*

As usual in calculi with ground values, the observational preorder can be defined by just one well-chosen observation; one possible choice is reduction to **top** at ϵ.

Lemma 1. *$M \lesssim N$ if and only if*

$$\forall C[\cdot]\ C[M] \Downarrow_\epsilon \mathbf{top} \Rightarrow C[N] \Downarrow_\epsilon \mathbf{top}.$$

Informally, this lemma says that terms are equivalent if and only if they generate errors in the same set of contexts.

It is worthwhile to compare our calculus with Abramsky's *lazy λ-calculus* [1]. Writing Ω for the looping term (*e.g.* $\Omega = ((\lambda x.(x\ x))\ \lambda x.(x\ x)))$, notice that Ω and $\lambda x.\Omega$ are observationally distinct. Indeed, taking

$$C[\cdot] = \mathbf{if}\ \cdot\ \mathbf{then\ tt\ else\ ff\ fi}$$

we have $C[\Omega] \Uparrow_\epsilon$, while $C[\lambda x.\Omega] \Downarrow_\epsilon$ **top**. On the other hand, as we shall see in Section 2, we introduce no explicit lifting in a sound and computationally adequate model. Thus, we believe that our calculus combines the most desirable characteristics of what Abramsky calls the *standard interpretation* of the λ-calculus with those of the lazy calculus. The fundamentally call-by-name nature of the construction is reflected in the syntax by the fact that the terms **top** and $\lambda x.$**top** are observationally equivalent (see the end of Section 2.2).

2 A Game Semantics for the Untyped Calculus

This section roughly outlines the semantic framework used for modelling untyped terms. As Game Semantics has been described before [2,11], and so has our particular framework [7,8], this section remains informal.

In Game Semantics, a term is represented by a *strategy*, the set of its behaviours in all possible contexts. A behaviour is modelled as a play between two players, Player, who represents the term under consideration, and Opponent, who represents its environment (the context it is in). The two players exchange tokens of information known as *moves* — one may think of these as (visible) actions in process calculi, or messages in message-passing object-oriented languages. By convention, Opponent plays first when modelling a call-by-name calculus.

Moves are structured into *components* which correspond to paths in the syntax tree of a type. For example, a strategy corresponding to a term of type **Bool** $\to$ **Bool** exchanges moves in components 0 (the left-hand-side of the arrow) and 1. Precisely, a move is of the form m_c, where m is one of q, the *question*, $a^{\mathbf{tt}}$, the *answer true*, or $a^{\mathbf{ff}}$, the *answer false*, and c, the *component* of the move, is a finite string over 0, 1, l, r. In addition, moves are decorated with justification pointers which, while absolutely necessary for the correction of the interpretation, are not essential for the ideas in this paper.

A *position* is an alternating sequence of moves — odd-ordered moves played by Opponent, even-ordered ones by Player. A *strategy* is a set of positions that specify the moves played by Player in response to a given sequence of moves from Opponent.

The main novelty of the formalism used in this work and introduced in [8,7] is that we allow strategies to refuse moves, which is used for modelling untrappable errors. Concretely, this is realised by allowing strategies to contain both even- and odd-length positions. In a spirit similar to that of Harmer [9, Chapter 4], even-length positions represent moves that are played by Player, while odd-length positions represent situations in which player loops.

Definition 2. *A set s of positions is*

- prefix-closed *if $p \cdot q \in s$ implies $p \in s$ for any positions p and q;*
- even-prefix-closed *if $p \cdot q \in s$ and $|p|$ even imply $p \in s$ for any positions p and q;*
- deterministic *if for any position $p \in s$, if $|p|$ is odd then, for any moves m and n,*

– $p \cdot m \in s$ *and* $p \cdot n \in s$ *imply* $m = n$; *and*
– $p \cdot m \in s$ *implies* $p \notin s$.

A strategy *is a non-empty even-prefix-closed deterministic set of positions.*

For any collection of positions A, we write Pref A for the prefix completion of A.

The even-prefix-closedness condition in this definition says that a strategy cannot mandate that Opponent should play at a given position: a strategy must allow for the situation in which Opponent never plays a move. As to the determinacy condition, it states that a strategy cannot mandate either playing two distinct moves or both playing and not playing a move at a given position. Taken together, even-prefix-closedness and determinacy imply that an odd-length position in a strategy cannot be extended (*i.e.* if $p \in s$ and $|p|$ is odd, then no $p \cdot q$ is in s): once a strategy has refused to play a move, the play will not proceed further.

In [7], we define a certain number of strategies. The strategy **top** consists of the single empty position ϵ; this strategy never accepts an Opponent's move. The strategy Ω consists of all positions of length 1; thus, it always accepts an initial move from Opponent, but never plays a move. The strategy **tt** consists of all even-length positions composed of alternations of the moves q and $a^{\mathbf{tt}}$; thus, it always accepts an initial question, and replies with the answer true (**ff** is analogous).

The class of strategies that copy moves between components are known as the *copy-cat* strategies; this class includes the identity **I**, the projections π_r and π_l, and, to a certain extent, the "if-then-else" strategy **ite**. In addition, we use a number of operations on strategies, including (functional) pairing $\langle \cdot, \cdot \rangle$, currying $\Lambda(\cdot)$, as well as the *injection* $K(\cdot)$ which "shifts" a strategy into component 1.

Composition of strategies s and t is performed by ranging over all behaviours in s and t, selecting those that are agree on a common component, and composing them, similarly to Baillot *et al.* [5]. However, we cannot just use their formalism, as we need to take into account *livelock*, or *infinite chattering*, the situation in which two strategies never disagree but never have positions that coincide. Indeed, suppose that when composing s with t, after the initial move is played in component 1 of t, both t and s keep playing in the common component. In this case, the two strategies would never ultimately reach agreement, and yet neither would ever play a move that is not accepted by the other.

Definition 3. *Given a natural integer* n*, we say that two positions* p *and* q agree at depth n *if* p *and* q *only contain moves within components* 1 *and* 0*, and the prefix of length* n *of* $p \upharpoonright 1$ *is equal to the prefix of same length of* $q \upharpoonright 0$ *(or* $p \upharpoonright 1 = q \upharpoonright 0$ *if both projections are of length smaller than* n*).*

Given two strategies s *and* t*, the strategy* $s;t$ *is the set of all positions* p *such that for any natural integer* n*, there exist positions* $q \in s$ *and* $q' \in t$ *such that* q *and* q' *agree at depth* n*,* $q \upharpoonright 0 = p \upharpoonright 0$ *and* $q' \upharpoonright 1 = p \upharpoonright 1$*.*

2.1 The Liveness Ordering

We now introduce an ordering — the *liveness ordering* $\preccurlyeq$ — which will model the observational preorder, the typing relation, and the subtyping relation (Lemmata 5 and 10 and Theorem 4), and is inspired by Abramsky's "back-and-forth inclusion relation" [2]. The definition of the liveness ordering is analogous to that of the observational preorder. Just like for terms M and N we have $M \lesssim N$ when M produces errors in less contexts than N, and N produces results in more contexts than M (Definition 1), we will want strategies s and t to satisfy $s \preccurlyeq t$ if and only if s accepts more positions and produces less positions than t when playing against any given opponent. We define $\preccurlyeq$ on prefix-closed sets of positions, and deduce a suitable definition for strategies from that.

For any non-empty position p, we write p_{-1} for the prefix of p of length $|p|-1$ (*i.e.* p without its last move).

Definition 4. *Given non-empty prefix-closed sets of positions A and B, we say that B is* more live *than A, or A is* safer *than B, and write $A \preccurlyeq B$, if*

- *for every position of odd length $q \in B$, if $q_{-1} \in A$ then $q \in A$; and*
- *for every position of even length $p \in A$ ($p \neq \epsilon$), if $p_{-1} \in B$, then $p \in B$.*

The definition of $\preccurlyeq$ may be paraphrased as follows. Given a prefix-closed collection of positions A, a position p is said to be *reachable* at A if $p_{-1} \in A$ or $p = \epsilon$. In order to have $A \preccurlyeq B$, the set of odd-length positions (positions ending in an Opponent's move) in A that are reachable at A needs to be a superset of the set of odd-length positions in B; and, dually, the set of even-length positions in B that are reachable at B should be a superset of the even-length positions in A. A clarification of the intuitions behind the liveness ordering may be found in [8, 7].

Theorem 1. *The relation $\preccurlyeq$ is a partial order on non-empty prefix-closed collections of positions.*

The definition of $\preccurlyeq$ above does not yield a transitive or antireflexive relation on arbitrary sets of positions. We may, however, extend $\preccurlyeq$ to all non-empty sets of positions by writing $A \preccurlyeq B$ whenever $\mathrm{Pref}(A) \preccurlyeq \mathrm{Pref}(B)$; while this only makes $\preccurlyeq$ into a preorder on arbitrary sets of positions, it does actually make it into a partial order on strategies.

Lemma 2. *If s and t are strategies, then $\mathrm{Pref}(s) = \mathrm{Pref}(t)$ implies $s = t$. The relation $\preccurlyeq$ is therefore a partial order on strategies.*

This property does depend on the fact that we have restricted ourselves to deterministic strategies.

2.2 Interpretation of the Calculus

We interpret a couple $\Gamma \vdash M$, where Γ is an (ordered) list of variables, and M a term such that $\mathrm{FV}(M) \subseteq \Gamma$. The interpretation is defined as follows.

$$
\begin{array}{rclrcl}
[\![x \vdash x]\!] & = & \mathbf{I} & [\![\Gamma \vdash \lambda x.M]\!] & = & \Lambda([\![\Gamma, x \vdash M]\!]) \\
[\![\Gamma, x \vdash x]\!] & = & \pi_r & [\![\Gamma \vdash (M\ N)]\!] & = & \langle [\![\Gamma \vdash M]\!], [\![\Gamma \vdash N]\!] \rangle; \mathbf{eval} \\
[\![\Gamma, y \vdash x]\!] & = & \pi_l; [\![\Gamma \vdash x]\!] & [\![\Gamma \vdash (M, N)]\!] & = & \langle [\![\Gamma \vdash M]\!], [\![\Gamma \vdash N]\!] \rangle \\
[\![\Gamma \vdash \mathbf{tt}]\!] & = & K(\mathbf{tt}) & [\![\Gamma \vdash \pi_l(M)]\!] & = & [\![\Gamma \vdash M]\!]; \pi_l \\
[\![\Gamma \vdash \mathbf{ff}]\!] & = & K(\mathbf{ff}) & [\![\Gamma \vdash \pi_r(M)]\!] & = & [\![\Gamma \vdash M]\!]; \pi_r \\
[\![\Gamma \vdash \mathbf{top}]\!] & = & \mathbf{top} & [\![\Gamma \vdash \mathbf{if}\ M\ \mathbf{then}\ \ N\ \ \mathbf{else}\ N'\ \mathbf{fi}]\!] & = & \\
 & & & & = & \langle [\![\Gamma \vdash M\ ]\!], \langle [\![\Gamma \vdash N]\!], [\![\Gamma \vdash N']\!] \rangle \rangle; \mathbf{ite}
\end{array}
$$

The notion of soundness that we use is somewhat complicated by the fact that we use a family of reduction relations. Given a component c, we say that two strategies s and t are equal at component c, and write $s =_c t$, when the sets of positions starting with q_c in s and t coincide.

Lemma 3. (Equational Soundness) *If $[\![\Gamma \vdash M]\!]$ is defined and $M \Downarrow_c N$, then $[\![\Gamma \vdash N]\!]$ is defined and $[\![\Gamma \vdash N]\!] =_{1 \cdot c} [\![\Gamma \vdash M]\!]$.*

The interpretation is also computationally adequate.

Lemma 4. (Computational Adequacy) *If $[\![\Gamma \vdash M]\!]$ is defined and there is no term N such that $M \Downarrow_\epsilon N$, then $[\![\Gamma \vdash M]\!] =_1 \bot$.*

In order to prove this property, we use a variant of Plotkin's method of formal approximation relations. We say that a family $\lhd_c$ of relations between strategies and terms, indexed by initial components, is a *family of formal approximation relations* when it satisfies a number of fairly natural properties that imply in particular that $s \lhd_\epsilon M$ implies $s =_1 \bot$ or $M \Downarrow_\epsilon$. We then show the existence of such a family, and that for any closed term M and initial component c, $[\![\Gamma \vdash M]\!] \lhd_c M$, which allows us to conclude by a standard argument.

Soundness, computational adequacy and Lemma 1 imply inequational soundness.

Lemma 5. (Inequational Soundness) *For any two terms M and N, if $[\![M]\!] \preccurlyeq [\![N]\!]$ then $M \lesssim N$.*

Inequational soundness can often be used for proving properties about the calculus itself. For example, as the terms **top** and $\lambda x.\mathbf{top}$ have the same interpretation, we may conclude that they are in fact observationally equivalent.

3 Type Assignment and Subtyping

In order to define a type assignment on our calculus, we assume the existence of a countable set of type variables $X, Y, \ldots$ and define the syntax of types as follows,

$$A, B ::= \mathbf{Bool} \mid \top \mid X \mid A \times B \mid A \to B \mid \mu X.C$$

where the type C is *guarded* in the type variable X. Thus, types consist of the ground type **Bool** of Booleans, the type $\top$ of all terms, type variables, product types, arrow types, and recursive types.

The set of types guarded in a type variable X is defined by the grammar

$$C, D ::= \mathbf{Bool} \mid \top \mid Y \mid A \times B \mid A \to B \mid \mu Y.C$$

In order to speak about subtyping of recursive types, we need a notion of *covariant* type. The set of types covariant in a type variable X is defined by the grammar

$$E, F ::= \mathbf{Bool} \mid \top \mid X \mid Y \mid E \times F \mid G \to E \mid \mu Y.E$$

where G is *contravariant* in X; the set of types contravariant in a type variable X is defined by the grammar

$$G, H ::= \mathbf{Bool} \mid \top \mid Y \mid G \times H \mid E \to G \mid \mu Y.G$$

An *environment* is a set of type variables and a map from variables to types. We use the letter E to range over environments, and write

$$X, Y, x : C, y : D$$

for the environment that specifies the free type variables X and Y, and maps x to C, y to D and all other type variables and variables to $\top$.

We use two kinds of *judgements*. A *subtyping judgement* is of the form $E \vdash A \leq B$ and specifies that in the environment E, the type A is a subtype of the type B; we write $E \vdash A = B$ for $E \vdash A \leq B$ and $E \vdash B \leq A$. A *typing judgement* is of the form $E \vdash M : A$ and states that in the environment E, the term M has type A.

The set of inference rules used for typing is given in Figures 3 and 4. Somewhat unusual is the fact that there are no explicit rules for the folding and unfolding of recursive types; these rules can in fact be derived from the penultimate rule in Fig. 4 and subsumption (the last rule in Fig. 3). The last rule in Fig. 4 is the subtyping rule proposed by Amadio and Cardelli [4].

3.1 Games and the Liveness Ordering

Types will be interpreted as games. A game is a set of positions that provide a specification that a strategy may or may not satisfy.

As we use the liveness ordering to interpret typing, a game A provides not only a specification for Player but also a specification for Opponent. A strategy s belongs to the game A if its behaviour satisfies the constraints expressed by A, but only as long as Opponent behaves according to A; Player's behaviour is otherwise unrestricted. Technically, this is expressed by the reachability condition in the definition of the liveness ordering.

Definition 5. *A* game *is a non-empty prefix-closed set of positions.*

$$E \vdash M : \top \qquad E, x : A \vdash x : A$$

$$\frac{E, x : A \vdash M : B}{E \vdash \lambda x.M : A \to B} \qquad \frac{E \vdash M : A \to B \quad E \vdash N : A}{E \vdash (M\ N) : B}$$

$$E \vdash \mathbf{tt} : \mathbf{Bool} \qquad E \vdash \mathbf{ff} : \mathbf{Bool}$$

$$\frac{E \vdash M : \mathbf{Bool} \quad E \vdash N : A \quad E \vdash P : A}{E \vdash \mathbf{if}\ M\ \mathbf{then}\ N\ \mathbf{else}\ P\ \mathbf{fi} : A}$$

$$\frac{E \vdash M : A \qquad E \vdash N : B}{E \vdash (M, N) : A \times B}$$

$$\frac{E \vdash M : A \times B}{E \vdash \pi_l(M) : A} \qquad \frac{E \vdash M : A \times B}{E \vdash \pi_r(M) : B}$$

$$\frac{E \vdash M : A \quad E \vdash A \leq B}{E \vdash M : B}$$

Fig. 3. Typing rules

We write $\mathcal{G}$ for the set of games.

The game $\top = \{\epsilon\}$ is the maximal element of the lattice of games. The game **Bool** of Booleans it is defined as the set of all interleavings of positions in Pref **tt** and Pref **ff**. The game $A \times B$ consists of the set of the injections of all positions in A in the component l, the injections of all positions in B in the component r, and all interleavings of such positions. Finally, the game $A \to B$ consists of all positions p entirely within components 0 and 1 such that $p \upharpoonright 0$ is an interleaving of positions in A and $p \upharpoonright 1$ is an interleaving of positions in B.

4 A Metric on Games

In order to solve recursive type equations, we use Banach's fixpoint theorem. We recall that a metric space is said to be *complete* when every Cauchy sequence has a limit. A map over a metric space (X, d) is said to be Lipschitz with constant $\lambda > 0$ when for all $x, y \in X$, $\mathrm{d}(f(x), f(y)) \leq \lambda \mathrm{d}(x, y)$. Such a map is said to be *nonexpanding* when $\lambda \leq 1$, and *contractive* when $\lambda < 1$.

Theorem 2. (Banach) *A contractive map f over a complete metric space has a unique fixpoint* $\mathrm{fix}(f)$.

In order to solve recursive type equations using Banach's theorem, we need to equip the set of games $\mathcal{G}$ with a metric that makes it into a complete space; furthermore, the metric should make all type constructors into contractive maps.

$$E \vdash A \leq A \qquad \frac{E \vdash A \leq B \quad E \vdash B \leq C}{E \vdash A \leq C}$$

$$E \vdash A \leq \top \qquad E \vdash \top \leq A \to \top$$

$$\frac{E \vdash A^{\bullet} \leq A \qquad E \vdash B \leq B^{\bullet}}{E \vdash A \to B \leq A^{\bullet} \to B^{\bullet}} \qquad \frac{E \vdash A \leq A^{\bullet} \qquad E \vdash B \leq B^{\bullet}}{E \vdash A \times B \leq A^{\bullet} \times B^{\bullet}}$$

$$E \vdash \mu X.B[X] = B[\mu X.B[X]]$$

$$\frac{E, X \vdash A[X] \leq B[X]}{E \vdash \mu X.A[X] \leq \mu X.B[X]} \quad (A,\ B \text{ covariant in } X)$$

Fig. 4. Subtyping rules

Games, being prefix-closed collections of sequences, may be seen as trees, so it would seem natural to equip $\mathcal{G}$ with the tree metric. Unfortunately, this simple approach does not yield enough contractive maps, failing in particular to make the product contractive. For this reason, we apply the tree metric method twice, once to components and once to positions.

Definition 6. *Let $p = m_0 \cdots m_{n-1}$ be a position, and M_p the set of moves m such that $p \cdot m$ is a position. For any move $m \in M_p$, the* weight *of m w.r.t. p is defined by $\mathrm{w}_p(m) = 2^{-|c|}$, where c is the component of m.*

The ultrametric d_p on M_p is defined, for distinct moves m, m', as $\mathrm{d}_p(m, m') = 2^{-|c|}$, where c is the longest common prefix of the components of m, m'.

We are now ready to define the metric on positions that will serve our needs. Given two distinct positions p and p', either one is the prefix of the other, in which case we will use the weight of the first differing move, or neither is a prefix of the other, in which case we use the distance between the first differing moves.

Definition 7. *Given a position $p = q \cdot m_n = m_0 \cdot m_1 \cdots m_{n-1} \cdot m_n$, the* weight *of p is defined as $\mathrm{w}(p) = 2^{-n}\mathrm{w}_q(m_n)$.*

The metric d on the set of positions is defined as follows. Given two distinct positions p, p', let $q = m_0 \cdots m_{n-1}$ be their longest common prefix. If

$$p = q \cdot m_n \cdot r, \qquad p' = q \cdot m'_n \cdot r',$$

then $\mathrm{d}(p,p') = 2^{-n}\mathrm{d}_q(m_n, m'_n)$. On the other hand, if

$$p = q \cdot m_n \cdot r, \qquad p' = q,$$

then $\mathrm{d}(p,p') = 2^{-n}\mathrm{w}_q(m_n)$.

Note that this metric does not induce the discrete topology on the set of positions; games, however, are closed with respect to it, and therefore we may still apply the Hausdorff formula to games.

Definition 8. *The metric* d *on the set of games is defined by the Hausdorff formula*

$$\mathrm{d}(A, B) = \max(\sup_{p \in A} \inf_{q \in B} \mathrm{d}(p, q), \sup_{p \in B} \inf_{q \in A} \mathrm{d}(p, q)).$$

As the space of positions is not complete, and games are not necessarily compact[1], we cannot take any of the properties of Hausdorff's metric for granted. However, a fairly standard proof shows that in fact d does have all the desired properties.

Theorem 3. *The space of games* $(\mathcal{G}, \mathrm{d})$ *is a complete ultrametric space.*

There is another property that we will need in order to prove soundness of typing: the fact that least upper bounds preserve the ordering in some cases. The following property is simple enough to prove directly and is sufficiently strong for our needs:

Lemma 6. *If* A *is a game, then the order ideal* $\{B \mid B \preccurlyeq A\}$ *is closed with respect to* d.

This is proved by considering a game $C \not\preccurlyeq A$. If A contains a position p such that $p \notin C$ but all strict prefixes of p are in C, we define the real number δ as the minimum of the weights of all prefixes of p, and show that for any $B \preccurlyeq A$, $\mathrm{d}(B, C) \geq \delta$. A similar argument applies in the case when $p \in C$, $p \notin A$ and all strict prefixes of p are in A.

This property implies the following one, which we will need in order to prove soundness:

Lemma 7. *Let* $f, g : \mathcal{G} \to \mathcal{G}$ *be monotone, contractive maps over games such that for any game* A, $f(A) \preccurlyeq g(A)$. *Then* $\mathrm{fix}(f) \preccurlyeq \mathrm{fix}(g)$.

Finally, as the metric was constructed *ad hoc*, it is a simple matter to show that all type constructors are contractive.

Lemma 8. *The maps over games* $\cdot \times \cdot$ *and* $\cdot \to \cdot$ *are contractive in all of their arguments.*

5 Interpreting Types

In order to interpret types, we need to give values to free type variables. A *type environment* is a map from type variables to games; we range over type environments with the Greek letter η. We write $\eta[X \backslash A]$ for the type environment

[1] Actually, they are in this case, but would no longer necessarily be so if we chose to use an infinite set of ground values.

that is equal to η except at X, which it maps to A, and interpret types as maps from type environments to types as follows:

$$\begin{aligned} [\![\mathbf{Bool}]\!]\eta &= \mathbf{Bool} & [\![A \times B]\!]\eta &= [\![A]\!]\eta \times [\![B]\!]\eta \\ [\![\top]\!]\eta &= \top & [\![A \to B]\!]\eta &= [\![A]\!]\eta \to [\![B]\!]\eta \\ [\![X]\!]\eta &= \eta(X) & [\![\mu X.A[X]]\!]\eta &= \mathrm{fix}(\lambda \mathcal{X}.[\![A]\!](\eta[X \backslash \mathcal{X}])) \end{aligned}$$

The well-foundedness of this definition is a consequence of the following lemma:

Lemma 9. *If A is a type, then*

(i) $[\![A]\!]$ is well-defined;
(ii) $[\![A]\!]$ is a pointwise nonexpanding map;
(iii) if A is guarded in X, then $[\![A]\!](\eta[X\backslash\mathcal{X}])$ is contractive in $\mathcal{X}$;
(iv) if A is covariant (resp. contravariant) in X, then $[\![A]\!](\eta[X\backslash\mathcal{X}])$ is monotone (resp. antimonotone) in $\mathcal{X}$.

The four properties are shown simultaneously by induction on the syntax of types. The only issue with part (i) is that of the existence and unicity of fixpoints, which is a consequence of part (iii) of the induction hypothesis, the fact that we restrict the fixpoint operator to guarded types, and Banach's fixpoint theorem. Parts (ii) and (iii) follow from Lemma 8 and the fact that d is an ultrametric, and part (iv) only depends on itself.

5.1 Soundness of Typing

The following lemma expresses the soundness of subtyping and is proved by induction on the derivation of $E \vdash A \leq A'$.

Lemma 10. (Soundness of subtyping) *Let E be an environment, and A and A' types such that $E \vdash A \leq A'$. Let η be a type environment; then $[\![A]\!]\eta \preccurlyeq [\![A']\!]\eta$.*

The main novelty in this lemma is the soundness of the last subtyping rule for the fixpoint operator; this is a consequence of Lemma 7.

Expressing the soundness of typing is slightly more involved, as we need to consider not only free type variables but also free variables.

Theorem 4. *Let E be a typing environment, M a term and A a type such that $E \vdash M : A$. Suppose that $E = X_1, \ldots X_n, x_1 : C_1, \ldots x_m : C_m$, and let $\Gamma = x_1, \ldots x_m$; let η be a typing environment, and C be the type $C = (\cdots(C_1 \times C_2) \times \cdots C_m)$. Then $[\![\Gamma \vdash M]\!] \preccurlyeq [\![C]\!]\eta \to [\![A]\!]\eta$.*

The usual statement of the *safety* of typing — that "well-typed terms cannot go wrong" — translates in our setting into the statement that terms that have a non-trivial type do not generate untrappable errors.

Corollary 1. (Safety of Typing) *If $\vdash M : A$, where M is a closed term and A a closed type such that $[\![A]\!]\varnothing \neq_\epsilon \top$, then it is not the case that $M \Downarrow_\epsilon$* **top**.

6 Conclusions and Further Work

In a previous work [7], we have shown how the addition of explicit untrappable errors to a simple λ-calculus with ground values induces a notion of subtyping, and have shown a sound Game Semantics model of the calculus with explicit errors and subtyping. In this paper, we have shown how an minor modification of the operational semantics makes our model computationally adequate. We believe that this calculus combines the best features of the standard and lazy semantics of the λ-calculus.

In addition, we have shown how the model supports recursive types by using fairly standard machinery, mainly a variant of the tree topology, and Banach's fixpoint theorem. By proving a property relating various order-theoretic and metric topologies, we have shown how our model validates the subtyping rule proposed by Amadio and Cardelli.

There is, however, an issue remaining. In [7], we have shown how the model supports bounded quantification. As we note in [8], we have been unable to make recursive types and quantifiers coexist in the same model. Indeed, while there is no problem with quantifying over fixpoints, there is no apparent reason why a least upper bound of contractive maps should itself be contractive; the issue is analogous to the well-known lack of properties of intersection with respect to Hausdorff's metric.

References

1. S. Abramsky. The lazy Lambda calculus. In D. Turner, editor, *Research Topics in Functional Programming*. Addison Wesley, 1990.
2. S. Abramsky. Semantics of interaction. In P. Dybjer and A. M. Pitts, editors, *Semantics and Logics of Computation*. Cambridge University Press, 1997.
3. S. Abramsky, R. Jagadeesan, and P. Malacaria. Full abstraction for PCF (extended abstract). In *Proc. TACS'94*, volume 789 of *Lecture Notes in Computer Science*, pages 1–15. Springer-Verlag, 1994.
4. R. M. Amadio and L. Cardelli. Subtyping recursive types. *ACM Transactions on Programming Languages and Systems*, 15(4):575–631, 1993.
5. P. Baillot, V. Danos, T. Ehrhard, and L. Regnier. Believe it or not, AJM's games model is a model of classical linear logic. In *Proceedings of the Twelfth International Symposium on Logic in Computer Science*. IEEE Computer Society Press, 1997.
6. R. Cartwright, P.-L. Curien, and M. Felleisen. Fully abstract models of observably sequential languages. *Information and Computation*, 111(2):297–401, 1994.
7. J. Chroboczek. Game Semantics and Subtyping In *Proceedings of the Fifteenth Annual IEEE Symposium on Logic in Computer Science*, Santa Barbara, California, June 2000.
8. J. Chroboczek. *Game Semantics and Subtyping*. Forthcoming PhD thesis, University of Edinburgh, 2000.
9. R. Harmer. *Games and Full Abstraction for Nondeterministic Languages*. PhD thesis, Imperial College, University of London, 1999.
10. J. M. E. Hyland and C.-H. L. Ong. On full abstraction for PCF: I, II and III. 1994.
11. G. McCusker. *Games and Full Abstraction for a Functional Metalanguage with Recursive Types*. PhD thesis, Imperial College, University of London, 1996.

Typing Lambda Terms in Elementary Logic with Linear Constraints

Paolo Coppola and Simone Martini

Dipartimento di Matematica e Informatica
Università di Udine
33100 Udine, Italy
{coppola,martini}@dimi.uniud.it

Abstract. We present a type inference algorithm for λ-terms in Elementary Affine Logic using linear constraints. We prove that the algorithm is correct and complete.

Introduction

The optimal reduction of λ-terms ([9]; see [3] for a comprehensive account and references) is a graph-based technique for normalization in which a redex is never duplicated. To achieve this goal, the syntax tree of the term is transformed into a graph, with an explicit node (a *fan*) expressing the sharing of two common subterms (these subterms are always variables in the initial translation). Giving correct reduction rules for these graphs is a surprisingly difficult problem, first solved in [8,7]. One of the main issues is to decide how to reduce two meeting fans, for which a complex machinery and new nodes have to be added (the *oracle*). There is large class of (typed) terms, however, for which this decision is very simple, namely the terms typeable in Elementary Logic, both in the Linear [6] (ELL) and the Affine [1] (EAL) flavor. Indeed, any proof-net for ELL or EAL may be (optimally) reduced with a simple check for the matching of fans. This fact was first observed in [1] and then exploited in [2] to obtain a certain complexity result on optimal reduction, where (following [10]) we also showed that EAL-typed λ-terms are powerful enough to encode arbitrary computations of elementary bounded Turing machines. We did not know, however, of any systematic way to derive EAL-types for λ-terms, a crucial issue if we want to exploit in an optimal reducer the added benefits of this class of terms. This is what we present in this paper.

Main contribution of the paper is a type inference algorithm (Section 2 and Appendix), assigning EAL-types (formulas) to *type-free* λ-terms. A typing derivation of a λ-term M in EAL consists of a *skeleton* – given by the derivation of a type for M in the simple type discipline – together with a *box assignment*, essential because EAL allows contraction only on boxed terms. The algorithm tries to introduce all possible boxes by collecting integer linear constraints during the exploration of the syntax tree of M. At the end, the integer solutions (if any) to the constraints give specific box assignments (i.e., EAL-derivations) for

S. Abramsky (Ed.): TLCA 2001, LNCS 2044, pp. 76–90, 2001.

M (for other approaches to the boxing of intuitionistic derivations, see [5,13]). Correctness and completeness of the algorithm are proved with respect to a natural deduction system for EAL, introduced in Section 2.1 together with terms annotating the derivations. For such term calculus we prove the main standard properties, including subject reduction.

1 Elementary Affine Logic

Elementary Affine Logic [1] (Figure 1) is a system with unrestricted weakening, where contraction is allowed only for modal formulas. There is only one *exponential* rule for the modality ! (*of-course*, or *bang*), which is introduced at once on both sides of the turnstile. Cut-elimination may be proved for EAL in a standard way.

$$\dfrac{}{A \vdash\ldots A}\ ax \qquad \dfrac{\Gamma \vdash\ldots A \quad A, \Delta \vdash\ldots B}{\Gamma, \Delta \vdash\ldots B}\ cut$$

$$\dfrac{\Gamma \vdash\ldots B}{\Gamma, A \vdash\ldots B}\ weak \qquad \dfrac{\Gamma, !A, !A \vdash\ldots B}{\Gamma, !A \vdash\ldots B}\ contr$$

$$\dfrac{\Gamma, A \vdash\ldots B}{\Gamma \vdash\ldots A \multimap B}\ {\multimap} R \qquad \dfrac{\Gamma \vdash\ldots A \quad B, \Delta \vdash\ldots C}{\Gamma, A \multimap B, \Delta \vdash\ldots C}\ {\multimap} L$$

$$\dfrac{A_1, \ldots, A_n \vdash\ldots B}{!A_1, \ldots, !A_n \vdash\ldots !B}\ !$$

Fig. 1. (Implicational) Elementary Affine Logic

A simple inspection of the rules of EAL shows that any λ-term with an EAL type has also a simple type[1]. Indeed, the simple type (and the corresponding derivation) is obtained by forgetting the exponentials, which must be present in an EAL derivation because of contraction.

The idea underlying our type inference algorithm is simple:

1. finding all "maximal decorations";
2. solving sets of linear constraints.

We informally present the main point with an example on the term $two \equiv \lambda xy.(x(x\ y))$. One (sequent) simple type derivation for two is:

$$\dfrac{\dfrac{\dfrac{\dfrac{\dfrac{\dfrac{}{w{:}\alpha\vdash w{:}\alpha} \quad \dfrac{}{y{:}\alpha\vdash y{:}\alpha}}{x{:}\alpha\to\alpha, y{:}\alpha\vdash (x\ y){:}\alpha} \quad \dfrac{}{z{:}\alpha\vdash z{:}\alpha}}{x{:}\alpha\to\alpha, x{:}\alpha\to\alpha, y{:}\alpha\vdash (x(x\ y)){:}\alpha}}{x{:}\alpha\to\alpha, x{:}\alpha\to\alpha\vdash \lambda y.(x(x\ y)){:}\alpha\to\alpha}}{x{:}\alpha\to\alpha\vdash \lambda y.(x(x\ y)){:}\alpha\to\alpha}}{\vdash \lambda xy.(x(x\ y)){:}(\alpha\to\alpha)\to\alpha\to\alpha}$$

[1] However there are simply typed terms not typeable in EAL, see [2].

If we change every $\rightarrow$ in $\multimap$, the previous derivation can be viewed as the skeleton of an EAL derivation. To obtain a full EAL derivation (provided it exists), we need to decorate this skeleton with exponentials, and to check that the contraction is performed only on exponential formulas.

Let's produce first a *maximal decoration* of the skeleton, interleaving n ! introduction rules after each logical rule. For example

$$\dfrac{\dfrac{}{w{:}\alpha \vdash w{:}\alpha} \quad \dfrac{}{y{:}\alpha \vdash y{:}\alpha}}{x{:}\alpha \multimap \alpha, y{:}\alpha \vdash (x\ y){:}\alpha}$$

becomes

$$\dfrac{\dfrac{\dfrac{}{w{:}\alpha \vdash w{:}\alpha}}{!^{n_1} w{:}\alpha \vdash !^{n_1} w{:}\alpha}\,!^{n_1} \quad \dfrac{\dfrac{}{y{:}\alpha \vdash y{:}\alpha}}{!^{n_2} y{:}\alpha \vdash !^{n_2} y{:}\alpha}\,!^{n_2}}{x{:}!^{n_2}\alpha \multimap !^{n_1}\alpha, y{:}!^{n_2}\alpha \vdash (x\ y){:}!^{n_1}\alpha}$$

where n_1 and n_2 are fresh variables. We obtain in this way a meta-derivation representing all EAL derivations with $n_1, n_2 \in \mathbb{N}$.

Continuing to decorate the skeleton of *two* (i.e. to interleave n_i ! rules) we obtain

$$\dfrac{\dfrac{\dfrac{\dfrac{\dfrac{\dfrac{\dfrac{\dfrac{}{w{:}\alpha \vdash w{:}\alpha}}{w{:}!^{n_1}\alpha \vdash w{:}!^{n_1}\alpha}\,!^{n_1} \quad \dfrac{\dfrac{}{y{:}\alpha \vdash y{:}\alpha}}{y{:}!^{n_2}\alpha \vdash y{:}!^{n_2}\alpha}\,!^{n_2}}{x{:}!^{n_2}\alpha \multimap !^{n_1}\alpha, y{:}!^{n_2}\alpha \vdash (x\ y){:}!^{n_1}\alpha}}{x{:}!^{n_3}(!^{n_2}\alpha \multimap !^{n_1}\alpha), y{:}!^{n_2+n_3}\alpha \vdash (x\ y){:}!^{n_1+n_3}\alpha}\,!^{n_3} \quad \dfrac{\dfrac{}{z{:}\alpha \vdash z{:}\alpha}}{z{:}!^{n_4}\alpha \vdash z{:}!^{n_4}\alpha}\,!^{n_4}}{x{:}!^{n_1+n_3}\alpha \multimap !^{n_4}\alpha, x{:}!^{n_3}(!^{n_2}\alpha \multimap !^{n_1}\alpha), y{:}!^{n_2+n_3}\alpha \vdash (x(x\ y)){:}!^{n_4}\alpha}}{x{:}!^{n_5}(!^{n_1+n_3}\alpha \multimap !^{n_4}\alpha), x{:}!^{n_3+n_5}(!^{n_2}\alpha \multimap !^{n_1}\alpha), y{:}!^{n_2+n_3+n_5}\alpha \vdash (x(x\ y)){:}!^{n_4+n_5}\alpha}\,!^{n_5}}{x{:}!^{n_5}(!^{n_1+n_3}\alpha \multimap !^{n_4}\alpha), x{:}!^{n_3+n_5}(!^{n_2}\alpha \multimap !^{n_1}\alpha) \vdash \lambda y.(x(x\ y)){:}!^{n_2+n_3+n_5}\alpha \multimap !^{n_4+n_5}\alpha}}{x{:}!^{n_5+n_6}(!^{n_1+n_3}\alpha \multimap !^{n_4}\alpha), x{:}!^{n_3+n_5+n_6}(!^{n_2}\alpha \multimap !^{n_1}\alpha) \vdash \lambda y.(x(x\ y)){:}!^{n_6}(!^{n_2+n_3+n_5}\alpha \multimap !^{n_4+n_5}\alpha)}\,!^{n_6}}{x{:}!^{n_5+n_6}(!^{n_1+n_3}\alpha \multimap !^{n_4}\alpha) \vdash \lambda y.(x(x\ y)){:}!^{n_6}(!^{n_2+n_3+n_5}\alpha \multimap !^{n_4+n_5}\alpha)}$$

The last rule – contraction – is correct in EAL iff the types of x are unifiable and banged. In other words iff the following constraints are satisfied:

$$n_1,n_2,n_3,n_4,n_5,n_6 \in \mathbb{N} \quad \wedge \quad n_5 = n_3 + n_5 \quad \wedge \quad n_1 + n_3 = n_2 \quad \wedge \quad n_4 = n_1 \quad \wedge \quad n_5 + n_6 \geq 1.$$

The second, third and fourth of these constraints come from unification; the last one from the fact that contraction is allowed only on exponential formulas. These constraints are equivalent to

$$n_1,n_5,n_6 \in \mathbb{N} \quad \wedge \quad n_3 = 0 \quad \wedge \quad n_1 = n_2 = n_4 \quad \wedge \quad n_5 + n_6 \geq 1.$$

Since clearly these constraints admit solutions, we conclude the decoration procedure obtaining

$$\dfrac{\dfrac{\vdots}{x{:}!^{n_5+n_6}(!^{n_1}\alpha \multimap !^{n_1}\alpha) \vdash \lambda y.(x(x\ y)){:}!^{n_6}(!^{n_1+n_5}\alpha \multimap !^{n_1+n_5}\alpha)}}{\vdash \lambda xy.(x(x\ y)){:}!^{n_5+n_6}(!^{n_1}\alpha \multimap !^{n_1}\alpha) \multimap !^{n_6}(!^{n_1+n_5}\alpha \multimap !^{n_1+n_5}\alpha)}$$

Thus *two* has EAL types $!^{n_5+n_6}(!^{n_1}\alpha \multimap !^{n_1}\alpha) \multimap !^{n_6}(!^{n_1+n_5}\alpha \multimap !^{n_1+n_5}\alpha)$, for any n_1, n_5, n_6 solutions of

$$n_1,n_5,n_6 \in \mathbb{N} \quad \wedge \quad n_5+n_6 \geq 1.$$

We may display the full derivation in a more manageable way, representing the skeleton with the syntax tree of the lambda term with edges labelled with types and adding boxes representing the ! introduction rules, as in Figure 2.

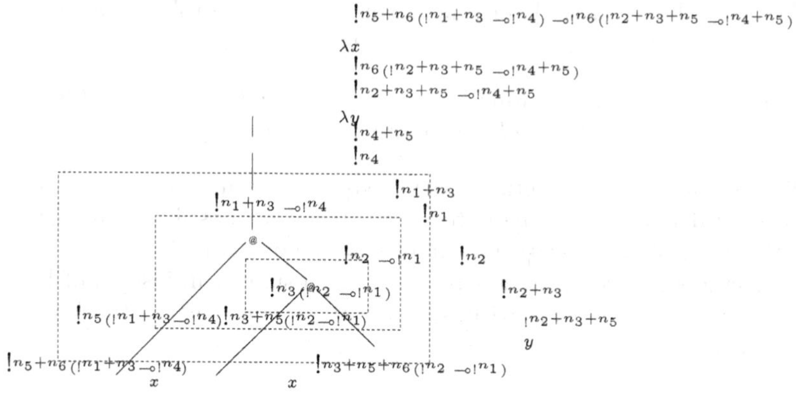

Fig. 2. Meta EAL type derivation of *two*.

Finally notice that at the beginning of this section, we started with "*one (sequent) derivation*" for *two* (there are other derivations, building in a different way the application $x(xy)$)). If that derivation had produced an unsolvable set of constraint, the procedure should restart with another derivation. To avoid this problem, our search for maximal decorations (i.e., the collection of constraints) is not performed on sequent derivations, but on the syntax tree of the term. However, the fact that multiple derivations for a term and principal type scheme are possible, will surface again. It may happen that a solution to a set of constraints corresponds to more than one derivation (a *superposition of derivations*), with non compatible box-assignments. In this case, Lemma 6 ensures that compatible box assignments may be found.

2 Type Inference

A class of types for an EAL-typeable term can be seen as a decoration of a simple type with a suitable number of boxes. The main contribution of the paper is an algorithm collecting integer constraints whose solutions corresponds to proper box assignments.

Definition 1. *A* general EAL-type Θ *is generated from the following grammar:*

$$\Theta ::= o|\Theta \multimap \Theta|!^{n_1+\cdots+n_k}\Theta,$$

where $n_1, \ldots, n_k$ *are variables ranging on integers* $\mathbb{Z}$.

Definition 2 (Type Synthesis Algorithm). *Given a simply typeable lambda term and its principal type scheme $M : \sigma$, the* type synthesis algorithm $\mathcal{S}(M : \sigma)$ *returns a triple $\langle \Theta, B, A \rangle$, where Θ is a general EAL-type, B is a base (i.e. a multi-set of pairs variable, general EAL-type) and A is a set of linear constraints.*

The algorithm $\mathcal{S}(M : \sigma)$ is defined in the Appendix. One of the crucial issues is the localization of the points where derivations may differ for the presence or absence of boxes around some subterms. This is the role of *critical points*, managed by the boxing procedure, $\mathcal{B}$ (see A.3).

Proposition 1 (Termination). *Let M be a simply typed term and let σ be its most general type. $\mathcal{S}(M : \sigma)$ always terminates with a triple $\langle \Theta, B, A \rangle$.*

The algorithm is exponential in the size of the λ-term, because to investigate all possible derivations we need to (try to) box all possible combinations of critical points (see the clauses for the product union, $\uplus$, in A.4).

Correctness and completeness of $\mathcal{S}$ are much simpler if, instead of EAL, we formulate proofs and results with reference to an equivalent natural deduction formulation.

2.1 NEAL

The natural deduction calculus (NEAL) for EAL in given in Figure 3, after [4, 1,12].

Lemma 1 (Weakening). *If $\Gamma \vdash .... A$ then $B, \Gamma \vdash .... A$.*

$$\frac{}{\Gamma, A \vdash .... A}\ ax \qquad \frac{\Gamma \vdash .... !A \quad \Delta, !A, !A \vdash .... B}{\Gamma, \Delta \vdash .... B}\ contr$$

$$\frac{\Gamma, A \vdash .... B}{\Gamma \vdash .... A \multimap B}\ (\multimap I) \qquad \frac{\Gamma \vdash .... A \multimap B \quad \Delta \vdash .... A}{\Gamma, \Delta \vdash .... B}\ (\multimap E)$$

$$\frac{\Delta_1 \vdash .... !A_1 \cdots \Delta_n \vdash .... !A_n \quad A_1, \ldots, A_n \vdash .... B}{\Gamma, \Delta_1, \ldots, \Delta_n \vdash .... !B}\ !$$

Fig. 3. Natural Elementary Affine Logic in sequent style notation

To annotate NEAL derivations, we use terms generated by the following grammar (*elementary terms*):

$$M ::= x \mid \lambda x.M \mid (M\ M) \mid !\,(M)\,[M/x, \ldots, M/x] \mid \|M\|^{M}_{x,x}$$

Observe that in $!\,(M)\,[M/x, \ldots, M/x]$, the $[M/x]$ is a kind of explicit substitution. To define ordinary substitution, define first the set of free variables of a term M, $\mathtt{FV}(M)$, inductively as follows:

- $\mathtt{FV}(x)=\{x\}$
- $\mathtt{FV}(\lambda x.M)=\mathtt{FV}(M)\setminus\{x\}$
- $\mathtt{FV}(M_1\ M_2)=\mathtt{FV}(M_1)\cup\mathtt{FV}(M_2)$
- $\mathtt{FV}(!(M)[M_1/x_1,...,M_n/x_n])=\bigcup_{i=1}^{n}\mathtt{FV}(M_i)$
- $\mathtt{FV}(\|M\|_{x_1,x_2}^{N})=(\mathtt{FV}(M)\setminus\{x_1,x_2\})\cup\mathtt{FV}(N)$

Ordinary substitution $N\{M/x\}$ of a term M for the free occurrences of x in N, is defined in the obvious way. The (pedantic) exponential cases are as follows:

1. $!(N)[P_1/x_1,...,P_n/x_n]\{M/x\}=!(N\{y_1/x_1\}\cdots\{y_n/x_n\}\{M/x\})[P_1\{M/x\}/y_1,...,P_n\{M/x\}/y_n]$ if $x\notin\{x_1,...,x_n\}$, where $y_1,...,y_n$ are all fresh variables;
2. $!(N)[P_1/x_1,...,P_n/x_n]\{M/x\}=!(N)[P_1\{M/x\}/x_1,...,P_n\{M/x\}/x_n]$ if $\exists i$ s.t. $x_i=x$;
3. $\|N\|_{y,z}^{P}\{M/x\}=\|N\{y'/y\}\{z'/z\}\{M/x\}\|_{y',z'}^{P\{M/x\}}$ if $x\notin\{y,z\}$, where y',z' are fresh variables;
4. $\|N\|_{y,z}^{P}\{M/x\}=\|N\|_{y,z}^{P\{M/x\}}$ if $x\in\{y,z\}$.

Elementary terms may be mapped to λ-terms, by forgetting the exponential structure:

- $x^*=x$
- $(\lambda x.M)^*=\lambda x.M^*$
- $(M_1\ M_2)^*=(M_1^*\ M_2^*)$
- $(!(M)[M_1/x_1,...,M_n/x_n])^*=M^*\{M_1^*/x_1,...,M_n^*/x_n\}$
- $(\|M\|_{x_1,x_2}^{N})^*=M^*\{N^*/x_1,N^*/x_2\}$

Definition 3 (Legal elementary terms). *The elementary terms are* legal *under the following conditions:*

1. *x is legal;*
2. *$\lambda x.M$ is legal iff M is legal;*
3. *$(M_1\ M_2)$ is legal iff M_1 and M_2 are both legal and $\mathtt{FV}(M_1)\cap\mathtt{FV}(M_2)=\emptyset$;*
4. *$!(M)[M_1/x_1,\dots,M_n/x_n]$ is legal iff M and M_i are legal for any i $1\le i\le n$ and $\mathtt{FV}(M)=\{x_1,\dots,x_n\}$ and $(i\neq j\Rightarrow\mathtt{FV}(M_i)\cap\mathtt{FV}(M_j)=\emptyset)$;*
5. *$\|M\|_{x,y}^{N}$ is legal iff M and N are both legal and $\mathtt{FV}(M)\cap\mathtt{FV}(N)=\emptyset$.*

Proposition 2. *If M is a legal term, then every free variable $x\in\mathtt{FV}(M)$ is linear in M.*

Note 1. From now on we will consider only legal terms.

Notation. *Let $\Gamma=\{x_1:A_1,\dots,x_n:A_n\}$ be a basis. $dom(\Gamma)=\{x_1,\dots,x_n\}$; $\Gamma(x_i)=A_i$; $\Gamma\restriction V=\{x:A|x\in V\wedge A=\Gamma(x)\}$.*

Legal terms are the ones induced by the Curry-Howard isomorphism applied to NEAL-derivations (see [11,12] for different approaches to Curry-Howard isomorphism for Linear and Light Linear Logic). The term assignment system is shown in Figure 4, where all bases in the premises of the contraction, $\multimap$ elimination and !-rule, have domains with empty intersection.

$$\frac{}{\Gamma, x : A \vdash x : A}\ ax \qquad \frac{\Gamma \vdash M :!A \quad \Delta, x :!A, y :!A \vdash N : B}{\Gamma, \Delta \vdash \|N\|^{M}_{x,y} : B}\ contr$$

$$\frac{\Gamma, x : A \vdash M : B}{\Gamma \vdash \lambda x.M : A \multimap B}\ (\multimap I) \qquad \frac{\Gamma \vdash M : A \multimap B \quad \Delta \vdash N : A}{\Gamma, \Delta \vdash (M\ N) : B}\ (\multimap E)$$

$$\frac{\Delta_1 \vdash M_1 :!A_1 \cdots \Delta_n \vdash M_n :!A_n \quad x_1 : A_1, \ldots, x_n : A_n \vdash N : B}{\Gamma, \Delta_1, \ldots, \Delta_n \vdash !(N)\,[M_1/x_1, \ldots, M_n/x_n] :!B}\ !$$

Fig. 4. Term Assignment System for Natural Elementary Affine Logic

Lemma 2. *1. If* $\Gamma \vdash \mathbf{....}\ M : A$ *then* $\mathtt{FV}(M) \subseteq dom(\Gamma)$*;*
2. if $\Gamma \vdash \mathbf{....}\ M : A$ *then* $\Gamma \restriction \mathtt{FV}(M) \vdash \mathbf{....}\ M : A$.

Lemma 3 (Substitution). *If* $\Gamma, x : A \vdash \mathbf{....}\ M : B$ *and* $\Delta \vdash \mathbf{....}\ N : A$ *and* $dom(\Gamma) \cap dom(\Delta) = \emptyset$ *then* $\Gamma, \Delta \vdash \mathbf{....}\ M\{N/x\} : B$.

Theorem 1 (Equivalence). $\Gamma \vdash_{EAL} A$ *if and only if* $\Gamma \vdash_{NEAL} A$.

Lemma 4 (Unique Derivation). *For any legal term* M *and formula* A*, if there is a valid derivation of the form* $\Gamma \vdash \mathbf{....}\ M : A$*, then such derivation is unique (up to weakening).*

Although we are not interested in this paper in the dynamics (i.e., normalization) of NEAL, a notion of reduction is needed to state and obtain our main result. We have first two *logical* reductions ($\rightarrow_\beta$ and $\rightarrow_{\mathbf{...}}$) corresponding to the elimination of principal cuts in EAL. The other five reductions are permutation rules, allowing contraction to be moved out of a term.

$$(\lambda x.M\ N) \qquad \rightarrow_\beta \qquad M\{N/x\}$$

$$\|N\|^{!(M)[M_1/x_1,\ldots,M_n/x_n]}_{x,y} \qquad \rightarrow_{\mathbf{...}}$$
$$\left\|\left\|N\{^{!(M)[x'_1/x_1,\ldots,x'_n/x_n]}/x\}\{^{!(M')[y'_1/y_1,\ldots,y'_n/y_n]}/y\}\right\|^{M_1}_{x'_1,y'_1} \cdots\right\|^{M_n}_{x'_n,y'_n}$$

$$!(M)[^{M_1}/x_1, \cdots, ^{!(N)[P_1/y_1,\ldots,P_m/y_m]}/x_i, \cdots, ^{M_n}/x_n] \qquad \rightarrow_{!-!}$$
$$!(M\{N/x_i\})[^{M_1}/x_1, \cdots, ^{P_1}/y_1, \cdots, ^{P_m}/y_m, \cdots ^{M_n}/x_n]$$

$$(\|M\|^{M_1}_{x_1,x_2}\ N) \qquad \rightarrow_{@-\bullet} \qquad \|(M\{x'_1/x_1, x'_2/x_2\}\ N)\|^{M_1}_{x'_1,x'_2}$$

$$!(M)[^{M_1}/x_1, \cdots, ^{\|M_i\|^{N}_{y,z}}/x_i, \cdots, ^{M_n}/x_n] \qquad \rightarrow_{!-c}$$
$$\left\|!(M)[^{M_1}/x_1, \cdots, ^{M_i\{y'/y,z'/z\}}/x_i, \cdots, ^{M_n}/x_n]\right\|^{N}_{y',z'}$$

$$\|M\|_{x_1,x_2}^{\|N\|_{y_1,y_2}^{P}} \quad \to_{\bullet-\bullet} \quad \left\| \|M\|_{x_1,x_2}^{N\{y_1'/y_1,y_2'/y_2\}} \right\|_{y_1',y_2'}^{P}$$

$$\lambda x.\, \|M\|_{y,z}^{N} \quad \to_{\lambda-\bullet} \quad \|\lambda x.M\|_{y,z}^{N} \text{ where } x \notin \mathtt{FV}(N)$$

where M' in the $\to_{\bullet\bullet\bullet}$-rule is obtained from M replacing all its free variables with fresh ones (x_i is replaced with y_i); x_1' and x_2' in the $\to_{@-\bullet}$-rule, y' and z' in the $\to_{!-c}$-rule and y_1', y_2' in the $\to_{\bullet-\bullet}$-rule are fresh variables.

Definition 4. *The reduction relation on legal terms* $\rightsquigarrow$ *is defined as the reflexive and transitive closure of the union of* $\to_\beta, \to_{\bullet\bullet\bullet}, \to_{!-!}, \to_{@-\bullet}, \to_{!-\bullet}, \to_{\bullet-\bullet}, \to_{\lambda-\bullet}$.

Proposition 3. *Let* $M \rightsquigarrow N$ *and* M *be a legal term, then* N *is a legal term.*

Proposition 4. *Let* $M \to_r N$ *where* r *is not* $\to_\beta$, *then* $M^* = N^*$.

Lemma 5. *Let* M *be in* $\{\mathsf{dup}, !-!, @-\mathsf{c}, !-\mathsf{c}, \mathsf{c}-\mathsf{c}, \lambda-\mathsf{c}\}$*-normal form, then*

1. *if* $R = \|N\|_{x,y}^{P}$ *is a subterm of* M, *then either* $P = (P_1\ P_2)$ *or* P *is a variable;*
2. *if* $R = !(N)\,[P_1/x_1, \ldots, P_k/x_k]$ *is a subterm of* M, *then for any* $i \in \{1, \ldots, k\}$ *either* $P_i = (Q_i\ S_i)$ *or* P_i *is a variable.*

Theorem 2 (Subject Reduction). *Let* $\Gamma \vdash_{\bullet\bullet\bullet\bullet} M : A$ *and* $M \rightsquigarrow N$, *then* $\Gamma \vdash_{\bullet\bullet\bullet\bullet} N : A$.

2.2 Properties of the Type Syntesis Algorithm

Lemma 6 (Superimposing of derivations). *Let* $\mathcal{S}(M : \sigma) = \langle \Theta, B, A \rangle$ *and let* A *be solvable. If there is a solution* X_1 *of* A *that instantiates two boxes belonging to two superimposed derivations that are not compatible, then there exists another solution* X_2 *where all the instantiated boxes belong to the same derivation.*

Moreover the instantiations Θ', B' *of* Θ, B *using* X_1 *and the instantiations* Θ'', B'' *of* Θ, B *using* X_2 *are identical.*

Proof. (sketch) We may think of boxes as levels; boxing a subterm can then be seen as raising that subterm, as in Figure 5, where also some types label the edges of the syntax tree of a simple term. In particular, the edge starting from the @-node and ending in x_0 has label $!^{n_2}(\alpha \multimap !^{n_1}(\beta \multimap \gamma))$ at level 0 (nearest to x_0) and has label $(\alpha \multimap !^{n_1}(\beta \multimap \gamma))$ at level n_2. This is the graphical counterpart of the !-rule

$$\frac{\ldots, x_0 : T, \ldots \vdash \ldots}{\ldots, x_0 : !^{n_2}T, \ldots \vdash \ldots}\ !^{n_2}$$

The complete decoration of Figure 5 can be produced in NEAL in two ways: by

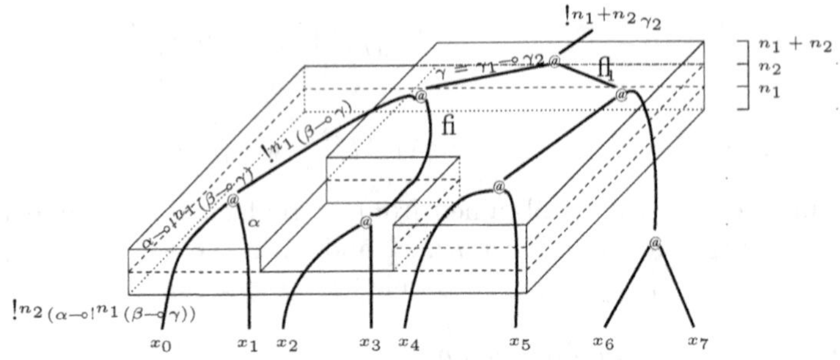

Fig. 5. Boxes can be viewed as levels.

the instantiation of

$$!^{n_2}\,((((x_0\ x_1)y)((x_4\ x_5)w)))\,[(x_2\ x_3)/y, (x_6\ x_7)/w]$$

and[2]

$$!^{n_1}\,(((z(x_2\ x_3))((x_4\ x_5)w)))\,[(x_0\ x_1)/z, (x_6\ x_7)/w],$$

which are boxes belonging to two different derivations. Graphically such an instantiation can be represented as in the first row of Figure 6, where incompatibility is evident by the fact that the boxes are not well stacked, in particular the rectangular one covers a hole. To have a correct EAL-derivation it is necessary to find the equivalent, well stacked configuration (that corresponds to the subsequent application of boxes from the topmost to the bottommost).

The procedure by which we find the well stacked box configuration is visualized in Figure 6. The reader may imagine the boxes subject to gravity (the passage from the first to the second row of Figure 6) and able to fuse each other when they are at the same level (the little square in the third row fuse with the solid at its left in the passage from the third to the fourth row).

The "gravity operator" corresponds to finding the minimal common subterm of all the superimposed derivations and it is useful for finding the correct order of application of the ! rule. The "fusion operator" corresponds to the elimination of a cut between two exponential formulas. Moreover, the final configuration of Figure 6 corresponds to a particular solution of the set of constraints produced by the type synthesis algorithm, that instantiates the following boxes:

$$!^{n_1}\,(!^{n_2-n_1}\,(!^{n_1}\,(((z\ w)((x_4\ x_5)t)))\,[(x_0\ x_1)/z])\,[(x_2\ x_3)/w])\,(x_6\ x_7)/t]$$

Finally, notice that during the procedure all types labelling the boundary edges of the lambda-term never changes, i.e. the instantiations of the term type (the label of the topmost edge) and the base types (the labels of the edges at the bottom) remain unchanged.

[2] The correct legal terms should have all free variable inside the square brackets. We omit to write variables when they are just renamed, for readability reasons (compare the first elementary term above with the correct one $!^{n_2}\,((((x_0\ x_1)y)((x_4\ x_5)w)))\,[\dot{x_0}/x_0, \dot{x_1}/x_1, (x_2\ x_3)/y, \dot{x_4}/x_4, \dot{x_5}/x_5, (x_6\ x_7)/w]$).

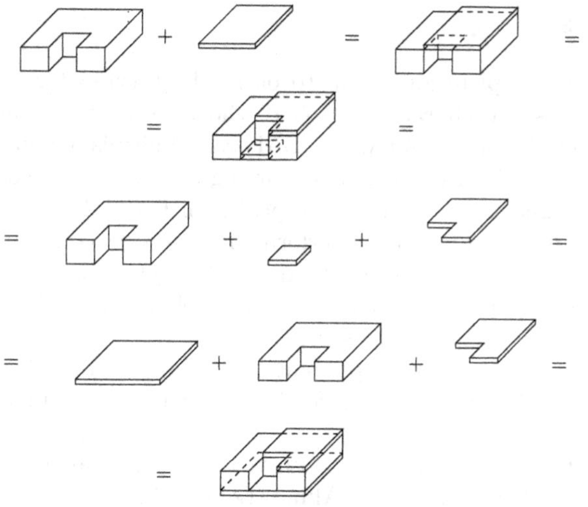

Fig. 6. Equivalences of boxes.

Theorem 3 (Soundness). *Let $\mathcal{S}(M:\sigma) = \langle \Theta, B, A\rangle$. For every X integer solution of A, and B', Θ' instantiations of B and Θ using X, there exists P elementary term such that $P^* = M$ and $B' \vdash P : \Theta'$ is derivable in NEAL.*

Proof. By induction on the structure of M, using the superimposing lemma.

Theorem 4 (Completeness). *Let $\Gamma \vdash\!\!\ldots\; P : \Psi$ and let P be in $\{!-!, @-\mathsf{c}, !-\mathsf{c}, \mathsf{c}-\mathsf{c}, \lambda-\mathsf{c}, \mathsf{dup}\}$-normal form with contraction only on variables ($\|R\|^{Q}_{x,y}$ is a subterm of P only if Q is a variable). Let $\mathcal{S}(P^* : \overline{\Psi}) = \langle \Theta, B, A\rangle$, where $\overline{\Psi}$ is the erasure of Ψ, i.e. the simple type obtained from Ψ erasing all ! and converting $\multimap$ in $\rightarrow$, then there exist X integer solution of A such that the instantiation B' of B using X is a subset of Γ and Ψ is the instantiation of Θ using X and $B' \vdash\!\!\ldots\; P : \Psi$.*

The request on the $\{!-!, @-\mathsf{c}, !-\mathsf{c}, \mathsf{c}-\mathsf{c}, \lambda-\mathsf{c}, \mathsf{dup}\}$-normal form is not a loss of generality, for the subject reduction lemma and Proposition 4. By Lemma 5, the only restriction is the exclusion of elementary terms with subterms of the form $\|R\|^{(Q_1\ Q_2)}_{x,y}$. In a sense, these terms "contract too much". Indeed, it could be the case that a term P is elementary thanks to the sharing of a β-redex (inside $(Q_1\ Q_2)$). However, the corresponding λ-term P^*, cannot share any redex – there is no sufficient syntax for this in the λ-calculus – hence P^* could be not elementary. As we discussed in the Introduction, our aim is to identify λ-terms that are reducible using optimal reduction without the *oracle* needed for the correct matching of fans. The NEAL terms excluded in the completeness theorem corresponds to EAL proof-nets which are not (the initial encoding of) λ-terms, since they contract an application.

Conclusions

We presented a complete algorithm to derive EAL-types for λ-terms. One of our main goals is the characterization of those terms that can be optimally reduced without the oracle, for which EAL-typeability is a sufficient condition. One should not see (N)EAL as a programming language; instead, it is a kind of intermediate language: if a λ-term is typeable in EAL, then we can compile it in a special manner with excellent performances during reduction, otherwise we compile it in the usual way, using the oracle. To get a more powerful (and, to a certain extent, flexible) language, a major development of this work would be the extension of our algorithm to second order EAL.

The same technique of this paper may be applied to Multiplicative Exponential Linear Logic. However, to treat dereliction, the number of constraints grows in an exponential way.

We believe techniques similar to those we used in this paper may be applied to type-inference for Light Linear (or Affine) Logic (LLL), a system characterizing polytime. A type-inference for LLL would be a uniform proof-technique to prove polynomiality of certain algorithms.

A puzzling open problem is whether there exist terms yielding constraints with only non integer solutions. Of course they have to be non EAL-typeable terms, in view of our completeness theorem. Our estensive experiments never produced such a scenario, yet we could not prove that the constraints have always integral solutions. Would there be any logical meaning for a term with a non integral number of boxes?

Acknowledgments. Harry Mairson provided useful criticism and comments on the form and the substance of the paper.

References

1. Andrea Asperti. Light affine logic. In *Proc. of the 13-th Annual IEEE Symposium on Logic in Computer Science (LICS '98)*, pages 300–308, Indianapolis, U.S.A., 1998.
2. Andrea Asperti, Paolo Coppola, and Simone Martini. (Optimal) duplication is not elementary recursive. In *ACM POPL'00*, pages 96–107, Boston, Massachusetts, January 19–21, 2000.
3. Andrea Asperti and Stefano Guerrini. *The Optimal Implementation of Functional Programming Languages*, volume 45 of *Cambridge Tracts in Theoretical Computer Science*. Cambridge University Press, 1998.
4. Nick Benton, Gavin Bierman, Valeria de Paiva, and Martin Hyland. A term calculus for intuitionistic linear logic. *TLCA'93*, volume 664 of *Lecture Notes in Computer Science*, pages 75–90, March 1993.
5. Vincent Danos, Jean-Baptiste Joinet and Harold Schellinx. On the linear decoration of intuitionistic derivations. *Archive for Mathematical Logic*, 33:387–412, 1995.
6. Jean-Yves Girard. Light linear logic. *Information and Computation*, 204:143–175, 1998.

7. Vinod K. Kathail. *Optimal Interpreters for Lambda-calculus Based Functional Programming Languages.* PhD thesis, MIT, May 1990.
8. John Lamping. An Algorithm for Optimal Lambda Calculus Reduction. In *ACM POPL '90*, pages 16–30, New York, NY, USA, 1990.
9. Jean-Jacques Lévy. Optimal reductions in the lambda-calculus. In Jonathan P. Seldin and J. Roger Hindley, editors, *To H. B. Curry: Essays on Combinatory Logic, Lambda Calculus and Formalism*, pages 159–191. Academic Press, London, 1980.
10. Harry G. Mairson. A simple proof of a theorem of Statman. *Theoretical Computer Science*, 103(2):387–394, September 1992.
11. Alberto Pravato and Luca Roversi. $\Lambda_!$ considered both as a paradigmatic language and a meta-language. In *Fifth Italian Conference on Theoretical Computer Science*, Salerno (Italy), 1995.
12. Luca Roversi. A Polymorphic Language which Is Typable and Poly-step. In *Proc. of the Asian Computing Science Conference (ASIAN'98)*, volume 1538 of *Lecture Notes in Computer Science*, pages 43 – 60, Manila (The Philippines), 1998.
13. Harold Schellinx. *The Noble Art of Linear Decorating.* PhD thesis, Institute for Logic, Language and Computation, University of Amsterdam, 1994.

A Appendix

In the following n, n_1, n_2 are always fresh variables, o is the base type. Moreover, we consider $!^{n_1}(!^{n_2}\Theta)$ syntactically equivalent to $!^{n_1+n_2}\Theta$. The list of free variable occurrences FVO is defined as follows:

1. $\mathtt{FVO}(x) = [x]$;
2. $\mathtt{FVO}(\lambda x.M) = \mathtt{FVO}(M) - x$;
3. $\mathtt{FVO}((M_1\ M_2)) = \mathtt{FVO}(M_1) + \mathtt{FVO}(M_2)$ (the concatenation of lists).

A.1 Unification: $\mathcal{U}$

$$\frac{}{\mathcal{U}\left(!^{\sum n_{i_1}}o,\ldots !^{\sum n_{i_h}}o\right)=\sum n_{i_1}-\sum n_{i_2}=0;\ldots \sum n_{i_{h-1}}-\sum n_{i_h}=0}\ \mathrm{u}o$$

$$\frac{\mathcal{U}(\Theta_{1_1},\ldots,\Theta_{1_h})=A_1 \qquad \mathcal{U}(\Theta_{2_1},\ldots,\Theta_{2_h})=A_2}{\mathcal{U}\left(!^{\sum n_{i_1}}(\Theta_{1_1}\multimap\Theta_{2_1}),\ldots,!^{\sum n_{i_h}}(\Theta_{1_h}\multimap\Theta_{2_h})\right)= \begin{array}{l}\sum n_{i_1}-\sum n_{i_2}=0;\\ \vdots \\ \sum n_{i_{h-1}}-\sum n_{i_h}=0;A_1;A_2\end{array}}\ \mathrm{u}\multimap$$

A.2 Contraction ($\mathcal{C}$) and Type Processing ($\mathcal{P}$)

$$\frac{}{\mathcal{C}(\Theta)=\emptyset}\ \mathrm{c}\emptyset \qquad \frac{\mathcal{U}(!^{n_1+\cdots+n_h}\Theta_1,\Theta_2,\ldots,\Theta_k)=A}{\mathcal{C}(!^{n_1+\cdots+n_h}\Theta_1,\ldots,\Theta_k)= \begin{array}{c}n_1+\cdots+n_h\geq 1\\ A\end{array}}\ \mathrm{c}$$

Where Θ_1 is either $\Gamma \multimap \Delta$ or o.

$$\frac{}{\mathcal{P}(o)=\langle !^n o, n\geq 0\rangle}\ \mathrm{p}o \qquad \frac{\mathcal{P}(\sigma)=\langle\Theta,A_1\rangle \quad \mathcal{P}(\tau)=\langle\Gamma,A_2\rangle}{\mathcal{P}(\sigma\to\tau)=\langle !^n(\Theta\multimap\Gamma), n\geq 0 \quad A_1 \quad A_2\rangle}\ \mathrm{p}\!\to$$

A.3 Boxing: $\mathcal{B}$

The boxing procedure superimposes all boxes due to the existence of critical points. Every time there are two possible EAL-derivations for the unique simple type, there is a critical point. For example during the type synthesis of

$$\frac{\vdots}{x{:}\alpha\to\alpha, x{:}\alpha\to\alpha, y{:}\alpha \vdash (x(x\ y)){:}\alpha}$$

we need to try all possible decorations of both derivations below

$$\frac{\dfrac{\vdots}{x{:}\alpha\to\alpha, y{:}\alpha\vdash(x\ y){:}\alpha} \quad \dfrac{}{z{:}\alpha\vdash z{:}\alpha}}{x{:}\alpha\to\alpha, x{:}\alpha\to\alpha, y{:}\alpha\vdash(x(x\ y)){:}\alpha} \qquad \frac{\dfrac{}{y{:}\alpha\vdash y{:}\alpha} \quad \dfrac{\vdots}{z{:}\alpha, x{:}\alpha\to\alpha\vdash(xz){:}\alpha}}{x{:}\alpha\to\alpha, x{:}\alpha\to\alpha, y{:}\alpha\vdash(x(x\ y)){:}\alpha}$$

In our graphical notation, the two decorations appear as follows (the one corresponding to the first derivation is at the left; the star indicates the critical point):

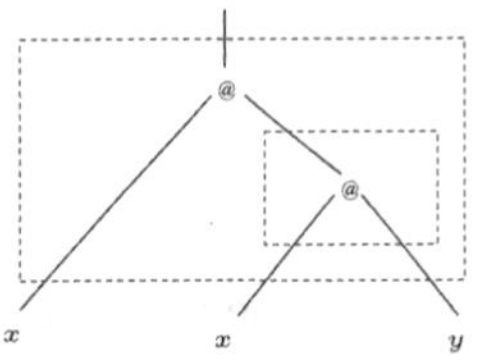

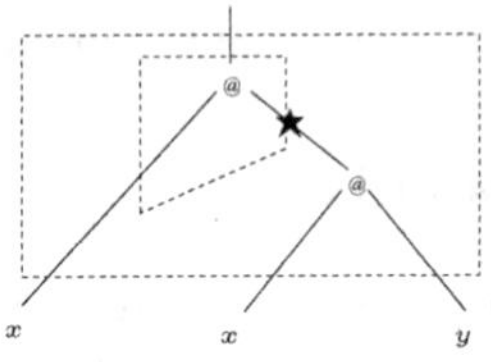

When $\mathcal{B}$ is called during the synthesis of $(x(x\ y))$, the base B is something like $\{x :!^{n_1}\alpha \multimap !^{n_2}\alpha, x :!^{n_6}(!^{n_3}\alpha \multimap !^{n_4}\alpha), y :!^{n_5+n_6}\alpha\}$, where the first x is the leftmost in the figure, the type Γ of $(x(x\ y))$ is $!^{n_2}\alpha$, the set of critical points *cpts* is $\{(n_4+n_6-n_1=0, [x,y])\}$ and the set of constraints A is $\{n_4+n_6-n_1=0; n_5-n_3=0\}$ (see inference rules in A.5).

At that stage of the type synthesis procedure, the decoration corresponds to the first one (the two constraints are needed for the correct type matching between the two occurrences of x and the respective arguments). $\mathcal{B}$ superimposes the second derivation, adding n_7 boxes as in the second figure, obtaining the superimposed decoration

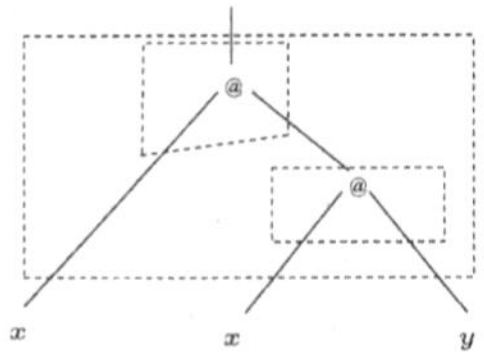

and modifying the base B in $\{x :!^{n_7}(!^{n_1}\alpha \multimap !^{n_2}\alpha), x :!^{n_6}(!^{n_3}\alpha \multimap !^{n_4}\alpha), y : !^{n_5+n_6}\alpha\}$ and the set of constraints A in $\{n_4+n_6-n_1-n_7=0; n_5-n_3=0\}$.

Definition 5. *A* slice *is a set of critical points, i.e. pairs (constraint, list of free variable occurrences) as in the following:*

$$sl = \{(A^{j_1}, [y_{1_1}, \ldots, y_{1_h}]), \ldots, (A^{j_k}, [y_{k_1}, \ldots, y_{k_h}])\}$$

A slice corresponds to a combination of critical points.

Notation. – *$sl(x)$ means that x is an element of every list of variables in $sl(x)$.*
- *$x \in sl$ if and only if there exists one element of sl whose list of variables contains x.*
- *$A^j \in sl$ if and only if there exists one element of sl whose constraint is A^j.*
- *Being A^j the constraint $\pm n_{j_1} \pm \cdots \pm n_{j_k} = 0$, $A^j - n$ corresponds to the constraint $\pm n_{j_1} \pm \cdots \pm n_{j_k} - n = 0$.*

$$\frac{}{\mathcal{B}(B,\Gamma,\emptyset,A)=\langle B,\Gamma,A\rangle}\ \mathrm{b}\emptyset$$

$$\frac{\mathcal{B}\left(B_1,!^n\Gamma,cpts,\begin{smallmatrix}A_2\\ n\geq 0\end{smallmatrix}\right)=\langle B,\Delta,A_1\rangle \quad B_1=\{x_i{:}!^n\Theta_i \text{ if } x_i\notin sl \vee x_i{:}\Theta_i \text{ if } x_i\in sl\}_i \quad A_2=(A^j \text{ if } A^j\notin sl \vee A^j-n \text{ if } A^j\in sl)_j}{\mathcal{B}(\{x_i{:}\Theta_i\}_i,\Gamma,\{sl\}\cup cpts,A)=\langle B,\Delta,A_1\rangle}\ \mathrm{b}sl$$

A.4 Product Union: ⋓

$$\frac{}{\emptyset \Cup X=X}\ \emptyset\Cup \qquad \frac{}{X\Cup\emptyset=X}\ \Cup\emptyset$$

$$\frac{\{sl_{2_1},\ldots,sl_{n_1}\}\Cup\{sl_{1_2},\ldots,sl_{n_2}\}=X}{\{sl_{1_1},\ldots,sl_{n_1}\}\Cup\{sl_{1_2},\ldots,sl_{n_2}\}=\{sl_{1_1},sl_{1_1}\cup sl_{1_2},\ldots,sl_{1_1}\cup sl_{n_2}\}\cup X}\ \Cup$$

A.5 Type Synthesis: $\mathcal{S}$

$$\frac{\mathcal{P}(\sigma)=\langle\Theta,A\rangle}{\mathcal{S}(x{:}\sigma)=\langle\Theta,x{:}\Theta,A,\emptyset\rangle}\ \mathrm{s}x$$

$$\frac{\mathcal{S}((M_1\ \ M_2){:}\tau)=\langle\Gamma,B\uplus\{x{:}\Theta_1,\ldots,x{:}\Theta_h\},A_1,cpts\cup\{sl_1(x),\ldots,sl_k(x)\}\rangle \quad \mathcal{C}(\Theta_1,\ldots,\Theta_h)=A_2}{\mathcal{S}(\lambda x.(M_1\ \ M_2){:}\sigma\to\tau)=\langle\Theta_1\multimap\Gamma,B,A_1;A_2,cpts\rangle}\ \mathrm{s}\lambda x@$$

Where $h \geq 1$.

$$\frac{cpts=cpts_1\cup\{(n-\sum n_i=0,\mathtt{FVO}(M_1\ \ M_2))\} \quad \mathcal{P}(\sigma)=\langle\Theta,A_2\rangle \quad \mathcal{S}((M_1\ \ M_2){:}\tau)=\langle !^{\sum n_i}\Gamma,B,A_1,cpts_1\rangle}{\mathcal{S}(\lambda x.(M_1\ \ M_2){:}\sigma\to\tau)=\left\langle \Theta\multimap !^n\Gamma,B,\begin{smallmatrix}A_1 & n\geq 0\\ A_2 & n-\sum n_i=0\end{smallmatrix},cpts\right\rangle}\ \mathrm{s}\lambda@$$

Where Γ is not exponential and x does not occur free in the body of the abstraction.

$$\frac{\mathcal{C}(!^n\Theta_1,\ldots,!^n\Theta_h)=A_2 \quad \mathcal{B}(B_1,\Gamma_1,\{sl_i(x)\}_i\cup cpts,A)=\langle B\uplus\{x{:}\Theta_1,\ldots,x{:}\Theta_h\},\Gamma,A_1\rangle \quad \mathcal{S}(M{:}\tau)=\langle\Gamma_1,B_1,A,\{sl_i(x)\}_i\cup cpts\rangle}{\mathcal{S}(\lambda x.M{:}\sigma\to\tau)=\langle\Theta_1\multimap !^n\Gamma,!^nB,A_1;A_2;n\geq 0,cpts\rangle}\ \mathrm{s}\lambda x$$

Where $h \geq 1$ and M is not an application.

$$\frac{\mathcal{P}(\sigma)=\langle\Theta,A_3\rangle \quad \mathcal{B}(B_1,\Gamma_1,cpts,A_1)=\langle B,\Gamma,A_2\rangle \quad \mathcal{S}(M{:}\tau)=\langle\Gamma_1,B_1,A_1,cpts\rangle}{\mathcal{S}(\lambda x.M{:}\sigma\to\tau)=\langle\Theta\multimap !^n\Gamma,!^nB,A_2;A_3;n\geq 0,cpts\rangle}\ \mathrm{s}\lambda$$

Where Γ is not exponential, M is not an application and x does not occur free in the body of the abstraction.

$$\frac{\begin{array}{l}\mathcal{B}(B_1\uplus !^{n_1}B_3,\Gamma_1,cpts_1\Cup cpts_2,A_1;A_3;A_4;n_1\geq 0)=\langle B,\Gamma,A\rangle\\ \mathcal{U}(!^{n_1}\Theta_3,\Theta_1)=A_4\\ \mathcal{B}(B_2,\Theta_2,cpts_2,A_2)=\langle B_3,\Theta_3,A_3\rangle\\ \mathcal{S}(N{:}\sigma)=\langle\Theta_2,B_2,A_2,cpts_2\rangle\\ \mathcal{S}(\lambda x.M{:}\sigma\to\tau)=\langle\Theta_1\multimap\Gamma_1,B_1,A_1,cpts_1\rangle\end{array}}{\mathcal{S}((\lambda x.M\ \ N){:}\tau)=\langle !^n\Gamma,!^nB,A;n\geq 0,cpts_1\Cup cpts_2\rangle}\ \mathrm{s}@\lambda$$

Where N is not an application.

$$\frac{\begin{array}{l}\mathcal{B}(B_1 \uplus !^{n_1} B_2, \Gamma_1, cpts_1 \Cup cpts_3, A_1; A_2; A_3; n_1 \geq 0) = \langle B, \Gamma, A\rangle \\ cpts_3 = cpts_2 \cup \{(A_3^1, \mathtt{FVO}((N_1\ \ N_2)))\} \\ \mathcal{U}(!^{n_1}\Theta_2, \Theta_1) = A_3 \\ \mathcal{S}((N_1\ \ N_2){:}\sigma) = \langle \Theta_2, B_2, A_2, cpts_2\rangle \\ \mathcal{S}(\lambda x.M{:}\sigma \to \tau) = \langle \Theta_1 \multimap \Gamma_1, B_1, A_1, cpts_1\rangle\end{array}}{\mathcal{S}((\lambda x.M\ \ (N_1\ \ N_2)){:}\tau) = \langle !^n\Gamma, !^n B, A; n \geq 0, cpts_1 \Cup cpts_3\rangle}\ \text{s@}\lambda\text{@}$$

$$\frac{\begin{array}{l}\mathcal{B}\big(B_1 \uplus !^{n_1} B_3, \Gamma_1, cpts_1 \Cup cpts_2, \sum n_{i_j} = 0; A_1; A_3; A_4; n_1 \geq 0\big) = \langle B, \Gamma, A\rangle \\ \mathcal{U}(!^{n_1}\Theta_3, \Theta_1) = A_4 \\ \mathcal{B}(B_2, \Theta_2, cpts_2, A_2) = \langle B_3, \Theta_3, A_3\rangle \\ \mathcal{S}(N{:}\sigma) = \langle \Theta_2, B_2, A_2, cpts_2\rangle \\ \mathcal{S}(x{:}\sigma \to \tau) = \langle !^{\sum n_{i_j}}(\Theta_1 \multimap \Gamma_1), B_1, A_1, cpts_1\rangle\end{array}}{\mathcal{S}((x\ \ N){:}\tau) = \langle !^n\Gamma, !^n B, A; n \geq 0, cpts_1 \Cup cpts_2\rangle}\ \text{s@}x$$

Where N is not an application.

$$\frac{\begin{array}{l}\mathcal{B}\big(B_1 \uplus !^{n_1} B_2, \Gamma_1, cpts_1 \Cup cpts_3, \sum n_{i_j} = 0; A_1; A_2; A_3; n_1 \geq 0\big) = \langle B, \Gamma, A\rangle \\ cpts_3 = cpts_2 \cup \{(A_3^1, \mathtt{FVO}((N_1\ \ N_2)))\} \\ \mathcal{U}(!^{n_1}\Theta_2, \Theta_1) = A_3 \\ \mathcal{S}((N_1\ \ N_2){:}\sigma) = \langle \Theta_2, B_2, A_2, cpts_2\rangle \\ \mathcal{S}(x{:}\sigma \to \tau) = \langle !^{\sum n_{i_j}}(\Theta_1 \multimap \Gamma_1), B_1, A_1, cpts_1\rangle\end{array}}{\mathcal{S}((x\ \ (N_1\ \ N_2)){:}\tau) = \langle !^n\Gamma, !^n B, A; n \geq 0, cpts_1 \Cup cpts_3\rangle}\ \text{s@}x\text{@}$$

$$\frac{\begin{array}{l}\mathcal{B}\big(B_1 \uplus !^{n_1} B_3, \Gamma_1, cpts_3 \Cup cpts_2, \sum n_{i_j} = 0; A_1; A_3; A_4; n_1 \geq 0\big) = \langle B, \Gamma, A\rangle \\ \mathcal{U}(!^{n_1}\Theta_3, \Theta_1) = A_4 \\ cpts_3 = cpts_1 \cup \{(\sum n_i = 0, \mathtt{FVO}((M_1\ \ M_2)))\} \\ \mathcal{B}(B_2, \Theta_2, cpts_2, A_2) = \langle B_3, \Theta_3, A_3\rangle \\ \mathcal{S}(N{:}\sigma) = \langle \Theta_2, B_2, A_2, cpts_2\rangle \\ \mathcal{S}((M_1\ \ M_2){:}\sigma \to \tau) = \langle !^{\sum n_{i_j}}(\Theta_1 \multimap \Gamma_1), B_1, A_1, cpts_1\rangle\end{array}}{\mathcal{S}((M_1\ \ M_2)\ \ N){:}\tau) = \langle !^n\Gamma, !^n B, A; n \geq 0, cpts_3 \Cup cpts_2\rangle}\ \text{s@@}$$

Where N is not an application.

$$\frac{\begin{array}{l}\mathcal{B}\big(B_1 \uplus !^{n_1} B_2, \Gamma_1, cpts_3 \Cup cpts_4, \sum n_{i_j} = 0; A_1; A_2; A_3; n_1 \geq 0\big) = \langle B, \Gamma, A\rangle \\ cpts_4 = cpts_2 \cup \{(A_3^1, \mathtt{FVO}((N_1\ \ N_2)))\} \\ \mathcal{U}(!^{n_1}\Theta_2, \Theta_1) = A_3 \\ cpts_3 = cpts_1 \cup \{(\sum n_i = 0, \mathtt{FVO}((M_1\ \ M_2)))\} \\ \mathcal{S}((N_1\ \ N_2){:}\sigma) = \langle \Theta_2, B_2, A_2, cpts_2\rangle \\ \mathcal{S}((M_1\ \ M_2){:}\sigma \to \tau) = \langle !^{\sum n_{i_j}}(\Theta_1 \multimap \Gamma_1), B_1, A_1, cpts_1\rangle\end{array}}{\mathcal{S}((M_1\ \ M_2)\ \ (N_1\ \ N_2)){:}\tau) = \langle !^n\Gamma, !^n B, A; n \geq 0, cpts_3 \Cup cpts_4\rangle}\ \text{s@@@}$$

A.6 Type Synthesis Algorithm: $\mathcal{S}$

$\mathcal{S}$ simply forgets the set of critical points and eventually contracts common variables in the base.

$$\frac{\mathcal{S}(M{:}\sigma = \langle \Theta, B, A, cpts\rangle}{\mathcal{S}(M{:}\sigma) = \langle \Theta, B, A\rangle}$$

If all variables in B are distinct. Otherwise

$$\frac{\begin{array}{c}\mathcal{C}(!^n\Theta_{1_h}, \ldots, !^n\Theta_{k_h}) = A_h \quad \ldots \quad \mathcal{C}(!^n\Theta_{1_1}, \ldots, !^n\Theta_{k_1}) = A_1 \\ \mathcal{S}(M{:}\sigma) = \big\langle \Theta, \big\{x_1{:}\Theta_{1_1}, \ldots, x_1{:}\Theta_{k_1}, \ldots, x_h{:}\Theta_{1_h}, \ldots, x_h{:}\Theta_{k_h}\big\}, A, cpts\big\rangle\end{array}}{\mathcal{S}(M{:}\sigma) = \big\langle !^n\Theta, \big\{x_1{:}!^n\Theta_{1_1}, \ldots, x_h{:}!^n\Theta_{1_h}\big\}, A; A_1; \ldots; A_h; n \geq 0\big\rangle}$$

Ramified Recurrence with Dependent Types

Norman Danner*

Department of Mathematics, University of California, Los Angeles
ndanner@member.ams.org

Abstract. We present a version of Gödel's system T in which the types are ramified in the style of Leivant and a system of dependent typing is introduced. The dependent typing allows the definition of recursively defined types, where the recursion is controlled by ramification; these recursively defined types in turn allow the definition of functions by repeated iteration. We then analyze a subsystem of the full system and show that it defines exactly the primitive recursive functions. This result supports the view that when data use is regulated (for example, by ramification), standard function constructions are intimately connected with standard type-theoretic constructions.

1 Introduction

Recently there has been a great deal of interest in characterizing function classes such as polynomial time with definitions that make no explicit mention of resources or bounds. This general area of study is referred to as *implicit computational complexity.* One of the field's major tools is the notion of *safe*, or *ramified* recurrence. The general idea is to use standard notions of definition by recurrence (such as primitive recursion) but to classify the arguments into different types. For example, in safe recurrence, function arguments are classified as either *normal* or *safe*. One thinks of the former as being arguments that have been predicatively defined; intuitively, one can use $g(x)$ in a normal position only if g has been completely defined as a function. The safe arguments are available for impredicatively defined data. In particular, when $f(x, y)$ is defined by a recursion $f(x, 0) = g(x)$ and $f(x, y + 1) = g(x, y, f(x, y))$, we have not completely defined f before using its output as an argument to g. Thus, this argument of g takes impredicatively defined data, and must therefore be safe.

This view has proven to be very fruitful. Bellantoni and Cook [5] characterize the polynomial-time computable functions using safe recursion with no ad-hoc initial functions such as the smash function ($(x, y) \mapsto 2^{|x|\cdot|y|}$) and without any explicit bounds on definition by recursion. Leivant [12] captures the same class using a more general ramification notion, for which functions computable in time n^k are defined using ramification levels $\leq k$. Passing out of the

* The author would like to thank Daniel Leivant and the referees for several helpful suggestions. The type derivations are typeset with Makoto Tatsuta's `proof.sty` package, version 3.0. The author was partially supported by National Science Foundation grant number DMS-9983726.

S. Abramsky (Ed.): TLCA 2001, LNCS 2044, pp. 91–105, 2001.

realm of feasibility, Covino et al. [7] add a scheme of "constructive diagonalization" to define classes $\mathcal{T}_\alpha$ for $\alpha < \varepsilon_0$ so that $\bigcup_n \mathcal{T}_n$ is exactly the polynomial time computable functions and $\bigcup_{\alpha<\varepsilon_0} \mathcal{T}_\alpha$ is exactly the elementary-time computable functions. The author and Pollett [8] add a safe minimization operator to Bellantoni-Cook's safe recursion to capture the partial multifunctions computable in nondeterministic polynomial time (see Selman [17] for details about this class). Bellantoni [4] uses a different safe minimization operator to capture the functions computable in polynomial time with oracles from the polynomial hierarchy. Finally, Leivant and Marion [14] use ramification combined with parameter substitution to characterize the polynomial-space computable functions.

In all of these characterizations, the authors use only type-1 functions. However, Bellantoni et al. [6] combine higher types with a notion of linearity to define a system the type-1 fragment of which is polynomial time. Hofmann [10,11] uses notions of modal types and linear function spaces to obtain a similar result. The result we are most interested in here, though, is that of Leivant [13], who extends the notion of ramification of [12] to all finite types in order to define a ramified variant of Gödel's System T of functionals [9] definable by primitive recursion in all finite types. Whereas the type-1 fragment of System T captures the functions provably recursive in Peano Arithmetic, the functions of Leivant's system are exactly those computable in elementary time.[1] The key idea in extending ramification to higher types is to understand it as specifying information about the use of data. In Leivant's notation, a functional f of type $\sigma \to \sigma$ is iterated by an object of type $\Omega\sigma$; i.e., if $g(0) = a$ and $g(x+1) = f(g(a))$, then g is given the type $\Omega\sigma \to \sigma$.

As we have just discussed, when data use is carefully regulated in this way, we are unable to define very complex functions (from a classical point of view, where elementary-time functions are considered simple). However, this fact opens up a very interesting line of inquiry. Once one passes to higher types, it is natural to ask about type-forming operations other than the function space type constructor. An obvious question is whether other standard constructions such as polymorphism and dependent types, when combined with (alternatively, restricted by) data ramification, can be used to define more complex functions. Specifically, two constructions predominate in the study of subrecursive hierarchies: iteration and diagonalization (see Rose [16]). Are there type constructors that correspond to these operations? In this paper, we consider iteration and show that it is intimately connected with dependent type formation. To do so, we add dependent types to Leivant's ramified version of System T and show that this extension allows us to define functions by iteration. In a subsystem of the extension, definition by iteration may be repeated any finite number of times, and so the resulting system provides a type-theoretic characterization of

[1] It is also interesting to note that when recursion is restricted to object type in System T, the resulting class is exactly the primitive recursive functions; the corresponding restriction in Leivant's system (adapted to binary words) yields the polynomial-time computable functions.

the primitive recursive functions using one of the standard constructs of type theory.

The plan of this paper is as follows. In the next section, we motivate the use of dependent types by considering the passage from the doubling function to exponentiation in Leivant's system. Section 3 is given over to formally defining the type system $\boldsymbol{\lambda}\mathbf{R}$ and its subsystem $\boldsymbol{\lambda}\mathbf{R}_0^0$; the latter is our concern in this paper. In Sections 4 and 5, we prove that the functions definable in $\boldsymbol{\lambda}\mathbf{R}_0^0$ are exactly the primitive recursive functions. We close in Section 6 with some speculation about using polymorphism to implement diagonalization.

2 Motivation: Why Dependent Types?

We motivate our use of dependent types by considering the passage from the function $\underline{\mathrm{dbl}}(x) =_{\mathrm{df}} 2x$ to $\exp(x) =_{\mathrm{df}} 2^{x+1} = \underline{\mathrm{dbl}}^x(2)$, where $f^0(y) = y$ and $f^{n+1}(y) = f(f^n(y))$. As noted above, in Leivant's system an argument used to iterate a function of type $\sigma \to \sigma$ is given type $\Omega\sigma$. As $\underline{\mathrm{dbl}}$ is obtained by iterating the successor function, which is represented by a term of type Nat$\to$Nat, we have that $\underline{\mathrm{dbl}}$ is represented by a term $\mathsf{dbl} : \Omega\mathrm{Nat} \to \mathrm{Nat}$. Since only functions with the same input and output type may be iterated, the doubling function cannot be iterated to obtain exponentiation. In Leivant's system, exp is instead obtained by iterating the double-application *functional* $(f, x) \mapsto f(f(x))$ starting with the successor; the corresponding term has type $\Omega(\mathrm{Nat} \to \mathrm{Nat}) \to \mathrm{Nat}$.

Semantically we understand types of the form $\Omega\sigma$ to be interpreted as the natural numbers (although there are models in which this is false). Thus we refer to Nat or any type of the form $\Omega\sigma$ as an *object* type. One can in fact define terms $\mathsf{dbl}_\sigma : \Omega\sigma \to \sigma$ for any object type σ to obtain terms $\mathsf{dbl}_{\Omega\sigma} : \Omega^2\sigma \to \Omega\sigma$, $\mathsf{dbl}_{\Omega^2\sigma} : \Omega^3\sigma \to \Omega^2\sigma$, etc. Further, the definitions of the various terms dbl_σ are all the same modulo the different types. Thus it seems that if one had a term that, given the numeral $\overline{n}$ as an argument, could "construct" the term $\mathsf{dbl}_{\Omega^n\mathrm{Nat}}$, one could then define a term exp representing exponentiation such that $\mathsf{exp}\overline{n} = \mathsf{dbl}_{\mathrm{Nat}}(\mathsf{dbl}_{\Omega\mathrm{Nat}}(\ldots(\mathsf{dbl}_{\Omega^{n-1}\mathrm{Nat}}\overline{2})\ldots))$.

Here the type(s) of dbl depend on the argument term; following Barendregt [3], this is the paradigm for *dependent types*. Thus we define a system of ramified recurrence with dependent types, the key feature of which is a recursor term $\hat{R}$ of type $\prod x^A.B. \to \prod x^A.B[\mathsf{s}x] \to B[x]. \to \Omega(\prod x^A.B) \to B[\mathsf{z}^A]$. Typically, we use $\hat{R}$ to iterate a term that can be given a recursively constructed type of the form $\Omega^{n+1}\sigma \to \Omega^n\sigma$. For example, we may define the type constructor Q in this system so that for an appropriately typed numeral $\overline{n}$, $Q\overline{n} = \Omega^n\mathrm{Nat}$. Furthermore, we can define a term $\mathsf{dbl} : \prod n.Q(\mathsf{s}n) \to Qn$ reflecting the uniformity of the definition of the doubling term in the usual ramified setting. The reduction rules for $\hat{R}$ will be defined in such a way that if we define exp as $\hat{R}\overline{2}\mathsf{dbl}$, we will have that $\mathsf{exp}\overline{n}$ expands as indicated in the previous paragraph.

We present the formalism $\boldsymbol{\lambda}\mathbf{R}$ following a style similar to [3]. Because of the dependence of types upon terms, we must first define a set of expressions, then provide inference rules for judging that a given expression is a kind, type, type constructor, or term. Here, we stratify the types into kinds T_j for $j \geq -1$. The kind T_{-1} consists of object types such as Nat and $\Omega(\text{Nat} \to \text{Nat})$; the intended interpretation of every element of T_{-1} is the natural numbers. For $j \geq 0$, T_{j+1} contains T_j as an element and is closed under products indexed by types of T_i for $i \leq j$.[2] T_0 is similar, but contains T_{-1} as a *subset*. The kinds consist of the T_j and are closed under type-indexed products. This allows us to name functions on types (e.g., the type constructor Q mentioned above represents a map from T_{-1} to T_{-1}). In Barendregt's notation, we are using elements of λP (dependent types) and $\lambda\underline{\omega}$ (functions at the type level).

3 The Dependent Typing System

As usual with systems of dependent types, since the types may depend on the terms, we first define a set of expressions, then give inference rules for assertions of the form $\Gamma \vdash A : B$ where Γ is a context and A and B are expressions. We posit a set of variables Var and define the set of *expressions* as:

$$x \in \text{Var} | s \in \{\Box, T_{-1}, T_0, T_1, \ldots\} | \text{Nat} | \Omega A | \prod x^A.B |$$
$$\mathsf{z}^A | \mathsf{s}^A | \mathsf{p}^A | \mathsf{c}^A | \tilde{R}_A | \hat{R}_A | \lambda x^A.E | EF$$

The *free variables* of an expression A, fv A, is defined as usual, as is the *substitution* of the term F for the variable x in A, $A[F/x]$. To reduce notation, if an expression such as $A[x]$ is mentioned during exposition, later occurrences of $A[F]$ will mean $A[F/x]$ unless otherwise noted; a similar convention will apply in derivations. The expression $A \to B$ is shorthand for $\prod x^A.B$ when $x \notin \text{fv}\, B$. We now define the inference rules. Throughout, s ranges over $\Box, T_{-1}, T_0, \ldots$. First we have two structural rules:

$$\frac{\Gamma \vdash A : s}{\Gamma, x : A \vdash x : A}\ \text{VarIntro} \qquad\qquad \frac{\Gamma \vdash A : B \quad \Gamma \vdash C : s}{\Gamma, x : C \vdash A : B}\ \text{Weak}$$

Next are the rules for assertions that a given expression names a kind or type:

$$\vdash \text{Nat} : T_{-1} \qquad \vdash T_j : \Box \qquad \vdash T_j : T_{j+1}$$
$$\frac{\Gamma \vdash A : s}{\Gamma \vdash \Omega A : T_{-1}} \qquad \frac{\Gamma \vdash A : T_{-1}}{\Gamma \vdash A : T_0}$$
$$\frac{\Gamma \vdash A : s' \quad \Gamma, w : A \vdash B[w] : s}{\Gamma \vdash \prod x^A.B[x] : s}$$

We allow the axiom $\vdash T_j : T_{j+1}$ only for $j \geq 0$. In the product formation rule, the pairs (s', s) may be (T_j, T_k) for $0 \leq j \leq k$, $(T_j, \Box)$ for any j, or $(\Box, \Box)$. Now we have the rules for assertions that a given expression names a type constructor or

[2] This restriction allows for a reasonably straightforward set-theoretic semantics; because of space, though, we do not pursue this topic here.

term.[3] Our system has constants for predecessor and conditional at every object type, implementing the "flat recurrence" of [13], as well as two recursors $\tilde{R}$ and $\hat{R}$, which we discuss more fully after defining the reduction rules (technically $\tilde{R}$ is an iterator, but we will ignore this distinction).

$$\frac{\Gamma \vdash A : T_{-1}}{\Gamma \vdash \mathsf{z}^A : A} \qquad \frac{\Gamma \vdash A : T_{-1}}{\Gamma \vdash \mathsf{s}^A : A \to A} \qquad \frac{\Gamma \vdash A : T_{-1}}{\Gamma \vdash \mathsf{p}^A : A \to A}$$

$$\frac{\Gamma \vdash A : T_{-1}}{\Gamma \vdash \mathsf{c}^A : A \to A \to A \to A}$$

$$\frac{\Gamma \vdash B : s}{\Gamma \vdash \tilde{R}_B : B \to (B \to B) \to \Omega B \to B}$$

$$\frac{\Gamma \vdash A : T_{-1} \quad \Gamma, w : A \vdash B[w] : s}{\Gamma \vdash \hat{R}_{\prod x^A . B[x]} : \prod x^A . B[x] \to \prod x^A . B[\mathsf{s}^A x] \to B[x]. \to \Omega(\prod x^A . B[x]) \to B[\mathsf{z}^A]}$$

$$\frac{\Gamma, w : A \vdash E[w] : B[w] \quad \Gamma \vdash \prod x^A . B[x] : s}{\Gamma \vdash \lambda x^A . E[x] : \prod x^A . B[x]} \qquad \frac{\Gamma \vdash E : \prod x^A . B \quad \Gamma \vdash F : A}{\Gamma \vdash EF : B[F/x]}$$

Finally, we allow a term to take any of its equivalent types:

$$\frac{\Gamma \vdash E : A \quad \Gamma \vdash A : T_j \quad \Gamma \vdash B : T_j \quad A = B}{\Gamma \vdash E : B}$$

where

$$(\lambda x^C . D)F = D[F/x]$$

$$\mathsf{p}^A(\mathsf{s}^A q) = q \qquad \mathsf{c}^A \mathsf{z}^A EF = E \qquad \mathsf{c}^A(\mathsf{s}^A q)EF = F$$

$$\tilde{R}_B EF\mathsf{z}^{\Omega B} = E \qquad \tilde{R}_B EF(\mathsf{s}^{\Omega B} q) = F(\tilde{R}_B EFq)$$

$$\hat{R}_C EF\mathsf{z}^{\Omega C} = E\mathsf{z}^A \qquad \hat{R}_C EF(\mathsf{s}^{\Omega C} q) = \hat{R}_C(\lambda w^A . Fw(E(\mathsf{s}^A w)))Fq$$

where $C = \prod x^A . B[x]$ for some A and B in the last line and the equations are extended to all expressions in the obvious way. While these equalities hold for all expressions, we are interested in them only as applied to type constructors and terms.

We now turn our attention to the two recursors. $\hat{R}$ was discussed in Section 2. Its purpose is to allow the iteration of a term such as dbl, which can be given the type $\prod m^{\Omega T_{-1}} . Q[\mathsf{s}m] \to Q[m]$, where $Q[\overline{n}^{\Omega T_{-1}}] = \Omega^n(\mathrm{Nat})$ and $\overline{n}^{\Omega T_{-1}} = \mathsf{s}^{\Omega T_{-1}}(\ldots \mathsf{s}^{\Omega T_{-1}} \mathsf{z}^{\Omega T_{-1}})$ (n times). The reduction rules are designed with this purpose in mind; they ensure that in the expansion of $\mathsf{dbl}\,\overline{m}\,\overline{n}$, the innermost dbl has type $Q[\mathsf{s}\overline{m}] \to Q[\overline{m}]$ and the outermost has type $Q[\mathsf{sz}] \to Q[\mathsf{z}]$. However, these very reduction rules prevent us from using $\hat{R}$ effectively to recursively

[3] The phrases "kind," "type constructor," etc. have no formal meaning here. Typically though, if we are considering an assertion $\Gamma \vdash A : B$, we call A a kind if $B = \Box$, a type if $B = T_j$ for some j, a type constructor if there is (at least implicitly) an assertion $\Delta \vdash B : \Box$, and a term if there is an assertion of $\Delta \vdash B : T_j$. Note that types are also type constructors under this reading.

build up types. For example, consider the type constructor Q just mentioned. An obvious approach would be to take $Q =_{\mathrm{df}} \hat{R}(\lambda x^{\mathrm{Nat}}.\mathrm{Nat})(\lambda x^{\mathrm{Nat}} u^{T_{-1}}.\Omega u)m$. We will need to assert that $Q[\mathsf{s}m] = \Omega(Q[m])$; however, our reduction rules for $\hat{R}$ tell us that $Q[\mathsf{s}m] = \hat{R}(\lambda w^{\mathrm{Nat}}.\Omega(\mathrm{Nat}))(\lambda xu.\Omega u)m$ (although for a numeral $\overline{m}$, we do have that $Q[\overline{m}] = \Omega^m(\mathrm{Nat})$, by applying several β-reductions after the $\hat{R}$-reductions). Thus we introduce the second iterator $\tilde{R}$ with corresponding reduction rules, which will let us define Q as $\tilde{R}(\mathrm{Nat})(\lambda u^{T_{-1}}.\Omega u)m$. In fact, $\tilde{R}$ ought to be a more symmetric version of $\hat{R}$ by taking its initial argument of type $B[\mathsf{z}]$ and a function argument of type $\prod x.B[x] \to B[\mathsf{s}x]$. Somewhat surprisingly, this poses non-trivial technical challenges related to the type of the recurrence argument. As we do not need the more symmetric version here, we sacrifice it for ease of exposition.

If we view the equations as reduction rules, with the expressions on the left reducing to the expressions on the right, then any derivation of an assertion typing an expression on the left is canonically translated into a derivation of an assertion typing the expression on the right. If Π and Σ are the corresponding derivations, we write $\Pi \to \Sigma$. To handle β-reduction, we need the usual substitution lemma:

Lemma 1. *Suppose that Π is a derivation of $\Gamma, x : A, \Delta[x] \vdash B[x] : C[x]$ and Σ is a derivation of $\Gamma \vdash F : A$. Then there is a derivation denoted $\Pi[\Sigma/x]$ of $\Gamma, \Delta[F] \vdash B[F] : C[F]$.*

Proof. Formally Π^* is defined by recursion on the height of Π. The only cases of interest are when the last rule of Π is one of the structural rules. If Π ends with a variable introduction, Π^* is Σ. Otherwise, suppose that Π ends with weakening and that the left premise of Π is Π_0. If the assumptions of the last line of Π_0 do not include $x : A$, then Π^* is Π_0; otherwise, it is Π_0^*, which is given inductively.

We denote this typing system $\boldsymbol{\lambda}\mathbf{R}$. Thus subsystem $\boldsymbol{\lambda}\mathbf{R}_n$ is obtained from $\boldsymbol{\lambda}\mathbf{R}$ by allowing T_j only for $j \leq n$ and removing the axiom $T_n : \Box$; thus we have just the first n "levels" of types, and cannot refer to the top level as an object. $\boldsymbol{\lambda}\mathbf{R}_n^k$ is obtained from $\boldsymbol{\lambda}\mathbf{R}_n$ by allowing recursor rules only when $s \in \{\Box, T_{-1}, T_0, \ldots, T_{k-1}\}$. For the remainder of this paper, we will be primarily concerned with the system $\boldsymbol{\lambda}\mathbf{R}_0^0$.

Although formally we cannot discuss types or terms independently of assertions, we shall often refer to, for example, a term $E[u^{T_{-1}}] : \sigma$. When we do so, we are implicitly claiming that there is a derivation of $\Gamma, u : T_{-1}, \Delta \vdash E : \sigma$. If no particular derivation is specified, then there is a straightforward one to which we are referring.

4 Representing Primitive Recursion

Throughout this section, u will denote a variable of type T_{-1} (i.e., we assume $u : T_{-1}$ occurs in the context of any relevant assertion). Given a type $A[u] : T_{-1}$

and initial type $C:T_{-1}$, we want a term P such that, when appropriately typed, $P\overline{n} = A[\ldots A[C]\ldots]$ (n times). The intended application is when $A = \Omega u$, so that $P\overline{n} = \Omega^n C$. Thus, given $C : T_{-1}$ and $A[u]$, define the term $P_C^A =_{\mathrm{df}} \tilde{R}_{T_{-1}} C(\lambda t^{T_{-1}}.A[t]) : \Omega T_{-1} \to T_{-1}$. By the reduction rules for $\tilde{R}$, we have that $P_C^A \mathsf{z} = C$ and $P_C^A(\mathsf{s}q) = A[P_C^A q]$; thus P_C^A builds up object types by "repeated application" of A. Furthermore, if the kinding of C and A are derivable in $\boldsymbol{\lambda}\mathbf{R}_0^0$, the same is true of P_C^A.

As we discussed in Section 2, if the output type of a representable function f is related in a suitably uniform manner to its input type and g is defined from f by iteration, then g can also be represented. The purpose of the present section is to formalize our earlier informal argument. First we define two notions of representability of a function f by a term F: (*open*) *representability* captures the most straightforward notion, in which the input and output types of F are object types; *uniform* representability captures the uniform dependence of the output type on the input type necessary for f to be iterated.

Definition 1. *Fix any $A:T_{-1}$. The* A-numeral $\overline{n}^A$ *is* $\mathsf{s}^A(\ldots\mathsf{s}^A\mathsf{z}^A)$ *and is typed in the obvious way.*

Definition 2. *Let f be a unary numeric function.*

1. f *is* openly represented *by the term $F[u^{T_{-1}}]$ if there is $B[u]:T_{-1}$ such that $F:B\to u$ and for all n we have $F\overline{n}^B = \overline{f(n)}^B$.*
2. f *is* represented *by F if there is $B[u]:T_{-1}$ such that $F:B[\mathrm{Nat}]\to \mathrm{Nat}$ and for all n we have $F\overline{n}^{B[\mathrm{Nat}]} = \overline{f(n)}^{\mathrm{Nat}}$.*
3. f *is* uniformly represented *by F if there is $B[u,m^{\Omega T_{-1}}] : T_{-1}$ such that $B[u,\mathsf{z}] = u$, $F:\prod m^{\Omega T_{-1}}.B[u,\mathsf{s}m]\to B[u,m]$, and for all m and n we have $F\overline{m}^{\Omega T_{-1}}\overline{n}^{B[\bullet m]} = \overline{f(n)}^{B[m]}$.*

Proposition 1. *Let f be a unary function.*

1. *f is openly represented by $F[u^{T_{-1}}]$ iff f is represented by $F[\mathrm{Nat}]$.*
2. *If f is uniformly represented, then f is openly represented.*

Proof. 1. The forward direction is immediate. For the reverse direction, let Π be any derivation of $\Gamma\vdash E:A$ and let u be a variable not occurring in Π. Obtain Σ from Π by weakening each leaf of Π to add $u:T_{-1}$ as a premise, then replace Nat throughout with u. It is easy to verify that Σ is a derivation of $u:T_{-1},\Gamma[u/\mathrm{Nat}]\vdash E[u/\mathrm{Nat}]:A[u/\mathrm{Nat}]$. In particular, if Π is a derivation of $\vdash F:B[\mathrm{Nat}]\to\mathrm{Nat}$, then Σ is a derivation of $u:T_{-1}\vdash F[u]:B[u]\to u$.

2. If f is uniformly represented by $F:\prod m^{\Omega T_{-1}}.B[\mathsf{s}m]\to B[m]$, then f is openly represented by $F\mathsf{z}$ because $B[\mathsf{z}]=u$.

Proposition 2.

1. *The constant zero and successor functions are represented in $\boldsymbol{\lambda}\mathbf{R}_0^0$.*

2. The $\boldsymbol{\lambda}\mathbf{R}_0^0$-representable functions are closed under composition.

Proof. (1) is immediate. We prove a special case of (2); the general case is similar. Suppose that the functions $f(x)$ and $g(x)$ are openly represented in $\boldsymbol{\lambda}\mathbf{R}_0^0$ by the terms $F : B[u] \to u$ and $G : C[u] \to u$, respectively. Define $G^* =_{\mathrm{df}} G[B/u]$; then $\lambda x^{C[B/u]}.F(G^* x) : C[B/u] \to u$ clearly represents $f \circ g$.

We also note that the $\hat{R}$-reductions behave as expected when an $\hat{R}$-term is applied to a numeral:

Lemma 2. *For any terms $E : \prod x^A.B$ and $F : \prod x^A.B[\mathsf{s}x] \to B[x]$ and any n, we have*

$$\hat{R}_{\prod x^A.B} E F \overline{n}^{\Omega(\prod x^A.B)} = F\mathsf{z}^A(F\overline{1}^A(\ldots F\overline{n-1}^A(E\overline{n}^A)\ldots)).$$

Proof. We prove the lemma by induction on n for all terms E and F:

$$\begin{aligned} \hat{R}EF\mathsf{z} &= E\mathsf{z}^A \\ \hat{R}EF(\mathsf{s}\overline{n}) &= \hat{R}(\lambda w^A.Fw(E(\mathsf{s}^A w)))F\overline{n} \\ &= F\mathsf{z}^A(F\overline{1}^A(\ldots F\overline{n-1}^A\left((\lambda w^A.Fw(E(\mathsf{s}^A w)))\,\overline{n}^A\right)\ldots)) \\ &= F\mathsf{z}^A(F\overline{1}^A(\ldots F\overline{n-1}^A(F\overline{n}^A(E\overline{n+1}^A))\ldots)). \end{aligned}$$

We now prove that the $\boldsymbol{\lambda}\mathbf{R}_0^0$-representable functions are closed under iteration.

Proposition 3. *Let f be a unary function.*

1. *If f is openly represented in $\boldsymbol{\lambda}\mathbf{R}_0^0$, then f is uniformly represented in $\boldsymbol{\lambda}\mathbf{R}_0^0$. In particular, by Prop. 1(2) f is openly represented iff it is uniformly represented.*
2. *If f is uniformly represented in $\boldsymbol{\lambda}\mathbf{R}_0^0$, then f' is openly represented in $\boldsymbol{\lambda}\mathbf{R}_0^0$, where $f'(0) = n$ and $f'(x+1) = f(f'(x))$ for some fixed n.*

Proof. 1. Suppose that $B[u] : T_{-1}$ and the function f is openly represented by the term $F[u] : B[u] \to u$. Fix a fresh variable $m : \Omega T_{-1}$ and consider the term $F^* =_{\mathrm{df}} F[P_u^B m/u] : B[P_u^B m] \to P_u^B m$. By the definition of P_u^B, we have that in fact $F^* : P_u^B(\mathsf{s}m) \to P_u^B m$. Furthermore, since only the types have changed, we also have that $F^* \overline{n}^{P_u^B(\cdot m)} = \overline{f(n)}^{P_u^B m}$, since $F\overline{n}^B = \overline{f(n)}^u$. Finally, $P_u^B \mathsf{z} = \tilde{R}u(\lambda t.B[t])\mathsf{z} = u$ by definition of P_u^B. Thus $\lambda m^{\Omega T_{-1}}.F^*$ uniformly represents f.

2. Suppose that f is uniformly represented by $F[u] : \prod m^{\Omega T_{-1}}.B[u, \mathsf{s}m] \to B[u, m]$. Define the term

$$F_0[u] =_{\mathrm{df}} \hat{R}_{\prod w^{\Omega T_{-1}}.B[u,w]}(\lambda y^{\Omega T_{-1}}.\overline{n}^{B[u,y]})F : \Omega(\prod w^{\Omega T_{-1}}.B[u,w]) \to B[u,\mathsf{z}].$$

By definition of uniform representability we have $B[u, \mathsf{z}] = u$, and therefore $F_0[u] : \Omega(\prod w.B[u,w]) \to u$. We now apply Lemma 2 to conclude that F_0 openly represents f':

$$\begin{aligned} F_0\overline{q}^B &= F\mathsf{z}^{\Omega T_{-1}}(\dots F\overline{q-1}^{\Omega T_{-1}}((\lambda y^{\Omega T_{-1}}.\overline{n}^{B[u,y]})\overline{q}^{\Omega T_{-1}})\dots) \\ &= F\mathsf{z}^{\Omega T_{-1}}(\dots F\overline{q-1}^{\Omega T_{-1}}(\overline{n}^{B[u,\overline{q}]})\dots) \\ &= \overline{f^q(n)}^{B[u,\bullet^{\Omega T_{-1}}]} \\ &= \overline{f^q(n)}^{u} \end{aligned}$$

Finally, we note that in each part, if the types and terms initially asserted are derivable in $\boldsymbol{\lambda}\mathbf{R}_0^0$, then so are all the types and terms involved in each step. Thus we conclude that the functions represented in $\boldsymbol{\lambda}\mathbf{R}_0^0$ are closed under definition by iteration.

Since primitive recursion is reducible to iteration (see Rose [16]), Propositions 2 and 3 combine to prove:

Theorem 1. *All primitive recursive functions are representable in* $\boldsymbol{\lambda}\mathbf{R}_0^0$.

As an example, we show how to represent the stacking function $\underline{\mathrm{stk}}(x)$ in $\boldsymbol{\lambda}\mathbf{R}_0^0$, where $\underline{\mathrm{stk}}(0) = 2$ and $\underline{\mathrm{stk}}(x+1) = \exp(\underline{\mathrm{stk}}(x))$. As before, we assume that u is a type variable of kind T_{-1}. We have already discussed one representation of the exponentiation function for which $\mathsf{exp} : \Omega(\mathrm{Nat} \to \mathrm{Nat}) \to \mathrm{Nat}$; the term Leivant defines in [13] can be easily adapted to this purpose. Applying Prop. 1(1) we assume that $u : T_{-1} \vdash \mathsf{exp} : \Omega(u \to u) \to u$. Prop. 3(1) gives us the uniform representation

$$\mathsf{exp}^* =_{\mathrm{df}} \lambda m^{\Omega T_{-1}}.(\mathsf{exp}[P_u^{\Omega(u\to u)}m/u]) : \prod m^{\Omega T_{-1}}.P_u^{\Omega(u\to u)}(\mathsf{s}m) \to P_u^{\Omega(u\to u)}m.$$

Now we apply Prop. 3(2) to obtain the open representation of $\underline{\mathrm{stk}}$:

$$\mathsf{stk} =_{\mathrm{df}} \hat{R}_{\prod w^{\Omega T_{-1}}.P_u^{\Omega(u\to u)}w}(\lambda y^{\Omega T_{-1}}.\overline{2}^{P_u^{\Omega(u\to u)}y})(\mathsf{exp}^*) : \\ \Omega(\prod w^{\Omega T_{-1}}.P_u^{\Omega(u\to u)}w) \to u.$$

Of course, this process can be repeated any finite number of times. Given that the iterate of any representable function can be represented, one wonders about defining a term for the iteration *functional*, formalizing this argument. This would allow us to pass out of primitive recursion by defining, for example, the Ackermann function. In Section 6, we analyze this idea further and give some indication why in fact we cannot do so in $\boldsymbol{\lambda}\mathbf{R}_0^0$.

5 Upper Bound on Representable Functions

To show that the functions representable in $\boldsymbol{\lambda}\mathbf{R}_0^0$ are exactly the primitive recursive functions, we translate derivations into an extension of Gödel's system T

that we call T^+. T^+ is conservative over T in the sense that if E is a T^+-term, then it is already a T-term. We then notice that the translations of the derivations for terms only use recurrence operators at object type, which implies that derivations of terms representing functions in $\boldsymbol{\lambda}\mathbf{R}_0^0$ are mapped to functions definable in T using recurrence at object type, and hence the images are primitive recursive functions. Since the translation will respect reduction/normalization, we may conclude that any function representable in $\boldsymbol{\lambda}\mathbf{R}_0^0$ is primitive recursive.

The main point to keep in mind for this translation is the following. In $\boldsymbol{\lambda}\mathbf{R}_0^0$, the type constructors ultimately return elements of T_{-1}. That is, if we consider a given type constructor as a function of numeric and type arguments that outputs a type, then regardless of the input, the output is a type of kind T_{-1}. This follows because no type of the form $\prod x^A.T_0$ can be formed in $\boldsymbol{\lambda}\mathbf{R}_0^0$. Since the intended semantics of any element of T_{-1} is the natural numbers, intuitively every type constructor is really a constant function. Thus we expand system T to include type constructor constants to which (derivations of) the type constructors of $\boldsymbol{\lambda}\mathbf{R}_0^0$ are mapped; the mapping of applications is translated in such a way as to implement this constant behavior.

Formally, we modify the usual system T by replacing the usual recursors with the following two recursors for every finite type σ:

$$\tilde{R}_\sigma : \sigma \to (\sigma \to \sigma) \to \mathrm{Nat} \to \sigma$$

$$\hat{R}_\sigma : (\mathrm{Nat} \to \sigma) \to (\mathrm{Nat} \to \sigma \to \sigma) \to \mathrm{Nat} \to \sigma$$

with the reductions

$$\tilde{R}EF\mathsf{z} = E \qquad \tilde{R}EF(\mathsf{s}q) = F(\tilde{R}EFq)$$

$$\hat{R}EF\mathsf{z} = E\mathsf{z} \qquad \hat{R}EF(\mathsf{s}q) = \hat{R}(\lambda w^{\mathrm{Nat}}.Fw(E(\mathsf{s}w)))Fq$$

We call the recursors $\tilde{R}_{\mathrm{Nat}}$ and $\hat{R}_{\mathrm{Nat}}$ *object-type* recursors. We also add constants p and c for predecessor and conditional with the corresponding reductions. Of course, all of these could be defined in the usual system T; by adding them we just simplify the details of the translation. It is well-known that if a function is definable in System T using only object-type recursors, then it is primitive recursive (see, for example, Feferman and Avigad [2]).

To define the system T^+, we will have two sorts of types. The *N-types* are the simple types over Nat. The *U-types* are defined as follows:

- U_0 is a U-type (U_0 can be taken to be the set of N-types).
- If ρ is an N-type and σ and τ are U-types, then $\rho \to \tau$ and $\sigma \to \tau$ are U-types.

The *type constructors* of T^+ consist of Nat and constants $\mathsf{f}_{\sigma\to\tau}$ for every U-type $\sigma \to \tau$. Finally, we define the *terms* of T^+ to consist of the usual terms of system T; we have variables x^σ only for N-types σ. Since the type constructors are not used in defining the terms of T^+, any term of T^+ is already a term of T. The reductions for T^+ consist of the usual β-reductions along with the $\tilde{R}$ and $\hat{R}$ reductions given above.

Table 1. Translation of $\boldsymbol{\lambda}\mathbf{R}_0^0$-judgements for kinds and type constructors.

$\dfrac{\overset{\Pi}{\Gamma \vdash T_{\cdot 1} : \Box}}{\Gamma, x : T_{\cdot 1} \vdash x : T_{\cdot 1}}$	$\rightsquigarrow$ Nat	$\dfrac{\overset{\Pi}{\Gamma \vdash A : s}}{\Gamma \vdash \Omega A : T_{\cdot 1}}$	$\rightsquigarrow$ Nat
$\dfrac{\overset{\Pi_0}{\Gamma \vdash A : T_j} \quad \overset{\Pi_1}{\Gamma \vdash C : s}}{\Gamma, x : C \vdash A : T_j}$	$\rightsquigarrow$ $\Pi_0^{\bullet}$	$\dfrac{\overset{\Pi}{\Gamma \vdash A : T_{\cdot 1}}}{\Gamma \vdash A : T_0}$	$\rightsquigarrow$ $\Pi^{\bullet}$
$\vdash \mathrm{Nat} : T_{\cdot 1}$	$\rightsquigarrow$ Nat	$\vdash T_{\cdot 1} : \Box$	$\rightsquigarrow$ Nat
$\dfrac{\overset{\Pi_0}{\Gamma \vdash A : s_0} \quad \overset{\Pi_1}{\Gamma, x : A \vdash B : s_1}}{\Gamma \vdash \prod x^A.B : s_1}$	$\rightsquigarrow$ $\Pi_0^{\bullet} \to \Pi_1^{\bullet}$		
$\dfrac{\overset{\Pi}{\Gamma \vdash A : \Box}}{\Gamma \vdash \tilde{R}_A : \cdots}$	$\rightsquigarrow$ ${}^{\bullet}\Pi'{}^{\bullet}\ (\Pi'{}^{\bullet}\ \Pi'){}^{\bullet}\ \mathrm{Nat}^{\bullet}\ \Pi'$		
$\dfrac{\overset{\Pi_0}{\Gamma \vdash A : T_{\cdot 1}} \quad \overset{\Pi_1}{\Gamma, w : A \vdash B : \Box}}{\Gamma \vdash \hat{R}_{\prod x^A.B} : \cdots}$	$\rightsquigarrow$ ${}^{\bullet}(\mathrm{Nat}^{\bullet}\ \Pi_1'){}^{\bullet}\ (\mathrm{Nat}^{\bullet}\ \Pi_1'{}^{\bullet}\ \Pi_1'){}^{\bullet}\ \mathrm{Nat}^{\bullet}\ \Pi_1'$		
$\dfrac{\overset{\Pi_0}{\Gamma, w : A \vdash E : B} \quad \overset{\Pi_1}{\prod x^A.B : \Box}}{\Gamma \vdash \lambda x^A.E : \prod x^A.B}$	$\rightsquigarrow$ ${}^{\bullet}\Pi_1'$		
$\dfrac{\overset{\Pi_0}{\Gamma \vdash C : \prod x^A.B} \quad \overset{\Pi_1}{\Gamma \vdash F : A}}{\Gamma \vdash CF : B[F/x]}$	$\rightsquigarrow$ $\begin{cases} \mathrm{Nat}, & \Pi_0^{\bullet} = {}^{\bullet}\sigma^{\bullet}\ \mathrm{Nat} \\ {}^{\bullet}\tau, & \Pi_0^{\bullet} = {}^{\bullet}\sigma^{\bullet}\ \tau, \tau \neq \mathrm{Nat} \end{cases}$		

We now give the translation of $\boldsymbol{\lambda}\mathbf{R}_0^0$-derivations, writing Π' for the translation of Π. In Table 1 we give the translations of derivations for kinds and type constructors and note that Π' is always a type or type constructor. In Table 2 we give the translations for derivations of terms, and note that Π' is always a term of the appropriate type. In the λ-abstraction clause, we write $\Pi_{1,0}$ to denote the left immediate subderivation of Π_1 as given in the inference rules of Section 3. To save space we write, for example, $\tilde{R} : \cdots$ instead of the actual type of $\tilde{R}$, which can be easily reconstructed from the derivation rules. We also write $\mathsf{z}|\cdots$ as shorthand for the rules for the constants z, s, p, and c.

The translation of the type equality rule in Table 2 is sufficient for $\boldsymbol{\lambda}\mathbf{R}_0^0$-derivations by the following lemma:

Lemma 3. *If in $\boldsymbol{\lambda}\mathbf{R}_0^0$ one can derive $\Gamma \vdash A : T_j$ and A is a redex, then $j = -1$.*

Proof. This is proved by verifying that for each possible type of redex, a derivation of $\Gamma \vdash A : T_j$ must include a subderivation of $\Gamma, \Delta \vdash T_j : \Box$ for some (pos-

Table 2. Translations of $\boldsymbol{\lambda}\mathbf{R}_0^0$-judgements for terms.

$\dfrac{\overset{\Pi}{\Gamma \vdash A : T_j}}{\Gamma, x : A \vdash x : A}$	$\leadsto$	$x^{\Pi'}$
$\dfrac{\overset{\Pi_0}{\Gamma \vdash E : A} \quad \overset{\Pi_1}{\Gamma \vdash C : s}}{\Gamma, x : C \vdash E : A}$	$\leadsto$	$\Pi_0^{\bullet}$
$\dfrac{\overset{\Pi}{\Gamma \vdash A : T_{-1}}}{\Gamma \vdash \bullet \vert \cdots}$	$\leadsto$	$\bullet \vert \cdots$
$\dfrac{\overset{\Pi}{\Gamma \vdash A : T_{-1}}}{\Gamma \vdash \tilde{R}_A : \cdots}$	$\leadsto$	$\tilde{R}_{\Pi'}$
$\dfrac{\overset{\Pi_0}{\Gamma \vdash A : T_{-1}} \quad \overset{\Pi_1}{\Gamma, w : A \vdash B : T_{-1}}}{\Gamma \vdash \hat{R}_{\prod x^A . B} : \cdots}$	$\leadsto$	$\hat{R}_{\Pi_1'}$
$\dfrac{\overset{\Pi_0}{\Gamma, w : A \vdash E : B} \quad \overset{\Pi_1}{\prod x^A . B : T_j}}{\Gamma \vdash \lambda x^A . E : \prod x^A . B}$	$\leadsto$	$\lambda x^{\Pi_{1,0}'} . \Pi_0^{\bullet}[x/w]$
$\dfrac{\overset{\Pi_0}{\Gamma \vdash E : \prod x^A . B} \quad \overset{\Pi_1}{\Gamma \vdash F : A}}{\Gamma \vdash EF : B[F/x]}$	$\leadsto$	$\Pi_0^{\bullet} \Pi_1^{\bullet}$
$\dfrac{\overset{\Pi_0}{\Gamma \vdash E : A} \quad \overset{\Pi_1}{\Gamma \vdash A : T_{-1}} \quad \overset{\Pi_2}{\Gamma \vdash B : T_{-1}} \quad A = B}{\Gamma \vdash E : B}$	$\leadsto$	$\Pi_0^{\bullet}$

sibly empty) Δ. But in $\boldsymbol{\lambda}\mathbf{R}_0^0$, the only possible derivation of this latter form is for $j = -1$.

Lemma 4. *If Π is an $\boldsymbol{\lambda}\mathbf{R}_0^0$-derivation of $\Gamma \vdash A : \prod x_1^{A_1} \cdots \prod x_m^{A_m} . T_{-1}$, then there are $\sigma_1, \ldots, \sigma_m$ such that*

$$\Pi' = \begin{cases} \mathrm{Nat}, & m = 0 \\ \mathsf{f}_{\sigma_1 \to \cdots \to \sigma_m \to \mathrm{Nat}}, & m > 0 \end{cases}$$

In particular, if Π is a derivation of $\Gamma \vdash A : T_{-1}$, then $\Pi' = \mathrm{Nat}$.

Proof. By induction on derivations of kinds and type constructors. In the case of λ-abstraction, one must expand out the right subderivation and notice that its translation is of the appropriate form.

The following is the key proposition; its proof is a straightforward induction on the definition of the reduction relation. Recall that if Π is a derivation of a typing of a redex, then Π can be canonically converted to a derivation Σ typing the contractum, and we write $\Pi \to \Sigma$.

Proposition 4. *If Π and Σ are $\boldsymbol{\lambda}\mathbf{R}_0^0$-derivations and $\Pi \to \Sigma$, then $\Pi' \to \Sigma'$.*

Given that the standard typing of a numeral $\overline{n}^A$ in $\boldsymbol{\lambda}\mathbf{R}_0^0$ is translated to the standard system T numeral $\overline{n}$, we conclude that if $F[u^{T_{-1}}]{:}A[u]{\to}u$ represents the function f in $\boldsymbol{\lambda}\mathbf{R}_0^0$ and Π_F is the derivation typing F, then Π' is a system T term of type Nat $\to$ Nat also representing f. Furthermore, since recursor judgements in term derivations are translated to recursors only when the expression is of type T_{-1} (and type constructors otherwise), all recursors in Π'_F use only object-type recursion. We conclude that f must be primitive recursive. Combined with the work of the previous section in which we showed that all primitive recursive functions are representable in $\boldsymbol{\lambda}\mathbf{R}_0^0$, we have just proved our main result:

Theorem 2. *The functions representable in $\boldsymbol{\lambda}\mathbf{R}_0^0$ are exactly the primitive recursive functions.*

Although space prevents us from giving a detailed example of the translation of the present section, we will briefly outline the result of applying it to the terms exp^* and stk defined at the end of Section 4. Assume that the derivation typing the term exp is translated to the System T^*-term $\mathsf{e}{:}\mathrm{Nat}{\to}\mathrm{Nat}$ which represents exp in system T. Roughly speaking, any derivation consists of "derivation leaves" that assign a kind to a type, which are used as immediate subderivations in Variable Introduction, Weakening, and Recursor rules. Following these rules are the more usual typing rules for terms. In analyzing the derivation of exp^*, all subderivations of types assign the kind T_{-1}, and hence each is translated to the type Nat (Lemma 4). The typing derivation for exp^*, then, is translated to the term $\mathsf{e}^* =_{\mathrm{df}} \lambda m^{\mathrm{Nat}}.\mathsf{e} : \mathrm{Nat} \to \mathrm{Nat} \to \mathrm{Nat}$. As m is not free in e, the extra argument is a dummy argument. The $\hat{R}$-rule used in the derivation of stk requires an immediate subderivation that $w : \Omega T_{-1} \vdash P_u^{\Omega(u \to u)} w : T_{-1}$; this derivation is translated to Nat, and so the $\hat{R}$-rule is translated to $\hat{R}_{\mathrm{Nat}}$. The derivation for stk itself is then translated to $\hat{R}_{\mathrm{Nat}}(\lambda y^{\mathrm{Nat}}.\mathsf{s}(\mathsf{sz}))\mathsf{e}^* : \mathrm{Nat} \to \mathrm{Nat}$, which represents the function $\underline{\mathrm{stk}}$ in System T.

6 Polymorphism and Diagonalization

To extend this formalism to a system that captures more than the primitive recursive functions, we would like to implement diagonalization. A standard way to do so is to define the iteration functional $\underline{\mathrm{It}}(f)(x) =_{\mathrm{df}} f^x(2)$ and then iterate $\underline{\mathrm{It}}$: if $f_0(y) = y + 1$ and $g(x) = \underline{\mathrm{It}}^x(f_0)(x)$, then g diagonalizes across generating functions for primitive recursion and is itself not primitive recursive. Summarizing the argument that the functions representable in $\boldsymbol{\lambda}\mathbf{R}_0^0$ are closed under iteration, we know that if f is openly represented by $F[u] : B[u] \to u$, then $\underline{\mathrm{It}}(f)$ is openly represented by

$$\hat{R}_{\prod w^{\Omega T_{-1}} P_u^B w}(\lambda y^{\Omega T_{-1}}.\overline{2}^{P_u^B y})(\lambda m^{\Omega T_{-1}}.F[P_u^B m]) : \Omega\left(\prod w^{\Omega T_{-1}} P_u^B w\right) \to u.$$

So it seems that to genericize $F[u]$ from an arbitrary but fixed term to a *variable*, we need to be able to change the value of the *type variable* u to $P_u^B m$. In other words, it appears that some form of polymorphism is necessary. This ties in very neatly with our contention that the main function-definition procedures correspond neatly with standard type-theoretic constructions.

Again using Barendregt's approach in [3], the product rule is a general enough mechanism for handling dependent products and polymorphism—all that must be adjusted is the allowable pairs (s, s'). Reading polymorphism as "terms depending on a type," one must allow product rules in which $(s, s') = (\Box, T_j)$. However, this naive approach is for full polymorphism, which we certainly wish to avoid here. Instead, let us consider the following version of the product rule:

$$\frac{\Gamma \vdash A : s_A \quad \Gamma, w : A \vdash B[w] : s_B}{\Gamma \vdash \prod x^A.B[x] : s}$$

Our current formalism $\boldsymbol{\lambda}\mathbf{R}$ allows the product rule for triples (s_A, s_B, s) of the form (T_j, T_k, T_k) for $0 \le j \le k$, $(T_j, \Box, \Box)$, and $(\Box, \Box, \Box)$. To allow for a controlled version of polymorphism in which impredicativity is disallowed, we now permit also triples of the form (T_k, T_j, T_k) for $k > j \ge 0$ and $(\Box, T_j, \Box)$. With such a rule, one can then define an iteration term

$$\lambda f^{\prod x^{T_{-1}}.B[x] \to x} \lambda x^{T_{-1}}.\hat{R}_{\prod w.Qw}(\lambda y^{\Omega T_{-1}} \overline{2}^{Qy})(\lambda m^{\Omega T_{-1}}.f(P_x^B m)) :$$
$$\prod x^{T_{-1}}.B[x] \to x. \to \prod x^{T_{-1}}.\Omega(\prod w.P_x^B w) \to x$$

where $Q = P_x^B$.

However, we still cannot iterate this term. Our recursor $\hat{R}$ is designed to handle the iteration of function(al)s with an input type that is "more complex" than the output type, and this term reverses the situation. This is actually more in line with the approach taken when adding dependent types to the unramified System T (see, for example, Avigad's [1], in which the recursor corresponding to our $\hat{R}$ has type $B[\mathsf{z}] \to \prod x^{\mathrm{Nat}}.B[x] \to B[\mathsf{s}x]. \to \prod x^{\mathrm{Nat}}.B$; also see Nelson [15]) and does not seem escapable. Intuitively, the iterate of a function must take "more complicated" inputs than the original function and thus the term representing the iterate is more complicated than the term representing the original function. However, as per our discussion of $\tilde{R}$ in Section 3, defining a recursor that allows iteration of function(al) whose output type is more complicated than its input type poses non-trivial technical challenges. We hope to report on research in this direction in a future paper.

References

1. J. Avigad. Predicative functionals and an interpretation of $\widehat{\mathrm{ID}}_{<\omega}$. *Ann. Pure Appl. Logic*, 92(1):1–34, 1998.
2. J. Avigad and S. Feferman. Gödel's functional ("Dialectica") interpretation. In S. Buss, editor, *Handbook of Proof Theory*, pages 337–405. North-Holland, Amsterdam, 1998.
3. H. Barendregt. Lambda calculi with types. In S. Abramsky, D. M. Gabbay, and T. Maibaum, editors, *Handbook of Logic in Computer Science, Vol. 2*, pages 117–309. Oxford University Press, Oxford, 1992.
4. S. Bellantoni. Predicative recursion and the polytime hierarchy. In *Feasible Mathematics II (Ithaca, NY, 1992)*, pages 15–29. Birkhäuser Boston, Boston, MA, 1995.
5. S. Bellantoni and S. Cook. A new recursion-theoretic characterization of the polytime functions. *Comput. Complexity*, 2(2):97–110, 1992.
6. S. Bellantoni, K.-H. Niggl, and H. Schwichtenberg. Higher type recursion, ramification and polynomial time. *Ann. Pure Appl. Logic*, 104(1–3):17–30, 2000.
7. E. Covino, G. Pani, and S. Caporaso. Extending the implicit computational complexity approach to the sub-elementary time-space classes. In *Algorithms and Complexity (Rome, 2000)*, pages 239–252. Springer, Berlin, 2000.
8. N. Danner and C. Pollett. Minimization and the class NPMV. In preparation.
9. K. Gödel. Über eine bisher noch nicht benützte Erweiterung des finiten Standpunktes. *Dialectica*, 12:280–287, 1958.
10. M. Hofmann. A mixed modal/linear lambda calculus with applications to Bellantoni-Cook safe recursion. In *Computer Science Logic (Aarhus, 1997)*, pages 275–294. Springer, Berlin, 1998.
11. M. Hofmann. Safe recursion with higher types and BCK-algebra. *Ann. Pure Appl. Logic*, 104(1-3):113–166, 2000.
12. D. Leivant. Ramified recurrence and computational complexity I: Word recurrence and poly-time. In *Feasible Mathematics II (Ithaca, NY, 1992)*, pages 320–343. Birkhäuser Boston, Boston, MA, 1995.
13. D. Leivant. Ramified recurrence and computational complexity III: Higher type recurrence and elementary complexity. *Ann. Pure Appl. Logic*, 96(1-3):209–229, 1999. Festschrift on the occasion of Professor Rohit Parikh's 60th birthday.
14. D. Leivant and J.-Y. Marion. Ramified recurrence and computational complexity II: Substitution and poly-space. In *Computer Science Logic (Kazimierz, 1994)*, pages 486–500. Springer, Berlin, 1995.
15. N. Nelson. Primitive recursive functionals with dependent types. In *Mathematical Foundations of Programming Semantics (Pittsburgh, PA, 1991)*, pages 125–143. Springer, Berlin, 1992.
16. H. E. Rose. *Subrecursion: functions and hierarchies.* Number 9 in Oxford Logic Guides. Oxford University Press, Oxford, 1984.
17. A. L. Selman. Much ado about functions. In *Eleventh IEEE Conference on Computational Complexity*, pages 198–212. 1996.

Game Semantics for the Pure Lazy λ-Calculus*

Pietro Di Gianantonio

Dipartimento di Matematica e Informatica, Università di Udine
via delle Scienze 206 I-33100 Udine Italy
e-mail: digianantonio@dimi.uniud.it

Abstract. In this paper we present a fully abstract game model for the pure lazy λ-calculus, i.e. the lazy λ-calculus without constants. In order to obtain this result we introduce a new category of games, the monotonic games, whose main characteristic consists in having an order relation on moves.

1 Introduction

The aim of this paper is to present a fully abstract model, based on game semantics, for the lazy λ-calculus. The λ-calculus we consider is the untyped one, with a lazy, call-by-name, reduction strategy. The model we construct lies in the category of monotonic games introduced in this paper. This new category is derived from the one defined by Abramsky Jagadeesan and Malacaria in [AJM94].

This paper is quite similar to the article [AM95a]. It has the same aims and uses a similar model, but there is also an important difference: the lazy λ-calculus considered in [AM95a] contains a constant C that, in the operational semantics, is able to perform a sequential test for convergence. The introduction of the constant C is essential in order to obtain a full definability result and, as a consequence, the full abstraction of the model. Similarly in [AO93], through syntactic methods, it was obtained a fully abstract model for the lazy λ-calculus extended with the constant C; while the problem of finding a fully abstract model for the pure lazy λ-calculus was left open.

In this paper we show that it is possible to have a fully abstract model for the pure lazy λ-calculus without constants. In order to obtain this result, we need to introduce a new category of games that we call monotonic games and indicate with $\mathcal{G}_{\mathcal{M}}$. The category $\mathcal{G}_{\mathcal{M}}$ differs from the more standard category of AJM-games $\mathcal{G}$ in several aspects. In $\mathcal{G}_{\mathcal{M}}$, moves are questions or answers and are ordered according to a notion of strength. Intuitively, a question a is stronger than a question b if it asks for more information. This means that if question a can receive an answer, then also b can receive an answer, or, from another point of view, a requires more work than b to be fulfilled. Similarly, an answer is stronger than another if it gives more information. Using this notion of strength, we impose some new restrictions on the way that a play can evolve and

* Research partially supported by Esprit Working Group TYPES and TMR Network LINEAR

S. Abramsky (Ed.): TLCA 2001, LNCS 2044, pp. 106–120, 2001.

in the way that a strategy can behave. Intuitively, we ask that a play proceeds with stronger and stronger questions, and that a strategy preserves the strength order relation. By these restrictions, in our game λ-model strategies are forced to behave as interpretations of λ-terms, and hence we have a fully complete and fully abstract model.

2 The Calculus

We define here the language λl, together with its operational semantics. Language λl is a lazy λ-calculus, its set of terms constructed from a set of variables $Var(\ni x)$ by the grammar:

$$M ::= x \mid MM \mid \lambda x.M$$

The operational semantics is given by a big-step reduction relation, $M \Downarrow N$, evaluating a term to a weak head normal form. The strategy of evaluation is lazy and call-by-name.

$$\frac{}{\lambda x.M \Downarrow \lambda x.M}$$

$$\frac{M \Downarrow \lambda x.P \qquad P[N/x] \Downarrow Q}{MN \Downarrow Q}$$

The above reduction strategy gives rise to a contextual pre-order ($\sqsubseteq_l$) on closed λ-terms (Λ^0) defined by:

$$M \sqsubseteq_l N \;\Leftrightarrow\; (\forall C[\;] \in \Lambda^0 \;.\; C[M] \Downarrow \Rightarrow C[N] \Downarrow)$$

We indicate with $\approx_l$ the equivalence relation induced by $\sqsubseteq_l$.
The following properties will be used:

- the lazy reduction strategy converges on any term M β-equivalent to a λ-abstraction;
- the relation $\approx_l$ is a $\lambda\beta$-theory.

The above properties follow immediately from the fact that there exist adequate models for the lazy λ-calculus (see [AO93,AM95a,BPR98]).

3 The Categories of Monotonic Games

In this section, we define the two categories of games employed in this article. These two categories are closely related to the categories $\mathcal{G}$ and $K_!(\mathcal{G})$ presented in [AJM94]. They are defined following similar patterns; essentially, they only differ with respect to the strength order relation on moves. We begin by giving the basic definitions.

As usual, we consider games between two participants: the *Player* and the *Opponent*. A play consists in an alternate sequence of moves, while each move consists in posing a question ($\in M^Q$) or giving an answer ($\in M^A$). Before giving the definition of games, we introduce the notation that will be used in the following.

- We use the metavariables A, B, C to range on games, the metavariables s, t, r, q to range on plays and the metavariables a, b, c to range on moves.
- The empty sequence is denoted by ϵ, concatenation of sequences is denoted by juxtaposition, the prefix relation between sequences is denoted by $\sqsubseteq$.
- Given a sequence s of moves in M and a subset M' of M, $s\lceil_{M'}$ denotes the subsequence of s formed by elements contained in M', and $|s|$ denotes the length of s.
- The function nl (nesting level) from plays to integers is defined as follows:

$$\mathsf{nl}(\epsilon) = 0$$
$$\mathsf{nl}(sa) = |s\lceil_{M^Q}| - |sa\lceil_{M^A}|$$

In a play questions and answers match like opened and closed parenthesis in an expression, the value $\mathsf{nl}(sa)$ gives the nested level of questions at which the move a lies in the sequence sa. Note that, in the above definition of $\mathsf{nl}(sa)$, the move a is "counted" only if it is an answer; as a consequence, the function nl has the same value on a question and on the corresponding answer.

Definition 1. *A game A is a tuple $(M_A, \lambda_A, \prec_A, P_A, \approx_A)$ where*

- *M_A is a set of* moves,
- *$\lambda_A : M_A \to \{O, P\} \times \{Q, A\}$ is the* labelling *function: it tells us if a move is taken by the Opponent or by the Player, and if it is a Question or an Answer. We can decompose λ_A into $\lambda_A^{OP} : M_A \to \{O, P\}$ and $\lambda_A^{QA} : M_A \to \{Q, A\}$ and put $\lambda_A = \langle \lambda_A^{OP}, \lambda_A^{QA} \rangle$. We denote by $^{-}$ the function which exchanges Player and Opponent, and Question and Answer i.e. $\overline{O} = P$, $\overline{P} = O$, $\overline{Q} = A$ and $\overline{A} = Q$. We also denote with $\overline{\lambda_A^{OP}}$ the function defined by $\overline{\lambda_A^{OP}}(a) = \overline{\lambda_A^{OP}(a)}$ and with $\overline{\lambda_A}$ the function $\langle \overline{\lambda_A^{OP}}, \lambda_A^{QA} \rangle$.*
- *$\prec_A \subseteq (M_A \times M_A)$ is a strict order relation on the set of moves.*
- *P_A is the set of* plays *of the game A, that is a non-empty and prefix-closed subset of the set $M_A^{\circledast}$, where $M_A^{\circledast}$ is the set of all sequences of moves which satisfy the following conditions:*
  - *$s = as' \Rightarrow \lambda_A(a) = OQ$, a play starts with a question made by the Opponent.*
  - *$s = rabt \Rightarrow \lambda_A^{OP}(a) = \overline{\lambda_A^{OP}(b)}$, Player and Opponent alternate.*
  - *$s = rt \Rightarrow \mathsf{nl}(r) \geq (0)$, it is possible to play an answer only if there exists a pending question.*

  - $s = qarbt \vee nl(qa) = nl(qarb) \Rightarrow a \prec_A b$, *a question is weaker than the corresponding answer, if an answer a is followed by a new question b, then b is stronger than a, and this condition recursively apply to nested moves.*
- $\approx_A$ *is an equivalence relation on* P_A *which satisfies the following properties:*
  - $s \approx_A \epsilon \;\Rightarrow\; s = \epsilon$
  - $sa \approx_A s'a' \;\Rightarrow\; s \approx_A s'$
  - $s \approx_A s' \;\wedge\; sa \in P_A \;\Rightarrow\; \exists a' .\, sa \approx_A s'a'$
  - $sa \approx_A s'a' \wedge sarb \approx_A s'a'r'b'$ $\Rightarrow\; ((a \prec_A b \Rightarrow a' \prec_A b') \wedge (b \prec_A a \Rightarrow b' \prec_A a'))$

Definition 2 (Strategies).
A strategy σ in a game A is a non-empty set of plays of even length such that $\sigma \cup dom(\sigma)$ is prefix-closed, where $dom(\sigma) = \{t \in P_A \mid \exists a\,.\, ta \in \sigma\}$.
A strategy can be seen as a set of rules which tells the Player which move to make after the last move by the Opponent.

In this paper we shall consider strategies that are *deterministic*, *history-free* and *monotone*. A strategy is history-free if it depends only on the *last* move by the Opponent; it is monotone if, in some particular cases, it respects the partial order $\prec$ (see bellow).

Before giving the definition of monotone strategy, we need to introduce two new concepts: the set of derived questions and the set of derived answers. The intuitive idea is the following: if, in a play s, a question a of the Opponent is followed by a question b of the Player, one can consider b an effect of question a, since in order to answer to a, the Player needs to know the answer to b. Moreover, if after receiving an answer c to b the Player asks a second question b', this means that the information given by c was not sufficient and new information is required; that is, also b' can be considered a direct consequence of a. The above argument can be repeated until a receives an answer, in this way defining the set of the derived questions of question a. Formally, the set of the derived questions of a question a in a play s is defined by:

$$\mathsf{drv}(s,a) = \{b \mid b \in M^Q,\; s = s'arbs'',\; \mathsf{nl}(r) = 0,\; \forall r' \subseteq r\,.\, nl(r') \geq 0\}$$

Similarly, one can associate to an answer a the set of answers generated thanks to the information given by a, and define the set of the derived answers of an answer a in play s

$$\mathsf{drv}(s,a) = \{b \mid b \in M^A,\; s = s'arbs'',\; \mathsf{nl}(r) = 0,\; \forall r' \subseteq r\,.\, nl(r) \leq 0\}$$

The function drv can be extended to strategies. Given a strategy σ and an Opponent move a, the set of moves derived from a in strategy σ is defined by:

$$\mathsf{drv}(\sigma,a) = \bigcup_{s \in \sigma} \mathsf{drv}(s,a)$$

Definition 3 (Deterministic, history-free and monotone strategies). *A strategy σ for a game A is deterministic if:*

$$sb, sc \in \sigma \ \Rightarrow \ b = c$$

The strategy σ is history-free if:

$$sab, t \in \sigma \ \wedge \ ta \in P_A \ \Rightarrow \ tab \in \sigma$$

The strategy σ is monotone if:

$$s \in \sigma \ \wedge \ sa \in P_A \ \wedge \ a' \prec_A a \wedge \mathit{drv}(\sigma, a') \neq \emptyset \\ \Rightarrow \ \exists b. \, (sab \in \sigma \ \wedge \ \forall b' \in \mathit{drv}(\sigma, a') \, . \, b' \prec_A b)$$

In the following we implicitly assume strategies to be deterministic, history free and monotone.

The condition of monotonicity requires that if a strategy σ reacts to a question a with another question b (or to an answer a with an answer b), then σ needs to react to any move stronger than a with a move that is stronger than b (and stronger that any other moves derived from a). The notion of derived moves is essential in order to assure that the composition of two monotonic strategies is a monotonic strategies.

The condition of monotonicity is quite strong. In particular, there are "few" finite monotone strategies: in general, a monotone strategy cannot be approximated by a chain of finite and monotone strategies. This shortage of finite strategies is necessary in order to have a full definability result. In game semantics the interpretation of a solvable λ-term is always an infinite object. In [KNO00, KNO99] the semantic interpretations of λ-terms are characterised as almost everywhere copy-cat strategies. On the game λ-models we are going to construct, the condition of monotonicity essentially forces the behaviour of strategies to be almost everywhere copy-cat strategies.

The equivalence relation on plays $\approx$ generates a relation $\precapprox$ and a partial equivalence relation $\approx$ on strategies in the following way.

Definition 4 (Order-enrichment). *Given strategies σ and τ we write $\sigma \precapprox \tau$ iff*

$$sab \in \sigma \wedge \ s' \in \tau \wedge \ sa \approx s'a' \Longrightarrow \exists b'.(s'a'b' \in \tau \ \wedge \ sab \approx s'a'b')$$

The relation $\approx$ on strategies is the reflexive closure of the relation $\precapprox$

It is easy to check that $\approx$ is a partial equivalence relation. It is not an equivalence since it might lack reflexivity. If σ is a strategy for a game A such that $\sigma \approx \sigma$, we write $\sigma : A$. It is also immediate that $\precapprox$ defines a partial order on the equivalence classes of strategies.

Definition 5 (Tensor product).
Given games A and B the tensor product *$A \otimes B$ is the game defined as follows:*

- $M_{A \otimes B} = M_A + M_B$;
- $\lambda_{A \otimes B} = [\lambda_A, \lambda_B]$;
- $\prec_{A \otimes B} = \prec_A \cup \prec_B$;
- $P_{A \otimes B} \subseteq M^{\circledast}_{A \otimes B}$ *is the set of plays, s, satisfying the projection condition:* $s \upharpoonright_{M_A} \in P_A$ *and* $s \upharpoonright_{M_B} \in P_B$ *(the projections on each component are plays for the games A and B respectively);*
- $s \approx_{A \otimes B} s'$ *iff* $s \upharpoonright_A \approx_A s' \upharpoonright_A \wedge s \upharpoonright_B \approx_B s' \upharpoonright_B \wedge \forall i.(s_i \in M_A \Leftrightarrow s'_i \in M_A)$.

Here + denotes disjoint union of sets, that is $A + B = \{in_l(a) \mid a \in A\} \cup \{in_r(b) \mid b \in B\}$, *and* $[-,-]$ *is the usual (unique) decomposition of a function defined on disjoint unions.*

One should notice that, differently from the standard definition of [AJM94], it is not necessarily to impose the Stack discipline, which says that in a play every answer must be in the same component game as the corresponding question. The stack discipline is forced by monotonicity condition on plays, in fact a question a and the corresponding answer b have the same nested level, therefore $a \prec_{A \otimes B} b$, and by the definition of $\prec_{A \otimes B}$, a and b lie in the same component. It is also useful to observe that if $sab \in P_{A \otimes B}$, and a, b are in different components then $\lambda^{QA}(a) = \lambda^{QA}(b)$. As a consequence, in a product game, only the Opponent can switch component, and this can happen only by reacting to a question of the Player with another question, or by giving an answer in the correct component.

Definition 6 (Unit). *The unit element for the tensor product is given by the empty game* $I = (\varnothing, \varnothing, \varnothing, \{\epsilon\}, \{(\epsilon, \epsilon)\})$.

Definition 7 (Linear implication). *Given games A and B the compound game* $A \multimap B$ *is defined as follows:*

- $M_{A \multimap B} = M_A + M_B$
- $\lambda_{A \multimap B} = [\overline{\lambda_A}, \lambda_B]$
- $\prec_{A \otimes B} = \prec_A \cup \prec_B$
- $P_{A \multimap B} \subseteq M^{\circledast}_{A \multimap B}$ *is the set of plays, s, which satisfy the Projection condition:* $s \upharpoonright_{M_A} \in P_A$ *and* $s \upharpoonright_{M_B} \in P_B$
- $s \approx_{A \multimap B} s'$ *iff* $s \upharpoonright_A \approx_A s' \upharpoonright_A \wedge s \upharpoonright_B \approx_B s' \upharpoonright_B \wedge \forall i\,.\,(s_i \in M_A \Leftrightarrow s'_i \in M_A)$

By repeating the arguments used for the tensor product, it is not difficult to see that in a "linear implication game" only the Player can switch component, and this can happen only by reacting to question of the Opponent with another question, or by giving an answer in the correct component.

Definition 8 (Exponential). *Given a game A the game !A is defined as follows:*

- $M_{!A} = \mathbb{N} \times M_A = \sum_{i \in \mathbb{N}} M_A$
- $\lambda_{!A}(\langle i, a \rangle) = \lambda_A(a)$
- $(i, a) \prec_{!A} (j, b)$ *iff* $i = j$ *and* $a \prec_A b$

- $P_{!A} \subseteq M_{!A}^{\circledast}$ *is the set of plays, s, which satisfy the conditions:* $\forall i \in \mathbb{N} . s\restriction_{A_i} \in P_{A_i}$
- $s \approx_{!A} s'$ *iff there exists a permutation of* α *on* $\mathbb{N}$ *such that:*
  - $\pi_1^*(s) = \alpha^*(\pi_1^*(s'))$
  - $\forall i \in \mathbb{N} . (\pi_2^*(s' \restriction_{\alpha(i)}) \approx \pi_2^*(s \restriction_i))$

  where α^* *denotes the pointwise extension of the function* α *to sequence of naturals.* π_1 *and* π_2 *are the projections of* $\mathbb{N} \times M_A$ *and* $s \restriction_i$ *is an abbreviation of* $s \restriction_{A_i}$.

Definition 9 (The category of games $\mathcal{G}_{\mathcal{M}}$).
The category $\mathcal{G}_{\mathcal{M}}$ *has as objects games and as morphisms, between games A and B, the equivalence classes, w.r.t. the relation* $\approx_{A \multimap B}$*, of deterministic, history-free and monotone strategies* $\sigma : A \multimap B$*. We denote the equivalence class of* σ *by* $[\sigma]$.

The identity for each game A is given by the (equivalence class) of the copy-cat *strategy, recursively defined as follows,*

$$id_A = \{sa'a'' \in P_{A \multimap A} \mid s \in id_A, \{a', a''\} = \{in_l(a), in_r(a)\}\} \cup \{\epsilon\}$$

Composition is given by the extension to equivalence classes of the following composition of strategies. Given strategies $\sigma : A \multimap B$ *and* $\tau : B \multimap C$, $\tau \circ \sigma : A \multimap C$ *is defined by*

$$\tau \circ \sigma = \{s\restriction_{(A,C)} \mid s \in (M_A + M_B + M_C)^* \ \& \ s\restriction_{(A,B)} \in \overline{\sigma}, s\restriction_{(B,C)} \in \overline{\tau}\}^{even}$$

where with S^{even} *denote the set of plays in S having even length.*

The correctness of the above definition follows, in part, from the correctness of the definition of AJM-games. In addition we need to prove that:

- id_A is a monotone strategy,
- the composition of two monotone strategies is again a monotone strategy.

The monotonicity of id_A follows immediately from the fact that for every pair of moves $a \in M_A^P$, $b \in M_A^O$, $\mathsf{drv}(id_A, \mathrm{in}_l(a)) = \{\mathrm{in}_r(a)\}$ and $\mathsf{drv}(id_A, \mathrm{in}_r(b)) = \{\mathrm{in}_l(b)\}$. The preservation of monotonicity by strategy composition follows easily from the fact that for every pair of strategies $\sigma : A \multimap B$ and $\tau : B \multimap C$ and for every pair of moves $a \in M_A^P$ and $c \in M_C^O$, if $a \in \mathsf{drv}(\tau \circ \sigma, c)$ then there exists a chain $b_1, \ldots b_{2n+1}$ such that $b_1 \in \mathsf{drv}(\tau, c)$, $a \in \mathsf{drv}(\sigma, b_{2n+1})$, $\forall i]in\{0 \ldots n\} . b_{2i} \in \mathsf{drv}(\sigma, b_{2i-1}), \quad b_{2i+1} \in \mathsf{drv}(\tau, b_{2i})$.

The constructions introduced in Definitions 5, 7 and 8 can be made to be functorial.

Definition 10. *Given two strategies* $\sigma : A \multimap B$ *and* $\sigma' : A' \multimap B'$ *the strategies* $\sigma \otimes \sigma' : (A \otimes A') \multimap (B \otimes B')$, $\sigma \multimap \sigma' : (A \multimap A') \multimap (B \multimap B')$, $!\sigma : !A \multimap !B$ *are recursively defined as follows:*

$$\sigma \otimes \sigma' = \{sab \in P_{(A \otimes A') \multimap (B \otimes B')} \mid s \in \sigma \otimes \sigma', \ sab\restriction_{M_A \cup M_B} \in \sigma, \ sab\restriction_{M_{A'} \cup M_{B'}} \in \sigma'\} \cup \{\epsilon\}$$

$$\sigma \multimap \sigma' = \{sab \in P_{(A \multimap A') \multimap (B \multimap B')} \mid s \in \sigma \multimap \sigma',\ sab\restriction_{M_A \cup M_B} \in \sigma,\ sab\restriction_{M_{A'} \cup M_{B'}} \in \sigma'\} \cup \{\epsilon\}$$

$$!\sigma = \{s \in P_{!A \multimap !B} \mid \forall i\ .\ s\restriction_{M_{A_i} \cup M_{B_i}} \in \sigma\}$$

It is not difficult to check that the above definitions are correct and that $\otimes$ and I indeed provide a categorical tensor product and its unit.

The category $\mathcal{G}_{\mathcal{M}}$ is monoidal closed, but not Cartesian closed. Analogously to what happens in AJM-games, a Cartesian closed category of games can be obtained by taking the co-Kleisli category $K_!(\mathcal{G}_{\mathcal{M}})$ over the co-monad $(!, \mathtt{der}, \delta)$, where for each game A the strategies $\mathtt{der}_A : !A \multimap A$ and $\delta_A : !A \multimap !!A$ are defined as follows:

- $\mathtt{der}_A = [\{s \in P_{!A \multimap A} \mid s\restriction_{(!A)_0} = s\restriction_A\}]$
- $\delta_A = [\{s \in P_{!A \multimap !!A} \mid s\restriction_{(!A)_{p(i,j)}} = s\restriction_{(!(!A)_i)_j}\}]$ where $p : \mathbb{N} \times \mathbb{N} \to \mathbb{N}$ is a pairing function

Hence one can easily see that the following definitions are well posed.

Definition 11 (A Cartesian closed category of games).
The category $K_!(\mathcal{G}_{\mathcal{M}})$ has as objects games and as morphisms between games A and B the equivalence classes of history-free strategies in the game $!A \multimap B$.

In order to give semantics to the lazy λ-calculus, it is necessary to define the lifting constructor.

Definition 12 (Lifting).
Given a game A, the lifted game $A_\perp$ is defined as follows:

- $M_{A_\perp} = M_A + \{\circ, \bullet\}$
- $\lambda_{A_\perp} = [\lambda_A, \{\circ \to OQ, \bullet \to PA\}]$
- $\prec_{A_\perp} = \prec_A \cup \{\langle b, a\rangle \mid a \in M,\ b \in \{\circ, \bullet\}\} \cup \{\langle \circ, \bullet\rangle\}$
- $P_{A_\perp} = \{\epsilon, \circ\} \cup \{\circ\bullet s \mid s \in P_A\}$
- $s \approx_{A_\perp} s'$ *iff* $s = s'$ *or*
  $s = \circ\bullet t$ *and* $s' = \circ\bullet t'$ *and* $t \approx_A t'$

Note that the above definition cannot be made functorial, at least not in a standard way. Given a strategy $\sigma : A \to B$, strategy $\sigma_\perp : A_\perp \to B_\perp$ is usually defined ([AM95a]) by:

$$\sigma_\perp = \{\circ_A \circ_B \bullet_B \bullet_A s \mid s \in \sigma\} \cup \{\epsilon, \circ_A \circ_B\}$$

In the category $\mathcal{G}_{\mathcal{M}}$, with the above definition, $\sigma_\perp$ is not necessarily a monotone strategy. In fact, the initial behaviour of $\sigma_\perp$ $(\circ_A \circ_B \bullet_B \bullet_A \in \sigma_\perp)$ imposes conditions on the future behaviour of the strategy that are not necessarily satisfied.

However, given any game A having a single initial move a, it is possible to define two strategies: $\mathsf{up}_A : A \multimap A_\perp$ and $\mathsf{dn}_A : A_\perp \multimap A$ as follows:

$$\mathsf{up}_A = \{\circ_{A_\perp} \bullet_{A_\perp} s \mid s \in id_A\} \cup \{\epsilon\}$$

$$\mathsf{dn}_A = \{a \circ_{A_\perp} \bullet_{A_\perp} s \in P_{!(A_\perp \multimap A} \mid as \in id_A\} \cup \{\epsilon, a \circ_{A_\perp}\}$$

It is not difficult to prove that the above strategies are well defined and that $\mathsf{dn}_A \circ \mathsf{up}_A \approx id_A$.

In order to define a model for the lazy λ calculus, the functoriality of the lifting constructor is not necessary, the existence of the strategies, dn_A and up_A suffices.

4 Solution of Recursive Game Equations

The categories of games $\mathcal{G}_\mathcal{M}$ and $K_!(\mathcal{G}_\mathcal{M})$ allow for the existence of *recursive* objects, *i.e.* objects that are fixed points of game constructors. We present the method proposed by Abramsky and McCusker ([AM95b]) for defining recursive games. This method allows to define *initial* fixed points for a large set of functors and it follows the pattern used for building initial fixed points in the context of information systems. First a complete partial order $\trianglelefteq$ on games is introduced.

Definition 13. *Let A, B be games, A is a* sub-game *of B ($A \trianglelefteq B$) if*

- $M_A \subseteq M_B$;
- $\lambda_A = \lambda_B \lceil_{M_A}$;
- $\prec_A = \prec_B \lceil_{M_A}$;
- $P_A = P_B \cap M_A^{\circledast}$;
- $s \approx_A s'$ *iff* $s \approx_B s'$ *and* $s \in P_A$.

One can easily see that the sub-game relation defines a complete partial order on games. Hence a game constructor F which is continuous with respect to $\trianglelefteq$ has a (minimal) fixed point $D = F(D)$ given by $\bigsqcup_{n\in\mathbb{N}} F^n(I)$. Notice that we have indeed an identity between D and $F(D)$ and that we do not need the game constructor F to be a functor; as a result we can also apply this method to the lifting game constructor.

One can easily see that the game constructors $\otimes$, $\multimap$, $!$, $(\ )_\perp$ and their compositions are continuous with respect to $\trianglelefteq$; therefore, the method applies to them.

5 Lazy λ-Models in $K_!(\mathcal{G}_\mathcal{M})$

A standard way to construct a model for the lazy λ-calculus consists in taking the initial fixed-point of the functor $F(D) = (D \to D)_\perp$ [AO93], [AM95a], [BPR98], [EHR92]. Here we use the same technique. We denote with D the least fixed-point of the game constructor: $F(A) = (A \to A)_\perp = (!A \multimap A)_\perp$ in the category of monotonic games ($D = \bigcup F^n(I)$). We denote with $\varphi :!D \multimap (!D \multimap D)$ the morphism $\mathsf{dn}_{!D \multimap D} \circ \mathsf{der}_D$ and with $\psi :!(!D \multimap D) \multimap D$ the morphism $\mathsf{up}_{!D\multimap D} \circ \mathsf{der}_{!D\multimap D}$.

The morphisms φ and ψ define a retraction between D and $!D \multimap D$ such that $\psi \circ \perp \not\approx_D \perp$, where with $\perp$ we indicate the smallest strategy $\{\epsilon\}$. It follows that the tuple $\mathsf{D} = \langle D, \varphi, \psi\rangle$ defines a categorical model of the lazy λ-calculus.

Definition 14. *The interpretation of a λ-term M (whose free variables are among the list $\Gamma = \{x_1, \ldots, x_n\}$) in the model $\mathtt{D} = \langle D, \varphi, \psi \rangle$ is strategy $[\![M]\!]_\Gamma$: $\overbrace{(!D \otimes \cdots \otimes !D)}^{|\Gamma|} \to D$ defined inductively as follows:*

$$[\![x_i]\!]_\Gamma = \pi_i^\Gamma;$$
$$[\![MN]\!]_\Gamma = ev \circ \langle (\varphi \circ [\![M]\!]_\Gamma) \ , \ [\![N]\!]_\Gamma \rangle;$$
$$[\![\lambda x.M]\!]_\Gamma = \psi \circ \Lambda(\ [\![M]\!]_{\Gamma,x});$$

where π_i^Γ are the canonical projection morphisms, ev and Λ denote "evaluation" and "abstraction" in the Cartesian closed category $K_!(\mathcal{G}_\mathcal{M})$.

It is useful to give some intuitive explanations concerning the plays in the game λ-model D. In game D the Opponent can be identified with the environment, while the Player can be identified with a program (λ-term) interrogated by the Opponent. A possible play s in game D proceeds as follows: the initial move of s is a request of the Opponent to know if the Player is a λ-abstraction; the Player fails to reply if it is a strongly unsolvable term, otherwise it answers (positively) to the question. After that, the play proceeds by a consecutive question of the Opponent asking if there is another λ-abstraction inside the first λ-abstraction. Again the Player may fail to answer, if it is an unsolvable term of order 1, or it may answer, if it contains two λ-abstractions. This time however, the Player can also pose a question to the Opponent; this happens if the Player is in the form $\lambda x \,.\, xM_1 \ldots M_m$. In this case, the Player contains a second λ-abstraction depending on value (behaviour) of x (the first argument passed by the Opponent). In particular, in the above case, the Player needs to check whether x contains $m+1$ λ-abstractions. In reaction to the questions of the Player, on the argument x, the Opponent can reply by posing questions on the arguments passed to x; in this case, the Player will answer according to the terms M_i. The plays can proceed with questions and answer in an arbitrary nested level, with the Opponent asking information on the deeper structure of the Player.

The order relation $\prec_D$ models the fact that consecutive questions, at the same nested level, ask for more and more λ-abstractions. The condition of monotonicity on plays models the fact that the questions posed by of a λ-term (Player) at nested level 1 always concern the argument appearing as head variable.

More formally, we present a set of results describing the theory induced by the game λ-model D. Since $\mathtt{D}$ is a λ model and since the interpretation of a λ-abstraction is never equivalent to strategy $\bot$, one immediately has:

Proposition 1. *For any pair of closed λ-terms M, N, if $M \Downarrow N$ then $[\![M]\!] \approx_D [\![N]\!] \neq \bot$.*

The proof of adequacy need to be more complex. In general sophisticated proof techniques, such as the computability method, the invariants relations or the approximation theorem, are needed to prove the adequacy of a model. In this case we can use a previous result concerning the games semantics of the untyped λ-calculus. In this way we are also able to characterize precisely the theory induced by D.

In [DGF00], a complete characterisation of the theories induced by game models in the category $\mathcal{G}$ of games and history free strategies has been carried out. In particular it has been shown that every categorical game model $\langle A, \varphi_A, \psi_A \rangle$, such that $\psi_A \circ \bot \neq \bot$, induces the theory $\mathcal{LT}$. In theory $\mathcal{LT}$ two terms are identified if and only if they have the same Lévy-Longo tree [L$\acute{7}$5, Lon83]. We briefly recall the definitions.

Definition 15. *Let* $\Sigma = \{\lambda x_1 \ldots x_n.\bot \mid n \in \mathbb{N}\} \bigcup \{T\} \bigcup \{\lambda x_1 \ldots x_n.y \mid n \in \mathbb{N}\}$, *with* $x_1, \ldots, x_n, y \in Var$

Lévy-Longo tree associated to λ-term M, $LLT(M)$, *is a* Σ*-labelled infinitary tree defined informally as follows:*

- $LLT(M) = T$ *if* M *is unsolvable of order* ∞, *that is for each natural numbers* n *there exists a λ-term* $\lambda x_1 \ldots x_n.M'$ *β-equivalent to* M.

- $LLT(M) = \lambda x_1 \ldots x_n.\bot$ *if* M *is unsolvable of order* n

- $LLT(M) = \lambda x_1 \ldots x_n.y$

$$
\begin{array}{c}
/ \quad \backslash \\
LLT(M_1) \ldots LLT(M_m)
\end{array}
$$

if M *is solvable and has principal head normal form* $\lambda x_1 \ldots x_n.yM_1 \ldots M_m$.

The arguments used in [DGF00] can be straightforwardly applied also to category $\mathcal{G}_\mathcal{M}$. In particular, through an application of the Approximation Theorem it is possible to derive that:

Proposition 2. *For any pair of closed λ-terms* M, N, *if* $LLT(M) \Downarrow LLT(N)$ *then* $[\![M]\!] \approx_D [\![N]\!]$.

Proposition 3 (Adequacy and Soundness). *For any pair of closed λ-terms* M, N*:*

– *if* $[\![M]\!] \not\approx_D \bot$ *then* $M \Downarrow$*;*
– *if* $[\![M]\!] \lessapprox_D [\![N]\!]$ *then* $M \sqsubseteq_l N$.

Proof. By Proposition 2, if a λ-term M is such that $[\![M]\!] \not\approx_D \bot$, then M is not strongly unsolvable, and the lazy reduction strategy converges on M (see Section 2). The second point is readily proved observing that denotational semantics is compositional and monotonic, therefore for every closed context $C[\ ]$, if $[\![M]\!] \lessapprox_D [\![N]\!]$, then $[\![C[M]]\!] \lessapprox_D [\![[N]]\!]$ and therefore $C[M] \Downarrow \Rightarrow C[N] \Downarrow$. □

5.1 Extensional Collapse

Theory $\mathcal{LT}$ is strictly weaker than theory λ_l; for example, the terms $\lambda x.xx$ and $\lambda x.x(\lambda y.xy)$ have different Lévy-Longo trees but they are equated in λ_l. In order to obtain a fully abstract model, we need to interpret λ-terms in the category $\mathcal{E}_\mathcal{M}$, defined as an extensional collapse of category $\mathcal{G}_\mathcal{M}$. We need to use the Sierpinski game, that is the game $I_\bot$.

Definition 16 (Intrinsic pre-order). *Give a game A, the intrinsic pre-order $\precsim_A$ on the strategies for A is defined by:*

$$\sigma_1 \precsim_A \sigma_2 \;\; \textit{iff} \;\; \forall \tau : A \multimap I_\perp \,.\, \tau \circ \sigma_1 \sqsubseteq_{I_\perp} \tau \circ \sigma_2$$

In the above expression we implicitly coerce the strategies in A into $I \multimap A$.

We indicate with $\simeq_A$ the partial equivalence relation induced by $\precsim_A$.

The category $\mathcal{E}_\mathcal{M}$ has as objects games and as morphism equivalence classes w.r.t. $\simeq$ of strategies.

It is not difficult to verify that intrinsic pre-order is preserved by all the categorical constructions presented above. Therefore, the category $\mathcal{E}_\mathcal{M}$ can be used in modeling the λ-calculus. In particular, the game λ-model $\mathtt{D}$ with the interpretation of Definition 14 gives rise to a λ-model also inside the category $\mathcal{E}_\mathcal{M}$.

6 Full-Abstraction

As usual, the proof of full-abstraction splits in two proofs.

Theorem 1 (Soundness). *For any pair of closed λ-terms M, N, we have:*

$$[\![M]\!] \precsim_D [\![N]\!] \;\Rightarrow\; M \sqsubseteq_l N$$

Proof. It is immediate to check that the only strategy $\simeq_D$-equivalent to $\perp$ is strategy $\perp$ itself. It follows that model $\mathtt{D}$ is adequate also in category $\mathcal{E}_\mathcal{M}$. By the compositionality of the interpretation, soundness follows immediately. □

In order to prove completeness, some preliminary results need to be presented.

Proposition 4. *The following properties hold in the game λ-model D:*

(i) *Every question has one only possible answer and every answer has one only possible consecutive question. Formally, for every pair of plays $sab, tab' \in P_D$ if $\lambda_D^{QA}(b) = \lambda_D^{QA}(b') = \overline{\lambda_D^{QA}(a)}$ then $b = b'$.*

(ii) *For every move $a \in M_D$, the set of predecessors of a, w.r.t. the order $\prec_D$, is a finite and linearly ordered set.*

(iii) *With respect to the order $\prec_D$, every question move has one successor, the corresponding answer; while every answer move a has infinitely many immediate successors which are the consecutive question at the same nested level, and an infinite number of questions at the next nested level (these questions are the initial moves of a subcomponent $!D$ of game D).*

(iv) *For every strategy $\sigma : D$ and for every move $a \in M_D$ the set $\mathsf{drv}(\sigma, a)$ is linearly order w.r.t. $\prec_D$*

Proof. The first three points can be proved by induction on the chain of games $F^n(I)$. Point (iv) follows from point (i). □

Lemma 1. *Any strategy $\alpha : D \to I_\perp$ can be extended to a strategy $\alpha_D : D \to D$ such that for any strategy $\sigma : D$, $\circ : \bullet \in \alpha \circ \sigma$ if and only if $\circ : \bullet \in \alpha_D \circ \sigma$*

Proof. Since game $D \to I_\perp$ is a sub-game of game $D \to D$, it is sufficient to extend α to a monotone strategy on game $D \to D$, which can be done incrementally. Let $\alpha_0, \dots, \alpha_n, \dots$ be an infinite chain of strategies constructed in the following way:

$$\begin{aligned}
&\alpha_0 = \alpha \\
&\alpha_{i+1} = \alpha_i \cup \{sab \in P_{D\to D} \mid s \in \alpha_i,\ sa \in P_{D\to D}, \\
&\qquad \textstyle\bigcup_{a' \in M^O, a' \prec a} \mathsf{drv}(\alpha_i, a') \neq \emptyset, \\
&\qquad b \text{ minimal upper bound in } M^P \text{ of } \textstyle\bigcup_{a' \in M^O, a' \prec a} \mathsf{drv}(\alpha_i, a')\}.
\end{aligned}$$

In the above definition the choice of the element b is not necessarily unique. In some cases the minimal upper bounds of $\bigcup_{a' \in M^O, a' \prec a} \mathsf{drv}(\alpha_i, a')$ can form a countable set: the initial questions or answers in the some subcomponent $!D$ of game D. In these cases, almost any possible choice gives rise to an equivalent (w.r.t. $\approx_D$) strategy; some care have to be taken when the move a is itself an initial question or answer in some other subcomponent $!D$ of game D, in which case it is sufficient to choose for b the same index as for the move a.

Strategy α_D is finally defined as $\alpha_D = \bigcup_{n \in N} \alpha_n$. □

It is interesting to observe that if one performs the above construction starting from strategy $\alpha = \{\epsilon,\ \circ\circ, \circ\circ\bullet\bullet\}$, one obtains a strategy $\alpha_D \approx_D id_D$.

Proposition 5 (Definability). *For any play s in game D such that $\mathsf{nl}(s) = 0$, there exists a closed λ-term M such that $s \in [\![M]\!]$ and $[\![M]\!] \lessapprox \sigma$ for any strategy σ with $s \in \sigma$.*

To the above proposition, we just give an informal and intuitive proof. A formal proof will require the introduction of several new concepts and will be more difficult to grasp.

We will associate to play s a Lévy-Longo tree or equivalently a λ-term that represents a Lévy-Longo tree. In order to do that, we decompose play s in several levels, each level determining a node of the Lévy-Longo tree.

We need to introduce some notation. Given a play t and an interval I of natural numbers, we denote with $t\lceil_I$ the subsequence of t formed by the moves whose nested level is a value in the interval I.

Sequence $s\lceil_{[0,1]}$ is a play contained in strategy σ. In fact, sequence $s\lceil_{[0,1]}$ describes the behaviour of strategy σ on the hypothesis that the Opponent answers immediately to questions posed by the Player (without posing nested questions). Since, in game D, the Opponent is always allowed to answer immediately to the questions of the Player, the sequence $s\lceil_{[0,1]}$ is a play. Since in $s\lceil_{[0,1]}$ the behaviour of the Player, in reaction to the last move of the Opponent, is the same that in s, and since σ is a history free strategy, it follows that $s\lceil_{[0,1]} \in \sigma$.

Play $s\lceil_{[0,1]}$ can be in one of the following forms:

– the Player always answers to the questions of the Opponent

– the Player answers for n times to the questions of the Opponent, then at the $n+1$ question q of the Opponent, the Player replays posing a question. In this case, the Player is in the position to make the second move in a game having form: $\underbrace{!D \multimap (\ldots(!D}_{n} \multimap D)\ldots)$ and it can choose to pose a question in one of the n instances of D staying on the left of an arrow. After that, the monotonicity condition on plays forces the Player to react to the answer of the Opponent by either posing a consecutive question in the same component either by giving an answer to question q. In all cases, after having posed m consecutive questions in the same component, the Player will answer to question q. The condition of monotonicity on strategies now forces play $s\restriction_{[0,1]}$ to proceed in only one possible way. At the consecutive question of the Opponent the Player needs to reply with a move that is stronger (w.r.t the $\prec_D$ order) to the last question posed by the Player (q_p). This implies that the Player needs to pose the question consecutive to question q_p. At the answer of the Opponent the Player, by the monotonicity condition on strategies, needs to reply with an answer (only one answer available). And the previous argument applies to all consecutive moves.

In the first case above, play s has all moves at the nested level 0, and it is possible to check that: $s \in [\![\lambda x_1 \ldots x_n.\Omega]\!]$ and that $[\![\lambda x_1 \ldots x_n.\Omega]\!] \lessapprox \sigma$. On the second case, it is possible to check that: $s\restriction_{[0,1]} \in [\![\lambda x_1 \ldots x_n.x_i \underbrace{\Omega \ldots \Omega}_{m}]\!]$. Moreover, it is possible to prove that $[\![\lambda x_1 \ldots x_n.x_i\Omega \ldots \Omega]\!] \lessapprox \sigma$; this can be done proving, by induction on the length of plays, that the monotonicity condition forces strategy σ to behave in a copy-cat way.

Sequence $s\restriction_{[2,\infty]}$ is a play in a game in the form $!D \multimap !D \ldots \multimap (!D\otimes \ldots \otimes !D)$, where the instances of D, on the left of the $\multimap$ arrow, denote variables that can be interrogated by the Player, and the instances of D on the right of the $\multimap$ arrow denote the arguments of the head-variable. The first move in $s\restriction_{[2,\infty]}$ is a question of the Opponent asking if one of the arguments of the head variable is a λ-abstraction and $s\restriction_{[2,3]}$ is a play in game $!D \multimap !D \ldots \multimap (!D \otimes \ldots \otimes !D)$ defining the external structure of one argument of the head-variable. It follows that there exists a λ-term $P_1 = \lambda x_1 \ldots x_n.x_i\Omega \ldots (\lambda x_{j1} \ldots x_{jn_j}.x_k\Omega \ldots \Omega) \ldots \Omega$ such that $s\restriction_{[0,3]} \in [\![P]\!]$ and $[\![P]\!] \lessapprox \sigma$.

The above analysis can be repeated for the consecutive levels of nested moves, each slice $s\restriction_{[2i,2i+1]}$ representing a play where the Opponent interrogates the Player in order to know the structure of some subterms of the Player. In this way play s determines a Lévy-Longo tree approximation of strategy σ at which s belongs.

From the above propositions one can finally conclude:

Theorem 2 (Completeness). *For any pair of closed λ-terms M, N, we have:*

$$M \sqsubseteq_l N \;\Rightarrow\; [\![M]\!] \lesssim_D [\![N]\!]$$

Proof. Suppose there exists a strategy α such that $\alpha \circ [\![M]\!] \neq \bot$, by Lemma 1, there exists a minimal strategy α_D such that $\alpha_D \circ [\![M]\!] \neq \bot$. Let $\circ s \bullet$ be the (initial) sequence of moves generated by in the interaction between the strategies α_D and $[\![M]\!]$. Play $\circ s \bullet$ is contained in strategy $\alpha_D : D \to D$, while play $t = \circ\bullet\circ s \bullet$ is contained in strategy $\mathsf{up} \circ \alpha_D : D$. Let P be the term defining the minimal strategy containing the play t. By a simple calculation it follows that $[\![PM]\!] \neq \bot$ and the following chain of implications is immediate: $[\![PM]\!] \neq \bot \Rightarrow PM \Downarrow \Rightarrow PN \Downarrow \Rightarrow [\![PN]\!] \neq \bot \Rightarrow \alpha_D \circ [\![N]\!] \neq \bot \Rightarrow \alpha \circ [\![N]\!] \neq \bot$. □

References

AJM94. S. Abramsky, R. Jagadeesan, and P. Malacaria. Full abstraction for pcf (extended abstract). In *Theoretical Aspects of Computer software. International Symposium TACS '94*, volume 789 of *LNCS*. Springer-Verlag, 1994.

AM95a. S. Abramsky and G. McCusker. Games and full abstraction for the lazy λ-calculus. In *Proceedings LICS '95*, 1995.

AM95b. S. Abramsky and G. McCusker. Games for recursive types. In I. C. Mackie C. L. Hankin and R. Nagarajan, editors, *Theory and Formal Methods of Computing 1994: Proceedings of the Second Imperial College Department of Computing Workshop on Theory and Formal Methods.* Imperial College Press, October 1995.

AO93. S. Abramsky and C.H.L. Ong. Full abstraction in the lazy λ-calculus. *Information and Computation*, 105:159–267, 1993.

BPR98. O. Bastonero, A. Pravato, and S Ronchi. Structures for lazy semantics. In *Programming Concepts and Methods, PROCOMET'98.* Chapman & Hall, 1998.

DGF00. P. Di Gianantonio and G. Franco. The fine structure of game lambda-models. In *Conference on the Foundation of Software Technology and Theoretical Computer Science (FSTTS '00)*, volume 1974 of *LNCS*, pages 429–441. Springer-Verlag, 2000.

EHR92. L. Egidi, F. Honsell, and S Ronchi. Operational, denotational and logical description: A case study. *Fundamenta Informaticae*, 16(2):149–170, 1992.

KNO99. A. D. Ker, H. Nickau, and C. H. L. Ong. A universal innocent game model for the Böhm tree lambda theory. In *Computer Science Logic: Proceedings of the 8th Annual Conference of the EACSL Madrid, Spain*, volume 1683 of *Lecture Notes in Computer Science*, pages 405–419. Springer-Verlag, September 1999.

KNO00. A. D. Ker, H. Nickau, and C. H. L. Ong. Innocent game models of untyped lambda calculus. To appear in *Theoretical Computer Science*, 2000.

L*ý*75. J.J. Lévy. An algebraic interpretation of λ-calculus and a labelled λ-calculus. In *Lambda Calculus and Computer Science*, volume 37 of *LNCS*, pages 147–165. Springer-Verlag, 1975.

Lon83. G. Longo. Set-theoretical models of λ-calculus: theories, expansions and isomorphisms. *Annals of Pure and Applied Logic*, 24:153–188, 1983.

Reductions, intersection types, and explicit substitutions

Dan Dougherty[†] and Pierre Lescanne,[‡]

[†]Department of Mathematics and Computer Science, Wesleyan University
Middletown, CT 06459 USA
E-mail: ddougherty@wesleyan.edu

[‡]Laboratoire de l'Informatique du Parallélisme, École Normale Supérieure de Lyon
46, Allée d'Italie, 69364 Lyon 07, FRANCE
E-mail: Pierre.Lescanne@ens-lyon.fr

1 Introduction

The λ-calculus plays a key role in the foundations of logic and of programming language design, and in the implementation of logics and languages as well. The foundation of λ-calculus itself is β-conversion, which relates the primitive notions of abstraction and application in terms of *substitution*. Classical λ-calculus treats substitution as an atomic operation, but in the presence of variable-binding substitution it is a complex operation to define and to implement. So a more careful analysis is required if one is to reason about the correctness of compilers, theorem provers, or proof-checkers. Furthermore the actual cost of performing substitution should be considered when reasoning about complexity of implementations.

Abadi, Cardelli, Curien, and Lévy [1] defined a calculus of *explicit substitutions* to serve as a more faithful model of implementations of the λ-calculus. Since then a variety of explicit substitutions calculi have been defined. The original motivation for the Abadi-Cardelli-Curien-Lévy calculus was pragmatic, but there is another point of view one may take on such a calculus, namely that making substitution explicit permits a more refined analysis of substitution than does classical λ-calculus. As historical context we note that in their book [12] Curry and Feys insist on the importance of substitution in logic in general and especially in the framework of λ-calculus. They write [page 6] that the synthetic theory of combinators "gives the ultimate analysis of substitutions in terms of a system of extreme simplicity. The theory of lambda-conversion is intermediate in character between synthetic theories and ordinary logics. Although its analysis is in some ways less profound—many of the complexities in regard to variables are still unanalyzed there—yet it is none the less significant; and it has the advantage of departing less radically from our intuition." From this point of view one can see an explicit substitution calculus as an improvement on both the system of combinators and the classical λ-calculus, since it is a system whose mechanics are first-order and as simple as those of combinatory logic yet which retains the same intensional character as

S. Abramsky (Ed.): TLCA 2001, LNCS 2044, pp. 121–135, 2001.

traditional λ-calculus. In particular we may view explicit substitution calculi as primary and see the classical λ-calculus as a subsystem of these systems, defined by a particular strategy of "eagerly" evaluating the substitution constructed by contracting a β-redex. In this way the study of explicit substitutions represents a deeper examination of the relationship between abstraction and application. This setting invites the programme of refining the results of the classical λ-calculus by finding proofs of their explicit-substitutions analogues *in the explicit substitutions system itself.* One can reasonably expect in this way to gain insight into the original λ-calculus. As a case study, in this paper we present a systematic study of the relation between normalization and types.

In many calculi of explicit substitutions, including the original Abadi, Cardelli, Curien, Lévy system, substitutions are first-class citizens and there is an algebraic/computational structure on the substitutions themselves, reflecting the fact that composition is a natural operation on substitutions. Melliès [17] made the somewhat surprising discovery that the presence of substitution-composition leads to the failure of strong normalization even for simply-typed terms. This suggests that it is useful to analyze the effect of making substitution explicit independently of studying composition of substitutions. Composition-free calculi of explicit substitutions have been studied in [16, 7, 4] among others.

Here we work in the composition-free calculus $\lambda\mathrm{x}$ (which uses names rather than de Bruijn indices) and the calculus $\lambda\mathrm{x}_{gc}$ obtained by adding explicit garbage-collection to $\lambda\mathrm{x}$.

Summary of results

Our main results concern the set of terms typable in various intersection-types disciplines. We show that in each of $\lambda\mathrm{x}$ and $\lambda\mathrm{x}_{gc}$ the terms which normalize by leftmost reduction and the terms which normalize by head reduction can each be characterized as the set of terms typable in a certain system. Our notions of leftmost- and head-reduction are non-deterministic, and our normalization theorems apply to any computations obeying these strategies. In this way we refine and strengthen these classical normalization theorems. See [18] where a similar issue is discussed. Surprisingly, the situation for the strongly normalizing terms diverges from the classical λ-calculus. For the natural generalization of the classical type system we prove that typable terms are strongly normalizing. But the converse fails: see Section 7.

In addition to their theoretical and methodological interest our results have consequences for the study of the implementation of functional programming languages. Recall that the theoretical foundation for the correctness of the standard evaluation strategy for functional languages is the classical theorem that leftmost reduction is normalizing (see for example [19] Prop. 2.4.12). When explicit substitutions are offered as a basis for an implementation one should define and analyze a corresponding notion of "leftmost" reduction. To our knowledge this analysis has not been previously done. The natural notion of leftmost reduction we define here is related to, but a refinement of, the classical notion; the non-determinism in leftmost reduction here corresponds to a choice between certain standard implementation strategies ([5]). The proofs we present here readily yield the results that a term is (strong-, leftmost-, or head-) normalizing iff it is so in the calculus extended by garbage-collection. Our results support the claim that garbage-collection is a very natural addition to the system, even from a purely theoretical point of view, since the resulting calculus has more convenient closure properties than the simple calculus (Lemma 2 is an example).

The intersection type systems we study are natural generalizations of the corresponding classical systems, and in fact the global structure of the proofs follow a standard paradigm (as in [3]). But the explicit reductions involving substitutions lead to combinatorial complications not arising in the traditional treatments and the proofs require some new techniques. The first result on strong normalization of calculi of explicit substitution was the so-called *preservation of strong normalization:* a pure (substitution-free) term is strongly normalizing under reduction in the presence of explicit substitutions if and only if it is strongly normalizing under β-reduction. We stress that, in keeping with our aim of treating the explicit substitutions calculus as logically prior to the traditional λ-calculus, we develop the machinery needed for direct proofs which do not depend on results from the theory of β-reduction.

This paper contains few proofs, but a full version with all the proofs is available at `http://www.ens-lyon.fr/~plescann/publications.html`

2 Terms and reduction strategies

In this section, we describe the terms of the calculus of explicit substitutions with explicit names λx and specify strategies of reduction toward normal forms, namely λx-reduction, head reduction, and leftmost reduction.

Actually the same set of terms can be described in many different ways which we call *taxonomies.*

Definition 1 (The basic taxonomy). *The set of terms with explicit substitutions $\Lambda\boldsymbol{x}$ is the set of terms M defined as follows:*

$$M, N ::= x \mid \lambda x.M \mid M\,N \mid M\langle x = N\rangle$$

The set of free variables of a term is defined just as for classical λ-calculus, with an additional clause ensuring that the free variables of $M\langle x = N\rangle$ are the same as the free variables of $(\lambda x.M)N$. In particular, x is bound in $M\langle x = N\rangle$. The set of free variables of a term M is written $FV(M)$, sometimes for simplicity we write $x \in M$ instead of $x \in FV(M)$.

We assume Barendregt's [2] convention, namely that *a variable does not occur free and bound in the same term.* For instance, we assume that x does not occur free in N in the term $M\langle x = N\rangle$. The rules we define further assume this convention and the reader should keep this fact in mind when reading them and certain forthcoming lemmas.

To describe the second taxonomy nicely it will be very convenient to have a notation to describe a term M on which is applied a sequence of closures $\langle z_1 = S_1\rangle$, ..., $\langle z_m = S_m\rangle$ then a sequence of applications of terms T_1,..., T_n. Such a term $M\langle z_1 = S_1\rangle ... \langle z_m = S_m\rangle T_1 ... T_n$ will be abbreviated as $M\langle \boldsymbol{z} = \boldsymbol{S}\rangle\, \boldsymbol{T}$.

Lemma 1 (The head form taxonomy). *Every term is of precisely one of the following forms:*

$\lambda x.B$	$(\lambda y.B)\langle x = A\rangle\langle \boldsymbol{z} = \boldsymbol{S}\rangle\, \boldsymbol{T}$
$(\lambda x.B)AT_1 \cdots T_n$ *with* $n \geq 0$	$(UV)\langle x = A\rangle\langle \boldsymbol{z} = \boldsymbol{S}\rangle\, \boldsymbol{T}$
$xT_1 \cdots T_n$ *with* $n \geq 0$	$x\langle x = A\rangle\langle \boldsymbol{z} = \boldsymbol{S}\rangle\, \boldsymbol{T}$
	$y\langle x = A\rangle\langle \boldsymbol{z} = \boldsymbol{S}\rangle\, \boldsymbol{T}$ *with* $x \not\equiv y$

Proof. Straightforward. ///

Following Barendregt [2] we distinguish between a set of rules defining a *notion of reduction* and a *reduction relation* induced by closing a notion of reduction under certain contexts. Sometimes the latter are called strategies and play a main role in evaluation of functional programming languages [5]. Some reductions are deterministic, which means that the structural rules determine a unique redex to be reduced. Others are non deterministic.

The following notion of reduction is due to R. Bloo and K. Rose [8, 21, 6]. The rules VarI and VarK, called respectively xv and xvgc by Rose, have been renamed here to recall the distinction between the classical λ_I and λ_K calculi.

Definition 2. *The notions of reduction* $\lambda\mathsf{x}$ *and* $\lambda\mathsf{x}_{gc}$ *are induced by the rules in Table 1: the notion of reduction* $\lambda\mathsf{x}$ *is obtained by deleting the rule* gc, *and the notion of reduction* $\lambda\mathsf{x}_{gc}$ *is obtained by deleting the rule* VarK.

The rule gc *is called "garbage collection", as it removes useless substitutions.*

(B)	$(\lambda x B)\, A$	$\rightarrow$	$B\langle x = A\rangle$	
(App)	$(MN)\langle x = A\rangle$	$\rightarrow$	$M\langle x = A\rangle\, N\langle x = A\rangle$	
(Abs)	$(\lambda y M)\langle x = N\rangle$	$\rightarrow$	$\lambda y M\langle x = N\rangle$	
(VarI)	$x\langle x = N\rangle$	$\rightarrow$	N	
(VarK)	$y\langle x = N\rangle$	$\rightarrow$	y	
(gc)	$M\langle x = A\rangle$	$\rightarrow$	M	if $x \notin FV(M)$

Table 1. The reduction rules.

Of course in the presence of rule gc we do not need rule VarK. On the other hand it is not the case that gc can be directly simulated by the other rules: consider the garbage-collection $x\langle x = y\rangle\langle v = w\rangle \longrightarrow x\langle x = y\rangle$. Rule gc has a different character from the other rules in the sense that it represents a more complex transformation than those of the other, atomic, substitution operations. On the other hand with an appropriate data structure for maintaining (the free variables in) terms it can be efficiently implemented and provides a tool to prevent memory leaks. And as Bloo and Rose have demonstrated it is quite convenient when reasoning about the formal properties of the calculus.

Notation. In the main technical development of this paper we will work exclusively with the full system $\lambda\mathrm{x}_{gc}$. So unless explicitly stated otherwise, phrases such as "reduction" refer to reduction in system $\lambda\mathrm{x}_{gc}$. At the end of the paper (Section 8) we will see that the results for the system not including garbage collection follow readily from the results for $\lambda\mathrm{x}_{gc}$.

Remark. As is well-known, $\lambda\mathrm{x}_{gc}$ has a critical pair, namely:

$$((\lambda x.M)N)\langle y = L\rangle$$

$$(\lambda x.M\langle y = L\rangle)\; N\langle y = L\rangle \quad \swarrow \qquad \searrow \quad M\langle x = N\rangle\langle y = L\rangle$$

Most of the difficulty in working with the system is due to this critical pair; this will be amply demonstrated in the sequel.

Definition 3 (Unrestricted reduction). *Unrestricted reduction (or $\lambda\mathrm{x}_{gc}$-reduction) allows a reduction rule to be applied in any context.*

A term M is strongly normalizing *if there is no infinite $\lambda\mathrm{x}_{gc}$-reduction starting from M. The set of strongly normalizing terms is denoted $\mathcal{SN}$.*

Definition 4 (Head reduction). Head reduction *is the closure of $\lambda\mathrm{x}_{gc}$ under the structural rules of Table 2.*

$$\frac{U \xrightarrow{h} U' \qquad U \text{ is not an abstraction}}{UV \xrightarrow{h} U'V} \qquad \frac{B \xrightarrow{h} B'}{\lambda x.B \xrightarrow{h} \lambda x.B'}$$

$$\frac{M \xrightarrow{h} M' \qquad M \text{ is not an abstraction}}{M\langle x = A\rangle \xrightarrow{h} M'\langle x = A\rangle}$$

Table 2. Head reduction

A term M is head normalizing *if there is no infinite head-reduction starting from M. The set of head normalizing terms is denoted $\mathcal{HN}$.*

A head normal form *is a term of the form $\lambda x_1..x_k.xA_1...A_n$ where x is a free variable or one of the x_i and $A_i \in \Lambda x$.*

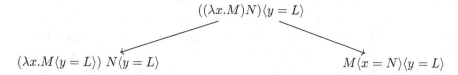

$$\frac{U \xrightarrow{l} U' \quad U \text{ is not an abstraction}}{UV \xrightarrow{l} U'V} \qquad \frac{B \xrightarrow{l} B'}{\lambda x.B \xrightarrow{l} \lambda x.B'}$$

$$\frac{M \xrightarrow{l} M' \quad M \text{ is not an abstraction}}{M\langle x = A\rangle \xrightarrow{l} M'\langle x = A\rangle} \qquad \frac{A_i \xrightarrow{l} A_i' \quad A_i \text{ is the leftmost non-n}}{xA_1...A_i...A_n \xrightarrow{l} xA_1...A_i'...A}$$

Table 3. Leftmost reduction

Definition 5 (Leftmost reduction). Leftmost reduction *is the closure of* λx_{gc} *under the structural rules in Table 3.*

A term M *is* leftmost normalizing *if there is no infinite leftmost reduction starting from* M. *The set of leftmost-normalizing terms is denoted* $\mathcal{LN}$.

Remark. Observe that in contrast to the classical notions both head reduction and leftmost reduction are nondeterministic strategies. Indeed both reductions out of the critical pair noted earlier count as head reductions. For example, let M be $((\lambda x.B)A)\langle y = C\rangle\langle \boldsymbol{z} = \boldsymbol{S}\rangle\boldsymbol{T}$. Then M can rewrite by leftmost reduction either to $P \equiv B\langle x = A\rangle\langle y = C\rangle\langle \boldsymbol{z} = \boldsymbol{S}\rangle\boldsymbol{T}$, or (in two steps) to $Q \equiv ((\lambda x.B\langle y = C\rangle)\ A\langle y = C\rangle)\langle \boldsymbol{z} = \boldsymbol{S}\rangle\boldsymbol{T}$. Then since $\lambda x.B\langle y = C\rangle$ is an abstraction Q leftmost-rewrites via rule B leading to $Q' \equiv B\langle y = C\rangle\langle x = A\langle y = C\rangle\rangle\langle \boldsymbol{z} = \boldsymbol{S}\rangle\boldsymbol{T}$.

3 Two fundamental Lemmas

To prove the main theorems of this paper, we need two very general lemmas which we present in this section. These lemmas aim at proving the following two facts (where $\mathcal{N}$ stands for $\mathcal{SN}$, $\mathcal{LN}$, or $\mathcal{HN}$).

$$M\langle x = N\rangle\langle y = L\rangle \in \mathcal{N} \text{ if } M\langle y = L\rangle\langle x = N\langle y = L\rangle\rangle \in \mathcal{N},$$

and

$$M\langle x = A\rangle \in \mathcal{N} \text{ if } M \in \mathcal{N} \text{ and } x \notin FV(M),$$

where, when $\mathcal{N}$ is $\mathcal{SN}$, we require $A \in \mathcal{SN}$ as well.

A remark on the classical Substitution Lemma

The Substitution Lemma of the classical λ-calculus [2] states a fundamental property of (implicit) substitutions, namely that, when x is not free in L

$$M[x := N][y := L] \equiv M[y := L][x := N[y := L]]$$

The two terms are syntactically identical above. When generalized to an explicit substitutions calculus the analogous statement is weakened to provable equality:

$$M\langle x = N\rangle\langle y = L\rangle = M\langle y = L\rangle\langle x = N\langle y = L\rangle\rangle$$

It is not hard to see that the two terms above can have quite different reduction behavior. In particular it is possible for the left-hand side to be $\mathcal{SN}$ while the right-hand side admits an infinite reduction. For instance, take $M \equiv z$, $N \equiv yy$ and $L \equiv \lambda u.uu$.

The Composition Lemma below states that if the right-hand side is $\mathcal{SN}$ then so is the left-hand side. So there is a fundamental asymmetry in this situation.

3.1 The Composition Lemma

Let us consider the following rule

$$M\langle x = N\rangle\langle y = L\rangle \;\rightarrow\; M\langle y = L\rangle\langle x = N\langle y = L\rangle\rangle$$

which we call *composition.* It abstracts the composition one finds in systems like $\lambda\sigma$ [1, 11] (namely the rule called *Map*) or in the extension $\lambda\mathrm{x}\|c$ of $\lambda\mathrm{x}$ [6, 7, 14, 15] (namely the rule $\xrightarrow[\|c]{}$ of [6], see also [21] page 75).

We would like to see that the converse of the composition rule preserves (strong, head, or leftmost) normalization. Unfortunately this rule does not commute in a nice way with reduction, essentially due to the duplication of substitutions in the App rule.

The following relation is a "bottom-up parallel extension" of the composition rule, which propagates and duplicates the applications of this rule inside terms. In particular, rule Cpabs pushes a substitution through an abstraction, Cpapp pushes through an application, and Cpclo pushes through a closure. The other rules make it a congruence.

Definition 6. *The relation $\Rrightarrow$ is given by the inductive definition indicated in Table 4.*

[Cref]	$M \Rrightarrow M$		
[Cabs]	$\dfrac{B \Rrightarrow B'}{\lambda x.B \Rrightarrow \lambda x.B'}$	[Cpabs]	$\dfrac{B\langle y = Q\rangle \Rrightarrow B^+}{(\lambda x.B)\langle y = Q\rangle \Rrightarrow \lambda x.B^+}$
[Capp]	$\dfrac{U \Rrightarrow U' \qquad V \Rrightarrow V'}{UV \Rrightarrow U'V'}$	[Cpapp]	$\dfrac{U\langle y = Q\rangle \Rrightarrow U^+ \qquad V\langle y = Q\rangle \Rrightarrow V^+}{(UV)\langle y = Q\rangle \Rrightarrow (U^+V^+)}$
[Cclo]	$\dfrac{B \Rrightarrow B' \qquad A \Rrightarrow A'}{B\langle z = A\rangle \Rrightarrow B'\langle z = A'\rangle}$	[Cpclo]	$\dfrac{M\langle y = Q\rangle \Rrightarrow M^+ \qquad P\langle y = Q\rangle \Rrightarrow P^+}{M\langle x = P\rangle\langle y = Q\rangle \Rrightarrow M^+\langle x = P^+\rangle}$

Table 4. The rules for $\Rrightarrow$

Lemma 2. *Let $\longrightarrow$ stand for either unrestricted reduction, leftmost reduction, or head reduction. If $M \longrightarrow M''$ and $M \Rightarrow M'$ then there is an M^* with $M'' \Rightarrow M^*$ and $M' \longrightarrow M^*$. Furthermore, if $M \longrightarrow M''$ is a B-step then in fact $M' \xrightarrow{+} M^*$ with at least one B-step.*

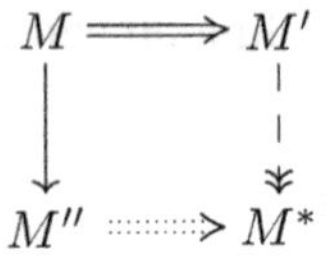

Proof. By induction over the definition of $M \Rightarrow M'$. ///

Corollary 1. *Suppose $M \Rightarrow M'$.*

- *If $M' \in \mathcal{HN}$ then $M \in \mathcal{HN}$.*
- *If $M' \in \mathcal{LN}$ then $M \in \mathcal{LN}$.*
- *If $M' \in \mathcal{SN}$ then $M \in \mathcal{SN}$.*

Proof. As is well-known, the set of rules of λx_{gc} other than the B-rule comprise a strongly normalizing rewrite system. So any infinite reduction out of M must involve infinitely many B-steps. With this observation each of the claims follows easily from Lemma 2. ///

In particular, suppose that if T' is obtained from T by the composition rule as defined at the beginning of this section. We conclude that if T' is strongly normalizing then T is strongly normalizing; similarly for leftmost and head normalization. These results will be crucial in the coming sections.

3.2 The Closure lemma

Now we want to prove that (head, leftmost, or strong) normalization is not affected by garbage-collection of normalizing terms. For technical reasons we state the result rather generally.

Definition 7. *A n-multi-context is a term with n holes in which we can insert n terms. If n is understood, we say a multi-context.*

If $C[\![\ldots,\ldots,\ldots]\!]$ is a multi-context and $M_1, \ldots, M_n$ are terms, then the insertions of those terms in $C[\![\ldots,\ldots,\ldots]\!]$ is denoted $C[\![M_1,\ldots,M_n]\!]$.

Lemma 3. *Let $C[\![\ldots]\!]$ be a multi-context, $A_1,\ldots,A_n$, and $M_1,\ldots,M_n$ be terms, with $x \notin FV(M_1), \ldots, x \notin FV(M_n)$.*

- *if $C[\![M_1,\ldots,M_n]\!] \in \mathcal{HN}$ then $C[\![M_1\langle x = A_1\rangle,\ldots,M_n\langle x = A_n\rangle]\!] \in \mathcal{HN}$.*
- *if $C[\![M_1,\ldots,M_n]\!] \in \mathcal{LN}$ then $C[\![M_1\langle x = A_1\rangle,\ldots,M_n\langle x = A_n\rangle]\!] \in \mathcal{LN}$.*
- *if $C[\![M_1,\ldots,M_n]\!] \in \mathcal{SN}$ and $A_i \in \mathcal{SN}$ for $1 \le i \le n$ then $C[\![M_1\langle x = A_1\rangle,\ldots,M_n\langle x = A_n\rangle]\!] \in \mathcal{SN}$.*

Proof. By induction on triples $(D, \boldsymbol{M}, \boldsymbol{A})$ where D is a term, $\boldsymbol{M}$ and $\boldsymbol{A}$ are multisets of terms. ///

Corollary 2. *Let* $M \equiv N\langle x = A\rangle\langle \boldsymbol{z} = \boldsymbol{S}\rangle \boldsymbol{T}$ *with* $x \notin FV(N)$ *and let* $M' \equiv N\langle \boldsymbol{z} = \boldsymbol{S}\rangle \boldsymbol{T}$,

- *If* $M' \in \mathcal{HN}$ *then* $M \in \mathcal{HN}$.
- *If* $M' \in \mathcal{LN}$ *then* $M \in \mathcal{LN}$.
- *If* $M' \in \mathcal{SN}$ *and* $A \in \mathcal{SN}$ *then* $M \in \mathcal{SN}$.

4 Saturated Sets

Definition 8. *A set* $\mathcal{S}$ *is* $\mathcal{X}$*-saturated (or saturated if there is no ambiguity about the set* $\mathcal{X}$*), if it is closed under the rules of inference in Table 5.*

$$\text{sat-B}\ \frac{B\langle x = A\rangle\ \boldsymbol{T}}{(\lambda x.B)A\ \boldsymbol{T}} \qquad \text{sat-I}\ \frac{A\langle \boldsymbol{z} = \boldsymbol{S}\rangle\ \boldsymbol{T}}{x\langle x = A\rangle\langle \boldsymbol{z} = \boldsymbol{S}\rangle\ \boldsymbol{T}}$$

$$\text{sat-Abs}\ \frac{(\lambda y.B\langle x = A\rangle)\ \langle \boldsymbol{z} = \boldsymbol{S}\rangle\ \boldsymbol{T}}{(\lambda y.B)\langle x = A\rangle\langle \boldsymbol{z} = \boldsymbol{S}\rangle\ \boldsymbol{T}} \qquad \text{sat-App}\ \frac{(U\langle x = A\rangle)(V\langle x = A\rangle)\ \langle \boldsymbol{z} = \boldsymbol{S}\rangle\ \boldsymbol{T}}{(UV)\langle x = A\rangle\ \langle \boldsymbol{z} = \boldsymbol{S}\rangle\ \boldsymbol{T}}$$

$$\text{sat-comp}\ \frac{M\langle y = Q\rangle\langle x = P\langle y = Q\rangle\rangle\langle \boldsymbol{z} = \boldsymbol{S}\rangle\ \boldsymbol{T}}{M\langle x = P\rangle\langle y = Q\rangle\langle \boldsymbol{z} = \boldsymbol{S}\rangle\ \boldsymbol{T}} \qquad \text{sat-gc}\ \frac{N\langle \boldsymbol{z} = \boldsymbol{S}\rangle\ \boldsymbol{T} \qquad A \in \mathcal{X} \text{ and } x \notin FV(N}{N\langle x = A\rangle\langle \boldsymbol{z} = \boldsymbol{S}\rangle\ \boldsymbol{T}}$$

Table 5. $\mathcal{X}$-saturated sets

Note that the set $\mathcal{X}$ occurs only in the rule sat-gc. In practice the set $\mathcal{X}$ will depend on the reduction we consider.

Definition 9 (Function space). *If* $\mathcal{A}$ *and* $\mathcal{B}$ *are sets of terms then* $\mathcal{A} \twoheadrightarrow \mathcal{B}$ *is* $\{M \mid \forall A \in \mathcal{A}, (MA) \in \mathcal{B}\}$.

The following are very easy consequences of the definition.

Lemma 4. *Let* $\mathcal{A}$ *and* $\mathcal{B}$ *be sets of terms.*

1. *If* $\mathcal{B}$ *is saturated then so is* $\mathcal{A} \twoheadrightarrow \mathcal{B}$
2. *If* $\mathcal{A}$ *and* $\mathcal{B}$ *are saturated then so is* $\mathcal{A} \cap \mathcal{B}$.

The major part of the technical difficulty in lifting the classical normalization proofs to our explicit substitutions setting is embodied in the next Lemma. In fact all of the work in the previous section was for the purpose of establishing these results.

Lemma 5.

- $\mathcal{SN}$ *is* $\mathcal{SN}$*-saturated.*

– *$\mathcal{HN}$ is Λx-saturated.*
– *$\mathcal{LN}$ is Λx-saturated.*

The notion of saturation is key to the proof of the Soundness Theorem below for the type systems. It is amusing to note that closure under sat-gc for $\mathcal{SN}$, $\mathcal{HN}$, and $\mathcal{LN}$ is used precisely in showing that the start rule below for typing variables is sound: in the standard λ-calculus this is a triviality but in our calculus we ultimately rely on the difficult argument embodied in Lemma 3.

5 Types and Soundness

Definition 10 (The system of type assignment $\mathcal{D}_\omega$). *Given an infinite set of type-variables and a distinguished type-constant ω the set of types is formed by closing the type-variables and ω under the operations $\sigma \to \tau$ and $\sigma \cap \tau$.*

A statement is an expression of the form $M\colon \tau$; where M is a term, the subject of the statement, and τ is a type. A basis is a set of statements with distinct variables as subjects. A judgment is a triple Γ, M, τ where Γ is a basis, M is a term, and τ is a type; the notion of a judgment's being derivable, denoted $\Gamma \vdash M\colon \tau$ is given by the rules of inference in Table 6.

We say that a term M is typable if there exists a Γ and a τ such that $\Gamma \vdash M\colon \tau$.

We identify two systems: the system $\mathcal{D}_\omega$ itself and the subsystem $\mathcal{D}$ obtained by omitting type ω and the rule ω-I.

$$\text{start}\ \frac{(x\colon \sigma) \in \Gamma}{\Gamma \vdash x\colon \sigma}$$

$$\to\text{I}\ \frac{\Gamma, x\colon \sigma \vdash M\colon \tau}{\Gamma \vdash \lambda x.M\colon \sigma \to \tau} \qquad \to\text{E}\ \frac{\Gamma \vdash M\colon \sigma \to \tau \qquad \Gamma \vdash N\colon \sigma}{\Gamma \vdash (MN)\colon \tau}$$

$$\cap\text{-I}\ \frac{\Gamma \vdash M\colon \sigma_1 \qquad \Gamma \vdash M\colon \sigma_2}{\Gamma \vdash M\colon \sigma_1 \cap \sigma_2} \qquad \cap\text{-E}\ \frac{\Gamma \vdash M\colon \sigma_1 \cap \sigma_2}{\Gamma \vdash M\colon \sigma_i}\ i \in \{1,2\}$$

$$\omega\text{-I}\ \ \Gamma \vdash M\colon \omega \qquad \text{cut}\ \frac{x\colon \sigma, \Gamma \vdash M\colon \tau \qquad \Gamma \vdash N\colon \sigma}{\Gamma \vdash M\langle x = N\rangle\colon \tau}$$

Table 6. Typing rules for $\mathcal{D}_\omega$

The form of the cut rule ensures that a closure $M\langle x = N\rangle$ has exactly the same typing behavior as the associated B-redex $(\lambda x.M)N$. That is, for every Γ and τ,

$$\Gamma \vdash M\langle x = N\rangle\colon \tau \qquad \text{iff} \qquad \Gamma \vdash (\lambda x.M)N\colon \tau.$$

Definition 11. *An interpretation $\mathcal{I}$ is a function from types to sets of terms obeying the following*

- $\mathcal{I}_\omega = \Lambda\boldsymbol{x}$ *(in system $\mathcal{D}_\omega$)*
- $\mathcal{I}_{\alpha\cap\beta} = \mathcal{I}_\alpha \cap \mathcal{I}_\beta$
- $\mathcal{I}_{\alpha\to\beta} = \mathcal{I}_\alpha \twoheadrightarrow \mathcal{I}_\beta$

Obviously an interpretation is completely determined by its value on the type variables. Suppose $\mathcal{I}$ is an interpretation and $\mathcal{X}$ is a set of terms such that $\mathcal{I}_t$ is $\mathcal{X}$-saturated for each type-variable t. Then $\mathcal{I}_\tau$ is $\mathcal{X}$-saturated for each type τ: for $\tau = \omega$ we observe that $\Lambda\mathbf{x}$ is itself $\mathcal{X}$-saturated, and for the other types we invoke Lemma 4.

Theorem 1 (Soundness). *Let $\mathcal{I}$ be an interpretation and let $\mathcal{X}$ be a class of terms such that $\mathcal{I}_t$ is $\mathcal{X}$-saturated for each type-variable t and such that $\mathcal{I}_\sigma \subseteq \mathcal{X}$ for each type σ.*

Suppose M is typable with type τ in either $\mathcal{D}$ or $\mathcal{D}_\omega$. Then $M \in \mathcal{I}_\tau$.

6 Normalization

Definition 12. *(See Cardone and Coppo [10]) A type is* proper *if it has no positive occurrence of ω,* antiproper *if it has no negative occurrence of ω, and* strictly proper *if it has no occurrence of ω.*

The trivial *types are determined by the following rules:*

- *ω is trivial.*
- *If σ is trivial and θ is any type, then $\theta \to \sigma$ is trivial.*
- *If σ and τ are trivial, then $\sigma \cap \tau$ is trivial.*

Head normalization. Consider the system $\mathcal{D}_\omega$; let "type" mean "type of $\mathcal{D}_\omega$" and let "typable" mean "typable in $\mathcal{D}_\omega$".

Definition 13. *Let $\mathcal{H}$ be the interpretation which maps each type variable to the set $\mathcal{HN}$ of head normalizing terms.*

Lemma 6. *$\mathcal{H}_\tau \subseteq \mathcal{HN}$ for each non-trivial type τ.*

Corollary 3. *If M is typable in $\mathcal{D}_\omega$ with a non-trivial type then M is head normalizing.*

Leftmost normalization. Consider the system $\mathcal{D}_\omega$; let "type" mean "type of $\mathcal{D}_\omega$" and let "typable" mean "typable in $\mathcal{D}_\omega$".

Definition 14. *Let $\mathcal{L}$ be the interpretation which maps each type variable to the set $\mathcal{LN}$ of leftmost-normalizing terms.*

Lemma 7. *$\mathcal{L}_\tau \subseteq \mathcal{LN}$ for each proper type τ.*

Corollary 4. *If M is typable in $\mathcal{D}_\omega$ with a proper type then M is leftmost normalizing.*

Strong normalization. Consider the system $\mathcal{D}$; let "type" mean "type of $\mathcal{D}$" and let "typable" mean "typable in $\mathcal{D}$.

Definition 15. *Let $\mathcal{S}$ be the interpretation which maps each type variable to the set $\mathcal{SN}$ of strongly normalizing terms.*

Lemma 8. *$\mathcal{S}_\tau \subseteq \mathcal{SN}$ for each type τ.*

Corollary 5. *If M is typable in $\mathcal{D}$ then M is strongly normalizing.*

7 Typings for normalizable terms

Proposition 1.

1. *If H is a head normal form then H is typable in system $\mathcal{D}_\omega$ with a non-trivial type.*
2. *If N is a normal form then N is typable in system $\mathcal{D}$.*

Theorem 2 (Subject Reduction). *In either of the systems $\mathcal{D}_\omega$ or $\mathcal{D}$: suppose $\Gamma \vdash M : \tau$ and $M \longrightarrow M_1$. Then $\Gamma \vdash M_1 : \tau$.*

Theorem 3 (Subject Expansion for $\mathcal{D}_\omega$). *In system $\mathcal{D}_\omega$: suppose $\Gamma \vdash M : \tau$ and $M_0 \longrightarrow M$. Then $\Gamma \vdash M_0 : \tau$.*

Corollary 6. *Suppose $\Gamma \vdash M : \tau$ in system $\mathcal{D}_\omega$ and $M \twoheadleftarrow\!\!\twoheadrightarrow M'$. Then $\Gamma \vdash M' : \tau$.*

Subject Expansion and system $\mathcal{D}$.

The Subject Expansion theorem plays a key role in deriving converses involving system $\mathcal{D}_\omega$ to the normalization results concerning head- and leftmost reduction (Corollaries 3 and 4 above). We present these converses in the next section.

It is well-known from the classical λ-calculus that the Subject Expansion theorem fails for system $\mathcal{D}$. But with some care (involving the potential erasing of non-typable terms) one can analyze β-expansion in order to derive a converse to the classical version of Corollary 5 and so obtain a characterization of the strongly normalizing classical terms.

It seems to be much more difficult to perform such an analysis for expansion in the calculus λx. In particular it is not the case that $\mathcal{D}$-typability is preserved by expansion even when the reduction-rule in question erases a strongly-normalizing subterm.

Example. Let D be the term $\lambda u.uu$ and M be the term $(\lambda x.(\lambda y.z)(xx))D$. D is $\mathcal{D}$-typable by $(t \cap (t \to t)) \to t$ but DD is not $\mathcal{SN}$ so M is not $\mathcal{SN}$. Now consider

$$M \longrightarrow (\lambda x.z\langle y = xx\rangle)D \longrightarrow M' \equiv z\langle y = xx\rangle\langle x = D\rangle \longrightarrow M'' \equiv z\langle x = D\rangle$$

M'' is $\mathcal{SN}$ and is easily seen to be $\mathcal{D}$-typable. But M' is $\mathcal{SN}$ yet not $\mathcal{D}$-typable.

It is not hard to see that M' is $\mathcal{SN}$. To see that M' is not $\mathcal{D}$-typable, first note that by Corollary 5 M cannot be typed since it is not $\mathcal{SN}$. But $(\lambda y.z)(xx)$ and $z\langle y = xx\rangle$ have exactly the same typing behavior in our system.

The reduction from M' to M'' witnesses the failure of Subject Expansion; the notable thing here is the innocuous nature of the erased subterm xx. The reduction $(\lambda y.z)(xx) \longrightarrow z\langle y = xx\rangle$, which reduces an inner B-redex, changes the behavior of the term w.r.t. to normalization. This should somewhat be translated into the typing system, which is not the case in $\mathcal{D}$.

The natural reaction to such an example is conclude that the type system $\mathcal{D}$ should be modified. But the terms M and M' above are related simply by an application of the B-rule. So if we are to have a type system which characterizes strong normalization it seems that we must abandon the property that closures $B\langle x = A\rangle$ have the same typing behavior as the associated B-redexes $(\lambda x.B)A$ (perhaps only in the case when x is not free in B). This would be a fundamental change in what seems to us to be the most natural generalization of the classical type system. We leave the search for a type system characterizing the strongly normalizing terms as a subject for future investigation.

8 Summary

In this section we summarize the results of this paper and also address the role of the garbage-collection rule gc in the development.

As suggested in the introduction one may view the rule gc as being somewhat out of character with the rest of the explicit substitutions program, since it does not really correspond to an *atomic* operation on terms. So it is natural to ask whether the relationships we have established between typings and reduction properties carries for the "pure" calculus without rule gc.

Since the pure calculus is a subsystem of the full (gc) calculus one direction of the relationship is immediate, but it is mildly surprising that the full equivalence between various normalization properties and typing properties can be established for the pure calculus with essentially no extra work. This is shown in the three theorems of this section.

Recall that head- and leftmost reduction are each non-deterministic and that when we speak of head- or leftmost-reduction below we mean *any* sequence of reduction steps obeying the given discipline.

Theorem 4 (Head normalization). *Let M be a closed term. The following are equivalent.*

1. *M is typable with a non-trivial type in system $\mathcal{D}_\omega$.*
2. *$M \in \mathcal{HN}$.*
3. *M is head-normalizing in the calculus λx (without garbage-collection).*
4. *M has a head normal form.*
5. *M is solvable, that is, there is an n and terms $X_1, \ldots X_n$ such that $MX_1 \cdots X_n = \lambda x.x$.*

Proof. We prove $1 \Rightarrow 2 \Rightarrow 3 \Rightarrow 4 \Rightarrow 1$ and $4 \Rightarrow 5 \Rightarrow 2$. ///

Theorem 5 (Leftmost normalization). *Let M be a closed term. The following are equivalent.*

1. *M is typable in system $\mathcal{D}_\omega$ with a type not involving ω.*
2. *M is typable with a proper type in system $\mathcal{D}_\omega$.*
3. *$M \in \mathcal{LN}$.*
4. *M is leftmost-normalizing in the calculus λx (without garbage-collection).*
5. *M has a normal form.*

Proof. We prove $1 \Rightarrow 2 \Rightarrow 3 \Rightarrow 4 \Rightarrow 5 \Rightarrow 1$. ///

It is worth emphasizing the fact that the implications 5 to 3 and 5 to 4 state that in λx and λx_{gc} leftmost reduction is a normalizing strategy.

Theorem 6 (Strong normalization). *Let M be a closed term.*

1. *$M \in \mathcal{SN}$ if and only if M is strongly normalizing in the calculus λx (without garbage-collection).*
2. *If M is typable in system $\mathcal{D}$ then $M \in \mathcal{SN}$.*
3. *If M is a pure term then $M \in \mathcal{SN}$ if and only if M is typable in system $\mathcal{D}$.*

As described in the previous section we do not have the implication "$M \in \mathcal{SN}$ implies M is typable in system $\mathcal{D}$." As is well-known this result holds for pure terms under β-reduction. It then follows from the preservation of strong normalization in λx that the result holds for pure terms under λx_{gc}-reduction. The problem of finding a reasonable type system characterizing the strongly normalizing terms in Λx is left as an open problem.

Acknowledgements The authors are grateful to Eduardo Bonelli, Nachum Dershowitz, Delia Kesner, Frédéric Lang, and Kristoffer Rose for many helpful discussions, and to several anonymous referees for improvements in style and substance.

References

1. M. Abadi, L. Cardelli, P.-L. Curien, and J.-J. Lévy. Explicit substitutions. *Journal of Functional Programming*, 1(4):375–416, 1991.
2. H. P. Barendregt. *The Lambda-Calculus, its syntax and semantics.* Studies in Logic and the Foundation of Mathematics. Elsevier, Amsterdam, 1984. Second edition.
3. H. P. Barendregt. Lambda calculi with types. In S. Abramsky, D. M. Gabby, and T. S. E. Maibaum, editors, *Handbook of Logic in Computer Science*, volume 2, chapter 2, pages 117–309. Oxford University Press, 1992.
4. Z. Benaissa, D. Briaud, P. Lescanne, and J. Rouyer-Degli. $\lambda\upsilon$, a calculus of explicit substitutions which preserves strong normalisation. *Journal of Functional Programming*, 6(5):699–722, September 1996.
5. Z. Benaissa, P. Lescanne, and K. H. Rose. Modeling sharing and recursion for weak reduction strategies using explicit substitution. In H. Kuchen and D. Swierstra, editors, *PLILP '96—8th Int. Symp. on Programming Languages: Implementation, Logics and Programs*, number 1140 in LNCS, pages 393–407, Aachen, Germany, September 1996. Springer-Verlag.

6. R. Bloo. *Preservation of Termination for Explicit Substitution.* PhD thesis, Technische Universiteit Eindhoven, 1997. IPA Dissertation Series 1997-05.
7. R. Bloo and J. H. Geuvers. Explicit substitution: on the edge of strong normalization. *Theoretical Computer Science*, 211:375 - 395, 1999.
8. R. Bloo and K. H. Rose. Preservation of strong normalisation in named lambda calculi with explicit substitution and garbage collection. In *CSN '95—Computing Science in the Netherlands*, pages 62–72, Utrecht, November 1995.
9. E. Bonelli. Perpetuality in a named lambda calculus with explicit substitutions and some applications. Technical Report RR 1221, LRI, University of Paris-Sud, 1999. to appear in MSCS, 2000.
10. F. Cardone and M. Coppo. Two extension of Curry's type inference system. In P. Odifreddi, editor, *Logic and Computer Science*, volume 31 of *APIC Series*, pages 19–75. Academic Press, New York, NY, 1990.
11. P.-L. Curien, T. Hardin, and J.-J. Lévy. Confluence properties of weak and strong calculi of explicit substitutions. *Journal of the ACM*, 43(2):362–397, March 1996.
12. H. B Curry and R. Feys. *Combinatory Logic I.* North-Holland, Amsterdam, 1958.
13. R. Di Cosmo and D. Kesner. Strong normalization of explicit substitutions via cut elimination in proof nets. In *LICS '97—Twelfth Annual IEEE Symposium on Logic in Computer Science*, pages 35–46. Warsaw U., IEEE, June 1997.
14. F. Kamareddine and A. Ríos. Relating the lambda-sigma and lambda-s styles of explicit substitutions. *Journal of Logic and Computation*, 10(3), 2000. Special issue on Type Theory and Term Rewriting.
15. F. Kamereddine and R.P. Nederpelt. On stepwise explicit substitutions. *International Journal of Foundations of Computer Science*, 4(3):197–240, 1993.
16. P. Lescanne. From $\lambda\sigma$ to $\lambda\upsilon$: a journey through calculi of explicit substitutions. In Hans-J. Boehm, editor, *POPL '94—21st Annual ACM Symposium on Principles of Programming Languages*, pages 60–69, Portland, Oregon, January 1994. ACM.
17. P.-A. Melliès. Typed λ-calculi with explicit substitution may not terminate. In M. Dezani, editor, *TLCA '95—Int. Conf. on Typed Lambda Calculus and Applications*, volume 902 of *LNCS*, pages 328–334, Edinburgh, Scotland, April 1995. Springer-Verlag.
18. P.-A. Melliès. Axiomatic rewriting theory III, a factorisation theorem in rewriting theory. In *Proceedings of the 7th Conference on Category Theory and Computer Science*, volume 1290 of *LNCS*, pages 49–68. Springer-Verlag, 1997.
19. J. C. Mitchell. *Foundations for Programming Languages.* MIT Press, Cambridge, MA, 1996.
20. E. Ritter. Characterising explicit substitutions which preserve termination. In *TLCA'99*, volume 1581 of *LNCS*. Springer-Verlag, 1999.
21. K.H. Rose. *Operational Reduction Models for Functional Programming Languages.* PhD thesis, DIKU, København, February 1996. DIKU report 96/1.

The Stratified Foundations as a Theory Modulo

Gilles Dowek

INRIA-Rocquencourt
B.P. 105, 78153 Le Chesnay Cedex, France.
Gilles.Dowek@inria.fr, http://logical.inria.fr/~dowek

Abstract. The *Stratified Foundations* are a restriction of naive set theory where the comprehension scheme is restricted to stratifiable propositions. It is known that this theory is consistent and that proofs strongly normalize in this theory. *Deduction modulo* is a formulation of first-order logic with a general notion of cut. It is known that proofs normalize in a theory modulo if it has some kind of many-valued model called a *pre-model*. We show in this paper that the Stratified Foundations can be presented in deduction modulo and that the method used in the original normalization proof can be adapted to construct a pre-model for this theory.

The *Stratified Foundations* are a restriction of naive set theory where the comprehension scheme is restricted to stratifiable propositions. This theory is consistent [8] while naive set theory is not and the consistency of the Stratified Foundations together with the extensionality axiom - the so-called *New Foundations* - is open.

The Stratified Foundations extend simple type theory and, like in simple type theory, proofs strongly normalize in The Stratified Foundations [2]. These two normalization proofs, like many, have some parts in common, for instance they both use Girard's reducibility candidates. This motivates the investigation of general normalization theorems that have normalization theorems for specific theories as consequences. The normalization theorem for deduction modulo [7] is an example of such a general theorem. It concerns theories expressed in *deduction modulo* [5] that are first-order theories with a general notion of cut. According to this theorem, proofs normalize in a theory in deduction modulo if this theory has some kind of many-valued model called a *pre-model*. For instance, simple type theory can be expressed in deduction modulo [5,6] and it has a pre-model [7,6] and hence it has the normalization property. The normalization proof obtained this way is modular: all the lemmas specific to type theory are concentrated in the pre-model construction while the theorem that the existence of a pre-model implies normalization is generic and can be used for any other theory in deduction modulo.

The goal of this paper is to show that the Stratified Foundations also can be presented in deduction modulo and that the method used in the original normalization proof can be adapted to construct a pre-model for this theory. The normalization proof obtained this way is simpler than the original one because

S. Abramsky (Ed.): TLCA 2001, LNCS 2044, pp. 136–150, 2001.

it simply uses the fact that proofs normalize in the Stratified Foundations if this theory has a pre-model, while a variant of this proposition needs to be proved in the original proof.

It is worth noticing that the original normalization proof for the Stratified Foundations is already in two steps, where the first is the construction of a so-called *normalization model* and the second is a proof that proofs normalize in the Stratified Foundations if there is such a normalization model. Normalization models are, more or less, pre-models of the Stratified Foundations. So, we show that the notion of normalization model, that is specific to the Stratified Foundations, is an instance of a more general notion that can be defined for all theories modulo, and that the lemma that the existence of a normalization model implies normalization for the Stratified Foundations is an instance of a more general theorem that holds for all theories modulo.

The normalization proof obtained this way differs also from the original one in other respects. First, to remain in first-order logic, we do not use a presentation of the Stratified Foundations with a binder, but one with combinators. To express the Stratified Foundations with a binder in first-order logic, we could use de Bruijn indices and explicit substitutions along the lines of [6]. The pre-model construction below should generalize easily to such a presentation. Second, our cuts are cuts modulo, while the original proof uses Prawitz' *folding-unfolding* cuts. It is shown in [4] that the normalization theorems are equivalent for the two notions of cuts, but that the notion of cut modulo is more general that the notion of folding-unfolding cut. Third, we use untyped reducibility candidates and not typed ones as in the original proof. This quite simplifies the technical details.

A last benefit of expressing the Stratified Foundations in deduction modulo is that we can use the method developed in [5] to organize proof search. The method obtained this way, that is an analog of higher-order resolution for the Stratified Foundations, is much more efficient than usual first-order proof search methods with the comprehension axioms, although it remains complete as the Stratified Foundations have the normalization property.

1 Deduction Modulo

1.1 Identifying Propositions

In deduction modulo, the notions of language, term and proposition are that of first-order logic. But, a theory is formed with a set of axioms Γ *and a congruence* $\equiv$ defined on propositions. Such a congruence may be defined by a rewrite systems on terms and on propositions (as propositions contain binders - quantifiers -, these rewrite systems are in fact *combinatory reduction systems* [9]). Then, the deduction rules take this congruence into account. For instance, the *modus ponens* is not stated as usual

$$\frac{A \Rightarrow B \quad A}{B}$$

as the first premise need not be exactly $A \Rightarrow B$ but may be only congruent to this proposition, hence it is stated

$$\frac{C \quad A}{B} \text{ if } C \equiv A \Rightarrow B$$

$$\frac{}{\Gamma \vdash_{\equiv} B} \text{ axiom if } A \in \Gamma \text{ and } A \equiv B$$

$$\frac{\Gamma, A \vdash_{\equiv} B}{\Gamma \vdash_{\equiv} C} \Rightarrow\text{-intro if } C \equiv (A \Rightarrow B)$$

$$\frac{\Gamma \vdash_{\equiv} C \quad \Gamma \vdash_{\equiv} A}{\Gamma \vdash_{\equiv} B} \Rightarrow\text{-elim if } C \equiv (A \Rightarrow B)$$

$$\frac{\Gamma \vdash_{\equiv} A \quad \Gamma \vdash_{\equiv} B}{\Gamma \vdash_{\equiv} C} \wedge\text{-intro if } C \equiv (A \wedge B)$$

$$\frac{\Gamma \vdash_{\equiv} C}{\Gamma \vdash_{\equiv} A} \wedge\text{-elim if } C \equiv (A \wedge B)$$

$$\frac{\Gamma \vdash_{\equiv} C}{\Gamma \vdash_{\equiv} B} \wedge\text{-elim if } C \equiv (A \wedge B)$$

$$\frac{\Gamma \vdash_{\equiv} A}{\Gamma \vdash_{\equiv} C} \vee\text{-intro if } C \equiv (A \vee B)$$

$$\frac{\Gamma \vdash_{\equiv} B}{\Gamma \vdash_{\equiv} C} \vee\text{-intro if } C \equiv (A \vee B)$$

$$\frac{\Gamma \vdash_{\equiv} D \quad \Gamma, A \vdash_{\equiv} C \quad \Gamma, B \vdash_{\equiv} C}{\Gamma \vdash_{\equiv} C} \vee\text{-elim if } D \equiv (A \vee B)$$

$$\frac{\Gamma \vdash_{\equiv} B}{\Gamma \vdash_{\equiv} A} \bot\text{-elim if } B \equiv \bot$$

$$\frac{\Gamma \vdash_{\equiv} A}{\Gamma \vdash_{\equiv} B} (x, A)\ \forall\text{-intro if } B \equiv (\forall x\ A) \text{ and } x \not\in FV(\Gamma)$$

$$\frac{\Gamma \vdash_{\equiv} B}{\Gamma \vdash_{\equiv} C} (x, A, t)\ \forall\text{-elim if } B \equiv (\forall x\ A) \text{ and } C \equiv [t/x]A$$

$$\frac{\Gamma \vdash_{\equiv} C}{\Gamma \vdash_{\equiv} B} (x, A, t)\ \exists\text{-intro if } B \equiv (\exists x\ A) \text{ and } C \equiv [t/x]A$$

$$\frac{\Gamma \vdash_{\equiv} C \quad \Gamma, A \vdash_{\equiv} B}{\Gamma \vdash_{\equiv} B} (x, A)\ \exists\text{-elim if } C \equiv (\exists x\ A) \text{ and } x \not\in FV(\Gamma B)$$

$$\frac{}{\Gamma \vdash_{\equiv} A} B \text{ Excluded middle if } A \equiv B \vee (B \Rightarrow \bot)$$

Fig. 1. Natural deduction modulo

All the rules of intuitionistic natural deduction may be stated in a similar way. Classical deduction modulo is obtained by adding the *excluded middle* rule (see figure 1).

For example, in arithmetic, we can define a congruence with the following rewrite system

$$0 + y \rightarrow y$$

$$S(x) + y \to S(x + y)$$
$$0 \times y \to 0$$
$$S(x) \times y \to x \times y + y$$

In the theory formed with a set of axioms Γ containing the axiom $\forall x\ x = x$ and this congruence, we can prove, in natural deduction modulo, that the number 4 is even

$$\dfrac{\dfrac{\dfrac{}{\Gamma \vdash_{\equiv} \forall x\ x = x}\ \text{axiom}}{\Gamma \vdash_{\equiv} 2 \times 2 = 4}\ (x, x = x, 4)\ \forall\text{-elim}}{\Gamma \vdash_{\equiv} \exists x\ 2 \times x = 4}\ (x, 2 \times x = 4, 2)\ \exists\text{-intro}$$

Substituting the variable x by the term 2 in the proposition $2 \times x = 4$ yields the proposition $2 \times 2 = 4$, that is congruent to $4 = 4$. The transformation of one proposition into the other, that requires several proof steps in usual natural deduction, is dropped from the proof in deduction modulo.

In this example, all the rewrite rules apply to terms. Deduction modulo permits also to consider rules rewriting atomic propositions to arbitrary ones. For instance, in the theory of integral domains, we have the rule

$$x \times y = 0 \to x = 0 \vee y = 0$$

that rewrites an atomic proposition to a disjunction.

Notice that, in the proof above, we do not need the axioms of addition and multiplication. Indeed, these axioms are now redundant: since the terms $0 + y$ and y are congruent, the axiom $\forall y\ 0 + y = y$ is congruent to the axiom of equality $\forall y\ y = y$. Hence, it can be dropped. Thus, rewrite rules replace axioms.

This equivalence between rewrite rules and axioms is expressed by the the *equivalence lemma* that for every congruence $\equiv$, we can find a theory $\mathcal{T}$ such that $\Gamma \vdash_{\equiv} A$ is provable in deduction modulo if and only if $\mathcal{T}\Gamma \vdash A$ is provable in ordinary first-order logic [5]. Hence, deduction modulo is not a true extension of first-order logic, but rather an alternative formulation of first-order logic. Of course, the provable propositions are the same in both cases, but the proofs are very different.

1.2 Model of a Theory Modulo

A *model* of a congruence $\equiv$ is a model such that if $A \equiv B$ then for all assignments, A and B have the same denotation. A *model* of a theory modulo $\Gamma, \equiv$ is a model of the theory Γ and of the congruence $\equiv$. Unsurprisingly, the completeness theorem extends to classical deduction modulo [3] and a proposition is provable in the theory $\Gamma, \equiv$ if and only if it is valid in all the models of $\Gamma, \equiv$.

1.3 Normalization in Deduction Modulo

Replacing axioms by rewrite rules in a theory changes the structure of proofs and in particular some theories may have the normalization property when expressed with axioms and not when expressed with rewrite rules. For instance, from the normalization theorem for first-order logic, we get that any proposition that is provable with the axiom $A \Leftrightarrow (B \wedge (A \Rightarrow \bot))$ has a normal proof. But if we transform this axiom into the rule $A \rightarrow B \wedge (A \Rightarrow \bot)$ (Crabbé's rule [1]) the proposition $B \Rightarrow \bot$ has a proof, but no normal proof.

We have proved a *normalization theorem*: proofs normalize in a theory modulo if this theory has a *pre-model* [7]. A pre-model is a many-valued model whose truth values are reducibility candidates, i.e. sets of proof-terms. Hence we first define proof-terms, then reducibility candidates and at last pre-models.

Definition 1 (Proof-term).
Proof-terms *are inductively defined as follows.*

$$\begin{aligned}\pi ::= \;& \alpha \\ & \mid \lambda\alpha\ \pi \mid (\pi\ \pi') \\ & \mid \langle \pi, \pi' \rangle \mid fst(\pi) \mid snd(\pi) \\ & \mid i(\pi) \mid j(\pi) \mid (\delta\ \pi_1\ \alpha\pi_2\ \beta\pi_3) \\ & \mid (botelim\ \pi) \\ & \mid \lambda x\ \pi \mid (\pi\ t) \\ & \mid \langle t, \pi \rangle \mid (exelim\ \pi\ x\alpha\pi')\end{aligned}$$

Each proof-term construction corresponds to an intuitionistic natural deduction rule: terms of the form α express proofs built with the axiom rule, terms of the form $\lambda\alpha\ \pi$ and $(\pi\ \pi')$ express proofs built with the introduction and elimination rules of the implication, terms of the form $\langle \pi, \pi' \rangle$ and $fst(\pi)$, $snd(\pi)$ express proofs built with the introduction and elimination rules of the conjunction, terms of the form $i(\pi), j(\pi)$ and $(\delta\ \pi_1\ \alpha\pi_2\ \beta\pi_3)$ express proofs built with the introduction and elimination rules of the disjunction, terms of the form $(botelim\ \pi)$ express proofs built with the elimination rule of the contradiction, terms of the form $\lambda x\ \pi$ and $(\pi\ t)$ express proofs built with the introduction and elimination rules of the universal quantifier and terms of the form $\langle t, \pi \rangle$ and $(exelim\ \pi\ x\alpha\pi')$ express proofs built with the introduction and elimination rules of the existential quantifier.

Definition 2 (Reduction). Reduction *on proof-terms is defined by the following rules that eliminate cuts step by step.*

$$(\lambda\alpha\ \pi_1\ \pi_2) \rhd [\pi_2/\alpha]\pi_1$$

$$fst(\langle \pi_1, \pi_2 \rangle) \rhd \pi_1$$

$$snd(\langle \pi_1, \pi_2 \rangle) \rhd \pi_2$$

$$(\delta\ i(\pi_1)\ \alpha\pi_2\ \beta\pi_3) \rhd [\pi_1/\alpha]\pi_2$$

$$(\delta\ j(\pi_1)\ \alpha\pi_2\ \beta\pi_3) \rhd [\pi_1/\beta]\pi_3$$

$$(\lambda x\ \pi\ t) \rhd [t/x]\pi$$

$$(exelim\ \langle t, \pi_1 \rangle\ \alpha x\pi_2) \rhd [t/x, \pi_1/\alpha]\pi_2$$

Definition 3 (Reducibility candidates). *A proof-term is said to be* neutral *if it is a proof variable or an elimination (i.e. of the form $(\pi\ \pi')$, $fst(\pi)$, $snd(\pi)$, $(\delta\ \pi_1\ \alpha\pi_2\ \beta\pi_3)$, $(botelim\ \pi)$, $(\pi\ t)$, $(exelim\ \pi\ x\alpha\pi')$), but not an introduction. A set R of proof-terms is a* reducibility candidate *if*

- *if $\pi \in R$, then π is strongly normalizable,*
- *if $\pi \in R$ and $\pi \rhd \pi'$ then $\pi' \in R$,*
- *if π is neutral and if for every π' such that $\pi \rhd^1 \pi'$, $\pi' \in R$ then $\pi \in R$.*

We write $\mathcal{C}$ for the set of all reducibility candidates.

Definition 4 (Pre-model). *A* pre-model $\mathcal{N}$ *for a language $\mathcal{L}$ is given by:*

- *a set N,*
- *for each function symbol f of arity n a function $\hat{f}$ from N^n to N,*
- *for each predicate symbol P a function $\hat{P}$ from N^n to $\mathcal{C}$.*

Definition 5 (Denotation in a pre-model). *Let $\mathcal{N}$ be a pre-model, t be a term and φ an assignment mapping all the free variables of t to elements of N. We define the object $[\![t]\!]_\varphi^\mathcal{N}$ by induction over the structure of t.*

- $[\![x]\!]_\varphi^\mathcal{N} = \varphi(x)$,
- $[\![f(t_1,\ldots,t_n)]\!]_\varphi^\mathcal{N} = \hat{f}([\![t_1]\!]_\varphi^\mathcal{N},\ldots,[\![t_n]\!]_\varphi^\mathcal{N})$.

Let A be a proposition and φ an assignment mapping all the free variables of A to elements of N. We define the reducibility candidate $[\![A]\!]_\varphi^\mathcal{N}$ by induction over the structure of A.

- *If A is an atomic proposition $P(t_1,\ldots,t_n)$ then $[\![A]\!]_\varphi^\mathcal{N} = \hat{P}([\![t_1]\!]_\varphi^\mathcal{N},\ldots,[\![t_n]\!]_\varphi^\mathcal{N})$.*
- *If $A = B \Rightarrow C$ then $[\![A]\!]_\varphi^\mathcal{N}$ is the set of proofs π such that π is strongly normalizable and whenever it reduces to $\lambda\alpha\ \pi_1$ then for every π' in $[\![B]\!]_\varphi^\mathcal{N}$, $[\pi'/\alpha]\pi_1$ is in $[\![C]\!]_\varphi^\mathcal{N}$.*
- *If $A = B \wedge C$ then $[\![A]\!]_\varphi^\mathcal{N}$ is the set of proofs π such that π is strongly normalizable and whenever it reduces to $\langle\pi_1,\pi_2\rangle$ then π_1 is in $[\![B]\!]_\varphi^\mathcal{N}$ and π_2 is in $[\![C]\!]_\varphi^\mathcal{N}$.*
- *If $A = B \vee C$ then $[\![A]\!]_\varphi^\mathcal{N}$ is the set of proofs π such that π is strongly normalizable and whenever it reduces to $i(\pi_1)$ (resp. $j(\pi_2)$) then π_1 (resp. π_2) is in $[\![B]\!]_\varphi^\mathcal{N}$ (resp. $[\![C]\!]_\varphi^\mathcal{N}$).*
- *If $A = \bot$ then $[\![A]\!]_\varphi^\mathcal{N}$ is the set of strongly normalizable proofs.*
- *If $A = \forall x\ B$ then $[\![A]\!]_\varphi^\mathcal{N}$ is the set of proofs π such that π is strongly normalizable and whenever it reduces to $\lambda x\ \pi_1$ then for every term t and every element a of N $[t/x]\pi_1$ is in $[\![B]\!]_{\varphi+a/x}^\mathcal{N}$.*
- *If $A = \exists x\ B$ then $[\![A]\!]_\varphi^\mathcal{N}$ is the set of proofs π such that π is strongly normalizable and whenever it reduces to $\langle t,\pi_1\rangle$ then there exists an element a in N such that π_1 is in $[\![B]\!]_{\varphi+a/x}^\mathcal{N}$.*

Definition 6. *A pre-model is said to be a* pre-model of a congruence $\equiv$ *if when* $A \equiv B$ *then for every assignment* φ, $[\![A]\!]_\varphi^\mathcal{N} = [\![B]\!]_\varphi^\mathcal{N}$.

Theorem 1 (Normalization). *[7] If a congruence* $\equiv$ *has a pre-model all proofs modulo* $\equiv$ *strongly normalize.*

2 The Stratified Foundations

2.1 The Stratified Foundations as a First-Order Theory

Definition 7. *(Stratifiable proposition)*
A proposition A *in the language* $\in$ *is said to be* stratifiable *if there exists a function* S *mapping every variable (bound or free) of* A *to a natural number in such a way that every atomic proposition of* A, $x \in y$ *is such that* $S(y) = S(x)+1$.

For instance, the proposition

$$\forall v\ (v \in x \Leftrightarrow v \in y) \Rightarrow \forall w\ (x \in w \Rightarrow y \in w)$$

is stratifiable (take, for instance, $S(v) = 4$, $S(x) = S(y) = 5$, $S(w) = 6$) but not the proposition

$$\forall v\ (v \in x \Leftrightarrow v \in y) \Rightarrow x \in y$$

Definition 8. *(The stratified comprehension scheme)*
For every stratifiable proposition A *whose free variables are among* $x_1, \ldots, x_n, x_{n+1}$ *we take the axiom*

$$\forall x_1\ \ldots\ \forall x_n\ \exists z\ \forall x_{n+1}\ (x_{n+1} \in z \Leftrightarrow A)$$

Definition 9. *(The skolemized stratified comprehension scheme)*
When we skolemize this scheme, we introduce for each stratifiable proposition A *in the language* $\in$ *and sequence of variables* $x_1, \ldots, x_n, x_{n+1}$ *such that the free variables of* A *are among* $x_1, \ldots, x_n, x_{n+1}$, *a function symbol* $f_{x_1,\ldots,x_n,x_{n+1},A}$ *and the axiom*

$$\forall x_1\ \ldots\ \forall x_n\ \forall x_{n+1}\ (x_{n+1} \in f_{x_1,\ldots,x_n,x_{n+1},A}(x_1, \ldots, x_n) \Leftrightarrow A)$$

2.2 The Stratified Foundations as a Theory Modulo

Now we want to replace the axiom scheme above by a rewrite rule, defining a congruence on propositions, so that the Stratified Foundations are defined as an axiom free theory modulo.

Definition 10. *(The rewrite system* $\mathcal{R}$*)*

$$t_{n+1} \in f_{x_1,\ldots,x_n,x_{n+1},A}(t_1, \ldots, t_n) \rightarrow [t_1/x_1, \ldots, t_n/x_n, t_{n+1}/x_{n+1}]A$$

Proposition 1. *The rewrite system $\mathcal{R}$ is confluent and terminating.*

Proof. The system $\mathcal{R}$ is an orthogonal combinatory reduction system, hence it is confluent [9].

For termination, if A is an atomic proposition we write $\|A\|$ for the number of function symbols in A and if A is a proposition containing the atomic propositions $A_1, \dots, A_p$ we write A° for the multiset $\{\|A_1\|, \dots, \|A_p\|\}$. We show that if a proposition A reduces in one step to a proposition B then $B^\circ < A^\circ$ for the multiset ordering.

If the proposition A reduces in one step to B, there is an atomic proposition of A, say A_1, that has the form $t_{n+1} \in f_{x_1,\dots,x_n,x_{n+1},C}(t_1, \dots, t_n)$ and reduces to $B_1 = [t_1/x_1, \dots, t_n/x_n, t_{n+1}/x_{n+1}]C$. Every atomic proposition b of B_1 has the form $[t_1/x_1, \dots, t_n/x_n, t_{n+1}/x_{n+1}]c$ where c is an atomic proposition of C. The proposition c has the form $x_i \in x_j$ for distinct i and j (since C is stratifiable) $x_i \in y$, $y \in x_i$ or $y \in z$. Hence b has the form $t_i \in t_j$ for distinct i and j, $t_i \in y$, $y \in t_i$ or $y \in z$ and $\|b\| < \|A_1\|$. Therefore $B^\circ < A^\circ$.

Proposition 2. *A proposition A is provable from the skolemized comprehension scheme if and only if it is provable modulo the rewrite system $\mathcal{R}$.*

2.3 Consistency

We want now to construct a model for the Stratified Foundations.

If $\mathcal{M}$ is a model of set theory we write M for the set of elements of the model, $\in_{\mathcal{M}}$ for the denotation of the symbol $\in$ in this model, $\wp_{\mathcal{M}}$ for the powerset in this model, etc. We write also $[\![A]\!]_\varphi^{\mathcal{M}}$ for the denotation of a proposition A for the assignment φ.

The proof of the consistency of the Stratified Foundations rests on the existence of a model of Zermelo's set theory, such that there is a bijection σ from M to M and a family v_i of elements of M, $i \in \mathbb{Z}$ such that

$$a \in_{\mathcal{M}} b \text{ if and only if } \sigma a \in_{\mathcal{M}} \sigma b$$

$$\sigma v_i = v_{i+1}$$

$$v_i \subseteq_{\mathcal{M}} v_{i+1}$$

$$\wp_{\mathcal{M}}(v_i) \subseteq_{\mathcal{M}} v_{i+1}$$

The existence of such a model is proved in [8].

Using the fact that $\mathcal{M}$ is a model of the axiom of extensionality, we prove that $a \subseteq_{\mathcal{M}} b$ if and only if $\sigma a \subseteq_{\mathcal{M}} \sigma b$, $\sigma\{a, b\}_{\mathcal{M}} = \{\sigma a, \sigma b\}_{\mathcal{M}}$, $\sigma\langle a, b\rangle_{\mathcal{M}} = \langle \sigma a, \sigma b\rangle_{\mathcal{M}}$, $\sigma\wp(a) = \wp(\sigma a)$, etc.

For the normalization proof, we will further need that $\mathcal{M}$ is an ω-model. We define $\overline{0} = \emptyset_{\mathcal{M}}$, $\overline{n+1} = \overline{n} \cup_{\mathcal{M}} \{\overline{n}\}_{\mathcal{M}}$. An ω-model is a model such that $a \in_{\mathcal{M}} \mathbb{N}_{\mathcal{M}}$ if and only if there exists n in $\mathbb{N}$ such that $a = \overline{n}$. The existence of such a model is proved in [8] (see also [2]).

Using the fact that $\mathcal{M}$ is a model of the axiom of extensionality, we prove that $\sigma\emptyset_{\mathcal{M}} = \emptyset_{\mathcal{M}}$ and then, by induction on n that $\sigma\overline{n} = \overline{n}$.

Notice that since $\wp_{\mathcal{M}}(v_i) \subseteq_{\mathcal{M}} v_{i+1}$, $\emptyset_{\mathcal{M}} \in_{\mathcal{M}} v_i$ and for all n, $\overline{n} \in_{\mathcal{M}} v_i$. Hence as the model is an ω-model $\mathbb{N}_{\mathcal{M}} \subseteq_{\mathcal{M}} v_i$.

In an ω-model, we can identify the set $\mathbb{N}$ of natural numbers with the set of objects a in $\mathcal{M}$ such that $a \in_{\mathcal{M}} \mathbb{N}_{\mathcal{M}}$. To each proof-term we can associate a natural number n (its Gödel number) and then the element $\overline{n}$ of $\mathcal{M}$. Proof-terms, their Gödel number and the encoding of this number in $\mathcal{M}$ will be identified in the following.

We are now ready to construct a model $\mathcal{U}$ for the Stratified Foundations. The base set is the set U of elements a of M such that $a \in_{\mathcal{M}} v_0$. The relation $\in_{\mathcal{U}}$ is defined by $a \in_{\mathcal{U}} b$ if and only if $a \in_{\mathcal{M}} \sigma b$. This permits to define the denotation of propositions built without Skolem symbols. To be able to define the denotation of Skolem symbols, we prove the following proposition.

Proposition 3. *For every stratifiable proposition A in the language $\in$ whose free variables are among $x_1, \ldots, x_n, x_{n+1}$ and for all $a_1, \ldots, a_n$ in U, there exists an element b in U such that for every a_{n+1} in U, $a_{n+1} \in_{\mathcal{M}} \sigma b$ if and only if $[\![A]\!]^{\mathcal{U}}_{a_1/x_1,\ldots,a_n/x_n,a_{n+1}/x_{n+1}} = 1$*

Proof. Let $|A|$ be the proposition defined as follows.

- $|A| = A$ if A is atomic,
- $|A \Rightarrow B| = |A| \Rightarrow |B|$, $|A \wedge B| = |A| \wedge |B|$, $|A \vee B| = |A| \vee |B|$, $|\bot| = \bot$,
- $|\forall x\ A| = \forall x\ ((x \in E_{S(x)}) \Rightarrow |A|)$,
- $|\exists x\ A| = \exists x\ ((x \in E_{S(x)}) \wedge |A|)$.

Notice that the free variables of $|A|$ are among $E_0, \ldots, E_m, x_1, \ldots, x_n, x_{n+1}$. Let

$$\varphi = a_1/x_1, \ldots, a_n/x_n, a_{n+1}/x_{n+1}$$

$$\psi = v_0/E_0, \ldots, v_m/E_m, \sigma^{k_1} a_1/x_1, \ldots, \sigma^{k_n} a_n/x_n, \sigma^{k_{n+1}} a_{n+1}/x_{n+1}$$

where $k_1 = S(x_1), \ldots, k_{n+1} = S(x_{n+1})$. We check, by induction over the structure of A, that if A is a stratifiable proposition in the language $\in$, then

$$[\![|A|]\!]^{\mathcal{M}}_{\psi} = [\![A]\!]^{\mathcal{U}}_{\varphi}$$

- If A is an atomic proposition $x_i \in x_j$, then $k_j = k_i + 1$, $[\![|A|]\!]^{\mathcal{M}}_{\psi} = 1$ if and only if $\sigma^{k_i} a_i \in_{\mathcal{M}} \sigma^{k_j} a_j$ if and only if $a_i \in_{\mathcal{M}} \sigma a_j$, if and only if $[\![A]\!]^{\mathcal{U}}_{\varphi} = 1$.
- if $A = B \Rightarrow C$ then $[\![|A|]\!]^{\mathcal{M}}_{\psi} = 1$ if and only if $[\![|B|]\!]^{\mathcal{M}}_{\psi} = 0$ or $[\![|C|]\!]^{\mathcal{M}}_{\psi} = 1$ if and only if $[\![B]\!]^{\mathcal{U}}_{\varphi} = 0$ or $[\![C]\!]^{\mathcal{U}}_{\varphi} = 1$ if and only if $[\![A]\!]^{\mathcal{U}}_{\varphi} = 1$.
- if $A = B \wedge C$ then $[\![|A|]\!]^{\mathcal{M}}_{\psi} = 1$ if and only if $[\![|B|]\!]^{\mathcal{M}}_{\psi} = 1$ and $[\![|C|]\!]^{\mathcal{M}}_{\psi} = 1$ if and only if $[\![B]\!]^{\mathcal{U}}_{\varphi} = 1$ and $[\![C]\!]^{\mathcal{U}}_{\varphi} = 1$ if and only if $[\![A]\!]^{\mathcal{U}}_{\varphi} = 1$.
- if $A = B \vee C$ then $[\![|A|]\!]^{\mathcal{M}}_{\psi} = 1$ if and only if $[\![|B|]\!]^{\mathcal{M}}_{\psi} = 1$ or $[\![|C|]\!]^{\mathcal{M}}_{\psi} = 1$ if and only if $[\![B]\!]^{\mathcal{U}}_{\varphi} = 1$ and $[\![C]\!]^{\mathcal{U}}_{\varphi} = 1$ if and only if $[\![A]\!]^{\mathcal{U}}_{\varphi} = 1$.
- $[\![|\bot|]\!]^{\mathcal{M}}_{\psi} = 0 = [\![\bot]\!]^{\mathcal{U}}_{\varphi}$.
- if $A = \forall x\ B$ then $[\![|A|]\!]^{\mathcal{M}}_{\psi} = 1$ if and only if for every c in M such that $c \in_{\mathcal{M}} v_k$, $[\![|B|]\!]^{\mathcal{M}}_{\psi+c/x} = 1$, if and only if for every e in U, $[\![|B|]\!]^{\mathcal{M}}_{\psi+\sigma^k e/x} = 1$ if and only if for every e in U, $[\![B]\!]^{\mathcal{U}}_{\varphi+e/x} = 1$ if and only if $[\![A]\!]^{\mathcal{U}}_{\varphi} = 1$.

– if $A = \exists x\ B$ then $[\![|A|]\!]^{\mathcal{M}}_{\psi} = 1$ if and only if there exists c in M such that $c \in_{\mathcal{M}} v_k$ and $[\![|B|]\!]^{\mathcal{M}}_{\psi+c/x} = 1$, if and only if there exists e in U such that $[\![|B|]\!]^{\mathcal{M}}_{\psi+\sigma^k e/x} = 1$ if and only if there exists e in U such that $[\![B]\!]^{\mathcal{U}}_{\varphi+e/x} = 1$ if and only if $[\![A]\!]^{\mathcal{U}}_{\varphi} = 1$.

Then, the model $\mathcal{M}$ is a model of the comprehension scheme. Hence, it is a model of the proposition

$$\forall E_0 \ \dots \ \forall E_m \ \forall x_1 \ \dots \ \forall x_n \ \forall y \ \exists z \ \forall x_{n+1} \ (x_{n+1} \in z \Leftrightarrow (x_{n+1} \in y \wedge |A|))$$

Thus, for all $a_1, ..., a_n$, there exists an object b_0 such that for all a_{n+1}

$$[\![(x_{n+1} \in z \Leftrightarrow (x_{n+1} \in y \wedge |A|))]\!]^{\mathcal{M}}_{\psi+v_{k_{n+1}}/y+b_0/z} = 1$$

We have $\sigma^{k_{n+1}} a_{n+1} \in_{\mathcal{M}} b_0$ if and only if $\sigma^{k_{n+1}} a_{n+1} \in_{\mathcal{M}} v_{k_{n+1}}$ and $[\![|A|]\!]^{\mathcal{M}}_{\psi} = 1$ thus $a_{n+1} \in_{\mathcal{M}} \sigma^{-k_{n+1}} b_0$ if and only if a_{n+1} is in U and $[\![A]\!]^{\mathcal{U}}_{\varphi} = 1$. We take $b = \sigma^{-(k_{n+1}+1)} b_0$. For all a_{n+1} in U, we have $a_{n+1} \in_{\mathcal{M}} \sigma b$ if and only if $[\![A]\!]^{\mathcal{U}}_{\varphi} = 1$.

Notice finally that $b_0 \in_{\mathcal{M}} \wp_{\mathcal{M}}(v_{k_{n+1}})$, thus $b_0 \in_{\mathcal{M}} v_{k_{n+1}+1}$, $b \in_{\mathcal{M}} v_0$ and hence b is in U.

Definition 11 (Jensen's model). *The model* $\mathcal{U} = \langle U, \in_{\mathcal{U}}, \hat{f}_{x_1,\dots,x_n,y,A} \rangle$ *is defined as follows. The base set is* U. *The relation* $\in_{\mathcal{U}}$ *is defined above. The function* $\hat{f}_{x_1,\dots,x_n,x_{n+1},A}$ *maps* $(a_1, \dots, a_n)$ *to an object* b *such that for all* a_{n+1} *in* U, $a_{n+1} \in_{\mathcal{M}} \sigma b$ *if and only if* $[\![A]\!]^{\mathcal{U}}_{a_1/x_1,\dots,a_n/x_n,a_{n+1}/x_{n+1}} = 1$.

Proposition 4. *The model* $\mathcal{U}$ *is a model of the Stratified Foundations.*

Proof. If A is a stratifiable proposition in the language $\in$, then

$$[\![t_{n+1} \in f_{x_1,\dots,x_n,x_{n+1},A}(t_1, \dots, t_n)]\!]^{\mathcal{U}}_{\varphi} = 1$$

if and only if

$$[\![t_{n+1}]\!]^{\mathcal{U}}_{\varphi} \in_{\mathcal{M}} \sigma \hat{f}_{x_1,\dots,x_n,x_{n+1},A}([\![t_1]\!]^{\mathcal{U}}_{\varphi}, \dots, [\![t_n]\!]^{\mathcal{U}}_{\varphi})$$

if and only if

$$[\![[t_1/x_1, \dots, t_n/x_n, t_{n+1}/x_{n+1}]A]\!]^{\mathcal{U}}_{\varphi} = 1$$

Hence, if $A \equiv B$ then A and B have the same denotation.

Corollary 1. *The Stratified Foundations are consistent.*

2.4 Normalization

We want now to construct a pre-model for the Stratified Foundations.

Let $u_i = v_{3i}$ and $\tau = \sigma^3$. The function τ is an automorphism of $\mathcal{M}$, $\tau u_i = u_{i+1}$, $u_i \subseteq_{\mathcal{M}} u_{i+1}$ and $\wp_{\mathcal{M}}(\wp_{\mathcal{M}}(\wp_{\mathcal{M}}(u_i))) \subseteq_{\mathcal{M}} u_{i+1}$.

As $\mathcal{M}$ is an ω-model of set theory, for each recursively enumerable relation R on natural numbers, there is an object r in $\mathcal{M}$ such that $R(a_1, \dots, a_n)$ if and only if $\langle a_1, \dots, a_n \rangle_{\mathcal{M}} \in_{\mathcal{M}} r$. In particular there is

- an object *Proof* such that $\pi \in_{\mathcal{M}} Proof$ if and only if π is (the encoding in $\mathcal{M}$ of the Gödel number of) a proof,
- an object *Term* such that $t \in_{\mathcal{M}} Term$ if and only if t is (the encoding of the Gödel number of) a term,
- an object *Subst* such that $\langle \pi, \pi_1, \alpha, \pi_2 \rangle_{\mathcal{M}} \in_{\mathcal{M}} Subst$ if and only if π, π_1 and π_2 are (encodings of Gödel numbers of) proofs, α is (the encoding of the Gödel number of) a proof variable and $\pi = [\pi_1/\alpha]\pi_2$,
- an object *Subst'* such that $\langle \pi, t, x, \pi_1 \rangle_{\mathcal{M}} \in_{\mathcal{M}} Subst'$ if and only if π and π_1 are (encodings of the Gödel numbers of) proofs, x is (the encoding of the Gödel number of) a term variable and t (the encoding of the Gödel number of) a term and $\pi = [t/x]\pi_1$,
- an object *Red* such that $\langle \pi, \pi_1 \rangle_{\mathcal{M}} \in_{\mathcal{M}} Red$ if and only if π and π_1 are (encodings of Gödel numbers of) proofs and $\pi \rhd^* \pi_1$,
- an object *Sn* such that $\pi \in_{\mathcal{M}} Sn$ if and only if π is (the encoding of the Gödel number of) a strongly normalizable proof,
- an object *ImpI* such that $\langle \pi, \alpha, \pi_1 \rangle_{\mathcal{M}} \in_{\mathcal{M}} ImpI$ if and only if π and π_1 are (encodings of Gödel numbers of) proofs, α is (the encoding of the Gödel number of) a proof variable and $\pi = \lambda\alpha\ \pi_1$,
- an object *AndI* such that $\langle \pi, \pi_1, \pi_2 \rangle_{\mathcal{M}} \in_{\mathcal{M}} AndI$ if and only if π, π_1 and π_2 are (encodings of Gödel numbers of) proofs and $\pi = \langle \pi_1, \pi_2 \rangle$,
- an object *OrI*1 (resp. *OrI*2) such that $\langle \pi, \pi_1 \rangle_{\mathcal{M}} \in_{\mathcal{M}} OrI1$ (resp. $\langle \pi, \pi_2 \rangle_{\mathcal{M}} \in_{\mathcal{M}} OrI2$) if and only if π and π_1 (resp. π and π_2) are (encodings of Gödel numbers of) proofs and $\pi = i(\pi_1)$ (resp. $\pi = j(\pi_2)$),
- an object *ForallI* such that $\langle \pi, \alpha, \pi_1 \rangle_{\mathcal{M}} \in_{\mathcal{M}} ForallI$ if and only π and π_1 are (encodings of Gödel numbers of) proofs, α is (the encoding of the Gödel number of) a proof variable, and $\pi = \lambda\alpha\pi_1$,
- an object *ExistsI* such that $\langle \pi, t, \pi_1 \rangle_{\mathcal{M}} \in_{\mathcal{M}} ExistsI$ if and only if π and π_1 are (encodings of Gödel numbers of) proofs, t is (the encoding of the Gödel number of) a term and $\pi = \langle t, \pi_1 \rangle$.

Notice also that, since $\mathcal{M}$ is a model of the comprehension scheme, there is an object Cr such that $\alpha \in_{\mathcal{M}} Cr$ if and only if α is a reducibility candidate (i.e. the set of objects β such that $\beta \in_{\mathcal{M}} \alpha$ is a reducibility candidate).

Definition 12 (Admissible). *An element α of M is said to* admissible *at level i if α is a set of pairs $\langle \pi, \beta \rangle_{\mathcal{M}}$ where π is a proof and β an element of u_i and for each β in u_i the set of π such that $\langle \pi, \beta \rangle_{\mathcal{M}} \in_{\mathcal{M}} \alpha$ is a reducibility candidate.*

Notice that if R is any reducibility candidate then the set $R \times_{\mathcal{M}} u_i$ is admissible at level i. Hence there are admissible elements at all levels.

Proposition 5. *There is an element A_i in M such that $\alpha \in_{\mathcal{M}} A_i$ if and only if α is admissible at level i.*

Proof. An element α of $\mathcal{M}$ admissible at level i if and only if

$$\alpha \in_{\mathcal{M}} \wp_{\mathcal{M}}(Proof \times_{\mathcal{M}} u_i) \\ \wedge \forall \beta\ (\beta \in_{\mathcal{M}} u_i \Rightarrow \exists C\ (C \in_{\mathcal{M}} Cr \wedge (\langle \pi, \beta \rangle_{\mathcal{M}} \in_{\mathcal{M}} \alpha \Leftrightarrow \pi \in_{\mathcal{M}} C)))$$

Hence, as $\mathcal{M}$ is a model of the comprehension scheme, there is an element A_i in M such that $\alpha \in_{\mathcal{M}} A_i$ if and only if α is admissible at level i.

Notice that $\alpha \in \tau A_i$ if and only if $\alpha \in A_{i+1}$. Hence as $\mathcal{M}$ is a model of the extensionality axiom, $\tau A_i = A_{i+1}$.

Notice, at last, that $A_i \subseteq_{\mathcal{M}} \wp_{\mathcal{M}}(Proof \times_{\mathcal{M}} u_i) \subseteq_{\mathcal{M}} \wp_{\mathcal{M}}(u_i \times_{\mathcal{M}} u_i) \subseteq_{\mathcal{M}} \wp_{\mathcal{M}}(\wp_{\mathcal{M}}(\wp_{\mathcal{M}}(u_i))) \subseteq_{\mathcal{M}} u_{i+1}$.

Proposition 6. *If $\beta \in_{\mathcal{M}} A_i$ and $\alpha \in_{\mathcal{M}} A_{i+1}$ then the set of π such that $\langle \pi, \beta \rangle \in_{\mathcal{M}} \alpha$ is a reducibility candidate.*

Proof. As $\alpha \in_{\mathcal{M}} A_{i+1}$ and $\beta \in_{\mathcal{M}} A_i \subseteq_{\mathcal{M}} u_{i+1}$, the set of π such that $\langle \pi, \beta \rangle \in_{\mathcal{M}} \alpha$ is a reducibility candidate.

We are now ready to construct a pre-model $\mathcal{N}$ of the Stratified Foundations. The base set of this pre-model is the set N of elements of M that are admissible at level 0. We take $\in_{\mathcal{N}} (\alpha, \beta) = \{\pi \mid \langle \pi, \alpha \rangle_{\mathcal{M}} \in_{\mathcal{M}} \tau\beta\}$. This permits to define the denotation of propositions built without Skolem symbols. To define the denotation of Skolem symbols, we prove the following proposition.

Proposition 7. *For every stratifiable proposition A in the language $\in$ whose free variables are among $x_1, \dots, x_n, x_{n+1}$ and for all $a_1, \dots, a_n$ in N, there exists an element b in N such that for every a_{n+1} in N, $\langle \pi, a_{n+1} \rangle_{\mathcal{M}} \in_{\mathcal{M}} \tau b$ if and only if π is in $[\![A]\!]^{\mathcal{N}}_{a_1/x_1, \dots, a_{n+1}/x_{n+1}}$.*

Proof. Let $|A|$ be the proposition (read p *realizes* A) defined as follows.

- $|x_i \in x_j| = \langle p, x_i \rangle \in x_j$,
- $|A \Rightarrow B| = p \in sn \wedge \forall q\ \forall w\ \forall r\ (\langle p, q \rangle \in red \wedge \langle q, w, r \rangle \in implI) \Rightarrow \forall s\ [s/p]|A| \Rightarrow \forall t\ \langle t, s, w, r \rangle \in subst \Rightarrow [t/p]|B|)$,
- $|A \wedge B| = p \in sn \wedge \forall q\ \forall r\ \forall s\ (((\langle p, q \rangle \in red \wedge \langle q, r, s \rangle \in andI) \Rightarrow [r/p]|A| \wedge [s/p]|B|)$,
- $|A \vee B| = p \in sn \wedge \forall q\ \forall r\ (((\langle p, q \rangle \in red \wedge \langle q, r \rangle \in orI1) \Rightarrow [r/p]|A|) \wedge \forall q\ \forall r\ (((\langle p, q \rangle \in red \wedge \langle q, r \rangle \in orI2) \Rightarrow [r/p]|B|)$,
- $|\bot| = p \in sn$,
- $|\forall x\ A| = p \in sn \wedge \forall q\ \forall w\ \forall r\ (\langle p, q \rangle \in red \wedge (\langle q, w, r \rangle \in forallI) \Rightarrow \forall x\ \forall y\ (x \in E_{S(x)} \wedge y \in term) \Rightarrow \forall s\ (\langle s, w, y, r \rangle \in subst' \Rightarrow [r/p, x/x]|A|))$,
- $|\exists x\ A| = p \in sn \wedge \forall q\ \forall t\ \forall r\ (\langle p, q \rangle \in red \wedge (\langle q, t, r \rangle \in existsI) \Rightarrow \exists x\ x \in E_{S(x)} \Rightarrow [r/p, x/x]|A|))$.

Notice that the free variables of $|A|$ are among $term, subst, subst', red, sn, implI, andI, orI1, orI2, forallI, existsI, p, E_0, \dots, E_m, x_1, \dots, x_n, x_{n+1}$. Let

$$\varphi = a_1/x_1, \dots, a_n/x_n, a_{n+1}/x_{n+1}$$

$$\psi = Term/term, Subst/subst, Subst'/subst', Red/red, Sn/sn,$$

$$ImplI/implI, AndI/andI, OrI1/orI1, OrI2/orI2, ForallI/forallI, ExistsI/existsI,$$

$$A_0/E_0, \dots, A_m/E_m, \tau^{k_1} a_1/x_1, \dots, \tau^{k_n} a_n/x_n, \tau^{k_{n+1}} a_{n+1}/x_{n+1}$$

We check, by induction over the structure of A, that if A is a stratifiable proposition in the language $\in$, then the set of proofs π such that $[\![|A|]\!]^{\mathcal{M}}_{\psi+\pi/p} = 1$ is $[\![A]\!]^{\mathcal{N}}_{\varphi}$.

- If A is an atomic proposition $x_i \in x_j$, then $k_j = k_i + 1$, we have $[\![|A|]\!]^{\mathcal{M}}_{\psi+\pi/p} = 1$ if and only if $\langle \pi, \tau^{k_i} a_i \rangle_{\mathcal{M}} \in_{\mathcal{M}} \tau^{k_j} a_j$ if and only if $\langle \tau^{k_i}\pi, \tau^{k_i} a_i \rangle_{\mathcal{M}} \in_{\mathcal{M}} \tau^{k_j} a_j$ if and only if $\tau^{k_i} \langle \pi, a_i \rangle_{\mathcal{M}} \in_{\mathcal{M}} \tau^{k_j} a_j$ if and only if $\langle \pi, a_i \rangle_{\mathcal{M}} \in_{\mathcal{M}} \tau a_j$ if and only if π is in $[\![A]\!]^{\mathcal{N}}_{\varphi}$.
- if $A = B \Rightarrow C$ then we have $[\![|A|]\!]^{\mathcal{M}}_{\psi+\pi/p} = 1$ if and only if π is strongly normalizable and whenever π reduces to $\lambda\alpha\ \pi_1$ then for all π' such that $[\![|B|]\!]^{\mathcal{M}}_{\psi+\pi'/p} = 1$ we have $[\![|C|]\!]^{\mathcal{M}}_{\psi+[\pi'/\alpha]\pi_1/p} = 1$ if and only if π is strongly normalizable and whenever π reduces to $\lambda x\ \pi_1$ then for all π' in $[\![B]\!]^{\mathcal{N}}_{\varphi}$, $[\pi'/\alpha]\pi_1$ is in $[\![C]\!]^{\mathcal{N}}_{\varphi}$ if and only if π is in $[\![A]\!]^{\mathcal{N}}_{\varphi}$.
- If $A = B \wedge C$ then we have $[\![A]\!]^{\mathcal{M}}_{\psi+\pi/p} = 1$ if and only if π is strongly normalizable and whenever π reduces to $\langle \pi_1, \pi_2 \rangle$ then $[\![B]\!]^{\mathcal{M}}_{\psi+\pi_1/p} = 1$ and $[\![C]\!]^{\mathcal{M}}_{\psi+\pi_2/p} = 1$ if and only if π is strongly normalizable and whenever π reduces to $\langle \pi_1, \pi_2 \rangle$ then π_1 is in $[\![B]\!]^{\mathcal{N}}_{\varphi}$ and π_2 is in $[\![C]\!]^{\mathcal{N}}_{\varphi}$ if and only if π is in $[\![A]\!]^{\mathcal{N}}_{\varphi}$.
- If $A = B \vee C$ then we have $[\![A]\!]^{\mathcal{M}}_{\psi+\pi/p} = 1$ if and only if π is strongly normalizable and whenever π reduces to $i(\pi_1)$ (resp. $j(\pi_2)$) then $[\![B]\!]^{\mathcal{M}}_{\psi+\pi_1/p} = 1$ (resp. $[\![C]\!]^{\mathcal{M}}_{\psi+\pi_2/p} = 1$) if and only if π is strongly normalizable and whenever π reduces to $i(\pi_1)$ (resp. $j(\pi_2)$) then π_1 is in $[\![B]\!]^{\mathcal{N}}_{\varphi}$ (resp. $[\![C]\!]^{\mathcal{N}}_{\varphi}$) if and only if π is in $[\![A]\!]^{\mathcal{N}}_{\varphi}$.
- If $A = \bot$ then $[\![A]\!]^{\mathcal{M}}_{\psi+\pi/p} = 1$ if and only if π is strongly normalizable if and only if π is in $[\![A]\!]^{\mathcal{N}}_{\varphi}$.
- if $A = \forall x\ B$, then $[\![|A|]\!]^{\mathcal{M}}_{\psi+\pi/p} = 1$ if and only if π is strongly normalizable and whenever π reduces to $\lambda x\ \pi_1$, for all term t and for all c in M such that $c \in_{\mathcal{M}} A_k$, $[\![|B|]\!]^{\mathcal{M}}_{\psi+c/x,[t/x]\pi_1/p} = 1$ if and only if π is strongly normalizable and whenever π reduces to $\lambda x\ \pi_1$, for all t and for all e in N, $[\![|B|]\!]^{\mathcal{M}}_{\psi+\tau^k e/x+[t/x]\pi_1/p} = 1$ if and only if π is strongly normalizable and whenever π reduces to $\lambda x\ \pi_1$, for all t and for all e in N, $[t/x]\pi_1$ is in $[\![B]\!]^{\mathcal{N}}_{\varphi+e/x}$ if and only if π is in $[\![A]\!]^{\mathcal{N}}_{\varphi}$.
- if $A = \exists x\ B$, then $[\![|A|]\!]^{\mathcal{M}}_{\psi+\pi/p} = 1$ if and only if π is strongly normalizable and whenever π reduces to $\langle t, \pi_1 \rangle$, there exists a c in M such that $c \in_{\mathcal{M}} A_k$ and $[\![|B|]\!]^{\mathcal{M}}_{\psi+c/x,[t/x]\pi_1/p} = 1$ if and only if π is strongly normalizable and whenever π reduces to $\langle t, \pi_1 \rangle$, there exists a e in N such that $[\![|B|]\!]^{\mathcal{M}}_{\psi+\tau^k e/x+[t/x]\pi_1/p} = 1$ if and only if π is strongly normalizable and whenever π reduces to $\langle t, \pi_1 \rangle$, there exists a e in N such that $[t/x]\pi_1$ is in $[\![B]\!]^{\mathcal{N}}_{\varphi+e/x}$ if and only if π is in $[\![A]\!]^{\mathcal{N}}_{\varphi}$.

Then, the model $\mathcal{M}$ is a model of the comprehension scheme. Hence, it is a model of the proposition

$$\forall E_0 \ldots \forall E_m \forall x_1 \ldots \forall x_n\ \exists z\ \forall p\ \forall x_{n+1}\ \langle p, x_{n+1} \rangle \in z \Leftrightarrow \langle p, x_{n+1} \rangle \in proof \times U \wedge |A|$$

Thus, for all $a_1, ..., a_n$, there exists an object b_0 such that for all a_{n+1}

$$[\![\langle p, x_{n+1} \rangle \in z \Leftrightarrow \langle p, x_{n+1} \rangle \in \mathbb{N}_{\mathcal{M}} \times U \wedge |A|]\!]^{\mathcal{M}}_{\psi+Proof/proof,b_0/z,u_{k_{n+1}+1}/U,\pi/p} = 1$$

We have $\langle\pi, \tau^{k_{n+1}} a_{n+1}\rangle_{\mathcal{M}} \in_{\mathcal{M}} b_0$ if and only if π is a proof, $\tau^{k_{n+1}} a_{n+1} \in_{\mathcal{M}} u_{k_{n+1}+1}$ and $[\![|A|]\!]^{\mathcal{M}}_{\psi+\pi/p} = 1$. Thus $\langle\pi, a_{n+1}\rangle_{\mathcal{M}} \in_{\mathcal{M}} \tau^{-k_{n+1}} b_0$ if and only if $a_{n+1} \in_{\mathcal{M}} u_1$ and π is in $[\![A]\!]^{\mathcal{N}}_{\varphi}$. We take $b = \tau^{-(k_{n+1}+1)} b_0$ and for all a_{n+1} in N we have $\langle\pi, a_{n+1}\rangle_{\mathcal{M}} \in_{\mathcal{M}} \tau b$ if and only if π is in $[\![A]\!]^{\mathcal{N}}_{\varphi}$. Finally, notice that b_0 is a set of pairs $\langle\pi, \beta\rangle_{\mathcal{M}}$ where π is a proof and β an element of $u_{k_{n+1}+1}$ and for each β in $u_{k_{n+1}+1}$ the set of π such that $\langle\pi, \beta\rangle_{\mathcal{M}} \in_{\mathcal{M}} b_0$ is $[\![|A|]\!]^{\mathcal{M}}_{\psi+\beta/x_{k_{n+1}}, \pi/p} = 1$, hence it is a reducibility candidate. Hence $b_0 \in_{\mathcal{M}} A_{k_{n+1}+1}$ and b is in N.

Definition 13 (Crabbé's pre-model).
The pre-model $\mathcal{N} = \langle N, \in_{\mathcal{N}}, \hat{f}_{x_1,\ldots,x_n,y,A}\rangle$ is defined as follows. The base set is N. The function $\in_{\mathcal{N}}$ is defined above. The function $\hat{f}_{x_1,\ldots,x_n,x_{n+1},A}$ maps $(a_1, \ldots, a_n)$ to the object b such that for all a_{n+1} in N, $\langle\pi, a_{n+1}\rangle_{\mathcal{M}} \in_{\mathcal{M}} \tau b$ if and only if π is in $[\![A]\!]^{\mathcal{N}}_{a_1/x_1,\ldots,a_n/x_n,a_{n+1}/x_{n+1}}$.

Proposition 8. *The pre-model $\mathcal{N}$ is a pre-model of the Stratified Foundations.*

Proof. If A is a stratifiable proposition in the language $\in$, then

$$\pi \text{ is in } [\![t_{n+1} \in f_{x_1,\ldots,x_n,x_{n+1},A}(t_1, \ldots, t_n)]\!]^{\mathcal{N}}_{\varphi}$$

if and only if

$$\langle\pi, [\![t_{n+1}]\!]^{\mathcal{N}}_{\varphi}\rangle_{\mathcal{M}} \in_{\mathcal{M}} \tau \hat{f}_{x_1,\ldots,x_n,x_{n+1},A}([\![t_1]\!]^{\mathcal{N}}_{\varphi}, \ldots, [\![t_n]\!]^{\mathcal{N}}_{\varphi})$$

if and only if

$$\pi \text{ is in } [\![[t_1/x_1, \ldots, t_n/x_n, t_{n+1}/x_{n+1}]A]\!]^{\mathcal{N}}_{\varphi}$$

Hence, if $A \equiv B$ then A and B have the same denotation.

Corollary 2. *All proofs strongly normalize in the Stratified Foundations.*

Remark 1. As already noticed in [2], instead of constructing the a pre-model of the Stratified Foundations within an automorphic ω-model of Zermelo's set theory, we could construct it within an ω-model of the Stratified Foundations. In such a model $\mathcal{U}$, we can define recursively enumerable relations, because the Stratified Foundations contains enough arithmetic and comprehension. Then we can take the sequence u_i to be the constant sequence equal to w where w is a universal set, i.e. a set such that $a \in_{\mathcal{U}} w$ for all element a of the model. Such an object obviously verifies $\wp_{\mathcal{U}}(\wp_{\mathcal{U}}(\wp_{\mathcal{U}}(w))) \subseteq_{\mathcal{U}} w$. In other words, we say that an element of U is admissible if it is a set of pairs $\langle\pi, \beta\rangle_{\mathcal{U}}$ where π is a proof and for each β in U, the set of π such that $\langle\pi, \beta\rangle \in_{\mathcal{U}} \alpha$ is a reducibility candidate. Proposition 6 becomes trivial, but we need to use the existence of a universal set to prove that there are admissible elements in the model and that there is a set A of admissible elements in the model. Hence, the difficult part in this pre-model construction (the part that would not go through for Zermelo's set theory for instance) is the construction of the base set.

Conclusion

In this paper, we have have shown that the Stratified Foundations can be expressed in deduction modulo and that the normalization proof for this theory be decomposed into two lemmas: one expressing that it has a pre-model and the other that proof normalize in this theory if it has a pre-model. This second lemma is not specific to the Stratified Foundations, but holds for all theories modulo. The idea of the first lemma is to construct a pre-model within an ω-model of the theory with the help of formal realizability. This idea does not seems to be specific to the Stratified Foundations either, but, its generality remains to be investigated. Thus, this example contributes to explore of the border between the theories modulo that have the normalization property and those that do not.

References

1. M. Crabbé. Non-normalisation de ZF. *Manuscript*, 1974.
2. M. Crabbé. Stratification and cut-elimination. *The Journal of Symbolic Logic*, 56:213–226, 1991.
3. G. Dowek. *La part du calcul.* Université de Paris 7, 1999. Mémoire d'habilitation.
4. G. Dowek, *About folding-unfolding cuts and cuts modulo. Journal of Logic and Computation* (to appear).
5. G. Dowek, Th. Hardin, and C. Kirchner. Theorem proving modulo. Rapport de Recherche 3400, Institut National de Recherche en Informatique et en Automatique, 1998. *Journal of Automated Reasoning* (to appear).
6. G. Dowek, Th. Hardin, and C. Kirchner. HOL-$\lambda\sigma$ an intentional first-order expression of higher-order logic. *Mathematical Structures in Computer Science*, 11, 2001, pp. 1-25.
7. G. Dowek and B. Werner. Proof normalization modulo. In *Types for proofs and programs 98*, volume 1657 of *Lecture Notes in Computer Science*, pages 62–77. Springer-Verlag, 1999.
8. R.B. Jensen. On the consistency of a slight (?) modification of Quine's new foundations. *Synthese*, 19, 1968-69, pp. 250–263.
9. J.W. Klop, V. van Oostrom, and F. van Raamsdonk. Combinatory reduction systems: introduction and survey. *Theoretical Computer Science*, 121:279–308, 1993.

Normalization by Evaluation for the Computational Lambda-Calculus

Andrzej Filinski

BRICS*, Department of Computer Science, University of Aarhus
andrzej@brics.dk

Abstract. We show how a simple semantic characterization of normalization by evaluation for the $\lambda_{\beta\eta}$-calculus can be extended to a similar construction for normalization of terms in the computational λ-calculus. Specifically, we show that a suitable *residualizing* interpretation of base types, constants, and computational effects allows us to extract a syntactic normal form from a term's denotation. The required interpretation can itself be constructed as the meaning of a suitable functional program in an ML-like language, leading directly to a practical normalization algorithm. The results extend easily to product and sum types, and can be seen as a formal basis for call-by-value type-directed partial evaluation.

1 Introduction

The basic idea of normalization by evaluation is to extract the normal form (with respect to some notion of conversion) of a term from its interpretation in a suitably chosen, quasi-syntactic denotational model of the conversion relation [5].

For instance, let us consider the interpretation of a pure, simply typed lambda-term E in a model where all base types are interpreted as the set Λ of well-formed lambda-terms, and function types are interpreted as full set-theoretic function spaces. Then it is fairly simple to (at least informally) construct for any type τ, a function $nf_\tau \in [\![\tau]\!] \to \Lambda$, such that for any closed term $\widetilde{E} : \tau$ in $\beta\eta$-long normal form, $nf_\tau([\![\widetilde{E}]\!]) =_\alpha \widetilde{E}$. We proceed as follows:

Let $\tau = \tau_1 \to \cdots \to \tau_n \to b$ $(n \geq 0)$, where each $\tau_i = \tau_{i1} \to \cdots \to \tau_{im_i} \to b_i$. Then $\widetilde{E} : \tau$ must be of the form $\lambda x_1. \cdots . \lambda x_n. x_i\, \widetilde{E}_1 \cdots \widetilde{E}_{m_i}$ where each $\widetilde{E}_j : \tau_{ij}$ is again in normal form. We can thus define nf_τ inductively as:

$$
\begin{aligned}
nf_\tau = \lambda f.\, LAM(v_1, \ldots, LAM(v_n,\\
&f\,(\lambda a_1. \cdots . \lambda a_{m_1}.\, APP(\cdots APP(VAR\, v_1, nf_{\tau_{11}}\, a_1) \cdots, nf_{\tau_{1m_1}}\, a_{m_1}))\\
&\qquad \vdots\\
&(\lambda a_1. \cdots . \lambda a_{m_n}.\, APP(\cdots APP(VAR\, v_n, nf_{\tau_{n1}}\, a_1) \cdots, nf_{\tau_{nm_n}}\, a_{m_n})))))
\end{aligned}
$$

where the v_i are "fresh" variable names, and we use VAR, LAM, and APP for constructing elements of Λ, to distinguish them from function abstraction and application in the set-theoretic model.

* Basic Research in Computer Science (www.brics.dk), funded by the Danish National Research Foundation.

S. Abramsky (Ed.): TLCA 2001, LNCS 2044, pp. 151–165, 2001.

Moreover, it is easy to see that $nf_\tau(\llbracket - \rrbracket)$ is in fact a normalization function: since $\beta\eta$-convertibility is sound for equality in all set-theoretic interpretations (and hence also in our chosen one), we have, for all terms E and E', that $E =_{\beta\eta} E'$ implies $\llbracket E \rrbracket = \llbracket E' \rrbracket$. So if $\widetilde{E}$ is the $\beta\eta$-long normal form of E, then $nf_\tau(\llbracket E \rrbracket) = nf_\tau(\llbracket \widetilde{E} \rrbracket) =_\alpha \widetilde{E}$.

Finally, if we can also construct a syntactic term $\mathsf{nf}_\tau : \tau \to \Lambda$ such that $nf_\tau \llbracket E \rrbracket = \llbracket \mathsf{nf}_\tau E \rrbracket$, then $\mathsf{nf}_\tau\, E$ is a closed term of base type Λ, and can thus be executed as a functional program. This gives us a very efficient executable *algorithm* for computing normal forms, and was indeed one of the motivations behind the construction [4,3]: we are reducing the general problem of term normalization to a special case for which we already have a good solution.

A natural question arises whether this semantic technique for normalization of lambda-terms is inherently tied to $\beta\eta$-conversion. Somewhat surprisingly, it is not: in the following, we show how the same idea can be used to normalize terms with respect to the computational lambda-calculus [16], where it also extends to product and – more notably – sum types. In fact, we can systematically extract the computational normal form of any pure, typed lambda-term from only its observable behavior in an imperative functional language such as ML.

Despite the relative simplicity of the construction, there are still a few technical details to nail down, even in the purely functional case. Accordingly, we will first present in Section 2 the normalization algorithm for a call-by-name setting, then show in Section 3 how it can be refined to call-by-value. In Section 4 we show how to further extend the normalizer with product and sum types, and in Section 5 we consider the relationship between normalization by evaluation and type-directed partial evaluation. Finally, Section 6 concludes and points out some directions for further work.

2 Normalization by Evaluation for Call by Name

The normalization construction sketched above, essentially due to Berger and Schwichtenberg [4], has been studied in many formulations [8], including more syntactic variants [2] as well as category-theoretic ones [1,6]. In the following, we present it in a call-by-name functional-programming setting [11]; this formulation extends particularly naturally to the call-by-value variant in the next section.

2.1 Language and Semantic Framework

Syntax. A *signature* Σ includes, first, a collection of base types b. The set of well-formed Σ-types τ is then given by the grammar

$$\tau ::= b \mid \tau_1 \to \tau_2$$

Further, Σ assigns Σ-types to a (possibly infinite) collection of constants c. Let x range over variable names, and Γ be a finite assignment of Σ-types to variables. Then the set of well-typed Σ-terms $\Gamma \vdash_\Sigma E : \tau$ is again given by the usual rules:

$$\frac{\Sigma(c) = \tau}{\Gamma \vdash_\Sigma c : \tau} \qquad \frac{\Gamma(x) = \tau}{\Gamma \vdash_\Sigma x : \tau} \qquad \frac{\Gamma, x{:}\,\tau_1 \vdash_\Sigma E : \tau_2}{\Gamma \vdash_\Sigma \lambda x^{\tau_1}.\, E : \tau_1 \to \tau_2} \qquad \frac{\Gamma \vdash_\Sigma E_1 : \tau_1 \to \tau_2 \quad \Gamma \vdash_\Sigma E_2 : \tau_1}{\Gamma \vdash_\Sigma E_1\, E_2 : \tau_2}$$

Finally, a Σ*-program* is a closed Σ-term of base type.

Semantics. For concreteness, and to accommodate the refinements in Section 5, we consider only a specific, domain-theoretic framework, but the results also adapt easily to a set-theoretic setting, by forgetting the order structure.

We work in the setting of (bottomless) cpos and (total) continuous functions. We also use the concept of a *monad* (more precisely, a *Kleisli triple*) $\mathcal{T} = (T, \eta, \star)$, where T maps cpos to cpos, $\eta_A \in A \to TA$ is the *unit* function, and $\star_{A,B} \in TA \times (A \to TB) \to TB$ is the *extension* operation written backwards, i.e., with the function last; this makes longer sequences of extensions easier to read. We omit the subscripts on units and extensions where they are clear from the context. A particularly important instance is the *lifting monad*, $\mathcal{T}_l$, where $T^l A = A_\perp = \{\iota a \mid a \in A\} \cup \{\perp\}$ with the usual ordering, $\eta^l a = \iota a$, $\perp \star^l f = \perp$, and $(\iota a) \star^l f = f a$.

An *interpretation* $\mathcal{I}$ of a signature Σ is a pair of functions $(\mathcal{B}, \mathcal{C})$. $\mathcal{B}$ assigns to every base type b in Σ, a cpo $\mathcal{B}(b)$. This assignment determines for any Σ-type τ, a pointed (i.e., containing a least element) cpo $[\![\tau]\!]^{\mathcal{I}}$ as follows:

$$[\![b]\!]^{\mathcal{I}} = T^l \mathcal{B}(b) \qquad\qquad [\![\tau_1 \to \tau_2]\!]^{\mathcal{I}} = [\![\tau_1]\!]^{\mathcal{I}} \to [\![\tau_2]\!]^{\mathcal{I}}$$

where $A \to B$ denotes the cpo of all continuous functions between A and B. We also give meaning to a type assignment Γ as a finite product:

$$[\![\Gamma]\!]^{\mathcal{I}} = \textstyle\prod_{x \in \mathrm{dom}\,\Gamma} [\![\Gamma(x)]\!]^{\mathcal{I}} = \{\rho \mid \forall x \in \mathrm{dom}\,\Gamma.\, \rho x \in [\![\Gamma(x)]\!]^{\mathcal{I}}\}$$

The function $\mathcal{C}$ assigns to every Σ-constant c an element $\mathcal{C}(c) \in [\![\Sigma(c)]\!]^{\mathcal{I}}$. Again this assignment extends to a full semantics of terms: for any $\Gamma \vdash_\Sigma E : \tau$, we define a continuous function $[\![E]\!]^{\mathcal{I}} \in [\![\Gamma]\!]^{\mathcal{I}} \to [\![\tau]\!]^{\mathcal{I}}$ in the usual way:

$$\begin{aligned} [\![c]\!]^{\cdot}\,\rho &= \mathcal{C}(c) & [\![\lambda x^\tau. E]\!]^{\cdot}\,\rho &= \lambda a. [\![E]\!]^{\cdot}\,(\rho[x \mapsto a]) \\ [\![x]\!]^{\cdot}\,\rho &= \rho x & [\![E_1\, E_2]\!]^{\cdot}\,\rho &= [\![E_1]\!]^{\cdot}\,\rho([\![E_2]\!]^{\cdot}\,\rho) \end{aligned}$$

Equivalence and normal forms. We say that two Σ-terms E and E' are semantically equivalent, written $\models E = E'$, if for all interpretations $\mathcal{I}$ of Σ, $[\![E]\!]^{\mathcal{I}} = [\![E']\!]^{\mathcal{I}}$. It is easy to see that if $E =_{\beta\eta} E'$ then $\models E = E'$. More generally, if *Int* is a subset of all possible interpretations of Σ (e.g., constraining the meanings of some of the constants), we write $\models^{Int} E = E'$ iff for all $\mathcal{I} \in Int$, $[\![E]\!]^{\mathcal{I}} = [\![E']\!]^{\mathcal{I}}$; we will return to this notion in Section 5.

Among the well-typed terms $\Gamma \vdash_\Sigma E : \tau$, we distinguish those in *normal* and *atomic* (also known as *neutral*) form:

$$\frac{\Gamma \vdash^{\mathrm{at}} E : b}{\Gamma \vdash^{\mathrm{nf}} E : b} \qquad \frac{\Gamma, x{:}\tau_1 \vdash^{\mathrm{nf}} E : \tau_2}{\Gamma \vdash^{\mathrm{nf}} \lambda x^{\tau_1}. E : \tau_1 \to \tau_2}$$

$$\frac{\Gamma(x) = \tau}{\Gamma \vdash^{\mathrm{at}} x : \tau} \qquad \frac{\Sigma(c) = \tau}{\Gamma \vdash^{\mathrm{at}} c : \tau} \qquad \frac{\Gamma \vdash^{\mathrm{at}} E_1 : \tau_1 \to \tau_2 \quad \Gamma \vdash^{\mathrm{nf}} E_2 : \tau_1}{\Gamma \vdash^{\mathrm{at}} E_1\, E_2 : \tau_2}$$

A *normalization function*, in the sense of Coquand and Dybjer [5] (but with a semantic notion of equivalence), then maps any term E to a normal-form term $norm(E)$, such that $\models norm(E) = E$, and such that for all E' with $\models E' = E$, $norm(E') = norm(E)$.

2.2 A Normalization Result

The traditional way of computing $norm(E)$ is by repeated β-reductions, possibly followed by η-expansions. However, we can also compute *norm* by a subtler, semantic method, *reduction-free normalization*.

Representing lambda-terms. Let $\mathbf{V}$ be a set (= discrete cpo) of explicitly typed variable names, and $\mathbf{E}$ a set suitable for representing lambda-terms, i.e., allowing us to define injective functions with mutually disjoint ranges,

$$CST \in \mathrm{dom}\,\Sigma \to \mathbf{E}, \quad VAR \in \mathbf{V} \to \mathbf{E}, \quad LAM \in \mathbf{V} \times \mathbf{E} \to \mathbf{E}, \quad APP \in \mathbf{E} \times \mathbf{E} \to \mathbf{E}$$

Then for any term E with variables from $\mathbf{V}$, we define its representation $\ulcorner E \urcorner \in \mathbf{E}$ in the obvious way, e.g., $\ulcorner \lambda x.\, E \urcorner = LAM\,(x, \ulcorner E \urcorner)$. Because of the injectivity and disjointness assumptions, for any $e \in \mathbf{E}$, there is *at most one* E such that $\ulcorner E \urcorner = e$; we need not require that all elements of $\mathbf{E}$ represent well-formed lambda-terms, let alone well-typed ones.

(We deliberately use a very concrete representation of terms, rather than a higher-level notion based on abstract syntax with binding constructs such as [12]. Our ultimate goal is to implement the normalization process as a simple functional program, without assuming potentially expensive operations, such as capture-avoiding substitution, as primitives.)

The task is now to construct an interpretation $\mathcal{I}_{\mathrm{r}}$ of Σ such that we can recover E's normal form from $[\![E]\!]^{\mathcal{I}_{\mathrm{r}}}$. We want to use the idea from the introduction, but need to account rigorously for "fresh" variable names. Freshness could be captured abstractly in a framework such as Fraenkel-Mostowski sets [14], but this again removes us a level from a direct implementation. Instead, we will explicitly generate non-clashing variable names. Perhaps the simplest scheme for doing so is through de Bruijn levels [2,11], but we adopt instead a scheme for generating "globally unique", gensym-style names using a monad, as it scales better to the constructions of the next section.

Auxiliary definitions. We first define the name-generation monad $\mathcal{T}_{\mathrm{g}}$. This is just a state-passing monad atop $\mathcal{T}_{\mathrm{l}}$; the state is the "next free index":

$$T^{\mathrm{g}}A = \mathbf{N} \to T^{\mathrm{l}}(A \times \mathbf{N}) \qquad \eta^{\mathrm{g}}\, a = \lambda i.\, \eta^{\mathrm{l}}\,(a, i) \qquad t \star^{\mathrm{g}} f = \lambda i.\, t\, i \star^{\mathrm{l}} \lambda(a, i').\, f\, a\, i'$$

With respect to T^{g}, we can define an effectful computation that generates a fresh name, and one that initializes the index within a delimited subcomputation:

$$\begin{aligned} new_\tau &\in T^{\mathrm{g}}\mathbf{V} & \qquad withct_A &\in T^{\mathrm{g}}A \to T^{\mathrm{l}}A \\ new_\tau &= \lambda i.\, \eta^{\mathrm{l}}\,(g_i^\tau, i+1) & \qquad withct_A\, t &= t\,0 \star^{\mathrm{l}} \lambda(a, i').\, \eta^{\mathrm{l}}\, a \end{aligned}$$

where the $g_i^\tau \in \mathbf{V}$ are assumed distinct for distinct i. Note that the codomain of $withct_A$ is simply $T^{\mathrm{l}}A$, i.e., $withct\, t$ represents a side-effect-free, purely functional computation, for any name-generating computation t.

The residualizing interpretation. We can now define a suitable residualizing interpretation $\mathcal{I}_r = (\mathcal{B}_r, \mathcal{C}_r)$. For $\mathcal{B}_r$ we take, for all base types b in Σ,

$$\mathcal{B}_r(b) = T^g\mathbf{E}$$

Formalizing the construction from the introduction, we further define, for any Σ-type τ, a pair of functions commonly called *reification* and *reflection*:

$$\begin{aligned}
\downarrow^\tau &\in [\![\tau]\!]^{\mathcal{I}_r} \to T^g\mathbf{E} \\
\downarrow^b &= \lambda\varepsilon.\,\varepsilon \\
\downarrow^{\tau_1\to\tau_2} &= \lambda f.\,new_{\tau_1} \star^g \lambda v.\downarrow^{\tau_2}\left(f\,(\uparrow_{\tau_1}(\eta^g\,(VAR\,v)))\right) \star^g \lambda e.\,\eta^g\,(LAM\,(v,e))
\end{aligned}$$

$$\begin{aligned}
\uparrow_\tau &\in T^g\mathbf{E} \to [\![\tau]\!]^{\mathcal{I}_r} \\
\uparrow_b &= \lambda\varepsilon.\,\varepsilon \\
\uparrow_{\tau_1\to\tau_2} &= \lambda\varepsilon.\,\lambda a.\uparrow_{\tau_2}\left(\varepsilon \star^g \lambda e.\downarrow^{\tau_1} a \star^g \lambda e'.\,\eta^g\,(APP\,(e,e'))\right)
\end{aligned}$$

(It may be helpful, on a first reading, to think of $\mathcal{T}_g$ as just the identity monad, and new_τ as "magically" generating fresh variable names; then the reification function simplifies to precisely the construction of *nf* sketched in Section 1.) Finally, we define the residualizing interpretation of constants by

$$\mathcal{C}_r(c) \in [\![\Sigma(c)]\!]^{\mathcal{I}_r}, \qquad \mathcal{C}_r(c) = \uparrow_{\Sigma(c)}(\eta^g\,(CST\,c))$$

The normalization function. To extract the syntactic normal form from the residualizing meaning of a term, we only need to supply a starting index for name generation. We can thus define an *extraction function*:

$$nf_\tau \in [\![\tau]\!]^{\mathcal{I}_r} \to T^l\mathbf{E}, \qquad nf_\tau = \lambda a.\,withct_{\mathbf{E}}\,(\downarrow^\tau a)$$

Finally, we define the (potentially partial) syntax-to-syntax function *norm* on closed terms $\vdash_\Sigma E : \tau$ by

$$norm(E) = \widetilde{E} \quad \text{iff} \quad nf_\tau([\![E]\!]^{\mathcal{I}_r}\,\emptyset) = \eta^l\,\ulcorner\widetilde{E}\urcorner$$

(We can find normal forms of open terms by explicitly lambda-abstracting over their free variables. The closed-term formulation leads to a particularly natural implementation, as sketched below.)

Theorem 1 (CBN semantic normalization). *Let $\vdash_\Sigma E : \tau$ be a closed Σ-term. Then (0) $\widetilde{E} = norm(E)$ is defined, (1) $\vdash^{nf} \widetilde{E} : \tau$, (2) $\models \widetilde{E} = E$, and (3) for all $\vdash_\Sigma E' : \tau$ such that $\models E' = E$, $norm(E') = \widetilde{E}$.*

If we already know that any Σ-term has a unique (up to α-conversion) $\beta\eta$-long normal form, the proof is fairly simple, using the argument sketched in the introduction. However, it is also possible to prove the theorem directly, using a suitable Kripke logical relation between the meanings of terms in the residualizing interpretation $\mathcal{I}_r$ and in an arbitrary one $\mathcal{I}$, with the base relation taken as the denotational meaning function. The details (for a more general setting, as sketched in Section 5) can be found in [11].

A normalization algorithm. The normalization function described above can be effectively computed as a program in any PCF-like functional language. The key idea is to express the residualizing semantic interpretation as a syntactic *realization* of all base types and constants in Σ in terms of types and terms of the programming-language signature Σ_{pl}, giving a substitution Φ^{r}, such that $[\![E\{\Phi^{\mathrm{r}}\}]\!]^{\mathcal{I}_{\mathrm{pl}}} = [\![E]\!]^{\mathcal{I}_{\mathrm{r}}}$. Likewise, for any Σ-type τ, we construct a term nf_τ such that $[\![\mathsf{nf}_\tau]\!]^{\mathcal{I}_{\mathrm{pl}}}\,\emptyset = nf_\tau$. Then for any closed Σ-term $\vdash_\Sigma E : \tau$, we can compute its normal form by evaluating the Σ_{pl}-program "$\mathsf{nf}_\tau\; E\{\Phi^{\mathrm{r}}\}$". Again, the details can be found in [11].

3 Normalization by Evaluation for Call by Value with Effects

We now refine the normalization result to a language based on Moggi's computational lambda-calculus λ_c [16], which provides a semantic framework for ML-like languages where "functions" may have effects such as mutating the state, performing input/output operations, or raising exceptions.

3.1 Language and Semantic Framework

Syntax and semantics. The syntax of types is the same as before. For terms, we also add a let-construct, with the usual typing rule:

$$\frac{\Gamma \vdash_\Sigma E_1 : \tau_1 \qquad \Gamma, x{:}\,\tau_1 \vdash_\Sigma E_2 : \tau_2}{\Gamma \vdash_\Sigma \mathbf{let}\; x = E_1 \;\mathbf{in}\; E_2 : \tau_2}$$

Now an interpretation $\mathcal{I}$ of a signature Σ consists a triple $(\mathcal{B}, \mathcal{T}, \mathcal{C})$. As before, $\mathcal{B}$ assigns cpos to base types of Σ. The new component $\mathcal{T} = (T, \eta, \star)$ is a monad used to model computational effects. These could be just divergence (modeled with the lifting monad), but also state, exceptions, continuations, etc.; the actual effectful operations are invoked through suitable constants from Σ.

We need to assume that $\mathcal{T}$ is layered atop $\mathcal{T}_1$ [10]; this amounts to requiring that TA is pointed for any A, and that $\lambda t.\, t \star f \in TA \to TB$ is strict for any $f \in A \to TB$. The CBV semantics of types is then given by:

$$[\![b]\!]^{\mathcal{I}}_{\mathrm{v}} = \mathcal{B}(b) \qquad\qquad [\![\tau_1 \to \tau_2]\!]^{\mathcal{I}}_{\mathrm{v}} = [\![\tau_1]\!]^{\mathcal{I}}_{\mathrm{v}} \to T[\![\tau_2]\!]^{\mathcal{I}}_{\mathrm{v}}$$

The meaning of a typing environment, $[\![\Gamma]\!]^{\mathcal{I}}_{\mathrm{v}}$, is a $(\mathrm{dom}\,\Gamma)$-indexed product of the meanings of the individual types, as before.

For the semantic function $\mathcal{C}$, we again require that for any $c \in \mathrm{dom}\,\Sigma$, $\mathcal{C}(c) \in [\![\Sigma(c)]\!]^{\mathcal{I}}_{\mathrm{v}}$. Then we define the meaning of a well-typed term $\Gamma \vdash_\Sigma E : \tau$ as a continuous function $[\![E]\!]^{\mathcal{I}}_{\mathrm{v}} \in [\![\Gamma]\!]^{\mathcal{I}}_{\mathrm{v}} \to T[\![\tau]\!]^{\mathcal{I}}_{\mathrm{v}}$, as follows:

$$\begin{aligned}
[\![c]\!]^{\cdot}_{\mathrm{v}}\,\rho &= \eta(\mathcal{C}(c)) & [\![\lambda x^\tau.\,E]\!]^{\cdot}_{\mathrm{v}}\,\rho &= \eta(\lambda a.\,[\![E]\!]^{\cdot}_{\mathrm{v}}\,(\rho[x \mapsto a])) \\
[\![x]\!]^{\cdot}_{\mathrm{v}}\,\rho &= \eta(\rho x) & [\![E_1\,E_2]\!]^{\cdot}_{\mathrm{v}}\,\rho &= [\![E_1]\!]^{\cdot}_{\mathrm{v}}\,\rho \star \lambda f.\,[\![E_2]\!]^{\cdot}_{\mathrm{v}}\,\rho \star \lambda a.\,f\,a
\end{aligned}$$
$$[\![\mathbf{let}\; x = E_1 \;\mathbf{in}\; E_2]\!]^{\cdot}_{\mathrm{v}}\,\rho = [\![E_1]\!]^{\cdot}_{\mathrm{v}}\,\rho \star \lambda a.\,[\![E_2]\!]^{\cdot}_{\mathrm{v}}\,(\rho[x \mapsto a])$$

Note that the **let**-construct appears redundant, because $[\![\mathbf{let}\; x = E_1 \;\mathbf{in}\; E_2]\!]^{\mathcal{I}}_{\mathrm{v}}\,\rho = [\![(\lambda x.\,E_2)\,E_1]\!]^{\mathcal{I}}_{\mathrm{v}}\,\rho$, but including it enables a nicer syntactic characterization of normal forms.

Equivalence and normal forms. Analogously to the CBN case, we write $\models_{\mathrm{v}} E = E'$ if for all $\mathcal{I}$, $[\![E]\!]^{\mathcal{I}}_{\mathrm{v}} = [\![E']\!]^{\mathcal{I}}_{\mathrm{v}}$. The shape of normal forms is now somewhat different, however: instead of normal and atomic forms, we have *normal values* and *normal computations*:

$$\frac{\Sigma(c) = b}{\Gamma \vdash^{\mathrm{nv}} c : b} \qquad \frac{\Gamma(x) = b}{\Gamma \vdash^{\mathrm{nv}} x : b} \qquad \frac{\Gamma, x{:}\,\tau_1 \vdash^{\mathrm{nc}} E : \tau_2}{\Gamma \vdash^{\mathrm{nv}} \lambda x^{\tau_1}.E : \tau_1 \to \tau_2}$$

$$\frac{\Gamma \vdash^{\mathrm{nv}} E : \tau}{\Gamma \vdash^{\mathrm{nc}} E : \tau} \qquad \frac{\Sigma(c) = \tau_1 \to \tau_2 \quad \Gamma \vdash^{\mathrm{nv}} E : \tau_1 \quad \Gamma, x{:}\,\tau_2 \vdash^{\mathrm{nc}} E^{\bullet} : \tau}{\Gamma \vdash^{\mathrm{nc}} \mathbf{let}\ x = c\,E\ \mathbf{in}\ E^{\bullet} : \tau}$$

$$\frac{\Gamma(x^{\bullet}) = \tau_1 \to \tau_2 \quad \Gamma \vdash^{\mathrm{nv}} E : \tau_1 \quad \Gamma, x{:}\,\tau_2 \vdash^{\mathrm{nc}} E^{\bullet} : \tau}{\Gamma \vdash^{\mathrm{nc}} \mathbf{let}\ x = x^{\bullet}\,E\ \mathbf{in}\ E^{\bullet} : \tau}$$

That is, a normal value is either a base-typed constant or variable, or of the form $\lambda x.\,\mathbf{let}\ x_1 = f_1\,V_1\ \mathbf{in}\ \cdots\ \mathbf{let}\ x_n = f_n\,V_n\ \mathbf{in}\ V$ where all the Vs are normal values, and each f_i is a function-typed constant or variable.

The set of normal-form terms is similar to Flanagan et al.'s *A-normal forms* [13]. However, their notion of A-reduction does not include even restricted β-conversion, so a term such as $(\lambda x.\,x)\,y$ is already A-normal. (Nor does it include η-like let-conversions: both $f\,x$ and $\mathbf{let}\ y = f\,x\ \mathbf{in}\ y$ are A-normal.) A much closer match is Ohori's language of *cut-free A-normal sequent proofs* [17], but still with one important difference: since we also care about uniqueness of normal forms, a variable is only a normal value in our sense if it is of base type; function-typed normal values must always be syntactic lambda-abstractions.

3.2 A Normalization Result

Term representations. Corresponding to the extended source syntax, we also assume given an additional constructor function $LET \in \mathbf{V} \times \mathbf{E} \times \mathbf{E} \to \mathbf{E}$, injective and with range disjoint from the others. The representation function for terms is also extended in the obvious way.

Residualizing monad. For constructing the residualizing interpretation, we now also need to pick a residualizing monad $\mathcal{T}_{\mathrm{r}}$. It is easy to see that we cannot simply use the lifting monad here, even if we only care about "purely functional" call-by-value languages, i.e., interpretations with $\mathcal{T}$ taken as lifting. The reason is that the two normal-form terms $E_1 = \lambda f^{b \to b}.\,\lambda x^b.\,\mathbf{let}\ y = f\,x\ \mathbf{in}\ x$ and $E_2 = \lambda f^{b \to b}.\,\lambda x^b.\,x$ are not semantically equivalent: for any $\mathcal{I}$ where $\mathcal{T}$ is the lifting monad, we only have $[\![E_1]\!]^{\mathcal{I}}_{\mathrm{v}} \sqsubseteq [\![E_2]\!]^{\mathcal{I}}_{\mathrm{v}}\,\rho$ (with the strictness of the inequality demonstrated by application of both sides to $\lambda a.\,\bot$). But $\ulcorner E_1 \urcorner \not\sqsubseteq \ulcorner E_2 \urcorner$, so there can be no *monotone* (let alone continuous) function $nf : [\![(b \to b) \to b \to b]\!]^{\mathcal{I}_{\mathrm{r}}} \to \mathbf{E}_{\bot}$ such that $nf([\![E_1]\!]^{\mathcal{I}_{\mathrm{r}}}\,\emptyset) = \eta^{1}\ulcorner E_1 \urcorner$ and $nf([\![E_2]\!]^{\mathcal{I}_{\mathrm{r}}}\,\emptyset) = \eta^{1}\ulcorner E_2 \urcorner$, if we require $\mathcal{I}_{\mathrm{r}}$ to use only the lifting monad to interpret computational effects.

Indeed, looking at the shape of normal computations, we see that the residualizing monad most allow us to register exactly where and when a function-typed constant or variable was applied, even if its return value is never used. We will show that it suffices that $\mathcal{T}_{\mathrm{r}}$ can be equipped with operations

$$bind_{\tau} \in \mathbf{E} \to T^{\mathrm{r}}\mathbf{V} \qquad \text{and} \qquad collect \in T^{\mathrm{r}}\mathbf{E} \to T^{\mathrm{g}}\mathbf{E}$$

satisfying the equational constraints

$$\begin{aligned} collect\,(\eta^{\mathrm{r}}\,e) &= \eta^{\mathrm{g}}\,e \\ collect\,(bind_{\tau}\,e \star^{\mathrm{r}} f) &= new_{\tau} \star^{\mathrm{g}} \lambda v.\,collect\,(f\,v) \star^{\mathrm{g}} \lambda e'.\,\eta^{\mathrm{g}}\,(LET\,(v,e,e')) \end{aligned}$$

These equations ensure that a $T^{\mathrm{r}}\mathbf{E}$-computation consisting of a sequence of calls to *bind* followed by returning a term has the effect of wrapping that term in a corresponding sequence of *LET*s:

$$\begin{aligned} &collect\,(bind_{\tau_1}\,e_1 \star^{\mathrm{r}} \lambda v_1.\cdots bind_{\tau_n}\,(e_n\,v_1\cdots v_{n-1}) \star^{\mathrm{r}} \lambda v_n.\,\eta^{\mathrm{r}}\,(e\,v_1\cdots v_n)) = \\ &\quad new_{\tau_1} \star^{\mathrm{g}} \lambda v_1.\cdots new_{\tau_n} \star^{\mathrm{g}} \lambda v_n.\,\eta^{\mathrm{g}}\,(LET\,(v_1,e_1,\cdots LET\,(v_n,e_n\,v_1\cdots v_{n-1},e\,v_1\cdots v_n))) \end{aligned}$$

To view T^{g}-computations as special cases of T^{r}-computations, we will also need a *monad morphism* from $\mathcal{T}_{\mathrm{g}}$ to $\mathcal{T}_{\mathrm{r}}$, i.e., a collection of functions $\gamma_A^{\mathrm{g,r}} \in T^{\mathrm{g}}A \to T^{\mathrm{r}}A$, such that $\gamma^{\mathrm{g,r}}\,(\eta^{\mathrm{g}}\,a) = \eta^{\mathrm{r}}\,a$ and $\gamma^{\mathrm{g,r}}\,(t \star^{\mathrm{g}} f) = \gamma^{\mathrm{g,r}}\,t \star^{\mathrm{r}} \lambda a.\,\gamma^{\mathrm{g,r}}\,(f\,a)$. We can construct a monad with these operations in several ways, notably including the following two:

The continuation monad with answer domain $T^{\mathrm{g}}\mathbf{E}$. We take:

$$\begin{aligned} T^{\mathrm{r}}A &= (A \to T^{\mathrm{g}}\mathbf{E}) \to T^{\mathrm{g}}\mathbf{E} & \gamma^{\mathrm{g,r}}\,t &= \lambda\kappa.\,t \star^{\mathrm{g}} \kappa \\ \eta^{\mathrm{r}}\,a &= \lambda\kappa.\,\kappa\,a & bind_{\tau}\,e &= \lambda\kappa.\,new_{\tau} \star^{\mathrm{g}} \lambda v.\,\kappa\,v \star^{\mathrm{g}} \lambda e'.\,\eta^{\mathrm{g}}\,(LET\,(v,e,e')) \\ t \star^{\mathrm{r}} f &= \lambda\kappa.\,t\,(\lambda a.\,f\,a\,\kappa) & collect\,t &= t\,\eta^{\mathrm{g}} \end{aligned}$$

This continuation-based wrapping of syntactic bindings was originally used for an "administrative-reduction free" continuation-passing transformation [9], and later adapted for a similar purpose in type-directed partial evaluation [7]. It is notable for also allowing an extension of CBV NBE to sum types (see Section 4).

The accumulation monad over the monoid of $(\mathbf{V}\times\mathbf{E})$*-lists.* Writing $[]$ for the empty list, $[-]$ for a singleton list, and @ for list concatenation, we take:

$$\begin{aligned} T^{\mathrm{r}}A &= T^{\mathrm{g}}(A \times (\mathbf{V}\times\mathbf{E})^{*}) & \gamma^{\mathrm{g,r}}\,t &= t \star^{\mathrm{g}} \lambda a.\,\eta^{\mathrm{g}}\,(a,[]) \\ \eta^{\mathrm{r}}\,a &= \eta^{\mathrm{g}}\,(a,[]) & bind_{\tau}\,e &= new_{\tau} \star^{\mathrm{g}} \lambda v.\,\eta^{\mathrm{g}}\,(v,[(v,e)]) \\ t \star^{\mathrm{r}} f &= t \star^{\mathrm{g}} \lambda(a,l).\,f\,a \star^{\mathrm{g}} \lambda(b,l').\,\eta^{\mathrm{g}}\,(b,l \mathbin{@} l') & collect\,t &= t \star^{\mathrm{g}} \lambda(e,l).\,\eta^{\mathrm{g}}\,(wrap\,l\,e) \end{aligned}$$

with the auxiliary function $wrap : (\mathbf{V}\times\mathbf{E})^{*} \to \mathbf{E} \to \mathbf{E}$ defined inductively as

$$wrap\,[]\,e = e \qquad wrap\,([(v,e')] \mathbin{@} l)\,e = LET\,(v,e',wrap\,l\,e)$$

This choice is a refinement of state-based TDPE [18]; see the end of this section for a brief account of the relationship between accumulation and state. Other constructions are also possible, such as accumulation with respect to the monoid $(\mathbf{E}\to\mathbf{E}, id, \circ)$.

Residualizing interpretation. For the residualizing interpretation, we again interpret all base types of Σ as syntactic lambda-terms; this time, however, we do not need to involve the name-generation monad yet, but simply take:

$$\mathcal{B}_{\mathrm{r}}(b) = \mathbf{E}$$

For any definition of $\mathcal{T}_\mathrm{r}$ satisfying the equational constraints on *bind* and *collect*, we can then define new reification and reflection functions:

$$\begin{aligned}
\downarrow^{\tau} &\in [\![\tau]\!]_\mathrm{v}^{\bullet,\mathrm{r}} \to T^\mathrm{g}\mathbf{E} \\
\downarrow^{b} &= \lambda e.\, \eta^\mathrm{g}\, e \\
\downarrow^{\tau_1 \bullet\ \tau_2} &= \lambda f.\, \mathit{new}_{\tau_1} \star^\mathrm{g} \lambda v.\, \mathit{collect}\, (\uparrow_{\tau_1} (\mathit{VAR}\, v) \star^\mathrm{r} \lambda a.\, f\, a \star^\mathrm{r} \lambda b.\, \gamma^{\mathrm{g,r}} (\downarrow^{\tau_2} b)) \star^\mathrm{g} \lambda e. \\
&\qquad \eta^\mathrm{g} (\mathit{LAM}\, (v, e))
\end{aligned}$$

$$\begin{aligned}
\uparrow_{\tau} &\in \mathbf{E} \to T^\mathrm{r} [\![\tau]\!]_\mathrm{v}^{\bullet,\mathrm{r}} \\
\uparrow_{b} &= \lambda e.\, \eta^\mathrm{r}\, e \\
\uparrow_{\tau_1 \bullet\ \tau_2} &= \lambda e.\, \eta^\mathrm{r}\, (\lambda a.\, \gamma^{\mathrm{g,r}} (\downarrow^{\tau_1} a) \star^\mathrm{r} \lambda e^{\bullet}.\, \mathit{bind}_{\tau_2} (\mathit{APP}\, (e, e^{\bullet})) \star^\mathrm{r} \lambda v.\, \uparrow_{\tau_2} (\mathit{VAR}\, v))
\end{aligned}$$

(The codomain of $\uparrow_\tau$ is $T^\mathrm{r} [\![\tau]\!]_\mathrm{v}^{\mathcal{I}_\mathrm{r}}$, rather than simply $[\![\tau]\!]_\mathrm{v}^{\mathcal{I}_\mathrm{r}}$, to accommodate the extensions in Section 4.) Note in particular how every construction of an *APP*-term is wrapped in a *bind*.

Finally, as for call by name, we interpret all Σ-constants as reflected *CST*-constructors:

$$\mathcal{C}_\mathrm{r}(c) \in [\![\Sigma(c)]\!]_\mathrm{v}^{\mathcal{I}_\mathrm{r}}, \qquad \mathcal{C}_\mathrm{r}(c) = a, \text{ where } \uparrow_{\Sigma(c)} (\mathit{CST}\, c) = \eta^\mathrm{r}\, a$$

($\mathcal{C}_\mathrm{r}(c)$ is well defined, because the reflection function factors through the injective η^r.) We also define the extraction function essentially as before:

$$\mathit{nf}_\tau \in [\![\tau]\!]_\mathrm{v}^{\mathcal{I}_\mathrm{r}} \to T^\mathrm{l}\mathbf{E}, \qquad \mathit{nf}_\tau = \lambda a.\, \mathit{withct}_\mathbf{E} (\downarrow^{\tau} a)$$

and the CBV normalization function for a closed value (constant or lambda-abstraction) E as

$$\mathit{norm}_\mathrm{v}(E) = \widetilde{E} \quad \text{iff} \quad \mathit{nf}_\tau\, a = \eta^\mathrm{l} \ulcorner \widetilde{E} \urcorner, \text{ where } [\![E]\!]_\mathrm{v}^{\mathcal{I}_\mathrm{r}}\, \emptyset = \eta^\mathrm{r}\, a.$$

(We can find the normal form of a non-value term by wrapping a dummy lambda-abstraction around it.)

Theorem 2 (CBV semantic normalization). *Let* $\vdash_\Sigma E : \tau$ *be a value, and take* $\widetilde{E} = \mathit{norm}_\mathrm{v}(E)$*. Then (0)* $\widetilde{E}$ *is defined, (1)* $\vdash^{\mathrm{nv}} \widetilde{E} : \tau$*, (2)* $\models_\mathrm{v} \widetilde{E} = E$*, and (3) if* $\models_\mathrm{v} E' = E$ *then* $\mathit{norm}_\mathrm{v}(E') = \widetilde{E}$*.*

The proof is similar to the CBN case, but using a pair of mutually inductively defined logical relations, one for values and one for computations. Very roughly, one again establishes that the residualizing and the arbitrary interpretations of all terms are related, and that for a pair of related values $(a, a') \in [\![\tau]\!]_\mathrm{v}^{\mathcal{I}_\mathrm{r}} \times [\![\tau]\!]_\mathrm{v}^{\mathcal{I}}$ the $\mathcal{I}$-meaning of $\downarrow^{\tau} a$ equals a'.

A normalization algorithm. Phrasing the normalization function as a functional program is a bit more complicated than for the CBN case. A typical CBV host language will have its own notion of effects, modeled by some monad $\mathcal{T}_\mathrm{pl}$, which is not likely to be exactly our residualizing monad $\mathcal{T}_\mathrm{r}$. However, much as we can embed the normalization algorithm into a host language with a signature much larger than what we need for the construction, we can realize both the residualizing and the name-generating monad through a uniform *effect-embedding* into a more general notion of effect in the host language [10].

For example, the accumulation-based choice of $\mathcal{T}_r$ can be easily (and more efficiently) implemented by passing around the bindings accumulated so far in a mutable state cell, rather than appending the bindings from both subcomputations in $\star$. The current name-generation index is naturally kept in another cell. An analogous, but somewhat more involved, construction also allows us to simulate $\mathcal{T}_r$ taken as a continuation monad, provided the host language provides both (higher-typed) state and first-class continuations, as found in Scheme or SML/NJ. The resulting implementation forms the basis of the CBV normalization algorithm used in the context of type-directed partial evaluation [7].

4 Structured Data Types

In this section, we consider normalization for the call-by-value language extended with product and sum types. (Adding products to the call-by-name language is trivial, but it does not appear possible to add even weak sum types, at least in the domain-theoretic semantics.)

Syntax. We extend the set of types by two new type constructors:

$$\tau ::= \cdots \mid \tau_1 \times \tau_2 \mid \tau_1 + \tau_2$$

(The generalizations to n-ary ($n \geq 0$) products and sums are completely straightforward and thus omitted.) The associated new terms are:

$$\frac{\Gamma \vdash_\Sigma E_1 : \tau_1 \quad \Gamma \vdash_\Sigma E_2 : \tau_2}{\Gamma \vdash_\Sigma (E_1, E_2) : \tau_1 \times \tau_2} \qquad \frac{\Gamma \vdash_\Sigma E : \tau_1 \times \tau_2 \quad \Gamma, x_1{:}\tau_1, x_2{:}\tau_2 \vdash_\Sigma E' : \tau}{\Gamma \vdash_\Sigma \mathbf{split}\,(E, x_1.\,x_2.\,E') : \tau}$$

$$\frac{\Gamma \vdash_\Sigma E : \tau_1}{\Gamma \vdash_\Sigma \mathbf{inl}\,(E) : \tau_1 + \tau_2} \qquad \frac{\Gamma \vdash_\Sigma E : \tau_2}{\Gamma \vdash_\Sigma \mathbf{inr}\,(E) : \tau_1 + \tau_2}$$

$$\frac{\Gamma \vdash_\Sigma E : \tau_1 + \tau_2 \quad \Gamma, x_1{:}\tau_1 \vdash_\Sigma E_1 : \tau \quad \Gamma, x_2{:}\tau_2 \vdash_\Sigma E_2 : \tau}{\Gamma \vdash_\Sigma \mathbf{case}\,(E, x_1.\,E_1, x_2.\,E_2) : \tau}$$

(Instead of **split**, we could have used explicit projections, but the characterization of normal forms becomes more uniform with **split**. In practice, the separate split-construct above is usually folded into pattern-matching let- and lambda-bindings.)

Semantics. The semantics of the type constructors is standard:

$$\begin{aligned}
[\![\tau_1 \times \tau_2]\!]^{\bullet}_{\mathrm{v}} &= [\![\tau_1]\!]^{\bullet}_{\mathrm{v}} \times [\![\tau_2]\!]^{\bullet}_{\mathrm{v}} = \{(a_1, a_2) \mid a_1 \in [\![\tau_1]\!]^{\bullet}_{\mathrm{v}}, a_2 \in [\![\tau_1]\!]^{\bullet}_{\mathrm{v}}\} \\
[\![\tau_1 + \tau_2]\!]^{\bullet}_{\mathrm{v}} &= [\![\tau_1]\!]^{\bullet}_{\mathrm{v}} + [\![\tau_2]\!]^{\bullet}_{\mathrm{v}} = \{\iota_1 a \mid a \in [\![\tau_1]\!]^{\bullet}_{\mathrm{v}}\} \cup \{\iota_2 a \mid a \in [\![\tau_2]\!]^{\bullet}_{\mathrm{v}}\}
\end{aligned}$$

as is the semantics of the associated terms:

$$\begin{aligned}
[\![(E_1, E_2)]\!]^{\bullet}_{\mathrm{v}}\, \rho &= [\![E_1]\!]^{\bullet}_{\mathrm{v}}\, \rho \star \lambda a_1.\, [\![E_2]\!]^{\bullet}_{\mathrm{v}}\, \rho \star \lambda a_2.\, \eta\,(a_1, a_2) \\
[\![\mathbf{split}\,(E, x_1.\,x_2.\,E')]\!]^{\bullet}_{\mathrm{v}}\, \rho &= [\![E]\!]^{\bullet}_{\mathrm{v}}\, \rho \star \lambda(a_1, a_2).\, [\![E']\!]^{\bullet}_{\mathrm{v}}\, (\rho[x_1 \mapsto a_1, x_2 \mapsto a_2])
\end{aligned}$$

$$[\![\mathbf{inl}\,(E)]\!]^{\bullet}_{\mathrm{v}}\,\rho = [\![E]\!]^{\bullet}_{\mathrm{v}}\,\rho \star \lambda a.\,\eta\,(\iota_1\,a)$$
$$[\![\mathbf{inr}\,(E)]\!]^{\bullet}_{\mathrm{v}}\,\rho = [\![E]\!]^{\bullet}_{\mathrm{v}}\,\rho \star \lambda a.\,\eta\,(\iota_2\,a)$$
$$[\![\mathbf{case}\,(E, x_1.\,E_1, x_2.\,E_2)]\!]^{\bullet}_{\mathrm{v}}\,\rho = [\![E]\!]^{\bullet}_{\mathrm{v}}\,\rho \star \lambda s.\begin{cases} [\![E_1]\!]^{\bullet}_{\mathrm{v}}\,(\rho[x_1 \mapsto a_1]) & \text{if } s = \iota_1\,a_1 \\ [\![E_2]\!]^{\bullet}_{\mathrm{v}}\,(\rho[x_2 \mapsto a_2]) & \text{if } s = \iota_2\,a_2 \end{cases}$$

Normal forms. With the addition of product and sum types, CBV normal forms exhibit a striking similarity with cut-free proofs in Gentzen-style intuitionistic sequent calculus, as also noted by Ohori [17]. In fact, we also get the usual sequent-calculus inconvenience of having to make arbitrary choices about the order in which we apply left-rules to decompose the types of variables. To keep normal forms unique, we choose to eliminate structured-type variables immediately as they are introduced, in a stack-like manner. (Immediate elimination leads to a slight anomaly for constants introduced by the signature: we can only allow Σ to declare constants of base and top-level-functional types. In the rare cases where we need, e.g., a sum-typed constant $c : \tau_1 + \tau_2$ in Σ, it can be provided as a function $c' : 1 \to \tau_1 + \tau_2$ where 1 is the zero-ary product type.)

We now have three mutually recursive notions of normality: normal values $\Gamma \vdash^{\mathrm{nv}} E : \tau$, normal computations $\Gamma \vdash^{\mathrm{nc}} E : \tau$, and normal bodies $\Gamma \mid \Theta \vdash^{\mathrm{nb}} E : \tau$, where Θ is an *ordered* list of typing assumptions:

$$\frac{\Sigma(c) = b}{\Gamma \vdash^{\mathrm{nv}} c : b} \qquad \frac{\Gamma(x) = b}{\Gamma \vdash^{\mathrm{nv}} x : b} \qquad \frac{\Gamma \mid x{:}\,\tau_1 \vdash^{\mathrm{nb}} E : \tau_2}{\Gamma \vdash^{\mathrm{nv}} \lambda x^{\tau_1}.\,E : \tau_1 \to \tau_2}$$

$$\frac{\Gamma \vdash^{\mathrm{nv}} E_1 : \tau_1 \quad \Gamma \vdash^{\mathrm{nv}} E_2 : \tau_2}{\Gamma \vdash^{\mathrm{nv}} (E_1, E_2) : \tau_1 \times \tau_2} \qquad \frac{\Gamma \vdash^{\mathrm{nv}} E : \tau_1}{\Gamma \vdash^{\mathrm{nv}} \mathbf{inl}\,(E) : \tau_1 + \tau_2} \qquad \frac{\Gamma \vdash^{\mathrm{nv}} E : \tau_2}{\Gamma \vdash^{\mathrm{nv}} \mathbf{inr}\,(E) : \tau_1 + \tau_2}$$

$$\frac{\Gamma \vdash^{\mathrm{nv}} E : \tau}{\Gamma \vdash^{\mathrm{nc}} E : \tau} \qquad \frac{\Sigma(c) = \tau_1 \to \tau_2 \quad \Gamma \vdash^{\mathrm{nv}} E : \tau_1 \quad \Gamma \mid x{:}\,\tau_2 \vdash^{\mathrm{nb}} E' : \tau}{\Gamma \vdash^{\mathrm{nc}} \mathbf{let}\ x = c\,E\ \mathbf{in}\ E' : \tau}$$

$$\frac{\Gamma(x') = \tau_1 \to \tau_2 \quad \Gamma \vdash^{\mathrm{nv}} E : \tau_1 \quad \Gamma \mid x{:}\,\tau_2 \vdash^{\mathrm{nb}} E' : \tau}{\Gamma \vdash^{\mathrm{nc}} \mathbf{let}\ x = x'\,E\ \mathbf{in}\ E' : \tau}$$

$$\frac{\Gamma \vdash^{\mathrm{nc}} E : \tau}{\Gamma \mid \ \vdash^{\mathrm{nb}} E : \tau} \qquad \frac{\Gamma, x{:}\,b \mid \Theta \vdash^{\mathrm{nb}} E : \tau}{\Gamma \mid \Theta, x{:}\,b \vdash^{\mathrm{nb}} E : \tau} \qquad \frac{\Gamma, x{:}\,\tau_1 \to \tau_2 \mid \Theta \vdash^{\mathrm{nb}} E : \tau}{\Gamma \mid \Theta, x{:}\,\tau_1 \to \tau_2 \vdash^{\mathrm{nb}} E : \tau}$$

$$\frac{\Gamma \mid \Theta, x_1{:}\,\tau_1, x_2{:}\,\tau_2 \vdash^{\mathrm{nb}} E : \tau}{\Gamma \mid \Theta, x{:}\,\tau_1 {\times} \tau_2 \vdash^{\mathrm{nb}} \mathbf{split}\,(x, x_1.\,x_2.\,E) : \tau} \qquad \frac{\Gamma \mid \Theta, x_1{:}\,\tau_1 \vdash^{\mathrm{nb}} E_1 {:}\, \tau \quad \Gamma \mid \Theta, x_2{:}\,\tau_2 \vdash^{\mathrm{nb}} E_2 {:}\, \tau}{\Gamma \mid \Theta, x{:}\,\tau_1 {+} \tau_2 \vdash^{\mathrm{nb}} \mathbf{case}\,(x, x_1.\,E_1, x_2.\,E_2) : \tau}$$

Note how newly-introduced variables are put into the quarantined context Θ, where their types are decomposed and the pieces migrate back into the ordinary context Γ. In particular, without the rules for product and sum types, the new definitions of normal values and computations agree exactly with the original ones.

Normalization by evaluation. We show only how the continuation-based residualizing interpretation can be extended. Products could be added to an accumulation-based interpretation without too much trouble, but sums apparently require the full power of applying a single continuation multiple times.

We assume the syntax-constructor functions for **E** are extended with functions *PAIR*, *SPLIT*, *INL*, *INR*, and *CASE* with the obvious types. Then we can define additional helper functions analogous to $bind_\tau$ from before:

$$\begin{aligned} bindp_{\tau_1,\tau_2} &\in \mathbf{E} \to T^{\mathrm{r}}(\mathbf{V}\times\mathbf{V}) \\ &= \lambda e.\,\lambda\kappa.\,new_{\tau_1} \star^{\mathrm{g}} \lambda v_1.\,new_{\tau_2} \star^{\mathrm{g}} \lambda v_2.\,\kappa(v_1,v_2) \star^{\mathrm{g}} \lambda e^{\bullet}.\,\eta^{\mathrm{g}}(SPLIT(e,v_1,v_2,e^{\bullet})) \\ binds_{\tau_1,\tau_2} &\in \mathbf{E} \to T^{\mathrm{r}}(\mathbf{V}+\mathbf{V}) \\ &= \lambda e.\,\lambda\kappa.\,new_{\tau_1} \star^{\mathrm{g}} \lambda v_1.\,new_{\tau_2} \star^{\mathrm{g}} \lambda v_2.\,\kappa(\iota_1 v_1) \star^{\mathrm{g}} \lambda e_1.\,\kappa(\iota_2 v_2) \star^{\mathrm{g}} \lambda e_2. \\ &\qquad \eta^{\mathrm{g}}(CASE(e,v_1,e_1,v_2,e_2)) \end{aligned}$$

and the corresponding cases for the reification and reflection functions, extending the ones from Section 2.2:

$$\begin{aligned} \downarrow^{\tau_1 \bullet \tau_2} &= \lambda(a_1,a_2).\downarrow^{\tau_1} a_1 \star^{\mathrm{g}} \lambda e_1.\downarrow^{\tau_2} a_2 \star^{\mathrm{g}} \lambda e_2.\,\eta^{\mathrm{g}}(PAIR(e_1,e_2)) \\ \downarrow^{\tau_1+\tau_2} &= \lambda s.\begin{cases} \downarrow^{\tau_1} a_1 \star^{\mathrm{g}} \lambda e_1.\,\eta^{\mathrm{g}}(INL\, e_1) & \text{if } s = \iota_1 a_1 \\ \downarrow^{\tau_2} a_2 \star^{\mathrm{g}} \lambda e_2.\,\eta^{\mathrm{g}}(INR\, e_2) & \text{if } s = \iota_2 a_2 \end{cases} \\ \uparrow_{\tau_1 \bullet \tau_2} &= \lambda e.\,bindp_{\tau_1,\tau_2}\, e \star^{\mathrm{r}} \lambda(v_1,v_2).\uparrow_{\tau_1}(VAR\, v_1) \star^{\mathrm{r}} \lambda a_1.\uparrow_{\tau_2}(VAR\, v_2) \star^{\mathrm{r}} \lambda a_2.\,\eta^{\mathrm{r}}(a_1,a_2) \\ \uparrow_{\tau_1+\tau_2} &= \lambda e.\,binds_{\tau_1,\tau_2}\, e \star^{\mathrm{r}} \lambda s.\begin{cases} \uparrow_{\tau_1}(VAR\, v_1) & \text{if } s = \iota_1 v_1 \\ \uparrow_{\tau_2}(VAR\, v_2) & \text{if } s = \iota_2 v_2 \end{cases} \end{aligned}$$

The residualizing interpretation is as before. (Note that we cannot give a residualizing interpretation to constants of top-level product or sum type, since the reflection function does *not* factor through η^{r} in these cases.) The normalization function and Theorem 2 remain the same. For the implementation, the realizations of *bindp*, *binds*, and the new clauses for reification and reflection in terms of continuation-manipulating primitives are straightforward.

5 Type-Directed Partial Evaluation

A primary application of normalization by evaluation is for type-directed partial evaluation (TDPE) [7]. The goal is to simplify a partially applied function of multiple arguments by propagating the values of the known arguments throughout the body of the function. Here, in addition to eliminating β-redexes, we also want to simplify occurrences of constants, such as arithmetic operations, when they are applied to literal values. In other words, we now want to normalize terms with *interpreted* base types and constants. There are in fact two natural ways to achieve this, both expressible in terms of the notion of constrained interpretations:

Offline TDPE. For offline TDPE [7, Section 3], we say that a program is *binding-time separated* if it is expressed over a signature Σ partitioned into a static and a dynamic part, Σ_{s} and Σ_{d}, each containing some type and term constants. The interpretation is likewise partitioned into $\mathcal{I}_{\mathrm{s}}$ and $\mathcal{I}_{\mathrm{d}}$. We then constrain the allowable interpretations so that $\mathcal{I}_{\mathrm{s}}$ is always the standard interpretation (i.e., int as $\mathbf{Z}$, $+$ as addition, fix as the domain-theoretic least fixed point, etc.), while the dynamic part remains completely unconstrained. That is, we consider the notion of *static equivalence*, $\models^{Int_{\mathrm{s}}} E = E'$ where $Int_{\mathrm{s}} = \{\mathcal{I} \mid \mathcal{I}|_{\Sigma_{\mathrm{s}}} = \mathcal{I}_{\mathrm{s}}\}$. When the static normal form $\widetilde{E}$ of a term E exists, it is then equivalent to E with respect to all interpretations of Σ_{d}. The residualizing interpretation of Σ, like any interpretation in Int_{s}, also uses the standard interpretation of Σ_{s}, and the syntax-reconstructing interpretation from Section 3.2 for $\mathcal{I}_{\mathrm{d}}$.

The separation allows us to realize the static part of the signature completely natively in terms of the corresponding construct of the programming language, and in fact we can use syntactic conveniences of the host language, such as pattern matching or **letrec**-forms directly for the static computations. It also becomes possible to self-apply the partial evaluator (the so-called second Futamura projection) [15].

Online TDPE. In the online variant [7, Section 4], like in general online partial evaluation, we do not annotate types, nor most occurrences of constants, with their binding times. Instead, the partial evaluator "opportunistically" propagates statically known data and performs reductions such as $2+3 \to 5$.

This corresponds to normalizing with respect to the set Int_{c} of interpretations $(\mathcal{B}, \mathcal{C})$ that satisfy constraints such as $\mathcal{C}(+)\,(\mathcal{C}(2), \mathcal{C}(3)) = \mathcal{C}(5)$, and possibly also additional ones, such as $\mathcal{C}(+)\,(x, \mathcal{C}(0)) = x$ for all x in $\mathcal{B}(\mathsf{int})$. Again, the standard interpretation of $\mathcal{C}(+)$ satisfies these constraints automatically. The residualizing one includes explicit checks for the reducible cases, to avoid constructing the corresponding redexes in the generated code. This formulation is similar to recent work on merging the reduction-free normalization of function abstraction and application with explicit reduction rules for constants [3].

The advantage of the online approach is that we do not have to explicitly separate the binding times in the source program, but correspondingly it becomes less predictable how much of the source program can be simplified at partial-evaluation time. An online normalizer also requires all primitive operations to explicitly check whether their arguments are literals (so the operation can be eliminated) or more general expressions (so the operation must remain in the normal form), slowing down the specialization process somewhat. Finally, fixed-point operators must still either be explicitly classified as static or dynamic, or need some ad hoc mechanism for deciding whether their unfolding equation should be applied.

In both cases, the normalization function may be partial, i.e., without part (0) of Theorems 1 and 2. This is unavoidable, since in the presence of recursion, some terms simply have no normal form. However, for the CBN case, one can still show that when it is defined, $\widetilde{E}$ satisfies parts (1-3), and also that whenever a $\widetilde{E}$ satisfying (1-3) exists, $norm(E)$ is defined [11]. The situation for CBV TDPE has not been fully analyzed yet, although it seems reasonable to conjecture an analogous result.

In any case, the semantic treatment of TDPE allows us to uniformly analyze the construction and state its correctness criterion independently of the details of its implementation. That is, we can think about partial evaluation in terms of normalization with respect to a class of interpretations, without worrying about whether the normalization is achieved through repeated reductions, or through reduction-less normalization by evaluation.

6 Conclusions and Future Work

We have seen that the same basic idea that allows us to compute normal forms of lambda-terms with respect to purely functional interpretations also allows us to compute such normal forms with respect to general computational interpretations. In both cases, we chose a "quasi-syntactic" interpretation of the types and constants, and in the latter, also a binding-accumulating monad as the interpretation of computational effects. Both variants of semantic normalization can be phrased as functional-program evaluation, although the construction is significantly more involved in the computational case.

An important application of the normalization construction is type-directed partial evaluation, which seeks to compute normal forms not with respect to *all* interpretations of a signature, but only with respect to a subset of those; different choices of such subsets lead to offline and online partial evaluation. The offline, call-by-name case is analyzed in isolation in an earlier paper [11], but the more general constrained-interpretation formulation presented here seems worth investigating further, especially since it also leads to natural TDPE formulations of other partial-evaluation concepts, such as polyvariant program-point specialization expressed as suitable constraints on the recursion operator.

Additionally, it should be possible to obtain a syntactic analog of the CBV normalization result, showing that the computed normal form is not only equivalent to the original term with respect to arbitrary set-, or domain-theoretic interpretations, but is provably equal to it using the axioms of the computational lambda-calculus [16]. However, it seems that the semantic characterization may be ultimately more convenient for reasoning about programs, since it seems to scale more directly to partially constrained interpretations, and especially static recursion.

Acknowledgments. The author wishes to thank Olivier Danvy and Peter Dybjer for many fruitful discussions of the topics presented here, and the anonymous TLCA'01 reviewers for numerous helpful suggestions (of which regrettably not all could be followed here, due to space constraints.)

References

1. Thorsten Altenkirch, Martin Hofmann, and Thomas Streicher. Categorical reconstruction of a reduction free normalization proof. In *Category Theory and Computer Science, 6th International Conference*, number 953 in Lecture Notes in Computer Science, 1995.
2. Ulrich Berger. Program extraction from normalization proofs. In M. Bezem and J. F. Groote, editors, *Typed Lambda Calculi and Applications*, number 664 in Lecture Notes in Computer Science, pages 91–106, Utrecht, The Netherlands, March 1993.
3. Ulrich Berger, Matthias Eberl, and Helmut Schwichtenberg. Normalization by evaluation. In *Prospects for Hardware Foundations (NADA)*, number 1546 in Lecture Notes in Computer Science, pages 117–137, 1998.

4. Ulrich Berger and Helmut Schwichtenberg. An inverse of the evaluation functional for typed λ-calculus. In *Proceedings of the Sixth Annual IEEE Symposium on Logic in Computer Science*, pages 203–211, Amsterdam, The Netherlands, July 1991.
5. Thierry Coquand and Peter Dybjer. Intuitionistic model constructions and normalization proofs. *Mathematical Structures in Computer Science*, 7:75–94, 1997.
6. Djordje Čubrić, Peter Dybjer, and Philip Scott. Normalization and the Yoneda embedding. *Mathematical Structures in Computer Science*, 8:153–192, 1998.
7. Olivier Danvy. Type-directed partial evaluation. In J. Hatcliff, T. Æ. Mogensen, and P. Thieman, editors, *Partial Evaluation – Practice and Theory; Proceedings of the 1998 DIKU Summer School*, number 1706 in Lecture Notes in Computer Science, pages 367–411. Springer-Verlag, Copenhagen, Denmark, July 1998.
8. Olivier Danvy and Peter Dybjer, editors. *Preliminary Proceedings of the APPSEM Workshop on Normalization by Evaluation.* Department of Computer Science, University of Aarhus, May 1998. BRICS Note NS-98-1.
9. Olivier Danvy and Andrzej Filinski. Representing control: A study of the CPS transformation. *Mathematical Structures in Computer Science*, 2(4):361–391, December 1992.
10. Andrzej Filinski. Representing layered monads. In *Proceedings of the 26th ACM SIGPLAN-SIGACT Symposium on Principles of Programming Languages*, pages 175–188, San Antonio, Texas, January 1999.
11. Andrzej Filinski. A semantic account of type-directed partial evaluation. In G. Nadathur, editor, *International Conference on Principles and Practice of Declarative Programming*, number 1702 in Lecture Notes in Computer Science, pages 378–395, Paris, France, September 1999.
12. Marcelo Fiore, Gordon Plotkin, and Daniele Turi. Abstract syntax and variable binding. In *Proceedings of the 14th Annual IEEE Symposium on Logic in Computer Science*, pages 193–202, Trento, Italy, July 1999.
13. Cormac Flanagan, Amr Sabry, Bruce F. Duba, and Matthias Felleisen. The essence of compiling with continuations. In *Proceedings of the SIGPLAN '93 Conference on Programming Language Design and Implementation*, pages 237–247, Albuquerque, New Mexico, June 1993.
14. Murdoch Gabbay and Andrew Pitts. A new approach to abstract syntax involving binders. In *Proceedings of the 14th Annual IEEE Symposium on Logic in Computer Science*, pages 214–224, Trento, Italy, July 1999.
15. Bernd Grobauer and Zhe Yang. The second Futamura projection for type-directed partial evaluation. In *ACM SIGPLAN Workshop on Partial Evaluation and Semantics-Based Program Manipulation*, pages 22–32. ACM Press, January 2000.
16. Eugenio Moggi. Computational lambda-calculus and monads. In *Proceedings of the Fourth Annual Symposium on Logic in Computer Science*, pages 14–23, Pacific Grove, California, June 1989. IEEE.
17. Atsushi Ohori. A Curry-Howard isomorphism for compilation and program execution. In J.-Y. Girard, editor, *Typed Lambda-Calculi and Applications*, volume 1581 of *Lecture Notes in Computer Science*, L'Aquila, Italy, April 1999.
18. Eijiro Sumii and Naoki Kobayashi. Online-and-offline partial evaluation: A mixed approach. In *ACM SIGPLAN Workshop on Partial Evaluation and Semantics-Based Program Manipulation*, pages 12–21. ACM Press, January 2000.

Induction Is Not Derivable in Second Order Dependent Type Theory

Herman Geuvers*

Department of Computer Science, University of Nijmegen, The Netherlands

Abstract. This paper proves the non-derivability of induction in second order dependent type theory ($\lambda P2$). This is done by providing a model construction for $\lambda P2$, based on a saturated sets like interpretation of types as sets of terms of a weakly extensional combinatory algebra. We give counter-models in which the induction principle over natural numbers is not valid. The proof does not depend on the specific encoding for natural numbers that has been chosen (like e.g. polymorphic Church numerals), so in fact we prove that there can not be an encoding of natural numbers in $\lambda P2$ such that the induction principle is satisfied. The method extends immediately to other data types, like booleans, lists, trees, etc.

In the process of the proof we establish some general properties of the models, which we think are of independent interest. Moreover, we show that the Axiom of Choice is not derivable in $\lambda P2$.

1 Introduction

In second order dependent type theory, $\lambda P2$, we can encode all kinds of inductive data types, like the types of natural numbers, lists, trees etcetera. This is usually done via the Böhm-Berarducci encoding (see [Girard et al. 1989] for a general exposition), which yields e.g. the well-known polymorphic Church numerals as interpretation of the natural numbers. This encoding already works for non-dependent second order type theory (the well-known polymorphic λ-calculus $\lambda 2$), but dependent types give the extra advantage that we can also state the *induction principle* for the inductive data types. For example, if nat is the type of polymorphic Church numerals with zero 0 and successor function succ, then the induction principle is represented by the type ind defined as

$$\mathsf{ind} := \Pi P{:}\mathsf{nat}\to\star.(P\mathsf{0})\to(\Pi y{:}\mathsf{nat}.(Py)\to(P(\mathsf{succ}\,y)))\to\Pi x{:}\mathsf{nat}.(Px).$$

Here, $\star$ denotes the 'kind' (universe) of all types, which captures both the sets ($\mathsf{nat}:\star$) and the propositions ($\mathsf{ind}:\star$). The induction principle for nat is said to be *derivable* in $\lambda P2$ if there is a closed term of type ind.

In this paper we show that the induction principle for nat is not derivable in $\lambda P2$. As a matter of fact, we prove something stronger: the non-derivability of

* email: herman@cs.kun.nl, fax: +31 24 3652525

S. Abramsky (Ed.): TLCA 2001, LNCS 2044, pp. 166–181, 2001.

induction does not depend on the specific choice of the encoding of the natural numbers: given any (closed) type N with $0 : N$ and $s : N \to N$, there can be no closed term of type $\Pi P{:}N\to\star.(P0)\to(\Pi y{:}N.(Py)\to(P(sy)))\to\Pi x{:}N.(Px)$. This rules out any 'smart' encoding of the natural numbers (like the N above) for which induction would be provable in $\lambda P2$. What a 'smart encoding' could possibly look like, see the small diversion below in 1.1.

It should be pointed out here that, of course, inductive reasoning can easily represented in $\lambda P2$ by 'relativizing' all statements about nat to the inductive natural numbers. If we let $\mathsf{Ind}\,x$ say that x is an 'inductive natural number', defined in $\lambda P2$ as follows,

$$\mathsf{Ind}\,x := \Pi P{:}\mathsf{nat}\to\star.(P\mathsf{O})\to(\Pi y{:}\mathsf{nat}.(Py)\to(P(\mathsf{succ}y)))\to(Px),$$

we can relativize $\Pi x{:}\mathsf{nat}.\varphi$ to $\Pi x{:}\mathsf{nat}.(\mathsf{Ind}\,x)\to\varphi$. Then one can reason by induction, just because all statements about nat are restricted to the inductive natural numbers. However, this does not give us an *inductive type* of natural numbers.

Our result extends immediately to other inductive data types, so induction is not derivable for any encoding of any inductive data type in $\lambda P2$. Also we show in this paper that the induction principle for one data type can not be derived from the induction principle for another data type. The results extend immediately to other systems like the Calculus of Constructions (without inductive types). In [Streicher 1991], also a non-derivability induction result is proved, using a realizability semantics, but only for one specific encoding of the natural numbers, as polymorphic Church numerals. Our proof of non-derivability uses a fairly simple model construction which originates from [Geuvers 1996] and [Stefanova and Geuvers 1996]. The model we construct has some similarities with the one used in [Berardi 1993] to justify encoding mathematics in the Calculus of Constructions. To establish our main result we construct a model in which the type that represents induction is empty.

Apart from the induction principle we also show the non-derivability of the Axiom of Choice.

1.1 Small Diversion: A Possible Smart Encoding of the Naturals

One may wonder whether there are other 'smarter' encodings of the natural numbers for which induction *is* provable. In this subsection we suggest a possible different encoding of the naturals. Our final result implies that induction is also non-derivable for this representation. Let us define

$$N := \exists x{:}\mathsf{nat}.(\mathsf{Ind}\,x),$$

with $\mathsf{Ind}\,x$ saying that x is an 'inductive natural number', defined as above. Now the 'inductivity' of the natural numbers is 'built in' in their encoding. ($\exists$ is defined in the well-known second order way: $\exists x{:}\sigma.\tau := \Pi\alpha{:}\star.(\Pi x{:}\sigma.\tau\to\alpha)\to\alpha$.) By using the definable $\exists$-elim and $\exists$-intro rules, it is now easy to define $\underline{\mathsf{O}}$, $\underline{\mathsf{succ}}$

for this encoding:

$$\underline{\mathsf{O}} := \lambda\alpha : \star.\lambda h{:}(\Pi x{:}\mathsf{nat}.(\mathsf{Ind}x)\to\alpha).h\mathsf{O}q_{\bullet}\ ,$$
$$\underline{\mathsf{succ}} := \lambda n{:}N.nN\big(\lambda x{:}\mathsf{nat}.\lambda p{:}(\mathsf{Ind}\,x).$$
$$\lambda\alpha{:}\star.\lambda h{:}(\Pi y{:}\mathsf{nat}.(\mathsf{Ind}y)\to\alpha).h(\mathsf{succ}\,x)(q_{\bullet\bullet\bullet\bullet}xp)\big),$$

where $q_{\bullet}$ and $q_{\bullet\bullet\bullet\bullet}$ are terms such that $q_{\bullet}$: $(\mathsf{Ind}\,\mathsf{O})$ and $q_{\bullet\bullet\bullet\bullet}$: $\Pi x{:}\mathsf{nat}.(\mathsf{Ind}\,x)\to(\mathsf{Ind}\,(\mathsf{succ}\,x))$. One may wonder whether the induction principle is derivable for the type N. It is not the case, which can intuitively be grasped from the fact that there is no 'coherence' among the possible proofs of $\mathsf{Ind}\,x$. (There are many possible proofs of $\mathsf{Ind}\,\mathsf{O}$, which are not all captured.)

2 Second Order Dependent Type Theory

The system of second order dependent type theory, $\lambda P2$, is an extension of the polymorphic λ-calculus with dependent types and it was first introduced in [Longo and Moggi 1988]. It can be seen as a subsystem of the Calculus of Constructions ([Coquand and Huet 1988], [Coquand 1990]), where the operations of forming type constructors are restricted to second order ones. (So, one can quantify over type constructors of kind $\sigma\to\star$, but one can not form type constructors of kind $(\sigma\to\star)\to\star$.) It can also be seen as an extension of the first order system λP, where quantification over type constructors has been added. For an extensive discussion on these systems and their relations, we refer to [Barendregt 1992] or [Geuvers 1993]. Here we just define the system $\lambda P2$ and give some initial motivation for it.

Definition 1. *The type system $\lambda P2$ is defined as follows. The set of* pseudo-terms, T, *is defined by*

$$\mathsf{T} ::= \star \mid \mathsf{Kind} \mid \mathrm{Var} \mid (\Pi\mathrm{Var}{:}\mathsf{T}.\mathsf{T}) \mid (\lambda\mathrm{Var}{:}\mathsf{T}.\mathsf{T}) \mid \mathsf{TT},$$

where Var *is a countable set of variables. On* T *we have the usual notion of β-reduction, $\longrightarrow_\beta$. We adopt from the untyped λ-calculus the conventions of denoting the transitive reflexive closure of $\longrightarrow_\beta$ by $\twoheadrightarrow_\beta$ and the transitive symmetric closure of $\twoheadrightarrow_\beta$ by $=_\beta$.*

The typing of terms is done under the assumption of specific types for the free variables that occur in the term. This is done in a context, *a finite sequence of declarations $\Gamma = v_1{:}T_1, \ldots, v_n{:}T_n$ (the v are variables and the T are pseudo-terms). Typing judgments are*

written as $\Gamma \vdash M : T$, with Γ a context and M and T pseudo-terms.

The deduction rules for $\lambda P2$ are as follows. (v ranges over Var, *s, s_1 and s_2 range over $\{\star, \mathsf{Kind}\}$ and M, N, T and U range over* T.*)*

$$(\mathit{axiom})\ \vdash \star : \mathsf{Kind} \quad (\mathit{var})\frac{\Gamma \vdash T : \star/\mathsf{Kind}}{\Gamma, v{:}T \vdash v : T} \quad (\mathit{weak})\frac{\Gamma \vdash T : \star/\mathsf{Kind} \qquad \Gamma \vdash M : U}{\Gamma, v{:}T \vdash M : U}$$

$$(\Pi)\frac{\Gamma\vdash T:s_1 \quad \Gamma,v{:}T\vdash U:s_2}{\Gamma\vdash \Pi v{:}T.U:s_2}\ \textit{if } (s_1,s_2)\neq(\mathsf{Kind},\mathsf{Kind})$$

$$(\lambda)\frac{\Gamma,v{:}T\vdash M:U \quad \Gamma\vdash \Pi v{:}T.U:s}{\Gamma\vdash \lambda v{:}T.M:\Pi v{:}T.U}$$

$$(app)\frac{\Gamma\vdash M:\Pi v{:}T.U \quad \Gamma\vdash N:T}{\Gamma\vdash MN:U[N/v]} \qquad (conv_\beta)\frac{\Gamma\vdash M:T \quad \Gamma\vdash U:s}{\Gamma\vdash M:U}\ \textit{if } T=_\beta U$$

In the rules (var) and (weak) it is always assumed that the newly declared variable is fresh, that is, it has not yet been declared in Γ. *For convenience, we split up the set* Var *into a set* $\mathrm{Var}^\star$, *the* object variables, *and* $\mathrm{Var}^{\bullet\bullet\bullet}$, *the* constructor variables. *Object variables will be denoted by* $x, y, z, \ldots$ *and constructor variables by* $\alpha, \beta, \ldots$. *In the rules (var) and (weak), we take the variable* v *out of* $\mathrm{Var}^\star$ *if* $s = \star$ *and out of* $\mathrm{Var}^{\bullet\bullet\bullet}$ *if* $s = \mathsf{Kind}$.

We call a pseudo-term M *well-typed* if there is a context Γ and another pseudo-term N such that either $\Gamma \vdash M : N$ or $\Gamma \vdash N : M$ is derivable. The well-typed terms can be split into the following disjoint subsets:

- $\{\mathsf{Kind}\}$,
- the set of *kinds*: terms A such that $\Gamma \vdash A : \mathsf{Kind}$ for some Γ; this includes $\star$. In $\lambda P2$ all kinds are of the form $\Pi x_1{:}\sigma_1 \ldots \Pi x_n{:}\sigma_n.\star$, with $\sigma_1, \ldots, \sigma_n$ *types* and $x_1, \ldots, x_n \in \mathrm{Var}^\star$.
- the set of *constructors*: terms of type a 'kind', i.e. terms P such that $\Gamma \vdash P : A$ for some kind A; this includes the *types*, terms of type $\star$.
  In $\lambda P2$ all constructors are of one of the following forms
  - $\alpha \in \mathrm{Var}^{\bullet\bullet\bullet}$,
  - Pt, with P a constructor and t an *object*,
  - $\lambda x{:}\sigma.P$, with σ a type, P a constructor, $x \in \mathrm{Var}^\star$,
  - $\Pi x{:}\sigma.\tau$, with σ and τ types, $x \in \mathrm{Var}^\star$,
  - $\Pi\alpha{:}A.\tau$, with A a kind, τ a type, $\alpha \in \mathrm{Var}^{\bullet\bullet\bullet}$.
- the *objects*: terms of type a 'type', i.e. terms M such that $\Gamma \vdash M : \sigma$ for some type σ. In $\lambda P2$ all objects are of one of the following forms
  - $x \in \mathrm{Var}^\star$,
  - qt, with q and t an objects,
  - qP, with P a constructor and q an object,
  - $\lambda x{:}\sigma.t$, with σ a type, t an object, $x \in \mathrm{Var}^\star$,
  - $\lambda\alpha{:}A.t$, with A a kind, t an object, $\alpha \in \mathrm{Var}^{\bullet\bullet\bullet}$.

Convention. We denote kinds by $A, B, C, \ldots$, types by $\sigma, \tau, \ldots$, constructors by $P, Q, \ldots$ and objects by $t, q, \ldots$.
If v is not free in U, we denote – as usual – $\Pi v{:}T.U$ by $T{\to}U$. In arrow types, we let brackets associate to the right, so $T{\to}T{\to}T$ denotes $T{\to}(T{\to}T)$. In application types, we let brackets associate to the left, so MNP denotes $(MN)P$.

Data types and formulas in $\lambda P2$. The well-known encoding of inductive data types in polymorphic λ-calculus extends immediately to $\lambda P2$. For the general procedure we refer to [Girard et al. 1989]. Here we give some examples. It is also standard that these inductive data types come together with the possibility of defining functions by iteration. We do not discuss the iteration scheme, as it is outside the scope of this paper. We do give, for each data type the associated induction principle. In this paper we show that the induction principle for natural numbers is not provable in $\lambda P2$. However the same method applies immediately to other data types, like the ones given below.

1. The natural numbers can be encoded by $\mathsf{nat} := \Pi\alpha{:}\star.\alpha{\to}(\alpha{\to}\alpha){\to}\alpha$, with zero and successor:

$$\begin{aligned} \mathsf{O} &:= \lambda\alpha{:}\star.\lambda x{:}\alpha.\lambda f{:}\alpha{\to}\alpha.x, \\ \mathsf{succ} &:= \lambda n{:}\mathsf{nat}.\lambda\alpha{:}\star.\lambda x{:}\alpha.\lambda f{:}\alpha{\to}\alpha.f(n\alpha x f). \end{aligned}$$

   The induction principle reads

$$\mathsf{ind...} := \Pi P{:}\mathsf{nat}{\to}\star.(P\mathsf{O}){\to}(\Pi y{:}\mathsf{nat}.(Py){\to}(P(\mathsf{succ}y))){\to}\Pi x{:}\mathsf{nat}.(Px).$$

2. The list over a given carrier type σ can be encoded by $\mathsf{list}_\sigma := \Pi\alpha{:}\star.\alpha{\to}(\sigma{\to}\alpha{\to}\alpha){\to}\alpha$, with empty list and 'cons' map:

$$\begin{aligned} \mathsf{nil} &:= \lambda\alpha{:}\star.\lambda x{:}\alpha.\lambda f{:}\sigma{\to}\alpha{\to}\alpha.x, \\ \mathsf{cons} &:= \lambda a{:}\sigma.\lambda l{:}\mathsf{list}_\sigma.\lambda\alpha{:}\star.\lambda x{:}\alpha.\lambda f{:}\sigma{\to}\alpha{\to}\alpha.fa(l\alpha x f). \end{aligned}$$

   As we are in $\lambda P2$, we can not define list as a type constructor $\mathsf{list} := \lambda\alpha{:}\star.\mathsf{list}_\alpha : \star{\to}\star$, simply because the kind $\star{\to}\star$ is not available in $\lambda P2$. For simplicity we write list for list_σ if the σ is clear from the context.
   The induction principle reads

$$\mathsf{ind....} := \Pi P{:}\mathsf{list}{\to}\star.(P\mathsf{nil}){\to}(\Pi a{:}\sigma.\Pi y{:}\mathsf{list}.(Py){\to}(P(\mathsf{cons}ay))){\to}\Pi x{:}\mathsf{list}.(Px).$$

3. The well-founded labeled trees of branching type τ and with labels in σ can be encoded by $\mathsf{tree}_{\tau\sigma} := \Pi\alpha{:}\star.(\sigma{\to}\alpha){\to}(\sigma{\to}(\tau{\to}\alpha){\to}\alpha){\to}\alpha$, with maps leaf and join (taking a label and a 'τ-sequence' of trees and returning a tree):

$$\begin{aligned} \mathsf{leaf} &:= \lambda a{:}\sigma.\lambda\alpha{:}\star.\lambda x{:}\sigma{\to}\alpha.\lambda f{:}\sigma{\to}(\tau{\to}\alpha){\to}\alpha.xa, \\ \mathsf{join} &:= \lambda a{:}\sigma.\lambda t{:}\tau{\to}\mathsf{tree}_{\tau\sigma}.\lambda\alpha{:}\star.\lambda x{:}\sigma{\to}\alpha.\lambda f{:}\sigma{\to}(\tau{\to}\alpha){\to}\alpha.fa(\lambda z{:}\tau.tz\alpha x f). \end{aligned}$$

   The remark about not being able to define $\mathsf{list} : \star{\to}\star$ also applies to tree. We omit the indices in tree if no confusion arises. The induction principle reads

$$\begin{aligned} \mathsf{ind....} := \Pi P{:}\mathsf{tree}{\to}\star.&(\Pi a{:}\sigma.(P(\mathsf{leaf}a))){\to} \\ &(\Pi a{:}\sigma.\Pi y{:}\tau{\to}\mathsf{tree}.(\Pi z{:}\tau.(P(yz))){\to}(P(\mathsf{join}ay))){\to}\Pi x{:}\mathsf{tree}.(Px). \end{aligned}$$

There is a *formulas-as-types* embedding from constructive second order predicate logic into $\lambda P2$.

3 Model Construction for $\lambda P2$

The model notion for $\lambda P2$ we give is not a general (categorical) one, but a description of a class of models, which is the same as in [Geuvers 1996]. It can be extended to a class of models for the Calculus of Constructions, which is done in [Stefanova and Geuvers 1996].

The models of $\lambda P2$ are built from *weakly extensional combinatory algebras* (weca for short). A *combinatory algebra* (ca for short) is a tuple $\mathcal{A} = \langle \mathbf{A}, \cdot, \mathbf{k}, \mathbf{s} \rangle$, with $\mathbf{A}$ a set, $\cdot$ a binary function from $\mathbf{A} \times \mathbf{A}$ to $\mathbf{A}$ (as usual denoted by infix notation), $\mathbf{k}, \mathbf{s} \in \mathbf{A}$ such that $(\mathbf{k}\cdot a)\cdot b = a$ and $((\mathbf{s}\cdot a)\cdot b)\cdot c = (a\cdot c)\cdot(b\cdot c)$. For $\mathcal{A}$ a combinatory algebra, the set of *terms over* $\mathcal{A}$, $\mathcal{T}(\mathcal{A})$, is defined by letting $\mathcal{T}(\mathcal{A})$ contain infinitely many variables $v_1, v_2, \ldots$ and distinct elements c_a for every $a \in \mathbf{A}$, and letting $\mathcal{T}(\mathcal{A})$ be closed under application (the operation $\cdot$). Given a term t and a valuation ρ, mapping variables to elements of $\mathbf{A}$, the *interpretation of t in* $\mathbf{A}$ *under* ρ, notation $[\![t]\!]^{\mathcal{A}}_{\rho}$, is defined in the usual way ($[\![c_a]\!]^{\mathcal{A}}_{\rho} = a$, $[\![MN]\!]^{\mathcal{A}}_{\rho} = [\![M]\!]^{\mathcal{A}}_{\rho} \cdot [\![N]\!]^{\mathcal{A}}_{\rho}$, etcetera). An important property of cas is that they are *combinatory complete*, i.e. if $t[v] \in \mathcal{T}(\mathcal{A})$ is a term with free variable v, then there is an element in $\mathbf{A}$, usually denoted by $\lambda^* v.t[v]$, such that $\forall x((\lambda^* v.t[v])\cdot x = t[x])$ in $\mathcal{A}$. (More technically, this means that $[\![(\lambda^* v.t[v]) \cdot x]\!]^{\mathcal{A}}_{\rho} = [\![t[x]]\!]^{\mathcal{A}}_{\rho}$ for all ρ.) A ca is *weakly extensional* if $[\![t_1]\!]^{\mathcal{A}}_{\rho(x:=a)} = [\![t_2]\!]^{\mathcal{A}}_{\rho(x:=a)}$ for all $a \in \mathbf{A}$ implies that $[\![\lambda^* x.t_1]\!]^{\mathcal{A}}_{\rho} = [\![\lambda^* x.t_2]\!]^{\mathcal{A}}_{\rho}$. In other words: a ca is weakly extensional if abstraction is a *function* on the weca $\langle \mathcal{T}(\mathcal{A}), \cdot, \mathbf{k}, \mathbf{s} \rangle$, i.e. if (in $\mathcal{T}(\mathcal{A})$) $t_1 = t_2$, then $\lambda^* x.t_1 = \lambda^* x.t_2$.

The need for *weakly extensional* cas comes from the fact that we want

$$M =_\beta N \Rightarrow (\![M]\!)_\rho = (\![N]\!)_\rho \text{ for all } \rho,$$

where $(\![-]\!)_\rho$ interprets pseudo-terms as elements of $\mathbf{A}$, using a valuation ρ for the free variables. Of course, $(\![-]\!)_\rho$ is close to $[\![-]\!]^{\mathcal{A}}_{\rho}$, except for the fact that now we also have to interpret abstraction: under $(\![-]\!)_\rho$, λ is interpreted as λ^*. [1]

Example 1. 1. A standard example of a weca is $\mathbf{\Lambda}$, consisting of the classes of open λ-terms modulo β-equality. So, $\mathbf{A}$ is just Λ/β and $[M] = [N]$ iff $M =_\beta N$. It is easily verified that this yields a weca.

2. Given a set of constants C, we define the weca $\mathbf{\Lambda}(C)$ as the equivalence classes of open λ_C-terms (i.e. lambda-terms over the constant set C) modulo βc-equality, where the c-equality rules says

$$cN =_c c \qquad \lambda v.c =_c c$$

for all $c \in C$ and $N \in \Lambda_C$.

3. Another example of a weca is $\mathbf{1}$, the *degenerate* weca where $\mathbf{A} = 1$, the one-element set. In this case $\mathbf{k} = \mathbf{s}$, which is usually not allowed in combinatory algebras, but note that we do allow it here.

[1] In general, for cas, $M = N \not\Rightarrow (\![M]\!)_\rho = (\![N]\!)_\rho$ (e.g. take combinatory logic and $M \equiv x$, $N \equiv Ix$). However, for wecas this implication holds.

The types of $\lambda P2$ will be interpreted as subsets of $\mathbf{A}$.

Definition 2. *A* polyset structure *over the weakly extensional combinatory algebra $\mathcal{A}$ is a collection $\mathcal{P} \subseteq \wp(\mathbf{A})$ such that*

1. $\mathbf{A} \in \mathcal{P}$,
2. *$\mathcal{P}$ is closed under arbitrary intersection $\bigcap$,*
3. *$\mathcal{P}$ is closed under* dependent products, *i.e. if $X \in \mathcal{P}$ and $F : X \to \mathcal{P}$, then $\Pi_{t \in X} F(t) \in \mathcal{P}$, where $\Pi_{t \in X} F(t)$ is defined as*

$$\{a \in \mathbf{A} \,|\, \forall t \in X (a \cdot t \in F(t))\}.$$

The elements of a polyset structure are called polysets. *If F is the constant function with value Y, we write $X \to Y$ instead of $\Pi_{t \in X} Y$.*

Example 2. 1. We obtain the *full polyset structure* over the weca $\mathcal{A}$ if we take $\mathcal{P} = \wp(\mathbf{A})$.
2. The *simple polyset structure* over the weca $\mathcal{A}$ is obtained by taking $\mathcal{P} = \{\emptyset, \mathbf{A}\}$. It is easily verified that this is a polyset structure.
3. Given the weca $\mathbf{\Lambda}(C)$ as defined in Example 1 (so C is a set of constants), we define the *polyset structure generated from C* by

$$\mathcal{P} := \{X \subseteq \mathbf{\Lambda}(C) \mid X = \emptyset \vee C \subseteq X\}.$$

To show that $\mathcal{P}$ is a polyset structure, the only interesting thing is to verify that $\mathcal{P}$ is closed under dependent product. So, let $X \in \mathcal{P}$ and $F : X \to \mathcal{P}$. We distinguish cases: if $X = \emptyset$, then $\Pi_{t \in X} F(t) = \mathbf{\Lambda}(C) \in \mathcal{P}$; if $F(t) = \emptyset$ for some $t \in X$, then $\Pi_{t \in X} F(t) = \emptyset \in \mathcal{P}$; in all other cases $C \subseteq \Pi_{t \in X} F(t)$, because for $c \in C$ and $t \in X$, $ct =_c c \in C \subseteq F(t)$, so $ct \in F(t)$.
4. Given the weca $\mathcal{A}$ and a set $C \subseteq \mathbf{A}$ such that $\forall a, b \in \mathbf{A}(a \cdot b \in C \Rightarrow a \in C$, we define the *power polyset structure of C* by

$$\mathcal{P} := \{X \subseteq \mathbf{A} \,|\, X \subseteq C \vee X = \mathbf{A}\}.$$

To check
that this is a polyset structure, one only has to verify that, for $X \in \mathcal{P}$ and $F : X \to \mathcal{P}$, $\Pi_{t \in X} F(t) \in \mathcal{P}$. This follows from an easy case distinction: $\forall t \in X (F(t) = \mathbf{A})$ or $\exists t \in X (F(t) \subseteq C)$.
An interesting instance of a power polyset structure is the one arising from $C = \text{HNF}$, the set of λ-terms with a head-normal-form, in the weca Λ/β.

The dependent product of a polyset structure will be used to interpret types of the form $\Pi x{:}\sigma.\tau$, where both σ and τ are types. The intersection will be used to interpret types of the form $\Pi \alpha{:}A.\sigma$, where σ is a type and A is a kind. To interpret kinds we need a *predicative structure.*

Definition 3. *For $\mathcal{P}$ a polyset structure, the* predicative structure over $\mathcal{P}$ *is the collection of sets $\mathcal{N}$ defined inductively by*

1. $\mathcal{P} \in \mathcal{N}$,
2. If $X \in \mathcal{P}$ *and* $\forall t \in X(F(t) \in \mathcal{N}$, *then* $\prod_{t \in X} F(t) \in \mathcal{N}$.

If F *is a constant function with value* $\mathcal{P}$, *we write* $X \to \mathcal{P}$ *in stead of* $\prod_{t \in X} \mathcal{P}$.

Definition 4. *If* $\mathcal{A}$ *is a combinatory algebra,* $\mathcal{P}$ *a polyset structure over* $\mathcal{A}$ *and* $\mathcal{N}$ *the predicative structure over* $\mathcal{P}$, *then we call the tuple* $\langle \mathcal{A}, \mathcal{P}, \mathcal{N} \rangle$ *a* $\lambda P2$-model.

The predicative structure over a polyset structure $\mathcal{P}$ is intended to give a domain of interpretation for the kinds. For example, if the type σ is interpreted as the polyset X, then the kind $\sigma \to \sigma \to \star$ is interpreted as $\prod_{t \in X} \prod_{q \in X} \mathcal{P}$, for which we usually write $X \to X \to \mathcal{P}$.

We now define three interpretation functions, one for kinds, $\mathcal{V}(-)$, that maps kinds to elements of $\mathcal{N}$, one for constructors (and types), $[\![-]\!]$, that maps constructors to elements of $\bigcup \mathcal{N}$ (and types to elements of $\mathcal{P}$, which is a subset of $\bigcup \mathcal{N}$) and one for objects, $(\![-]\!)$, that maps objects to elements of the combinatory algebra $\mathcal{A}$. All these interpretations are parametrized by *valuations*, assigning values to the free variables (declared in the context).

Let in the following $\mathcal{M} = \langle \mathcal{A}, \mathcal{P}, \mathcal{N} \rangle$ be a $\lambda P2$-model: $\mathcal{A} = \langle \mathbf{A}, \cdot, \mathbf{k}, \mathbf{s} \rangle$ is a combinatory algebra, $\mathcal{P}$ is a polyset structure over $\mathcal{A}$ and $\mathcal{N}$ is the predicative structure over the polyset structure $\mathcal{P}$.

Definition 5. *A* constructor variable valuation *is a map* ξ *from* $\mathrm{Var}^{\star\star}$ *to* $\bigcup \mathcal{N}$. *An* object variable valuation *is a map* ρ *from* $\mathrm{Var}^{\star}$ *to* $\mathbf{A}$.

Definition 6. *For* ρ *an object variable valuation, we define the map* $(\![-]\!)^{\mathcal{M}}_{\rho}$ *from the set of objects to* $\mathbf{A}$ *as follows. (We leave the model* $\mathcal{M}$ *implicit.)*

$$
\begin{aligned}
(\![x]\!)_{\rho} &:= \rho(x), \\
(\![tq]\!)_{\rho} &:= (\![t]\!)_{\rho} \cdot (\![q]\!)_{\rho}, \text{ if } q \text{ is an object}, \\
(\![tQ]\!)_{\rho} &:= (\![t]\!)_{\rho}, \text{ if } Q \text{ is a constructor}, \\
(\![\lambda x{:}\sigma.t]\!)_{\rho} &:= \lambda^* v.(\![t]\!)_{\rho(x:=v)}, \text{ if } \sigma \text{ is a type}, \\
(\![\lambda \alpha{:}A.t]\!)_{\rho} &:= (\![t]\!)_{\rho}, \text{ if } A \text{ is a kind}.
\end{aligned}
$$

Definition 7. *For* ρ *an object variable valuation and* ξ *a constructor variable valuation, we define the maps* $\mathcal{V}(-)^{\mathcal{M}}_{\xi\rho}$ *and* $[\![-]\!]^{\mathcal{M}}_{\xi\rho}$ *respectively from kinds to* $\mathcal{N}$ *and from constructors to* $\bigcup \mathcal{N}$ *as follows. (We leave the model* $\mathcal{M}$ *implicit.)*

$$
\begin{aligned}
\mathcal{V}(\star)_{\xi\rho} &:= \mathcal{P}, \\
\mathcal{V}(\Pi x{:}\sigma.B)_{\xi\rho} &:= \prod_{t \in [\![\sigma]\!]_{\xi\rho}} \mathcal{V}(B)_{\xi\rho(x:=t)}, \\
[\![\alpha]\!]_{\xi\rho} &:= \xi(\alpha),
\end{aligned}
$$

$$[\![\Pi\alpha{:}A.\tau]\!]_{\xi\rho} := \bigcap_{a\in\mathcal{V}(A)_{\xi\rho}} [\![\tau]\!]_{\xi(\alpha:=a)\rho}, \text{ if } A \text{ is a kind,}$$
$$[\![\Pi x{:}\sigma.\tau]\!]_{\xi\rho} := \Pi_{t\in[\![\sigma]\!]_{\xi\rho}} [\![\tau]\!]_{\xi\rho(x:=t)}, \text{ if } \sigma \text{ is a type,}$$
$$[\![Pt]\!]_{\xi\rho} := [\![P]\!]_{\xi\rho}((\![t]\!)_\rho),$$
$$[\![\lambda x{:}\sigma.P]\!]_{\xi\rho} := \lambda t \in [\![\sigma]\!]_{\xi\rho}.[\![P]\!]_{\xi\rho(x:=t)}.$$

Note that $\mathcal{V}(A)_{\xi\rho}$ and $[\![P]\!]_{\xi\rho}$ may be undefined. For example, in the definition of $[\![Pt]\!]_{\xi\rho}$, $(\![t]\!)_\rho$ may not be in the domain of $[\![P]\!]_{\xi\rho}$, in the definition of $[\![\Pi x{:}\sigma.\tau]\!]_{\xi\rho}$, $[\![\sigma]\!]_{\xi\rho}$ may not be a polyset and in the definition of $\mathcal{V}(\Pi x{:}\sigma.B)_{\xi\rho}$, $[\![\sigma]\!]_{\xi\rho}$ may not be defined. From the Soundness Theorem (1) it will follow that, under certain natural conditions for ξ and *rho*, $\mathcal{V}(A)_{\xi\rho}$ and $[\![P]\!]_{\xi\rho}$ are well-defined.

Definition 8. *For Γ a $\lambda P2$-context, ρ an object variable valuation and ξ a constructor variable valuation, we say that ξ, ρ* fulfills *Γ, notation $\xi, \rho \models \Gamma$, if for all $x \in \mathrm{Var}^\star$ and $\alpha \in \mathrm{Var}^{\star\star}$, $x : \sigma \in \Gamma \Rightarrow \rho(x) \in [\![\sigma]\!]_{\xi\rho}$ and $\alpha : A \in \Gamma \Rightarrow \xi(\alpha) \in \mathcal{V}(A)_{\xi\rho}$.*

It is (implicit) in the definition that $\xi\rho \models \Gamma$ only if for all declarations $x{:}\sigma \in \Gamma$, $[\![\sigma]\!]_{\xi\rho}$ is defined (and similarly for $\alpha{:}A \in \Gamma$).

Definition 9. *The notion of* truth in a $\lambda P2$-model, *notation $\models^{\mathcal{M}}$ and of* truth, *notation $\models$ are defined as follows. For Γ a context, t an object, σ a type, P a constructor and A a kind of $\lambda P2$,*

$$\Gamma \models^{\mathcal{M}} t : \sigma \text{ if } \forall \xi, \rho[\xi, \rho \models \Gamma \Rightarrow (\![t]\!)_\rho \in [\![\sigma]\!]_{\xi\rho}],$$
$$\Gamma \models^{\mathcal{M}} P : A \text{ if } \forall \xi, \rho[\xi, \rho \models \Gamma \Rightarrow [\![P]\!]_{\xi\rho} \in \mathcal{V}(A)_{\xi\rho}].$$

Quantifying over the class of all $\lambda P2$-models, we define, for M an object or a constructor of $\lambda P2$,

$$\Gamma \models M : T \text{ if } \Gamma \models^{\mathcal{M}} M : T \text{ for all } \lambda P2\text{-models } \mathcal{M}.$$

Soundness states that if a judgment $\Gamma \vdash M : T$ is derivable, then it is true in all models. It is proved 'model-wise', by induction on the derivation in $\lambda P2$.

Theorem 1 (Soundness). *For Γ a context, M an object or a constructor and T a type or a kind of $\lambda P2$,*

$$\Gamma \vdash M : T \Rightarrow \Gamma \models M : T.$$

Example 3. Let $\mathcal{A}$ be a weca.

1. The *full $\lambda P2$-model* over $\mathcal{A}$ is $\mathcal{M} = \langle \mathcal{A}, \mathcal{P}, \mathcal{N} \rangle$, where $\mathcal{P}$ is the full polyset structure over $\mathcal{A}$ (as defined in Example 2).
2. The *simple $\lambda P2$-model* over $\mathcal{A}$ is $\mathcal{M} = \langle \mathcal{A}, \mathcal{P}, \mathcal{N} \rangle$, where $\mathcal{P}$ is the simple polyset structure over $\mathcal{A}$. (So $\mathcal{P} = \{\emptyset, \mathbf{A}\}$.)
3. The simple $\lambda P2$-model over the degenerate $\mathcal{A}$ is also called the *proof-irrelevance model* or *PI-model* for $\lambda P2$.
4. For C a set of constants, the *$\lambda P2$-model generated from C* is defined by $\mathcal{M} = \langle \mathbf{\Lambda}(C), \mathcal{P}, \mathcal{N} \rangle$, where $\mathcal{P}$ is the polyset structure generated from C.

4 Non-derivability Results in $\lambda P2$

We now show that the induction-principle is not derivable in $\lambda P2$ by constructing a counter-model. We first introduce some notation and then we study some specific models and their properties.

In a logical model, validity of a formula φ means that the interpretation of φ is true in the model. In a type theoretical model, we call a type *valid* if its interpretation is nonempty. This conforms with the 'formulas-as-types' embedding from PRED2 to $\lambda P2$, where a formula is interpreted as the type of its proofs. (Hence, a formula is provable iff its associated type is nonempty.)

Definition 10. *For $\mathcal{M}$ a $\lambda P2$-model, Γ a context, σ a type in Γ and ξ, ρ valuations such that $\xi, \rho \models \Gamma$, we say that σ is* valid in $\mathcal{M}$ under ξ, ρ, *notation $\mathcal{M}, \xi, \rho \Vdash^{\lambda P2} \sigma$, if*

$$[\![\sigma]\!]^{\mathcal{M}}_{\xi\rho} \neq \emptyset.$$

In case the model $\mathcal{M}$ is clear from the context, we omit it. Similarly we omit ξ and/or ρ if they are clear from the context or if the specific choice of ξ or ρ is irrelevant (e.g. in case of a closed type σ).

So, to prove the non-derivability of ind in $\lambda P2$, we are looking for a $\lambda P2$-model $\mathcal{M}$ such that

$$\mathcal{M} \not\Vdash^{\lambda P2} \mathsf{ind}.$$

Definition 11. *A $\lambda P2$-model $\mathcal{M}$ is* consistent *if $\emptyset \in \mathcal{P}$.*

For a $\lambda P2$-model, being consistent is equivalent to saying that $[\![\bot]\!] = \emptyset$, because $[\![\bot]\!]$ is the minimal element (w.r.t. $\subseteq$) of $\mathcal{P}$. Here, $\bot$ is defined as usual as $\Pi\alpha{:}\star.\alpha$.

Note that the polyset structures of Example 2 all yield a consistent $\lambda P2$-model.

Convention 12 *From now on we only discuss consistent $\lambda P2$-models.*

Definition 13. *In a $\lambda P2$-model $\mathcal{M} = \langle \mathcal{A}, \mathcal{P}, \mathcal{N} \rangle$ we define the 'connectives' $\bot$, $\neg$, $\wedge$, $\vee$ and $\exists$ as follows. ($X, Y \in \mathcal{P}$, $F : X \to \mathcal{P}$ and $Y_i \in \mathcal{P}$ for all $i \in I$; as in types, we let brackets associate to the right.)*

$$\bot := \textstyle\bigcap_{Z\in\mathcal{P}} Z, \qquad \neg X := X \to \bot,$$

$$X \wedge Y := \textstyle\bigcap_{Z\in\mathcal{P}} (X \to Y \to Z) \to Z, \qquad X \vee Y := \textstyle\bigcap_{Z\in\mathcal{P}} (X \to Z) \to (Y \to Z) \to Z,$$

$$\exists_{x\in X} F(x) := \textstyle\bigcap_{Z\in\mathcal{P}} (\Pi_{x\in X} F(x) \to Z) \to Z, \quad \exists_{i\in I} Y_i := \textstyle\bigcap_{Z\in\mathcal{P}} (\bigcap_{i\in I} Y_i \to Z) \to Z.$$

Note that, due to the assumptions on a polyset structure, these are all elements of $\mathcal{P}$.

Remark 1. The definition of $\exists_{i\in I}Y_i$ is close to the *union*. If we define the elements F and G of the weca $\mathbf{A}$ by $F := \lambda^* x.xI$ and $G := \lambda^* xh.hx$ (where I denotes the identity in $\mathbf{A}$: $I := \mathbf{skk}$), then $F \in \exists_{i\in I}Y_i \to \bigcup_{i\in I} Y_i$ and $G \in \bigcup_{i\in I} Y_i \to \exists_{i\in I}Y_i$ even with $F \circ G = I$[2]. Note however, that $\bigcup_{i\in I} Y_i$ need not be an element of $\mathcal{P}$[3], but we do have $\exists_{i\in I}Y_i = \emptyset \Leftrightarrow \bigcup_{i\in I} Y_i = \emptyset$.

Lemma 1. *The following holds in arbitrary (consistent) $\lambda P2$-models $\mathcal{M}$.*

$$\neg X = \emptyset \Leftrightarrow X \neq \emptyset, \tag{1}$$

$$X \to Y \neq \emptyset \Leftrightarrow \textit{if } X \neq \emptyset \textit{ then } Y \neq \emptyset, \tag{2}$$

$$X \wedge Y \neq \emptyset \Leftrightarrow X \neq \emptyset \textit{ and } Y \neq \emptyset, \tag{3}$$

$$X \vee Y \neq \emptyset \Leftrightarrow X \neq \emptyset \textit{ or } Y \neq \emptyset, \tag{4}$$

$$\exists_{x\in X} F(x) \neq \emptyset \Leftrightarrow \exists t \in X(F(t) \neq \emptyset), \tag{5}$$

$$\exists_{i\in I} Y_i \neq \emptyset \Leftrightarrow \exists i \in I(Y_i \neq \emptyset), \tag{6}$$

$$\Pi_{x\in X} F(x) \neq \emptyset \Rightarrow \forall t \in X(F(t) \neq \emptyset), \tag{7}$$

$$\bigcap_{i\in I} Y_i \neq \emptyset \Rightarrow \forall i \in I(Y_i \neq \emptyset). \tag{8}$$

Proof. We reason classically in the meta-theory of the models (otherwise $\Leftarrow$ in (2) and $\Rightarrow$ in (4)-(6) are problematic).
(1) follows immediately from $\bot = \emptyset$ (i.e. the consistency of the $\lambda P2$-model).
For (2), $\Rightarrow$ is immediate. For $\Leftarrow$, we distinguish cases: if $X \neq \emptyset$, then $Y \neq \emptyset$, say $q \in Y$, and hence $\lambda^* x.q \in X \to Y$; if $X = \emptyset$, then $\lambda^* x.x \in X \to Y$. For (3), $\Rightarrow$: $M \in X \wedge Y$, then $M\mathbf{k} \in X$ and $M(\mathbf{ki}) \in Y$ (where $\mathbf{i}$ is the identity in the weca, $\mathbf{i} := \mathbf{skk}$). $\Leftarrow$: if $M_1 \in X$, $M_2 \in Y$, then $\lambda^* h.hM_1M_2 \in X \wedge Y$.
For (4), $\Rightarrow$: let $M \in X \vee Y$ and suppose $X = Y = \emptyset$. Then $Maa \in \emptyset$ ($a \in \mathcal{A}$ arbitrary), contradiction. So $X \neq \emptyset$ or $Y \neq \emptyset$ $\Leftarrow$: if $M \in X$, then $\lambda^* hg.hM \in X \vee Y$ and similarly for $M \in Y$.
For (5), $\Rightarrow$: let $M \in \exists_{x\in X} F(x)$ and suppose $\forall x \in X(F(x) = \emptyset)$. Then $M(\lambda^* x.\lambda^* y.y) \in \emptyset$, contradiction, so $\exists x \in X(F(x) \neq \emptyset)$. $\Leftarrow$: If $q \in F(t)$ for certain $t \in X$, then $\lambda^* h.htq \in \exists_{x\in X} F(x)$.
(6) follows from Remark 1 and (7) and (8) are immediate.

Remark 2. The reverse implications in Lemma 1, cases (7) and (8), do not hold in general. A counterexample can be found by looking at the full polyset structure over $\mathbf{A} = \mathbf{\Lambda}$. Define $F : \mathbf{A} \to \mathcal{P}$ by $F(t) = \mathbf{\Lambda} \setminus \{t\}$. Then $F(t) \neq \emptyset$ for all $t \in \mathbf{\Lambda}$. Now suppose $M \in \Pi_{x\in X} F(x)$. Then $Mt \neq t$ for all $t \in \mathbf{\Lambda}$, but this is not possible, since M has a fixed point. This contradicts the reverse implication of (7). If we consider $\bigcap_{x\in \mathbf{A}} F(x)$, we immediately find a counterexample to the reverse implication of (8).

[2] In a weca $\mathbf{A}$, composition is defined as usual by $a \circ b := \lambda^* x.a \cdot (b \cdot x)$.

[3] The example $\mathcal{P}$s of Example 2 are all closed under arbitrary union and at this moment we don 't know of any $\mathcal{P}$ that is *not* closed under unions. However, Definition 2 does not a priori require a $\mathcal{P}$ to be closed under union.

Lemma 2. *For a simple $\lambda P2$-model over $\mathcal{A}$ the reverse implications in Lemma 1, cases (7) and (8), hold. Similarly for a $\lambda P2$-model generated from a set C.*

Proof. Case (8) is immediate: $\bigcap_{i \in I} Y_i$ can only be empty if one of the Y_i is empty. For (7), if for all $t \in X$, $Ft \neq \emptyset$, then there is an element q such that $\forall t \in X(q \in Ft)$ (this is a peculiar feature of these models) and hence $\lambda^* x.q \in \Pi_{t \in X} Ft$.

Lemma 3. *All $\lambda P2$-models satisfy classical logic, i.e.*

$$\neg\neg X \to X \neq \emptyset$$

for all $X \in \mathcal{P}$ in all $\lambda P2$-models.

Proof. We reason classically in the models, using Lemma 1. Let $X \in \mathcal{P}$. If $X \neq \emptyset$, say $t \in X$, then $\neg\neg X \to X \neq \emptyset$, because e.g. $\lambda^* x.t \in \neg\neg X \to X$. If $X = \emptyset$, then $\neg X = \mathbf{\Lambda}$, so $\neg\neg X = \emptyset$, so $\neg\neg X \to X = \mathbf{\Lambda}$.

Remark 3. It is not the case that $\cap_{X \in \mathcal{P}} \neg\neg X \to X \neq \emptyset$ in all $\lambda P2$-models. In fact we have the following.

1. In the full $\lambda P2$-model over $\mathbf{\Lambda}$, $\cap_{X \in \mathcal{P}} \neg\neg X \to X = \emptyset$.
2. In simple $\lambda P2$-models or models generated by some C, $\cap_{X \in \mathcal{P}} \neg\neg X \to X \neq \emptyset$.

The first is proved by defining $X_i = \{x_i\}$ for all $i \in \mathbb{N}$ (with, of course all x_i different). Then $\neg\neg X_i = \mathbf{\Lambda}$. Now, suppose $M \in \cap_{X \in \mathcal{P}} \neg\neg X \to X$. Then for any $N \in \mathbf{\Lambda}$, we find that $\forall i \in \mathbb{N}(MN \in X_i)$, i.e. $MN =_\beta x_i$ for all i, which is not possible, as MN contains only finitely many free variables.
The second is proved by noticing that, in these models there is an element P such that $X \neq \emptyset \Rightarrow P \in X$. Hence $\lambda^* x.P \in \cap_{X \in \mathcal{P}} \neg\neg X \to X$, following the reasoning in the proof of Lemma 3.

Equality is defined in $\lambda P2$ using Leibniz equality: for $\sigma : \star$, $M, N : \sigma$

$$M =_\sigma N \ := \ \Pi P{:}\sigma \to \star.(PM) \to (PN).$$

In case the type is clear from the context, we often do not write it as a subscript in the Leibniz equality. The notion of 'Proof-Irrelevance', meaning that for any type σ, all terms of type σ are equal, is defined by PI $:= \Pi\alpha{:}\star.\Pi x, y{:}\alpha.x =_\alpha y$.

Lemma 4. *Given a $\lambda P2$-model $\mathcal{M}$, a type σ and terms $M, N : \sigma$, we have*

$$\mathcal{M}, \xi, \rho \models M =_\sigma N \Leftrightarrow (\![M]\!)_\rho = (\![N]\!)_\rho.$$

Proof. $\Rightarrow$: Suppose $\cap_{Q \in [\![\sigma]\!] \to \mathcal{P}} Q(\![M]\!)_\rho \to Q(\![N]\!)_\rho \neq \emptyset$. Take Q such that $Qx \neq \emptyset$ iff $x = (\![M]\!)_\rho$ Then it is the case that $Q(\![N]\!)_\rho \neq \emptyset$, hence $(\![M]\!)_\rho = (\![N]\!)_\rho$.
$\Leftarrow$: If $(\![M]\!)_\rho = (\![N]\!)_\rho$, then $Q(\![M]\!)_\rho = Q(\![N]\!)_\rho$, so $\lambda^* x.x \in \cap_{Q \in [\![\sigma]\!] \to \mathcal{P}} Q(\![M]\!)_\rho \to Q(\![N]\!)_\rho$.

Corollary 1. *$\mathcal{M} \models PI \Leftrightarrow \mathcal{M}$ is the PI-model.*

In this paper we focus especially on the induction principle for (an arbitrary encoding of) the natural numbers. We therefore characterize when a $\lambda P2$-model satisfies induction for the natural numbers.

Definition 14. *Given a closed $\lambda P2$-type N and closed terms $0 : N$ and $S : N \rightarrow N$, we define the type $\mathsf{ind}_{N,0,S}$ by*

$$\Pi P{:}N \rightarrow \star . P0 \rightarrow (\Pi x{:}N.Px \rightarrow P(Sx)) \rightarrow \Pi x{:}N.Px.$$

Lemma 5. *For $\mathcal{M} = \langle \mathcal{A}, \mathcal{P}, \mathcal{N} \rangle$ a $\lambda P2$-model,*

$$\mathcal{M} \models \mathsf{ind}_{N,0,S} \;\Rightarrow\; [\![N]\!] = \{S^n 0 \mid n \in \mathbb{N}\}$$

If, moreover, the test-for-zero and the predecessor function are definable on the type N in the model $\mathcal{M}$, then also

$$[\![N]\!] = \{S^n 0 \mid n \in \mathbb{N}\} \;\Rightarrow \mathcal{M} \;\models \mathsf{ind}_{N,0,S}.$$

Proof. For simplicity, we denote the interpretations of N, 0 and S in the model just by N, 0 and S. Suppose $\mathcal{M} \models \mathsf{ind}_{N,0,S}$. Then

$$\bigcap_{Q \in N \rightarrow \mathcal{P}} Q0 \rightarrow (\Pi_{t \in N} Qt \rightarrow Q(St)) \rightarrow \Pi_{t \in N} Qt \neq \emptyset.$$

Let X be some non-empty element of $\mathcal{P}$. Define $Q : N \rightarrow \mathcal{P}$ as follows: $Qt = X$ if $t = S^n 0$ for some $n \in \mathbb{N}$ and $Qt = \emptyset$ otherwise. Then $Q0 \neq \emptyset$ and $\Pi_{t \in N} Qt \rightarrow Q(St) \neq \emptyset$, hence $\Pi_{t \in N} Qt \neq \emptyset$, say $M \in \Pi_{t \in N} Qt$. Now, suppose $q \in N$ with $q \neq S^n 0$ (for all $n \in \mathbb{N}$). Then $Qq = \emptyset$ but also $Mq \in Qq$, contradiction. So all $q \in N$ are of the form $S^n 0$.

For the reverse implication, suppose that the test-for-zero and the predecessor function are definable in the model and suppose that $N = \{S^n 0 \mid n \in \mathbb{N}\}$. To prove that $\bigcap_{Q \in N \rightarrow \mathcal{P}} Q0 \rightarrow (\Pi_{t \in N} Qt \rightarrow Q(St)) \rightarrow \Pi_{t \in N} Qt \neq \emptyset$, let $Q \in N \rightarrow \mathcal{P}$ arbitrary and let $Z \in Q0$, $F \in \Pi_{t \in N} Qt \rightarrow Q(St)$. We are looking for an element of $\Pi_{t \in N} Qt$, which is given by an H which is a solution to

$$Hx = \text{if } \mathrm{Zero}(x) \text{ then } Z \text{ else } F(x-1)(H(x-1)).$$

This

can be obtained by taking for H a fixed point of $\lambda^* hx.$if $\mathrm{Zero}(x)$ then Z else $F(x-1)(h(x-1))$. Note that we need the test-for-zero and predecessor to be able to define this H.

Theorem 2. *Induction over the natural numbers is not derivable in $\lambda P2$ for any type N and terms $0 : N$, $S : N \rightarrow N$.*

Proof. In the simple $\lambda P2$-model over $\mathbf{\Lambda}$ (see Example 3), the interpretation of N is $\mathbf{\Lambda}$. So, using the Lemma, we conclude that $\mathsf{ind}_{N,0,S}$ is not valid in the model and hence $\mathsf{ind}_{N,0,S}$ is not inhabited in $\lambda P2$.

As can be observed from the proof, the non-derivability of induction in $\lambda P2$ is not caused by the fact that the logic of $\lambda P2$ is constructive. logic. Note that, taking the PI-model in the proof of the Theorem does not work, because then $[\![N]\!] = 1 = \{S^n 0 \mid n \in \mathbb{N}\}$, so we do not obtain a counterexample.

The arguments of Lemma 5 and Theorem 2 also apply to other data types like lists and trees and even to a finite data type like the booleans. So, induction is not derivable for any data type.

Remark 4. It is in general not the case in $\lambda P2$ that the induction principle for one data type (say the natural numbers) implies the induction principle for another data type (say booleans). For a counterexample consider the context $\Gamma = N : \star, 0 : N, S : N \to N, h : \mathsf{ind}_{N,0,S}$ and the $\lambda P2$-model $\langle \mathbf{\Lambda}(C), \mathcal{P}, \mathcal{N} \rangle$, where $C = \{S^n(0) \mid n \in \mathbb{N}\}$ (so the $S^n(0)$ are considered as constants) and $\mathcal{P}$ is the polyset structure generated from C. (See Example 2.)

Now, take valuations ξ and ρ with $\xi(N) = C$, $\rho(0) = 0$, $\rho(S) = S$ and $\rho(h) = \lambda^* z f x.0$. Then $\rho(h) \in [\![\mathsf{ind}_{N,0,S}]\!]_{\xi\rho}$:

$$\lambda^* z f x.0 \in \bigcap_{Q \in C \to \mathcal{P}} Q0 \to (\Pi_{t \in C} Qt \to Q(St)) \to \Pi_{t \in C} Qt,$$

because for $Q \in C \to \mathcal{P}$, $Z \in Q0$, $G \in \Pi_{t \in C} Qt \to Q(St)$ and $t \in C$, we find that $t = S^n(0)$ (def of C) and for all $n \in \mathbb{N}$, $Q(S^n(0)) \neq \emptyset$ (induction on n, using Z and G), so $0 \in Qt$. We conclude that $\xi, \rho \models \Gamma$.

So, $\mathcal{M}, \xi, \rho \models \mathsf{ind}_{N,0,S}$. On the other hand, for any closed type B (the 'booleans') with closed terms $T : B$ and $F : B$, $[\![B]\!] \supsetneq \{(\![F]\!), (\![T]\!)\}$, so induction over booleans is not valid.

One may wonder what happens with the counterexample in the proof of Theorem 2 if we add induction over natural numbers to $\lambda P2$ as a primitive concept, together with the associated reduction rules. Let's take a closer look at this situation.

We extend $\lambda P2$ with a type constant N and term constants $0 : N$, $S : N \to N$, $R : \Pi P{:}N \to \star.(P0) \to (\Pi y{:}N.Py \to P(Sy)) \to \Pi x{:}N.(Px)$. Furthermore we add reduction rules

$$RPzf0 \longrightarrow_r z \quad \text{and} \quad RPzf(Sx) \longrightarrow_r fx(RPzfx).$$

To make a model of this extension of $\lambda P2$ we have to give an interpretation to the constants in such a way that the equality rule for R is preserved. For $\mathbf{\Lambda}$ (that we used in the counter-model of 2), this can be achieved by adding primitive constants 0, S and R to Λ, with the reduction rules

$$Rzf0 \longrightarrow_r z \quad \text{and} \quad Rzf(Sx) \longrightarrow_r fx(Rzfx).$$

Let's denote this extension of λ-calculus (it is a weca) by $\mathbf{\Lambda}^+$. (So we interpret 0 by 0, S by S and R by R.) Now consider the simple $\mathbf{\Lambda}^+$-model determined by the polyset structure $\{\emptyset, \mathbf{\Lambda}\}$ and notice that it is *not* a model of this $\lambda P2$ extension, because $\mathsf{ind}_{N,0,S}$ is empty in this model (so we can not interpret R).

We give one more non-derivability result in $\lambda P2$, based on our models.

Lemma 6. *There are closed types σ, τ and a relation $R : \sigma \to \tau \to \star$ in $\lambda P2$ for which the Axiom of Choice, $(\Pi x{:}\sigma.\exists y{:}\tau.Rxy) \to (\exists f{:}\sigma \to \tau.\Pi x{:}\sigma.Rx(fx))$, is not derivable.*

Proof. The counterexample is similar to the one in Remark 2. Take $\sigma = \tau = \mathsf{nat}$ and $Rxy := x \neq_{\dots} y$ and consider the simple $\lambda P2$-model over $\mathbf{A} = \mathbf{\Lambda}$. Now $\mathcal{M} \models \Pi x{:}\sigma.\exists y{:}\tau.Rxy$, because this is equivalent to (using Lemmas 1 and 4) $\forall t \in \mathbf{\Lambda} \exists q \in \mathbf{\Lambda}(t \neq_\beta q)$. On the other hand, $\mathcal{M} \not\models \exists f{:}\sigma \to \tau.\Pi x{:}\sigma.Rx(fx)$, because this is equivalent to the statement $\exists g \in \mathbf{\Lambda} \forall t \in \mathbf{\Lambda}(gt \neq_\beta t)$, which is not possible, because every element of $\mathbf{\Lambda}$ has a fixed point.

The proof of non-derivability of the Axiom of Choice bears a strong similarity to a proof in [Barendregt 1973], credited originally to Scott, showing that classical Combinatory Logic extended with the Axiom of Choice is inconsistent.

Acknowledgments. Thanks to the referees for pointing out some mistakes in the original manuscripts and suggesting several improvements. Furthermore I want to thank Thierry Coquand for raising the question of derivability of induction in $\lambda P2$ and for some valuable discussions on the topic.

References

[Barendregt 1973] H.P. Barendregt, Combinatory Logic and the Axiom of Choice, in *Indagationes Mathematicae*, vol. 35, nr. 3, pp. 203 – 221.

[Barendregt 1992] H.P. Barendregt, Typed lambda calculi. In *Handbook of Logic in Computer Science*, eds. Abramski et al., Oxford Univ. Press.

[Berardi 1993] S. Berardi, Encoding of data types in Pure Construction Calculus: a semantic justification, in *Logical Environments*, eds. G. Huet and G. Plotkin, Cambridge University Press, pp 30–60.

[Coquand 1990] Th. Coquand, Metamathematical investigations of a calculus of constructions. In *Logic and Computer Science*, ed. P.G. Odifreddi, APIC series, vol. 31, Academic Press, pp 91-122.

[Coquand and Huet 1988] Th. Coquand and G. Huet, The calculus of constructions, *Information and Computation*, 76, pp 95-120.

[Geuvers 1993] J.H. Geuvers, *Logics and Type systems*, Ph.D. Thesis, University of Nijmegen, Netherlands.

[Geuvers 1996] J.H. Geuvers, Extending models of second order logic to models of second order dependent type theory, *Computer Science Logic*, Utrecht, eds. D. van Dalen and M. Bezem, LNCS 1258, 1997, pp 167–181.

[Hyland and Ong 1993] J.M.E. Hyland and C.-H. L. Ong, Modified realizability toposes and strong normalization proofs. In *Typed Lambda Calculi and Applications*, Proceedings, eds. M. Bezem and J.F. Groote, LNCS 664, pp. 179–194, Springer-Verlag, 1993.

[Girard et al. 1989] J.-Y. Girard, Y. Lafont and P. Taylor, *Proofs and types*, Camb. Tracts in Theoretical Computer Science 7, Cambridge University Press.

[Longo and Moggi 1988] G. Longo and E. Moggi, Constructive Natural Deduction and its "Modest" Interpretation. Report CMU-CS-88-131.

[Stefanova and Geuvers 1996] M. Stefanova and J.H. Geuvers, A Simple Model Construction for the Calculus of Constructions, in *Types for Proofs and Programs*, Int. Workshop, Torino, eds. S. Berardi and M. Coppo, LNCS 1158, 1996, pp. 249–264.

[Streicher 1991] T. Streicher, Independence of the induction principle and the axiom of choice in the pure calculus of constructions, *TCS* 103(2), pp 395 - 409.

Strong Normalization of Classical Natural Deduction with Disjunction

Philippe de Groote

LORIA UMR n° 7503 – INRIA
Campus Scientifique, B.P. 239
54506 Vandœuvre lès Nancy Cedex – France
e-mail: Philippe.de.Groote@loria.fr
phone: +33 3 83 59 30 32
fax: +33 3 83 27 83 19

Abstract. We introduce $\boldsymbol{\lambda\mu}^{\cdot\ \cdots}$, an extension of Parigot's $\lambda\mu$-calculus where disjunction is taken as a primitive. The associated reduction relation, which includes the permutative conversions related to disjunction, is Church-Rosser, strongly normalizing, and such that the normal deductions satisfy the subformula property. From a computer science point of view, $\boldsymbol{\lambda\mu}^{\cdot\ \cdots}$ may be seen as the core of a typed CBN functional language featuring product, coproduct, and control operators.

1 Introduction

During this last decade, several authors have investigated the relation existing between functional control operators, on the one hand, and normalization procedures for classical logic, on the other hand. This research originated in Griffin's observation [13] that Felleisen's control operator $\mathcal{C}$ [8,9] may be typed with the classical tautology $\neg\neg\alpha \to \alpha$.

Griffin's discovery resulted in lot of work aiming at extending the Curry-Howard correspondance [14] to the case of classical logic [1,2,4,6,11,15,17,18,19, 26]. On the proof-theoretic side, the problem consists in defining a classical natural deduction system together with an appropriate proof normalization procedure such that the resulting normal deductions satisfy the subformula property. This ensures that proof normalization may be interpreted as an evaluation process.

In fact, it appears *a posteriori* that the line of research opened by Griffin may be traced back to Prawitz [23] who shows how to normalize deductions in the presence of the following classical absurdity rule:

$$\frac{\begin{array}{c}[\neg\alpha]\\ \bot\end{array}}{\alpha}\ (\bot_c)$$

In order to obtain the subformula property, Prawitz restricts the use of Rule $\bot_c$ to the case where α is atomic. Then he shows how to transform any deduction in order to fulfill this requirement. For instance, any application of Rule $\bot_c$ whose conclusion is an implication may be transformed as follows:

S. Abramsky (Ed.): TLCA 2001, LNCS 2044, pp. 182–196, 2001.

$$\dfrac{\begin{array}{c}[\neg(\alpha\to\beta)]\\ \vdots\ \Pi_1\\ \bot\end{array}}{\alpha\to\beta} \quad\text{reduces to}\quad \dfrac{\dfrac{\begin{array}{c}\dfrac{\dfrac{\overset{(1)}{\neg\beta}\qquad \dfrac{\overset{(2)}{\alpha\to\beta}\quad\overset{(3)}{\alpha}}{\beta}}{\bot}}{\neg(\alpha\to\beta)}\,{\scriptstyle(2)}\\ \vdots\ \Pi_1\\ \bot\end{array}}{\beta}\,{\scriptstyle(1)}}{\alpha\to\beta}\,{\scriptstyle(3)} \tag{1}$$

It is worth noting that this reduction, which dates back to 1965, is strongly related to Felleisen's operator $\mathcal{C}$. Indeed it corresponds precisely to the following rewriting rule:

$$\mathcal{C}\,(\lambda k.\,M) \to \lambda n.\,\mathcal{C}\,(\lambda k.\,M[k{:=}\lambda f.\,k\,(f\,n)]),$$

which is reminiscent of one of Felleisen's rules [9].

Prawitz gives similar rules for conjunction and the universal quantifier. For disjunction, however, there is a problem. Indeed, restricting the application of Rule $\bot_c$ to the case where its conclusion is atomic would imply, in the presence of disjunction, the existence of a universal decision procedure. A similar problem arises with the existential quantifier. Consequently Prawitz does not take disjunction and existential quantification as primitives.

A way of circumventing this problem is to observe that reductions akin to (1) are only needed when the conclusion of Rule $\bot_c$ is the principal formula of an elimination rule. In the case of implication, this idea gives rise to the following reduction scheme:

$$\dfrac{\dfrac{\begin{array}{c}[\neg(\alpha\to\beta)]\\ \vdots\ \Pi_1\\ \bot\end{array}}{\alpha\to\beta}\qquad \begin{array}{c}\vdots\ \Pi_2\\ \alpha\end{array}}{\beta} \quad\text{reduces to}\quad \dfrac{\begin{array}{c}\dfrac{\dfrac{\overset{(1)}{\neg\beta}\qquad \dfrac{\overset{(2)}{\alpha\to\beta}\quad\begin{array}{c}\vdots\ \Pi_2\\ \alpha\end{array}}{\beta}}{\bot}}{\neg(\alpha\to\beta)}\,{\scriptstyle(2)}\\ \vdots\ \Pi_1\\ \bot\end{array}}{\beta}\,{\scriptstyle(1)} \tag{2}$$

which is used by both [30] and [26], and which amounts, modulo some additional β-contraction steps, to Parigot's μ-reduction [19]. Applying the same idea to the case of disjunction yields the following figure:

$$\dfrac{\dfrac{\begin{array}{c}[\neg(\alpha\vee\beta)]\\ \vdots\ \Pi_1\\ \bot\end{array}}{\alpha\vee\beta} \quad \begin{array}{c}[\alpha]\\ \vdots\ \Pi_2\\ \gamma\end{array} \quad \begin{array}{c}[\beta]\\ \vdots\ \Pi_3\\ \gamma\end{array}}{\gamma} \quad \text{reduces to} \quad \begin{array}{c} \dfrac{\dfrac{\overset{(1)}{\neg\gamma} \quad \dfrac{\overset{(2)}{\alpha\vee\beta}\quad \begin{array}{c}[\alpha]\\ \vdots\ \Pi_2\\ \gamma\end{array}\quad \begin{array}{c}[\beta]\\ \vdots\ \Pi_3\\ \gamma\end{array}}{\gamma}}{\bot}}{\neg(\alpha\vee\beta)}\,{\scriptstyle(2)} \\ \vdots\ \Pi_1 \\ \dfrac{\bot}{\gamma}\,{\scriptstyle(1)}\end{array} \tag{3}$$

which corresponds precisely to one of the reduction rules proposed in [26]. Reduction (3), however, is not sufficient to solve the problems related to disjunction. Indeed, in order to obtain the subformula property, one also needs the so-called permutative conversions [31]. Consequently, normalization procedures for full classical logic [29,30] might be rather intricate since they involve three kinds of reduction steps:

- the usual detour conversions of intuitionistic natural deduction,
- conversions related to Rule $\bot_c$,
- permutative conversions related to disjunction.

In the present paper, we revisit this problem from a type-theoretic point of view. Our main contribution is the definition of an extension of the $\lambda\mu$-calculus ($\boldsymbol{\lambda\mu}^{\rightarrow\wedge\vee\bot}$) such that:

(a) intuitionistic disjunction—i.e., coproduct—is taken as a primitive;
(b) normal deductions satisfy the subformula property;
(c) the reduction relation is defined by means of local reduction steps;[1]
(d) the reduction relation is proven to be strongly normalizing;
(e) the reduction relation is Church-Rosser;
(f) the reduction relation is defined at the untyped level;[2]
(g) the reduction relation satisfies the subject reduction property.

Properties (e), (f), and (g) are of special interest from a programming language perspective. Property (e) ensures the unicity of the normal forms, which allows the reduction relation to be considered as the core of an operational semantics. Properties (f) and (g), on the other hand, allow the typing information to be ignored at run time.

[1] By a *local* reduction step, we mean a rewriting rule that is compatible with the term formation rules, i.e., a rewriting rule $s \rightarrow t$ such that $C[s] \rightarrow C[t]$ independently of the shape of the context C.

[2] From a proof-theoretic point of view, it means that the proof normalization steps are defined on the shape of the derivations, according to the inference rules that are used and without any proviso on the formulas that are introduced or eliminated.

To the best of our knowledge, there is no *classical* extension of the simply typed λ-calculus (or equivalently, no normalisation procedure for classical natural deduction) that enjoys all of the above properties. In particular, neither (d) nor (e) are satisfied by [29]. The handling of the permutative conversions in [30] does not satisfy (c) and (e), while the handling of classical negation does not satisfy (f). Finally, in [26], the extension of λ_{Δ} that satifies (a) does not satisfy (b).

In a series of paper [25,27,28], Ritter, Pym, and Wallen introduce an extension of the $\lambda\mu$-calculus that also features disjunction as a primitive. Their system is rather different from ours because they take as primitive a classical form of disjunction that amounts to $\neg A \rightarrow B$. Nevertheless, in [25], Pym and Ritter give a brief account of another extension of the $\lambda\mu$-calculus with an intuitionistic disjunction. However, the reduction rules they give are not sufficient to guarantee that the normal proofs satisfy the subformula property.

In order to prove that our system is strongly normalizable, we use the method that we have introduced in [7]. This yields several auxiliary results that have independent interest:

- We reduce, by finitary means, the strong normalization of classical logic to the strong normalization of intuitionistic implicative logic. Consequently, when combined with an arithmetizable proof of the strong normalization of the simply typed λ-calculus, our method yields a completely arithmetizable proof of the strong normalization of classical propositional logic.
- We prove the strong normalization of Parigot's μ-reduction (together with similar reduction relations) on the *untyped terms*. This result confirms that the reduction relations related to the classical absurdity rule are structural reductions, akin to permutative conversions, which do not carry any real computational content.
- We provide $\boldsymbol{\lambda\mu}^{\rightarrow\wedge\vee\perp}$ with a *continuation passing style* semantics that may be used to construct denotational models of classical logic with disjunction as a primitive.
- Our CPS-translation, since it is defined on the untyped $\lambda\mu$-terms, may be raised to the second order as in [5]. This yields a new proof of the strong normalization of Parigot's original second-order $\lambda\mu$-calculus. In contrast to [20, 21], this proof does not require any extension of the reducibility candidate method since it consists in reducing the problem to the strong normalization of Girard's system F [12].

2 Classical Propositional Logic as an Extension of the $\boldsymbol{\lambda\mu}$-Calculus

The types of $\boldsymbol{\lambda\mu}^{\rightarrow\wedge\vee\perp}$ are the formulas of propositional logic built upon a finite set of atomic types, using the connectives $\rightarrow$, $\vee$, and $\wedge$. The set of atomic types contains the constant $\perp$ that stands for absurdity, and negation is defined in the usual way: $\neg\alpha \equiv \alpha \rightarrow \perp$.

The untyped terms of $\boldsymbol{\lambda\mu}^{\rightarrow\wedge\vee\bot}$ (that we will call $\lambda\mu$-terms, for short) are built upon two disjoint alphabets of variables $\mathcal{X}$ and $\mathcal{A}$, according to the following grammar:

$$\begin{aligned}\mathcal{T} ::= \mathcal{X} \mid \lambda\mathcal{X}.\,\mathcal{T} \mid \mathcal{T}\,\mathcal{T} \mid \langle\mathcal{T},\mathcal{T}\rangle \mid \pi_1\,\mathcal{T} \mid \pi_2\,\mathcal{T} \mid \\ \iota_1\,\mathcal{T} \mid \iota_2\,\mathcal{T} \mid \delta(\mathcal{T},\mathcal{X}.\,\mathcal{T},\mathcal{X}.\,\mathcal{T}) \mid \mu\mathcal{A}.\,\mathcal{T} \mid \mathcal{A}\,\mathcal{T}\end{aligned}$$

The elements of $\mathcal{X}$ are called λ-variables, and these of $\mathcal{A}$ are called μ-variables. We let letters from the end of the alphabet (x, y, z) range over $\mathcal{X}$, and letters from the beginning of the alphabet (a, b, c) range over $\mathcal{A}$. δ and μ are binding operators. Any free occurrence of x in N and any free occurrence of y in O is bound in the term $\delta(M, x.\,N, y.\,O)$. Similarly, any free occurrence of a in M is bound in $\mu a.\,M$. We consider that some implicit convention (e.g., [3, p. 26]) prevents clashes between free and bound variables. We write "=", without subscript, for the relation of α-conversion, and we let $M[x{:=}N]$ denote the usual capture-avoiding substitution.

We define an antecedent to be a set of declarations of the form $x{:}\alpha$ where x is a λ-variable, α is a type, and where all the declared variables are distinct. Similarly, we define a succedent to be a set of declarations of μ-variables obeying the same constraints. The typing system of $\boldsymbol{\lambda\mu}^{\rightarrow\wedge\vee\bot}$ is given by means of sequents of the form $\Gamma \vdash M : \alpha;\, \Delta$, where Γ is an antecedent, M is a $\lambda\mu$-term, α is a type, and Δ is a succedent. The typing rules are the following:

$$x : \alpha, \Gamma \vdash x : \alpha;\, \Delta$$

$$\frac{x : \alpha, \Gamma \vdash M : \beta;\, \Delta}{\Gamma \vdash \lambda x.\,M : \alpha \rightarrow \beta;\, \Delta}\ (\rightarrow\text{-I}) \qquad \frac{\Gamma \vdash M : \alpha \rightarrow \beta;\, \Delta \quad \Gamma \vdash N : \alpha;\, \Delta}{\Gamma \vdash M\,N : \beta;\, \Delta}\ (\rightarrow\text{-E})$$

$$\frac{\Gamma \vdash M : \alpha;\, \Delta \quad \Gamma \vdash N : \beta;\, \Delta}{\Gamma \vdash \langle M, N\rangle : \alpha \wedge \beta;\, \Delta}\ (\wedge\text{-I})$$

$$\frac{\Gamma \vdash M : \alpha \wedge \beta;\, \Delta}{\Gamma \vdash \pi_1\,M : \alpha;\, \Delta}\ (\wedge\text{-E1}) \qquad \frac{\Gamma \vdash M : \alpha \wedge \beta;\, \Delta}{\Gamma \vdash \pi_2\,M : \beta;\, \Delta}\ (\wedge\text{-E2})$$

$$\frac{\Gamma \vdash M : \alpha;\, \Delta}{\Gamma \vdash \iota_1\,M : \alpha \vee \beta;\, \Delta}\ (\vee\text{-I1}) \qquad \frac{\Gamma \vdash M : \beta;\, \Delta}{\Gamma \vdash \iota_2\,M : \alpha \vee \beta;\, \Delta}\ (\vee\text{-I2})$$

$$\frac{\Gamma \vdash M : \alpha \vee \beta;\, \Delta \quad x : \alpha, \Gamma \vdash N : \gamma;\, \Delta \quad y : \beta, \Gamma \vdash O : \gamma;\, \Delta}{\Gamma \vdash \delta(M, x.\,N, y.\,O) : \gamma;\, \Delta}\ (\vee\text{-E})$$

$$\frac{\Gamma \vdash M : \bot;\, \Delta, a : \alpha}{\Gamma \vdash \mu a.\,M : \alpha;\, \Delta}\ (\textsc{muabs}) \qquad \frac{\Gamma \vdash M : \alpha;\, \Delta, a : \alpha}{\Gamma \vdash a\,M : \bot;\, \Delta, a : \alpha}\ (\textsc{name})$$

The one-step reduction relation of $\boldsymbol{\lambda\mu}^{\rightarrow\wedge\vee\bot}$ is defined as the union of three different one-step reduction relations: the relation of detour-reduction ($\rightarrow_D$), which corresponds to the usual detour conversions of intuitionistic logic; the relation of δ-reduction ($\rightarrow_\delta$), which corresponds to the permutative conversions related to the elimination rule of disjunction; the relation of μ-reduction ($\rightarrow_\mu$), which corresponds to conversions that are proper to classical logic. Following Barendregt [3], we write "$\stackrel{+}{\rightarrow}_X$" and "$\twoheadrightarrow_X$" to denote, repectively, the transitive closure and the transitive, reflexive closure of a reduction relation "$\rightarrow_X$".

Definition 1. *(detour-reduction)*

(a) $(\lambda x.\, M)\, N \rightarrow_D M[x{:=}N]$
(b) $\pi_1\,\langle M, N\rangle \rightarrow_D M$
(c) $\pi_2\,\langle M, N\rangle \rightarrow_D N$
(d) $\delta(\iota_1\, M, x.\, N, y.\, O) \rightarrow_D N[x{:=}M]$
(e) $\delta(\iota_2\, M, x.\, N, y.\, O) \rightarrow_D O[y{:=}M]$ ■

Definition 2. *(δ-reduction)*

(a) $\delta(M, x.\, N, y.\, O)\, P \rightarrow_\delta \delta(M, x.\, N\, P, y.\, O\, P)$
(b) $\pi_1\, \delta(M, x.\, N, y.\, O) \rightarrow_\delta \delta(M, x.\, \pi_1\, N, y.\, \pi_1\, O)$
(c) $\pi_2\, \delta(M, x.\, N, y.\, O) \rightarrow_\delta \delta(M, x.\, \pi_2\, N, y.\, \pi_2\, O)$
(d) $\delta(\delta(M, x.\, N, y.\, O), u.\, P, v.\, Q) \rightarrow_\delta \delta(M, x.\, \delta(N, u.\, P, v.\, Q), y.\, \delta(O, u.\, P, v.\, Q))$
■

In order to define the different basic μ-reduction steps, we must first introduce a notion of structural substitution. Let $C[\,]$ be a context, (i.e., a $\lambda\mu$-term with a hole). The structural substitution $M[a * := C[*]]$ is inductively defined as follows:

$$
\begin{aligned}
&x^* = x\\
&(\lambda x.\, M)^* = \lambda x.\, (M^*)\\
&(M\, N)^* = M^*\, N^*\\
&\langle M, N\rangle^* = \langle M^*, N^*\rangle\\
&(\pi_i\, M)^* = \pi_i\, (M^*)\\
&(\iota_i\, M)^* = \iota_i\, (M^*)\\
&\delta(M, x.\, N, y.\, O)^* = \delta(M^*, x.\, N^*, y.\, O^*)\\
&(\mu b.\, M)^* = \mu b.\, (M^*) \quad \text{if } a \neq b\\
&(\mu a.\, M)^* = \mu a.\, M\\
&(b\, M)^* = b\, (M^*) \quad \text{if } a \neq b\\
&(a\, M)^* = C[M^*]
\end{aligned}
$$

where M^* stands for $M[a * := C[*]]$.

Definition 3. *(μ-reduction)*

(a) $(\mu a.\, M)\, N \rightarrow_\mu \mu a.\, M[a * := a\, (*\, N)]$
(b) $\pi_1\, (\mu a.\, M) \rightarrow_\mu \mu a.\, M[a * := a\, (\pi_1\, *)]$
(c) $\pi_2\, (\mu a.\, M) \rightarrow_\mu \mu a.\, M[a * := a\, (\pi_2\, *)]$

(d) $\delta(\mu a.\, M, x.\, N, y.\, O) \rightarrow_{\mu} \mu a.\, M[a * := a\, \delta(*, x.\, N, y.\, O)]$

where $a \in \mathrm{FV}(M)$. ■

In the above definition, we stipulate that a must occur free in M. Nevertheless, the above reductions also make sense when the μ-abstraction is vacuous. In this case, they correspond to the $\perp$-reductions of intuitionistic logic [31], which are also needed for the subformula property. These $\perp$-reductions may therefore be seen as particular cases of μ-reductions. However, for technical reasons, we prefer to keep them separate.

Definition 4. *($\perp$-reduction)*

(a) $(\mu a.\, M)\, N \rightarrow_{\perp} \mu a.\, M$
(b) $\pi_1\, (\mu a.\, M) \rightarrow_{\perp} \mu a.\, M$
(c) $\pi_2\, (\mu a.\, M) \rightarrow_{\perp} \mu a.\, M$
(d) $\delta(\mu a.\, M, x.\, N, y.\, O) \rightarrow_{\perp} \mu a.\, M$

where $a \notin \mathrm{FV}(M)$. ■

The next proposition ensures that the above relations of reduction, which are defined at the untyped level, correspond indeed to proof-theoretic conversions.

Proposition 1. (Subject reduction) *Let M, N, α, Γ, and Δ be such that $\Gamma \vdash M : \alpha;\ \Delta$ and $M \rightarrow_{D\delta\mu\perp} N$. Then $\Gamma \vdash N : \alpha;\ \Delta$.* □

We end this section by giving a characterization of the normal terms, from which we derive the subformula property. Consider the following grammar:

$$\begin{aligned} \mathcal{P} &::= \lambda \mathcal{X}.\, \mathcal{P} \mid \langle \mathcal{P}, \mathcal{P} \rangle \mid \iota_1\, \mathcal{P} \mid \iota_2\, \mathcal{P} \mid \delta(\mathcal{Q}, \mathcal{X}.\, \mathcal{P}, \mathcal{X}.\, \mathcal{P}) \mid \mu \mathcal{A}.\, \mathcal{P} \mid \mathcal{Q} \\ \mathcal{Q} &::= \mathcal{X} \mid \mathcal{Q}\mathcal{P} \mid \pi_1\, \mathcal{Q} \mid \pi_2\, \mathcal{Q} \mid \mathcal{A}\mathcal{P} \end{aligned}$$

we say that a $\lambda\mu$-term that conforms to $\mathcal{P}$ is P-canonical (or simply, canonical). Similarly, a $\lambda\mu$-term that conforms to $\mathcal{Q}$ is said to be Q-canonical.

Lemma 1. *Let M, α, Γ, and Δ be such that*

$$\Gamma \vdash M : \alpha;\ \Delta \qquad (*)$$

and let Π be the derivation of $(*)$.

(a) *If M is Q-canonical then every type occurring in Π is either $\perp$, or a subformula of a type occurring in Γ or Δ.*
(b) *If M is P-canonical then every type occurring in Π is either $\perp$, or a subformula of a type occurring in Γ or Δ, or a subformula of α.* □

Lemma 2. *Let M, α, Γ, and Δ be such that $\Gamma \vdash M : \alpha;\ \Delta$. If M is $D\delta\mu\perp$-normal then M is canonical.* □

We get the subformula property as a direct consequence of these two lemmas.

Proposition 2. *Let M, α, Γ, and Δ be such that*

$$\Gamma \vdash M : \alpha;\, \Delta \qquad (*)$$

If M is $D\delta\mu\bot$-normal then every type occurring in the derivation of $(*)$ is either $\bot$, or a subformula of a type occurring in Γ or Δ, or a subformula of α. □

3 Strong Normalisation of the Structural Reductions

In this section, we prove that the untyped $\lambda\mu$-terms are strongly normalizable with respect to the reduction relation induced by the structural reduction steps (i.e., δ, μ, and $\bot$). To this end, we provide the $\lambda\mu$-terms with a norm that strictly decreases under the relation of $\delta\mu\bot$-reduction. This norm is adapted from the norm that we introduced in [7]. Nevertheless it is more involved because we have to accommodate the structural substitutions of the μ-reductions.

Definition 5. *The norm $|\cdot|$ assigned to the $\lambda\mu$-terms is inductively defined as follows:*

(a) $|x| = 1;$
(b) $|\lambda x.\, M| = |M|$
(c) $|M\, N| = |M| + \#M \times |N|$
(d) $|\langle M, N\rangle| = |M| + |N|$
(e) $|\pi_1\, M| = |M| + \#M$
(f) $|\pi_2\, M| = |M| + \#M$
(g) $|\iota_1\, M| = |M|$
(h) $|\iota_2\, M| = |M|$
(i) $|\delta(M, x.\, N, y.\, O)| = |M| + \#M \times (|N| + |O|)$
(j) $|\mu a.\, M| = |M|$
(k) $|a\, M| = |M|$

where:

(a) $\#x = 1;$
(b) $\#\lambda x.\, M = 1$
(c) $\#M\, N = \#M$
(d) $\#\langle M, N\rangle = 1$
(e) $\#\pi_1\, M = \#M$
(f) $\#\pi_2\, M = \#M$
(g) $\#\iota_1\, M = 1$
(h) $\#\iota_2\, M = 1$
(i) $\#\delta(M, x.\, N, y.\, O) = 2 \times \#M \times (\#N + \#O)$
(j) $\#\mu a.\, M = \lfloor M \rfloor_a + 1$
(k) $\#a\, M = 1$

and where:

(a) $\lfloor x \rfloor_a = 0$;
(b) $\lfloor \lambda x.\, M \rfloor_a = \lfloor M \rfloor_a$
(c) $\lfloor M\, N \rfloor_a = \lfloor M \rfloor_a + \#M \times \lfloor N \rfloor_a$
(d) $\lfloor \langle M, N \rangle \rfloor_a = \lfloor M \rfloor_a + \lfloor N \rfloor_a$
(e) $\lfloor \pi_1\, M \rfloor_a = \lfloor M \rfloor_a$
(f) $\lfloor \pi_2\, M \rfloor_a = \lfloor M \rfloor_a$
(g) $\lfloor \iota_1\, M \rfloor_a = \lfloor M \rfloor_a$
(h) $\lfloor \iota_2\, M \rfloor_a = \lfloor M \rfloor_a$
(i) $\lfloor \delta(M, x.\, N, y.\, O) \rfloor_a = \lfloor M \rfloor_a + \#M \times (\lfloor N \rfloor_a + \lfloor O \rfloor_a)$
(j) $\lfloor \mu b.\, M \rfloor_a = \lfloor M \rfloor_a$
(k) $\lfloor a\, M \rfloor_a = \lfloor M \rfloor_a + \#M$
(l) $\lfloor b\, M \rfloor_a = \lfloor M \rfloor_a$ ■

This norm is strictly positive and compatible with the term formation rules.

Lemma 3. *Let $C[\,]$ be any context.*

(a) *If $\#M \geq \#N$ and $|M| > |N|$ then $|C[M]| > |C[N]|$;*
(b) *If $\#M \geq \#N$ and $\lfloor M \rfloor_a \geq \lfloor N \rfloor_a$ then $\#C[M] \geq \#C[N]$ and $\lfloor C[M] \rfloor_a \geq \lfloor C[N] \rfloor_a$.* □

The next proposition may be easily established by adapting the proof given in [7].

Proposition 3. *If $S \to_\delta T$ then $|S| > |T|$, for any $\lambda\mu$-terms S, T.* □

We end this section by sketching the proof that the norm of Definition 5 decreases under the relation of $\mu\bot$-reduction.

Lemma 4. *Let M, N be $\lambda\mu$-terms and let $M^* \equiv M[a\,{*}{:=}a\,({*}\,N)]$. If $a \notin \mathrm{FV}(N)$ then:*

(a) $\lfloor M \rfloor_a = \lfloor M^* \rfloor_a$;
(b) $\lfloor M \rfloor_b + \lfloor M \rfloor_a \times \lfloor N \rfloor_b = \lfloor M^* \rfloor_b$;
(c) $|M| + \lfloor M \rfloor_a \times |N| = |M^*|$. □

Lemma 5. *Let S, T be $\lambda\mu$-terms. If $S \to_{\mu\bot} T$ then:*

(a) $\#S \geq \#T$;
(b) $\lfloor S \rfloor_b \geq \lfloor T \rfloor_b$.

Proof. We establish the property as a consequence of Lemma 3 (b), by proving Inequations (a) and (b) for each basic reduction step. We give only the first case, and leave the other ones to the reader.

(a)
$$\begin{aligned}
\#S &= \#(\mu a.\, M)\, N \\
&= \#\mu a.\, M \\
&= \lfloor M \rfloor_a + 1 \\
&= \lfloor M[a\,{*}{:=}a\,({*}\,N)] \rfloor_a + 1 \qquad \textit{by Lemma 4 (a)} \\
&= \#\mu a.\, M[a\,{*}{:=}a\,({*}\,N)] \\
&= \#T
\end{aligned}$$

(b) $$\begin{aligned} \lfloor S \rfloor_b &= \lfloor (\mu a.\, M)\, N \rfloor_b \\ &= \lfloor \mu a.\, M \rfloor_b + \#\mu a.\, M \times \lfloor N \rfloor_b \\ &= \lfloor M \rfloor_b + (\lfloor M \rfloor_a + 1) \times \lfloor N \rfloor_b \\ &\geq \lfloor M \rfloor_b + \lfloor M \rfloor_a \times \lfloor N \rfloor_b \\ &= \lfloor M[a * := a\,(*\, N)] \rfloor_b \qquad \textit{by Lemma 4 (b)} \\ &= \lfloor \mu a.\, M[a * := a\,(*\, N)] \rfloor_b \\ &= \lfloor T \rfloor_b \end{aligned}$$

□

Proposition 4. *If $S \to_{\mu\bot} T$ then $|S| > |T|$, for any $\lambda\mu$-term S, T.*

Proof. *We establish the property as a consequence of Lemma 3 (a) and Lemma 5, by proving it for each basic reduction step. We give only the first case, and leave the other ones to the reader.*

$$\begin{aligned} |S| &= |(\mu a.\, M)\, N| \\ &= |\mu a.\, M| + \#\mu a.\, M \times |N| \\ &= |M| + (\lfloor M \rfloor_a + 1) \times |N| \\ &> |M| + \lfloor M \rfloor_a \times |N| \\ &= |M[a * := a\,(*\, N)]| \qquad \textit{by Lemma 4 (c)} \\ &= |\mu a.\, M[a * := a\,(*\, N)]| \\ &= |T| \end{aligned}$$

□

As a direct consequence of Propositions 3 and 4, we obtain that any $\lambda\mu$-term is strongly $\delta\mu\bot$-normalizable.

4 Postponement of the $\bot$-Reductions

All the right-hand sides of the rules of Definition 4 are of the same form: $\mu a.\, M$, where $a \notin \mathrm{FV}(M)$. One easily checks that there is no critical pair between this form and any left-hand side of the rules of Definitions 1, 2 and 3. Consequently, the following proposition may be established easily.

Proposition 5. *Let $R \in \{D, \delta, \mu\}$, and let L, M and N be three $\lambda\mu$-terms. If $L \to_\bot M \to_R N$ then there exists a $\lambda\mu$-term O such that $L \overset{+}{\to}_R O \twoheadrightarrow_\bot N$.* □

As a consequence of this proposition together with the strong normalization of the $\bot$-reductions, we have that any infinite sequence of $D\delta\mu\bot$-reduction steps may be turned into an infinite sequence of $D\delta\mu$-reduction steps.

5 Negative Translation and CPS-Simulation

In this section, we adapt to $\boldsymbol{\lambda\mu}^{\rightarrow\wedge\vee\bot}$ the negative translation and the CPS-simulation given in [7].

Definition 6. *The negative translation* $\overline{\alpha}$ *of any type* α *is defined as* $\overline{\alpha} = {\sim}{\sim}\alpha^\circ$ *where* ${\sim}\alpha = \alpha \rightarrow o$ *for some distinguished atomic type* o *(that is not used elsewhere), and where:*

(a) $\bot^\circ = o$
(b) $a^\circ = a$
(c) $(\alpha \rightarrow \beta)^\circ = \overline{\alpha} \rightarrow \overline{\beta}$
(d) $(\alpha \wedge \beta)^\circ = {\sim}(\overline{\alpha} \rightarrow {\sim}\overline{\beta})$
(e) $(\alpha \vee \beta)^\circ = {\sim}\overline{\alpha} \rightarrow {\sim}{\sim}\overline{\beta}$ ■

Definition 7. *The CPS-translation* $\overline{M}$ *of any* $\lambda\mu$*-term* M *is inductively defined as follows:*

(a) $\overline{x} = \lambda k.\, x\, k$
(b) $\overline{\lambda x.\, M} = \lambda k.\, k\, (\lambda x.\, \overline{M})$
(c) $\overline{(M\, N)} = \lambda k.\, \overline{M}\, (\lambda m.\, m\, \overline{N}\, k)$
(d) $\overline{\langle M, N\rangle} = \lambda k.\, k\, (\lambda p.\, p\, \overline{M}\, \overline{N})$
(e) $\overline{\pi_1\, M} = \lambda k.\, \overline{M}\, (\lambda p.\, p\, (\lambda i.\, \lambda j.\, i\, k))$
(f) $\overline{\pi_2\, M} = \lambda k.\, \overline{M}\, (\lambda p.\, p\, (\lambda i.\, \lambda j.\, j\, k))$
(g) $\overline{\iota_1\, M} = \lambda k.\, k\, (\lambda i.\, \lambda j.\, i\, \overline{M})$
(h) $\overline{\iota_2\, M} = \lambda k.\, k\, (\lambda i.\, \lambda j.\, j\, \overline{M})$
(i) $\overline{\delta(M, x.\, N, y.\, O)} = \lambda k.\, \overline{M}\, (\lambda m.\, m\, (\lambda x.\, \overline{N}\, k)\, (\lambda y.\, \overline{O}\, k))$
(j) $\overline{\mu a.\, M} = \lambda a.\, \overline{M}\, (\lambda k.\, k)$
(k) $\overline{a\, M} = \lambda k.\, \overline{M}\, a$

where k*,* m*,* p*,* i *and* j *are fresh variables.* ■

The next proposition states that these two translations commute with the typing relation.

Proposition 6. *Let* M*,* α*,* Γ*, and* Δ *be such that* $\Gamma \vdash M : \alpha; \Delta$ *Then* $\overline{M}$ *is a* λ*-term of the simply typed* λ*-calculus, typable with type* $\overline{\alpha}$ *under the set of declarations* $\overline{\Gamma}, {\sim}\Delta^\circ$. □

The translation of Definition 7 does not allow the detour-reduction steps to be simulated by β-reduction. This is due to the so-called administrative redexes [22]. In order to circumvent this problem, we introduce the following modified translation.

Definition 8. *The modified CPS-translation* $\overline{\overline{M}}$ *of any* $\lambda\mu$*-term* M *is defined as:*

$$\overline{\overline{M}} = \lambda k.\, (M : k)$$

where k *is a fresh variable, and where the infix operator ":" obeys the following definition:*

(a) $x : K = x\,K$
(b) $\lambda x.\,M : K = K\,(\lambda x.\,\overline{\overline{M}})$
(c) $(M\,N) : K = M : \lambda m.\,m\,\overline{\overline{N}}\,K$
(d) $\langle M, N\rangle : K = K\,(\lambda p.\,p\,\overline{\overline{M}}\,\overline{\overline{N}})$
(e) $\pi_1\,M : K = M : \lambda p.\,p\,(\lambda i.\,\lambda j.\,i\,K)$
(f) $\pi_2\,M : K = M : \lambda p.\,p\,(\lambda i.\,\lambda j.\,j\,K)$
(g) $\iota_1\,M : K = K\,(\lambda i.\,\lambda j.\,i\,\overline{\overline{M}})$
(h) $\iota_2\,M : K = K\,(\lambda i.\,\lambda j.\,j\,\overline{\overline{M}})$
(i) $\delta(M, x.\,N, y.\,O) : K = M : \lambda m.\,m\,(\lambda x.\,(N : K))\,(\lambda y.\,(O : K))$
(j) $\mu a.\,M : K = (M : \lambda k.\,k)\,[a{:=}K]$ *if* $a \in \mathrm{FV}(M)$
(k) $\mu a.\,M : K = (\lambda a.\,(M : \lambda k.\,k))\,K$ *if* $a \notin \mathrm{FV}(M)$
(l) $a\,M : K = M : a$

where m, p, i and j are fresh variables. ∎

This modified CPS-translation is consistent with the translation of Definition 7 in the sense of the following lemma.

Lemma 6. *Let M be a $\lambda\mu$-term. Then:*

(a) $\overline{M} \twoheadrightarrow_\beta \overline{\overline{M}}$,
(b) $\overline{M}\,K \twoheadrightarrow_\beta M : K$, *for any λ-term K.* □

As a consequence of this lemma, we obtain the following proposition.

Proposition 7. *Let M, α, Γ, and Δ be such that $\Gamma \vdash M : \alpha;\,\Delta$. Then $\overline{\overline{M}}$ is a λ-term of the simply typed λ-calculus, typable with type $\overline{\alpha}$ under the set of declarations $\overline{\Gamma}, {\sim}\Delta^{\circ}$.* □

We now prove that the modified CPS-translation of Definiton 8 simulates the relation of detour-reduction by strict β-reduction, and the relations of δ- and μ-reduction by equality. We first state a few technical lemmas.

Lemma 7. *Let M and N be $\lambda\mu$-terms and K be a simple λ-term such that $x \notin \mathrm{FV}(K)$. Then $(M : K)[x{:=}\overline{\overline{N}}] \twoheadrightarrow_\beta (M[x{:=}N]) : K$.* □

Lemma 8. *Let M, N, and O be $\lambda\mu$-terms, and K be a simple λ-term such that $a \notin \mathrm{FV}(K)$. Then:*

(a) $(M : K)[a{:=}\lambda m.\,m\,\overline{\overline{N}}\,a] \twoheadrightarrow_\beta (M[a\,{*}{:=}\,a\,({*}\,N)]) : K$,
(b) $(M : K)[a{:=}\lambda p.\,p\,(\lambda i.\,\lambda j.\,i\,a)] \twoheadrightarrow_\beta (M[a\,{*}{:=}\,a\,(\pi_1\,{*})]) : K$,
(c) $(M : K)[a{:=}\lambda p.\,p\,(\lambda i.\,\lambda j.\,j\,a)] \twoheadrightarrow_\beta (M[a\,{*}{:=}\,a\,(\pi_2\,{*})]) : K$,
(d) $(M : K)[a{:=}\lambda m.\,m\,(\lambda x.\,(N : a))\,(\lambda y.\,(O : a))]$
$\twoheadrightarrow_\beta (M[a\,{*}{:=}\,a\,\delta({*}, x.\,N, y.\,O)]) : K$. □

Lemma 9. *Let M and N be two $\lambda\mu$-terms and let $C[\,]$ be any context. Then, for any simple λ-term K:*

(a) *if* $M : K \xrightarrow{+}_\beta N : K$ *then* $C[M] : K \xrightarrow{+}_\beta C[N] : K$,
(b) *if* $M : K = N : K$ *then* $C[M] : K = C[N] : K$. □

The next two lemmas concern the simulation of the relations of detour- and δ-reduction. Their proofs may be found in [7]

Lemma 10. *Let* S *and* T *be two* $\lambda\mu$*-terms. If* $S \to_D T$ *then* $\overline{\overline{S}} \xrightarrow{+}_\beta \overline{\overline{T}}$. □

Lemma 11. *Let* S *and* T *be two* $\lambda\mu$*-terms. If* $S \to_\delta T$ *then* $\overline{\overline{S}} = \overline{\overline{T}}$. □

It remains to prove that the μ-reductions are interpreted as equality.

Lemma 12. *Let* S *and* T *be two* $\lambda\mu$*-terms such that* $S \to_\mu T$*. Then* $\overline{\overline{S}} = \overline{\overline{T}}$.

Proof. *We prove that, for any simple* λ*-term* K,

$$S : K = T : K \qquad (*)$$

from which the property follows.

Equation $(*)$ *may be established as a consequence of Lemma 9 (b) by proving that it holds for each basic reduction step. We give only the first case, and leave the other ones to the reader.*

$$\begin{aligned} S : K &= (\mu a.\, M)\, N : K \\ &= \mu a.\, M : \lambda m.\, m\, \overline{\overline{N}}\, K \\ &= (M : \lambda k.\, k)[a{:=}\lambda m.\, m\, \overline{\overline{N}}\, K] \\ &= (M : \lambda k.\, k)[a{:=}\lambda m.\, m\, \overline{\overline{N}}\, a][a{:=}K] \\ &= (M[a * {:=}\, a\, (* N)] : \lambda k.\, k)[a{:=}K] \qquad \text{by Lemma 8 (a)} \\ &= \mu a.\, M[a * {:=}\, a\, (* N)] : K \\ &= T \end{aligned}$$

□

6 Strong Normalization

We are now in a position of proving the main proposition of this paper.

Proposition 8. (Strong Normalization) *Any well-typed* $\lambda\mu$*-term of* $\boldsymbol{\lambda\mu}^{\to\wedge\vee\bot}$ *is strongly normalizable with respect to the relation of* $D\delta\mu\bot$*-reduction.*

Proof. Suppose it is not the case. Then, by Proposition 4 and 5, there would exist an infinite sequence of D- and $\delta\mu$*-reduction steps starting from a typable term (say,* M*) of* $\boldsymbol{\lambda\mu}^{\to\wedge\vee\bot}$*. If this infinite sequence contains infinitely many D-reduction steps, there must exist, by Lemmas 10, 11 and 12 an infinite sequence of* β*-reduction steps starting from* $\overline{\overline{M}}$*. But this, by Proposition 7, would contradict the strong normalization of the simply typed* λ*-calculus. Hence the infinite sequence may contain only a finite number of D-reduction steps. But then, it would contain an infinite sequence of consecutive* $\delta\mu$*-reduction steps, which is impossible by Propositions 3 and 4.* □

7 Confluence of the Reductions

We prove the Church-Rosser property by establishing the local confluence of the reductions.

Lemma 13. *Let M, N, O be $\lambda\mu$-terms such that $M \rightarrow_{D\delta\mu} N$ and $M \rightarrow_{D\delta\mu} O$ then there exists a $\lambda\mu$-term P such that $N \twoheadrightarrow_{D\delta\mu} P$ and $M \twoheadrightarrow_{D\delta\mu} P$.* □

Proposition 9. (Church-Rosser Property) *Let M, N, O be typable $\lambda\mu$-terms such that $M \twoheadrightarrow_{D\delta\mu} N$ and $M \twoheadrightarrow_{D\delta\mu} O$ then there exists a $\lambda\mu$-term P such that $N \twoheadrightarrow_{D\delta\mu} P$ and $M \twoheadrightarrow_{D\delta\mu} P$.*

Proof. A consequence of Proposition 8, Lemma 13, and Newman lemma. □

References

1. F. Barbanera and S. Berardi. Extracting constructive content from classical logic via control-like reductions. In M. Bezem and J.F. Groote, editors, *Proceedings of the International Conference on Typed Lambda Calculi and Applications, TLCA'93*, volume 664 of *Lecture Notes in Computer Science*, pages 45–59. Springer Verlag, 1993.
2. F. Barbanera and S. Berardi. A symmetric lambda-calculus for "classical" program extraction. In M. Hagiya and J.C. Mitchell, editors, *Proceedings of the International Symposium on Theoretical Aspects of Computer Software*, volume 789 of *Lecture Notes in Computer Science*, pages 494–515. Springer Verlag, 1994.
3. H.P. Barendregt. *The lambda calculus, its syntax and semantics.* North-Holland, revised edition, 1984.
4. R. Constable and C. Murthy. Finding computational content in classical proofs. In G. Huet and G. Plotkin, editors, *Logical Frameworks*, pages 341–362. Cambridge University Press, 1991.
5. Ph. de Groote. A CPS-translation of the $\lambda\mu$-calculus. In S. Tison, editor, *19th International Colloquium on Trees in Algebra and Programming, CAAP'94*, volume 787 of *Lecture Notes in Computer Science*, pages 85–99. Springer Verlag, 1994.
6. Ph. de Groote. A simple calculus of exception handling. In M. Dezani-Ciancaglini and G. Plotkin, editors, *Second International Conference oń Typed Lambda Calculi and Applications, TLCA'95*, volume 902 of *Lecture Notes in Computer Science*, pages 201–215. Springer Verlag, 1995.
7. Ph. de Groote. On the strong normalisation of natural deduction with permutation-conversions. In *10th International Conference on Rewriting Techniques and Applications, RTA'99*, volume 1631 of *Lecture Notes in Computer Science*, pages 45–59. Springer Verlag, 1999.
8. M. Felleisen, D.P. Friedman, E. Kohlbecker, and B. Duba. A syntactic theory of sequential control. *Theoretical Computer Science*, 52:205–237, 1987.
9. M. Felleisen and R. Hieb. The revised report on the syntactic theory of sequential control and state. *Theoretical Computer Science*, 102:235–271, 1992.
10. G. Gentzen. *Recherches sur la déduction logique (Untersuchungen über das logische schliessen).* Presses Universitaires de France, 1955. Traduction et commentaire par R. Feys et J. Ladrière.

11. J.-Y. Girard. A new constructive logic: Classical logic. *Mathematical Structures in Computer Science*, 1:255–296, 1991.
12. J.-Y. Girard, Y. Lafont, and P. Taylor. *Proofs and Types*, volume 7 of *Cambridge Tracts in Theoretical Computer Science*. Cambridge University Press, 1989.
13. T. G. Griffin. A formulae-as-types notion of control. In *Conference record of the seventeenth annual ACM symposium on Principles of Programming Languages*, pages 47–58, 1990.
14. W.A. Howard. The formulae-as-types notion of construction. In J. P. Seldin and J. R. Hindley, editors, *to H. B. Curry: Essays on Combinatory Logic, Lambda Calculus and Formalism*, pages 479–490. Academic Press, 1980.
15. J.-L. Krivine. Classical logic, storage operators and second order λ-calculus. *Annals of Pure and Applied Logic*, 68:53–78, 1994.
16. A. Meyer and M. Wand. Continuation semantics in typed lambda-calculi (summary). In R. Parikh, editor, *Logics of Programs*, volume 193 of *Lecture Notes in Computer Science*, pages 219–224. Springer Verlag, 1985.
17. C. R. Murthy. An evaluation semantics for classical proofs. In *Proceedings of the sixth annual IEEE symposium on logic in computer science*, pages 96–107, 1991.
18. C. R. Murthy. A computational analysis of Girard's translation and LC. In *Proceedings of the seventh annual IEEE symposium on logic in computer science*, pages 90–101, 1992.
19. M. Parigot. $\lambda\mu$-Calculus: an algorithmic interpretation of classical natural deduction. In A. Voronkov, editor, *Proceedings of the International Conference on Logic Programming and Automated Reasoning*, volume 624 of *Lecture Notes in Artificial Intelligence*, pages 190–201. Springer Verlag, 1992.
20. M. Parigot. Strong normalization for second order classical natural deduction. In *Proceedings of the eighth annual IEEE symposium on logic in computer science*, pages 39–46, 1993.
21. M. Parigot. Proofs of strong normalisation for second order classical natural deduction. *Journal of Symbolic Logic*, 62(4):1461–1479, 1997.
22. G. D. Plotkin. Call-by-name, call-by-value and the λ-calculus. *Theoretical Computer Science*, 1:125–159, 1975.
23. D. Prawitz. *Natural Deduction, A Proof-Theoretical Study.* Almqvist & Wiksell, Stockholm, 1965.
24. D. Prawitz. Ideas and results in proof-theory. In J.E. Fenstad, editor, *Proceedings of the Second Scandinavian Logic Symposium*, pages 237–309. North-Holland, 1971.
25. D. Pym and E. Ritter. On the semantics of classical disjunction. *Journal of Pure and Applied Algebra*, To appear.
26. N.J. Rehof and M.H. Sørensen. The λ_{Δ}-calculus. In M. Hagiya and J.C. Mitchell, editors, *Proceedings of the International Symposium on Theoretical Aspects of Computer Software, TACS'94*, pages 516–542. Lecture Notes in Computer Science, 789, Springer Verlag, 1994.
27. E. Ritter, D. Pym, and L. Wallen. On the intuitionistic force of classical search. *Theoretical Computer Science*, 232:299–333, 2000.
28. E. Ritter, D. Pym, and L. Wallen. Proof-terms for classical and intuitionistic resolution. *Journal of Logic and Computation*, 10(2):173–207, 2000.
29. J. Seldin. On the proof theory of the intermediate logic MH. *Journal of Symbolic Logic*, 51(3):626–647, 1986.
30. G. Stålmarck. Normalization theorems for full first-order classical natural deduction. *Journal of Symbolic Logic*, 56(1):129–149, 1991.
31. A. Troelstra and D. van Dalen. *Constructivism in Mathematics*, volume II. North-Holland, 1988.

Partially Additive Categories and Fully Complete Models of Linear Logic

Esfandiar Haghverdi*

Department of Mathematics
University of Pennsylvania
Philadelphia, PA 19104
esfan@math.upenn.edu

Abstract. We construct a new class of models for linear logic. These models are constructed on partially additive categories using the *Int* construction of Joyal, Street and Verity and double glueing construction of Hyland and Tan. We prove full completeness for MLL+MIX in these models.

1 Introduction

Partially Additive Categories (PACs) were introduced by Manes and Arbib [27] to provide an algebraic semantics for programming languages. They have also been used to provide a categorical model for the Geometry of Interaction (GoI) interpretation [15, 16]. PACs are also closely related to iteration theories and categories with fixed-point operations [15]. Mascari and Pedicini have used partially additive categories to provide a structured dynamics for algorithm executions inspired by GoI [10, 28]. Their approach is completely different from what we will discuss in this paper. In this paper, we construct a new class of models for the multiplicative fragment of linear logic. These models are constructed on partially additive categories using the *Int* construction of Joyal, Street and Verity and the double glueing construction of Hyland and Tan. We prove full completeness for these models.

The contributions of this work are: (1) Using PACs to construct models of multiplicative linear logic and thus opening new directions relating GoI type semantics to denotational semantics and iteration and fixed-point theories to linear logic models, (2) Introducing new examples of traced symmetric monoidal, compact closed and eventually *-autonomous categories via the compact closure and glueing constructions, (3) Providing a new class of fully complete models for MLL + MIX in a fashion that is in the spirit of a unified approach to full completeness problem.

The paper is organized as follows: Section 2 contains a brief introduction to traced symmetric monoidal categories and the Int construction. In section 3 we briefly discuss PACs and give some examples. A short introduction to

* Research supported in part by Natural Sciences and Engineering Research Council of Canada.

S. Abramsky (Ed.): TLCA 2001, LNCS 2044, pp. 197–216, 2001.

full completeness and a brief discussion of functorial polymorphism and free monoidal categories form the bulk of Section 4. We recall the double glueing construction in Section 5 and present the main theorems of the paper in Section 6. Section 7 contains some concluding remarks and thoughts on future research directions.

We have omitted many proofs due to lack of space. We have been cautious to do this in such a way that the coherence and consistency of the text is not jeopardized. The reader is referred to author's PhD thesis [15] for all the proofs that are not included in this paper.

2 Traced Monoidal Categories

Definition 2.1. *A* traced symmetric monoidal category *is a symmetric monoidal category* $(\mathbb{C}, I, \otimes, \sigma)$ *with a family of functions* $Tr^U_{X,Y} : \mathbb{C}(X \otimes U, Y \otimes U) \to \mathbb{C}(X, Y)$, *called a* trace, *subject to the following axioms:*

- **Natural** *in* X, $Tr^U_{X,Y}(f)g = Tr^U_{X',Y}(f(g \otimes 1_U))$ *where* $f : X \otimes U \to Y \otimes U$, $g : X' \to X$,
- **Natural** *in* Y, $gTr^U_{X,Y}(f) = Tr^U_{X,Y'}((g \otimes 1_U)f)$ *where* $f : X \otimes U \to Y \otimes U$, $g : Y \to Y'$,
- **Dinatural** *in* U, $Tr^U_{X,Y}((1_Y \otimes g)f) = Tr^{U'}_{X,Y}(f(1_X \otimes g))$ *where* $f : X \otimes U \to Y \otimes U'$, $g : U' \to U$,
- **Vanishing (I,II)**, $Tr^I_{X,Y}(f) = f$ *and* $Tr^{U \otimes V}_{X,Y}(g) = Tr^U_{X,Y}(Tr^V_{X \otimes U, Y \otimes U}(g))$ *for* $f : X \otimes I \to Y \otimes I$ *and* $g : X \otimes U \otimes V \to Y \otimes U \otimes V$,
- **Superposing**, $g \otimes Tr^U_{X,Y}(f) = Tr^U_{W \otimes X, Z \otimes Y}(g \otimes f)$, *for* $f : X \otimes U \to Y \otimes U$ *and* $g : W \to Z$,
- **Yanking**, $Tr^U_{U,U}(\sigma_{U,U}) = 1_U$.

2.1 Int Construction

The Int construction was introduced by Joyal, Street and Verity in [21]. It is used to construct a free tortile monoidal category from a given traced balanced monoidal category. In this paper we will work with traced symmetric monoidal categories and hence in this case the main result of [21] reads as follows.

Theorem 2.2 (Joyal, Street & Verity). [1] *Suppose* $\mathbb{C}$ *is a traced symmetric monoidal category and* $\mathbb{D}$ *is a compact closed category. Then there exists a compact closed category* $Int\,\mathbb{C}$ *such that for all traced monoidal functors* $F : \mathbb{C} \to \mathbb{D}$, *there exists a symmetric monoidal functor* $K : Int\,\mathbb{C} \to \mathbb{D}$ *which is unique up to monoidal natural isomorphism with the property* $KN \cong F$, *where* $N : \mathbb{C} \to Int\,\mathbb{C}$ *is the full faithful inclusion functor.*

Let $\mathbb{C}$ be a traced symmetric monoidal category. The category $Int\,\mathbb{C}$ is defined as follows:

[1] Note that this is the version of the original theorem for the case of symmetric monoidal categories.

- Objects: Pairs of objects from $\mathbb{C}$, e.g. (A^+, A^-) where A^+ and A^- are objects of $\mathbb{C}$.
- Arrows: An arrow $f : (A^+, A^-) \to (B^+, B^-)$ in $Int\,\mathbb{C}$ is $f : A^+ \otimes B^- \to B^+ \otimes A^-$ in $\mathbb{C}$.
- Identity: $1_{(A^+,A^-)} = 1_{A^+ \otimes A^-}$.
- Composition: Given $f : (A^+, A^-) \to (B^+, B^-)$ and $g : (B^+, B^-) \to (C^+, C^-)$, $gf : (A^+, A^-) \to (C^+, C^-)$ is given by:
$gf = Tr^{B^-}_{A^+ \otimes C^-, C^+ \otimes A^-}((1_{C^+} \otimes \sigma_{B^-,A^-})(g \otimes 1_{A^-})(1_{B^+} \otimes \sigma_{A^-,C^-})(f \otimes 1_{C^-})$
$(1_{A^+} \otimes \sigma_{C^-,B^-}))$
- Tensor: $(A^+, A^-) \otimes (B^+, B^-) = (A^+ \otimes B^+, B^- \otimes A^-)$ and for $f : (A^+, A^-) \to (B^+, B^-)$ and $g : (C^+, C^-) \to (D^+, D^-)$, $f \otimes g$ is defined to be the following composite: $A^+ \otimes C^+ \otimes D^- \otimes B^- \stackrel{\sigma \otimes \sigma}{\longrightarrow} C^+ \otimes A^+ \otimes B^- \otimes D^- \stackrel{1 \otimes f \otimes 1}{\longrightarrow} C^+ \otimes B^+ \otimes A^- \otimes D^- \stackrel{\sigma \otimes \sigma}{\longrightarrow} B^+ \otimes C^+ \otimes D^- \otimes A^- \stackrel{1 \otimes g \otimes 1}{\longrightarrow} B^+ \otimes D^+ \otimes C^- \otimes A^-$
- Unit: (I, I).

Proposition 2.3. *Let $\mathbb{C}$ be a traced symmetric monoidal category , $Int\,\mathbb{C}$ is a compact closed category. Moreover, $N : \mathbb{C} \to Int\,\mathbb{C}$ with $N(A) = (A, I)$ and $N(f) = f$ is a full and faithful embedding.*

Proof. (Sketch) This is just a specialisation of the proof that appears in [21] and we will not repeat it here. However we give the main morphisms of the closed structure. For any two objects (A^+, A^-) and (B^+, B^-) in $Int\,\mathbb{C}$, $\sigma_{(A^+,A^-),(B^+,B^-)} =_{def} \sigma_{A^+,B^+} \otimes \sigma_{A^-,B^-}$. The left dual of (A^+, A^-), $(A^+, A^-)^* = (A^-, A^+)$. The unit is given by $\eta_{(A^+,A^-)} : (I, I) \to (A^+, A^-) \otimes (A^+, A^-)^* =_{def} 1_{A^+ \otimes A^-}$ and counit is $\epsilon_{(A^+,A^-)} : (A^+, A^-)^* \otimes (A^+, A^-) \to (I, I) =_{def} 1_{A^- \otimes A^+}$. The internal homs are given by $(A^+, A^-) \multimap (B^+, B^-) = (B^+, B^-) \otimes (A^+, A^-)^* = (B^+ \otimes A^-, A^+ \otimes B^-)$. □

3 Partially Additive Categories

In this section we recall the definitions of partially additive monoids and categories enriched over such monoids, partially additive categories. Partially additive categories were defined and used by Manes and Arbib to provide an algebraic semantics for programming languages [27]. Our interest in partially additive categories is primarily due to the fact that they provide a canonical construction for trace and composition in geometry of interaction categories, together with unique decomposition property of morphisms (see Proposition 3.6 and the following discussion).

Definition 3.1. *A* partially additive monoid *is a pair (M, Σ), where M is a nonempty set and Σ is a partial function which maps countable families in M to elements of M (we say that $\{x_i\}_{i \in I}$ is* summable *if $\sum_{i \in I} x_i$ is defined)*[2] *subject to the following axioms:*

[2] Throughout, "countable" means finite or denumerable. All index sets are countable.

1. Partition-Associativity Axiom. *If* $\{x_i\}_{i\in I}$ *is a countable family and if* $\{I_j\}_{j\in J}$ *is a (countable) partition of* I *, then* $\{x_i\}_{i\in I}$ *is summable if and only if* $\{x_i\}_{i\in I_j}$ *is summable for every* $j \in J$ *and* $\sum_{i\in I_j} x_i$ *is summable for* $j \in J$. *In that case,* $\sum_{i\in I} x_i = \sum_{j\in J}(\sum_{i\in I_j} x_i)$.
2. Unary Sum Axiom. *Any family* $\{x_i\}_{i\in I}$ *in which* I *is a singleton is summable and* $\sum_{i\in I} x_i = x_j$ *if* $I = \{j\}$.
3. Limit Axiom. *If* $\{x_i\}_{i\in I}$ *is a countable family and if* $\{x_i\}_{i\in F}$ *is summable for every finite subset* F *of* I *then* $\{x_i\}_{i\in I}$ *is summable.*

We observe the following facts about partially additive monoids:

(i) Axiom 1 implies that every subfamily of a summable family is summable.
(ii) Axioms 1 and 2 imply that the empty family is summable. We denote $\sum_{i\in\emptyset} x_i$ by 0, which is an additive identity for summation. In fact, 0 is a countable additive identity.
(iii) Axiom 1 implies the obvious equations of commutativity and associativity for the sum (when defined). More generally, $\sum_i x_{\varphi(i)}$ is defined for any permutation φ of I whenever $\sum_i x_i$ exists and $\sum_i x_{\varphi(i)} = \sum_i x_i$; just consider the partition $\{\varphi(j)\}_{j\in I}$.
(v) There are no additive inverses. Indeed, let $\{x_i\}_{i\in I}$ be a summable family with $\sum_{i\in I} x_i = 0$ and set $y = \sum_{j\in I-\{i\}} x_j$ for some $i \in I$. Then, $y + x_i = 0$ and $x_i = x_i + (y + x_i) + (y + x_i) + \cdots = (x_i + y) + (x_i + y) + \cdots = 0$. Thus, $x_i = 0$ for all $i \in I$.

A doubly indexed family f in a partially additive monoid M, $f : I \times J \to M$, is denoted as $\{f_{ij}\}_{i\in I, j\in J}$ or simply $\{f_{ij}\}$ if the index sets are clear from the context. Such a family is summable iff $\sum_{i\in I}(\sum_{j\in J} f_{ij})$ exists and in that case $\sum_{i,j} f_{ij} = \sum_{i\in I}(\sum_{j\in J} f_{ij})$. It follows, using Axiom 1, that for a summable family $\{f_{ij}\}$, $\sum_{i\in I}(\sum_{j\in J} f_{ij}) = \sum_{j\in J}(\sum_{i\in I} f_{ij})$. Here are some examples of partially additive monoids.

Example 3.2. 1. $M = \mathbf{PInj}(X, Y)$, the set of partial injective functions from X to Y. A family $\{f_i\}_{i\in I} \in \mathbf{PInj}(X, Y)$ is said to be summable iff $Dom(f_i) \cap Dom(f_j) = \emptyset$ and $Codom(f_i) \cap Codom(f_j) = \emptyset$ for all $i \neq j$. In that case, $(\sum_i f_i)(x) = f_j(x)$ if $x \in Dom(f_j)$ for some $j \in I$ and undefined, otherwise.
2. $M = \mathbf{Pfn}(X, Y)$, the set of partial functions from X to Y. A family $\{f_i\}_{i\in I} \in M$ is summable iff $Dom(f_i) \cap Dom(f_j) = \emptyset$ for all $i \neq j$. In that case, $(\sum_i f_i)(x) = f_j(x)$ if $x \in Dom(f_j)$ for some $j \in I$ and undefined, otherwise. We denote this partially additive monoid by $(\mathbf{Pfn}(X, Y), \Sigma^{di})$.
3. $M = \mathbf{Rel}_+(X, Y)$, the set of relations from a set X to a set Y. Any family $\{R_i\}_{i\in I} \in \mathbf{Rel}_+(X, Y)$ is summable with $\sum_i R_i = \bigcup_{i\in I} R_i$.

Definition 3.3. *The category of partially additive monoids,* **PAMon**, *has as objects partially additive monoids* (M, Σ). *Its arrows* $(M, \Sigma) \xrightarrow{f} (M', \Sigma')$ *are maps from* M *to* M' *which preserve the sum. Composition and identities are inherited from* **Set**.

Observe that **PAMon** has finite products: given (M_1, Σ_1) and (M_2, Σ_2), their product is $(M_1 \times M_2, \Sigma)$ where $\sum_{i\in I}(x_i, y_i) = (\sum_1 x_i, \sum_2 y_i)$ for all summable families $\{(x_i, y_i)\}_{i\in I}$ in $M_1 \times M_2$. The zero object $\mathbf{0}$ is $(\{0\}, \sum)$ in which all families are summable, with sum equal to 0. In particular, **PAMon** is a symmetric monoidal category with product as the tensor.

A **PAMon**-category $\mathbb{C}$ is a category enriched in **PAMon**; that is, the homsets are enriched with an additive structure such that composition distributes over addition from left and right. More specifically, for all $f : W \to X, h : Y \to Z$ and for all summable families $\{g_i\}_{i\in I}$ in $\mathbb{C}(X, Y)$, $\{g_i f\}_{i\in I}$ and $\{hg_i\}_{i\in I}$ are also summable and $(\sum_{i\in I} g_i)f = \sum_{i\in I} g_i f$ and $h(\sum_{i\in I} g_i) = \sum_{i\in I} hg_i$.

Note that such categories have non-empty homsets and automatically have zero morphisms, namely $0_{XY} : X \to Y = \sum_{i\in\emptyset} f_i$ for $f_i \in \mathbb{C}(X, Y)$.

Notation: We will use $+$ for the addition operation on the homsets. We use $\oplus$ for coproduct.

Definition 3.4. *Let $\mathbb{C}$ be a* **PAMon**-*category with countable coproducts $\bigoplus_{i\in I} X_i$. We define* quasi projections *$\rho_j : \bigoplus_{i\in I} X_i \to X_j$ for all $j \in I$ as follows: $\rho_j in_k = 1_{X_j}$ if $k = j$ and $0_{X_k X_j}$ otherwise. Note that ρ_j exists for all $j \in I$ since $\mathbb{C}$ has zero morphisms.*

Definition 3.5. *A* partially additive category *$\mathbb{C}$ is a* **PAMon**-*category with countable coproducts which satisfies the following axioms:*

1. Compatible Sum Axiom*: If $\{f_i\}_I \in \mathbb{C}(X, Y)$ is a countable family and there exists $f : X \to I.Y$ such that $\rho_i f = f_i$ for all $i \in I$, (we say the f_i are* compatible*), then $\sum_{i\in I} f_i$ exists.*
2. Untying Axiom*: If $f + g : X \to Y$ exists then so does $in_1 f + in_2 g : X \to Y \oplus Y$.*

The dual of a partially additive category is a **PAMon**-category with countable products which satisfies the dual of the above axioms.

PACs enjoy the following properties: (i) the unique decomposition property, (ii) the existence of the iteration (dagger) operation, and (iii) the uniqueness of the additive structure. We will be more explicit regarding the first two properties.

Proposition 3.6 (Manes and Arbib[27]). *Given $f : X \to \bigoplus_{i\in I} Y_i$ in a partially additive category. There exists a unique family $f_i : X \to Y_i$ with $f = \sum_{i\in I} in_i f_i$, namely, $f_i = \rho_i f$.*

Corollary 3.7. *Given $f : \bigoplus_{j\in J} X_j \to \bigoplus_{i\in I} Y_i$ in a partially additive category, there exists a unique family $\{f_{ij}\}_{i\in I, j\in J} : X_j \to Y_i$ with $f = \sum_{i\in I, j\in J} in_i f_{ij} \rho_j$, namely, $f_{ij} = \rho_i f in_j$.*

Proof. In any PAC, $\sum_{i\in I} in_i \rho_i$ exists and $\sum_{i\in I} in_i \rho_i = 1 : \bigoplus_{i\in I} X_i \to \bigoplus_{i\in I} X_i$. To see this, note that by the theorem above $1_{\bigoplus_I X_i}$ can be uniquely written as $1 = \sum_{i\in I} in_i \rho_i 1 = \sum_{i\in I} in_i \rho_i$.

Now let $f = 1_{\bigoplus_I Y_i} f 1_{\bigoplus_J X_j} = (\sum_i in_i \rho_i) f (\sum_j in_j \rho_j) = (\sum_i in_i \rho_i f)(\sum_j in_j \rho_j) = \sum_{ij} in_i \rho_i f in_j \rho_j = \sum_{ij} in_i f_{ij} \rho_j$. For uniqueness, suppose there is another family

$\{g_{kl}\}_{k\in I, l\in J}$ such that $f = \sum_{kl} in_k g_{kl} \rho_l$. Then $f_{ij} = \rho_i f in_j = \sum_{kl} \rho_i in_k g_{kl} \rho_l in_j = g_{ij}$ for all i, j. □

Based on this proposition, every morphism $f : \bigoplus_J X_j \to \bigoplus_I Y_i$ can be represented by its components. When I and J are finite, we will use the corresponding matrices to represent morphisms, for example f above with $|I| = m$ and $|J| = n$ is represented by an $m \times n$ matrix (f_{ij}). It also follows that the composition of morphisms in a PAC with finite coproducts in their domain and codomain corresponds to matrix multiplication of their matricial representations.

Remark 3.8. Note that although any morphism $f : \bigoplus_J X_j \to \bigoplus_I Y_i$ can be represented by the unique family $\{f_{ij}\}_{i\in I, j\in J}$ of its components, the converse is not necessarily true, that is to say given a family $\{f_{ij}\}$ there may not be a morphism $f : \bigoplus_J X_j \to \bigoplus_I Y_i$ satisfying $f = \sum_{ij} in_i f_{ij} \rho_j$. However, in case such an f exists it will be unique.

Theorem 3.9 (Manes and Arbib[27]). *Given a map $f : X \to Y \oplus X$ in a partially additive category. The sum $f^\dagger = \sum_{n=0}^{\infty} f_1 f_2^n : X \to Y$ exists, where $f_1 : X \to Y$ and $f_2 : X \to X$ are the components of f.*

$f^\dagger$ is called the *iterate* (or *dagger*) of f. We define a family of operations $\dagger_{X,Y} : \mathbb{C}(X, Y \oplus X) \to \mathbb{C}(X, Y)$ which to each f associates its iterate $f^\dagger$. This operation induces a trace operator on $\mathbb{C}$ as follows:

Proposition 3.10 ([15]). *Every partially additive category is a traced symmetric monoidal category, where given $f : X \oplus U \to Y \oplus U$,*

$$Tr^U_{X,Y} = f_{11} + \sum_{n\in\omega} f_{12} f_{22}^n f_{21}$$

and f_{ij} are the components of f.

There are many interesting connections between trace operation and the iteration and fixed-point operators. However, we will refrain from discussing these for lack of space, for more details see [15, 18]. We give a few illustrative examples, for more examples see [15].

3.1 Examples

1. Consider the category **Pfn** of sets and partial functions. Recall that a partial function from X to Y is a function from $Dom(f) \subseteq X$ to Y. Given $f : X \to Y$ and $g : Y \to Z$, $gf : X \to Z$ is defined by $Dom(gf) = \{x \in X | x \in Dom(f), f(x) \in Dom(g)\}$ and $(gf)(x) = g(f(x))$ for $x \in Dom(gf)$. The additive structure was given in Example 3.2, the zero morphism $0_{XY} : X \to Y$ is the everywhere undefined partial function. **Pfn** has countable coproducts given by disjoint union.
2. Consider the category $\mathbf{Rel}_+$ of sets and binary relations. Given morphisms $R : X \to Y$ and $S : Y \to Z$ the composition $SR : X \to Z$ is given by the usual relational product and identity morphisms are identity relations. We have

already defined the additive structure for the homsets in Example 3.2. Hence, for any X and Y, $\mathbf{Rel}_+(X, Y)$ is a partially additive monoid with all families summable and the zero morphism is the empty relation ($\emptyset \subseteq X \times Y$). Note that $\mathbf{Rel}_+$ has countable coproducts given by the disjoint union.

3. Consider the category **SRel** of stochastic relations with measurable spaces $(X, \mathcal{F}_X)$ as objects and stochastic kernels as arrows. An arrow $f : (X, \mathcal{F}_X) \to (Y, \mathcal{F}_Y)$ is a map $f : X \times \mathcal{F}_Y \to [0, 1]$ such that $f(\cdot, B) : X \to [0, 1]$ is a bounded measurable function for fixed $B \in \mathcal{F}_Y$ and $f(x, \cdot) : \mathcal{F}_Y \to [0, 1]$ is a subprobability measure (i.e., σ-additive, set function, $f(x, \emptyset) = 0$ and $f(x, Y) \leq 1$). The identity morphism $1_X : (X, \mathcal{F}_X) \to (X, \mathcal{F}_X)$ is $1_X : X \times \mathcal{F}_X \to [0, 1]$ and is defined by $1_X(x, A) = \delta(x, A) = 1$ if $x \in A$ and 0, if $x \notin A$.

For A fixed, $\delta(x, A)$ is the characteristic function of A and for x fixed, $\delta(x, A)$ is the Dirac distribution. Finally, composition is defined as follows: given $f : (X, \mathcal{F}_X) \to (Y, \mathcal{F}_Y)$ and $g : (Y, \mathcal{F}_Y) \to (Z, \mathcal{F}_Z)$, $gf : (X, \mathcal{F}_X) \to (Z, \mathcal{F}_Z)$ is given by $gf(x, C) = \int_Y g(y, C)\, f(x, dy)$.

SRel was proven to be a PAC jointly by Panangaden [29] and the author. It was shown to be a traced symmetric monoidal category directly (without using partially additive structure) in [1].

4 Full Completeness: A Brief Introduction

Traditional completeness theorems are with respect to provability, whereas full completeness is with respect to proofs. This can be best explained in a categorical model [24]. Let $\mathbb{M}$ be a categorical model of the formulas and proofs of a logic $\mathcal{L}$. This means that $\mathbb{M}$ is a category with an appropriate structure such that formulas of $\mathcal{L}$ are interpreted as objects in $\mathbb{M}$ and proofs Π in $\mathcal{L}$ of entailments $A \vdash B$ are interpreted by morphisms $[\![\Pi]\!] : [\![A]\!] \to [\![B]\!]$. Finally convertibility of proofs in $\mathcal{L}$ with respect to cut-elimination is soundly modeled by the equations between morphisms holding in $\mathbb{M}$. Traditional completeness theorems assert that $\mathbb{M}([\![A]\!], [\![B]\!]) \neq \emptyset$ implies $A \vdash B$ is provable in the logic $\mathcal{L}$ (= truth implies provability.)

We say that $\mathbb{M}$ is *fully complete* for $\mathcal{L}$ if for all formulas A, B of $\mathcal{L}$, every morphism f in $\mathbb{M}([\![A]\!], [\![B]\!])$ is the denotation of some proof Π of $A \vdash B$ in $\mathcal{L}$: $f = [\![\Pi]\!]$. This amounts to asking that the unique free functor (with respect to any interpretation of the generators) $[\![-]\!] : \mathbb{F} \to \mathbb{M}$ be full. Here $\mathbb{F}$ is the free category generated by the logic $\mathcal{L}$. Thus, full completeness establishes a tight connection between syntax and semantics compared to completeness. This connection can be made even stronger by requiring that the functor $[\![-]\!]$ be faithful too. In other word, a full faithful completeness theorem asserts that every morphism in $\mathbb{M}([\![A]\!], [\![B]\!])$ is the denotation of a unique proof of $A \vdash B$.

The term "full completeness" was coined by Abramsky and Jagadeesan in [2] where they also proved full completeness for a game semantics of Multiplicative Linear Logic with the MIX rule (MLL + MIX). This was followed by a series of papers which established full completeness results for a variety of models with respect to various versions of MLL [20, 8, 7, 25, 26]. Recently Abramsky and Mellies [3] introduced a new concurrent form of game semantics for linear logic and proved a full completeness theorem for Multiplicative-Additive Linear Logic for this semantics. In this paper we will be mainly concerned with MLL and hence we will not further discuss this latter work.

The idea that dinatural transformations could provide a semantics for proofs of a logical system was first introduced in [4] in the programme called "functorial polymorphism" (see below). In this setting a formula is interpreted by a multivariant functor and a dinatural transformation between multivariant functors provides the interpretation for proofs. The problem with dinatural transformations is that they do not compose in general to give a dinatural transformation. Girard, Scedrov and P. Scott [14] showed that a dinatural interpretation in the framework of cartesian closed categories is sound with respect to intuitionistic logic without the cut rule. Based on these ideas and results R. Blute and P. Scott [8] proved a full completeness theorem for MLL + MIX in the category of reflexive topological vector spaces and a full completeness theorem for the multiplicative fragment of Yetter's *cyclic linear logic* (CyLL) with the MIX rule [7]. See also [17].

There has also been a considerable body of work on full completeness theorems for MLL in *-autonomous categories constructed from compact closed or symmetric monoidal closed categories. Devarajan, Hughes, Plotkin and Pratt [12] prove a full completeness theorem for MLL without MIX interpreted over binary logical transformations of Chu spaces over a two-letter alphabet $\mathbf{Chu}(\mathbf{Set}, \mathbf{2})$.

Another important approach to full completeness theorems was introduced in Tan's PhD thesis [31]. Hyland and Tan introduced the double glueing construction which given a compact closed category $\mathbb{C}$ constructs a *-autonomous category $\mathbf{G}\mathbb{C}$. The setting is the proofs as dinatural transformations paradigm. This work is especially important as it has initiated a systematic approach to the full completeness theorems for such categorical models. Explicitly, Tan defines a *compact closed full completeness* and reduces the full completeness problem for $\mathbf{G}\mathbb{C}$ to compact closed full completeness for $\mathbb{C}$. The lifting of compact closed full completeness to $\mathbf{G}\mathbb{C}$ establishes the desired full completeness result. Tan studies several examples: $\mathbf{Rel}_\times$, $\mathbf{FDVec}$, a category of Conway games and topological vector spaces. She proves full completeness for MLL + MIX in these categories (with the exception of $\mathbf{Rel}_\times$ where she has the result for MLL without MIX). However, the passage from compact closed full completeness to full completeness of $\mathbf{G}\mathbb{C}$ is not completely algorithmic, that is each case requires a different treatment.

In section 6, we will construct categorical models for MLL+MIX based on partially additive categories and prove full completeness theorems for such models. The semantic setting we will be using is the functorial polymorphism of [4]. More explicitly, we start with a partially additive category $\mathbb{D}$ and use the *Int* construction of Joyal, Street and Verity (equivalently the Geometry of Interaction construction $\mathcal{G}$ of Abramsky [1,15]), to get a compact closed category $\mathit{Int}\,\mathbb{D}$. We next prove compact closed full completeness for this category. Finally, applying the double glueing construction of Hyland and Tan, we construct a *-autonomous category $\mathbf{G}(Int\,\mathbb{D})$, which is a model of MLL + MIX. Finally, we prove full completeness for MLL + MIX in $\mathbf{G}(Int\,\mathbb{D})$ by lifting compact closed full completeness in $Int\,\mathbb{D}$. This approach works for all partially additive categories in a uniform way. In this paper we will only be concerned with full completeness for *unit-free* formulas.

4.1 Functorial Polymorphism

Functorial polymorphism will be the semantic setting we use in our categorical models. *Functorial polymorphism* introduced in [4], provides a general categorical framework for parametric polymorphic lambda calculus. In this setting, types are represented by multivariant functors and terms by certain multivariant, i.e. *dinatural* transformations. Applications of this framework for proving full completeness theorems for fragments of linear logic can be found in [8, 7, 17, 31, 12, 3, 15].

Dinatural Interpretation for MLL.

Definition 4.1. *Let $\mathbb{C}$ be a category and $F, G : \mathbb{C}^n \times (\mathbb{C}^{op})^n \to \mathbb{C}$ be multivariant functors. We write $\underline{X}$ for the list $X_1, X_2, \cdots, X_n$. A* dinatural trans- formation *$\rho : F \stackrel{..}{\longrightarrow} G$ is a family of $\mathbb{C}$-morphisms $\rho = \{\rho_{\underline{X}} : F(\underline{X}, \underline{X}) \to G(\underline{X}, \underline{X}) | \underline{X}$ a list of objects in $\mathbb{C}\}$ satisfying (for all $f_i : X_i \to Y_i$):*

$$G(\underline{f}, 1_{\underline{X}})\rho_{\underline{X}}F(1_{\underline{X}}, \underline{f}) = G(1_{\underline{Y}}, \underline{f})\rho_{\underline{Y}}F(\underline{f}, 1_{\underline{Y}}).$$

It is well known that, a model of MLL consists of a *-autonomous category $\mathbb{C}$ [30]. Following the methods of functorial polymorphism, we interpret formulas of MLL as multivariant functors over such a category $\mathbb{C}$, using the operations $(F \otimes G)(\underline{A}, \underline{B}) = F(\underline{A}, \underline{B}) \otimes G(\underline{A}, \underline{B})$ and $F^{\perp}(\underline{A}, \underline{B}) = (F(\underline{B}, \underline{A}))^{\perp}$ on n-ary multivariant functors $F, G : \mathbb{C}^n \times (\mathbb{C}^{op})^n \to \mathbb{C}$. Here $\underline{A}$ and $\underline{B}$ are lists of objects in $\mathbb{C}$ that occur co- and contravariantly respectively.

Let $\varphi(\alpha_1, \cdots, \alpha_n)$ be an MLL formula built from the literals $\alpha_1, \cdots, \alpha_n$ and $\alpha_1^{\perp}, \cdots, \alpha_n^{\perp}$. To each such formula we associate its *interpretation* $[\![\varphi(\alpha_1, \cdots, \alpha_n)]\!] : \mathbb{C}^n \times (\mathbb{C}^{op})^n \to \mathbb{C}$ as follows:

1. If $\varphi(\alpha_1, \cdots, \alpha_n) \equiv \alpha_i$, then $[\![\varphi]\!](\underline{A}, \underline{B}) = A_i$, the covariant projection functor onto the ith component of $\underline{A}$. We denote this functor by Π_i.
2. If $\varphi(\alpha_1, \cdots, \alpha_n) \equiv \alpha_i^{\perp}$, then $[\![\varphi]\!](\underline{A}, \underline{B}) = B_i^{\perp}$, the linear negation of the contravariant projection onto the ith component of $\underline{B}$, denoted $\Pi_i^{\perp}$.
3. If $\varphi = \varphi_1 \otimes \varphi_2$, then $[\![\varphi]\!] = [\![\varphi_1]\!] \otimes [\![\varphi_2]\!]$
4. If $\varphi = \varphi_1^{\perp}$, then $[\![\varphi]\!] = [\![\varphi_1]\!]^{\perp}$.

The connective ⅋ is defined by De Morgan duality.

We say that a functor is *definable* if it is the interpretation of a formula in the logic or equivalently it is an interpretation of an object in the free category representing the logic. A proof Π of $\vdash \Gamma$ is interpreted as a dinatural transformation from the constant $\mathbf{1}$ functor, $\mathcal{K}_1$, to the multivariant functor $[\![\Gamma]\!]$.

Remark 4.2. A formula $\varphi(\alpha_1, \cdots, \alpha_n)$ in MLL is an object $F(\underline{X}) = F(X_1, \cdots, X_n)$ in the free *-autonomous category $\mathcal{F}^*(\{X_1, \cdots, X_n\})$ generated on n objects. $F(\underline{X})$ is built from $X_1, \cdots, X_n$ and $X_1^\perp, \cdots, X_n^\perp$ using tensor and par products.[3] A proof Π of $\vdash \varphi(\alpha_1, \cdots, \alpha_n)$ in MLL is a morphism from the unit of tensor, **1**, to the object $F(\underline{X})$ in $\mathcal{F}^*(\{X_1, \cdots, X_n\})$.

4.2 Coherence and Free Monoidal Categories

The logical approach to coherence initiated by Lambek in [23] via the equivalence between the deductions in a deductive system and morphisms in a free category, can be used to describe morphisms in a free category. For example, a morphism in the free *-autonomous category (without units), can be interpreted as a proof-net [5]. As proof nets are graphs satisfying a correctness criterion, they may be used to determine the existence of morphisms in various free monoidal categories [6]. We assume familiarity with the free compact closed and *-autonomous categories generated on a set of objects. The reader can find a lucid presentation of these constructions in [31]. See also [15].

Definition 4.3. *For any object A in a traced symmetric monoidal category $\mathbb{C}$ we define the* dimension *of A to be the endomorphism $dim(A) = Tr^A_{I,I}(1_I \otimes 1_A) : I \to I$.*

For a compact closed category we have: $dim(A) : I \xrightarrow{\eta_A} A \otimes A^* \xrightarrow{\sigma_{A,A^*}} A^* \otimes A \xrightarrow{\epsilon_A} I$.

In $\mathbf{Rel}_\times$, $dim(\emptyset) = 0_I$, the empty relation, and $dim(A) = 1_I$ for all $A \neq \emptyset$. In **FDVec**, $dim(V)$ is the dimension of the vector space V. In a partially additive category, $dim(A) = 1_I = 0_I$ for all A, since I is the zero object. An object A with $dim(A) = 1_I$ is said to have *trivial* dimension. So in a PAC all objects have trivial dimensions. The Kelly-Mac Lane graph [22] of $dim(A)$, for A in a compact closed category is a "loop" passing through A and A^*.

We now describe the morphisms in the free compact closed category $\mathcal{F}(\mathcal{A})$. For this purpose let $\mathcal{F}_1(\mathcal{A})$ be the free compact closed category generated on a set of objects $\mathcal{A} = \{A_1, \cdots, A_n\}$ with trivial dimension. Now suppose $F(A_1, \cdots, A_n)$ and $G(A_1, \cdots, A_n)$ are objects in $\mathcal{F}_1(\mathcal{A})$ built from $A_1, \cdots A_n$, $A_1^*, \cdots, A_n^*$, called *literals* using $\otimes$.

A morphism in $\mathcal{F}_1(\mathcal{A})$ from $F(A_1, \cdots, A_n)$ to $G(A_1, \cdots, A_n)$ is described by pairing the occurrences of literals in the objects (formulas) F and G as follows: (i) Each literal occurrence is paired with precisely one other literal occurrence, (ii) An occurrence of A_i (in F, say) may be paired with either an occurrence of A_i^* in the same formula (F), or with another occurrence of A_i in the other formula (G), (iii) An occurrence of A_i^* may be paired with either an occurrence of A_i in the same formula, or with another occurrence of A_i^* in the other formula.

Now a morphism $F(A_1, \cdots, A_n) \to G(A_1, \cdots, A_n)$ in $\mathcal{F}(\mathcal{A})$ is a morphism in $\mathcal{F}_1(\mathcal{A})$ tensored with finitely many maps of the form $dim(A_i) : I \to I$.

[3] We sometimes use $F(\underline{X}, \underline{X})$ to denote $F(\underline{X})$, in particular when we want to emphasize the functoriality of F.

Let $F(\underline{X}) = F(X_1, \cdots, X_n)$ be a unit-free object built from $X_1, \cdots, X_n$ and $X_1^\perp, \cdots, X_n^\perp$ by tensor and par connectives. $F(\underline{X})$ corresponds to a formula in MLL and a proof $\vdash F(\underline{X})$ in MLL has a categorical interpretation in $\mathcal{F}^*(\underline{X})$ as a morphism $\mathbf{1} \to F(\underline{X})$ and conversely a morphism $\mathbf{1} \to F(\underline{X})$ in $\mathcal{F}^*(\underline{X})$ is the categorical representation of a proof of $\vdash F(\underline{X})$ in MLL. Therefore, MLL proof nets can be regarded as a graphical description of morphisms in the free *-autonomous category [5]. Finally, note that the free *-autonomous category supporting the MIX rule merely requires the addition of the unary MIX morphism $m : \perp \to \mathbf{1}$ and the necessary coherence equations.

5 A Double Glueing Construction

The double glueing construction we recall here is due to Tan and Hyland. Given a compact closed category, this construction produces a *-autonomous category which makes distinction between tensor and par products. The motivation for this construction lies in the work of Loader [25] on Linear Logical Predicates (LLP). See also Hasegawa [19] for a more abstract treatment and generalisations of glueing construction. The presentation here follows [31].

Let $\mathbb{C} = (\mathbb{C}, \otimes, I, (-)^*)$ be a compact closed category. Let H denote the covariant hom functor $\mathbb{C}(I, -) : \mathbb{C} \to \mathbf{Set}$ and K denote the contravariant functor $\mathbb{C}(-, I) \cong \mathbb{C}(I, (-)^*) : \mathbb{C}^{op} \to \mathbf{Set}$. Define a new category, $\mathbf{G}\mathbb{C}$ the glueing category of $\mathbb{C}$, whose objects are triples $\mathcal{A} = (|\mathcal{A}|, \mathcal{A}_s, \mathcal{A}_t)$ where

- $|\mathcal{A}|$ is an object of $\mathbb{C}$
- $\mathcal{A}_s \subseteq H(|\mathcal{A}|) = \mathbb{C}(I, A)$, is a set of *points* of A,
- $\mathcal{A}_t \subseteq K(|\mathcal{A}|) = \mathbb{C}(A, I) \cong \mathbb{C}(I, A^*)$ is a set of *copoints* of A.

A morphism $f : \mathcal{A} \to \mathcal{B}$ in $\mathbf{G}\mathbb{C}$ is a morphism $f : |\mathcal{A}| \to |\mathcal{B}|$ in $\mathbb{C}$ such that $Hf : \mathcal{A}_s \to \mathcal{B}_s$ and $Kf : \mathcal{B}_t \to \mathcal{A}_t$. Given $f : \mathcal{A} \to \mathcal{B}$ and $g : \mathcal{B} \to \mathcal{C}$ in $\mathbf{G}\mathbb{C}$, the composite $gf : \mathcal{A} \to \mathcal{C}$ is induced by the morphism gf in $\mathbb{C}$. The identity morphism on $\mathcal{A}$ is given by the identity morphism on $|\mathcal{A}|$ in $\mathbb{C}$.

We will denote the underlying object of $\mathcal{A}$ by A, etc. Given objects $\mathcal{A}$ and $\mathcal{B}$ we define the tensor product as follows:

- $| \mathcal{A} \otimes \mathcal{B} | = A \otimes B$
- $(\mathcal{A} \otimes \mathcal{B})_s = \{\sigma \otimes \tau \mid \sigma \in \mathcal{A}_s, \tau \in \mathcal{B}_s\}$,
- $(\mathcal{A} \otimes \mathcal{B})_t = \mathbf{G}\mathbb{C}(\mathcal{A}, \mathcal{B}^\perp)$.

where given $\mathcal{A}$, $\mathcal{A}^\perp = (A^*, \mathcal{A}_t, \mathcal{A}_s)$. We define $\mathcal{A} \multimap \mathcal{B} = (\mathcal{A} \otimes \mathcal{B}^\perp)^\perp$ and $\mathcal{A} \parr \mathcal{B} = (\mathcal{A}^\perp \otimes \mathcal{B}^\perp)^\perp$.

Proposition 5.1 (Tan). *For any compact closed category $\mathbb{C}$, $\mathbf{G}\mathbb{C}$ is a *-autonomous category with tensor $\otimes$ as above and unit $\mathbf{1} = (I, \{id_I\}, \mathbb{C}(I, I))$.*

Remark 5.2. Note that $\mathbf{G}\mathbb{C}$ is a nontrivial categorical model of MLL. That is, the tensor and par products are always distinct. For example, $(I, \emptyset, \emptyset) \otimes (I, \emptyset, \emptyset) = (I, \emptyset, \mathbb{C}(I, I))$ while $(I, \emptyset, \emptyset) \parr (I, \emptyset, \emptyset) = (I, \mathbb{C}(I, I), \emptyset)$.

Proposition 5.3 (Tan). $\mathbf{G}\mathbb{C}$ *supports the MIX rule iff* $\mathbb{C}(I, I) = \{1_I\}$.

In a logical setting one can think of an object $\mathcal{A}$, as an object A in $\mathbb{C}$ together with a collection of proofs of A (the collection $\mathcal{A}_s$) and a collection of disproofs or refutations of A (the collection $\mathcal{A}_t$.)

Proposition 5.4 (Tan). *The forgetful functor* $U : \mathbf{G}\mathbb{C} \to \mathbb{C}$ *preserves the *-autonomous structure of* $\mathbf{G}\mathbb{C}$. *Furthermore, it has a right adjoint* $R : \mathbb{C} \to \mathbf{G}\mathbb{C}$, *specified by* $RA = (A, \mathbb{C}(I, A), \emptyset)$ *and a left adjoint* $L : \mathbb{C} \to \mathbf{G}\mathbb{C}$, *specified by* $LA = (A, \emptyset, \mathbb{C}(I, A^*))$.

5.1 Approaching Full Completeness

The full completeness problem for MLL in our setting amounts to the following: Given a *-autonomous category $\mathbb{C}$ and a dinatural transformation $\rho : \mathcal{K}_{\mathbf{1}} \to [\![F]\!]$ where $\mathbf{1}$ is the unit of tensor and $[\![F]\!]$ is a definable multivariant functor, we would like to prove that ρ is induced by (is a denotation of) a morphism $\mathbf{1} \to F(\underline{X}, \underline{X})$ in the free *-autonomous category on n objects $\{X_1, \cdots, X_n\}$. We will be working with unit-free formulas and thus such a morphism is described by the proof net of the formula F.

The novelty in Loader and Tan's work included the approach to this problem using *-autonomous categories which are the glueing of compact closed categories. That is, $\mathbf{G}\mathbb{C}$ with $\mathbb{C}$ a compact closed category. Now, there is a forgetful functor $U : \mathbf{G}\mathbb{C} \to \mathbb{C}$ as we saw in the previous section. The idea is that a dinatural transformation $\rho : \mathcal{K}_{\mathbf{1}} \to F$ in $\mathbf{G}\mathbb{C}$ induces a dinatural transformation $U\rho : \mathcal{K}_I \to UF$ in the underlying compact closed category $\mathbb{C}$ and is completely determined by it. Note that UF simply consists of tensor products. Full completeness for a compact closed category, is defined in the same way, that is a dinatural $\mathcal{K}_I \to F$ must be the denotation of a morphism $I \to F(\underline{X}, \underline{X})$ in the free compact closed category on n objects. Therefore, the full completeness problem for a certain class of *-autonomous categories (those that are glueings of compact closed categories) is reduced to: (1) Proving full completeness for the underlying compact closed category, (2) Lifting the result to the *-autonomous category. We proceed by recalling the necessary formal definitions and theorems from [31].

Definition 5.5. *Let* $\mathbb{C}$ *be a compact closed category. Then* $\mathbb{C}$ *satisfies* compact closed full completeness *if every dinatural transformation* $\rho : \mathcal{K}_I \to [\![F]\!]$ *(with* $[\![F]\!] : \mathbb{C}^n \times (\mathbb{C}^{op})^n \to \mathbb{C}$*), is induced by a morphism* $I \to F(\underline{X}, \underline{X})$ *in the free compact closed category on* n *objects* $X_1, \cdots, X_n$.

Proposition 5.6 (Tan). *Let* $\mathbb{C}$ *be a compact closed category, let* $F : \mathbb{C}^n \times (\mathbb{C}^{op})^n \to \mathbb{C}$ *be a multivariant functor such that* $F(\underline{A}, \underline{A}) \cong A_{\mu_1} \otimes \cdots \otimes A_{\mu_l} \otimes A^*_{\lambda_1} \otimes \cdots \otimes A^*_{\lambda_m}$, *(where* $\mu_i, \lambda_i \in \{1, \ldots, n\}$ *for all* i*) and let* σ *be a collection of morphisms* $\sigma_{\underline{A}} : I \to F(\underline{A}, \underline{A})$ *in* $\mathbb{C}$. *Define* $F^-(\underline{A}) = A_{\lambda_1} \otimes \cdots \otimes A_{\lambda_m}$ *and* $F^+(\underline{A}) = A_{\mu_1} \otimes \cdots \otimes A_{\mu_l}$, *so that each* $\sigma_{\underline{A}}$ *is canonically equivalent to a morphism* $\tilde{\sigma}_{\underline{A}} : F^-(\underline{A}) \to F^+(\underline{A})$. *Then,* σ *is a dinatural transformation in* $\mathbb{C}$ *iff* $\tilde{\sigma}$ *is a natural transformation in* $\mathbb{C}$.

In view of this observation we can redefine compact closed full completeness as:

> A compact closed category $\mathbb{C}$ satisfies compact closed full completeness if every natural transformation $[\![F^-]\!] \to [\![F^+]\!]$ (with $[\![F^-]\!], [\![F^+]\!] : \mathbb{C}^n \to \mathbb{C}$), is induced by a morphism $F^-(\underline{X}) \to F^+(\underline{X})$ in the free compact closed category on n objects.

Theorem 5.7 (Tan). *Suppose that we have a multivariant functor $F : (\mathbf{G}\mathbb{C})^n \otimes (\mathbf{G}\mathbb{C}^{op})^n \to \mathbf{G}\mathbb{C}$, such that $\rho : \mathcal{K}_1 \to F$ is a dinatural transformation in $\mathbf{G}\mathbb{C}$. If $\mathcal{A}_1, \ldots, \mathcal{A}_n, \mathcal{B}_1, \ldots, \mathcal{B}_n$, objects in $\mathbf{G}\mathbb{C}$ are such that $U\mathcal{A}_i = U\mathcal{B}_i$ for all i, then $U\rho_{\underline{\mathcal{A}}} = U\rho_{\underline{\mathcal{B}}}$.*

6 Full Completeness in PAC-based Models

In this section we use PACs to construct models of MLL. Recall that PACs are traced symmetric monoidal categories. For any PAC $\mathbb{D}$, $Int\,\mathbb{D}$ is a compact closed category and hence $\mathbf{G}(Int\,\mathbb{D})$ is a *-autonomous category. In this way we get a class of models for MLL+MIX, which we show are fully complete for MLL+MIX. Our models support the MIX rule: $Int\,\mathbb{D}(I,I) = Int\,\mathbb{D}((I,I),(I,I)) = \mathbb{D}(I,I) = \{1_I\}$. Hereafter, $\mathbb{D}$ denotes a PAC and $\mathbb{C}$ denotes $Int\,\mathbb{D}$.

6.1 Compact Closed Full Completeness

Definition 6.1. *A sequent Γ is* balanced *if each propositional atom α occurs the same number of times as does its linear negation $\alpha^\perp$. The* length *of a sequent Γ is the number of occurrences of literals in Γ.*

If Γ has length p, then we can speak of the position where each literal occurs, numbered 1 to p. If Γ is balanced, and hence p is even, then we can specify the axiom links of a cut-free proof structure associated with Γ by a map $\varphi : \{1,\cdots,p\} \to \{1,\cdots,p\}$ such that φ is a fixed-point-free involution *and if a propositional atom α occurs in position i, then there is an occurrence of $\alpha^\perp$ in position $\varphi(i)$. Thus a cut-free proof structure can be specified as (Γ,φ), where Γ is a balanced sequent of length p and φ is a fixed-point-free involution on $\{1,\cdots,p\}$ specifying the axiom links.*

Let $F(\underline{X},\underline{X})$ be a formula of length p generated by $X_1,\ldots X_n$, $X_1^* \ldots X_n^*$ using $\otimes$. $F(\underline{X},\underline{X})$ induces a multivariant functor $[\![F]\!] : \mathbb{C}^n \times (\mathbb{C}^{op})^n \to \mathbb{C}$, which we will refer to as F. Also let $\sigma : \mathcal{K}_I \to F$ be a dinatural transformation from constant I functor to F. We can canonically transform σ into a natural transformation $\tilde{\sigma} : F^- \to F^+$. Suppose that the component of σ at $\underline{A}$ is given by $\sigma_{\underline{A}} : I \to F(\underline{A},\underline{A})$ where $F(\underline{A},\underline{A}) = A_{\xi_1}^{\zeta_1} \otimes \cdots \otimes A_{\xi_p}^{\zeta_p}$ with $\xi_i \in \{1,\cdots,n\}$ and $\zeta_i \in \{1,*\}$, (A_i^1 is read as A_i). Also let $N = \{i|\zeta_i = *\}$ and $P = \{i|\zeta_i = 1\}$. The component of $\tilde{\sigma}$ at $\underline{A}$ is of the form $\tilde{\sigma}_{\underline{A}} : F^-(\underline{A}) \to F^+(\underline{A})$ where $F^-(\underline{A}) = A_{\lambda_1} \otimes \cdots \otimes A_{\lambda_m}$ and $F^+(\underline{A}) = A_{\mu_1} \otimes \cdots \otimes A_{\mu_l}$ with $\lambda_i, \mu_i \in \{1,2,\cdots,n\}$. Therefore $|N| = m$ and $|P| = l$.

Lemma 6.2. *Let $\tilde{\sigma} : F^- \to F^+$ be a natural transformation as above. Then each type variable that occurs in F^- must also occur in F^+. Moreover it must occur in F^+ with the same multiplicity.*

Proof. (idea) The proof of this lemma is prohibitingly long to include here. However, the main idea can be sketched as follows. One starts with assuming the negation of the conclusion, the naturality conditions written out in the category $\mathbb{C}$, then give rise to systems of recursive equations in the components of $\tilde{\sigma}$. These equations are then proven to be inconsistent by double induction on the number of equations and summands in each equation. □

Proposition 6.3. *Let $F(\underline{X}, \underline{X})$ be an MLL formula of length p generated by $X_1, \ldots X_n$, $X_1^* \ldots X_n^*$ using $\otimes$. Let $[\![F]\!] : \mathbb{C}^n \times (\mathbb{C}^{op})^n \to \mathbb{C}$ be the induced multivariant functor on $\mathbb{C}$. If $\sigma : \mathcal{K}_I \to [\![F]\!]$ is a dinatural transformation, then F is balanced.*

Theorem 6.4. *$\tilde{\sigma} : F^- \to F^+$ is a permutation on the tensor factors, i.e. $\tilde{\sigma}_{\underline{A}} : F^-(\underline{A}) \to F^+(\underline{A})$ is of the form $\begin{bmatrix} \mathbf{B}_1 & \mathbf{0} \\ \mathbf{0} & \mathbf{B}_4 \end{bmatrix}$ where $\mathbf{B}_1$ and $\mathbf{B}_4$ are permutation matrices and $\mathbf{B}_4 = \mathbf{B}_1^t$ and the permutation $\delta \in S_m$ induced by $\mathbf{B}_1$ satisfies $\mu_i = \lambda_{\delta(i)}$ for $i = 1, \cdots, m$. Here $(-)^t$ denotes matrix transposition obtained by reflection across* antidiagonal *elements.*

Proof. Let $\tilde{\sigma} : F^- \to F^+$ be a natural transformation in $\mathbb{C}$, that is the following diagram commutes:

$$\begin{array}{ccc} A_{\lambda_1} \otimes \cdots \otimes A_{\lambda_m} & \xrightarrow{\tilde{\sigma}_{\underline{A}}} & A_{\mu_1} \otimes \cdots \otimes A_{\mu_m} \\ {\scriptstyle f_{\lambda_1} \otimes \cdots \otimes f_{\lambda_m}} \downarrow & & \downarrow {\scriptstyle f_{\mu_1} \otimes \cdots \otimes f_{\mu_m}} \\ B_{\lambda_1} \otimes \cdots \otimes B_{\lambda_m} & \xrightarrow{\tilde{\sigma}_{\underline{B}}} & B_{\mu_1} \otimes \cdots \otimes B_{\mu_m} \end{array}$$

Hence, we have $Tr^{A^-_{\mu_m} \otimes \cdots \otimes A^-_{\mu_1}}(M) = (f_{\mu_1} \otimes \cdots \otimes f_{\mu_m})\tilde{\sigma}_{\underline{A}} = \tilde{\sigma}_{\underline{B}}(f_{\lambda_1} \otimes \cdots \otimes f_{\lambda_m}) = Tr^{B^-_{\lambda_m} \otimes \cdots \otimes B^-_{\lambda_1}}(M')$ where M and M' are given below.
M:

$$\begin{array}{ccccccccc|c}
A^+_{\lambda_1} & \cdots & A^+_{\lambda_m} & B^-_{\mu_m} & \cdots & B^-_{\mu_1} & A^-_{\mu_m} & \cdots & A^-_{\mu_1} & \\ \hline
f^{\mu_1}_{11}\tilde{\sigma}_{1,1} & \cdots & f^{\mu_1}_{11}\tilde{\sigma}_{1,m} & & & f^{\mu_1}_{12} & f^{\mu_1}_{11}\tilde{\sigma}_{1,m+1} & \cdots & f^{\mu_1}_{11}\tilde{\sigma}_{1,2m} & B^+_{\mu_1} \\
\vdots & & \vdots & \mathbf{0} & \iddots & \mathbf{0} & \vdots & & \vdots & \vdots \\
f^{\mu_m}_{11}\tilde{\sigma}_{m,1} & \cdots & f^{\mu_m}_{11}\tilde{\sigma}_{m,m} & f^{\mu_m}_{12} & & & f^{\mu_m}_{11}\tilde{\sigma}_{m,m+1} & \cdots & f^{\mu_m}_{11}\tilde{\sigma}_{m,2m} & B^+_{\mu_m} \\
\tilde{\sigma}_{m+1,1} & \cdots & \tilde{\sigma}_{m+1,m} & & & & \tilde{\sigma}_{m+1,m+1} & \cdots & \tilde{\sigma}_{m+1,2m} & A^-_{\lambda_m} \\
\vdots & & \vdots & & \mathbf{0} & & \vdots & & \vdots & \vdots \\
\tilde{\sigma}_{2m,1} & \cdots & \tilde{\sigma}_{2m,m} & & & & \tilde{\sigma}_{2m,m+1} & \cdots & \tilde{\sigma}_{2m,2m} & A^-_{\lambda_1} \\
f^{\mu_m}_{21}\tilde{\sigma}_{m,1} & \cdots & f^{\mu_m}_{21}\tilde{\sigma}_{m,m} & f^{\mu_m}_{22} & & & f^{\mu_m}_{21}\tilde{\sigma}_{m,m+1} & \cdots & f^{\mu_m}_{21}\tilde{\sigma}_{m,2m} & A^-_{\mu_m} \\
\vdots & & \vdots & \mathbf{0} & \ddots & \mathbf{0} & \vdots & & \vdots & \vdots \\
f^{\mu_1}_{21}\tilde{\sigma}_{1,1} & \cdots & f^{\mu_1}_{21}\tilde{\sigma}_{1,m} & & & f^{\mu_1}_{22} & f^{\mu_1}_{21}\tilde{\sigma}_{1,m+1} & \cdots & f^{\mu_1}_{21}\tilde{\sigma}_{1,2m} & A^-_{\mu_1}
\end{array}$$

M':

$$
\begin{array}{ccccccccc|c}
A^+_{\lambda_1} & \cdots & A^+_{\lambda_m} & B^-_{\mu_m} & \cdots & B^-_{\mu_1} & B^-_{\lambda_m} & \cdots & B^-_{\lambda_1} & \\
\hline
\tilde{\sigma}_{1,1}f_{11}^{\lambda_1} & \cdots & \tilde{\sigma}_{1,m}f_{11}^{\lambda_m} & \tilde{\sigma}_{1,m+1} & \cdots & \tilde{\sigma}_{1,2m} & \tilde{\sigma}_{1,m}f_{12}^{\lambda_m} & \cdots & \tilde{\sigma}_{1,1}f_{12}^{\lambda_1} & B^+_{\mu_1} \\
\vdots & & \vdots & \vdots & & \vdots & \vdots & & \vdots & \vdots \\
\tilde{\sigma}_{m,1}f_{11}^{\lambda_1} & \cdots & \tilde{\sigma}_{m,m}f_{11}^{\lambda_m} & \tilde{\sigma}_{m,m+1} & \cdots & \tilde{\sigma}_{m,2m} & \tilde{\sigma}_{m,m}f_{12}^{\lambda_m} & \cdots & \tilde{\sigma}_{m,1}f_{12}^{\lambda_1} & B^+_{\mu_m} \\
 & & f_{21}^{\lambda_m} & & & & f_{22}^{\lambda_m} & & & A^-_{\lambda_m} \\
\mathbf{0} & \iddots & \mathbf{0} & & \mathbf{0} & & \mathbf{0} & \ddots & \mathbf{0} & \vdots \\
f_{21}^{\lambda_1} & & & & & & & & f_{22}^{\lambda_1} & A^-_{\lambda_1} \\
\tilde{\sigma}_{m+1,1}f_{11}^{\lambda_1} & \cdots & \tilde{\sigma}_{m+1,m}f_{11}^{\lambda_m} & \tilde{\sigma}_{m+1,m+1} & \cdots & \tilde{\sigma}_{m+1,2m} & \tilde{\sigma}_{m+1,m}f_{12}^{\lambda_m} & \cdots & \tilde{\sigma}_{m+1,1}f_{12}^{\lambda_1} & B^-_{\lambda_m} \\
\vdots & & \vdots & \vdots & & \vdots & \vdots & & \vdots & \vdots \\
\tilde{\sigma}_{2m,1}f_{11}^{\lambda_1} & \cdots & \tilde{\sigma}_{2m,m}f_{11}^{\lambda_m} & \tilde{\sigma}_{2m,m+1} & \cdots & \tilde{\sigma}_{2m,2m} & \tilde{\sigma}_{2m,m}f_{12}^{\lambda_m} & \cdots & \tilde{\sigma}_{2m,1}f_{12}^{\lambda_1} & B^-_{\lambda_1}
\end{array}
$$

By instantiating at $f_{kl}^{\lambda_i} = f_{kl}^{\mu_i} = 0$ for $k, l = 1, 2$ for all $i = 1, \cdots, m$ and using $Tr(M) = Tr(M')$, we conclude that $\tilde{\sigma}_{\underline{A}}$ is of the form: $\begin{bmatrix} \mathbf{B}_1 & \mathbf{0} \\ \mathbf{0} & \mathbf{B}_4 \end{bmatrix}$ where

$$
\mathbf{B}_1 = \begin{array}{cccc|c}
A^+_{\lambda_1} & A^+_{\lambda_2} & \cdots & A^+_{\lambda_m} & \\
\hline
\tilde{\sigma}_{1,1} & \tilde{\sigma}_{1,2} & \cdots & \tilde{\sigma}_{1,m} & A^+_{\mu_1} \\
\tilde{\sigma}_{2,1} & \tilde{\sigma}_{2,2} & \cdots & \tilde{\sigma}_{2,m} & A^+_{\mu_2} \\
\vdots & \vdots & \cdots & \vdots & \vdots \\
\tilde{\sigma}_{m,1} & \tilde{\sigma}_{m,2} & \cdots & \tilde{\sigma}_{m,m} & A^+_{\mu_m}
\end{array}
$$

$$
\mathbf{B}_4 = \begin{array}{cccc|c}
A^-_{\mu_m} & A^-_{\mu_{m-1}} & \cdots & A^-_{\mu_1} & \\
\hline
\tilde{\sigma}_{m+1,m+1} & \tilde{\sigma}_{m+1,m+2} & \cdots & \tilde{\sigma}_{m+1,2m} & A^-_{\lambda_m} \\
\tilde{\sigma}_{m+2,m+1} & \tilde{\sigma}_{m+2,m+2} & \cdots & \tilde{\sigma}_{m+2,2m} & A^-_{\lambda_{m-1}} \\
\vdots & \vdots & \cdots & \vdots & \vdots \\
\tilde{\sigma}_{2m,m+1} & \tilde{\sigma}_{2m,m+2} & \cdots & \tilde{\sigma}_{2m,2m} & A^-_{\lambda_1}
\end{array}
$$

Next, let $A^+_{\lambda_i} = A^-_{\lambda_i}$ and $B^+_{\lambda_i} = B^-_{\lambda_i}$ and f^{λ_i} be the twist, i.e. $f_{11}^{\lambda_i} = f_{22}^{\lambda_i} = 0$ and $f_{12}^{\lambda_i} = f_{21}^{\lambda_i} = 1$ for all $i = 1, \cdots, m$. Similarly for A_{μ_i}, B_{μ_i} and f^{μ_i} for $i = 1, \cdots, m$. We get the following system of equations:

System I:

$$
NN' = \begin{bmatrix}
\tilde{\sigma}_{1,m} & \cdots & \tilde{\sigma}_{1,1} \\
\tilde{\sigma}_{2,m} & \cdots & \tilde{\sigma}_{2,1} \\
\vdots & \cdots & \vdots \\
\tilde{\sigma}_{m,m} & \cdots & \tilde{\sigma}_{m,1}
\end{bmatrix}
\begin{bmatrix}
\tilde{\sigma}_{m+1,m+1} & \cdots & \tilde{\sigma}_{m+1,2m} \\
\tilde{\sigma}_{m+2,m+1} & \cdots & \tilde{\sigma}_{m+2,2m} \\
\vdots & \cdots & \vdots \\
\tilde{\sigma}_{2m,m+1} & \cdots & \tilde{\sigma}_{2m,2m}
\end{bmatrix}
= antidiag(1, 1, \cdots, 1)
$$

that is, we have $R_i^N C_j^{N'} = 1$ for $j = m - i + 1$ and 0 else, for $i = 1, \cdots, m$.

System II:

$$PP' = \begin{bmatrix} \tilde{\sigma}_{m+1,m+1} & \cdots \tilde{\sigma}_{m+1,2m} \\ \tilde{\sigma}_{m+2,m+1} & \cdots \tilde{\sigma}_{m+2,2m} \\ \vdots & \cdots \vdots \\ \tilde{\sigma}_{2m,m+1} & \cdots \tilde{\sigma}_{2m,2m} \end{bmatrix} \begin{bmatrix} \tilde{\sigma}_{m,1} & \cdots \tilde{\sigma}_{m,m} \\ \tilde{\sigma}_{m-1,1} & \cdots \tilde{\sigma}_{m-1,m} \\ \vdots & \cdots \vdots \\ \tilde{\sigma}_{1,1} & \cdots \tilde{\sigma}_{1,m} \end{bmatrix} = antidiag(1, 1, \cdots, 1)$$

that is $R_i^P C_j^{P'} = 1$ for $j = m - i + 1$ and 0 else, for $i = 1, \cdots, m$.

The rest of the proof consists of several steps:

Step 1: We show that $\mathbf{B}_4 = \mathbf{B}_1^t$, that is, $\tilde{\sigma}_{i,j} = \tilde{\sigma}_{2m+1-j,2m+1-i}$ for $i, j = m+1, m+2, \cdots, 2m$. Note that $(-)^t$ denotes the matrix transposition which is obtained by reflection across the *antidiagonal* entries.

• *case 1*: $\tilde{\sigma}_{i,j} = 1, \tilde{\sigma}_{2m+1-j,2m+1-i} = 0$. Note that $R_{j-m}^P C_k^{P'} = 0$ for all $k = 1, \cdots, m$, $k \neq 2m - j + 1$. $\tilde{\sigma}_{i,j} = 1$ implies that $\tilde{\sigma}_{2m-j+1,k} = 0$ for $k \neq 2m - i + 1$, however $\tilde{\sigma}_{2m-j+1,2m-i+1} = 0$ is given. Therefore, $R_{j-m}^{P'} = 0$ and $R_{2m-j+1}^N = R_{j-m}^{P'} = 0$ giving $R_{2m-j+1}^N C_{j-m}^{N'} = 0$, a contradiction.

• *case 2*: $\tilde{\sigma}_{i,j} = 0, \tilde{\sigma}_{2m+1-j,2m+1-i} = 1$. Note that $R_{2m-i+1}^N C_k^{N'} = 0$ for all $k = 1, \cdots, m$, $k \neq i - m$. $\tilde{\sigma}_{2m-j+1,2m-i+1} = 1$ implies $\tilde{\sigma}_{i,k} = 0$ for all $k \neq j$, but $\tilde{\sigma}_{i,j} = 0$ is given, hence $R_{i-m}^{N'} = 0$, and $R_{i-m}^P = R_{i-m}^{N'} = 0$ giving $R_{i-m}^P C_{2m-i+1}^{P'} = 0$, a contradiction.

Hence $\mathbf{B}_4 = \mathbf{B}_1^t$.

Step 2: There are no all-zero rows or columns in $\mathbf{B}_1$ or $\mathbf{B}_4$.

The ith row of $\mathbf{B}_1$ is equal to R_i^N in reverse order, hence it cannot be all zero since $R_i^N C_{m-i+1}^{N'} = 1$. Also the jth column of $\mathbf{B}_1$ is equal to $C_j^{P'}$ in reverse order and hence it cannot be all zero since $R_{m+1-j}^P C_j^{P'} = 1$.

The statement is trivially true for $\mathbf{B}_4$ as $\mathbf{B}_4 = \mathbf{B}_1^t$.

Step 3: In $\mathbf{B}_1$ and $\mathbf{B}_4$ every row and column has exactly one 1. Suppose any two elements on the ith row of $\mathbf{B}_1$ are 1; $\tilde{\sigma}_{i,j} = \tilde{\sigma}_{i,k} = 1$ for $k \neq j$ with $i, j, k \in \{1, \cdots, m\}$. The ith row of $\mathbf{B}_1 = R_{m-i+1}^{P'}$ and hence $C_{m-i+1}^P = 0$. For example, suppose $\tilde{\sigma}_{1,1} = \tilde{\sigma}_{1,m} = 1$, then using system II we see that all the elements on the last column of P are zero by just using the fact that $\tilde{\sigma}_{1,1} = 1$. Also $\tilde{\sigma}_{2m,2m} = 0$ because $\tilde{\sigma}_{1,m} = 1$. Note that $C_{m-i+1}^P = C_{m-i+1}^{\mathbf{B}_4} = 0$, a contradiction.

Also let any two elements on the jth column of $\mathbf{B}_1$ be both 1; $\tilde{\sigma}_{i,j} = \tilde{\sigma}_{k,j} = 1$ for $i \neq k$ with $i, j, k \in \{1, \cdots, m\}$. The jth column of $\mathbf{B}_1 = C_{m-j+1}^N$ and hence $R_{m-j+1}^{N'} = 0$, and as $R_{m-j+1}^{\mathbf{B}_4} = R_{m-j+1}^{N'}$, we get a contradiction.

As $\mathbf{B}_4 = \mathbf{B}_1^t$, the statement follows for $\mathbf{B}_4$.

Therefore, $\mathbf{B}_1$ and $\mathbf{B}_4$ are permutation matrices. Let $\delta \in S_m$ be the permutation induced by $\mathbf{B}_1$, that is $\delta(i) = j$ iff $\tilde{\sigma}_{i,j} = 1$. Then we have $\tilde{\sigma}_{i,j} : A_{\lambda_j}^+ \to A_{\mu_i}^+ = 1$ and $\tilde{\sigma}_{2m+1-j,2m+1-i} : A_{\mu_i}^- \to A_{\lambda_j}^- = 1$ and thus $A_{\lambda_{\delta(i)}} = A_{\lambda_j} = (A_{\lambda_j}^+, A_{\lambda_j}^-) = (A_{\mu_i}^+, A_{\mu_i}^-) = A_{\mu_i}$ for all $i = 1, \cdots, m$. □

We can view the natural transformation $\tilde{\sigma}_{\underline{A}} : F^-(\underline{A}) \to F^+(\underline{A})$ as matching $A_{\lambda_{\delta(i)}}$ to A_{μ_i} for all $i = 1, \cdots, m$. Hence we have:

Corollary 6.5 (Full Completeness in $\mathbb{C}$). *Every natural transformation $\tilde{\sigma} : F^- \to F^+$ in $\mathbb{C}$ is induced by a unique morphism $F^-(\underline{X}) \to F^+(\underline{X})$ in the free compact closed category on n objects $X_1, \cdots, X_n$ with trivial dimension.*

Note that all objects in $\mathbb{C}$ have trivial dimension, i.e. $dim(A) = 1_I$ for all objects A in $\mathbb{C}$, since $\mathbb{C}(I,I) = \mathbb{D}(I,I) = \{1_I\}$. Therefore, the restriction on dimension can be removed: it is tensoring with finitely many 1_I maps, which have no effect. Thus

Corollary 6.6 (Full Completeness in $\mathbb{C}$). *Every natural transformation $\tilde{\sigma} : F^- \to F^+$ in $\mathbb{C}$ is induced by a morphism $F^-(\underline{X}) \to F^+(\underline{X})$ in the free compact closed category on n objects $X_1, \cdots, X_n$.*

Theorem 6.7. *Suppose that σ is a dinatural transformation in $\mathbb{C}$ from $\mathcal{K}_I$ to the multivariant functor F. Then there exists a fixed-point-free involution φ on $\{1, \cdots, p\}$ such that $\xi_{\varphi(i)} = \xi_i, \zeta_{\varphi(i)} \neq \zeta_i$.*

In view of this theorem we see that σ determines a *unique* set of axiom links and hence a unique MLL proof structure for the formula F. We will show that this proof structure is indeed a proof net. That is, we need to check the Danos-Regnier correctness criterion. However, as $\mathbb{C}(I,I) = \mathbb{D}(I,I) = \{1_I\}$, $\mathbf{G}\mathbb{C}$ satisfies the MIX rule and hence we need only check the acyclicity condition [11, 13].

6.2 Full Completeness in $\mathbf{G}\mathbb{C}$

Given a dinatural transformation $\rho : \mathcal{K}_1 \to F$ in $\mathbf{G}\mathbb{C}$, we have the specification of a unique proof structure because we have the formula F and the axiom links are given by the fixed-point free involution φ induced by the dinatural transformation $U\rho$ in the underlying compact closed category $\mathbb{C} = Int(\mathbb{D})$. We show that this proof structure is indeed a proof net. For this purpose we only need to prove acyclicity as our category $\mathbf{G}\mathbb{C}$ satisfies the MIX rule.

Lemma 6.8. *Let $F(\underline{X}, \underline{X}) = F_1(\underline{X}, \underline{X}) \otimes F_2(\underline{X}, \underline{X})$ be an object in the free compact closed category on n objects $X_1, \cdots, X_n$ with trivial dimension and $\Gamma : I \to F(\underline{X}, \underline{X})$ be a morphism. Suppose also that the induced fixed-point free involution φ does not make a matching between formulas in F_1 and those in F_2, then $\Gamma = \Gamma_1 \otimes \Gamma_2$ where $\Gamma_1 : I \to F_1(\underline{X}, \underline{X})$ and $\Gamma_2 : I \to F_2(\underline{X}, \underline{X})$.*

Theorem 6.9 (Acyclicity). *Suppose that ρ is a dinatural transformation in $\mathbf{G}\mathbb{C}$ from the constant functor $\mathcal{K}_1$ to F. Consider the unique proof structure associated with ρ. Then for any DR-switching, the associated DR-graph is acyclic.*

Proof. Suppose that for a certain DR-switching, the associated DR-graph contains a cycle. Express the shortest cycle as lower connected pairs $(a_1, b_1), \cdots, (a_r, b_r)$ where $\varphi(b_i) = a_{i+1}$ for all $i \in \mathbb{Z}_r$. Recall that a lower connected pair in a proof structure is a pair of formulas that are connected with paths not traversing any axiom links [25, 26]. Using the weak distributivity natural transformations, binary MIX morphisms and associativity and commutativity natural transformations for par and tensor [9], we transform the given dinatural transformation ρ into $\tilde{\rho} : \mathcal{K}_1 \to \tilde{F}$ such that the cycle is preserved, where $\tilde{\rho}_{\underline{\mathcal{A}}} : \mathbf{1} \to \tilde{F}(\underline{\mathcal{A}}, \underline{\mathcal{A}})$ and

$$\tilde{F}(\underline{\mathcal{A}}, \underline{\mathcal{A}}) = \Gamma_{\underline{\mathcal{A}}} \parr (\mathcal{A}^{\zeta_{a_1}}_{\xi_{a_1}} \otimes \mathcal{A}^{\zeta_{b_1}}_{\xi_{b_1}}) \parr \cdots \parr (\mathcal{A}^{\zeta_{a_r}}_{\xi_{a_r}} \otimes \mathcal{A}^{\zeta_{b_r}}_{\xi_{b_r}})$$

The procedure is as follows (see also [31] and [2]):

- If a fragment of F has the form $A \otimes (B \parr C)$, and A and B are lower connected, then the switching must have assigned *left* to the par-link in question. In this case, we compose ρ with a natural transformation built from w^L_{ABC},
- If A and C are lower connected, then the switching must have assigned *right* to the par-link. In this case, we compose with a natural transformation built from w^R_{ABC},
- We apply binary MIX, commutativity and associativity, whenever necessary to separate out each lower connected pair.

Consider the test object $\mathcal{A} = (A, \{0_{A^-A^+}\}, \{0_{A^+A^-}\})$ where $A = (A^+, A^-) \neq (I, I)$ and $A^+ = A^-$. Hence $\mathcal{A}^\perp = \mathcal{A}$. Put $\mathcal{A}_i = \mathcal{A}$ for $i = 1, \cdots, n$. In what follows there is no need to put ζ_- superscripts as $\mathcal{A} = \mathcal{A}^\perp$, however we have included these for clarity.

Notice that $U(\tilde{\rho_{\underline{A}}}) = f_1 \otimes f_2$, $f_1 : I \to \Gamma_{\underline{A}}$ and $f_2 : I \to (A^{\zeta_{a_1}} \otimes A^{\zeta_{b_1}}) \otimes \cdots \otimes (A^{\zeta_{a_r}} \otimes A^{\zeta_{b_r}})$, because the part in $\tilde{F}(\underline{A}, \underline{A})$ consisting of par product of tensored pairs is closed under the axiom link matchings induced by any dinatural transformation. Therefore, we have that f_2 must lift to a morphism in $\mathbf{G}\mathbb{C}$ from $\mathbf{1}$ to $(\mathcal{A}^{\zeta_{a_1}} \otimes \mathcal{A}^{\zeta_{b_1}}) \parr \cdots \parr (\mathcal{A}^{\zeta_{a_r}} \otimes \mathcal{A}^{\zeta_{b_r}})$.
Hence $f_2 \in ((\mathcal{A}^{\zeta_{a_1}} \otimes \mathcal{A}^{\zeta_{b_1}}) \parr \cdots \parr (\mathcal{A}^{\zeta_{a_r}} \otimes \mathcal{A}^{\zeta_{b_r}}))_s$.

$((\mathcal{A}^{\zeta_{a_1}} \otimes \mathcal{A}^{\zeta_{b_1}}) \parr \cdots \parr (\mathcal{A}^{\zeta_{a_r}} \otimes \mathcal{A}^{\zeta_{b_r}}))_s =$
$((\mathcal{A}^{\zeta_{a_1}} \otimes \mathcal{A}^{\zeta_{b_1}})^\perp \otimes \cdots \otimes (\mathcal{A}^{\zeta_{a_r}} \otimes \mathcal{A}^{\zeta_{b_r}})^\perp)_t =$
$\mathbf{G}\mathbb{C}((\mathcal{A}^{\zeta_{a_1}} \otimes \mathcal{A}^{\zeta_{b_1}})^\perp \otimes \cdots \otimes (\mathcal{A}^{\zeta_{a_{r-1}}} \otimes \mathcal{A}^{\zeta_{b_{r-1}}})^\perp, (\mathcal{A}^{\zeta_{a_r}} \otimes \mathcal{A}^{\zeta_{b_r}}))$

Now consider $((\mathcal{A}^{\zeta_{a_1}} \otimes \mathcal{A}^{\zeta_{b_1}})^\perp \otimes \cdots \otimes (\mathcal{A}^{\zeta_{a_{r-1}}} \otimes \mathcal{A}^{\zeta_{b_{r-1}}})^\perp)_s = \{\sigma_1 \otimes \cdots \otimes \sigma_{r-1} \mid \sigma_i \in (\mathcal{A}^{\zeta_{a_i}} \otimes \mathcal{A}^{\zeta_{a_i}})^\perp_s\}$.

Notice that $(\mathcal{A}^{\zeta_{a_i}} \otimes \mathcal{A}^{\zeta_{b_i}})^\perp_s = (\mathcal{A}^{\zeta_{a_i}} \otimes \mathcal{A}^{\zeta_{b_i}})_t = \mathbf{G}\mathbb{C}(\mathcal{A}, \mathcal{A})$ and hence $1_A \in (\mathcal{A}^{\zeta_{a_i}} \otimes \mathcal{A}^{\zeta_{b_i}})^\perp_s$ and therefore $\underbrace{1_A \otimes \cdots \otimes 1_A}_{r-1 \ times} \in ((\mathcal{A}^{\zeta_{a_1}} \otimes \mathcal{A}^{\zeta_{b_1}})^\perp \otimes \cdots \otimes (\mathcal{A}^{\zeta_{a_{r-1}}} \otimes \mathcal{A}^{\zeta_{b_{r-1}}})^\perp)_s$. On the other hand, $(\mathcal{A}^{\zeta_{a_r}} \otimes \mathcal{A}^{\zeta_{b_r}})_s = \{0\}$. Now by definition, $f_2 \circ \alpha \in (\mathcal{A}^{\zeta_{a_r}} \otimes \mathcal{A}^{\zeta_{b_r}})_s = \{0\}$ for all $\alpha \in ((\mathcal{A}^{\zeta_{a_1}} \otimes \mathcal{A}^{\zeta_{b_1}})^\perp \otimes \cdots \otimes (\mathcal{A}^{\zeta_{a_{r-1}}} \otimes \mathcal{A}^{\zeta_{b_{r-1}}})^\perp)_s$ and hence $f_2 = 0$ which yields a contradiction because such a morphism cannot induce any axiom links. □

Theorem 6.10 (Full completeness in $\mathbf{G}\mathbb{C}$). *Every dinatural transformation in $\mathbf{G}\mathbb{C}$ from the constant functor $\mathcal{K}_1$ to the multivariant functor F is the denotation of a unique cut-free proof in MLL+MIX of the formula F, and is therefore induced by a unique morphism $\mathbf{1} \to F(\underline{X}, \underline{X})$ in the free *-autonomous category supporting the MIX rule, on n objects $X_1, X_2, \cdots, X_n$.*

We conclude this section by stating a negative result for the class of categories $\mathbf{G}\mathbb{C}$ with $\mathbb{C} = Int(\mathbb{D})$ and $\mathbb{D}$ a PAC. Suppose we choose to use the traditional categorical semantics framework [24]. That is, formulas of MLL are objects in $\mathbf{G}\mathbb{C}$ and proofs are morphisms. Then we show that $\mathbf{G}\mathbb{C}$ fails to be fully complete for MLL.

Theorem 6.11. *Let $\mathbb{D}$ be a PAC and $\mathbb{C} = Int(\mathbb{D})$, interpret the formulas of MLL as objects in $\mathbf{G}\mathbb{C}$ and the proofs as morphisms. Then, $\mathbf{G}\mathbb{C}$ is not fully complete for MLL.*

Proof. Let $\mathcal{A} = (A, \mathcal{A}_s, \mathcal{A}_t)$ be an object in $\mathbf{G}\mathbb{C}$ with $A = (A^+, A^-)$ and $A^+ = A^-$. Also, let $\mathcal{A}_s = \mathcal{A}_t = \{1_{A^+}\}$. Note that $\mathcal{A} \otimes \mathcal{A}^\perp$ is not an MLL provable formula. We show that there exists a map $f : \mathbf{1} \to \mathcal{A} \otimes \mathcal{A}^\perp$. Let $f : I \to A \otimes A^*$ be $1_{A^+ \otimes A^+}$. Recall that $\mathbf{1}_s = \{1_I\}$ and therefore $f\alpha = f \in (\mathcal{A} \otimes \mathcal{A}^\perp)_s$ for all $\alpha \in \mathbf{1}_s$. Recall that $(\mathcal{A} \otimes \mathcal{A}^\perp)_t = \mathbf{G}\mathbb{C}(\mathcal{A}, \mathcal{A})$ and hence $(\mathcal{A} \otimes \mathcal{A}^\perp)_t \neq \emptyset$. To conclude the proof, we need to show that $\beta f = 1_I$ for all $\beta \in (\mathcal{A} \otimes \mathcal{A}^\perp)_t$, but $\beta f : I \to I$ in $\mathbb{C}$ and $\mathbb{C}(I, I) = \{1_I\}$, therefore $\beta f = 1_I$ and thus $\beta f \in \mathbf{1}_t$ for all $\beta \in (\mathcal{A} \otimes \mathcal{A}^\perp)_t$. □

7 Conclusion and Future Work

We have shown how to construct models of MLL, i.e. *-autonomous categories based on PACs. We made use of the Int and double glueing constructions to get such models. We also proved that such models are fully complete for MLL+MIX. The techniques we have used are general enough to allow us to prove that for a traced Unique Decomposition Category [15, 16] $\mathbb{D}$, $\mathbf{G}(Int\mathbb{D})$ is fully complete for MLL + MIX. We have not included this result due to lack of space, for details see [15].

A major problem is to extend full completeness to different fragments of linear logic (e.g. additives, exponentials.) Game-theoretical models have recently become a major tool in this area. We intend to unify these models with the ones studied in this paper and author's thesis [15]. Current results by several research groups appear amenable to an abstract and axiomatic approach which is yet to be developed. First steps towards such an approach were taken by Hyland, Abramsky and their colleagues and students. Recently Abramsky and Mellies announced a novel game-theoretic full completeness result for the multiplicative and additive fragment of linear logic. Current work aims to give new non-game-theoretic fully complete models for the multiplicative and additive fragment of linear logic.

References

1. Abramsky, S.: Retracing Some Paths in Process Algebra. In CONCUR 96, SLNCS **1119** (1996) 1–17.
2. Abramsky, S. and Jagadeesan, R.: Games and full completeness for Multiplicative Linear Logic. *J. of Symbolic Logic* **59** (1994) 543–574.
3. Abramsky, S. and Mellies, P.: Concurrent Games and Full Completeness. In *Proc. of 14th LICS* (1999) 431–442.
4. Bainbridge, E.S., Freyd, P., Scedrov, A. and Scott, P.J.: Functorial Polymorphism. *Theor. Comp. Science*, **70** (1990) 35–64.
5. Blute, R.F.: Linear Logic, Coherence and Dinaturality. *Theor. Computer Science* **115** (1993) 3–41.
6. Blute, R.F., Cockett, J.R.B., Seely, R.A.G. and Trimble, T.H.: Natural deduction and coherence for weakly distributive categories. *Jour. Pure and Applied Algebra* **113** (1996) 229–296.

7. Blute, R.F. and Scott, P.J.: The Shuffle Hopf Algebra and Noncommutative Full Completeness. *Journal of Symbolic Logic* **63** (4) (1998) 1413–1436.
8. Blute, R.F. and Scott, P.J.: Linear Läuchli Semantics. *Annals of Pure and Applied Logic* **77** (1996) 101–142.
9. Cockett, J.R.B. and Seely, R.A.G.: Weakly Distributive Categories. *Journal of Pure and Applied Algebra* **114** (1997) 133–173.
10. Costantini, F., Mascari, G. and Pedicini M.: Dynamics of Algorithms. *Technical Report* **n.6** (1996).
11. Danos, V. and Regnier, L.: The Structure of Multiplicatives. *Arch. Math. Logic* **28** (1989) 181–203.
12. Devarajan, P., Hughes, D., Plotkin, G. and Pratt, V.: Full completeness of the multiplicative linear logic of Chu spaces. In *Proc. of the 14th LICS*, (1999) 234–243.
13. Fleury, A. and Rétoré, C.: The MIX Rule, *Math. Struct. in Comp. Science* **4** (1994) 273–285.
14. Girard, J.Y., Scedrov, A. and Scott, P.J.: Normal forms and cut-free proofs as natural transformations. In *Logic from Computer Science*, (1991) 217–241.
15. Haghverdi, E.: *A Categorical Approach to Linear Logic, Geometry of Interaction and Full Completeness*, PhD Thesis, University of Ottawa, 2000. Available from `http:// www.math.upenn.edu/~esfan`.
16. Haghverdi, E.: Unique decomposition categories, Geometry of Interaction and combinatory logic, *Math. Struct. in Comp. Science*, vol 10 (2000) 205–231.
17. Hamano, M.: Pontrajagin Duality and Full Completeness for MLL. *Math. Struct. in Comp. Science*, vol 10 (2000) 231–259.
18. Hasegawa, M.: Recursion from Cyclic Sharing : Traced Monoidal Categories and Models of Cyclic Lambda Calculus, TLCA'97, *SLNCS* **1210** (1997) 196–213.
19. Hasegawa, M.: *Categorical glueing and logical predicates for models of linear logic*, Manuscript (1998).
20. Hyland, M and Ong, L.: *Full completeness for multiplicative linear logic without the mix-rule*, electronically distributed manuscript, (1993).
21. Joyal, A., Street, R. and Verity, D.: Traced Monoidal Categories. *Math. Proc. Camb. Phil. Soc.* **119** (1996) 447–468.
22. Kelly, G.M. and Mac Lane S. (1971), Coherence in closed categories. *Jour. Pure and Applied Algebra*, **1** (1) (1971) 97–140.
23. Lambek, J.: Deductive systems and categories II. *Springer Lecture Notes in Mathematics* **87** (1969).
24. Lambek, J. and Scott, P.J.: *Introduction to higher order categorical logic*. Cambridge University Press, (1986).
25. Loader, R.: Linear Logic, totality and full completeness. In *Proc. LICS* (1994).
26. Loader, R.: *Models of Lambda Calculi and Linear Logic : Structural, equational and proof-theoretic characterisations.* PhD Thesis, Oxford, (1994).
27. Manes, E.G. and Arbib, M.A.: *Algebraic Approaches to Program Semantics.* Springer-Verlag (1986).
28. Mascari, G. and Pedicini M.: Types and Dynamics in Partially Additive Categories. *Idempotency*, Ed. J. Gunawardena, (1995).
29. Panangaden, P.: *Stochastic Techniques in Concurrency.* Lecture Notes, BRICS, (1997).
30. Seely, R.A.G.: Linear logic, *-autonomous categories and cofree coalgebras. *Categories in Computer Science and Logic*, Contemp. Math. **92**. AMS (1989).
31. Tan, A.M.: *Full Completeness for Models of Linear Logic.* PhD thesis, Cambridge, (1997).

Distinguishing Data Structures and Functions: The Constructor Calculus and Functorial Types

C. Barry Jay

University of Technology, Sydney
cbj@it.uts.edu.au

Abstract. The expressive power of functional programming can be improved by identifying and exploiting the characteristics that distinguish data types from function types. Data types support generic functions for equality, mapping, folding, etc. that do not apply to functions. Such generic functions require case analysis, or pattern-matching, where the branches may have incompatible types, e.g. products or sums. This is handled in the *constructor calculus* where *specialisation* of program *extensions* is governed by constructors for data types. Typing of generic functions employs polymorphism over functors in a *functorial type system.* The expressive power is greatly increased by allowing the functors to be polymorphic in the number of arguments they take, i.e. in their *arities.* The resulting system can define and type the fundamental examples above. Some basic properties are established, namely subject reduction, the Church-Rosser property, and the existence of a practical type inference algorithm.

1 Introduction

Generic programming [BS98,Jeu00] applies the key operations of the Bird-Meertens style, such as mapping and folding to a general class of data structures that includes initial algebra types for lists and trees. Such operations are at the heart of data manipulation, so that any improvement here can have a major impact on the size of programs and the cost of their construction. Most treatments of generic programming either focus on the semantics [MFP91], or use type information to drive the evaluation [JJ97,Jan00,Hin00]. Functorial ML (FML) [JBM98] showed how evaluation could be achieved parametrically, without reference to types, but was unable to *define* generic functions and so had to represent them as combinators. Such definitions require a better understanding of *data structures* that demonstrates why pairing builds them but lambda-abstraction does not. The usual approach, based on introduction-elimination rules for types, does not do so as it derives both pairing and lambda-abstraction from introduction rules. As data structures are built using *constructors*, the challenge is to account for them in a new way. This is done in the *constructor calculus.*

Generic programs are polymorphic in the choice of structure used to hold the data. A second challenge is to represent this polymorphism within a type system. This requires an account of *data types* that demonstrates why the product of

S. Abramsky (Ed.): TLCA 2001, LNCS 2044, pp. 217–239, 2001.

two data types is a data type but their function type is not. The *functorial type system* will represent a typical data type as the application of a functor F (representing the structure) to a type (or tuple of types) X representing the data. Functor applications are the fundamental operations for constructing data types, in the sense that function types are fundamental for constructing program types. Quantification over functors will capture polymorphism in the structure.

This last statement is a slight simplification. Different functors take different numbers of type arguments, and produce different numbers of results. This information is captured by giving functors *kinds* of the form $m \to n$ where m and n are the *arities* of the arguments and results. Further, a typical functor is built from a variety of functors, all of different kinds. It follows that a typical generic function cannot be defined for functors of one kind only, but must be polymorphic in arities, too. The inability to quantify over arities was the biggest drawback of FML, whose primitive constants came in families indexed by arities.

In its basic form, the resulting system supports a large class of concrete data types whose terms, built from the given (finite) set of constructors, can be handled by generic operations. In practice, however, programmers need to define their own (abstract) data types. If these contribute new constructors then it is not at all clear how generic functions can be applied to them without additional coding. For example, when a new data type is defined in Haskell [Aa97] then the various fragments of code required for mapping, etc. are added by the programmer. This is better than re-defining the whole function but it is still something less than full genericity.

```
> -| > datatype tree(a, b) = ••• a | •••• b : tree(a, b) : tree(a, b);;
> -| > •• tr = •••• 3.1 (••• 4) (••• 5);;
tr : (tree :: 2 → 1)(••, ••••)
•••• 3.1 (••• 4) (••• 5)
> -| > •• f x = x + 1;;
f : •• → ••
> -| > •• g y = y * . 3.0;;
g : •••• → ••••
> -| > •• tr2 = • ••2 f g tr;;
tr2 : (tree :: 2 → 1)(••, ••••)
•••• 9.3 (••• 5) (••• 6)
> -| > •• tr3 = ••• tr tr;;
tr3 : (tree :: 2 → 1)(••, ••••)
•••• 6.2 (••• 8) (••• 10)
```

Fig. 1. Examples of generic programming in FISh2

The solution adopted here is to create the abstract data structures by *tagging* the underlying concrete data structures with the appropriate names. Since naming can be treated in a uniform fashion, we are able to apply existing generic programs to novel datatypes. Figure 1 contains a (tidied) session from the implementation of FISh2 language that illustrates some of these ideas. Lines beginning with >-|> and ending with ;; are input by the programmer. The others are responses from the system. A datatype of binary trees is declared. This introduces a new functor *tree* which takes two arguments. $tr : tree(\mathsf{int}, \mathsf{float})$ is a small example of such a tree. $tr2$ is obtained by mapping the functions f and g over tr. The generic function map2 is a specialised form of the generic function map whose type is given in (3) in Section 1.2. Note that even though trees are a new kind of data structure the mapping algorithm works immediately, without any further coding. The session concludes with an application of the generic addition function plus which is able to handle any data structure containing any kind of numerical data such as integers and floats.

```
••• •••••• x y =
  • •••• (x, y) • •••
    ••,•• → ••••
  | (x0, x1), (y0, y1) → (••••• x0 y0) && (••••• x1 y1)
  | ••• x0, ••• y0 → ••••• x0 y0
  | ••• x0, ••• y0 → ••••• x0 y0
  | _ → •••••
```

Fig. 2. Equality by generic patterns

1.1 The Constructor Calculus

Consider a generic equality function. Intuitively, two data structures are equal if they are built using the same constructor and the corresponding constructor arguments are equal. Figure 2 presents a fragment of pseudo-code which employs the desired style for just three kinds of data structure. un is the unique value of unit type, (x_0, x_1) is the pairing pair x_0 x_1 of x_0 and x_1 and inl and inr are the left and right coproduct inclusions (&& is the conjunction of booleans). These are not actually primitives of the constructor calculus but are familiar terms that will serve here to illustrate the principles. The actual program for equality is given in Figure 6.

Some such algorithm is supported by the equality types in Standard ML [MT91]. It is not, however, typable as a program in ML because the patterns for un, pair and inl have incompatible types. Generic pattern-matching must be able to branch on *any* constructor, of any type. This requirement generates a cascade of challenges for the construction of the terms themselves, and more especially for the type derivation rules.

Generic pattern-matching can be represented by iterating a particular form of case analysis called *function extension*

$$\text{under } c \text{ apply } f \text{ else } g$$

where c is a constructor, f is the *specialisation function* applied if the argument is built using c and g is the *default function.* Its application to a term t may be written as under c apply f else g to t. For example, the equality defined in Figure 2 can be de-sugared to a series of extensions that ends with

$$\begin{array}{l}\text{under inr apply } \lambda x_0.\text{under inr apply } \lambda y_0.(\text{equal } x_0\ y_0) \text{ else } \lambda y.\text{false} \\ \qquad\qquad \text{else } \lambda x, y.\text{false}.\end{array}$$

The specialisation rule is

$$\text{under } c \text{ apply } f \text{ else } g \text{ to } c\ t_0\ \ldots\ t_{n-1} > f\ \ t_0\ \ldots\ t_{n-1} \tag{1}$$

where n is the number of arguments taken by the constructor c. The *default* rule is

$$\text{under } c \text{ apply } f \text{ else } g \text{ to } \ t > g\ t \quad \text{if } t \text{ cannot be constructed by } c. \tag{2}$$

It applies if t is constructed by some other constructor or is an explicit function. Unlike most other approaches to generic programming, e.g. [ACPR95,Hin00], evaluation does not require explicit type information.

A type for this extension is given by a type for the default function $g : T \to T'$. If extension were like a standard case analysis or its underlying conditional then the same type constraints would also suffice for f. However, f need only be applied to terms t constructed by c. If c has given type scheme

$$c : \forall \Delta_c.T_0 \to \ldots \to T_{n-1} \to T_n$$

then specialisation to f is possible whenever T_n and T have been unified by some substitution, without loss of generality their most general unifier v. Hence f must have type $v(T_0 \to \ldots \to T_{n-1} \to T')$.

For example, the type of equal is the type of its ultimate default function

$$\lambda x.\lambda y.\text{false} : X \to Y \to \text{bool}$$

where bool is a type representing booleans, say $1+1$. The various specialisations take different types. For un it is $Y \to$ bool (as un takes no arguments). For pair : $X_0 \to X_1 \to X_0 * X_1$ it is $X_0 \to X_1 \to Y \to$ bool. For inl : $X_0 \to X_0 + X_1$ it is $X_0 \to Y \to$ bool.

Several points emerge from this discussion. First, constructors have an associated type scheme which must be principal (most general) for specialisation to preserve typing. Second, the type derivation rules rely on the existence of most general unifiers. Third, the definition of generic equality employs *polymorphic recursion*, e.g. recursive calls to equal are instantiated to product and coproduct types etc.

Several conclusions can be drawn from these observations. The need for most general unifiers is not an onerous restriction in practice, but their existence cannot be guaranteed if type schemes $\forall X.T$ are considered to be types, as in system F [GLT89]. Hence type schemes and types must be kept in separate classes. In other words, data types cannot here be reduced to functions and quantification. Also, the presence of polymorphic recursion means that not every term will have a principal type scheme [Hen93]. As constructors are required to have them it follows that constructors must be distinguishable from terms in general. Concerning type inference, we shall see that there is powerful and practical type inference algorithm, which only requires types to be given when *defining* generic functions (which is probably a good thing to do anyway) but not when applying them.

1.2 Functorial Types

Now let us consider the the data types. It has long been recognised that data types can be understood semantically as the application of a functor to a type [Ga77]. Very briefly, a functor $F : \mathcal{C} \to \mathcal{D}$ between categories $\mathcal{C}$ and $\mathcal{D}$ sends each arrow (or function) $f : X \to Y$ of $\mathcal{C}$ to an arrow $Ff : FX \to FY$ of $\mathcal{D}$ in a way that preserves composition of arrows and their identities. Ff is the mapping of f relative to F. There have been several approaches to representing functors in programming languages starting with CHARITY [CF92]. Basically, they can be represented either as type constructors or treated as a new syntactic class similar to the types.

The former approach is less radical, and can be incorporated into existing languages relatively easily, e.g. Haskell supports a type class of functors. It does, however, have several limitations. First, there are type constructors which are not functors. Hence, many operations, such as mapping, cannot be applied to an arbitrary type constructor. For example, the type constructor that takes X to $X \to X$ is *contravariant* in the first occurence of the type X so that to produce a function from $X \to X$ to $Y \to Y$ would require a function from Y to X as well as one from X to Y.

Second, and more fundamental, is the difficulty of determining where the boundary between the structure and the data lies. If the function f is to be mapped across a term of type GFX then it is not clear if f is to be applied to values of type X or to values of type FX. This can only be resolved by explicit type information at the point of application, which could be quite onerous in practice.

A third problem concerns handling data structures that contain several kinds of data. This would be easy if one could first define `map1` for functors of one argument in isolation, and then `map2` for functors of two arguments, etc. but the presence of inductive data types like lists and trees make it necessary to handle simultaneously functors of arbitrary arity. For example, `map1` f applied to a list `cons` h t introduces `map2` $(f, \texttt{map1}\ f)\ (h, t)$. In the simplest cases the problem can be avoided by providing a function `pmap` for mapping over polynomials in two variables [Jay95a,JJ97] but one can easily construct examples which require mapping of three or more functions.

The alternative approach, of introducing functors as a new syntactic class, was introduced in FML. Now mapping is always defined, and functor composition is explicit, so that $(GF)X$ and $G(FX)$ are distinct types with distinct behaviour under mapping. Unfortunately, the system required explicit arity constants for functors and combinators which was onerous for programming. Now the functorial type system supports arity variables and polymorphism in the arities of functors, as well as in the functors themselves. For example, the binary product functor is replaced by a finite product functor $P :: m \to 1$ where m is an arbitrary arity. In general, a functor F has kind $m \to n$ where m and n are both arities. When $m \to n$ is $0 \to 1$ then F is a type. When m is 0 then F is an n-tuple of types. The same arity polymorphism appears in terms, e.g. the family of mapping combinators map^m of FML have been replaced by a single generic function

$$\mathsf{map} : \forall n.\forall F :: n \to 1, X :: n, Y :: n.\ P(X \to Y) \to FX \to FY \tag{3}$$

which is polymorphic in the choice of functor F and its arity n as well as the argument types represented by the tuples X and Y. Kind inference means that it is rarely necessary to specify the kinds explicitly.

There is an ongoing tension between the functors, as representatives of data structures, and types, as representative of executable programs. Of course, function types are not data types: we cannot define a meaningful equality or mapping over them. (If we treat lambda-binding as a constructor then we derive mere syntactic equality of functions). So we must consider how to relate a system of functors, for building data types, with a type system designed to support programming with functions. In FML the functors and types are kept in separate syntactic classes. Here, the need for variables that represent tuples of types (of variable arity) drives us to regard both types and tuples of types as special kinds of functors. Note, however, that only types will have associated terms. That is, if t is a term whose type is the functor F then $F :: 0 \to 1$.

So the tension has shifted to the status of the functor of functions $F \to G$. When X and Y are types then of course $X \to Y$ is a type. More generally, if X and Y are n-tuples of types then so is $X \to Y$ as when typing map above. Category theory is able to provide some guidance for the general situation. The appropriate notion of arrow between functors is a natural transformation. A *natural transformation* $\alpha : F \to G$ between functors $F, G : \mathcal{C} \to \mathcal{D}$ is given by a family of arrows $\alpha_X : FX \to GX$ indexed by the objects of $\mathcal{C}$ such that for each arrow $f : X \to Y$ of $\mathcal{C}$ we have $Gf.\alpha_X = \alpha_Y.Ff$. This is a kind of parametricity condition [Rey85].

The definition of the exponential, or function object, in a category can be generalised to define an object in $\mathcal{D}$ that represents the natural transformations from F to G [Jay96]. Thus, if $F, G :: m \to n$ are functors in our system then $F \to G : 0 \to n$ is the *functor of functions* from F to G. When m is not necessarily 0 we may call this the *functor of natural transformations* from F to

G and describe its terms similarly. Note that the function functor never takes any arguments. If it did then we would run into the contravariance problem again. There is a certain similarity between the type $F \to G$ and the type scheme $\forall X.FX \to GX$. However, the type of map shows that $F \to G$ may appear within types where type schemes would not be allowed.

When X and Y are types then terms of type $X \to Y$ can be built by lambda-abstraction in the usual way. This will not work for arbitrary functors $F, G :: m \to n$ as in general there are no terms of type F or G to be manipulated. However, applying the *finite product functor* $P :: n \to 1$ yields the type $P(F \to G) :: 0 \to 1$ which may have terms given by tuples of functions, as in the first argument to map above.

At this point we are able to answer the two original questions. The data types are the types built from functors that do not employ the function type constructor. The constructors are the introduction symbols for these functors. They are able to support the extension mechanism that is used to define generic functions.

1.3 Additional Constant Functors

We shall consider two additional features that improve the expressive power of the constructor calculus. The first provides machinery necessary to support functors for abstract data types, like *tree* in Figure 1. The second is introduction of *datum types* for integers, floats, etc. Both must be introduced in a way that supports generic programming.

1.4 Contents of the Paper

The focus of this paper is to introduce the machinery necessary for this approach to generic programming with enough examples to illustrate its power. However, we shall also prove a number of standard results: the existence of most general unifiers; that reduction preserves typing; and reduction is Church-Rosser. Also, the calculus supports a powerful type inference algorithm.

The structure of the paper is as follows. Section 2 introduces the arities and kinds. Section 3 introduces the functors and types and their constructors. Some simple examples of functors, including lists and binary trees will be produced along the way. Section 4 introduces the full term language, including the extensions. Section 5 introduces the reduction rules and establishes that reduction satisfies subject reduction and is Church-Rosser. Section 6 introduces a constructor for creating exceptions (as when taking the head of an empty list) which can then be handled by extensions. Section 7 provides examples of generic functions, including programs for equality, mapping and folding. Section 8 develops an effective type inference algorithm. Section 9 introduces tagged terms for abstract data types. Section 10 introduces the datum types. Section 11 draws conclusions and looks to future work.

2 Kinds

Tuples of types will be characterised by their *arities*. The absence of any types is represented by the arity 0, a single type by the arity 1. The pairing of an m-tuple and an n-tuple of types will have arity (m, n). Hence the *arities* are generated by

$$m, n ::= a \mid 0 \mid 1 \mid (m, n)$$

where a is an arity variable. We will informally denote $(1, 1)$ by 2. Note, however, that 3 would be ambiguous as $(2, 1)$ and $(1, 2)$ are distinct arities (as are $(m, 0)$ and m). The importance of this distinction is that we will be able to index types within a tuple by a sequence of lefts and rights instead of by an integer, and so will only need a pair of constructors instead of an infinite family.

The *kinds* (meta-variable k) are used to characterise the functors. They are of the form $m \to n$ where m and n are arities. If F is a functor that acts on m-tuples of types and produces n-tuples of types then it has kind $m \to n$.

3 Functors and Their Constructors

A single syntactic class represents both functors in general and types. Each functor F has an associated kind k, written $F :: k$. The *types* are defined to be those functors T whose kind is $T :: 0 \to 1$. We shall use the meta-variables F, G and H for functors and T for types. When $F :: 0 \to n$ then it is a *tuple of types* and we may write its kinding as $F :: n$. If all the functors involved in an expression are types we may omit their kinds altogether. The *type schemes* (meta-variable S) are obtained by quantifying types with respect to both arity variables and kinded functor variables. The functors and raw type schemes are formally introduced in Figure 3. Let us introduce them informally first.

The functors (stripped of their kinds) and type schemes are given by

$$\begin{aligned} F, G, T &::= X \mid P \mid C \mid K \mid (F, G) \mid L \mid R \mid G\,F \mid \mu F \mid F \to G \\ S &::= T \mid \forall X :: k.S \mid \forall a.S \end{aligned}$$

X represents a functor variable. The *finite product* functor P has kind $P :: m \to 1$ for any arity m. When m is 0 then P is a type, namely the *unit type*. Its constructor is

$$\mathsf{intrU} : P.$$

Unfortunately, the constructors do not yet have descriptive names. When m is 1 then P is the unary product. Its constructor is

$$\mathsf{intrE} : \forall X.\ X \to PX.$$

When m is (p, q) then its constructor is

$$\mathsf{intrF} : \forall m, n. \forall X :: m, Y :: n.\ PX \to PY \to P(X, Y).$$

Thus, the usual, binary pairing is given by $\mathsf{pair}\ x\ y = \mathsf{intrF}\ (\mathsf{intrE}\ x)\ (\mathsf{intrE}\ y)$. The intrE's convert raw data into simple data structures (one-tuples) which are then combined using intrF. We may write (x, y) for $\mathsf{pair}\ x\ y$ from now on.

The *finite coproduct* functor $C :: m \to 1$ is dual to the product. When m is 0 then C is the empty type, and has no constructor. When m is 1 then the constructor is

$$\mathsf{intrC} : \forall X.\ X \to CX.$$

When m is (p, q) then the coproduct has two inclusions

$$\begin{aligned}\mathsf{intrA} &: \forall m, n. \forall X :: m, Y :: n.\ CX \to C(X, Y)\\ \mathsf{intrB} &: \forall m, n. \forall X :: m, Y :: n.\ CY \to C(X, Y).\end{aligned}$$

The usual inclusions to the binary coproduct may thus be written as $\mathsf{inl}\ x = \mathsf{intrA}\ (\mathsf{intrC}\ x)$ and $\mathsf{inr}\ y = \mathsf{intrB}\ (\mathsf{intrC}\ y)$

The functors P and C convert tuples of types into types. Now we must consider how to build the former. First we have empty tuples of types, constructed by the *kill* functor $K :: m \to 0$. It is used to convert a type T into a "constant functor" that ignores its argument. Its constructor is

$$\mathsf{intrK} : \forall m. \forall X :: 1, Y :: m.\ X \to (XK)Y.$$

For example, the empty list is built using $\mathsf{intrK}\ \mathsf{intrU} : (PK)(A, X)$ where A is the type of the list entries and X is the list type itself.

If $F :: p \to m$ and $G :: p \to n$ are functors able to act on the same arguments then their pairing is $(F, G) :: p \to (m, n)$. There are no constructors for pairs of functors as they are not types. Rather, we shall have to adapt the constructors to handle situations in which functor pairing is relevant.

Corresponding to functor pairing we have left and right functor projections $L :: (m, n) \to m$ and $R :: (m, n) \to n$ with constructors[1]

$$\begin{aligned}\mathsf{intrL} &: \forall m, n. \forall F :: m \to 1, X :: m, Y :: n.\ FX \to (FL)(X, Y)\\ \mathsf{intrR} &: \forall m, n. \forall F :: n \to 1, X :: m, Y :: n.\ FY \to (FR)(X, Y).\end{aligned}$$

They are used to introduce "dummy" functor arguments. For example, to build $\mathsf{leaf}\ x : tree(A, B)$ from some term $x : A$ we begin with

$$\mathsf{intrL}\ (\mathsf{intrL}\ (\mathsf{intrE}\ x)) : ((PL)L)\ ((A, B), tree(A, B))$$

to convert it into a data structure built from data of type A, B and $tree(A, B)$. Application of intrE introduces a functor application which supports the two application of intrL. Note that were F to be elided from the type of intrL then the outermost application above would fail.

The kinding of L and R is made possible by the way the arities are structured. For example, we have $LL(2, 1) \to 1$. By contrast, in FML arities are given by

[1] In earlier drafts there were four constructors here. The original •• ••• and •• ••• have been dropped and their names taken by the other two.

natural numbers which must then be used to index the projection functors such as $\Pi_0^3 :: 3 \to 1$. This indexing then leaks into the term languages, with onerous results.

If $F :: m \to n$ and $G :: n \to p$ are functors then $GF :: m \to p$ is their *composite* functor. When F is a type or tuple of types then we may speak of applying G to F. Composition associates to the right, so that GFX is to be read as $G(FX)$. The associated constructor is

$$\mathsf{intrG} : \forall m.\forall G :: 1 \to 1, F :: m \to 1, X :: m.\ G(FX) \to (GF)X.$$

The restriction on the kind of G is a consequence of the tension between functors and types discussed in the introduction. It is necessary to be able to define functions like map in Section 7. We also need a constructor for handling composites involving pairs of functors, namely

$$\mathsf{intr}H : \forall m.\forall H :: 2 \to 1, F :: m \to 1, G :: m \to 1.X :: m.\ H(FX, GX) \to (H(F,G))X.$$

With the structure available so far we can construct arbitrary polynomial functors. Now let us consider their initial algebras. For example, lists with entries of type A are often described as a solution to the domain isomorphism

$$X \cong 1 + A * X$$

given by $\mu_X.1 + A * X$. Here μ_X indicates that the smallest solution to the domain isomorphism is sought, i.e. the *inductive type* or *initial algebra* for the functor F where $F(A, X) \cong 1 + A * X$. We can represent such an F as follows. $A * X$ is just $P(A, X)$ and 1 becomes P which becomes $(PK)(A, X)$. Thus $F = C(PK, P) :: (1, 1) \to 1$. Now we must represent the initial algebra construction. Instead of introducing a type variable X only to bind it again, we adopt the convention that it is the second argument to the functor that represents the recursion variable. That is, if $F :: (m, n) \to n$ then $\mu F :: m \to n$. For example, $\mathsf{list}_p = \mu C(PK, P) :: 1 \to 1$ is a functor for lists. The corresponding constructor is

$$\mathsf{intrl} : \forall m.\forall F :: (m, 1) \to 1, X :: m.\ F(X, (\mu F)X) \to (\mu F)X.$$

Binary trees can be represented by the functor $\mu\ (PL)L + (PR)L * (PR * PR)$ called $tree_c$ where $(PL)L$ represents the leaf data, $(PR)L$ represents the node data and PR represents the sub-trees.

Let us consider functions between functors, or natural transformations. If $F :: m \to n$ and $G :: m \to n$ are functors of the same kind then $F \to G :: 0 \to n$ is their *function functor*. If X and Y are types then we can build lambda-terms $\lambda(x : X).(t : Y) : X \to Y$ in the usual way but λx is not a data constructor in the formal sense employed here.

We can recover a type from $F \to G :: 0 \to n$ by applying the product functor $P :: n \to 1$ to get $P(F \to G)$ (as appears in the type for map). Can we build any terms of such types? If $f_i : X_i \to Y_i$ for $i = 0, 1$ then there is a pair

$$(f_0, f_1) :: P(X_0 \to Y_0, X_1 \to Y_1)$$

whose type is structurally different to $P((X_0, X_1) \to (Y_0, Y_1))$. The solution adopted is to exploit semantic insights and assert the type identity

$$(X_0, X_1) \to (Y_0, Y_1) = (X_0 \to Y_0, X_1 \to Y_1).$$

Arities (m, n)

$$\frac{}{A \vdash ::: a}\ a \text{ in } A \qquad \frac{}{A \vdash ::: 0} \qquad \frac{}{A \vdash ::: 1} \qquad \frac{A \vdash ::: m \quad A \vdash ::: n}{A \vdash ::: (m, n)}$$

Functor contexts(Δ)

$$\frac{A \vdash}{A; \vdash} \qquad \frac{\Delta \vdash ::: m \quad \Delta \vdash ::: n}{\Delta, X :: m \to n \vdash}\ n \text{ is not a pair, } X \notin \mathrm{dom}(\Delta)$$

Functors and types (F, G, T)

$$\frac{\Delta \vdash}{\Delta \vdash X :: m \to n}\ \Delta(X) = m \to n \qquad \frac{\Delta \vdash F :: m \to n \quad \Delta \vdash G :: m \to n}{\Delta \vdash F \to G :: 0 \to n}$$

$$\frac{\Delta \vdash ::: m}{\Delta \vdash P :: m \to 1} \qquad \frac{\Delta \vdash ::: m}{\Delta \vdash C :: m \to 1}$$

$$\frac{\Delta \vdash ::: m}{\Delta \vdash K :: m \to 0} \qquad \frac{\Delta \vdash F :: p \to m \quad \Delta \vdash G :: p \to n}{\Delta \vdash (F, G) :: p \to (m, n)}$$

$$\frac{\Delta \vdash ::: m \qquad \Delta \vdash ::: n}{\Delta \vdash L :: (m, n) \to m} \qquad \frac{\Delta \vdash ::: m \qquad \Delta \vdash ::: n}{\Delta \vdash R :: (m, n) \to n}$$

$$\frac{\Delta \vdash F :: m \to n \quad \Delta \vdash G :: n \to p}{\Delta \vdash GF :: m \to p} \qquad \frac{\Delta \vdash F :: (m, n) \to n}{\Delta \vdash \mu F :: m \to n}$$

Type schemes (S)

$$\frac{\Delta \vdash T :: 0 \to 1}{\Delta \vdash T} \qquad \frac{A; \Delta \vdash \quad A, m; \Delta \vdash S}{A; \Delta \vdash \forall m.S} \qquad \frac{\Delta, F :: m \to n \vdash S}{\Delta \vdash \forall F :: m \to n.S}$$

Fig. 3. The functorial type system

A *functor context* $A; \Delta$ is given by an arity context A and a finite sequence Δ of distinct *functor variables* X with assigned kinds $m \to n$ where n is not of the form (p, q). This restriction arises because type inference for program extensions requires fine control over the effects of substitutions. The key case is when an arity variable n is replaced by a pair (p, q) and there is a functor variable $X :: m \to n$. To ensure that the arity substitution has achieved its full effect X must be replaced by (Y, Z) for some fresh variables $Y :: m \to p$ and $Z :: m \to q$. Write $\mathrm{dom}(\Delta)$ for the set of functor variables appearing in Δ.

We have the following judgement forms concerning functor contexts, functors and constructors. $A; \Delta \vdash$ asserts that $A; \Delta$ is a well-formed functor context. $A; \Delta \vdash F :: k$ asserts that F is a well-formed functor of kind k in functor context $A; \Delta$. $A; \Delta \vdash S$ asserts that S is a well-formed type scheme in functor context $A; \Delta$. The judgement $\vdash c : S$ asserts that the constructor c has type scheme S. We shall often write the context as Δ leaving the arity context A implicit, and may write $\Delta \vdash ::: m$ when $A \vdash ::: m$.

The free and bound variables of a functor or scheme are defined in the usual way. Type schemes are defined to be equivalence classes of well-formed raw type schemes under α-conversion of bound variables.

A *functor substitution* σ is given by an arity substitution σ_a and a partial function σ_f from functor variables to functors. Let $A; \Delta$ and $A'; \Delta'$ be well-formed functor contexts. Define $\sigma : A; \Delta \to A'; \Delta'$ if $\sigma_a : A \to A'$ and $\mathrm{dom}(\Delta)$ is contained in the domain of σ_f and further if $\Delta(X) = k$ then $\Delta' \vdash \sigma_a \sigma_f X :: \sigma_a k$. That is, σ preserves kinds. Note that if n is an arity variable such that $\sigma_a(n) = (p, q)$ and $X : m \to n$ then $\sigma_f(X)$ must be some pair (F, G) because the variable X cannot have kind $m \to (p, q)$ in Δ'. The *image* of σ is the set of arity variables and the set of functor variables that are free in arities (respectively functors) of the form σu where u is a variable in the domain of σ. The action of such a σ extends homomorphically to any expression that is well-formed in context Δ (including those to be defined below). Composition is defined as for arity substitutions.

Lemma 1. *If $\Delta \vdash J$ has a derivation and $\sigma : \Delta \to \Delta'$ is a functor substitution then $\Delta' \vdash \sigma J$ also has a derivation.*

Proof. By induction on the structure of J.

The most general unifier $\mathcal{U}(F, G)$ of a pair of functors is defined as usual.

Theorem 1. *Let $\Delta \vdash F :: k$ and $\Delta \vdash G :: k'$ be well-formed functors. If F and G have a unifier then they have a most general unifier.*

Proof. The proof is not quite standard. Note that if arity substitution causes a functor variable X to have kind $m \to (p, q)$ then X must be replaced by a pair of functor variables. Also, when unifying $F_0 \to F_1$ and (G_0, G_1) then let X_0, X_1, X_2 and X_3 be fresh variables and unify F_0 with (X_0, X_1) and F_1 with (X_2, X_3) and G_0 with $X_0 \to X_2$ and G_1 with $X_1 \to X_3$.

3.1 Denotational Semantics

The denotational semantics of these syntactic functors is defined as follows. Let $\mathcal{D}$ be a cartesian closed *locos* [Coc90,Jay95b]. These include all toposes that have a natural numbers object, such as **Set** or the effective topos [Hyl82], and also categories used in domain theory such as ω-complete partial orders. They can be thought of as a minimal setting in which both lists and functions are definable.

The denotational semantics of the type schemes has not yet been developed but should prove amenable to the methods developed for FML[JBM98].

3.2 Kind Inference

A *kinding* for an unkinded functor F is a functor context $A;\Delta$ and a kind k such that $A;\Delta \vdash F :: k$. It is a *principal kinding* for F if for any other kinding $A';\Delta' \vdash F :: k'$ there is an arity substitution $\sigma : A \to A'$ such that $\sigma k = k'$.

Theorem 2. *If an unkinded functor F has a kinding then it has a principal kinding.*

Proof. The kind inference algorithm $\mathcal{W}$ follows Milner's algorithm [Mil78].

Algorithm $\mathcal{W}$ takes a four-tuple (A, Δ, F, k) and tries to produce an arity substitution $\sigma : A \to A'$ such that $A';\sigma\Delta \vdash F :: \sigma k$. We initialise the choices of A, Δ and k with fresh variables as follows.

Assign each functor variable X in F a kind whose source and target are fresh kind variables. Let A be some sequence of these kind variables and Δ be some sequence of the kinded functor variables created above. Let $k = m \to n$ be another fresh kind.

The algorithm proceeds by induction on the structure of F. The proofs that σ produces a principal kind follows the same pattern. Here are the cases.

1. F is a functor variable X. Then $\sigma = \mathcal{U}(\Delta(X), k)$.
2. F is (G, H). Let n_0 and n_1 be a pair of fresh arity variables. Let

$$\begin{aligned} &\upsilon : A \to A' = \mathcal{U}(n, (n_0, n_1)) \\ &\sigma_1 : A' \to A'' = \mathcal{W}(A', \upsilon\Delta, G, \upsilon(m \to n_0)) \qquad \text{and} \\ &\sigma_2 : A'' \to A''' = \mathcal{W}(A'', \sigma_1\upsilon\Delta, H, \sigma_1\upsilon(m \to n_1)). \end{aligned}$$

   Then σ is $\sigma_2\sigma_1\upsilon$.
3. F is μG. Then $\sigma = \mathcal{W}(A, \Delta, G, (m, n) \to n)$.
4. F is $F_0 \to F_1$. Let m and n be fresh arity variable and let

$$\begin{aligned} &\upsilon : A \to A' = \mathcal{U}(k, 0 \to n) \\ &\sigma_1 : A' \to A'' = \mathcal{W}(A', \upsilon\Delta, F_0, m \to n) \qquad \text{and} \\ &\sigma_2 : A'' \to A''' = \mathcal{W}(A'', \sigma_1\upsilon\Delta, F_1, m \to n). \end{aligned}$$

   Then σ is $\sigma_2\sigma_1\upsilon$.
5. F is a constant of kind k'. Then σ is $\mathcal{U}(k, k')$.

4 Terms

The raw terms are given by

$$t ::= x \mid t\,t \mid \lambda x.t \mid \mathsf{let}\ x = t\ \mathsf{in}\ t \mid \mathsf{fix}(\lambda x.t) \mid c \mid \mathsf{under}\ c\ \mathsf{apply}\ t\ \mathsf{else}\ t.$$

x is a variable. The application, λ-abstraction and let-construct all take their standard meanings. Let $g.f$ be notation for $\lambda x.g(f\ x)$. The fixpoint construct supports fixpoints with respect to a type scheme instead of a type, i.e. *polymorphic recursion.* We shall often use explicit recursion to represent fixpoints. For example, a declaration of the form $f\ x = t$ where f is free in t stands for $\mathsf{fix}\ (\lambda f.\lambda x.t)$. A term of the form $c\ t_0\ \ldots\ t_{n-1}$ where c is a constructor taking n arguments is *constructed by* c. The lambda-abstractions, extensions and partially applied constants are collectively known as *explicit functions.*

A *term context* Γ is a finite sequence of term variables with assigned type schemes. A *context* $A; \Delta; \Gamma$ consists of a functor context $A; \Delta$ and a term context Γ whose type schemes are all well-formed with respect to $A; \Delta$. The set of *free functor (respectively, arity) variables* of Γ is given by the union of the sets of free functor (respectively arity) variables in the type schemes assigned to the term variables in Γ.

The *closure* $\mathsf{closure}(\Gamma, T)$ of a type T with respect to a term context Γ is given by quantifying T with respect to those of its free arity and functor variables which are not free in Γ (the order of variables will not prove to be significant).

We have the following judgement forms concerning term contexts and terms. $\Delta; \Gamma \vdash$ asserts that $\Delta; \Gamma$ is a well-formed context. $\Delta; \Gamma \vdash t : T$ asserts that t is a term of type T in the context $\Delta; \Gamma$. The type derivation rules for terms are given in Figure 4. Let us consider them now in turn.

If x is a term variable in Γ then it can be treated as a term typed by any instantiation of its type scheme, as given by a substitution σ from its bound variables to the functor context. Similarly, any constructor c can be treated as a term. The rules for application, lambda-abstraction and let-construction are standard.

The main premise for polymorphic recursion equips the recursion variable x with a type *scheme* whose instantiation will provide the type of the resulting term. This means that x may take on many different instantiations of this scheme in typing the recursion body, e.g. equal may act on many different kinds of structures. Of course, this will limit the power of type inference, because of the wide range of possible type schemes that can produce a given type.

The default behaviour of a term $\mathsf{under}\ c\ \mathsf{apply}\ t_1\ \mathsf{else}\ t_2$ is that of t_2. Hence any type for the whole extension must be a type $T \to T'$ for t_2. In a standard case analysis or conditional, t_1 would be required to have the same type as t_2 but here different cases may have different types, depending on whether the argument is a pair, an inclusion (of coproduct type), a list, etc. That is, each case must be typed in a context that includes local type information that is not relevant to the overall type. More precisely, if the extension is applied to a term constructed by c then its type will be an instantiation of both the result type T_n of c and the argument type T of t_2. In other words, it is an instantiation of the most general unifier υ of T_n and T. Thus when the specialisation is invoked the extension need only have type $\upsilon(T \to T')$ and so t_1 (which acts on the arguments of c) must have type $\upsilon(T_0 \to \ldots \to T_{n-1} \to T')$.

Term Contexts (Γ)

$$\frac{\Delta \vdash}{\Delta;\ \vdash} \qquad \frac{\Delta;\Gamma \vdash \quad \Delta \vdash S}{\Delta;\Gamma,x:S \vdash}\ x \notin \mathrm{dom}(\Gamma)$$

Terms (t)

$$\frac{\Delta;\Gamma \vdash}{\Delta;\Gamma \vdash x:\sigma T}\ \begin{array}{l}\Gamma(x) = \forall\Delta^{\bullet}.T\\ \sigma:\Delta^{\bullet} \to \Delta\end{array} \qquad \frac{\vdash c:\forall\Delta_c.T}{\Delta;\Gamma \vdash c:\sigma T}\ \sigma:\Delta_c \to \Delta$$

$$\frac{\Delta;\Gamma \vdash t:T_1 \to T_2 \quad \Delta;\Gamma \vdash t_1:T_1}{\Delta;\Gamma \vdash t\ t_1:T_2} \qquad \frac{\Delta;\Gamma,x:T_1 \vdash t:T_2}{\Delta;\Gamma \vdash \lambda x.t:T_1 \to T_2}$$

$$\frac{\Delta;\Gamma \vdash \quad \Delta,\Delta_1;\Gamma \vdash t_1:T_1 \quad \Delta;\Gamma,x:\forall\Delta_1.T_1 \vdash t_2:T_2}{\Delta;\Gamma \vdash \bullet\bullet\bullet\ x = t_1 \bullet\bullet\ t_2:T_2}$$

$$\frac{\Delta;\Gamma \vdash \quad \Delta,\Delta_1;\Gamma,x:\forall\Delta_1.T \vdash t:T}{\Delta;\Gamma \vdash \bullet\bullet(\lambda x.t):\sigma T}\ \sigma:\Delta_1 \to \Delta$$

$$\frac{\begin{array}{c}\vdash c:\forall\Delta_c.T_0 \to \ldots \to T_n \quad \Delta;\Gamma \vdash t_2:T \to T^{\bullet}\\ \upsilon = \mathcal{U}(T_n,T):\Delta\Delta_c \to \Delta^{\bullet} \quad \Delta^{\bullet};\upsilon\Gamma \vdash t_1:\upsilon(T_0 \to \ldots T_{n\bullet 1} \to T^{\bullet})\end{array}}{\Delta;\Gamma \vdash \bullet\bullet\bullet\bullet\bullet\ c \bullet\bullet\bullet\bullet\ t_1 \bullet\bullet\bullet\ t_2:T \to T^{\bullet}}$$

Fig. 4. Terms of the constructor calculus

It is unusual for unifiers to appear in type derivation rules, as opposed to type *inference*. Further, the substitution in the premises does not appear in the conclusion and so type inference will have to backtrack to remove its effects from the final result (see Section 8).

Free and bound term variables are defined in the usual way. A *term substitution* σ is a partial function from term variables to terms. If Δ is a functor context and Γ and Γ' are term contexts then $\sigma : \Delta;\Gamma \to \Delta;\Gamma'$ if $\Delta;\Gamma$ and $\Delta;\Gamma'$ are well-formed and for each term variable x in Γ we have $\Delta;\Gamma' \vdash \sigma x : \Gamma(x)$. A *term* is an equivalence class of raw terms under substitution for bound variables.

Lemma 2. *If $\Delta;\Gamma \vdash J$ has a derivation and $\sigma : \Delta;\Gamma \to \Delta;\Gamma'$ is a term substitution then $\Delta;\Gamma' \vdash \sigma J$ also has a derivation.*

Proof. By induction on the structure of the derivation of J.

5 Evaluation

The basic reduction rules are represented by the relation $>$ in Figure 5. A *reduction* $t \to t'$ is given by the application of a basic reduction to a sub-term. All of the reduction rules are standard except those for extensions. Reduction of extensions amounts to deciding whether to specialise or not. A term t *cannot be constructed by* a constructor c if it is an explicit function, or is constructed by some constructor other than c.

$$(\lambda x.t_2)\ t_1 > t_2\{t_1/x\}$$
$$\bullet\bullet\bullet\ x = t_1 \bullet\bullet\ t_2 > t_2\{t_1/x\}$$
$$\bullet\bullet(\lambda x.t) > (\lambda x.t) \bullet\bullet(\lambda x.t)$$
$$\bullet\bullet\bullet\bullet\bullet\ c \bullet\bullet\bullet\bullet\ f \bullet\bullet\bullet\ g \bullet\bullet\ c\ t_1\ \ldots t_n > f\ t_1\ \ldots t_n$$
$$\bullet\bullet\bullet\bullet\bullet\ c \bullet\bullet\bullet\bullet\ f \bullet\bullet\bullet\ g \bullet\bullet\ t > g\ t \text{ if } t \text{ cannot be constructed by } c$$

Fig. 5. Evaluation rules

Theorem 3 (subject reduction). *Reduction preserves typing.*

Proof. The only novel cases are in specialisation. Consider a reduction

$$\mathsf{under}\ c\ \mathsf{apply}\ f\ \mathsf{else}\ g\ \mathsf{to}\ c\ t_0\ \ldots t_{n-1} > f\ t_0\ \ldots t_{n-1}$$

and a type derivation for the left-hand side. Let c have type scheme $\forall \Delta_c.T_0 \to \ldots \to T_n$ and g have derived type $T \to T'$. It follows that the argument $c\ t_0\ \ldots\ t_{n-1}$ must have type T and that T must be $\sigma(T_n)$ for some substitution σ on Δ_c. Hence σ factors through the most general unifier of T_n and T by some substitution ρ. Hence $f : \sigma(T_0 \to \ldots T_{n-1} \to T')$ by Lemma 2 applied to ρ. Now the right-hand side of the reduction has type $\sigma T'$ which is T' as σ only acts on Δ_c.

Theorem 4. *Reduction is Church-Rosser.*

Proof. The rules for specialisation can be viewed as a finite family of rules, one for each constructor c. Then all the reduction rules are left-linear and non-overlapping, so we can apply Klop's general result [Klo80].

6 Exceptions

Any account of data types must address the issue of missing data. For example, taking the head of an empty list. In imperative languages this often results in a void pointer. In pointer-free languages like ML or Java such problems are addressed by introducing exceptions. These flag problems during evaluation and may change the flow of control in ways that are difficult to specify and understand. Exceptions may be caught and handled using any arguments to the exception as parameters.

Exceptions arise here when an extension is applied to an argument of the wrong form, either constructed by the wrong constructor or an explicit function. The solution is to add one more constructor

$$\mathsf{exn} : \forall X, Y.X \to Y$$

which represents an exception that carries an argument. As exn is a constructor it can be handled using the extension mechanism without additional machinery. The only issue is that the exception may produce a function, rather than data. This would be bad programming style but can be handled by introducing one additional evaluation rule

$$\mathsf{exn}\ s\ t > \mathsf{exn}\ s$$

Rewriting is still confluent because the left-hand side is not a term constructed by exn as it is applied to two arguments, not one.

7 Examples

Let us begin with a generic equality relation. We can use $C(P, P)$ to represent a type bool of booleans with true = inl intrU and false = inr intrU and let && be an infix form of conjunction, defined by nested extensions. In the generic test for equality it is not necessary that the arguments have the same type, though this will be the case if they are indeed equal. So the type for equality is $X \to Y \to$ bool. The program is given in Figure 6. Each pattern in the program corresponds to one extension. For example,

$$\mid \mathsf{intrE}\ x \to \mathsf{under\ intrE\ apply\ equal}\ x\ \mathsf{else}\ \lambda y.\mathsf{false}$$

represents under intrE apply $\lambda x.$under intrE apply equal x else $\lambda y.$false else while the final pattern of the form $\mid \_ \to t$ represents the default function $\lambda x.t$. Obviously, the algorithm is quite independent of the nature of the individual constructors, one of the reasons that equality can sometimes be treated by ad hoc methods.

Before tackling more complex examples like mapping, we shall require a little more infrastructure. In particular, we shall require eliminators corresponding to the constructors. Happily, these can be defined by extensions. For example,

$$\mathsf{elimE} = \mathsf{under\ intrE\ apply}\ \lambda x.x\ \mathsf{else\ exn\ intrE} : PX \to X$$

is the eliminator corresponding to intrE. If applied to some intrE t then it returns t. Otherwise an exception results.

In general each constructor has eliminators corresponding to the number of its arguments. intrU has no eliminator. Those for intrF are given by

$$\mathsf{elimF0} = \mathsf{under\ intrF\ apply}\ \lambda x, y.x\ \mathsf{else\ exn}\ (\mathsf{intrF}, 0) : P(X, Y) \to PX$$
$$\mathsf{elimF1} = \mathsf{under\ intrF\ apply}\ \lambda x, y.y\ \mathsf{else\ exn}\ (\mathsf{intrF}, 1) : P(X, Y) \to PY$$

```
(•••••: X → Y → ••••) z =
  • •••• z • •••
    •• ••• → ••••• •• ••• •••• •••• •••• λy.•• •••
  | •• ••• x → ••••• •• ••• •••• ••••• x •••• λy.•• •••
  | •• ••• x0 x1 → ••••• •• ••• •••• λy0, y1.(••••• x0 y0) && (••••• x1 y1) ••••
            λy.•• •••
  | •• ••• x → ••••• •• ••• •••• ••••• x •••• λy.•• •••
  | •• ••• x → ••••• •• ••• •••• ••••• x •••• λy.•• •••
  | •• ••• x → ••••• •• ••• •••• ••••• x •••• λy.•• •••
  | •• ••• x → ••••• •• ••• •••• ••••• x •••• λy.•• •••
  | •• ••• x → ••••• •• ••• •••• ••••• x •••• λy.•• •••
  | •• ••• x → ••••• •• ••• •••• ••••• x •••• λy.•• •••
  | •• ••• x → ••••• •• ••• •••• ••••• x •••• λy.•• •••
  | •• ••• x → ••••• •• ••• •••• ••••• x •••• λy.•• •••
  | •• ••• x → ••••• •• ••• •••• ••••• x •••• λy.•• •••
  | _ → λx, y.•• •••
```

Fig. 6. Defining equality by extension

We can define the usual projections for pairs by $\mathsf{fst} = \mathsf{elimE.elimF0}$ and $\mathsf{snd} = \mathsf{elimE.elimF1}$. If $\mathsf{intrZ} : T \to T'$ is any other constructor then its eliminator is

$$\mathsf{elimZ} = \mathsf{under\ intrZ\ apply}\ \lambda x.x\ \mathsf{else\ exn\ intrZ} : T' \to T.$$

The algorithm for mapping is fairly complex. Recall that the type for mapping is

$$\mathsf{map} : \forall m.\forall F :: m \to 1, X :: m, Y :: m.P(X \to Y) \to FX \to FY$$

If m is 1 and $f : X \to Y$ is an ordinary function then $\mathsf{intrE}\ f : P(X \to Y)$ is the corresponding one-tuple and $\mathsf{map}\ (\mathsf{intrE}\ f)$ has type $FX \to FY$. When applied to $\mathsf{intrE}\ x$ then semantically the expected result is

$$\mathsf{map}\ (\mathsf{intrE}\ f)\ (\mathsf{intrE}\ x) = \mathsf{intrE}\ (f\ x)$$

as x is of the type to which f is to be applied, and then intrE is applied to create a one-tuple, having the same structure as the original argument.

If m is (m_0, m_1) then we need a pair of functions $f_i : PX_i \to PY_i$ for $i = 0, 1$. When applied to a term of the form $\mathsf{intrF}\ x_0\ x_1$ then $x_i : PX_i$ and we get

$$\mathsf{map}\ (\mathsf{intrF}\ f_0\ f_1)\ (\mathsf{intrF}\ x_0\ x_1) = \mathsf{intrF}\ (\mathsf{map}\ f_0\ x_0)\ (\mathsf{map}\ f_1\ x_1).$$

Note that it is necessary to *map* f_0 and f_1 across x_0 and x_1. Putting these two rules together for pairs yields

$$\mathsf{map}\ (f_0, f_1)\ (x_0, x_1) = (f_0\ x_0, f_1\ x_1).$$

The program in Figure 7 represents such semantic equations within extensions, but replaces the explicit structures given to the functions by the appropriate eliminators for tuples. Note how exceptions carry both the mapping function and the argument as a pair. In particular, if z has evaluated to an exception then that is nested within the exception generated by the mapping. This gives detailed account of where the error has occurred which can be handled in sophisticated ways. We can customise map for any particular arity as in

$$\mathsf{map1}\ f = \mathsf{map}\ (\mathsf{intrE}\ f) : (X \to Y) \to FX \to FY$$
$$\mathsf{map2}\ f\ g = \mathsf{map}\ (f, g) : (X_0 \to Y_0) \to (X_1 \to Y_1) \to F(X_0, X_1) \to F(Y_0, Y_1).$$

```
(• •• : P(X → Y) → F X → F Y) f z =
  • •••• z • •••
    •• ••• x → •• ••• (•••• • f x)
   | •• ••• x y → •• ••• (• •• (•••• •0 f) x) (• •• (•••• •1 f) y)
   | •• ••• x → •• ••• (•••• • f x)
   | •• ••• x → •• ••• (•••• •0 f x)
   | •• ••• y → •• ••• (•••• •1 f y)
   | •• ••• x → •• ••• x
   | •• ••• x → •• ••• (• •• (•••• •0 f) x)
   | •• ••• y → •• ••• (• •• (•••• •1 f) y)
   | •• ••• x → •• ••• (• •• (•• ••• (• •• f)) x)
   | •• ••• x → •• ••• (• •• (• •• f, • •• f) x)
   | •• ••• x → •• ••• (• •• (f, • •• f) x)
   | _ → ••• (• •• f, z)
```

Fig. 7. Definition of • ••

Using map we can define the operation

$$\mathsf{induct} : \forall F :: 2 \to 1, X :: 1, Y :: 1.(F(X, Y) \to Y) \to (\mu F)X \to Y$$

associated with initial algebras for functors of kind $2 \to 1$ by

$$\mathsf{induct}\ f = f.(\mathsf{map2}\ (\lambda x.x)(\mathsf{induct}\ f)).\mathsf{elimI}.$$

This definition can be adapted to functors $F :: (n, 1) \to 1$ for any fixed n by replacing map2 $(\lambda x.x)$ by map_n applied to $n - 1$ copies of the identity function.

The most familiar example of foldleft takes a function $f : X \to Y \to X$ an $x : X$ and a list $[y_0, y_1, \ldots, y_n]$ and produces $f\ (\ldots(f\ x\ y_0)\ldots)\ y_n$. In general we must consider a functor which takes more than one argument. For example, to fold over $F(Y_0, Y_1)$ where Y_0 and Y_1 are types we need two functions $F_i : X \to Y_i \to X$. These can be combined by case analysis to give a function $X \to Y_0 + Y_1 \to X$. In general, we can let $Y :: n$ be a tuple of types and use a function $X \to CY \to X$. Hence the type of foldleft is as defined in Figure 8. When the data structure holds only one kind of data then we can employ

$$\mathsf{foldleft1} : (X \to Y \to X) \to X \to FY \to X$$

defined by foldleft1 f = foldleft $(\lambda u, v.\ f\ u\ (\mathsf{elimC}\ v))$.

```
(•• •• •• •• : (X → C(Y :: n) → X) → X → FY → X) x f y =
• •••• y • •••
 | •• ••• y0 → f x (•• ••• y0)
 | •• ••• y0 y1 → •• •• •• •• (λu, v.f u (•• ••• v)) (•• •• •• •• (λu, v.f u (•• ••• v)) x y0) y1
 | •• ••• y0 → f x (•• ••• y0)
 | •• ••• y0 → f x (•• ••• y0)
 | •• ••• y1 → f x (•• ••• y1)
 | •• ••• y0 → x
 | •• ••• y0 → •• •• •• •• (λu, v.f u (•• ••• v)) x y0
 | •• ••• y1 → •• •• •• •• (λu, v.f u (•• ••• v)) x y1
 | •• ••• y0 → •• •• •• •• (λu, v.•• •• •• •• f u (• ••• • v)) x y0
 | •• ••• y0 → •• •• •• •• (λu.•••• ((•• •• •• •• f u).• ••• •) ((•• •• •• •• f u).• ••• •) x y0
 | •• ••• y0 → •• •• •• •• (λu.•••• (f u) ((•• •• •• •• f u).• ••• •) x y0
 | _ → ••• (•• •• •• •• f, y)
```

Fig. 8. Definition of •• •• •• ••

8 Type Inference

The use of polymorphic recursion and extension mean that not every term has a principal type. The issues for polymorphic recursion are already well explored, e.g. [Hen93]. When inferring a type for $\mathsf{fix}(\lambda x.t)$ there are many choices of type scheme which could produce the necessary type. A similar problem arises with extensions.

Theorem 5. *Type inference is correct: if* $\mathcal{W}(\Delta; \Gamma, t, T) = \sigma : \Delta \to \Delta'$ *then* $\Delta'; \sigma\Gamma \vdash t : \sigma T$.

Proof.

9 Abstract Datatypes

This section addresses the creation of types by users, such as the tree functor defined in Figure 1. It is easy to introduce new functors and constants – the challenge is to support them in existing generic programs. For example, if we introduce new constants leaf and node for the user-defined trees then the generic mapping algorithm in Figure 7 will not be able to handle them. Of course, we could write new patterns for the new constants but this defeats the purpose of genericity.

The solution begins with the observation that the user-defined functors are always isomorphic to existing "concrete" functors, e.g. *tree* is isomorphic to $tree_c$. The point of creating a new functor is to create a new, separate class of data structures distinguished from the others by their *names*. So we shall introduce a single new constructor called intrT whose arguments will be a name and a data structure, each of which can be handled in a uniform way. For example, let leaf_c and node_c be the concrete versions of leaf and node. Then tree_name : $tree_c \rightarrow tree$ is used to name trees so that we can define

$$\mathsf{leaf} = \mathsf{intrT}\ \mathsf{tree\_name}\ \mathsf{leaf}_c.$$

intrT has type $\forall F :: m \rightarrow 1, G :: m \rightarrow 1, X :: m.\ (F \rightarrow G) \rightarrow (FX \rightarrow GX)$. That is, it takes a natural transformation $r : F \rightarrow G$ (the name) to form a *tag* intrT r which when applied to a term $t : FX$ produces a *tagged term* of type GX.

Now all tags can be treated in a uniform way. For example, mapping over tagged terms is given by

$$\mathsf{map}\ f\ (\mathsf{intrT}\ r\ t) = \mathsf{intrT}\ r\ (\mathsf{map}\ f\ t).$$

10 Datum Types

```
(•••• : X → X → X) x y =
  • •••• (x, y) • •••
    (••• x0, ••• y0) → ••• (•••••••• ••• x0 y0)
  | (•••• x0, •••• y0) → •••• (•••••••• •••• x0 y0)
  | (••••• , ••••• ) → •••••
  | (••••• x0, ••••• y0) → ••••• .(•••• x0 y0)
  | (••••• x0 x1, ••••• y0 y1) → ••••• (•••• x0 y0) (•••• x1 y1)
  | ...
```

Fig. 9. Generic addition

11 Conclusions

The constructor calculus with its functorial type system is able to define and type a wide range of generic functions. The type system is based on a class of functors which is similar to that of FML, but is supported by a system of polymorphic kinds that eliminates most of the need for explicit arities. Program extension is the truly novel contribution of the paper. It shows how generic programs can be incorporated within the typed lambda-calculus by giving a type derivation rule for extensions. Evaluation does not require type information, so there is no need for programmers to supply it except to aid the type inference mechanism, when defining complex functions.

All of the ideas in this paper have been tested during development of the programming language FISH2. In particular, all of the generic programs in the paper have been created, type-checked and evaluated therein. The ability to create such new and powerful programs shows the expressive power of this addition to the typed lambda-calculus.

The focus of this paper has been to demonstrate the expressive power of this approach in a purely functional setting. The successor to this paper will show how to add imperative features to the calculus so that we may define generic programs such as

$$\mathsf{assign} : \forall X.\mathsf{loc}X \to X \to \mathsf{comm}$$

where $\mathsf{loc}\ X$ represents a location for a value of type X and comm is a type of commands. In the process we will gain some insights into the optimisation techniques for reducing the execution overhead of generic programs, again based on our new understanding of constructors.

Acknowledgements. The Université de Paris VII and the University of Edinburgh each hosted me for a month while working on this material, the latter by EPSRC visiting fellowship No. GR/M36694. I would like to thank my hosts Pierre-Louis Curien and Don Sannella for making these visits so pleasant and productive. This work has also profited from discussions with many other colleagues, especially Manuel Chakravarty, Peter Dybyer, Martin Hofmann, Gabi Keller, Hai Yan Lu, Eugenio Moggi, Jens Palsberg, Gilles Peskine and Bob Tennent.

References

[Aa97] L Augustsson and all. Report on the functional programming language Haskell: version 1.4. Technical report, University of Glasgow, 1997.

[ACPR95] M. Abadi, L. Cardelli, B.C. Pierce, and D. Rémy. Dynamic typing in polymorphic languages. *Journal of Functional Programming*, 5(1):111–130, 1995.

[BS98] R. Backhouse and T. Sheard, editors. *Workshop on Generic Programming: Marstrand, Sweden, 18th June, 1998*. Chalmers University of Technology, 1998.

[CF92] J.R.B. Cockett and T. Fukushima. About Charity. Technical Report 92/480/18, University of Calgary, 1992.

[Coc90] J.R.B. Cockett. List-arithmetic distributive categories: locoi. *Journal of Pure and Applied Algebra*, 66:1–29, 1990.

[Ga77] J.A. Goguen and all. Initial algebra semantics and continuous algebras. *Journal of the Association for Computing Machinery*, 24:68–95, 1977.

[GLT89] J-Y. Girard, Y. Lafont, and P. Taylor. *Proofs and Types*. Tracts in Theoretical Computer Science. Cambridge University Press, 1989.

[Hen93] F. Henglein. Type inference with polymorphic recursion. *ACM Trans. on Progr. Lang. and Sys.*, 15:253–289, 1993.

[Hin00] R. Hinze. A new approach to generic functional programming. In *Proceedings of the 27th Annual ACM SIGPLAN-SIGACT Symposium on Principles of Programming Languages, Boston, Massachusetts, January 19-21, 2000*, 2000.

[Hyl82] J.M.E. Hyland. The effective topos. In A.S. Troelstra and D. van Dalen, editors, *The L.E.J. Brouwer Centenary Symposium*. North Holland, 1982.

[Jan00] P. Jansson. *Functional Polytypic Programming*. PhD thesis, Chalmers University, 2000.

[Jay95a] C.B. Jay. Polynomial polymorphism. In R. Kotagiri, editor, *Proceedings of the Eighteenth Australasian Computer Science Conference: Glenelg, South Australia 1–3 February, 1995*, volume 17, pages 237–243. A.C.S. Communications, 1995.

[Jay95b] C.B. Jay. A semantics for shape. *Science of Computer Programming*, 25:251–283, 1995.

[Jay96] C.B. Jay. Data categories. In M.E. Houle and P. Eades, editors, *Computing: The Australasian Theory Symposium Proceedings, Melbourne, Australia, 29–30 January, 1996*, volume 18, pages 21–28. Australian Computer Science Communications, 1996. ISSN 0157–3055.

[Jay00] C.B. Jay. Distinguishing data structures and functions: the constructor calculus and functorial types. `http://www-staff.it.uts.edu.au/~cbj/Publications/constructors.ps`, 2000.

[JBM98] C.B. Jay, G. Bellè, and E. Moggi. Functorial ML. *Journal of Functional Programming*, 8(6):573–619, 1998.

[Jeu00] J. Jeuring, editor. *Proceedings: Workshop on Generic Programming (WGP2000): July 6, 2000, Ponte de Lima, Portugal.* Utrecht University, UU-CS-2000-19, 2000.

[JJ97] P. Jansson and J. Jeuring. PolyP - a polytypic programming language extension. In *POPL '97: The 24th ACM SIGPLAN-SIGACT Symposium on Principles of Programming Languages*, pages 470–482. ACM Press, 1997.

[Klo80] J.W. Klop. *Combinatory Reduction Systems.* PhD thesis, Mathematical Center Amsterdam, 1980. Tracts 129.

[MFP91] E. Meijer, M. Fokkinga, and R. Paterson. Functional programming with bananas, lenses, envelopes and barbed wire. In J. Hughes, editor, *Procceding of the 5th ACM Conference on Functional Programming and Computer Architecture*, volume 523 of *Lecture Notes in Computer Science*, pages 124–44. Springer Verlag, 1991.

[Mil78] R. Milner. A theory of type polymorphism in programming. *JCSS*, 17, 1978.

[MT91] R. Milner and M. Tofte. *Commentary on Standard ML.* MIT Press, 1991.

[Rey85] J. Reynolds. Types, abstraction, and parametric polymorphism. In R.E.A. Mason, editor, *Information Processing '83.* North Holland, 1985.

The Finitely Generated Types of the λ-Calculus

Thierry Joly

Équipe Preuves, Programmes et Systèmes, Université Paris VII,
2, place Jussieu, 75251 Paris cedex 05, France
joly@logique.jussieu.fr

Abstract. We answer a question raised by Richard Statman (cf. [8]) concerning the simply typed λ-calculus (having o as only ground type): Is it possible to generate from a finite set of combinators all the closed terms of a given type ? (By combinators we mean closed λ-terms of any types).
Let us call *complexity* of a λ-term t the least number of distinct variables required for its writing up to α-equivalence. We prove here that a type T can be generated from a finite set of combinators iff there is a constant bounding the complexity of every closed normal λ-term of type T. The types of rank $\leqslant 2$ and the types $A_1 \to (A_2 \to \dots (A_n \to o))$ such that for all $i = 1, \dots, n$: $A_i = o$, $A_i = o \to o$ or $A_i = (o \to (o \to \dots (o \to o))) \to o$, are thus the only inhabited finitely generated types.

1 Introduction

We consider here the simply typed λ-calculus *à la Church* whose only atomic type is o. Let us introduce some general notations and definitions about it, before going into our subject.

Notation 1. *$A^n \to B$ and ${}^nA$ are the types inductively defined for all $n \in \mathbb{N}$ by: $A^0 \to B = B$, $A^{n+1} \to B = A \to (A^n \to B)$ and ${}^0A = A$, ${}^{n+1}A = {}^nA \to A$.*

A type $A_1 \to (A_2 \to \dots (A_n \to B))$ will be more simply written: $A_1, A_2, \dots, A_n \to B$.

Definition 2. *The rank $rk(T)$ of any type T is inductively defined by: $rk(o) = 0$ and $rk(A \to B) = \max(rk(A) + 1, rk(B))$.*

Every λ-term t will be considered up to α-equivalence and the relation $t \approx_\alpha u$ will be denoted by: $t = u$.

Notation 3. *For any set $\mathcal{C}$ of λ-terms, let $[\mathcal{C}]$ denote the set of λ-terms built from those of $\mathcal{C}$ using applications only.*

Definition 4. *We say that a set S of closed λ-terms is* finitely generated *if there is a finite set $\mathcal{C}$ of typed closed λ-terms such that every λ-term of S is $\beta\eta$-equivalent to some λ-term of $[\mathcal{C}]$.*

S. Abramsky (Ed.): TLCA 2001, LNCS 2044, pp. 240–252, 2001.

Definition 5. *Let us say that a simple type A is* finitely generated *if the set of the closed λ-terms of type A is finitely generated according to the previous definition.*

Our concern is the following question: What are the finitely generated types? This problem was first considered by R. Statman in [8], where the type ${}^3o, o \to o$ was shown not to be finitely generated inside the typed λI-calculus and where the similar statement about typed λK-calculus was conjectured. We will give a proof of the latter (see Section 2 below), which is a key step of the present work.

The type $\mathbb{L} = {}^2o, (o^2 \to o) \to o$ was given as a first example of non finitely generated type in [5]. This example was established as follows:
The type $\mathbb{L}$ allows the coding of the pure (i.e. type-free) closed λ-terms. For every pure λ-term t whose free variables are $x_1, \ldots, x_k$, let $|t|_{la}$ be the λ-term of type o taking its free variables among $l^{{}^2o}, a^{o^2 \to o}, x_1^o, \ldots, x_k^o$ and defined inductively by:

$$\begin{aligned} |x|_{la} &= x^o \\ |t_1 t_2|_{la} &= a^{o^2 \to o}\, |t_1|_{la}\, |t_2|_{la} \\ |\lambda x.t|_{la} &= l^{{}^2o}\, \lambda x^o.|t|_{la} \ . \end{aligned} \tag{1}$$

Definition 6. *If t is a pure closed λ-term, then $|t|_{la}^o$ is a λ-term having no free variable except $l^{{}^2o}, a^{o^2 \to o}$ and the code of t is the closed normal term of type $\mathbb{L}$:*

$$|t|^{\mathbb{L}} =_{def} \lambda l^{{}^2o} a^{o^2 \to o}.|t|_{la}^o \ . \tag{2}$$

Proposition 1. *Given any type A for which there is at least one closed λ-term t^A, we can construct a closed λ-term* $\mathrm{Decode}_A^{\mathbb{L}[T/o] \to A}$ *such that for any closed β-normal λ-term t^A (t being its underlying pure λ-term):*

$$\mathrm{Decode}_A\,(|t|^{\mathbb{L}}[T/o]) =_{\beta\eta} t^A \ . \tag{3}$$

Proof. Cf [4] (Lemma 3). □

From this proposition, it follows that if the type $\mathbb{L}$ had been finitely generated, then every type A having closed λ-terms t^A would have been finitely generated. Indeed, if $\mathcal{C}$ is a set of closed λ-terms generating $\mathbb{L}$, then the set $\mathcal{C}_A = \{t[T/o]\,;\, t \in \mathcal{C}\} \cup \{\mathrm{Decode}_A\}$ where T and Decode_A are as in Proposition 1 generates the type A. Moreover, as was proved in [5], it could have been decided out of the computable sets $\mathcal{C}_A$ whether a point in a full type structure over a finite ground domain is λ-definable or not. But this contradicts R. Loader's famous undecidability result given in [6], hence:

Proposition 2. *The type $\mathbb{L} = {}^2o, (o^2 \to o) \to o$ is not finitely generated.*

Proof. Cf [5] (p. 3, Proposition 4). □

It was also noted in [5] that Proposition 2 can be given the following equivalent form:

Proposition 3. *For any set C of closed* pure *λ-terms and any integer k, there is a closed pure normal λ-term t having no combinatorial representation $M \in [C]$ that reduces to t in less than k developments. (Recall that a development of a λ-term u is a β-reduction sequence starting with u and in which only residuals of redexes in u are reduced, see e.g. [1], Chapter 11.)*

Proof. Cf [5] (p. 4, Proposition 6). □

We will now consider the finite generation problem for *any* type A. Let us first remark that the question is of interest only if the type A has closed λ-terms. One may easily check that every type is either a classical tautology or a formula equivalent to o, and that there are closed λ-terms of the type A iff A is a tautology (this would be obviously false if we were given several ground types):

Definition 7. *Let us say that a type A is* inhabited *whenever the two following equivalent statments hold:*

· *There is at least one closed λ-term of the type A,*
· *A is a classical tautology.*

It turns out of the present work that an inhabited type A is finitely generated iff $rk(A) \leqslant 2$ or A is of the form $A_1, \ldots, A_n \to o$, where for all $i = 1, \ldots, n$: $A_i = o$ or $A_i = (o^k \to o) \to o$ for some $k \in \mathbb{N}$. The finitely generated types can also be given a nicer characterization as follows:

Definition 8. *Let us call* complexity $c(t)$ *of a λ-term t the least number of distinct variables required for its writing up to α-equivalence.*

Remark. If k is the maximal number of the free variables of a subterm of t, then $k \leqslant c(t) \leqslant k+1$. Indeed, the first inequality is obvious from the definitions of its members, and $c(t) \leqslant k+1$ is easily checked by induction on t.

Definition 9. *Let us say that an inhabited type A has a* bounded complexity *if there is an integer k such that for any closed normal λ-term t of type A: $c(t) \leqslant k$.*

We have then:

Theorem 4. *An inhabited type is finitely generated if and only if it has a bounded complexity.*

One way of this equivalence comes at once from:

Proposition 5. *For any inhabited type A and any integer k, the set $S_{A,k}$ of the closed β-normal λ-terms t of type A such that $c(t) \leqslant k$ is finitely generated.*

Proof. According to Proposition 1, we only have to prove that, for any integer k, the set $\Lambda_k = \{|t|^{\mathbb{L}}\,;\, t$ is a pure closed λ-term and $c(t) \leqslant k\}$ is finitely generated. Indeed, if C is a finite set of closed typed λ-terms generating Λ_k, then $\{\mathrm{Decode}_A^{\mathbb{L}[T/o]\to A}\} \cup \{M[T/o]\,;\, M \in C\}$, where Decode_A is as in 1, is a finite set generating $S_{A,k}$.

Now, let $\boldsymbol{x} = (x_0, x_1, \ldots, x_k)$ and let $\Lambda_k^{\boldsymbol{x}}$ be the set of the pure λ-terms of complexity $\leqslant k$ whose free variables are among $\boldsymbol{x}$. According to the remark following Definition 8, every element of $\Lambda_k^{\boldsymbol{x}}$ can be built from the variables $\boldsymbol{x}$ with the help of the functions $\mathrm{App} : (\Lambda_k^{\boldsymbol{x}})^2 \to \Lambda_k^{\boldsymbol{x}}$, $\mathrm{L}_i : \Lambda_k^{\boldsymbol{x}} \to \Lambda_k^{\boldsymbol{x}}$ $(1 \leqslant i \leqslant k)$ defined by: $\mathrm{App}(t, u) = tu$, $\mathrm{L}_i(t) = \lambda x_i.t$. Every element t of $\Lambda_k^{\boldsymbol{x}}$ can be represented by the *closed* λ-term of type $o^{k+1} \to \mathbb{L}$:

$$\rho(t) = \lambda x_1^o \ldots x_k^o l^{{}^2 o} a^{o^2 \to o}.|t|_{la}^o \ , \tag{4}$$

where $|t|_{la}^o$ is defined by (1). Moreover, the set $\{\rho(t); t \in \Lambda_k^{\boldsymbol{x}}\}$ is generated by the finite set $\{\rho(x_i); 0 \leqslant i \leqslant k\} \cup \{\mathcal{L}_i; 0 \leqslant i \leqslant k\} \cup \{\mathcal{A}pp\}$, where:

$$\mathcal{A}pp = \lambda z_1^{o^{k+1} \to \mathbb{L}} z_2^{o^{k+1} \to \mathbb{L}} \lambda \boldsymbol{x} la.a(z_1 \boldsymbol{x} la)(z_2 \boldsymbol{x} la) \ , \tag{5}$$

$$\mathcal{L}_i = \lambda z^{o^{k+1} \to \mathbb{L}} \lambda \boldsymbol{x} la.l(\lambda x_i.z \boldsymbol{x} la) \ , \tag{6}$$

since we have: $\mathcal{A}pp\, \rho(t)\rho(u) =_\beta \rho(\mathrm{App}(t, u))$ and $\mathcal{L}_i\, \rho(t) =_\beta \rho(\mathrm{L}_i(t))$. At last, by applying the closed λ-term:

$$\mathcal{R}^{(o^{k+1} \to \mathbb{L}) \to \mathbb{L}} = \lambda z^{o^{k+1} \to \mathbb{L}} \lambda la.z \underbrace{(l\lambda x.x) \ldots (l\lambda x.x)}_{k+1} \tag{7}$$

to every term $\rho(t)$, we obtain its representation $|t|^{\mathbb{L}}$. Λ_k is therefore finitely generated. □

Hence, every type of bounded complexity is finitely generated, but the converse is not so easy to establish and the rest of this paper is actually devoted to a rather lengthy proof of it. Since there are finitely generated sets S such that the complexities $c(t)$ of the terms $t \in S$ are not bounded by any constant k, any proof of the existence of such a constant in the particular case where S is also the set of the inhabitants of some type A should somehow take into account the latter hypothesis. This leads us here to discard the non suitable types A through a case study. It is nevertheless quite clear that further ideas for a direct proof of Theorem 4 are still missing.

2 The Monster Type Is Not Finitely Generated

The next step of our investigation is to prove that the type $\mathbb{M} = {}^3 o, o \to o$, nicknamed "Monster Type" by R. Statman, is not finitely generated. This was conjectured in [8] (example 5, p. 90).

2.1 The Type $\mathbb{P} = ((o \to o) \to o) \to o, ((o \to o)^2 \to o) \to o, o \to o$

One presents now a new coding of the pure closed λ-terms into the type:

$$\mathbb{P}_w = {}^3 w, (({}^1 w)^2 \to w) \to w, w \to w \ , \quad \text{where } w = ({}^1 o)^2 \to o \ . \tag{8}$$

For any pure λ-term t whose free variables are $x_1, \ldots, x_k$, let $\lceil t \rceil_{la}^{\overset{1}{w}}$ be the λ-term taking its free variables among $l^{\overset{3}{w}}, a^{((\overset{1}{w})^2 \to w) \to w}, x_1^{\overset{1}{w}}, \ldots, x_k^{\overset{1}{w}}$ and defined inductively by:

$$\begin{aligned} \lceil x \rceil_{la} &= x^{\overset{1}{w}} \\ \lceil t_1 t_2 \rceil_{la} &= \lambda \epsilon^w . a^{((\overset{1}{w})^2 \to w) \to w} (\lambda z_1^{\overset{1}{w}} z_2^{\overset{1}{w}} . \lceil t_1 \rceil_{la} (z_1 (\lceil t_2 \rceil_{la} (z_2 \epsilon)))) \\ \lceil \lambda x . t \rceil_{la} &= \lambda \epsilon^w . l^{\overset{3}{w}} (\lambda x^{\overset{1}{w}} . \lceil t \rceil_{la} \epsilon) \ . \end{aligned} \tag{9}$$

Definition 10. *If t is a pure closed λ-term, then $\lceil t \rceil_{la}^{\overset{1}{w}}$ is a λ-term having no free variable but $l^{\overset{3}{w}}, a^{((\overset{1}{w})^2 \to w) \to w}$ and the code of t is the closed normal term of type $\mathbb{P}_w$:*

$$\lceil t \rceil^{\mathbb{P}_w} =_{def} \lambda l^{\overset{3}{w}} a^{((\overset{1}{w})^2 \to w) \to w} \epsilon^w . \lceil t \rceil_{la} \epsilon \ . \tag{10}$$

Lemma 6. *There are closed λ-terms $\iota_1^{o \to w}$, $\iota_2^{o \to w}$ such that: $s^w f_1^{\overset{1}{o}} f_2^{\overset{1}{o}} =_{\beta\eta} f_i u^o$, if $s^w = \iota_i u^o$ (in other words, w behaves as the sum type: $w = o \oplus o$).*

Proof. Take: $\iota_i^{o \to w} = \lambda z^o f_1^{\overset{1}{o}} f_2^{\overset{1}{o}} . f_i z$, $i = 1, 2$. □

Lemma 7. *There are a closed λ-term $C^{\overset{1}{w}}$ and, for any pure variable x, a λ-term $V_x^{\overset{1}{w}}$ whose only free variable is x^o such that for any u^o:*

$$C(\iota_1 u^o) =_{\beta\eta} \iota_2 u^o \qquad C(\iota_2 u^o) =_{\beta\eta} \iota_1 u^o \tag{11}$$
$$V_x(\iota_1 u^o) =_{\beta\eta} \iota_1 x^o \qquad V_x(\iota_2 u^o) =_{\beta\eta} \iota_2 u^o \ . \tag{12}$$

Proof. Take: $C^{\overset{1}{w}} = \lambda s^w f_1^{\overset{1}{o}} f_2^{\overset{1}{o}} . s f_2 f_1$ and $V_x^{\overset{1}{w}} = \lambda s^w f^{\overset{1}{o}} . s(\lambda d^o . f x^o)$. □

Lemma 8. *There are λ-terms $L^{\overset{3}{w}}$ and $A^{((\overset{1}{w})^2 \to w) \to w}$ having no free variable but $l^{\overset{2}{o}}$, $a^{o^2 \to o}$, ϵ^o such that for any pure λ-term t whose free variables are $x_1, \ldots, x_n$:*

$$\begin{aligned} \lceil t \rceil_{la} [L/l, A/a, V_{x_i}/x_i^{\overset{1}{w}}]_{1 \leqslant i \leqslant n} (\iota_1 u^o) &=_{\beta\eta} \iota_1 |t|_{la}^o \\ \lceil t \rceil_{la} [L/l, A/a, V_{x_i}/x_i^{\overset{1}{w}}]_{1 \leqslant i \leqslant n} (\iota_2 u^o) &=_{\beta\eta} \iota_2 u^o \end{aligned} \tag{13}$$

Proof. One shows that the λ-terms $L = \lambda c^{\overset{2}{w}} f^{\overset{1}{o}} . c V_\epsilon (\lambda d^o . f (l^{\overset{2}{o}} (\lambda x^o . c V_x \mathbf{I}^{\overset{1}{o}} \mathbf{I}^{\overset{1}{o}})))$, $A = \lambda c^{(\overset{1}{w})^2 \to w} f^{\overset{1}{o}} . c \mathbf{I}^{\overset{1}{w}} \mathbf{I}^{\overset{1}{w}} (\lambda d^o . f (a^{o^2 \to o} (c C (\lambda d^w . \iota_2 \epsilon^o) \mathbf{I}^{\overset{1}{o}} \mathbf{I}^{\overset{1}{o}}) (c C (\lambda d^w . \iota_1 \epsilon^o) \mathbf{I}^{\overset{1}{o}} \mathbf{I}^{\overset{1}{o}})))$, where $\mathbf{I}^{\overset{1}{T}} = \lambda z^T . z^T$, suit by induction on t.

$\underline{t = x}$. This case follows at once from (12).

$\underline{t = \lambda x . t_0}$. Let $\tau^{\overset{1}{w}} = \lceil t_0 \rceil_{la} [L/l, A/a, V_{x_i}/x_i^{\overset{1}{w}}]_{1 \leqslant i \leqslant n}$. By the induction hypothesis, we have: $\tau[V_x/x](\iota_1 u^o) =_{\beta\eta} \iota_1 |t_0|_{la}^o$, $\tau[V_\epsilon/x](\iota_1 u^o) =_{\beta\eta} \iota_1 |t_0[\epsilon/x]|_{la}^o$ and $\tau[V_x/x](\iota_2 u^o) =_{\beta\eta} \tau[V_\epsilon/x](\iota_2 u^o) =_{\beta\eta} \iota_2 u^o$ (we may suppose that the variables

$x^{\overset{1}{w}}$ and x^o are not free in u^o). Then the λ-term $\theta = \lceil t \rceil_{la}[L/l, A/a, V_{x_i}/x_i^{\overset{1}{w}}]_{1\leqslant i\leqslant n}$ $= \lambda\epsilon^w.L(\lambda x^{\overset{1}{w}}.\tau\epsilon)$ is such that:

$$
\begin{aligned}
\theta(\iota_1 u^o) &=_\beta L(\lambda x^{\overset{1}{w}}.\tau(\iota_1 u^o))\\
&=_\beta \lambda f^{\overset{1}{o}}.(\lambda x^{\overset{1}{w}}.\tau(\iota_1 u^o))V_\epsilon(\lambda d^o.f(l^{\overset{2}{o}}(\lambda x^o.(\lambda x^{\overset{1}{w}}.\tau(\iota_1 u^o))V_x\mathbf{II})))\\
&=_\beta \lambda f^{\overset{1}{o}}.\tau[V_\epsilon/x](\iota_1 u^o)(\lambda d^o.f(l^{\overset{2}{o}}(\lambda x^o.\tau[V_x/x](\iota_1 u^o)\mathbf{II})))\\
&=_{\beta\eta} \lambda f^{\overset{1}{o}}.\iota_1|t_0[\epsilon/x]|^o_{la}(\lambda d^o.f(l^{\overset{2}{o}}(\lambda x^o.\iota_1|t_0|^o_{la}\mathbf{II}))) && \text{(by ind. hyp.)}\\
&=_\eta \lambda f_1^{\overset{1}{o}} f_2^{\overset{1}{o}}.\iota_1|t_0[\epsilon/x]|^o_{la}(\lambda d^o.f_1(l^{\overset{2}{o}}(\lambda x^o.\iota_1|t_0|^o_{la}\mathbf{II})))f_2\\
&=_{\beta\eta} \lambda f_1^{\overset{1}{o}} f_2^{\overset{1}{o}}.(\lambda d^o.f_1(l^{\overset{2}{o}}(\lambda x^o.\mathbf{I}|t_0|^o_{la})))|t_0[\epsilon/x]|^o_{la} && \text{(by Lemma 6)}\\
&=_\beta \iota_1(l^{\overset{2}{o}}(\lambda x^o.\mathbf{I}|t_0|^o_{la})) =_\beta \iota_1(l^{\overset{2}{o}}(\lambda x^o.|t_0|^o_{la})) = \iota_1|t|^o_{la}
\end{aligned}
$$

$$
\begin{aligned}
\theta(\iota_2 u^o) &=_\beta L(\lambda x^{\overset{1}{w}}.\tau(\iota_2 u^o))\\
&=_\beta \lambda f^{\overset{1}{o}}.(\lambda x^{\overset{1}{w}}.\tau(\iota_2 u^o))V_x(\lambda d^o.f(l^{\overset{2}{o}}(\lambda x^o.(\lambda x^{\overset{1}{w}}.\tau(\iota_2 u^o))V_x\mathbf{II})))\\
&=_\beta \lambda f^{\overset{1}{o}}.\tau[V_x/x](\iota_2 u^o)(\lambda d^o.f(l^{\overset{2}{o}}(\lambda x^o.\tau[V_x/x](\iota_2 u^o)\mathbf{II})))\\
&=_{\beta\eta} \lambda f^{\overset{1}{o}}.\iota_2 u^o(\lambda d^o.f(l^{\overset{2}{o}}(\lambda x^o.\iota_2 u^o\mathbf{II}))) && \text{(by ind. hyp.)}\\
&=_\eta \lambda f_1^{\overset{1}{o}} f_2^{\overset{1}{o}}.\iota_2 u^o(\lambda d^o.f_1(l^{\overset{2}{o}}(\lambda x^o.\iota_2 u^o\mathbf{II})))f_2\\
&=_{\beta\eta} \lambda f_1^{\overset{1}{o}} f_2^{\overset{1}{o}}.f_2 u^o =_\beta \iota_2 u^o && \text{(by Lemma 6)}
\end{aligned}
$$

<u>$t = t_1 t_2$</u>. Let $\tau_i^{\overset{1}{w}} = \lceil t_i \rceil_{la}[L/l, A/a, V_{x_j}/x_j^{\overset{1}{w}}]_{1\leqslant j\leqslant n}$, $i = 1, 2$. By induction hypothesis: $\tau_i(\iota_1 u^o) =_{\beta\eta} \iota_1|t_i|^o_{la}$ and $\tau_i(\iota_2 u^o) =_{\beta\eta} \iota_2 u^o$ for any u^o ($i = 1, 2$). Hence, one gets by Lemma 7:

$$
\begin{array}{ll}
\tau_1(C(\tau_2(\iota_1 u^o))) =_{\beta\eta} \iota_2|t_2|^o_{la} & \tau_1(\tau_2(\iota_1 u^o)) =_{\beta\eta} \iota_1|t_1|^o_{la}\\
\tau_1(C(\tau_2(\iota_2 u^o))) =_{\beta\eta} \iota_1|t_1|^o_{la} & \tau_1(\tau_2(\iota_2 u^o)) =_{\beta\eta} \iota_2 u^o\ .
\end{array}
\tag{14}
$$

Then, $\theta = \lceil t \rceil_{la}[L/l, A/a, V_{x_i}/x_i^{\overset{1}{w}}]_{1\leqslant i\leqslant n} = \lambda\epsilon^w.A(\lambda z_1^{\overset{1}{w}} z_2^{\overset{1}{w}}.\tau_1(z_1(\tau_2(z_2\epsilon))))$ is s.t.:

$$
\begin{aligned}
\theta(\iota_1 u^o) &=_\beta A(\lambda z_1^{\overset{1}{w}} z_2^{\overset{1}{w}}.\tau_1(z_1(\tau_2(z_2(\iota_1 u^o)))))\\
&=_\beta \lambda f^{\overset{1}{o}}.\tau_1(\tau_2(\iota_1 u^o))(\lambda d^o.f(a^{o^2\to o}(\tau_1(C(\tau_2(\iota_2\epsilon)))\mathbf{II})(\tau_1(C(\tau_2(\iota_1\epsilon)))\mathbf{II})))\\
&=_{\beta\eta} \lambda f^{\overset{1}{o}}.\iota_1|t_1|^o_{la}(\lambda d^o.f(a^{o^2\to o}(\iota_1|t_1|^o_{la}\mathbf{II})(\iota_2|t_2|^o_{la}\mathbf{II})))\\
&=_\eta \lambda f_1^{\overset{1}{o}} f_2^{\overset{1}{o}}.\iota_1|t_1|^o_{la}(\lambda d^o.f_1(a^{o^2\to o}(\iota_1|t_1|^o_{la}\mathbf{II})(\iota_2|t_2|^o_{la}\mathbf{II})))f_2\\
&=_{\beta\eta} \lambda f_1^{\overset{1}{o}} f_2^{\overset{1}{o}}.(\lambda d^o.f_1(a^{o^2\to o}(\mathbf{I}|t_1|^o_{la})(\mathbf{I}|t_2|^o_{la})))|t_1|^o_{la} && \text{(by Lemma 6)}\\
&=_\beta \lambda f_1^{\overset{1}{o}} f_2^{\overset{1}{o}}.f_1(a^{o^2\to o}|t_1|^o_{la}|t_2|^o_{la}) =_\beta \iota_1|t|^o_{la}
\end{aligned}
$$

$$
\begin{aligned}
\theta(\iota_2 u^o) &=_\beta A(\lambda z_1^{\overset{1}{w}} z_2^{\overset{1}{w}}.\tau_1(z_1(\tau_2(z_2(\iota_2 u^o)))))\\
&=_\beta \lambda f^{\overset{1}{o}}.\tau_1(\tau_2(\iota_2 u^o))(\lambda d^o.f(a^{o^2\to o}(\tau_1(C(\tau_2(\iota_2\epsilon)))\mathbf{II})(\tau_1(C(\tau_2(\iota_1\epsilon)))\mathbf{II})))\\
&=_{\beta\eta} \lambda f^{\overset{1}{o}}.\iota_2 u^o(\lambda d^o.f(a^{o^2\to o}(\iota_1|t_1|^o_{la}\mathbf{II})(\iota_2|t_2|^o_{la}\mathbf{II})))\\
&=_\eta \lambda f_1^{\overset{1}{o}} f_2^{\overset{1}{o}}.\iota_2 u^o(\lambda d^o.f_1(a^{o^2\to o}(\iota_1|t_1|^o_{la}\mathbf{II})(\iota_2|t_2|^o_{la}\mathbf{II})))f_2\\
&=_{\beta\eta} \lambda f_1^{\overset{1}{o}} f_2^{\overset{1}{o}}.f_2 u^o =_\beta \iota_2 u^o && \text{(by Lemma 6)}
\end{aligned}
$$

□

Proposition 9. *There is a closed λ-term Θ of the type $\mathbb{P}_w \to \mathbb{L}$ such that for any closed pure λ-term t:* $\Theta\lceil t \rceil^{\mathbb{P}_w} =_{\beta\eta} |t|^{\mathbb{L}}$.

Proof. By Lemma 8, there are terms $L^{{}^3w}$, $A^{(({}^1w)^2 \to w) \to w}$ whose free variables are $l^{{}^2o}$, $a^{o^2 \to o}$, ϵ^o s.t. for any closed pure λ-term t: $\lceil t \rceil_{la}[L/l, A/a](\iota_1 \epsilon^o) =_{\beta\eta} \iota_1 |t|^o_{la}$. Moreover, since ϵ^o never occurs in $\iota_1 |t|^o_{la}$, one gets: $\lceil t \rceil_{la}[L'/l, A'/a](\iota_1 E^o) =_{\beta\eta} \iota_1 |t|^o_{la}$, where $E^o = l^{{}^2o}\mathbf{I}^{{}^1o}$, $L' = L[E^o/\epsilon^o]$, $A' = A[E^o/\epsilon^o]$. Hence, one may choose the closed term: $\Theta = \lambda x^{\mathbb{P}_w} l^{{}^2o} a^{o^2 \to o}.xL'A'\mathbf{I}^{{}^1o}\mathbf{I}^{{}^1o}$. □

Proposition 10. *Every closed λ-term of type $\mathbb{L}$ is $\beta\eta$-equivalent to a λ-term of the form $|t|^{\mathbb{L}}$ for some closed pure λ-term t.*

Proof. Cf [3] (Chapter 1, p. 11, Proposition 1.1). □

Proposition 11. *The type $\mathbb{P} = {}^3o, (({}^1o)^2 \to o) \to o, o \to o$ is not finitely generated.*

Proof. Suppose for a contradiction that the type $\mathbb{P}$ is finitely generated. The set $\{\lceil t \rceil^{\mathbb{P}_w} ; t$ is a pure closed λ-term$\}$ would be finitely generated since each of its elements is of the form: $t^{\mathbb{P}}[w/o]$. By adding the λ-term Θ defined in Proposition 9 as a generator, the set $\{|t|^{\mathbb{L}} ; t$ is a pure closed λ-term$\}$ would then be also finitely generated. Therefore, according to Proposition 10, the type $\mathbb{L}$ itself would be finitely generated, but this contradicts Proposition 2. □

2.2 The Type $\mathbb{M} = (((o \to o) \to o) \to o), o \to o$

As remarked in [3] (Chapter 1, p. 27), the type $\mathbb{M}$ allows the coding of the closed pure λ-terms such that in every subterm of the form tu, t is a variable. These β-normal λ-terms will be called small λ-terms, more formally:

Definition 11. *The* small *λ-terms are the pure λ-terms inductively defined by:*

- *any λ-variable is a small term,*
- *if x is a variable and t a small term then xt is a small term,*
- *if t is a small term then $\lambda x.t$ is a small term.*

Let $x \mapsto \underline{x}^{{}^1o}$ be an embedding of the set of pure λ-variables into that of the λ-variables of type ${}^1o$. For any small λ-term t whose free variables are $x_1, \ldots, x_k$, let $\|t\|^o_{le}$ be the λ-term taking its free variables among $l^{{}^3o}, e^o, \underline{x_1}^{{}^1o}, \ldots, \underline{x_k}^{{}^1o}$ and defined inductively by:

$$\begin{aligned} \|x\|^o_{le} &= \underline{x}^{{}^1o} e^o \\ \|xt\|^o_{le} &= \underline{x}^{{}^1o} \|t\|^o_{le} \\ \|\lambda x.t\|^o_{le} &= l^{{}^3o} \lambda \underline{x}^{{}^1o}.\|t\|^o_{le} \end{aligned} \tag{15}$$

Definition 12. *If t is a closed small λ-term, then $\|t\|^o_{le}$ is a λ-term having no free variable except $l^{{}^3o}, e^o$ and the code of t is the normal closed term of type $\mathbb{M}$:*

$$\|t\|^{\mathbb{M}} =_{def} \lambda l^{{}^3o} e^o.\|t\|^o_{le} . \tag{16}$$

Lemma 12. *The set* $\{|t|^{\mathbb{L}}\,;\, t$ is a *small* closed λ-term$\}$ *is not finitely generated.*

Proof. It may be checked that the underlying form t of every normal closed term $t^{\mathbb{P}}$ is a small λ-term. Since we have moreover: $\mathrm{Decode}_{\mathbb{P}}^{\mathbb{L}[T/o]\to\mathbb{P}}(|t|^{\mathbb{L}}[T/o]) =_{\beta\eta} t^{\mathbb{P}}$, where $\mathrm{Decode}_{\mathbb{P}}$ is as in Proposition 1, it follows from Proposition 11 that the set $\{|t|^{\mathbb{L}}[T/o]\,;\, t$ is a small closed λ-term$\}$ cannot be finitely generated and we conclude at once. □

Proposition 13. *There is a closed term* $C^{\mathbb{M}[S/o]\to\mathbb{L}}$ *such that for every small closed λ-term* t*:* $C(\|t\|^{\mathbb{M}}[S/o]) =_{\beta\eta} |t|^{\mathbb{L}}$.

Proof. Let $S = {}^{1}o, o \to o$, $\iota_1^{o\to S} = \lambda z^o f_1^{{}^1o} f_2^o.f_1 z$ and $\iota_2^S = \lambda f_1^{{}^1o} f_2^o.f_2$. For any terms $t_1^{{}^1o}$, t_2^o, u^o, we then have: $(\iota_1 u)t_1 t_2 =_{\beta\eta} t_1 u$ and $\iota_2 t_1 t_2 =_{\beta\eta} t_2$ (S behaves as the sum type: $o \oplus \bot$). For any pure λ-variable x, let $\mathcal{V}_x^{{}^1S} = \lambda v^S.\iota_1(v(a^{o^2\to o}x^o)x^o)$. At last, let $\mathcal{L}^{{}^3S} = \lambda v^{{}^2S}.\iota_1(l^{{}^2o}\lambda x^o.v\mathcal{V}_x\mathbf{I}(l^{{}^2o}\mathbf{I}))$, where $\mathbf{I} = \lambda z^o.z^o$. Note that the only free variables of $\mathcal{L}$ are $l^{{}^2o}$ and $a^{o^2\to o}$. We will prove by induction on a small pure λ-term t whose free variables are $x_1, \ldots, x_n$ (with possibly $n = 0$) that the term $\tau = \|t\|_{le}[S/o, \mathcal{L}/l, \iota_2/e, \mathcal{V}_{x_i}/\underline{x}_i]_{1\leqslant i\leqslant n}$ is β-equivalent to $\iota_1|t|_{la}^o$: $\underline{t = x}$. In this case: $\tau = \mathcal{V}_x\iota_2 =_\beta \iota_1(\iota_2(a^{o^2\to o}x^o)x^o) =_\beta \iota_1(x^o) = \iota_1|t|_{la}^o$.
$\underline{t = xt_0}$. By induction hypothesis, $\tau = \mathcal{V}_x(\|t_0\|_{le}[S/o, \mathcal{L}/l, \iota_2/e, \mathcal{V}_{x_i}/\underline{x}_i]_{1\leqslant i\leqslant n}) =_\beta \mathcal{V}_x(\iota_1|t_0|^o) =_\beta \iota_1(\iota_1|t_0|^o(a^{o^2\to o}x^o)x^o) =_\beta \iota_1(a^{o^2\to o}x^o|t_0|^o) = \iota_1|t|_{la}^o$.
$\underline{t = \lambda x.t_0}$. Let $u^S = \|t_0\|_{le}[S/o, \mathcal{L}/l, \iota_2/e, \mathcal{V}_{x_i}/\underline{x}_i]_{1\leqslant i\leqslant n}$. By induction hypothesis, $u^S[\mathcal{V}_x/\underline{x}^{{}^1S}] =_\beta \iota_1|t_0|^o$, hence: $\tau = (\mathcal{L})\lambda\underline{x}^{{}^1S}.u^S =_\beta \iota_1(l(\lambda x^o.(\lambda\underline{x}^{{}^1S}.u^S)\mathcal{V}_x\mathbf{I}(l\mathbf{I}))) =_\beta \iota_1(l(\lambda x^o.u^S[\mathcal{V}_x/\underline{x}]\mathbf{I}(l\mathbf{I}))) =_\beta \iota_1(l(\lambda x^o.(\iota_1|t_0|^o)\mathbf{I}(l\mathbf{I}))) =_\beta \iota_1(l(\lambda x^o.\mathbf{I}|t_0|^o)) =_\beta \iota_1|t|^o$.

In particular, we have for every small closed λ-term t: $\|t\|^{\mathbb{M}}[S/o]\mathcal{L}\iota_2\mathbf{I}(l\mathbf{I}) =_\beta \|t\|_{le}[S/o, \mathcal{L}/l, \iota_2/e]\mathbf{I}(l\mathbf{I}) =_\beta \iota_1|t|^o\mathbf{I}(l\mathbf{I}) =_\beta \mathbf{I}|t|^o =_\beta |t|^o$. It follows that we may take: $C = \lambda v^{\mathbb{M}[S/o]} l^{{}^2o} a^{o^2\to o}.v\mathcal{L}\iota_2\mathbf{I}(l\mathbf{I})$. □

Proposition 14. *The set* $\{\|t\|^{\mathbb{M}}\,;\, t$ is a *small* closed λ-term$\}$ *is not finitely generated.*

Proof. Indeed, otherwise the set $\{\|t\|^{\mathbb{M}}[S/o]\,;\, t$ is a small closed λ-term$\}$ would be finitely generated and by Proposition 13, $\{|t|^{\mathbb{L}}\,;\, t$ is a small closed λ-term$\}$ would then be also finitely generated, contradicting Lemma 12. □

Corollary 15. *The type* $\mathbb{M} = {}^{3}o, o\to o$ *is not finitely generated.*

3 Some Other Non Finitely Generated Types

3.1 The Type $\mathbb{L}' = ((o\to o), o\to o), o\to o$

The type $\mathbb{L}'$ allows also a coding of every pure closed λ-term. For any pure λ-term t whose free variables are $x_1, \ldots, x_k$, let $\lfloor t\rfloor_{ae}'$ be the λ-term of type o

taking its free variables among $a^{{}^1o,o\to o}, e^o, x_1^o, \ldots, x_k^o$ and defined inductively by:

$$\begin{aligned} \lfloor x \rfloor'_{ae} &= x^o \\ \lfloor t_1 t_2 \rfloor'_{ae} &= a^{{}^1o,o\to o}(\lambda d^o.\lfloor t_1 \rfloor'_{ae})\, \lfloor t_2 \rfloor'_{ae} \\ \lfloor \lambda x.t \rfloor'_{ae} &= a^{{}^1o,o\to o}(\lambda x^o.\lfloor t \rfloor'_{ae})e^o \ . \end{aligned} \tag{17}$$

Definition 13. *If t is a pure closed λ-term, then $\lfloor t \rfloor'_{ae}$ is a λ-term having no free variable except $a^{{}^1o,o\to o}, e^o$ and the code of t in the type $\mathbb{L}'$ is the closed normal term:*

$$\lfloor t \rfloor^{\mathbb{L}'} =_{def} \lambda a^{{}^1o,o\to o} e^o.\lfloor t \rfloor'_{ae} \ . \tag{18}$$

Proposition 16. *There is a closed term $C^{\mathbb{L}'[S/o]\to\mathbb{L}}$ such that for all pure closed λ-terms t: $C(\lfloor t \rfloor^{\mathbb{L}'}[S/o]) =_{\beta\eta} |t|^{\mathbb{L}}$.*

Sketched Proof. Let S, $\iota_1^{o\to S}$ and ι_2^S be as in the proof of Proposition 13, i.e. such that: $(\iota_1 u)t_1 t_2 =_{\beta\eta} t_1 u$, $\iota_2 t_1 t_2 =_{\beta\eta} t_2$ (S behaves as the sum type: $o \oplus \perp$). Let $\mathcal{A}^{{}^1S,S\to S} = \lambda x^{{}^1S} y^S.\iota_1(y(\lambda d^o.a^{o^2\to o}(x\iota_2\mathbf{I}(l\mathbf{I}))(y\mathbf{I}(l\mathbf{I})))(l^{{}^2o}(\lambda z^o.x(\iota_1 z^o)\mathbf{I}(l\mathbf{I}))))$. Note that the only free variables of $\mathcal{A}$ are $l^{{}^2o}$ and $a^{o^2\to o}$. We can prove by induction on a small pure λ-term t whose free variables are $x_1, \ldots, x_n$ (with possibly $n = 0$) that the term $\lfloor t \rfloor'_{ae}[S/o, \mathcal{A}/a, \iota_2/e, \iota_1 x_i^o/x_i]_{1\leqslant i\leqslant n}$ is β-equivalent to $\iota_1 |t|^o_{la}$.

In particular, if t is any small closed λ-term then $\lfloor t \rfloor^{\mathbb{L}'}[S/o]\mathcal{A}\iota_2\mathbf{I}(l\mathbf{I}) =_\beta |t|^o_{la}$, so that we can take: $C = \lambda v^{\mathbb{L}'[S/o]} l^{{}^2o} a^{o^2\to o}.v\mathcal{A}\iota_2\mathbf{I}(l\mathbf{I})$. □

Hence, by Proposition 2 and the latter:

Proposition 17. *The type $\mathbb{L}'$ is not finitely generated.*

3.2 The Type $\mathbb{L}'' = ((o\to o)^2 \to o)\to o$

What has just been done for the type $\mathbb{L}'$ can be reproduced for the type $\mathbb{L}'' = (({}^1o)^2 \to o) \to o$: For any pure λ-term t whose free variables are $x_1, \ldots, x_k$, let $\lfloor t \rfloor''_a$ be the λ-term of type o taking its free variables among $a^{({}^1o)^2\to o}, x_1^o, \ldots, x_k^o$ and defined inductively by:

$$\begin{aligned} \lfloor x \rfloor''_a &= x^o \\ \lfloor t_1 t_2 \rfloor''_a &= a^{({}^1o)^2\to o}(\lambda d^o.\lfloor t_1 \rfloor''_a)\,(\lambda d^o.\lfloor t_2 \rfloor''_a) \\ \lfloor \lambda x.t \rfloor''_a &= a^{({}^1o)^2\to o}(\lambda x^o.\lfloor t \rfloor''_a)(\lambda x^o.x^o) \ . \end{aligned} \tag{19}$$

Definition 14. *If t is a pure closed λ-term, then $\lfloor t \rfloor''_a$ is a λ-term having no free variable except $a^{({}^1o)^2\to o}$ and the code of t in the type $\mathbb{L}''$ is the closed normal λ-term:*

$$\lfloor t \rfloor^{\mathbb{L}''} =_{def} \lambda a^{({}^1o)^2\to o}.\lfloor t \rfloor''_a \ . \tag{20}$$

It can be checked at once that the closed λ-term:

$$C' = \lambda v^{((^1o)^2\to o)\to o} a^{^1o,o\to o} e^o . v(\lambda x^{^1o} y^{^1o}.ax(ye)) \tag{21}$$

satisfies for all pure closed terms t: $C'\lfloor t\rfloor^{\mathbb{L}''} =_{\beta\eta} \lfloor t\rfloor^{\mathbb{L}'}$. By composing C' with the term C of Proposition 16, we get a closed λ-term C'' s.t. for all closed t: $C''\lfloor t\rfloor^{\mathbb{L}''} =_{\beta\eta} |t|^{\mathbb{L}}$; hence by Propositions 2 and 10:

Proposition 18. *The type $\mathbb{L}''$ is not finitely generated.*

3.3 The Type $\mathbb{M}' = (((o\to o), o\to o)\to o)\to o$

Let $x \mapsto \underline{x}^{^1o}$ be an embedding of the set of pure λ-variables into that of the λ-variables of type ^1o. For any small λ-term t whose free variables are $x_1, \ldots, x_k$, let $\llbracket t\rrbracket'_l$ be the λ-term taking its free variables among $l^{(^1o,o\to o)\to o}, z^o, \underline{x}_1, \ldots, \underline{x}_k$ and defined inductively by:

$$\begin{aligned} \llbracket x\rrbracket'_l &= \underline{x}z^o \\ \llbracket xt\rrbracket'_l &= \underline{x}\llbracket t\rrbracket'_l \\ \llbracket \lambda x.t\rrbracket'_l &= l^{(^1o,o\to o)\to o}\lambda \underline{x}z^o.\llbracket t\rrbracket'_l \ . \end{aligned} \tag{22}$$

Definition 15. *If t is a closed small λ-term, then $\llbracket t\rrbracket'_l$ is a λ-term having no free variable except $l^{(^1o,o\to o)\to o}$ and the code of t in the type $\mathbb{M}'$ is the normal closed λ-term:*

$$\llbracket t\rrbracket^{\mathbb{M}'} =_{def} \lambda l^{(^1o,o\to o)\to o}.\llbracket t\rrbracket^o_l \ . \tag{23}$$

Proposition 19. *There is a closed term $D^{\mathbb{M}'\to\mathbb{M}}$ such that for every small closed λ-term t: $D\llbracket t\rrbracket^{\mathbb{M}'} =_{\beta\eta} \|t\|^{\mathbb{M}}$.*

Sketched Proof. Let $\mathcal{L}^{(^1o,o\to o)\to o} = \lambda x^{^1o,o\to o}.l^{^3o}(\lambda z^{^1o}.xze)$. Note that the only free variables of $\mathcal{L}$ are $l^{^3o}$ and e^o. We can prove by induction on a small pure λ-term t that the term $\llbracket t\rrbracket'_l[\mathcal{L}/l, e^o/z^o]$ is β-equivalent to $\|t\|^o_{le}$.

In particular, if t is any small closed λ-term, then z^o is not free in $\llbracket t\rrbracket'_l$; hence we have $\llbracket t\rrbracket^{\mathbb{M}'}\mathcal{L} =_\beta \|t\|^o_{le}$, and we can take: $D = \lambda v^{\mathbb{M}'} l^{^3o} e^o.v\mathcal{L}$. □

Hence, by Proposition 14 and the latter:

Proposition 20. *The type $\mathbb{M}'$ is not finitely generated.*

3.4 The Type $\mathbb{M}'' = ((((o\to o)\to o)\to o)\to o)\to o$

Let $x \mapsto \underline{x}^{^2o}$ be an embedding of the set of pure λ-variables into that of the λ-variables of type ^2o. For any small λ-term t whose free variables are $x_1, \ldots, x_k$, let $\llbracket t\rrbracket''_l$ be the λ-term taking its free variables among $l^{^4o}, \underline{x}_1, \ldots, \underline{x}_k$ and defined inductively by:

$$\begin{aligned} \llbracket x\rrbracket''_l &= \underline{x}\lambda z^o.z^o \\ \llbracket xt\rrbracket''_l &= \underline{x}\lambda d^o.\llbracket t\rrbracket''_l \\ \llbracket \lambda x.t\rrbracket''_l &= l^{^4o}\lambda \underline{x}^{^2o}.\llbracket t\rrbracket''_l \ . \end{aligned} \tag{24}$$

Definition 16. *If t is a closed small λ-term, then $\llbracket t \rrbracket_l''$ is a λ-term having no free variable except $l^{\overset{4}{o}}$ and the code of t in the type $\mathbb{M}''$ is the normal closed λ-term:*

$$\llbracket t \rrbracket^{\mathbb{M}''} =_{def} \lambda l^{\overset{4}{o}}.\llbracket t \rrbracket_l'' \ . \tag{25}$$

Proposition 21. *There is a closed term $D^{\mathbb{M}''\to\mathbb{M}}$ such that for every small closed λ-term t: $D\llbracket t \rrbracket^{\mathbb{M}''} =_{\beta\eta} \|t\|^{\mathbb{M}}$.*

Sketched Proof. Let $\mathcal{L}^{\overset{4}{o}} = \lambda x^{\overset{3}{o}}.l^{\overset{3}{o}}(\lambda z^{\overset{1}{o}}.x(\lambda s^{\overset{1}{o}}.z(se^o)))$. Note that the only free variables of $\mathcal{L}$ are $l^{\overset{3}{o}}$ and e^o. We can prove by induction on a small pure λ-term t whose free variables are $x_1, \ldots, x_n$ (with possibly $n = 0$) that the term $\llbracket t \rrbracket_l''[\mathcal{L}/l, \lambda s^{\overset{1}{o}}.\underline{x}_i^{\overset{1}{o}}(se^o)/\underline{x}_i^{\overset{2}{o}}]_{1\leqslant i\leqslant n}$ is β-equivalent to $\|t\|_{le}^o$.

In particular, if t is any small closed λ-term then $\llbracket t \rrbracket^{\mathbb{M}''}\mathcal{L} =_\beta \|t\|_{le}^o$, so that we can take: $D = \lambda v^{\mathbb{M}''} l^{\overset{2}{o}} e^o.v\mathcal{L}$. □

Hence, by Proposition 14 and the latter:

Proposition 22. *The type $\mathbb{M}''$ is not finitely generated.*

4 The Types of Unbounded Complexity Are Not Finitely Generated

We will now reduce the case of every (inhabited) type of unbounded complexity to the previous ones ($\mathbb{L}, \mathbb{L}', \mathbb{L}'', \mathbb{M}, \mathbb{M}', \mathbb{M}''$) with the help of:

Definition 17. *Let $\succcurlyeq$ be the least (binary) reflexive and transitive relation on the set of types $\Lambda_o^{\to}$ such that for any types A, B, C:*

$$\begin{aligned} & A\to B \succcurlyeq B \ , \\ & A, B\to C \succcurlyeq B, A\to C \ , \\ & \textit{If } A \succcurlyeq B, \ \textit{then } C\to A \succcurlyeq C\to B \textit{ and } A\to C \succcurlyeq B\to C \ . \end{aligned} \tag{26}$$

Proposition 23. – de'Liguoro, Piperno, Statman (1992, cf [2], p. 464, Lemma 3.5) *If we have $A \succcurlyeq B$ for some types A, B, then there are (possibly open) λ-terms $\tau^{B\to A}, \theta^{A\to B}$, such that for all λ-terms u^B: $\theta(\tau u^B) =_{\beta\eta} u^B$.*

This was actually proved in [2] for a relation $\trianglerighteq$ larger than $\succcurlyeq$, defined in the same way but with the additional clause: $(A\to o)\to o \trianglerighteq A$ *(for any type A)*.

Corollary 24. *If A, B are inhabited types such that $A \succcurlyeq B$ and if A is finitely generated, then B is also finitely generated.*

Indeed, if $A = A_1, \ldots, A_m \to o$ and $B = B_1, \ldots, B_n \to o$ are inhabited types such that $A \succcurlyeq B$, then there are λ-terms $\theta^{A \to B} =_{\beta\eta} \lambda x^A x_1^{B_1} \ldots x_m^{B_m}.r^o$ and $\tau^{B \to A} =_{\beta\eta} \lambda x^B x_1^{A_1} \ldots x_n^{A_n}.r'^o$ s.t. for all u^B: $\theta(\tau u^B) =_{\beta\eta} u^B$. Since A (resp. B) is inhabited, there is a term κ^o whose free variables are among $x_1^{A_1}, \ldots, x_n^{A_n}$ (resp. among $x_1^{B_1}, \ldots, x_m^{B_m}$); hence, we may replace every free variable of τ (resp. of θ) with a λ-term of the form $\lambda d_1 \ldots d_k.\kappa^o$ in order to get a closed λ-term τ_0 (resp. a closed λ-term θ_0). Note that we have still for all u^B: $\theta_0(\tau_0 u^B) =_{\beta\eta} u^B$. Now, for all closed u^B, the closed λ-term $t^A = \tau_0 u^B$ is s.t. $\theta_0 t^A =_{\beta\eta} u^B$; hence, if $\mathcal{C}$ is a finite set generating the type A, then the set $\mathcal{C} \cup \{\theta_0\}$ generates B.

Recall that the signs of subtype occurrences in a type are inductively defined as follows: the only occurence of T in T is positive; if $T \neq A \to B$, the positive (resp. negative) occurrences of T in $A \to B$ are the negative (resp. positive) occurrences of T in A and the positive (resp. negative) occurrences of T in B.

Definition 18. – **Statman (1980, cf [7], p. 512)** *We say that a type is* small *if it has no negative occurrence of a subtype of the form $A, B \to C$ and* large *if it is not small.*

Proposition 25. *If A is an inhabited large type of rank 3, then $A \succcurlyeq \mathbb{L}$, $A \succcurlyeq \mathbb{L}'$ or $A \succcurlyeq \mathbb{L}''$.*

Proof. Since A is supposed to be large and of rank 3, we must have $A = A_1, \ldots, A_n \to o$, with for some i, j: $rk(A_i) = 2$ and $A_j = B_1, \ldots, B_m \to o$, $m \geqslant 2$.

- If $i \neq j$, then $A_i \succcurlyeq {}^2o$ and $A_j \succcurlyeq o^2 \to o$, hence $A \succcurlyeq \mathbb{L}$.
- If $i = j$, then $A_i \succcurlyeq {}^1o, o \to o$.
  - If A_i is inhabited, then $n \geqslant 2$, otherwise A would not be a classical tautology and would have no inhabitant; thus $A \succcurlyeq \mathbb{L}'$.
  - If A_i is not inhabited, then for all $1 \leqslant k \leqslant m$, B_k is inhabited, $B_k \succcurlyeq {}^1o$; hence, $A_i \succcurlyeq ({}^1o)^2 \to o$ and $A \succcurlyeq \mathbb{L}''$.

□

Proposition 26. *If A is an inhabited type of rank $\geqslant 4$, then $A \succcurlyeq \mathbb{M}$, $A \succcurlyeq \mathbb{M}'$ or $A \succcurlyeq \mathbb{M}''$.*

Proof. Suppose that $A = A_1, \ldots, A_n \to o$ is an inhabited type of rank $\geqslant 4$.

- If $n \geqslant 2$, then $A \succcurlyeq \mathbb{M}$.
- If $n = 1$, let $A_1 = B_1, \ldots, B_m \to o$, i s.t. $rk(B_i) \geqslant 2$, $B_i = C_1, \ldots, C_k \to o$ and j s.t. $rk(C_j) \geqslant 1$. Since A is inhabited, B_i must be a classical tautology i.e. an inhabited type.
  - If C_j is inhabited, then we must have $k \geqslant 2$, since B_i is inhabited. Therefore, $B_i \succcurlyeq {}^1o, o \to o$ and $A \succcurlyeq \mathbb{M}'$.
  - If C_j is not inhabited, then $rk(C_j) \geqslant 2$ and $A \succcurlyeq \mathbb{M}''$.

□

Proposition 27. *If an inhabited type A has not a bounded complexity, then at least one of the following relations holds: $A \succcurlyeq \mathbb{L}$, $A \succcurlyeq \mathbb{L}'$, $A \succcurlyeq \mathbb{L}''$, $A \succcurlyeq \mathbb{M}$, $A \succcurlyeq \mathbb{M}'$, $A \succcurlyeq \mathbb{M}''$.*

Proof. Let $A = A_1, \dots, A_n \to o$ be any inhabited type of unbounded complexity. If $rk(A) \geqslant 4$, then we have from Proposition 26: $A \succcurlyeq \mathbb{M}$, $A \succcurlyeq \mathbb{M}'$ or $A \succcurlyeq \mathbb{M}''$. If $rk(A) \leqslant 3$, let $t = \lambda x_1^{A_1} \dots x_n^{A_n}.\tau^o$ be a closed normal λ-term of type A with complexity $c(t) \geqslant n+2$. There must be at least one A_i $(1 \leqslant i \leqslant n)$ of rank >1, otherwise τ^o could not contain any abstraction and consequently could not have other variable occurrences than those of $x_1^{A_1}, \dots, x_n^{A_n}$. Moreover, since $rk(A) \leqslant 3$, all the bounded variables of τ^o have the type o, and some A_i must then be of the form $B, C \to D$, otherwise τ^o would necessarily have the form: $x_{i_1}(\lambda \boldsymbol{z}_1 x_{i_2}(\lambda \boldsymbol{z}_2 \dots x_{i_k}(\lambda \boldsymbol{z}_k.z^o)))$ and we would have: $c(t) \leqslant n+1$. A is therefore a large type of rank 3 and we get by Proposition 25: $A \succcurlyeq \mathbb{L}$, $A \succcurlyeq \mathbb{L}'$ or $A \succcurlyeq \mathbb{L}''$. □

Proof of Theorem 4. Let A be any inhabited type.
If A has a bounded complexity, then A is finitely generated by Proposition 5. If A has not a bounded complexity, then according to Propositions 27, 2, 17, 18, 15, 20 and 22, there is a non finitely generated type B ($B = \mathbb{L}, \mathbb{L}', \mathbb{L}'', \mathbb{M}, \mathbb{M}'$ or $\mathbb{M}''$) such that $A \succcurlyeq B$. It follows from Corollary 24 that A is not finitely generated. □

References

1. Barendregt, H.P.: The λ-Calculus. Vol. 103 of Studies in Logic and the Foundations of Mathematics. 2nd edition. North Holland, Amsterdam (1984)
2. de'Liguoro, U., Piperno, A., Statman, R.: Retracts in simply typed λβη-calculus. IEEE Symposium on Logic in Computer Science (1992) 461–469
3. Joly, T.: Codages, séparabilité et représentation de fonctions en λ-calcul simplement typé et dans d'autres systèmes de types. Thèse de Doctorat, Université Paris VII, Jan. 2000
4. Joly, T.: Constant time parallel computations in λ-calculus. Theoretical Computer Science B (to appear)
5. Joly, T.: Non finitely generated types & λ-terms combinatoric representation cost. C.R.A.S. of Paris, Série I **331** (2000) 581–586
6. Loader, R.: The Undecidability of λ-Definability. Church Memorial Volume (to appear)
7. Statman, R.: On the existence of closed terms in the typed λ-calculus I. In "To Curry: Essays on combinatory logic, lambda-calculus and formalism", Hindley and Seldin, eds., Academic Press (1980) 511–534
8. Statman, R.: Logical Relations and the Typed λ-calculus. Information & Computation **65** (1985) 85–97

Deciding Monadic Theories of Hyperalgebraic Trees

Teodor Knapik[1], Damian Niwiński[2*], and Paweł Urzyczyn[2**]

[1] Dept. de Mathématiques et Informatique, Université de la Réunion, BP 7151, 97715 Saint Denis Messageries Cedex 9, Réunion
knapik@univ-reunion.fr

[2] Institute of Informatics, Warsaw University
ul. Banacha 2, 02–097 Warszawa, Poland
{niwinski,urzy}@mimuw.edu.pl

Abstract. We show that the monadic second–order theory of any infinite tree generated by a higher–order grammar of level 2 subject to a certain syntactic restriction is decidable. By this we extend the result of Courcelle [6] that the MSO theory of a tree generated by a grammar of level 1 (algebraic) is decidable. To this end, we develop a technique of representing infinite trees by infinite λ–terms, in such a way that the MSO theory of a tree can be interpreted in the MSO theory of a λ–term.

Introduction

In 1969, Rabin [13] proved decidability of the monadic second–order (MSO) theory of the full n–ary tree, which is perhaps one of the most widely applied decidability results. There are several ways in which Rabin's Tree Theorem can be extended. One possibility is to consider a more general class of structures obtained in tree–like manner, i.e., by unwinding some initial structure. The decidability of the MSO theory of the unwound structure then relies on the decidability of the MSO theory of the initial structure, and the "regularity" of the unwinding process (see [16]). Another direction, which we will pursue here, is to remain with trees but consider more sophisticated modes of generation than unwinding.

To this end, it is convenient to rephrase Rabin's Tree Theorem for *labeled* trees, as follows: The MSO theory of any regular tree is decidable. Here, a labeled tree is seen as a logical structure with additional monadic predicates corresponding to the labels, and a tree is *regular* if it has only a finite number of non–isomorphic subtrees. An equivalent definition of regularity says that a tree is generated by a (deterministic) regular tree grammar, which gives rise to further generalizations. Indeed, Courcelle [6] proved that the MSO theory of any tree generated by an *algebraic* (or, context–free) tree grammar is also decidable. However nothing general is known about the MSO theories of trees generated

* Partly supported by KBN Grant 8 T11C 027 16.
** Partly supported by KBN Grant 8 T11C 035 14.

S. Abramsky (Ed.): TLCA 2001, LNCS 2044, pp. 253–267, 2001.

by higher order grammars, although the expressive power of such grammars was extensively studied by Damm [7] in the early eighties. This is the question we address in the present paper.

It is plausible to think that any tree generated by a higher–order grammar (see Section 4 below) has decidable MSO theory, this however is only a conjecture. At present, we are able to show decidability of the MSO theory of trees generated by grammars of level 2 satisfying some additional condition restricting occurrences of individual parameters in scope of functional ones. This however properly extends the aforementioned result of Courcelle [6].

Our method makes use of the idea of the *infinitary* λ*–calculus*, already considered by several authors [9,4]. Here we view infinite λ–terms as infinite trees additionally equipped with edges from bound variables to their binders. In course of a possibly infinite sequence of β–reductions, these additional edges may disappear, and the result is a tree consisting of constant symbols only. We show that the MSO theory of the resulting tree can be reduced to the MSO theory of the original λ–term, viewed as an appropriate logical structure (Theorem 2 below). Let us stress that the reduction is not a mere interpretation (which seems to be hardly possible). Instead, we use the μ–calculus as an intermediate logic, and an intermediate structure obtained by folding the tree. In order to interpret the MSO theory of the folded tree in the MSO theory of the λ–term, we use techniques similar to Caucal [5], combined with an idea originated from Geometry of Interaction and the theory of optimal reductions. Namely, we consider deformations of regular paths in λ–terms and push–down store computations along these paths.

The motivation behind the of MSO theories is that the MSO theory of a λ–term should be easier to establish than that of the tree after reduction. This is indeed the case of grammars satisfying our restriction. More specifically, for each such grammar, we are able to construct an infinite λ–term which is essentially an algebraic tree, and whose result of β–reduction is precisely the tree generated by the grammar. Hence, by the aforementioned Courcelle's theorem, we get our decidability result (Theorem 4 below).

Let us mention that the interest in deciding formal theories of finitely presentable infinite structures has grown among the verification community during last decade (see, e.g., [11] and references therein). In particular, a problem related to ours was addressed by H. Hungar, who studied graphs generated by some specific higher–order graph grammars. He showed [8] decidability of the monadic second–order theory (S1S) of *paths* of such graphs (not the full MSO theory of graphs).

1 Preliminaries

Types. We consider a set of types T constructed from a unique *basic* type $\mathbf{0}$. That is $\mathbf{0}$ is a type and, if τ_1, τ_2 are types, so is $(\tau_1 \to \tau_2) \in$ T. The operator $\to$ is assumed to associate to the right. Note that each type is of the form $\tau_1 \to \cdots \to \tau_n \to \mathbf{0}$, for some $n \geq 0$. A type $\mathbf{0} \to \cdots \to \mathbf{0}$ with $n+1$ occurrences of $\mathbf{0}$ is also written $\mathbf{0}^n \to \mathbf{0}$. The level $\ell(\tau)$ of a type τ is defined by $\ell(\tau_1 \to \tau_2) =$

$\max(1 + \ell(\tau_1), \ell(\tau_2))$, and $\ell(\mathbf{0}) = 0$. Thus $\mathbf{0}$ is the only type of level 0 and each type of level 1 is of the form $\mathbf{0}^n \to \mathbf{0}$ for some $n > 0$. A type $\tau_1 \to \cdots \to \tau_n \to \mathbf{0}$ is *homogeneous* (where $n \geq 0$) if each τ_i is homogeneous and $\ell(\tau_1) \geq \ell(\tau_2) \geq \ldots \geq \ell(\tau_n)$. For example $((\mathbf{0} \to \mathbf{0}) \to \mathbf{0}) \to (\mathbf{0} \to \mathbf{0}) \to (\mathbf{0} \to \mathbf{0} \to \mathbf{0}) \to \mathbf{0} \to \mathbf{0}$ is homogeneous, but $\mathbf{0} \to (\mathbf{0} \to \mathbf{0}) \to \mathbf{0}$ is not.

Higher–order terms. A *typed alphabet* is a set Γ of symbols with types in T. Thus Γ can be also presented as a T–indexed family $\{\Gamma_\tau\}_{\tau \in \mathrm{T}}$, where Γ_τ is the set of all symbols of Γ of type τ. We let the *type level* $\ell(\Gamma)$ of Γ be the supremum of $\ell(\tau)$, such that Γ_τ is nonempty. A *signature* is a typed alphabet of level 1.

Given a typed alphabet Γ, the set $T(\Gamma) = \{T(\Gamma)_\tau\}_{\tau \in \mathrm{T}}$ of *applicative terms* is defined inductively, by

(1) $\Gamma_\tau \subseteq T(\Gamma)_\tau$; (2) if $t \in T(\Gamma)_{\tau_1 \to \tau_2}$ and $s \in T(\Gamma)_{\tau_1}$ then $(ts) \in (\Gamma)_{\tau_2}$.

Note that each applicative term can be presented in a form $Zt_1 \ldots t_n$, where $n \geq 0$, $Z \in \Gamma$, and $t_1, \ldots, t_n$ are applicative terms. We say that a term $t \in T(\Gamma)_\tau$ is of type τ, which we also write $t : \tau$. We adopt the usual notational convention that application associates to the left, i.e. we write $t_0 t_1 \ldots t_n$ instead of $(\cdots((t_0 t_1) t_2) \cdots) t_n$.

Trees. The free monoid generated by a set X is written X^* and the empty word is written ε. The length of word $w \in X^*$ is denoted by $|w|$. A *tree* is any nonempty prefix–closed subset T of X^* (with ε considered as the *root*). If $u \in T$, $x \in X$, and $ux \in T$ then ux is an *immediate successor* of u in T. For $w \in T$, the set $T.w = \{v \in X^* : wv \in T\}$ is the *subtree* of T induced by w. Note that $T.w$ is also a tree, and $T.\varepsilon = T$.

Now let Σ be a signature and let $T \subseteq \omega^*$, where ω is the set of natural numbers, be a tree. A mapping $t : T \to \Sigma$ is called a Σ*–tree* provided that if $t(w) : \mathbf{0}^k \to \mathbf{0}$ then w has exactly k immediate successors which are $w1, \ldots, wk$ (hence w is a leaf whenever $t(w) : \mathbf{0}$). The set of Σ–trees is written $T^\infty(\Sigma)$.

If $t : T \to \Sigma$ is a Σ–tree, then T is called the *domain* of t and denoted by $T = \mathfrak{Dom}\ t$. For $v \in \mathfrak{Dom}\ t$, the *subtree* of t induced by v is a Σ–tree $t.v$ such that $\mathfrak{Dom}\ t.v = (\mathfrak{Dom}\ t).v$, and $t.v(w) = t(vw)$, for $w \in \mathfrak{Dom}\ t.v$. It is convenient to organize the set $T^\infty(\Sigma)$ into an algebra over the signature Σ, where for each $f \in \Sigma_{\mathbf{0}^n \to \mathbf{0}}$, the operation associated with f sends an n–tuple of trees $t_1, \ldots, t_n$ onto the unique tree t such that $t(\varepsilon) = f$ and $t.i = t_i$, for $i \in [n]$. (The notation $[n]$ abbreviates $\{1, \ldots, n\}$). Finite trees in $T^\infty(\Sigma)$ can be also identified with applicative terms of type $\mathbf{0}$ over the alphabet Σ in the usual manner.

We introduce a concept of limit. For a Σ–tree t, let $t{\restriction}n$ be its truncation to the level n, i.e., the restriction of the function t to the set $\{w \in \mathfrak{Dom}\ t : |w| \leq n\}$. Suppose $t_0, t_1, \ldots$ is a sequence of Σ–trees such that, for all k, there is an m, say $m(k)$, such that, for all $n, n' \geq m(k)$, $t_n{\restriction}k = t_{n'}{\restriction}k$. (This is a Cauchy condition in a suitable metric space of trees.) Then the *limit* of the sequence t_n, in symbols $\lim t_n$, is a Σ–tree t which is the set–theoretical union of the functions $t_n{\restriction}m(n)$ (understanding a function as a set of pairs).

Types as trees. Types in T can be identified with finite (unlabeled) binary trees. More specifically, we use the set of directions $\{p, q\}$, and let *tree*$(\tau_1 \to \tau_2)$ be the unique tree such that $tree(\tau_1 \to \tau_2).p = tree(\tau_1)$, $tree(\tau_1 \to \tau_2).q = tree(\tau_2)$ and $tree(\mathbf{0}) = \{\varepsilon\}$. In the sequel we will not make notational distinction between τ and $tree(\tau)$.

Monadic second–order logic. Let R be a *relational vocabulary*, i.e., a set of relational symbols, each r in R given with an arity $\rho(r) > 0$. The formulas of *monadic second order (MSO) logic* over vocabulary R use two kinds of variables : *individual variables* $x_0, x_1, \ldots$, and *set variables* $X_0, X_1, \ldots$. Atomic formulas are $x_i = x_j$, $r(x_{i_1}, \ldots, x_{i_{\rho(r)}})$, and $X_i(x_j)$. The other formulas are built using propositional connectives $\vee, \neg$, and the quantifier $\exists$ ranging over both kinds of variables. (The connectives $\wedge, \Rightarrow$, etc., as well as the quantifier $\forall$ are introduced in the usual way as abbreviations.) A formula without free variables is called a *sentence*. Formulas are interpreted in relational structures over the vocabulary R, which we usually present by $\mathbf{A} = \langle A, \{r^{\mathbf{A}} : r \in R\}\rangle$, where A is the *universe* of $\mathbf{A}$, and $r^{\mathbf{A}} \subseteq A^{\rho(r)}$ is a $\rho(r)$–ary relation on A. A *valuation* is a mapping v from the set of variables (of both kinds), such that $v(x_i) \in A$, and $v(X_i) \subseteq A$. The *satisfaction* of a formula φ in $\mathbf{A}$ under the valuation v, in symbols $\mathbf{A}, v \models \varphi$ is defined by induction on φ in the usual manner. The *monadic second–order theory* of $\mathbf{A}$ is the set of all MSO sentences satisfied in $\mathbf{A}$.

Let Σ be a signature and suppose that the maximum of the arities of symbols in Σ exists and equals m_Σ. A tree $t \in T^\infty(\Sigma)$ can be viewed as a logical structure $\mathbf{t}$, over the vocabulary $R_\Sigma = \{p_f : f \in \Sigma\} \cup \{d_i : 1 \le i \le m_\Sigma\}$, with $\rho(p_f) = 1$, and $\rho(d_i) = 2$:

$$\mathbf{t} = \langle \mathcal{D}om\, t,\ \{p_f^{\mathbf{t}} : f \in \Sigma\} \cup \{d_i^{\mathbf{t}} : 1 \le i \le m_\Sigma\}\rangle.$$

The universe of $\mathbf{t}$ is the domain of t, and the predicate symbols are interpreted by $p_f^{\mathbf{t}} = \{w \in \mathcal{D}om\, t : t(w) = f\}$, for $f \in \Sigma$, and $d_i^{\mathbf{t}} = \{(w, wi) : wi \in \mathcal{D}om\, t\}$, for $1 \le i \le m_\Sigma$. We refer the reader to [15] for a survey of the results on monadic second–order theory of trees.

2 Infinitary λ–Calculus

We will identify infinite λ–terms with certain infinite trees. More specifically, we fix a finite signature Σ and let $\Sigma^\perp = \Sigma \cup \{\perp\}$, where $\perp$ is a fresh symbol of type $\mathbf{0}$. All our finite and infinite terms, called *λ–trees* are simply typed and may involve constants from $\Sigma^\perp$, and variables from a fixed countably infinite set. In fact, we only consider λ–trees of types of level at most 1.

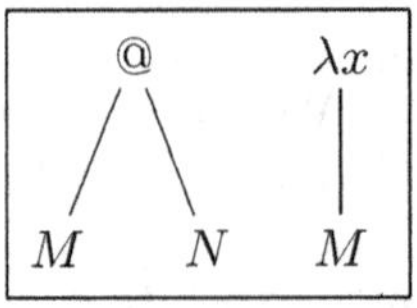

Fig. 1. Application and abstraction

Let Σ° be an infinite alphabet of level 1, consisting of a binary function symbol @, all symbols from $\Sigma^\perp$ as individual constants, regardless of their actual types, infinitely many individual variables as individual constants, unary function symbols λx for all variables x. The set of all *λ–trees* (over a signature Σ) is the greatest set of Σ°–trees, given together with their *types*, such that the following conditions hold.

- Each variable x is a λ–tree of type $\mathbf{0}$.
- Each function symbol $f \in \Sigma^{\perp}$ of type τ is a λ–tree of type τ.
- Otherwise each λ–tree is of type of level at most 1 and is either an *application* (MN) or an *abstraction* $(\lambda x.M)$ (see Fig. 1).
- If a λ–tree P of type τ is an application (MN) then M is a λ–tree of type $\mathbf{0} \to \tau$, and N is a λ–tree of type $\mathbf{0}$.
- If a λ–tree P of type τ has the form $(\lambda x.M)$, then $\tau = \mathbf{0} \to \sigma$, and M is a λ–tree of type σ.

Strictly speaking, the above is a co–inductive definition of the two–argument relation "M is a λ–tree of type τ". Formally, a λ–tree can be presented as a pair (M, τ), where M is a Σ°–tree, and τ is its type satisfying the conditions above. Whenever we talk about a "λ–tree" we actually mean a λ–tree together with its type.

Let M be a λ–tree and let x be a variable. Each node of M labeled x is called an *occurrence* of x in M. An occurrence of x is *bound* (resp. *free*), iff it has an (resp. no) ancestor labeled λx. The *binder* of this occurrence of x is the closest of all such ancestors λx. A variable x is *free* in a λ–tree M iff it has a free occurrence in M. The (possibly infinite) set of all free variables of M will be denoted by $FV(M)$. A λ–tree M with $FV(M) = \varnothing$ is called *closed*.

Definition 1. We call a λ–tree M *boundedly typed* if the set of types of all subterms of M is finite.

Clearly, ordinary λ–terms can be seen as a special case of λ–trees, and the notion of a free variable in a λ–tree generalizes the notion of a free variable in a λ–term. The *n–th approximant* of a λ–tree M, denoted $M{\restriction}n$ is defined by induction as follows:

- $M{\restriction}0 = \perp$, for all M;
- $(MN){\restriction}(n{+}1) = (M{\restriction}n)(N{\restriction}n)$
- $(\lambda x.M){\restriction}(n{+}1) = \lambda x(M{\restriction}n)$

That is, the n–th approximant is obtained by replacing all subtrees rooted at depth n by the constant $\perp$.

We denote by $M[x := N]$ the result of substitution of all free occurrences of x in M by N. The definition of the substitution of λ–trees is similar to that for ordinary λ–terms. (An α–conversion of some subterms of M may be necessary in order to avoid the capture of free variables of N.)

A *redex* in a λ–tree M is a subtree of the form $(\lambda x.P)Q$. The *contractum* of such a redex is of course $P[x := Q]$. We write $M \to_{\beta} N$ iff N is obtained from M by replacing a redex by its contractum. Note that infinite λ–trees, even simply typed, may have infinite reduction sequences, due to infinitely many redexes.

2.1 Paths in λ–Graphs

To each λ–tree M, we associate a *λ–graph* $G(M)$. Some edges of $G(M)$ are oriented and labeled either p or q. Other edges are non–oriented and unlabeled. To construct $G(M)$ we start with M where (for technical reasons) we add an additional node labeled "c" above each application, i.e., above any @–node.

(We refer to a node labeled by a symbol σ as to a σ–node.) We add an edge, oriented downward and labeled q from each c–node to the corresponding @–node. We also add an edge oriented upward and labeled p, form each bound occurrence of a variable to its binder. Since the α–equivalent λ–trees may be identified, without loss of information, we can replace all labels λx just by "λ" and all bound occurrences of variables by a uniform label "v". In addition we assign labels and orientation to the following existing edges. Each edge connecting the argument of an application with the corresponding @–node is labeled p and oriented upward. Each edge connecting the body of an abstraction with the corresponding λ–node is labeled q and oriented upward. Different cases of nodes and edges of $G(M)$ are depicted on Fig. 2.

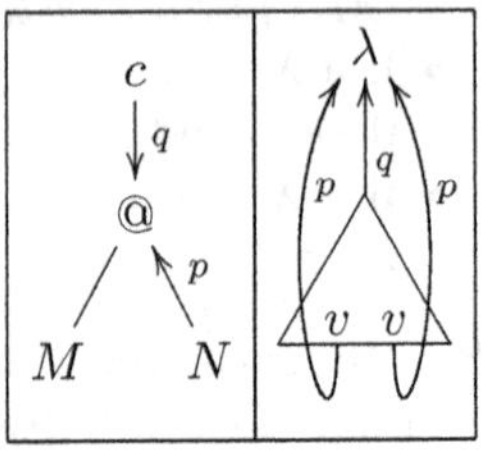

Fig. 2. Labeled application and abstraction

Each subterm N of M corresponds in an obvious way to a subgraph of $G(M)$, which will be denoted $G(N)$. Observe that nodes corresponding to free variables of N are connected to λ–nodes outside of $G(N)$, and these connecting edges are not part of $G(N)$. Each of these graphs $G(N)$ has a distinguished *entry* node (drawn at the top). The entry node of an abstraction is the appropriate λ–node, the entry node of an application is the additional c–node (not the @–node) and the entry node of a variable is the v–node itself. If confusion does not arise, we will write $\alpha := \alpha_N$ to mean that α is an entry node of a subterm N in M. Note that each node in $G(M)$, except for @–nodes, is an entry node of $G(N)$, for some subterm N of M.

Following the ideas of the Geometry of Interaction, see [2,3], we will now consider paths in λ–graphs. A sequence of adjacent edges in a λ–graph (possibly including the extra arcs from variable nodes to their binders), is called a *straight path* provided there is no backtracking, i.e., no edge is taken twice in a row and two edges connecting two different variable nodes with a λ–node which is their common binder may not directly follow one another. From now on by a *path* we always mean a straight path. Note that a path can pass an oriented edge forward (obeying the orientation) or backward (against the orientation).

Following [14], we will now consider a certain stateless pushdown automaton $\mathcal{P}$ walking on paths in a λ–graph. Informally, $\mathcal{P}$ moves along edges of a path Π, and each time it traverses a labeled edge forward or backward, a pushdown store (pds, for short) operation is performed (no pds operation if there is no label). Whenever we follow an arrow labeled p, we push p on the pds, and in order to traverse such an arrow backward, we must pop p from the pds. We proceed in an analogous way when we use edges labeled by q, forward or backward. In particular, in order to take a p–arrow backward, the top of the pds must be p, and similarly for q. This can be described more formally, considering that $\mathcal{P}$ works with the input alphabet $\{p\uparrow, p\downarrow, q\uparrow, q\downarrow, |\}$ ($|$ for a nonlabeled edge). A *configuration* of $\mathcal{P}$ is defined as a pair of the form (α, w) consisting of a node α and a word w representing the pds contents (top at the left). Note that once the path Π is fixed, the behaviour of $\mathcal{P}$ is fully determined by the initial contents of the pds.

An important property of $\mathcal{P}$ is that the contents of the pds in a configuration (α, w) may contain some information about the type of the subterm N whose entry node is α (i.e., $\alpha = \alpha_N$). Note that, according to our representation of types as trees, the subtree $\tau.w$ of a type τ is again a type.

Lemma 1. *Assume that $\mathcal{P}$ can move from (α_N, w) to (α_P, v), and that $N:\tau$ and $P:\sigma$. Suppose that $\tau.w$ is defined. Then $\sigma.v$ is defined and $\tau.w = \sigma.v$. Similarly, if $\sigma.v$ is defined then so is $\tau.w$ and the equality holds.*

In other words, the pds of $\mathcal{P}$ can be seen as a pointer to a certain subtype of the type of currently visited node. A crucial consequence of this fact is that if our λ–tree is boundedly typed (see Definition 1) then $\mathcal{P}$ can essentially be replaced by a finite automaton, if we only consider computations beginning with configurations (α_N, w), such that $\tau.w$ is defined, where τ is the type of N. Indeed, by Lemma 1, the type pointed to by the pds during the whole computation is $\tau' = \tau.w$. Now, if there are altogether only finitely many types in use, the type τ' can occur in all these types in a finite number of positions only. This means that there is only a finite number of possible values of the pds (uniformly bounded for a boundedly typed λ–term). Then we can convert the pds contents of $\mathcal{P}$ into states of a *finite* automaton. Thus we can compare computations of both automata in an obvious way.

We summarize above considerations in the following.

Proposition 1. *Suppose a λ-tree M is boundedly typed. There is a deterministic finite automaton $\mathcal{A}$ whose set of states is a subset of $\{p,q\}^*$, and whose computations along paths in $G(M)$ coincide with the computations of $\mathcal{P}$. Whenever $\mathcal{P}$ can traverse a path in $G(M)$, $\mathcal{A}$ can do it as well.*

It will now be convenient to define a computation path (of $\mathcal{P}$) more formally. A *computation path* Π in a λ-graph M is a finite or infinite sequence of configurations $(\alpha_0, w_0), (\alpha_1, w_1), \ldots$, such that the corresponding sequence of edges $(\alpha_0, \alpha_1), (\alpha_1, \alpha_2), \ldots$ forms a straight path (if Π is infinite, we mean that each initial segment is straight), and $w_0, w_1, \ldots$ are consecutive contents of the pds in $\mathcal{P}$'s computation along this path. Note that we allow a *trivial* computation path consisting of a single configuration (α_0, w_0) (no edges). A computation path is *maximal* if it is infinite or finite but cannot be extended to a longer computation path.

Now suppose a computation path Π ends in a configuration (α_N, w). If Π is nontrivial, there are two ways in which Π may reach the last configuration: It either comes from outside of $G(N)$ (i.e., the last but one node is not in $G(N)$) or from inside of $G(N)$. We will call Π *South–oriented* in the first case, and *North–oriented* in the second (as Π comes "from above" or "from below", respectively). If Π is trivial, and hence consists only of (α_N, w), we will qualify it as South–oriented if N is an application or a signature symbol of arity 0, and North–oriented if N is a signature symbol of arity > 0 (we do not care for other cases). Now, let $N : \tau$. We will say that Π is *properly ending* if the type $\tau.w$ is defined and equals $\mathbf{0}$, and moreover

- if Π is South–oriented then $w = q^n$, for some $n \geq 0$,
- if Π is North–oriented then $w = q^n p$, for some $n \geq 0$.

We are ready to state a lemma that will be crucial for further applications (see [10] for a proof).

Lemma 2. *Let M be a closed λ–tree of type $\mathbf{0}$.*

(1) *There is exactly one maximal computation path Π starting from configuration (α_M, ε). If Π is finite then it must end in a configuration (α_g, q^n), for some signature symbol g of arity n.*
(2) *Let $\alpha = \alpha_f$, be a node of M, where f is a signature symbol of arity $k > 0$, and let $i \in [k]$. There is exactly one maximal computation path Π starting from configuration $(\alpha_f, q^{i-1}p)$. If Π is finite then it must end in a configuration (α_g, q^n), for some signature symbol g of arity n.*

2.2 Derived Trees

An infinite tree in $T^{\infty}(\Sigma^{\perp})$ can be viewed as an infinite λ–term in a natural way if we read $f(t_1, \ldots, t_k)$ as the nested application $(\ldots((f t_1) t_2) \ldots t_k)$.

Conversely, we will show a method to derive a tree in $T^{\infty}(\Sigma^{\perp})$ from a closed boundedly typed λ–tree M of type $\mathbf{0}$. Intuitively, this will be a tree to which M eventually evaluates, after performing all β–reductions. However, not all branches of reduction need converge, and therefore we will sometimes put $\perp$ instead of a signature symbol.

A tree $t_M : \mathfrak{D}\mathrm{om}\, t_M \to \Sigma^{\perp}$ will be defined along with a partial mapping I_M from $\mathfrak{D}\mathrm{om}\, t_M$ to $G(M)$ in such a way that the label of $I_M(w)$ in $G(M)$ coincides with $t_M(w)$. More specifically, the domain of I_M will be $\mathfrak{D}\mathrm{om}_{+} t_M = \{w \in \mathfrak{D}\mathrm{om}\, t_M : t_M(w) \neq \perp\}$. At first, consider the maximal computation path Π in $G(M)$ starting from (α_M, ε) (cf. Lemma 2). If it is infinite, we let $t_M = \perp$ and $I_M = \varnothing$. Otherwise, again by Lemma 2, Π ends in a node labeled by a signature symbol. We call this node the *source of* M and denote it by s_M. We let $I_M : \varepsilon \mapsto s_M$, and $t_M(\varepsilon) = f$, where f is the label of s_M. Now suppose I_M and t_M are defined for a node w, and $t_M(w) = g$, $I_M(w) = \alpha = \alpha_g$, where g is a signature symbol of arity k. For each $i = 1, \ldots, k$, we consider the maximal computation path Π_i in $G(M)$ starting from $(\alpha, q^{i-1}p)$. By Lemma 2, the path Π_i is well defined, and if it is finite, the last node, say α_i, is labeled by a signature symbol, say g_i. Then we define t_M and (possibly) I_M for the k successors of w, by $t_M(wi) = g_i$ and $I_M(wi) = \alpha_i$, if Π_i is finite, and $t_M(wi) = \perp$, otherwise.

For Sect. 3, we also need an extension I^*_M of I_M defined on the whole $\mathfrak{D}\mathrm{om}\, t$. Without loss of generality, we can assume that the root ε of $G(M)$ is not labeled by a signature symbol.[1] We let $I^*_M(w) = I_M(w)$ if $I_M(w)$ is defined and $I^*_M(w) = \varepsilon$ otherwise.

[1] Otherwise $G(M)$ consists of a single node labeled by a constant, and all the results in the sequel become trivial. We choose ε for concreteness, but any other node not in $I_M(\mathfrak{D}\mathrm{om}_{+} t_M)$ could be used instead.

2.3 Beta Reduction

We will now examine how a β–reduction step applied to a λ–tree M may affect a path Π in $G(M)$. Suppose M_1 is obtained from M by a β–reduction replacing a redex $(\lambda y.A)B$ by $A[y := B]$. In order to transform $G(M)$ into $G(M_1)$ one finds out all the variable nodes bound by the λ–node at the top of $G(\lambda y.A)$. Then the subgraph $G((\lambda y.A)B)$ is replaced by $G(A)$ where all such variable nodes are replaced by copies of the whole graph $G(B)$. We consider only the case when y

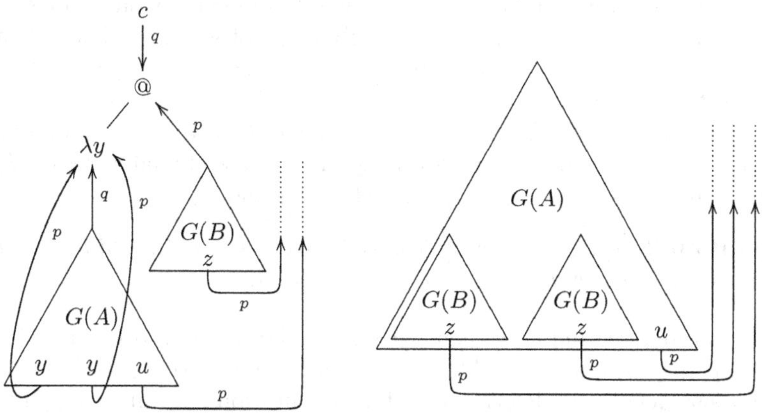

(a) Before reduction (b) After reduction

Fig. 3.

is free in A, the other case is easier and left to the reader. Fig. 3(a) presents a redex, and Fig. 3(b) shows its contractum. Consider a weakly regular path Π_1 in M_1. We define a path Π in M as follows:

- Outside of the redex the path is unchanged.
- The same for portions of the path within $G(A[y := B])$ but outside of $G(B)$.
- Each copy of $G(B)$ in $G(A[y := B])$ is represented in the redex by the single argument $G(B)$. Every portion of the path Π_1 that goes through any of the multiple $G(B)$'s in $G(A[y := B])$ is replaced by an identical portion going through $G(B)$ in the redex.
- Whenever Π_1 enters $G(A[y := B])$ through its top node, the path Π enters $G((\lambda y.A)B)$ through its top node (a q–arrow), then goes to $G(\lambda y.A)$ and takes the q–arrow backward to enter $G(A)$. (Note that there is no other choice.)
- Whenever Π_1 enters $G(B)$ through its top node the path Π reaches a v–node within $G(A)$. Then it must go to the argument $G(B)$, traversing p and then p backward.
- Whenever Π_1 enters or leaves $G(A[y := B])$ through a variable (i.e., via an edge between a variable node to its binder), the path Π enters or leaves the redex through the corresponding variable. (Note that no variable free in B can be bound in $A[y := B]$.)

In this case we say that Π_1 is a *deformation* of Π.

Suppose now that $M \to_\beta N$. Then every node β of $G(N)$ may be seen as obtained from a node α of $G(M)$. We say that β is an *offset* of α. This notion should be obvious up to one specific point: the entry node to a contractum of a redex (the entry to $G(A)$ at Fig. 3(b)) should be considered an offset of the entry node of the body of the abstraction (the entry to $G(A)$ at Fig. 3(a)) and *not* of the entry node of the redex or of the abstraction.

In this way we can say that a node of $G(M)$ may have one or more offsets or no offset at all, but each node of $G(N)$ is an offset of exactly one node of $G(M)$. In addition, a variable or constant may only be an offset of an identical variable or constant. It should be obvious that the type associated to a node and to its offset must be the same.

Let a path Π_1 in $G(N)$ from node β_1 to node β_2 be a deformation of a path Π in $G(M)$. Let α_1 and α_2 be respectively the initial and final nodes of Π. Then of course β_1 and β_2 are respectively offsets of α_1 and α_2.

Proposition 2. *Let M and N be closed λ–trees of level $\mathbf{0}$, and let t_M and t_N be the respective derived trees. If $M \to_\beta N$ then $t_M = t_N$.*

Proof. Consider the inductive construction of the tree t_N, as described in Section 2.2. It may be readily established that each computation path in $G(N)$ used in this construction is a deformation of a computation path in $G(M)$. If the former is infinite, the latter must be infinite too. If the former reaches a signature symbol f, so must the latter, because α_f must be an offset of α_f. Hence the result follows by induction on the levels of the tree t_N. □

3 Moving between MSO Theories

Let M be a closed boundedly typed λ–tree of level $\mathbf{0}$. We are going to show that the MSO theory of the derived tree t_M can be interpreted in the MSO theory of $G(M)$, viewed as a specific logical structure $\mathbf{G}_M$ defined below. We shall see that both structures are bisimilar in the usual process–theoretic sense (see e.g. Definition 6.3.10 of [1]). By composing several well–known facts about the propositional modal μ–calculus (see e.g. [1,12]), we can establish the following.

Proposition 3. *There is a recursive mapping ρ of MSO sentences such that for every MSO sentence φ, every tree $t \in T^\infty(\Sigma)$ and every countable structure $\mathbf{A}$ which is bisimilar to $\mathbf{t}$, the following holds:* $\mathbf{t} \models \varphi \iff \mathbf{A} \models \rho(\varphi)$.

We define $\Sigma^\bullet := \Sigma \cup \{@, \lambda, c, v\}$, where the symbol @ is binary, λ and c are unary, and v, as well as all symbols from Σ are 0–ary. Recall that $G(M)$ is a tree over $\Sigma^\bullet$ additionally equipped with the edges from v–nodes to their binders (λ–nodes). Let us denote the domain of $G(M)$ by W. We consider the structure $\mathbf{G}_M = \langle W, \{p_f^{\mathbf{G}_M} \mid f \in \Sigma^\bullet\} \cup \{d_i^{\mathbf{G}_M} \mid 1 \le i \le m_{\Sigma^\bullet}\} \cup \{E^{\mathbf{G}_M}\}\rangle$ where $p_f^{\mathbf{G}_M}$ and $d_i^{\mathbf{t}}$ are defined as in Section 1, and $(u, w) \in E^{\mathbf{G}_M}$, whenever $G(M)(u) = v$, $G(M)(w) = \lambda$, and there is an edge in $G(M)$ from u to w.

We now define the structure $\mathbf{I}_M$ over the same vocabulary as $\mathbf{t}_M$, of universe $I^*_M(\mathfrak{Dom}\ t_M)$ by letting, for $f \in \Sigma^\perp$, we let $p_f^{\mathbf{I}_M} = \{I^*_M(w) : t_M(w) = f\}$, and, for $1 \le i \le m_\Sigma$, $d_i^{\mathbf{I}_M} = \{(I^*_M(w), I^*_M(wi)) : wi \in \mathfrak{Dom}\ t_M\}$. Clearly I^*_M is an epimorphism from $\mathbf{t}_M$ onto $\mathbf{I}_M$, and moreover $(I^*_M(w), I^*_M(v)) \in d_i^{\mathbf{I}_M}$ implies $(w, v) \in d_i^{\mathbf{t}_M}$, and $I^*_M(w) \in p_f^{\mathbf{I}_M}$ implies $w \in p_f^{\mathbf{t}_M}$. This follows that $\mathbf{t}_M$ and $\mathbf{I}_M$ are bisimilar as transition systems.

The next lemma allows to accomplish the interpretation of the MSO theory of $\mathbf{t}_M$ in that of $\mathbf{G}_M$.

Lemma 3. *For any MSO sentence φ, one can construct effectively an MSO sentence ψ such that $\mathbf{I}_M \models \varphi$ if and only if $\mathbf{G}_M \models \psi$.*

Proof. It is enough to interpret the structure $\mathbf{I}_M$ in the MSO theory of $\mathbf{G}_M$. Since the universe of $\mathbf{I}_M$ is already a subset of the universe of $G(M)$, it is enough to write MSO formulas, say $Uni(x)$, $P_f(x)$, for $f \in \Sigma^\perp$, and $D_i(x, y)$, for $1 \le i \le m_\Sigma$, defining the relations $I^*_M(\mathfrak{Dom}\ t_M)$, $p_f^{\mathbf{I}_M}$, and $d_i^{\mathbf{I}_M}$, respectively.

The formulas $P_f(x)$ are obvious. To write formulas $D_i(x, y)$, the key point is to express a property "there is a finite computation path Π starting from configuration $(\alpha_f, q^{i-1}p)$ (where f is a signature symbol of arity k), and ending in a node labeled by a signature symbol". Note that such a path must be maximal, and hence, by Lemma 2, there is at most one such path. Moreover, by Proposition 1, this computation can be carried by a finite automaton. (It follows easily from Lemma 2 that the computation empties pds at least once.) Therefore, it is routine to express the desired property by the known techniques, see Caucal [5]. The existence of an infinite computation path can be expressed by negation of the existence of finite maximal paths. The argument for expressing a computation path starting from (α_M, ε) is similar. This allows to write formulas $D_i(x, y)$. Using formulas $P_f(x)$ and $D_i(x, y)$, it is routine to write the desired formula $Uni(x)$. □

By combining Proposition 3 and Lemma 3, we get the following result.

Theorem 2. *Let M be a closed boundedly typed λ–tree of type $\mathbf{0}$, and let $t_M \in T^\infty(\Sigma^\perp)$ be the tree derived from M. Then the MSO theory of $\mathbf{t}_M$ is reducible to the MSO theory of $\mathbf{G}_M$, that is, there exists a recursive mapping of sentences $\varphi \mapsto \varphi'$ such that $\mathbf{t}_M \models \varphi$ iff $\mathbf{G}_M \models \varphi'$.*

4 Grammars

We now fix two disjoint typed alphabets, $N = \{N_\tau\}_{\tau \in \mathrm{T}}$ and $\mathrm{X} = \{\mathrm{X}_\tau\}_{\tau \in \mathrm{T}}$ of *nonterminals* and *variables* (or *parameters*), respectively. A *grammar* is a tuple $\mathcal{G} = (\Sigma, V, S, E)$, where Σ is a signature, $V \subseteq N$ is a finite set of nonterminals, $S \in V$ is a *start symbol* of type $\mathbf{0}$, and E is a set of equations of the form $F z_1 \ldots z_m = w$, where $F : \tau_1 \to \cdots \to \tau_m \to \mathbf{0}$ is a nonterminal in V, z_i is a variable of type τ_i, and w is an applicative term in $T(\Sigma \cup V \cup \{z_1 \ldots z_m\})$.

We assume that for each $F \in V$, there is exactly one equation in E with F occurring on the left hand side. Furthermore, we make a *proviso* that each nonterminal in a grammar has a homogeneous type, and that if $m \geq 1$ then $\tau_m = \mathbf{0}$. This implies that each nonterminal of level > 0 has at least one parameter of level 0 (which need not, of course, occur at the right–hand side). The *level* of a grammar is the highest level of its nonterminals.

In this paper, we are interested in grammars as generators of Σ–trees. First, for any applicative term t over $\Sigma \cup V$, let $t^\perp$ be the result of replacing in t each nonterminal, together with its arguments, by $\perp$. (Formally, $t^\perp$ is defined by induction: $f^\perp = f$, for $f \in \Sigma$, $X^\perp = \perp$, for $X \in V$, and $(sr)^\perp = (s^\perp r^\perp)$ if $s^\perp \neq \perp$, otherwise $(sr)^\perp = \perp$.) It is easy to see that if t is an applicative term (over $\Sigma \cup V$) of type $\mathbf{0}$ then $t^\perp$ is an applicative term over $\Sigma^\perp$ of type $\mathbf{0}$. Recall that applicative terms over $\Sigma^\perp$ of type $\mathbf{0}$ can be identified with finite trees.

We will now define the single–step rewriting relation $\to_{\mathcal{G}}$ among the terms over $\Sigma \cup V$. Informally speaking, $t \to_{\mathcal{G}} t'$ whenever t' is obtained from t by replacing some occurrence of a nonterminal F by the right–hand side of the appropriate equation in which all parameters are in turn replaced by the actual arguments of F. Such a replacement is allowed only if F occurs as a head of a subterm of type $\mathbf{0}$. More precisely, the relation $\to_{\mathcal{G}} \subseteq T(\Sigma \cup V) \times T(\Sigma \cup V)$ is defined inductively by the following clauses.

- $Ft_1 \ldots t_k \to_{\mathcal{G}} t[z_1{:=}t_1, \ldots, z_k{:=}t_k]$ if there is an equation $Fz_1 \ldots z_k = t$ (with $z_i : \rho_i$, $i = 1, \ldots, k$), and $t_i \in T(\Sigma \cup V)_{\rho_i}$, for $i = 1, \ldots, k$.
- If $t \to_{\mathcal{G}} t'$ then $(st) \to_{\mathcal{G}} (st')$ and $(tq) \to_{\mathcal{G}} (t'q)$, whenever the expressions in question are applicative terms.

A *reduction* is a finite or infinite sequence $t_0 \to_{\mathcal{G}} t_1 \to_{\mathcal{G}} \ldots$ of terms in $T(\Sigma \cup V)$. We define the relation $t \twoheadrightarrow_{\mathcal{G}}^{\infty} t'$, where t is an applicative term in $T(\Sigma \cup V)$ and t' is a tree in $T^\infty(\Sigma^\perp)$, by

- t' is a finite tree, and there is a finite reduction sequence $t = t_0 \to_{\mathcal{G}} \ldots \to_{\mathcal{G}} t_n = t'$, or
- t' is infinite, and there is an infinite reduction sequence $t = t_0 \to_{\mathcal{G}} t_1 \to_{\mathcal{G}} \ldots$ such that $t' = \lim t_n^\perp$.

To define a unique tree produced by the grammar, we recall a standard *approximation ordering* on $T^\infty(\Sigma^\perp)$: $t' \sqsubseteq t$ if $\mathfrak{D}\mathrm{om}\, t' \subseteq \mathfrak{D}\mathrm{om}\, t$ and, for each $w \in \mathfrak{D}\mathrm{om}\, t'$, $t'(w) = t(w)$ or $t'(w) = \perp$. (In other words, t' is obtained from t by replacing some of its subtrees by $\perp$.) Then we let $[\![\mathcal{G}]\!] = \sup\{t \in T^\infty(\Sigma^\perp) : S \twoheadrightarrow_{\mathcal{G}}^{\infty} t\}$. It is easy to see that, by the Church–Rosser property of our grammar, the above set is directed, and hence $[\![\mathcal{G}]\!]$ is well defined since $T^\infty(\Sigma^\perp)$ with the approximation ordering is a cpo. Furthermore, it is routine to show that if an infinite reduction $S = t_0 \to_{\mathcal{G}} t_1 \to_{\mathcal{G}} \ldots$ is *fair*, i.e., any occurrence of a nonterminal symbol is eventually rewritten, then its result $t' = \lim t_n^\perp$ is $[\![\mathcal{G}]\!]$.

From grammar terms to λ–trees. Given a grammar $\mathcal{G}$, we define a map $\daleth_{\mathcal{G}}$ of $T(\Sigma \cup V \cup \mathrm{X})$ into the set of λ–trees (over Σ) such that

(1) $\daleth_{\mathcal{G}}(t) = f$, if t is a function symbol $f \in \Sigma$,
(2) $\daleth_{\mathcal{G}}(t) = x$, if t is a variable $x \in \mathrm{X}_{\mathbf{0}}$,
(3) $\daleth_{\mathcal{G}}(t) = \lambda x'_1 \dots x'_n . \daleth_{\mathcal{G}}(r[\phi_1{:=}t_1, \dots, \phi_m{:=}t_m, x_1{:=}x'_1, \dots, x_n{:=}x'_n])$, where the variables $x'_1, \dots, x'_n$ are chosen so that no x'_i occurs free in any of t_j, if $t = Ft_1 \dots t_m$, $F\phi_1 \dots \phi_m x_1 \dots x_n = r$ is an equation of $\mathcal{G}$ and $\mathrm{type}(\phi_i) = \mathrm{type}(t_i)$ for $i \in [m]$.
(4) $\daleth_{\mathcal{G}}(t) = \daleth_{\mathcal{G}}(t_1)\daleth_{\mathcal{G}}(t_2)$, if $t = t_1 t_2$ where $t_1 : \mathbf{0} \to \tau$ and $t_2 : \mathbf{0}$.

It is a routine exercise to prove that $\daleth_{\mathcal{G}}$ is well defined. To this end one may first define (co–inductively) an appropriate relation, establish its functionality and show that it corresponds to the above definition of $\daleth_{\mathcal{G}}$.

We have the following characterization of $[\![\mathcal{G}]\!]$ in terms of operation $\daleth_{\mathcal{G}}$ and derivation of trees from λ–graphs (see [10] for a proof).

Proposition 4. *Let* $M = \daleth_{\mathcal{G}}(S)$. *Then* $t_M = [\![\mathcal{G}]\!]$.

5 Decidability Result

By Proposition 4 and Theorem 2, the decision problem of the MSO theory of a tree generated by a grammar reduces to that of the graph $G(\daleth_{\mathcal{G}}(S))$. We are now interested in generating, in a sense, the last graph by a grammar of level 1. Note, however, that the underlying tree structure of $G(M)$ does not keep the complete information about the tree M. Indeed, while converting a λ–tree M into a graph $G(M)$ we have replaced (possibly) infinite number of labels λx_i and x_i, by only two labels, λ and v, at the expense of introducing "back edges". One might expect that these back edges are MSO definable in the underlying tree structure of $G(M)$, but it is not always the case. A good situation occurs if in part (3) of the definition of $\daleth_{\mathcal{G}}$ we need not to rename the bound variables (i.e., we can take $x'_i := x_i$, for $i = 1, \dots, m$).

Definition 3. Let $\mathcal{G}$ be a grammar of level 2. We call the grammar *unsafe* if there are two equations (not necessarily distinct) $F\phi_1 \dots \phi_m x_1 \dots x_n = r$ and $F'\phi'_1 \dots \phi'_{m'} x'_1 \dots x'_{n'} = r'$ (where the ϕ's are of level 1 and the x's of level 0) such that r has a subterm $F't_1, \dots t_{m'}$, such that some variable x_i occurs free in some term t_j. Otherwise the grammar is *safe*. (Note that in the above, x_i may occur in arguments of F' of type $\mathbf{0}$, but not in those of level 1.)

It is easy to see that if a grammar is safe then in the definition on $\daleth_{\mathcal{G}}(S)$ we are not obliged to introduce any new variables.

Let $\mathcal{G} = (\Sigma, V, S, E)$ be a safe grammar of level 2. We may assume that the parameters of type $\mathbf{0}$ occurring in distinct equations are different. Let $\mathrm{X}^{\mathcal{G}_0} = \{x_1, \dots, x_L\}$ be the set of all parameters of type $\mathbf{0}$ occurring in grammar $\mathcal{G}$. We define an *algebraic* (i.e., level 1) grammar $\mathcal{G}^\alpha = (\Sigma^\alpha, V^\alpha, S^\alpha, E^\alpha)$ as follows.

First we define a translation α of (homogeneous) types of level 2 to types of level 1 that maps $(\mathbf{0}^k \to \mathbf{0})$ to $\mathbf{0}$, and $(\mathbf{0}^{k_1} \to \mathbf{0}) \to \dots \to (\mathbf{0}^{k_m} \to \mathbf{0}) \to \mathbf{0}^\ell \to \mathbf{0}$ to $\mathbf{0}^m \to \mathbf{0}$. We will denote $\alpha(\tau)$ by τ^α. Let $\Sigma^\alpha = \Sigma \cup \{@, c, \lambda x_1, \dots, \lambda x_L, x_1, \dots, x_L\}$, where all symbols from Σ as well as (former) parameters x_i are considered constant, the symbol @ is binary, and the symbol c as well as all symbols λx_i are unary. Now, for a typed term $r:\tau$ over signature Σ, we define a term $r^\alpha : \tau^\alpha$ over Σ^α, as follows:

- F^α, for a variable $F:\tau$, is a fresh variable of type τ^α,
- $s^\alpha = s$ for each parameter of G (thus parameters of level 0 become constants, and parameters of level 1 change their types to $\mathbf{0}$),
- if $r = Ft_1 \dots t_m$ then $r^\alpha = F^\alpha t_1^\alpha \dots t_m^\alpha$, whenever F is a nonterminal of type $(\mathbf{0}^{k_1} \to \mathbf{0}) \to \dots \to (\mathbf{0}^{k_m} \to \mathbf{0}) \to \mathbf{0}^\ell \to \mathbf{0}$,
- if $r = (ts)$ with $s:\mathbf{0}$ then $r^\alpha = c(@t^\alpha s^\alpha)$.

Now $E^\alpha := \{F^\alpha \phi_1 \dots \phi_m = \lambda x_1 \dots \lambda x_n . r^\alpha \mid F\phi_1 \dots \phi_m x_1 \dots x_n = r \in E\}$ (where the ϕ's are of level 1 and the x's of level 0) and $V^\alpha = \{F^\alpha : F \in V\}$ which completes the definition of $\mathcal{G}^\alpha$.

Now let $t^\alpha = [\![\mathcal{G}^\alpha]\!]$ be the tree over $\Sigma^{\alpha\perp}$ generated by $\mathcal{G}^\alpha$, and let $\mathbf{t}^\alpha$ be the logical structure associated with it. We transform $\mathbf{t}^\alpha$ into a structure $\mathbf{t}_0^\alpha$ over the vocabulary $\{p_f : f \in \Sigma^\bullet\} \cup \{d_i \,|\, 1 \le i \le m_{\Sigma^\alpha}\} \cup \{E\}$ as follows. The universe remains the same as in $\mathbf{t}^\alpha$ as well as the interpretation of symbols d_i and p_f, for f different from λx_i and x_i. Furthermore, $w \in p_\lambda^{\mathbf{t}_0^\alpha}$ whenever $w \in p_{\lambda x_i}^{\mathbf{t}^\alpha}$, and $w \in p_v^{\mathbf{t}_0^\alpha}$ whenever $w \in p_{x_i}^{\mathbf{t}^\alpha}$, for some $x_i \in \mathrm{X}^{\mathcal{G}_0}$. Finally, we let $(u, w) \in E^{\mathbf{t}_0^\alpha}$ whenever w is binder of u, i.e., $t^\alpha(u) = x_i$, $t^\alpha(w) = \lambda x_i$, and w is the closest ancestor of u labeled by λx_i.

Lemma 4. *The structure $\mathbf{t}_0^\alpha$ is MSO definable in the structure $\mathbf{t}^\alpha$.*

Furthermore we claim the following.

Lemma 5. *Let grammars $\mathcal{G}$ and $\mathcal{G}^\alpha$ be as above, and let $M = \daleth_{\mathcal{G}}(S)$. Then the structure $\mathbf{t}_0^\alpha$ coincides with the structure $\mathbf{G}_M$ defined for M as in Section 3.*

We conclude by the main result of the paper.

Theorem 4. *Let $\mathcal{G}$ be a safe grammar of level 2. Then the monadic theory of $[\![\mathcal{G}]\!]$ is decidable.*

Proof. Since the tree $\mathbf{t}^\alpha$ is algebraic, its MSO theory is decidable, by the result of Courcelle [6]. Let $M = \daleth_{\mathcal{G}}(S)$. By Lemmas 5 and 4, the MSO theory of $\mathbf{G}_M$ is decidable. By Proposition 4, $[\![\mathcal{G}]\!] = t_M$. It is easy to see that, by construction, M is boundedly typed. Hence the result follows from Theorem 2. □

Example 1. Let f, g, c be signature symbols of arity 2,1,0, respectively. Consider a grammar of level 2 with nonterminals $S:\mathbf{0}$, and $F, G : (\mathbf{0} \to \mathbf{0}) \to \mathbf{0} \to \mathbf{0}$, and equations

$$S = Fgc \qquad F\varphi x = f\left(F(G\varphi)\,(\varphi x)\right) x \qquad G\psi y = \psi\,(\psi y)$$

It is easy to see that this grammar generates a tree t with $\mathfrak{Dom}\, t = \{1^n 2^m : m \le 2^n\}$, such that $t(1^n) = f$, $t(1^n 2^{2^n}) = c$, and $t(w) = g$ otherwise. Since $\mathfrak{Dom}\, t$ considered as a language is not context-free, the tree t is not algebraic (see [6]). Since the grammar is safe, the decidability of the MSO of $\mathbf{t}$ follows from Theorem 4.

References

1. A. Arnold and D. Niwiński. *Rudiments of μ–calculus.* Elsevier, 2001.
2. A. Asperti, V. Danos, C. Laneve, and L. Regnier. Paths in the lambda-calculus. In *Proc. 9th IEEE Symp. on Logic in Comput. Sci.*, pages 426–436, 1994.
3. A. Asperti and S. Guerrini. The optimal implementation of functional programming languages. In *Cambridge Tracts in Theoretical Computer Science*, volume 45. Cambridge University Press, 1998.
4. A. Berarducci and M. Dezani-Ciancaglini. Infinite lambda-calculus and types. *Theoret. Comput. Sci.*, 212:29–75, 1999.
5. D. Caucal. On infinite transition graphs having a decidable monadic second–order theory. In F. M. auf der Heide and B. Monien, editors, *23th International Colloquium on Automata Languages and Programming*, LNCS 1099, pages 194–205, 1996. A long version will appear in TCS.
6. B. Courcelle. The monadic second–order theory of graphs IX: Machines and their behaviours. *Theoretical Comput. Sci.*, 151:125–162, 1995.
7. W. Damm. The IO– and OI–hierarchies. *Theoretical Comput. Sci.*, 20(2):95–208, 1982.
8. H. Hungar. Model checking and higher-order recursion. In L. P. M. Kutyłowski and T. Wierzbicki, editors, *Mathematical Foundations of Computer Science 1999*, LNCS 1672, pages 149–159, 1999.
9. J. R. Kennaway, J. W. Klop, M. R. Sleep, and F. J. de Vries. Infinitary lambda calculus. *Theoret. Comput. Sci.*, 175:93–125, 1997.
10. T. Knapik, D. Niwiński, and P. Urzyczyn. Deciding monadic theories of hyperalgebraic trees, 2001. `http://www.univ-reunion.fr/~knapik/publications/`
11. O. Kupferman and M. Vardi. An automata-theoretic approach to reasoning about infinite-state systems. In *Computer Aided Verification, Proc. 12th Int. Conference*, Lecture Notes in Computer Science. Springer-Verlag, 2000.
12. D. Niwiński. Fixed points characterization of infinite behaviour of finite state systems. *Theoretical Comput. Sci.*, 189:1–69, 1997.
13. M. O. Rabin. Decidability of second-order theories and automata on infinite trees. *Trans. Amer. Soc*, 141:1–35, 1969.
14. Z. Spławski and P. Urzyczyn. Type fixpoints: iteration vs. recursion. In *Proc. 4th ICPF*, pages 102–113. ACM, 1999.
15. W. Thomas. Languages, automata, and logic. In G. Rozenberg and A. Salomaa, editors, *Handbook of Formal Languages*, volume 3, pages 389–455. Springer-Verlag, 1997.
16. I. Walukiewicz. Monadic second-order logic on tree-like structures. In C. Puech and R. Reischuk, editors, *Proc. STACS '96*, pages 401–414. Lect. Notes Comput. Sci. 1046, 1996.

A Deconstruction of Non-deterministic Classical Cut Elimination

J. Laird

COGS, University of Sussex, UK
e-mail: jiml@cogs.susx.ac.uk

Abstract. This paper shows how a symmetric and non-deterministic cut elimination procedure for a classical sequent calculus can be faithfully simulated using a non-deterministic choice operator to combine different 'double-negation' translations of each cut. The resulting interpretation of classical proofs in a λ-calculus with non-deterministic choice leads to a simple proof of termination for cut elimination.

1 Introduction

The problem faced when analysing classical logic from a computational perspective is that the very wildness which makes it hard to understand is also what makes it interesting — its symmetries cannot be disentangled from the non-deterministic behaviour of cut elimination, and this non-determinism is a formidable obstacle to proof theoretical, semantic, or computational interpretations of classical proofs.

Previous analyses, such as [10, 16, 11, 7] have often resolved this problem by 'controlling' the non-determinism of classical cuts out of existence by predetermining their behaviour, either systematically or by attaching additional information to proofs. This permits an interpretation in terms of a deterministic system such as double negation translation [13], control operators [14] or 'linear decoration' [7]. Although these interpretations have generated many key insights (ingredients of the approach described here) they seem inevitably to lose some proof-theoretic content because only a limited selection of the possible cut elimination behaviours can be pre-determined (see [2, 17] and the discussion in Section 1.2 below). On the other hand, more general symmetric and non-deterministic cut elimination and normalisation procedures have been *described* via rewriting systems for term-annotations of classical proofs [16, 2, 17] but without comparable analysis in terms of simpler or more well-behaved logics. The work reported here is an attempt to make a connection between these two approaches by further examination of the choices encountered in classical cut elimination, leading to a "deconstructive" translation into intuitionistic logic plus non-determinism.

1.1 Contribution and Organization of the Paper

The primary objective of this paper is to describe a 'double-negation' translation on propositional formulas and proofs of **LK** which is sound with respect

S. Abramsky (Ed.): TLCA 2001, LNCS 2044, pp. 268–282, 2001.

$$\frac{}{\alpha,\alpha^\perp}\text{AXIOM} \qquad \frac{\Gamma,A \quad \Gamma,A^\perp}{\Gamma}\text{CUT}$$

$$\frac{\Gamma}{\Gamma,A}\text{WK} \qquad \frac{\Gamma,A,A}{\Gamma,A}\text{ CON}$$

$$\frac{\Gamma,A \quad \Gamma,B}{\Gamma,A\wedge B}\text{ AND} \qquad \frac{\Gamma,A_i}{\Gamma,A_1\vee A_2}\text{ OR}_i\text{: } i=1,2$$

Table 1. Additive **LK** with multiset sequents

to a simple and symmetric cut elimination protocol, and thereby preserves non-determinism and makes it explicit. The proof theoretical dividend of this translation is immediate — a simple proof of strong normalisation for cut elimination (moreover, one which extends readily to second order). Less directly, the translation suggests new semantic and computational interpretations of classical logic.

The remainder of Section 1 consists of a discussion of classical cut elimination and non-determinism. In Section 2 a cut elimination procedure for additive (propositional) **LK** is formally defined by reduction of annotating terms of the symmetric λ-calculus, λ^{Sym}, of Barbanera and Berardi [2]. Section 3 describes $\mathbf{LK}_\uparrow$, a version of **LK** which incorporates a non-logical rule into proofs which determines how cuts will be eliminated. This allows the non-determinism inherent in cut elimination to be presented in terms of a choice between different proofs of the same formula. Proofs of $\mathbf{LK}_\uparrow$ can be annotated with terms of an ordinary simply-typed λ-calculus as a form of "double-negation-translation". In Section 4, **LK** proofs (represented as λ^{Sym} terms) are translated into the λ-calculus extended with an erratic choice operator, by combining different $\mathbf{LK}_\uparrow$ disambiguations of each cut. Soundness of the translation implies strong-normalisation of the cut elimination procedure. There is also a stronger soundness result; the normal forms of the translation of a term which annotates a proof are precisely the translations of the terms which annotate its cut free forms.

1.2 Cut Elimination and Non-determinism

Definition 1 (Sequent Calculus LK). *Formulas of the propositional calculus are given in 'negation normal form' — they are generated from a set of literals (atoms, $\alpha, \beta, \ldots$ and negated atoms $\alpha^\perp, \beta^\perp, \ldots$) by conjunction and disjunction:*

$$A ::= \alpha \mid \alpha^\perp \mid A \wedge A \mid A \vee A$$

Negation is defined by involutivity and deMorgan duality: i.e.
$(\alpha^\perp)^\perp = \alpha,\ \ (A \vee B)^\perp = A^\perp \wedge B^\perp\ (A \wedge B)^\perp = A^\perp \vee B^\perp$.
Sequents are multisets of formulas derived according to the rules in table 1.

Cut elimination [8] proceeds by transforming each *logical cut* in which both cut formulas are main conclusions of the last logical rule, into a cut on two of

the immediate subformulas, and transforming non-logical, or *commuting cuts* into logical cuts by commutation — moving the cut rule up the proof tree by successively commuting it with the rules above it on the right and left hand side (and duplicating or erasing proofs when moving past $\wedge$-introductions, and structural rules in which the cut formula is main).
The non-determinism which arises in the additive version of **LK** considered here comes from the 'structural dilemma' described in [7]. If neither cut formula is the main conclusion of the last logical rule, then it is necessary to choose which branch to commute the cut up first. Because of the presence of structural rules, this choice has important consequences. A well known example is the observation of Lafont [12] that *any* two proofs of the same formula can be merged (using *Cut*) into a proof which has cut free forms derived from *both* of the original proofs.

Definition 2. *Given proofs* $\pi, \pi' \vdash \Gamma$*, form* π **or** $\pi' \vdash \Gamma$ *as follows:*

$$\dfrac{\dfrac{\begin{matrix}\pi\\ \vdots\\ \Gamma\end{matrix}}{\Gamma, A}\,WK \qquad \dfrac{\begin{matrix}\pi'\\ \vdots\\ \Gamma\end{matrix}}{\Gamma, A^{\perp}}\,WK}{\Gamma}\,CUT$$

Commute the cut up the left branch, and the right branch is discarded, and vice-versa. So if π, π' are cut free, then there are two cut free forms of π **or** π' — π and π'. Thus any congruence (for instance, a denotational equivalence) generated by such a cut elimination procedure must equate π and π' and hence be trivial. However, this is consistent with an idea underlying the geometry of interaction [9], and game semantics [4, 1, 5], that cut elimination is a *process* analogous to computation. Non-determinism is both a standard property of computational processes and a key feature of many important algorithms and the physical systems on which programs are run. But can a connection between non-deterministic computation and non-deterministic cut elimination be established? Computationally, **or** corresponds to an erratic 'choice operator' [15], showing in principal that many non-deterministic algorithms may be extracted from proofs. A more difficult question is whether there are *natural* proofs which have computational content which is non-deterministic. Coquand [5, 6] has described examples of symmetric classical existence proofs from which two different witnesses can be extracted by different double negation translations, but much work remains to be done.

The structural dilemma can be seen as a problem of *ambiguity* — classical proofs do not carry sufficient information to determine their cut elimination behaviour, and so additional information must either be supplied with the proofs, or during the cut elimination process. The most comprehensive and systematic attempt to describe and analyse the former option is the **LKtq** calculus (and refinements) proposed by Danos, Joinet and Schellinx [7]. This 'disambiguates' proofs by attaching one of two complementary *colours*, t and q to each formula, to determine how to eliminate any commuting cuts (the details of *how* are not important here). This annotation permits a confluent cut elimination procedure to be defined which is preserved by a translation of coloured formulas and their proofs into linear logic (linear decoration). So, for example, the two colourings of the cut formula A in the proof π **or** π' correspond to the two possible cut elimination strategies; discard π or discard π'.

However, as observed by Urban and Bierman [17], predetermining the behaviour of the *cuts* of a proof by annotating its *formulas* places a somewhat arbitrary restriction on cut elimination behaviour. Reducing a cut on a formula C can generate multiple sub-cuts on its subformulas. If the subformulas carry extra information such as colouring, then that will also be copied, and hence each subcut on the same subformula must be reduced in the same way. So (as shown with an example in [17]) allowing the structural dilemma to be resolved afresh at each cut allows more normal forms to be reached. Moreover, reduction by colouring cut formulas lacks the following transitivity property: say that π colour-reduces to π' whenever there is a colouring of π, π' as π_c, π'_c such that π_c reduces to π'_c; then it is not the case that if π colour-reduces to π' and π' colour-reduces to π'' then π colour reduces to π'', as the following example shows. Suppose $\pi \vdash \Gamma, A$, $\lambda \vdash \Gamma, B$ and $\rho \vdash \Gamma, A \wedge B$ are distinct (cut free) proofs. Consider the following proof of $\Gamma, A \wedge B$.

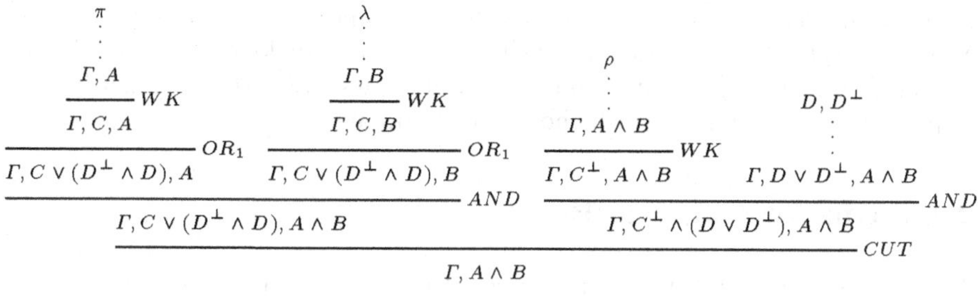

$$
\cfrac{
\cfrac{
\cfrac{
\cfrac{\pi \atop \vdots \atop \Gamma, A}{\Gamma, C, A}\ WK
}{\Gamma, C \vee (D^\perp \wedge D), A}\ OR_1
\qquad
\cfrac{
\cfrac{\lambda \atop \vdots \atop \Gamma, B}{\Gamma, C, B}\ WK
}{\Gamma, C \vee (D^\perp \wedge D), B}\ OR_1
}{\Gamma, C \vee (D^\perp \wedge D), A \wedge B}\ AND
\qquad
\cfrac{
\cfrac{\rho \atop \vdots \atop \Gamma, A \wedge B}{\Gamma, C^\perp, A \wedge B}\ WK
\qquad
\cfrac{D, D^\perp \atop \vdots}{\Gamma, D \vee D^\perp, A \wedge B}
}{\Gamma, C^\perp \wedge (D \vee D^\perp), A \wedge B}\ AND
}{\Gamma, A \wedge B}\ CUT
$$

This reduces to:

$$
\cfrac{
\cfrac{
\cfrac{\pi \atop \vdots \atop \Gamma, A}{\Gamma, C, A \wedge B, A}\ WK
\qquad
\cfrac{\rho \atop \vdots \atop \Gamma, A \wedge B}{\Gamma, C^\perp, A \wedge B, A}\ WK
}{\Gamma, A \wedge B, A}\ CUT
\qquad
\cfrac{
\cfrac{\lambda \atop \vdots \atop \Gamma, B}{\Gamma, C, A \wedge B, B}\ WK
\qquad
\cfrac{\rho \atop \vdots \atop \Gamma, A \wedge B}{\Gamma, C^\perp, A \wedge B, B}\ WK
}{\Gamma, A \wedge B, B}\ CUT
}{
\cfrac{\Gamma, A \wedge B, A \wedge B}{\Gamma, A \wedge B}\ CON
}\ AND
$$

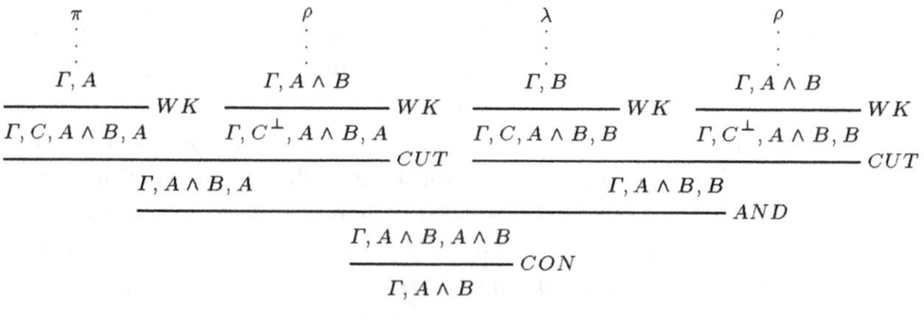

In the original proof, all of the occurrences of C must receive the same colour, — i.e. the structural dilemma must be resolved in the same way on each of the subcuts, hence there are only two cut free forms (having garbage-collected structural rules):

$$\dfrac{\begin{matrix}\pi & \lambda\\ \vdots & \vdots\\ \Gamma, A & \Gamma, B\end{matrix}}{\Gamma, A \wedge B}\,AND \qquad\qquad \begin{matrix}\rho\\ \vdots\\ \Gamma, A \wedge B\end{matrix}$$

But the cut formula C can be coloured differently (i.e. the structural dilemma can be resolved differently) in the two cuts, yielding another two cut free forms:

$$\dfrac{\dfrac{\dfrac{\begin{matrix}\pi\\ \vdots\\ \Gamma, A\end{matrix}}{\Gamma, A \wedge B, A}\,WK \quad \dfrac{\begin{matrix}\rho\\ \vdots\\ \Gamma, A \wedge B\end{matrix}}{\Gamma, A \wedge B, B}\,WK}{\Gamma, A \wedge B, A \wedge B}\,AND}{\Gamma, A \wedge B}\,CON \qquad \dfrac{\dfrac{\dfrac{\begin{matrix}\rho\\ \vdots\\ \Gamma, A \wedge B\end{matrix}}{\Gamma, A \wedge B, A}\,WK \quad \dfrac{\begin{matrix}\lambda\\ \vdots\\ \Gamma, B\end{matrix}}{\Gamma, A \wedge B, B}\,WK}{\Gamma, A \wedge B, A \wedge B}\,AND}{\Gamma, A \wedge B}\,CON$$

This example does not rely on any specific property of colouring; any attempt to determine cut elimination by adding information to formulas will encounter the same problem. Indeed, because of the large bounds on the number of cuts which may be generated in the course of cut elimination of a proof any fully general disambiguation will have to carry a similar amount of information. Non-determinism can be seen as a way of achieving the right level of generality by allowing the choice of how to reduce cuts to be made during cut elimination. Many of the choices introduced may be trivial in that they do not lead to different normal forms or shorter cut elimination. But representing them explicitly is a step towards determining which choices really do matter.

2 Terms for Classical Proofs

A system for annotatating classical proofs with appropriate terms will be used to formally define both the symmetric cut elimination protocol and the translation into intuitionistic proofs with non-determinism. The symmetric λ-calculus [2] has been adopted for this rôle, so that λ^{Sym} normalisation of annotations corresponds to cut elimination of proofs. This choice of annotation for classical proofs should be seen as a practical decision motivated by the simplicity of the typing-judgements and reduction system for λ^{Sym} (stressed in [2]). It is possible, however, to adapt the translation to other calculi for classical sequent proofs such as Urban and Bierman's [17], or annotation with the $\lambda\mu$-calculus [16].

The *types* of λ^{Sym} are the propositional formulas, together with a distinguished type $\bot$. The formal definition of λ^{Sym} terms is given in [2]. A restricted subset of these terms (the (proof) *annotations*) is defined by assignment to sequent proofs in table 2 — a proof of $\Gamma = A, B, \ldots$ is annotated with a term-in-context $\Gamma^{\bot} \vdash t : \bot$, where $\Gamma^{\bot}$ is a context of variables of negated types or *names* $a : A^{\bot}, b : B^{\bot}, \ldots$. This is similar to Urban and Bierman's calculus for classical sequents [17] — the main difference is that the single operator $\bigstar$ ("symmetric application") is used both to annotate the cut rule itself and also in

$$\frac{}{a{:}\alpha, b{:}\alpha^{\perp} \vdash a \bigstar b{:}\perp}\ \text{AXIOM} \qquad \frac{\Gamma, a{:}A^{\perp} \vdash s{:}\perp \quad \Gamma, b{:}A \vdash t{:}\perp}{\Gamma \vdash \lambda a.s \bigstar \lambda b.t{:}\perp}\ \text{CUT}$$

$$\frac{\Gamma \vdash t{:}\perp}{\Gamma, a{:}A^{\perp} \vdash t{:}\perp}\ \text{WK} \qquad \frac{\Gamma, a{:}A^{\perp}, b{:}A^{\perp} \vdash t{:}\perp}{\Gamma, c{:}A^{\perp} \vdash t[c/a][c/b]{:}\perp}\ \text{CON}$$

$$\frac{\Gamma, a{:}A^{\perp} \vdash s{:}\perp \quad \Gamma, b{:}B^{\perp} \vdash t{:}\perp}{\Gamma, c{:}(A \wedge B)^{\perp} \vdash c \bigstar \langle \lambda a.s, \lambda b.t \rangle{:}\perp}\ \text{AND} \quad \frac{\Gamma, a{:}A_i^{\perp} \vdash t{:}\perp}{\Gamma, b{:}(A_1 \vee A_2)^{\perp} \vdash b \bigstar \mathtt{in}_i(\lambda a.t){:}\perp}\ OR_i$$

Table 2. Derivation of proof-annotations (λ^{Sym} terms assigned to **LK** proofs)

the introduction rules for the logical connectives. This is because the purpose of the annotations for the introduction rules is to determine their behaviour under cut elimination, and names are simply place-holders for which subsequent cut formulas can be substituted.

2.1 Cut Elimination

The cut elimination procedure described in the introduction is implemented by reduction rules for λ^{Sym}.

Definition 3. *Cut reduction by rewriting of annotations:*

$$\lambda a.t \bigstar s \longrightarrow t[s/a] \qquad t \bigstar \lambda b.s \longrightarrow s[t/b]$$
$$\mathtt{in}_i(t) \bigstar \langle s_1, s_2 \rangle \longrightarrow t \bigstar s_i \qquad \langle s_1, s_2 \rangle \bigstar \mathtt{in}_i(t) \longrightarrow s_i \bigstar t$$

The first pair of rules implement structural cut-reduction by directly transporting copies of the proof of one cut formula to each of the points where the other was introduced, the latter two implement logical cut-reduction. So, for example, if we have annotations $s, t : \perp$, we can form the non-deterministic merging

$$s \ \mathbf{or}\ t = \lambda a\colon A.s \bigstar \lambda b\colon A^{\perp}.t \quad (a, b \notin FV(s) \cup FV(t))$$

such that $s\ \mathbf{or}\ t \longrightarrow s$ and $s\ \mathbf{or}\ t \longrightarrow t$.

Each full cut reduction step corresponds to three different operations on proofs — commute and copy the cut up one branch, commute and copy up the other branch, and reduce all resulting logical cuts. One consequence is that the price to pay for the simplicity of the rewriting system is that it lacks a kind of subject reduction property — the set of proof annotations is not closed under λ^{Sym} reduction. For example, $\lambda a.(a \bigstar b) \bigstar \lambda c.(d \bigstar c) \longrightarrow (\lambda c.d \bigstar c) \bigstar b$. However, the important point is that the notion of proof annotations is closed under *normalisation*. To show this, a larger set of *reachable* terms — which is closed under reduction — can be defined by including the intermediate logical and structural cut reduction steps as term-assignment rules.

Definition 4. *Define the reachable terms by adding the following additional formation rules to those given in table 2:*

$$\frac{\Gamma,a{:}A_1^{\perp}\vdash s{:}\bot \quad \Gamma,b{:}A_2^{\perp}\vdash t{:}\bot, \quad \Gamma,c{:}A_i\vdash r{:}\bot}{\Gamma\vdash \mathtt{in}_i(\lambda c.r)\bigstar\langle\lambda a.s,\lambda b.t\rangle{:}\bot, \quad \langle\lambda a.s,\lambda b.t\rangle\bigstar\mathtt{in}_i(\lambda c.r){:}\bot}$$

$$\frac{\Gamma,a{:}\alpha^{\perp}\vdash t{:}\bot}{\Gamma,a'{:}\alpha^{\perp}\vdash a'\bigstar\lambda a.t{:}\bot} \quad \frac{\Gamma,a{:}\alpha\vdash t{:}\bot}{\Gamma,a'{:}\alpha\vdash\lambda a.t\bigstar a'{:}\bot}$$

$$\frac{\Gamma,b{:}A_1\vee A_2\vdash t{:}\bot \quad \Gamma,c{:}A_i^{\perp}\vdash s{:}\bot,}{\Gamma\vdash\lambda b.t\bigstar\mathtt{in}_i(\lambda c.s){:}\bot}, \quad \frac{\Gamma,a{:}A^{\perp}\vdash s{:}\bot \quad \Gamma b{:}B^{\perp}\vdash t{:}\bot \quad \Gamma,c{:}A\wedge B\vdash r{:}\bot}{\Gamma\vdash\lambda c.r\bigstar\langle\lambda a.s,\lambda b.t\rangle}$$

Proposition 1. *Let t be an annotation. If s is a normal form of t then it is the annotation of a cut free proof.*

Proof. We establish the following two facts:

- If r is a reachable term and $r \to r'$, then r' is a reachable term.
- A reachable term is in normal form if and only if it is the annotation of a cut free proof.

3 $\mathbf{LK}_{\uparrow}$ — a Confluent Calculus

The basic idea of the translation of λ^{Sym}-annotated proofs will be to give two 'different disambiguations' for each cut (or potential cut), corresponding to the two different choices represented by the structural dilemma, and then to use a non-deterministic choice operator to combine them. The first step is to settle on a means of determining the behaviour of proofs under cut elimination by adding information. One possibility is to attach this information to *formulas* — this is the solution proposed in the **LKtq** calculus [7]. But combining proofs of different formulas non-deterministically is difficult when it comes to co-ordinating the choices of colouring for cut formulas.

There is another option. The structural dilemma can be resolved by allowing different cut elimination behaviours to be captured as *different proofs* of the *same formula*. This is achieved here by adding information to proofs in the form of a new, non-logical rule — *'lifting'* — which converts a formula which appears as the main conclusion of a logical rule or axiom into one which can be used as the main premiss of a logical rule or cut. Different cut elimination behaviours for the same **LK** proof are obtained by varying *where* the liftings are included. This allows the structural dilemma to be expressed as a simple erratic choice between two proofs of the same formula — i.e. λ-terms of the same type.

Definition 5. *The* types *of* $\mathbf{LK}_{\uparrow}$ *(represented as $C, D, \ldots$) consist of the propositional formulas $(A, B, \ldots)$, together with* lifted *formulas $\lfloor A\rfloor, \lfloor B\rfloor, \ldots)$*

$$C ::= A \mid \lfloor A\rfloor$$

The rules for the sequent calculus $\mathbf{LK}_{\uparrow}$ *are given in table 3.*

Note that any proof of $\mathbf{LK}_{\uparrow}$ has (as its 'skeleton' [7]) a corresponding **LK** proof of the same sequent which omits all of the liftings. The cut elimination procedure

$$\frac{}{\alpha,\alpha^{\perp}}\ \text{AXIOM} \qquad \frac{\Gamma,A}{\Gamma,\lfloor A\rfloor}\ \text{LIFT}$$

$$\frac{\Gamma,A \quad \Gamma,A^{\perp}}{\Gamma}\ \text{CUT} \qquad \frac{\Gamma,\lfloor A\rfloor \quad \Gamma,\lfloor A^{\perp}\rfloor}{\Gamma}\ \lfloor\text{CUT}\rfloor$$

$$\frac{\Gamma}{\Gamma,C}\ \text{WK} \qquad \frac{\Gamma,C,C}{\Gamma,C}\ \text{CON}$$

$$\frac{\Gamma,\lfloor A\rfloor \quad \Gamma,\lfloor B\rfloor}{\Gamma,A\wedge B}\ \text{AND} \qquad \frac{\Gamma,\lfloor A_i\rfloor}{\Gamma,A_1\vee A_2}\ OR_i :\ i=1,2$$

Table 3. Classical sequent calculus with lifting, $\mathbf{LK}_{\uparrow}$

for $\mathbf{LK}_{\uparrow}$ proofs is similar to that given for **LK**, except that *polarities* (together with the lifting rule) will be used to resolve the structural dilemma. The polarity of a formula is simply determined by its outermost connective.

Definition 6. Positive *formulas are defined:* $P ::= \alpha^{\perp} \mid A \vee B$. Negative *formulas are defined:* $N ::= \alpha \mid A \wedge B$.

The key feature for cut elimination is therefore that at each cut, one cut formula is positive and the other cut formula is negative. Thus the structural dilemma can be expressed in terms of polarities; each commuting cut can be reduced by commuting it up the branch containing the negative cut formula first, or up the branch containing the positive cut formula first. Lifting allows both of these possibilities to be combined without non-determinism.

- A structural cut between two lifted formulas reduces by transporting the proof of the lifted *negative* formula to each of the points where the *positive* formula was lifted. Each of these intermediate cuts reduces by transporting the proof of the newly lifted positive formula to the point where the negative formula was lifted.
- A "logical lifted cut" (i.e. a cut in which the last rule on both sides is the lifting of the cut formulas) reduces directly to a cut between the unlifted formulas.
- A structural cut between unlifted formulas reduces first by transporting the proof of the *positive* formula to each point where the *negative* formula was introduced, and then reducing each intermediate cut by transporting the proofs of the freshly introduced negative formulas to the points where the positive formula was introduced, creating logical cuts.

Thus different cut elimination protocols for **LK** can be simulated in $\mathbf{LK}_{\uparrow}$ by choices of *where* (not whether) to include liftings, as will be shown by using non-determinism to simulate symmetric cut elimination.

3.1 Assigning λ-terms to $\mathbf{LK}_\uparrow$ Proofs

The formal representation of the cut elimination procedure is analogous to the symmetric calculus. Terms of the simply-typed λ-calculus with pairing are assigned to proofs of $\mathbf{LK}_\uparrow$ and cut elimination is implemented by normalisation. An alternative would be a translation into proofs of linear logic, which could also be used to analyse the "logical dilemma" [7] encountered in cut elimination of proofs with multiplicative logical rules.

The term-calculus has a distinguished (empty) ground type $\mathbf{0}$ corresponding to the 'response type' of a cps translation; the other ground types are the atomic propositions of **LK**. Representing $T \Rightarrow \mathbf{0}$ as $\neg T$, we give a translation of positive formulas as negated types, negative formulas as doubly negated types.

Definition 7. *Each* $\mathbf{LK}_\uparrow$*-type* C *is interpreted as a* λ*-type* $[\![C]\!]$ *as follows:*

- $[\![\alpha]\!] = \neg\neg\alpha$, $[\![\alpha^\perp]\!] = \neg\alpha$
- $[\![\lfloor A \rfloor]\!] = \neg[\![A^\perp]\!]$
- $[\![A \wedge B]\!] = \neg\neg(\neg\neg[\![A]\!] \times \neg\neg[\![B]\!])$
- $[\![A \vee B]\!] = \neg(\neg\neg[\![A^\perp]\!] \times \neg\neg[\![B^\perp]\!])$

Note that for all positive formulas, $[\![P^\perp]\!] = \neg[\![P^\perp]\!]$ and $[\![\lfloor P \rfloor]\!] = \neg\neg[\![P]\!]$, whereas for all negative formulas, $[\![N]\!] = \neg[\![N^\perp]\!]$ and $[\![\lfloor N \rfloor]\!] = [\![N]\!]$.

Definition 8 (Annotation of $\mathbf{LK}_\uparrow$ proofs). *For each* $\mathbf{LK}_\uparrow$ *formula* A, *define a* λ*-type* $\overline{A}$ *such that* $\neg\overline{A} = [\![A]\!]$*: i.e.* $\overline{\alpha} = \neg\alpha$, $\overline{\alpha^\perp} = \alpha$, $\overline{\lfloor A \rfloor} = [\![A^\perp]\!]$, $\overline{A \wedge B} = \neg(\neg\neg[\![A]\!] \times \neg\neg[\![B]\!])$, $\overline{A \vee B} = \neg\neg[\![A^\perp]\!] \times \neg\neg[\![B^\perp]\!]$
Proofs of $\mathbf{LK}_\uparrow$ *sequents* $C, D, \ldots$ *are annotated with lambda terms-in-context:* $x : \overline{A}, y : \overline{B}, \ldots \vdash u : \mathbf{0}$ *as defined in table 4 (λ-terms are denoted* $u, v, w, \ldots$ *to distinguish them from* λ^{Sym}*-terms).*
(The operation $\lceil \_ \rceil$ *is defined such that if* $v : [\![\lfloor A \rfloor]\!]$, *then* $\lceil v \rceil : \neg\neg[\![A]\!]$*:*
$\lceil v : [\![\lfloor P \rfloor]\!] \rceil = v$, $\lceil v : [\![N]\!] \rceil = \lambda x : [\![N]\!] \Rightarrow \mathbf{0}.x\ v$.*)*

$$\frac{}{x{:}\overline{\alpha^\perp}, y{:}\overline{\alpha} \vdash y\ x{:}\mathbf{0}}\ \text{AXIOM} \qquad \frac{\Gamma, x{:}\overline{P} \vdash v{:}\mathbf{0}}{\Gamma, y{:}\overline{\lfloor P \rfloor} \vdash y\ \lambda x.v{:}\mathbf{0}},\ \frac{\Gamma, x{:}\overline{N} \vdash v{:}\mathbf{0}}{\Gamma, x{:}\overline{\lfloor N \rfloor} \vdash v{:}\mathbf{0}}\ \text{LIFT}$$

$$\frac{\Gamma, x{:}\overline{P} \vdash u{:}\mathbf{0}, \quad \Gamma, y{:}\overline{P^\perp} \vdash v{:}\mathbf{0}}{\Gamma \vdash (\lambda y.v)\ \lambda x.u{:}\mathbf{0}}\ \text{CUT} \qquad \frac{\Gamma, x{:}\overline{\lfloor P \rfloor} \vdash u{:}\mathbf{0}, \quad \Gamma, y{:}\overline{\lfloor P^\perp \rfloor} \vdash v{:}\mathbf{0}}{\Gamma \vdash (\lambda x.u)\ \lambda y.v{:}\mathbf{0}}\ \lfloor \text{CUT} \rfloor$$

$$\frac{\Gamma \vdash v{:}\mathbf{0}}{\Gamma, x{:}\overline{C} \vdash v{:}\mathbf{0}}\ \text{WK} \qquad \frac{\Gamma, x{:}\overline{C}, y{:}\overline{C} \vdash v{:}\mathbf{0}}{\Gamma, z{:}\overline{C} \vdash v[z/x, z/y]{:}\mathbf{0}}\ \text{CON}$$

$$\frac{\Gamma, x{:}\overline{\lfloor A \rfloor} \vdash u{:}\mathbf{0} \quad \Gamma, y{:}\overline{\lfloor B \rfloor} \vdash v{:}\mathbf{0}}{\Gamma, z{:}\overline{(A \wedge B)} \vdash z\ \langle \lceil \lambda x.u \rceil, \lceil \lambda y.v \rceil \rangle{:}\mathbf{0}}\ \text{AND} \qquad \frac{\Gamma, x{:}\overline{\lfloor A_i \rfloor} \vdash u{:}\mathbf{0}}{\Gamma, y{:}\overline{(A_1 \vee A_2)} \vdash \pi_i(y)\ \lambda x.u{:}\mathbf{0}}\ \text{OR}_i$$

Table 4. λ-calculus annotation of $\mathbf{LK}_\uparrow$

Definition 9. *Cut elimination for* $\mathbf{LK}_\uparrow$ *is by* $\beta\pi$*-reduction of annotations:*

$$\lambda x.v\ u \longrightarrow_{\beta\pi} v[u/x] \qquad \pi_i(\langle v_1, v_2\rangle) \longrightarrow_{\beta\pi} v_i$$

Note that this implements the procedure for $\mathbf{LK}_\uparrow$ cut elimination informally described above. As for λ^{Sym}-annotation of **LK** proofs, the set of annotations is not closed under reduction, but is closed under normalisation, which can be shown by defining a set of reachable terms and observing that a reachable term is a normal form if and only it annotates a cut free proof.

Proposition 2. *Cut elimination for* $\mathbf{LK}_\uparrow$ *is confluent and strongly normalising.*

Proof. This is a direct consequence of strong normalisation and confluence of the simply-typed λ-calculus with pairing.

4 Simulating Non-deterministic Cut Elimination

The target language for the translation of λ^{Sym}-annotated **LK** proofs is a simple extension of the λ-calculus with non-deterministic choice.

Definition 10. *The* λ^+*-calculus is the simply-typed* λ*-calculus with pairing, augmented with the following rule of 'superposition'; in computational terms, a kind of "erratic choice" construct [15]:*

$$\frac{\Gamma \vdash u : T \Rightarrow \mathbf{0} \quad \Gamma \vdash v : T \Rightarrow \mathbf{0}}{\Gamma \vdash u + v : T \Rightarrow \mathbf{0}}$$

The choice expressed by $u+v$ is between *functions* rather than arguments, so it is resolved when it calls (is applied to) another procedure, rather than vice-versa. In other words, reduction of superposition is *lazy.* This is because non-determinism in the classical sequent proofs is incorporated in the logical rules as well as the cut rule, but it is only when the logical rules are unpacked, in the course of cut elimination, that the choice can be resolved. Thus cut free proofs — and terms in normal form — should retain the possibility of non-deterministic behaviour.

Definition 11. *Superposition reduction:*

$$(u_1 + u_2)\ v \longrightarrow^+ u_i\ v \quad (i = 1, 2)$$

The union of (the compatible closure of) this reduction with $\beta\pi$ reduction on λ^+ will be written $\longrightarrow$, and the transitive reflexive closure, $\twoheadrightarrow$. The following is a straightforward inference from strong normalisation of the simply-typed λ-calculus.

Proposition 3. $(\lambda^+, \longrightarrow)$ *is strongly normalising.*

4.1 Translating the Structural Dilemma

Both of the choices presented by the (polarized version of the) structural dilemma for a commuting cut can be simulated by different (and in a sense, canonical) choices of where to lift the cut formulas in an associated $\mathbf{LK}_{\uparrow}$ proof. If the positive cut formula was lifted *immediately after introduction*, then according to the $\mathbf{LK}_{\uparrow}$ cut elimination protocol, the cut is eliminated by commutation up the branch containing the negative cut formula first. If the positive cut formula was lifted *immediately before the cut* then the cut is eliminated by commutation up the the branch containing the positive cut formula first. (Note that where the negative formula was lifted has no effect).

LKtq can therefore be translated into $\mathbf{LK}_{\uparrow}$ by using the colours to determine where the positive formula should be lifted. However, using the typed 'superposition' operator, we can also introduce explicit choices between these two possible liftings of the positive cut formula, and hence simulate the structural dilemma and λ^{Sym} reduction of proofs. So we shall now give a translation of λ^{Sym} annotations into λ^{+} such that the normal forms of each annotation and its translate are the same.

(Where it matters) negative names (i.e λ^{Sym}-variables of *positive type*) will be represented as $m, n, \ldots$, and positive names as $p, q, \ldots$. We shall assume that each positive name $p : P$ is associated with a complementary pair of fresh names $p_{\downarrow}, p_{\uparrow} : P$, with the intended meaning that an introduction named with $p_{\uparrow}$ is translated as an introduction followed by a lifting, but an introduction named with $p_{\downarrow}$ is not. The binding of p is translated as a choice between substituting $p_{\uparrow}$ for p, translating, and then binding $p_{\uparrow}$ *or* substituting $p_{\downarrow}$ for p, translating, binding $p_{\downarrow}$ and lifting the result.

Definition 12. *The translation of propositional formulas is as in definition 7. The translation of annotations assumes a bijective correspondence between names $n : N, p_{\downarrow}, p_{\uparrow} : P$, and λ^{+}-variables $x_n : \overline{N}, x_{p\downarrow} : \overline{P}, x_{p\uparrow} : \neg\neg\overline{P}$.*

- $[\![p_{\downarrow}\bigstar n]\!] = x_n\ x_{p\downarrow}$,
  $[\![p_{\uparrow}\bigstar n]\!] = x_{p\uparrow}\ \lambda y.x_n\ y$
- $[\![\lambda n.t]\!] = \lambda x_n.[\![t]\!]$,
  $[\![\lambda p.t]\!] = \lambda x_{p\uparrow}.[\![t[p_{\uparrow}/p]]\!] + (\lambda y.y\ \lambda x_{p\downarrow}.[\![t[p_{\downarrow}/p]]\!])$
- $[\![n\bigstar\langle s_1, s_2\rangle]\!] = x_n\ \langle\lceil[\![s_1]\!]\rceil, \lceil[\![s_2]\!]\rceil\rangle$,
- $[\![p_{\downarrow}\bigstar\mathtt{in}_i(s)]\!] = \pi_i(x_{p\downarrow})\ [\![s]\!]$,
  $[\![p_{\uparrow}\bigstar\mathtt{in}_i(s)]\!] = x_{p\uparrow}\ \lambda x.\pi_i(x)\ [\![s]\!]$
- $[\![\lambda p.s\bigstar\lambda n.t]\!] = [\![\lambda n.t\bigstar\lambda p.s]\!] = [\![\lambda p.s]\!]\ [\![\lambda n.t]\!]$

E.g. $[\![s\ \mathbf{or}\ t]\!] = (\lambda x_{a\uparrow}.[\![s]\!] + \lambda z.z\ \lambda x_{a\downarrow}.[\![s]\!])\ \lambda x_b.[\![t]\!]$.
$(\lambda x_{a\uparrow}.[\![s]\!] + \lambda z.z\ \lambda x_{a\downarrow}.[\![s]\!])\ \lambda x_b.[\![t]\!] \longrightarrow \lambda x_{a\uparrow}.[\![s]\!]\ \lambda x_b.[\![t]\!] \longrightarrow [\![s]\!]$
$(\lambda x_{a\uparrow}.[\![s]\!] + \lambda z.z\ \lambda x_{a\downarrow}.[\![s]\!])\ \lambda x_b.[\![t]\!] \longrightarrow (\lambda z.z\ \lambda x_{a\downarrow}.[\![s]\!])\ \lambda x_b.[\![t]\!] \twoheadrightarrow [\![t]\!]$.
In order to formally relate λ^{Sym} reduction of annotations to λ^{+} reduction of their translations (and derive the termination result) it is necessary to extend Definition 12 to a translation of the *reachable* λ^{Sym} terms into λ^{+}.

Definition 13. *Translation of reachable terms:*

- $\llbracket \lambda a.t\bigstar b \rrbracket = \llbracket b\bigstar \lambda a.t \rrbracket = \lambda x_a.\llbracket t \rrbracket\ x_b,$
- $\llbracket \lambda p.r\bigstar \langle s,t\rangle \rrbracket = \lambda x_{p\downarrow}.\llbracket r \rrbracket\ \langle \lceil \llbracket s \rrbracket \rceil, \lceil \llbracket t \rrbracket \rceil \rangle$
- $\llbracket \lambda n.s\bigstar \mathtt{in}_i(t) \rrbracket = \lambda x_n.\llbracket s \rrbracket\ (\lambda x.\pi_i(x)\ \llbracket t \rrbracket)$
- $\llbracket \mathtt{in}_i(r)\bigstar \langle s,t\rangle \rrbracket = \llbracket \langle s,t\rangle \bigstar \mathtt{in}_i(r) \rrbracket = \pi_i(\langle \lceil \llbracket s \rrbracket \rceil, \lceil \llbracket t \rrbracket \rceil \rangle)\ \llbracket r \rrbracket$

Proposition 4. *For any reachable terms s, t: $t \longrightarrow s$ implies $\llbracket t \rrbracket \twoheadrightarrow \llbracket s \rrbracket$.*

Proof. is by observing that λ^{Sym} redexes of annotations translate to a series of redexes of λ^+, e.g. $\llbracket \lambda p.t\bigstar \lambda n.s \rrbracket =$
$(\lambda x_{p\uparrow}.\llbracket t[p_\uparrow/p] \rrbracket + \lambda z.z\ \lambda x_p.\llbracket t[p_\downarrow/p] \rrbracket)\ \llbracket \lambda n.s \rrbracket \longrightarrow \lambda x_{p\uparrow}.\llbracket t[p_\uparrow/p] \rrbracket\ \llbracket \lambda n.s \rrbracket$
$\longrightarrow \llbracket t[p_\uparrow/p] \rrbracket[\llbracket \lambda n.s \rrbracket/x_{p\uparrow}] = \llbracket t[\lambda n.s/p] \rrbracket$ (proof by structural induction). And $(\lambda x_p.\llbracket t[p_\uparrow/p] \rrbracket + \lambda z.z\ \lambda x_{p\downarrow}.\llbracket t[p_\downarrow/p] \rrbracket)\ \lambda y_n.\llbracket s \rrbracket \longrightarrow (\lambda z.z\ \lambda x_{p\downarrow}.\llbracket t[p_\downarrow/p])\ \lambda y_n.\llbracket s \rrbracket$
$\longrightarrow \lambda y_n.\llbracket s \rrbracket\ \lambda x_p.\llbracket t[p_\downarrow/p] \rrbracket \longrightarrow \llbracket s \rrbracket[\lambda x_{p\downarrow}.\llbracket t[p_\downarrow/p] \rrbracket/y_n] = \llbracket s[\lambda p.t/n] \rrbracket$.
$\llbracket \mathtt{in}_i(t)\bigstar \langle s_1,s_2\rangle \rrbracket = \pi_i(\langle \lceil \llbracket s_1 \rrbracket \rceil, \lceil \llbracket s_2 \rrbracket \rceil \rangle)\ \llbracket t \rrbracket \longrightarrow \llbracket s_i \rrbracket\ \llbracket t \rrbracket = \llbracket s_i\bigstar t \rrbracket$ if s_i has a positive type, or $(\lambda z.z\ \llbracket s_i \rrbracket)\ \llbracket t \rrbracket \longrightarrow \llbracket t \rrbracket\ \llbracket s_i \rrbracket = \llbracket s_i\bigstar t \rrbracket$ if s_i has a negative type.

Corollary 1. *The cut elimination procedure for* **LK** *given by λ^{Sym} annotation and reduction is strongly normalising.*

This is of course already a corollary of the fact that λ^{Sym} reduction is terminating [2]. Proof of that result, however, is given by 'a non-trivial version of Tait and Girard's [reducibility candidates] method'. Proving strong normalisation by translation has the desirable consequence of limiting the use of such logically complex methods to a few well known intuitionistic cases (this is significant for the extension to second-order **LK**).

The following proposition and its corollary strengthen the soundness result given in Proposition 4 by establishing that the translation is *faithful*; λ^+-normalisation of translated annotations implements cut elimination for **LK** precisely.

Proposition 5. *If u is a λ^+ term such that $u \twoheadrightarrow_{\beta\pi} \llbracket s \rrbracket$ for some reachable term s and v is a λ^+ term such that $u \twoheadrightarrow v$, then there exists a reachable term t such that $v \twoheadrightarrow_{\beta\pi} \llbracket t \rrbracket$ and $s \twoheadrightarrow t$.*

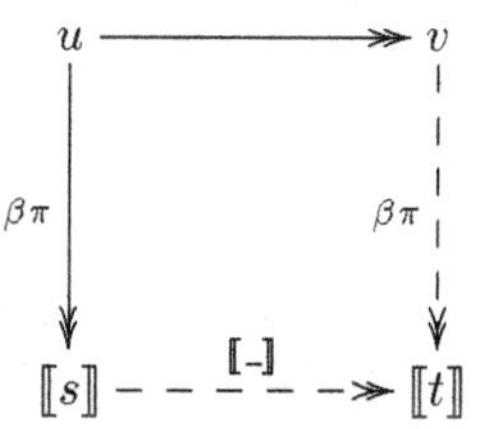

Corollary 2. *For any term M of λ^{Sym} term or λ^+, let $Nf(M)$ be the set of normal forms of M. Then for any annotation s, $Nf(\llbracket s \rrbracket) = \{\llbracket t \rrbracket \mid t \in Nf(s)\}$.*

To prove Proposition 5, we first establish it in a restricted form.

Lemma 1. *If s is a reachable term, and u is a λ^+-term such that $\llbracket s \rrbracket \longrightarrow u$, then there is a reachable term t such that $s \twoheadrightarrow t$ and $u \twoheadrightarrow_{\beta\pi} \llbracket t \rrbracket$.*

Next we prove that a form of the Church-Rosser theorem holds for λ^+ (using a method which is a straightforward variant of the standard proof of that theorem — see e.g. [3]).

Lemma 2. *Suppose u is a λ^+ term such $u \twoheadrightarrow v$ and $u \twoheadrightarrow_{\beta\pi} v'$, then there exists a λ^+-term w such that $v \twoheadrightarrow_{\beta\pi} w$ and $v' \longrightarrow w$.*

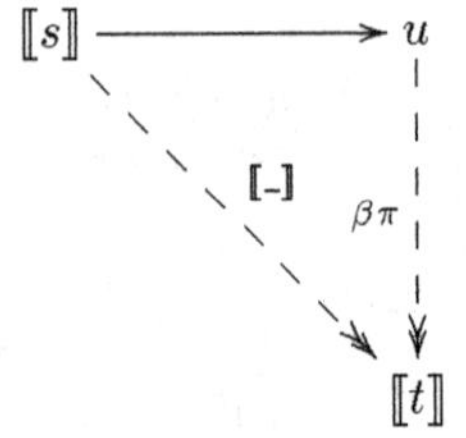

Lemma 1

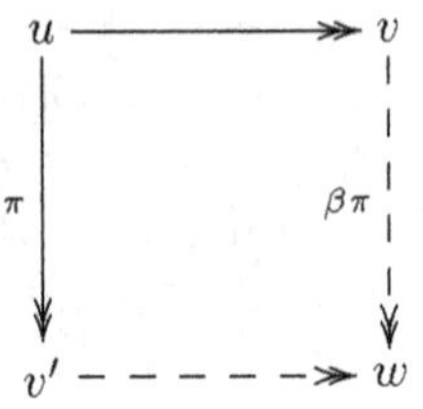

Lemma 2

Now Proposition 5 can be proved by induction on the maximum number of steps required to reduce u to normal form. There are two cases to consider, $u = [\![s]\!]$ and $u \neq [\![s]\!]$.

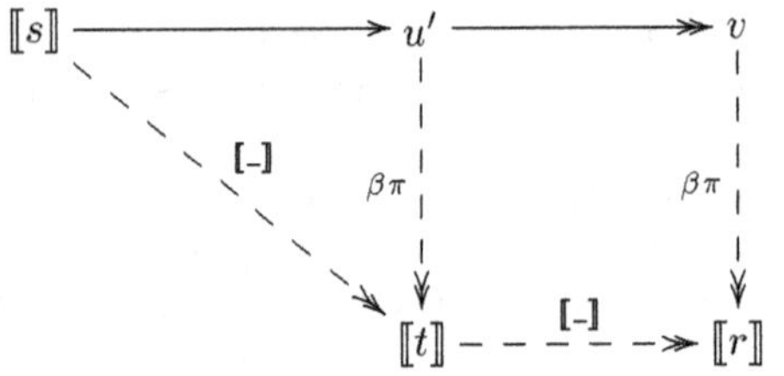

Case $u = [\![s]\!]$

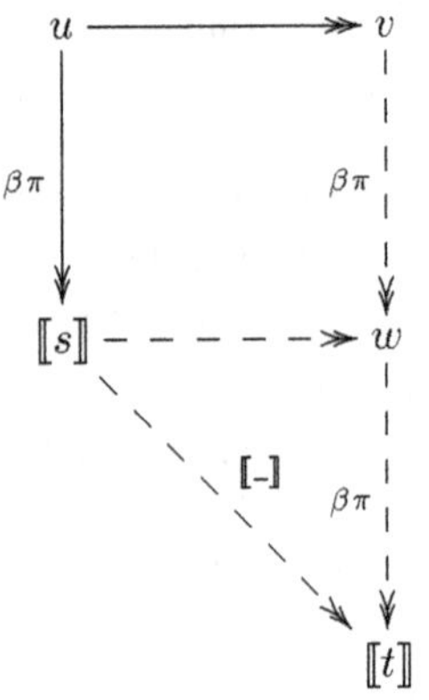

Case $u \neq [\![s]\!]$

$u = [\![s]\!]$ Either $u = v$, or $u \longrightarrow u' \twoheadrightarrow v$. Then by Lemma 1, there exists a reachable t such that $u' \twoheadrightarrow_{\beta\pi} [\![t]\!]$ and $s \twoheadrightarrow t$. The induction hypothesis now applies, as u' reduces to normal form in fewer steps than u, and so there is some reachable r such that $v \twoheadrightarrow_{\beta\pi} [\![w]\!]$ and $t \twoheadrightarrow r$ and $s \twoheadrightarrow r$ as required.

$u \neq [\![s]\!]$ Suppose $u \twoheadrightarrow_{\beta\pi} [\![s]\!]$ and $u \longrightarrow v$. Then by the restricted Church-Rosser property (Lemma 2), there exists w such that $[\![s]\!] \longrightarrow w$ and $v \twoheadrightarrow_{\beta\pi} w$. Now the induction hypothesis applies, as $[\![s]\!]$ normalises in fewer steps than u, so there exists a reachable t such that $s \twoheadrightarrow t$ and $w \twoheadrightarrow_{\beta\pi} [\![t]\!]$ and hence $u \twoheadrightarrow_{\beta\pi} [\![t]\!]$ as required.

5 Further Directions

Several avenues for further research exist which are beyond the scope of this paper, notably the extension to second order logic and the use of lifting and non-determinism to give denotational models of classical proofs.

It is relatively straightforward to extend the work outlined here to *second-order*. Using the second order symmetric λ-calculus, sequent rules for the introduction of the quantifiers can be annotated as follows:

$$\frac{\Gamma, a{:}A(X)^{\perp} \vdash t{:}\perp}{\Gamma, b{:}(\forall X.A(X))^{\perp} \vdash b \bigstar \Lambda X.\lambda a.t{:}\perp}\ \forall\text{-intro} \qquad \frac{\Gamma, a{:}A(T)^{\perp} \vdash t{:}\perp}{\Gamma, b{:}(\exists X.A(X))^{\perp} \vdash b \bigstar \langle T, \lambda a.t\rangle{:}\perp}\ \exists\text{-intro}$$

with cut elimination implemented by the rules:

$$\Lambda X.s \bigstar \langle T, t\rangle \longrightarrow s[T/X] \bigstar t \qquad \langle T, t\rangle \bigstar \Lambda X.s \longrightarrow s[T/X] \bigstar t$$

$\mathbf{LK}_{\uparrow}$ extends naturally to second order, and can be annotated with terms of System F with (a representation of) pairing, by extending the double negation translation:

$[\![\forall X.A(X)]\!] = \neg\neg\forall X.\neg\neg[\![A(X)]\!]$, $[\![\exists X.A(X)]\!] = \neg\forall X.\neg\neg[\![A(X)^{\perp}]\!]$.

This yields a translation of classical second-order proofs into System F with superposition. Since the latter is strongly normalising (easy to prove as a corollary of strong normalisation for System F), this shows that cut elimination for second order **LK** is strongly normalising without the need for a new and logically complex proof.

The analysis of classical proofs described so far has been wholly syntactic, but it suggests a recipe for a *semantics of classical proofs* in any model of the λ^{+}-calculus. However, there are a number of issues to adress if the translation of classical proofs into intuitionistic logic is to yield a reasonable semantic construction. For example, the interpretation given here is very liberal with the double-negation operator. This is necessary to represent the many cut elimination strategies available in classical logic compared with the intuitionistic case; for instance $\vee, \wedge$ are not associative — $[\![A \wedge (B \wedge C)]\!]$ is not isomorphic to $[\![(A \wedge B) \wedge C]\!]$ — and this faithfully reflects the fact that $A \wedge (B \wedge C)$ and $(A \wedge B) \wedge C$ behave differently as cut formulas because of the different order of introduction of the disjunctions. However, associativity would seem to be a minimal requirement of the logical connectives as semantic operators. To regain it, the double negation translation can be restricted so that it includes liftings — and hence cut choices — only when polarities *alternate*: e.g. let $[\![\lfloor N \rfloor]\!] = [\![N]\!]$, $[\![\lfloor P \rfloor]\!] = \neg\neg[\![P]\!]$, $[\![A \wedge B]\!] = [\![\lfloor A \rfloor]\!] \times [\![\lfloor B \rfloor]\!]$, $[\![A \vee B]\!] = \neg([\![A^{\perp} \wedge B^{\perp}]\!])$.

Then $[\![A \wedge (B \wedge C)]\!] = [\![\lfloor A \rfloor]\!] \times ([\![\lfloor B \rfloor]\!] \times [\![\lfloor C \rfloor]\!]) \cong ([\![\lfloor A \rfloor]\!] \times [\![\lfloor B \rfloor]\!]) \times [\![\lfloor C \rfloor]\!] = [\![(A \wedge B) \wedge C]\!]$. By contrast, $[\![A \wedge (B \vee C)]\!] = [\![\lfloor A \rfloor]\!] \times \neg\neg\neg([\![B^{\perp} \wedge C^{\perp}]\!])$.

This leads to the question of whether it is possible to give a 'most general' cut elimination procedure for classical proofs, and whether it is possible to translate it into a constructive framework. The translation given here is not *fully complete* — because more diverse behaviours exist in the target language than in the symmetric λ-calculus — is it possible to capture all cut elimination behaviours with a full completeness result?

Acknowledgements

Most of the work reported here was undertaken at the LFCS, University of Edinburgh, as part of the UK EPSRC project 'Foundational Structures for Computing Science'. I would like to thank Christian Urban for helpful discussions, and the anonymous referees for their comments. I used Paul Taylor's prooftree package.

References

1. S. Abramsky, R. Jagadeesan. Games and full completeness for multiplicative linear logic. *Journal of Symbolic Logic*, 59:543–574, 1994.
2. F. Barbanera, and S. Berardi. A symmetric lambda calculus for classical program extraction. *Information and Computation*, 125:103–117, 1996.
3. H. P. Barendregt. *The Lambda Calculus, its Syntax and Semantics, revised edition.* North-Holland, 1984.
4. A. Blass. A game semantics for linear logic. *Annals of Pure and Applied logic*, 56:183 – 220, 1992.
5. T. Coquand. A semantics of evidence for classical arithmetic. *Journal of Symbolic Logic*, 60:325–337, 1995.
6. T. Coquand. Computational content of classical logic. In A. Pitts and P. Dybjer, editor, *Semantics and Logics of Computation.* Cambridge University Press, 1997.
7. V. Danos, J.-B. Joinet and H. Schellinx. A new deconstructive logic, linear logic. *Journal of Symbolic Logic*, 62(3), 1996.
8. G. Gentzen. Unterschungen über das logische Schließen I and II. *Mathematische Zeitschrift*, 39:176–210, 405–431, 1935.
9. J.-Y. Girard. Towards a geometry of interaction. In *Categories in Computer Science and Logic*, volume 92 of *Contemporary Mathematics*, 1989.
10. J.-Y. Girard. A new constructive logic: classical logic. *Mathematical structures in Computer Science*, 1:259–256, 1991.
11. J.-Y. Girard. On the unity of logic. *Annals of Pure and Applied Logic*, 59:201–217, 1993.
12. J.-Y. Girard, P. Taylor and Y. Lafont. *Proofs and Types.* Cambridge University Press, 1990.
13. K. Gödel. On intuitionistic arithmetic and number theory. In M. Davis, editor, *The undecidable*, pages 75–81. Raven Press, 1965.
14. T. Griffin. A formulae-as-types notion of control. In *Proc. ACM Conference on the Principles of Programming Languages*, pages 47 – 58. ACM Press, 1990.
15. M. Hennessey, and E. Ashcroft. Parameter passing and non-determinism. In *Conference report of the ninth ACM Conference on Theory of Computing*, pages 306–311, 1977.
16. M. Parigot. $\lambda\mu$ calculus: an algorithmic interpretation of classical natural deduction. In *Proc. International Conference on Logic Programming and Automated Reasoning*, pages 190–201. Springer, 1992.
17. C. Urban, and G. Bierman. Strong normalisation of cut-elimination in classical logic. In *Proceedings of TLCA '99*, number 1581 in LNCS, pages 356 – 380. Springer, 1999.

A Token Machine for Full Geometry of Interaction
(Extended Abstract)

Olivier Laurent

Institut de Mathématiques de Luminy
163, avenue de Luminy - case 907
13288 MARSEILLE cedex 09 FRANCE
olaurent@iml.univ-mrs.fr

Abstract. We present an extension of the Interaction Abstract Machine (IAM) [10, 4] to full Linear Logic with Girard's Geometry of Interaction (GoI) [6]. We propose a simplified way to interpret the additives and the interaction between additives and exponentials by means of *weights* [7]. We describe the interpretation by a token machine which allows us to recover the usual • ••• case by forgetting all the additive information.

The Geometry of Interaction (GoI), introduced by Girard [5], is an interpretation of proofs (programs) by bideterministic automata, turning the global cut elimination steps (β-reduction) into local transitions [10, 4]. Because of its local feature, the GoI has proved to be a useful tool for studying the theory and implementation of optimal reduction of the λ-calculus [8, 2]. It is also strongly connected to some work on games semantics for Linear Logic and PCF ([1, 3] for example). Maybe the most exciting use of the locality of GoI is the current work, aiming at using it for implementing some parallel execution schemes [11].

Although the GoI has been very present in various works, its most popular version only deals with the MELL fragment of Linear Logic (which is sufficient for encoding the λ-calculus though). Girard proposed an extension of GoI to the additives [6], but his solution is quite technical (it makes an important use of an equivalence relation and entails a complex interpretation of the exponentials), which is probably the reason why it had not the same success as the MELL case. A first difficulty is to manage the "non linearity" of the additive cut elimination step (see below). A subtler problem comes from the interaction between additives and exponentials; in particular we will see that the weakening rule becomes very tricky to handle in presence of additives. Note that this is linked to the problem (still open at the time of this writing) of finding a good proof-net syntax for full LL.

In this paper, although we don't claim to give the final word on the GoI for LL, we propose an interpretation with an abstract machine, based on additive *weights* [7]. This greatly simplifies Girard's interpretation since we don't have to work up to isomorphism. We hope that this will prove to be a determining step towards the extension to additives of optimal reduction, games, ...

S. Abramsky (Ed.): TLCA 2001, LNCS 2044, pp. 283–297, 2001.

Additive linearity. A naive look at the usual &/⊕ cut elimination step of MALL shows that this reduction step is not so "linear": the sub-proof π_2 is completely erased.

$$\dfrac{\dfrac{\dfrac{\pi_1}{\vdash \Gamma, A} \quad \dfrac{\pi_2}{\vdash \Gamma, B}}{\vdash \Gamma, A \,\&\, B}\,\& \qquad \dfrac{\dfrac{\pi_3}{\vdash \Delta, A^\perp}}{\vdash \Delta, A^\perp \oplus B^\perp}\,\oplus_1}{\vdash \Gamma, \Delta}\,cut \quad \to \quad \dfrac{\dfrac{\pi_1}{\vdash \Gamma, A} \quad \dfrac{\pi_3}{\vdash \Delta, A^\perp}}{\vdash \Gamma, \Delta}\,cut$$

This is too drastic to be interpreted by the very local approach of GoI. To "linearize" this cut elimination step, Girard introduced the $\flat$-rules [6], in the calculus LL$^\flat$, which allow to keep the proof π_2 after reduction but marked with a $\flat$ symbol.

Soundness. By the modification of the additive reduction in LL$^\flat$, we get the preservation of the GoI by $\flat$-reduction. Although an LL$^\flat$-normal form is not an LL-normal form (it still has some $\flat$-rules), one can easily extract the latter from the former. This extraction procedure erases parts of the proof that had been memorized along the $\flat$-reduction thus it doesn't respect the GoI interpretation. This is why we will only show a soundness result for LL-proofs of $\vdash 1 \oplus 1$.

Proofs of $\vdash 1 \oplus 1$ give an encoding of booleans since there are exactly two normal proofs of this sequent in LL. The restriction to these boolean results is very drastic but sufficient, from a computational point of view, to distinguish different results (see [5] for a longer discussion).

Without clearly decomposing LL cut elimination into these two steps ($\flat$-reduction and extraction), the study of the modifications of the GoI interpretation during LL-reduction would be very complicated and the results very difficult to express and to prove. Moreover this precise analysis allows us to introduce a parallel version of the automaton which leads to simpler results in both steps (Propositions 1 and 2). This parallel approach may probably also be used to define GoI for the system LLP [9] for classical logic which contains generalized structural rules.

Additives and weakening. The other main technical (and complicated) point is the interaction between additives and exponentials, in particular the interaction with weakening (or $\perp$). According to its erasing behavior, the usual interpretation of a weakened formula is empty. In an additive setting, this idea leads to an inconsistency:

$$\dfrac{\dfrac{\dfrac{\dfrac{\dfrac{}{\vdash 1}\,1}{\vdash 1 \oplus 1}\,\oplus_1}{\vdash 1 \oplus 1, \perp}\,\perp \quad \dfrac{\dfrac{\dfrac{}{\vdash 1}\,1}{\vdash 1 \oplus 1}\,\oplus_2}{\vdash 1 \oplus 1, \perp}\,\perp}{\vdash 1 \oplus 1, \perp \,\&\, \perp}\,\& \qquad \dfrac{\dfrac{}{\vdash 1}\,1}{\vdash 1 \oplus 1}\,\oplus_i}{\vdash 1 \oplus 1}\,cut$$

there is no way to know if the $\oplus_1$ proof is "attach" to the left or to the right part of the &, if $\perp$ is empty. Thus the GoI interpretation of this proof doesn't depend on the value of i which is crucial since it determinates the boolean corresponding to the normal form. To solve this problem, we have to modify the weakening rule by attaching the weakened formula to a formula in the context (which corresponds to encoding $\bot$ by $\exists\alpha(\alpha \otimes \alpha^\bot)$, see [6]) ensuring that an explicit information in the GoI interpretation indicates which $\oplus$ is in the left and in the right.

Sequent calculus vs. proof-nets. The idea of GoI comes from the geometric representation of proofs given by proof-nets [7]. However, the technology of proof-nets for additives is not completely satisfactory, in particular because there is no good cut elimination procedure. Moreover using proof-nets would require a definition of $\flat$-proof-nets. For these reasons, we will interpret proofs in sequent calculus and prove our results for this interpretation but it is easy to define the interpretation of proof-nets while not talking about cut elimination.

The presentation is done in three distinct steps: first the MALL case, then we add the constants and eventually we obtain the full case by adding the exponentials. In this way, it is easier to see the modularity of the construction and to see which part of the interpretation corresponds to which subpart of Linear Logic. By forgetting the adequate constructions, we can easily obtain GoI for various fragments of LL, in particular we recover the usual IAM [4] for MELL.

1 Sequent Calculus MALL$^\flat$

To give the interpretation of proofs, we have to be very precise about the distinct occurrences of formulas. This is why we introduce annotations with indexes in the rules of the sequent calculus.

1.1 Usual MALL Sequent Calculus

$$\frac{}{\vdash A, A^\bot}\,ax \qquad \frac{\vdash \Gamma_1, A \qquad \vdash A^\bot, \Delta_1}{\vdash \Gamma, \Delta}\,cut$$

$$\frac{\vdash \Gamma_1, A \qquad \vdash \Delta_1, B}{\vdash \Gamma, \Delta, A \otimes B}\,\otimes \qquad \frac{\vdash \Gamma_1, A, B}{\vdash \Gamma, A \parr B}\,\parr$$

$$\frac{\vdash \Gamma_1, A \qquad \vdash \Gamma_2, B}{\vdash \Gamma, A \,\&\, B}\,\& \qquad \frac{\vdash \Gamma_1, A}{\vdash \Gamma, A \oplus B}\,\oplus_1 \qquad \frac{\vdash \Gamma_1, B}{\vdash \Gamma, A \oplus B}\,\oplus_2$$

1.2 $\flat$-Rules

We have to introduce a new symbol $\flat$, for marking some "partial" sequents in proofs, this is not a formula and thus no connective can be applied on it.

$$\frac{}{\vdash \Gamma, \flat}\,\flat \qquad \frac{\vdash \Gamma_1, \Delta_1 \qquad \vdash \Gamma_2, \flat \qquad \vdash \Delta_2, \flat}{\vdash \Gamma, \Delta}\,s^\flat$$

The two $\flat$-premises of the $s^\flat$-rule are used to memorize some sub-proofs through the additive reduction step (see Sect. 1.4).

A proof of a sequent containing the symbol $\flat$ is a kind of partial proof where some sub-proof is missing.

Definition 1 (Weight). *Given a set of* elementary weights, *i.e., boolean variables, a* basic weight *is an elementary weight p or a negation of an elementary weight $\bar{p}$ and a* weight *is a product (conjunction) of basic weights.*

As a convention, we use 1 for the empty product and 0 for a product where p and $\bar{p}$ appear. We also replace $p.p$ by p and $\bar{\bar{p}}$ by p. With this convention we say that the weight w depends on *p when p or $\bar{p}$ appears in w.*

We use the notations $w(p)$ (resp. $w(\bar{p})$) if p (resp. $\bar{p}$) appears in w and $w(\not{p})$ if w doesn't depend on p. The product of weights is denoted by $w.w'$.

We will consider weighted proofs, i.e., with a basic weight associated to each &-rule and to each $s^\flat$-rule. These two kinds of rules are called *sided rules*. For a &-rule, the sub-proof of the left (resp. right) premise is called its left (resp. right) side and for a $s^\flat$-rule the sub-proof of $\vdash \Gamma_1, \Delta_1$ is the left side and the sub-proofs of $\vdash \Gamma_2, \flat$ and $\vdash \Delta_2, \flat$ are the right side.

A weight describes a choice for the &-rules of one of their two premises. It corresponds to the notion of *additive slice* [7], that is the multiplicative proofs obtained by projecting each & on one of its sides.

Definition 2 (Correct weighting). *A weighted proof has a* correct weighting *when two sided rules have a basic weight corresponding to the same elementary weight only if they are in the left side and in the right side of a same sided rule (i.e., an elementary weight never appears twice in the same additive slice of a proof).*

1.3 The $\natural$-Translation

We are only interested in proofs of LL sequents (without $\flat$), $\flat$-rules are used as an intermediary step for the interpretation. This is why in the sequel we will consider only $\mathsf{LL}^\flat$ proofs of LL sequents.

There exists an easy way to transform such an $\mathsf{LL}^\flat$ proof π into an LL one $\pi^\natural$ called the *$\natural$-translation*: for LL-rules just change nothing and for each $s^\flat$-rule erase the right side and connect the left side to the conclusion.

1.4 Cut Elimination

For the $\mathsf{LL}^\flat$ sequent calculus, the cut elimination procedure is the usual one except for the additive step:

$$\dfrac{\dfrac{\dfrac{\pi_1}{\vdash \Gamma_1, A} \qquad \dfrac{\pi_2}{\vdash \Gamma_2, B}}{\vdash \Gamma_3, A \,\&\, B}\,\& \qquad \dfrac{\dfrac{\pi_3}{\vdash \Delta_1, A^\perp}}{\vdash \Delta_2, A^\perp \oplus B^\perp}\,\oplus_1}{\vdash \Gamma, \Delta}\,cut$$

$$\downarrow$$

$$\dfrac{\dfrac{\dfrac{\pi_1}{\vdash \Gamma_1, A} \quad \dfrac{\pi_3^1}{\vdash \Delta_1^1, A^\perp}}{\vdash \Gamma_3, \Delta_2}\,cut \qquad \dfrac{\dfrac{\pi_2}{\vdash \Gamma_2, B} \quad \dfrac{}{\vdash B^\perp, \flat}\,\flat}{\vdash \Gamma_4, \flat}\,cut \qquad \dfrac{\dfrac{\pi_3^2}{\vdash \Delta_1^2, A^\perp} \quad \dfrac{}{\vdash A, \flat}\,\flat}{\vdash \Delta_3, \flat}\,cut}{\vdash \Gamma, \Delta}\,s^\flat$$

For such a cut elimination step between a &-rule and a $\oplus_i$-rule, we can define a canonical weighting for the new proof from the one on the initial proof by associating to the $s^\flat$-rule the basic weight p if $i = 1$ and $\bar{p}$ if $i = 2$ where p is the basic weight of the &-rule.

Due to this modified reduction step, MALL is a sub-system of $\mathsf{MALL}^\flat$ which is not stable by reduction.

Remark 1. This new additive step is now "really" linear if we consider sub-proofs with their additive weight: before reduction we have $p.\pi_1 + \bar{p}.\pi_2 + \pi_3$ and after $p.\pi_1 + \bar{p}.\pi_2 + p.\pi_3^1 + \bar{p}.\pi_3^2$. Notice that the $\flat$-premises of the $s^\flat$-rule are crucial for this purpose: one for $\bar{p}.\pi_2$ and the other one for $\bar{p}.\pi_3$.

We also have to define new commutative steps for the $s^\flat$-rule:

$$\dfrac{\dfrac{\dfrac{\pi_1}{\vdash \Gamma_1, \Delta_1, C_1} \quad \dfrac{\pi_2}{\vdash \Gamma_2, C_2, \flat} \quad \dfrac{\pi_3}{\vdash \Delta_2, \flat}}{\vdash \Gamma_3, \Delta_3, C}\, s^\flat \qquad \dfrac{\pi_4}{\vdash \Sigma_1, C^\perp}}{\vdash \Gamma, \Delta, \Sigma}\, \mathit{cut}$$

$$\downarrow$$

$$\dfrac{\dfrac{\dfrac{\pi_1}{\vdash \Gamma_1, \Delta_1, C_1} \quad \dfrac{\pi_4^1}{\vdash \Sigma_1, C_1^\perp}}{\vdash \Gamma_3, \Delta_3, \Sigma_3}\, \mathit{cut} \qquad \dfrac{\dfrac{\pi_2}{\vdash \Gamma_2, C_2, \flat} \quad \dfrac{\pi_4^2}{\vdash \Sigma_2, C_2^\perp}}{\vdash \Gamma_4, \Sigma_4, \flat}\, \mathit{cut} \qquad \dfrac{\pi_3}{\vdash \Delta_2, \flat}}{\vdash \Gamma, \Delta, \Sigma}\, s^\flat$$

and the corresponding one for a cut on a formula in Δ.

For the other cut elimination steps, the new weighting is easy to define; when a sub-proof is duplicated, we preserve the same basic weights in the two copies.

We will now always consider proofs with correct weightings, noting that correctness is preserved by reduction.

Definition 3 (Quasi-normal form). *A proof in* $\mathsf{LL}^\flat$ *is said to be in* quasi-normal form *if it cannot be reduced by any step described above.*

Remark 2. A proof in quasi-normal form contains only cuts in which at least one of the two occurrences of the cut formula has been introduced by a $\flat$ axiom rule and used only in *cut*-rules.

It is possible to define a general cut elimination procedure as in [6] for $\mathsf{LL}^\flat$, but it would be more complicated and useless because we can remark that the $\natural$-translation of a proof of an LL sequent in quasi-normal form is a normal proof in LL.

2 The Interaction Abstract Machine

We now define the Interaction Abstract Machine (IAM) for $\mathsf{MALL}^\flat$. Forgetting the additive informations gives back the multiplicative IAM [4].

2.1 Tokens and Machine

Definition 4 (Token). *For the multiplicative-additive case, a* token *is a tuple* (m, a, w) *where* m *and* a *are stacks (*ε *will denote the empty stack) built on letters* $\{g, d\}$ *(Girard's notations corresponding to the french* gauche *and* droite*) and* w *is a weight.*

Definition 5 (Abstract machine). *A* state *of the* machine M_π *associated to the proof* π *of* $\mathsf{LL}^\flat$ *is* $F^\uparrow(m, a, w)$ *or* $F^\downarrow(m, a, w)$ *or* $\emptyset$ *where* F *is an occurrence of a formula appearing in the proof and the arrow indicates if the token* (m, a, w) *is going upwards or downwards.* $\emptyset$ *means that the machine stops.*

The transitions of M_π *through the rules of* π *are described in Figs. 1 and 2.* Γ *(resp.* Δ*) is used for one of the formulas of the multiset* Γ *(resp.* Δ*), but the same before and after the transition. If the result of a transition contains a weight* $w = 0$*, we consider it as* $\emptyset$*.*

Remark 3. In Fig. 2, changing the transition in the case $A \,\&\, B^\uparrow(m, g.a, w(\not{p}))$ (resp. $A \,\&\, B^\uparrow(m, d.a, w(\not{p})))$ into $A \,\&\, B^\uparrow(m, g.a, w(\not{p})) \rightarrow A^\uparrow(m, a, w.p)$ (resp. $A \,\&\, B^\uparrow(m, d.a, w(\not{p})) \rightarrow B^\uparrow(m, a, w.\bar{p})$) would make no difference since this p (resp. $\bar{p}$) information is also added when going down through the &-rule.

2.2 Properties of the Machine

Definition 6 (Partial function on tokens). *Let* π *be a proof and* A *one of its conclusions, we define the partial function* f_π *by:*

$$f_\pi(A, (m, a, w)) = \begin{cases} (B, (m', a', w')) & \text{if the computation on } A^\uparrow(m, a, w) \text{ ends} \\ & \text{by } B^\downarrow(m', a', w') \text{ whith } B \text{ conclusion of } \pi \\ \uparrow & \text{otherwise} \end{cases}$$

The partial function f_π is undefined in two cases: either if the machine stops inside the proof or if the execution doesn't terminate.

Lemma 1. *Let* π *be a proof, and* w' *and* w_0 *two weights s.t.* $w'.w_0 \neq 0$.
$f_\pi(A, (m, a, w)) = (B, (m', a', w')) \Rightarrow f_\pi(A, (m, a, w.w_0)) = (B, (m', a', w'.w_0))$

Theorem 1 (Soundness). *If* π *is a proof in* $\mathsf{MALL}^\flat$ *whose quasi-normal form is* π_0 *then for each pair formula-token* j*:*

- *if* $f_\pi(j) = \uparrow$ *then* $f_{\pi_0}(j) = \uparrow$
- *if* $f_{\pi_0}(j) = j'$ *then* $f_\pi(j) = j'$
- *if* $f_\pi(A, (m, a, w)) = (B, (m', a', w'))$ *and* $f_{\pi_0}(A, (m, a, w)) = \uparrow$ *then there exists* w_0 *s.t.* $f_{\pi_0}(A, (m, a, w.w_0)) = (B, (m', a', w'.w_0))$ *with* $w'.w_0 \neq 0$.

Moreover, if the execution in M_π *is infinite, it is infinite in* M_{π_0}*.*

$$
\begin{array}{ll}
\multicolumn{2}{c}{ax} \\
A^{\uparrow}(m,a,w) & \rightarrow A^{\perp\downarrow}(m,a,w) \\
A^{\perp\uparrow}(m,a,w) & \rightarrow A^{\downarrow}(m,a,w)
\end{array}
\qquad
\begin{array}{ll}
\multicolumn{2}{c}{cut} \\
A^{\downarrow}(m,a,w) & \rightarrow A^{\perp\uparrow}(m,a,w) \\
A^{\perp\downarrow}(m,a,w) & \rightarrow A^{\uparrow}(m,a,w) \\
\Gamma^{\uparrow}(m,a,w) & \rightarrow \Gamma_1^{\uparrow}(m,a,w) \\
\Delta^{\uparrow}(m,a,w) & \rightarrow \Delta_1^{\uparrow}(m,a,w) \\
\Gamma_1^{\downarrow}(m,a,w) & \rightarrow \Gamma^{\downarrow}(m,a,w) \\
\Delta_1^{\downarrow}(m,a,w) & \rightarrow \Delta^{\downarrow}(m,a,w)
\end{array}
$$

$$
\begin{array}{ll}
\multicolumn{2}{c}{\otimes} \\
A\otimes B^{\uparrow}(g.m,a,w) & \rightarrow A^{\uparrow}(m,a,w) \\
A\otimes B^{\uparrow}(d.m,a,w) & \rightarrow B^{\uparrow}(m,a,w) \\
A\otimes B^{\uparrow}(\varepsilon,a,w) & \rightarrow \emptyset \\
A^{\downarrow}(m,a,w) & \rightarrow A\otimes B^{\downarrow}(g.m,a,w) \\
B^{\downarrow}(m,a,w) & \rightarrow A\otimes B^{\downarrow}(d.m,a,w) \\
\Gamma^{\uparrow}(m,a,w) & \rightarrow \Gamma_1^{\uparrow}(m,a,w) \\
\Delta^{\uparrow}(m,a,w) & \rightarrow \Delta_1^{\uparrow}(m,a,w) \\
\Gamma_1^{\downarrow}(m,a,w) & \rightarrow \Gamma^{\downarrow}(m,a,w) \\
\Delta_1^{\downarrow}(m,a,w) & \rightarrow \Delta^{\downarrow}(m,a,w)
\end{array}
\qquad
\begin{array}{ll}
\multicolumn{2}{c}{\parr} \\
A\parr B^{\uparrow}(g.m,a,w) & \rightarrow A^{\uparrow}(m,a,w) \\
A\parr B^{\uparrow}(d.m,a,w) & \rightarrow B^{\uparrow}(m,a,w) \\
A\parr B^{\uparrow}(\varepsilon,a,w) & \rightarrow \emptyset \\
A^{\downarrow}(m,a,w) & \rightarrow A\parr B^{\downarrow}(g.m,a,w) \\
B^{\downarrow}(m,a,w) & \rightarrow A\parr B^{\downarrow}(d.m,a,w) \\
\Gamma^{\uparrow}(m,a,w) & \rightarrow \Gamma_1^{\uparrow}(m,a,w) \\
\Gamma_1^{\downarrow}(m,a,w) & \rightarrow \Gamma^{\downarrow}(m,a,w)
\end{array}
$$

Fig. 1. Identity and multiplicative groups.

$$
\begin{array}{ll}
\multicolumn{2}{c}{\&} \\
A\,\&\,B^{\uparrow}(m,g.a,w(p)) & \rightarrow A^{\uparrow}(m,a,w(p)) \\
A\,\&\,B^{\uparrow}(m,g.a,w(\not p)) & \rightarrow A^{\uparrow}(m,a,w(\not p)) \\
A\,\&\,B^{\uparrow}(m,g.a,w(\bar p)) & \rightarrow \emptyset \\
A\,\&\,B^{\uparrow}(m,d.a,w(\bar p)) & \rightarrow B^{\uparrow}(m,a,w(\bar p)) \\
A\,\&\,B^{\uparrow}(m,d.a,w(\not p)) & \rightarrow B^{\uparrow}(m,a,w(\not p)) \\
A\,\&\,B^{\uparrow}(m,d.a,w(p)) & \rightarrow \emptyset \\
A\,\&\,B^{\uparrow}(m,\varepsilon,w) & \rightarrow \emptyset \\
A^{\downarrow}(m,a,w) & \rightarrow A\,\&\,B^{\downarrow}(m,g.a,w.p) \\
B^{\downarrow}(m,a,w) & \rightarrow A\,\&\,B^{\downarrow}(m,d.a,w.\bar p) \\
\Gamma^{\uparrow}(m,a,w(p)) & \rightarrow \Gamma_1^{\uparrow}(m,a,w(p)) \\
\Gamma^{\uparrow}(m,a,w(\bar p)) & \rightarrow \Gamma_2^{\uparrow}(m,a,w(\bar p)) \\
\Gamma^{\uparrow}(m,a,w(\not p)) & \rightarrow \emptyset \\
\Gamma_1^{\downarrow}(m,a,w) & \rightarrow \Gamma^{\downarrow}(m,a,w.p) \\
\Gamma_2^{\downarrow}(m,a,w) & \rightarrow \Gamma^{\downarrow}(m,a,w.\bar p)
\end{array}
\qquad
\begin{array}{ll}
\multicolumn{2}{c}{\oplus_1} \\
A\oplus B^{\uparrow}(m,g.a,w) & \rightarrow A^{\uparrow}(m,a,w) \\
A\oplus B^{\uparrow}(m,d.a,w) & \rightarrow \emptyset \\
A\oplus B^{\uparrow}(m,\varepsilon,w) & \rightarrow \emptyset \\
A^{\downarrow}(m,a,w) & \rightarrow A\oplus B^{\downarrow}(m,g.a,w) \\
\Gamma^{\uparrow}(m,a,w) & \rightarrow \Gamma_1^{\uparrow}(m,a,w) \\
\Gamma_1^{\downarrow}(m,a,w) & \rightarrow \Gamma^{\downarrow}(m,a,w)
\end{array}
$$

$$
\begin{array}{ll}
\multicolumn{2}{c}{\flat} \\
\Gamma^{\uparrow}(m,a,w) & \rightarrow \emptyset
\end{array}
\qquad
\begin{array}{ll}
\multicolumn{2}{c}{s^{\flat}} \\
\Gamma^{\uparrow}(m,a,w(p)) & \rightarrow \Gamma_1^{\uparrow}(m,a,w(p)) \\
\Gamma^{\uparrow}(m,a,w(\bar p)) & \rightarrow \Gamma_2^{\uparrow}(m,a,w(\bar p)) \\
\Gamma^{\uparrow}(m,a,w(\not p)) & \rightarrow \emptyset \\
\Delta^{\uparrow}(m,a,w(p)) & \rightarrow \Delta_1^{\uparrow}(m,a,w(p)) \\
\Delta^{\uparrow}(m,a,w(\bar p)) & \rightarrow \Delta_2^{\uparrow}(m,a,w(\bar p)) \\
\Delta^{\uparrow}(m,a,w(\not p)) & \rightarrow \emptyset \\
\Gamma_1^{\downarrow}(m,a,w) & \rightarrow \Gamma^{\downarrow}(m,a,w.p) \\
\Gamma_2^{\downarrow}(m,a,w) & \rightarrow \Gamma^{\downarrow}(m,a,w.\bar p) \\
\Delta_1^{\downarrow}(m,a,w) & \rightarrow \Delta^{\downarrow}(m,a,w.p) \\
\Delta_2^{\downarrow}(m,a,w) & \rightarrow \Delta^{\downarrow}(m,a,w.\bar p)
\end{array}
$$

Fig. 2. Additive and $\flat$ groups. (p is the basic weight associated to the &-rule or to the $s^{\flat}$-rule and the $\oplus_2$ is easy to define from $\oplus_1$)

The introduction of the weight w_0 corresponds to the transformation of π_3 into $p.\pi_3 + \bar{p}.\pi_3$ during additive cut elimination (see Remark 1). Before reduction we don't need any information about p to go in π_3 but after reduction we have to know if we go to $p.\pi_3$ or to $\bar{p}.\pi_3$.

Proof. We have to prove that for each step of cut elimination the theorem is true and then by an easy induction on the length of a normalization we obtain the result.

We suppose that the cut-rule which we are eliminating is the last rule of the proof π and we obtain a proof π'. If it is not the case, we just have to remark that adding the same new rules at the end of π and π' is correct with respect to the interpretation.

We only consider the case of the additive cut elimination step (figure in Sect. 1.4) which is the most important one, the others are left to the reader.

We use the notation $j = (\Gamma, t)$ or $s = \Gamma^{\uparrow}(m, a, w)$ to say that the formula we are talking about is in the multiset Γ (idem for Δ, ...). Moreover $f(\Gamma, t) = (\Gamma, t')$ doesn't necessarily mean that the formula is the same before and after the computation.

Let p be the basic weight associated to the &-rule. We study the different possible cases for j:

- if $j = (\Gamma, (m, a, w(p)))$, we look at the sequence $s_1, s_2, \dots$ (resp. $s'_1, s'_2, \dots$) of the states $F^{\uparrow}(m, a, w)$ in the conclusions of the sub-proofs π_1, π_2 and π_3 (resp. π_1, π_2, π_3^1 and π_3^2) during the computation of M_{π} (resp. $M_{\pi'}$) on the state associated to j. In fact these states will always be in the conclusions of π_1 and π_3 (resp. π_1 and π_3^1) with s_1 in Γ_1, more precisely:
  - if $s_{2i+1} = F^{\uparrow}(m, a, w)$ with $F = \Gamma_1$ or A, $s_{2i+2} = A^{\perp\uparrow}(m', a', w')$ with $f_{\pi_1}(F, (m, a, w)) = (A, (m', a', w'))$ or s_{2i+2} doesn't exist;
  - if $s_{2i} = A^{\perp\uparrow}(m, a, w)$, $s_{2i+1} = A^{\uparrow}(m', a', w')$ with $f_{\pi_3}(A^{\perp}, (m, a, w)) = (A^{\perp}, (m', a', w'))$ or s_{2i+1} doesn't exist.

  The same facts occur for the s'_i by replacing π_3 with π_3^1 so we have $\forall i, s_i = s'_i$. If $s_1, s_2, \dots$ is infinite, $s'_1, s'_2, \dots$ too. There are two different reasons for $s_1, s_2, \dots$ to be finite, if s_n is the last state and is in the conclusions of π_k: either the evaluation of the corresponding machine M_{π_k} is infinite (or undefined at a step) on s_n and the same thing occurs in π' or f_{π_k} on this state gives a result j' in the context and $f_{\pi}(j) = j'$ (idem in π').
- if $j = (\Gamma, (m, a, w(\bar{p})))$, either $f_{\pi_2}(j) = (\Gamma, t')$ and $f_{\pi}(j) = (\Gamma, t') = f_{\pi'}(j)$ or $f_{\pi_2}(j) = (B, t')$ and $f_{\pi}(j) = \uparrow = f_{\pi'}(j)$;
- if $j = (\Gamma, (m, a, w(\not{p})))$, $f_{\pi}(j) = \uparrow$ and $f_{\pi'}(j) = \uparrow$;
- if $j = (\Delta, (m, a, w(p)))$, similar to the $(\Gamma, (m, a, w(p)))$ case;
- if $j = (\Delta, (m, a, w(\bar{p})))$, either $f_{\pi_3}(j) = (\Delta, t')$ and $f_{\pi}(j) = (\Delta, t') = f_{\pi'}(j)$ or $f_{\pi_3}(j) = (A^{\perp}, t')$ and $f_{\pi}(j) = \uparrow = f_{\pi'}(j)$;
- if $j = (\Delta, (m, a, w(\not{p})))$, $f_{\pi'}(j) = \uparrow$ but $f_{\pi}(j)$ may be defined with $f_{\pi}(j) = (F, (m', a', w'))$ in this situation we have by Lemma 1 (noting that $w'.p \neq 0$ for a correct weighting) and by applying the case $j = (\Delta, (m, a, w(p)))$, $f_{\pi'}(\Delta, (m, a, w.p)) = (F, (m', a', w'.p)) = f_{\pi}(\Delta, (m, a, w.p))$. This case is

very important because it is characteristic of the fact that f_π and $f_{\pi'}$ may differ. □

Corollary 1. *If w is a weight s.t. for all elementary weight p of π, $p \in w$ or $\bar{p} \in w$ then $f_\pi(F, (m, a, w)) = f_{\pi_0}(F, (m, a, w))$.*

Theorem 2 (Termination). *Let π be a proof, A a conclusion of π and t a token, the execution of the machine on $A^\uparrow(t)$ terminates.*

Proof. Let π_0 be a quasi-normal form of π. By Theorem 1, we have to prove that the execution of the machine associated to π_0 on $A^\uparrow(t)$ terminates.

In π_0, if the execution never uses the transition of a cut formula, either it stops in a transition or it goes up to an axiom and then down to a conclusion so it terminates. Moreover, by the definition of a quasi-normal form, the *cut*-rules appearing in π_0 are of the form:

$$\dfrac{\dfrac{\vdots}{\vdash \Gamma, A} \qquad \dfrac{\dfrac{\dfrac{}{\vdash A^\perp, \flat}\ \flat}{\vdots}\ cut}{}}{\vdash \Gamma, \flat}\ cut$$

and if the execution uses the transition on A in such a *cut*-rule, it stops in the $\flat$-rule. Thus the evaluation is always finite in a quasi-normal form. □

2.3 The Parallel IAM

In order to complete some transitions on which the IAM stops, we can introduce a parallel version of the machine for which states are formal sums of states of the IAM with 0 for the empty sum. To define the parallel machine M_π^p associated to a proof π, we modify some particular transitions and we replace $\emptyset$ by 0:

$$\&$$
$$\Gamma^\uparrow(m, a, w(\not p)) \to \Gamma_1^\uparrow(m, a, w.p) + \Gamma_2^\uparrow(m, a, w.\bar{p})$$

$$s^\flat$$
$$\Gamma^\uparrow(m, a, w(\not p)) \to \Gamma_1^\uparrow(m, a, w.p) + \Gamma_2^\uparrow(m, a, w.\bar{p})$$
$$\Delta^\uparrow(m, a, w(\not p)) \to \Delta_1^\uparrow(m, a, w.p) + \Delta_2^\uparrow(m, a, w.\bar{p})$$

We denote by f_π^p the partial function associated to this machine and defined like f_π (Definition 6). To simplify the results (formal sums of pairs formula-token), we use the following rewriting rule:

$$(A, (m, a, w.p)) + (A, (m, a, w.\bar{p})) \quad \to \quad (A, (m, a, w))$$

Proposition 1 (Parallel soundness). *If π is a proof whose quasi-normal form is π_0 then $f_\pi^p = f_{\pi_0}^p$.*

When a weight information is missing, the parallel machine tries all the possibilities thus it doesn't need any starting information. This is why the requirement of an additional weight w_0 in Theorem 1 disappears.

3 Adding the Constants

3.1 Rules and Machine

$$\frac{}{\vdash 1}\,1 \qquad \frac{\vdash \Gamma_1, A_1}{\vdash \Gamma, A, \bot}\,\bot \qquad \frac{\vdash \Gamma_1, \flat}{\vdash \Gamma, \top}\,\top$$

As explained in the introduction, we have to modify the $\bot$-rule by distinguishing a particular formula in the context.

We can extend the $\natural$-translation without any loss of its properties by replacing each $\top$-rule by the usual one $\frac{}{\vdash \Gamma, \top}$ and by erasing everything above it.

For the multiplicative constants, the cut elimination is as usual. For the additive constants, we obtain:

$$\frac{\dfrac{\vdash \Gamma_2, A_1, \flat}{\vdash \Gamma_1, A, \top_1}\,\top \qquad \vdash \Delta_1, A^\perp}{\vdash \Gamma, \Delta, \top}\,cut \quad \rightarrow \quad \frac{\dfrac{\vdash \Gamma_2, A, \flat \qquad \vdash \Delta_2, A^\perp}{\vdash \Gamma_1, \Delta_1, \flat}\,cut}{\vdash \Gamma, \Delta, \top}\,\top$$

We extend the notion of token by using the letters $\{g, d, \Uparrow, \Downarrow\}$ for the multiplicative stack and we add new transitions for the added rules (Fig. 3).

$$\begin{array}{lll}
\quad 1 & \quad \bot & \\
1^\uparrow(m,a,w) \rightarrow 1^\downarrow(m,a,w) & A^\uparrow(m,a,w) \rightarrow \bot^\downarrow(\Uparrow.m,a,w) & \\
 & A_1^\downarrow(m,a,w) \rightarrow \bot^\downarrow(\Downarrow.m,a,w) & \\
 & \bot^\uparrow(\Uparrow.m,a,w) \rightarrow A_1^\uparrow(m,a,w) & \\
\quad \top & \bot^\uparrow(\Downarrow.m,a,w) \rightarrow A^\downarrow(m,a,w) & \\
\top^\uparrow(m,a,w) \rightarrow \top^\downarrow(m,a,w) & \bot^\uparrow(m,a,w) \rightarrow \emptyset & m \neq \Uparrow.m', \Downarrow.m' \\
\Gamma^\uparrow(m,a,w) \rightarrow \Gamma_1^\uparrow(m,a,w) & \Gamma^\uparrow(m,a,w) \rightarrow \Gamma_1^\uparrow(m,a,w) & \\
\Gamma_1^\downarrow(m,a,w) \rightarrow \Gamma^\downarrow(m,a,w) & \Gamma_1^\downarrow(m,a,w) \rightarrow \Gamma^\downarrow(m,a,w) &
\end{array}$$

Fig. 3. Constant group.

Theorem 1.a (Soundness *continued*). *The Theorem 1 is still true in* $\mathsf{MALL}^\flat$ *with constants.*

3.2 Computation of Booleans

We want to compute results for the usual cut elimination procedure of LL. As already explained, we have to restrict ourselves to the particular case of proofs of $\vdash 1 \oplus 1$ that give a notion of booleans.

Lemma 2. *If* π *is a proof of* $\vdash 1$*, there exists* w *s.t.* $f_\pi(1, (\varepsilon, \varepsilon, w)) = (1, (\varepsilon, \varepsilon, w))$.

Lemma 3. *If* π *is a proof of* $\vdash 1 \oplus 1, \flat$ *then, for any* j, $f_\pi(j) = \uparrow$.

Theorem 3. *If* π *is a proof of* $\vdash 1 \oplus 1$ *whose quasi-normal form is* π_0 *then* $\pi_0^\natural = \dfrac{\dfrac{}{\vdash 1}\,1}{\vdash 1 \oplus 1}\,\oplus_i$ *and*

– *either there exists* w *s.t.* $f_\pi(1 \oplus 1, (\varepsilon, g, w)) = (1 \oplus 1, (\varepsilon, g, w))$ *and* $i = 1$
– *or there exists* w *s.t.* $f_\pi(1 \oplus 1, (\varepsilon, d, w)) = (1 \oplus 1, (\varepsilon, d, w))$ *and* $i = 2$.

Proof. We suppose that $i = 1$ and we make an induction on π_0.

– If the last rule is $\oplus_k$ it must be a $\oplus_1$ by normalization; we apply the Lemma 2 to the premise which gives us a weight w s.t. $f_\pi(1 \oplus 1, (\varepsilon, g, w)) = (1 \oplus 1, (\varepsilon, g, w))$. Moreover for any weight w', $f_\pi(1 \oplus 1, (\varepsilon, d, w')) = \uparrow$.
– If the last rule is $s^\flat$, let p be its basic weight. We can apply the induction hypothesis to the sub-proof π_0' of $\vdash 1 \oplus 1$ and we obtain a weight w s.t. $f_{\pi_0'}(1 \oplus 1, (\varepsilon, g, w)) = (1 \oplus 1, (\varepsilon, g, w))$ so $f_{\pi_0}(1 \oplus 1, (\varepsilon, g, w.p)) = (1 \oplus 1, (\varepsilon, g, w.p))$. Moreover for any weight w', $f_{\pi_0'}(1 \oplus 1, (\varepsilon, d, w')) = \uparrow$ thus, by Lemma 3, we also have $f_{\pi_0}(1 \oplus 1, (\varepsilon, d, w')) = \uparrow$.

Finally we conclude by Theorem 1. □

We cannot assume that the weight w is empty for the evaluation of f_π because, for some proofs, $f_\pi(A, (m, a, 1)) = \uparrow$ for any A, m and a (see the proof of $\vdash 1 \oplus 1$ in the introduction, for example).

The parallel machine gives a solution for this problem since it doesn't require any initial weight information. The weight may be built dynamically thus starting with 1 is sufficient.

Proposition 2. *If* π *is a proof of* $\vdash 1 \oplus 1$ *whose quasi-normal form is* π_0 *then* $\pi_0^\natural = \dfrac{\dfrac{}{\vdash 1}\,1}{\vdash 1 \oplus 1}\,\oplus_i$ *and either* $f_\pi^p(1 \oplus 1, (\varepsilon, g, 1)) \neq 0$ *and* $i = 1$ *or* $f_\pi^p(1 \oplus 1, (\varepsilon, d, 1)) \neq 0$ *and* $i = 2$.

4 Exponentials

We have now to generalize some of our definitions of Sect. 1 to deal with the following exponential rules. The interpretation is the one defined by Danos and Regnier [4], accommodated with the additives and extended to the $?w$-rule.

4.1 Sequent Calculus

$$\frac{\vdash \Gamma, A}{\vdash ?\Gamma, !A}\,! \qquad \frac{\vdash \Gamma_1, A}{\vdash \Gamma, ?A}\,?d$$

$$\frac{\vdash \Gamma_1, B_1}{\vdash \Gamma, B, ?A}\,?w \qquad \frac{\vdash \Gamma_1, ?A_1, ?A_2}{\vdash \Gamma, ?A}\,?c \qquad \frac{\vdash \Gamma_1, ??A}{\vdash \Gamma, ?A}\,??$$

The formula B in the context of the $?w$-rule is used for the same purpose as in the $\bot$-rule. We use a *functorial* promotion and a *digging* rule ($??$-rule) instead of the usual promotion because it allows us to decompose precisely the GoI.

Definition 7 (Weight, *Definition 1 continued*). *A* copy address c *is a word built on the letters* $\{g, d\}$.

A basic weight *is now a pair of an elementary weight* p *(or its negation* $\bar{p}$*) and a copy address* c*, and is denoted by* p_c *(*$\bar{p}_c$*).*

In order to deal with the erasing of sub-proofs by the weakening cut elimination step, we will only consider proofs with no ? in conclusions. To prove the preservation of the interpretation by reduction, we can restrict cut elimination to the particular strategy reducing only exponential cuts with no context in the !-rule.

In the $?c$ cut elimination step, we obtain two copies π_1^1 and π_1^2 of the proof π_1 of $\vdash\, !A$. In π_1^1 (resp. π_1^2), we replace all the basic weights p_c by $p_{g.c}$ (resp. $p_{d.c}$).

4.2 Extending the Machine

Definition 8 (Exponential informations).

- Exponential signatures σ *and* exponential stacks s *are defined by:*

$$\begin{aligned} \sigma &::= \Box \mid g.\sigma \mid d.\sigma \mid \ulcorner\sigma\urcorner.\sigma \mid [s]^{\Uparrow} \mid [s]^{\Downarrow} \\ s &::= \varepsilon \mid \sigma.s \end{aligned}$$

We will use the notation $[s]$ *to talk about both* $[s]^{\Uparrow}$ *and* $[s]^{\Downarrow}$.

- *The* copy address $\widetilde{\sigma}$ *of an exponential signature* σ *is defined by:* $\widetilde{\Box} = \varepsilon$, $\widetilde{[s]} = \varepsilon$, $\widetilde{g.\sigma} = g.\widetilde{\sigma}$, $\widetilde{d.\sigma} = d.\widetilde{\sigma}$ *and* $\widetilde{\ulcorner\sigma'\urcorner.\sigma} = \widetilde{\sigma'}.\widetilde{\sigma}$.
- *The* copy address *of an exponential stack is:* $\widetilde{\varepsilon} = \varepsilon$ *and* $\widetilde{\sigma.s} = \widetilde{\sigma}.\widetilde{s}$.
- *We define the predicate* weak() *on signatures by:*
  - weak($\Box$) = *false and* weak($[s]$) = *true*
  - weak($g.\sigma$) = weak(σ) *and* weak($d.\sigma$) = weak(σ)
  - weak($\ulcorner\sigma'\urcorner.\sigma$) = weak($\sigma'$)

The weak() predicate tells if the leaf of the exponential branch described by σ is a $?w$-rule.

Definition 9 (Token, *Definition 4* continued). *For the full case, a* token *is a tuple* (m, a, w, b, s) *where* b *and* s *are exponential stacks. Moreover the language of* m *is extended to* $\{g, d, \Uparrow, \Downarrow, |\}$ *and the language of* a *is extended to* $\{g, d, |\}$ *(*$\Uparrow$*,* $\Downarrow$ *and* $|$ *are only used for* $\bot$ *and* $?w$*).*

Definition 10 (Type of a token). *The* type *of a token* (m, a, w, b, s) *in a formula of a proof is the pair* $(|b| - d, |s| - n)$ *where* $|.|$ *is the length of a stack,* d *is the depth of the formula in the proof (i.e., the number of !-rules below it) and* n *is the number of exponential connectives in the scope of which the subformula described by* m *and* a *is (without looking at the right of any* $|$*,* $\Uparrow$ *or* $\Downarrow$ *symbol).*

In the transitions of the machine defined in Fig. 2, we replace everywhere p by $p_{\widetilde{b}.c}$ since we have to take into account the stack b and to look at the dependency with respect to $p_{\widetilde{b}.c}$, for example:

$$A \,\&\, B^{\uparrow}(m, g.a, w(p_{\widetilde{b}.c})) \to A^{\uparrow}(m, a, w(p_{\widetilde{b}.c}))$$

For the constants (Sect. 3.1), we have to refine the $\bot$-transitions (Fig. 4).

The new transitions of the token machine for exponential rules are described in Fig. 5. Some transitions are implicit to simplify the description: if no transition appears for a state $F^{\updownarrow}(m, a, w, b, s)$, it just corresponds to the transition $F^{\updownarrow}(m, a, w, b, s) \to \emptyset$.

$$\begin{array}{lll} & \bot & \\ A^{\uparrow}(m,a,w,b,s) & \to \bot^{\downarrow}(\Uparrow.m, |.a, w, b, s) & \\ A_1^{\downarrow}(m,a,w,b,s) & \to \bot^{\downarrow}(\Downarrow.m, |.a, w, b, s) & \\ \bot^{\uparrow}(\Uparrow.m, |.a, w, b, s) & \to A_1^{\uparrow}(m,a,w,b,s) & \\ \bot^{\uparrow}(\Downarrow.m, |.a, w, b, s) & \to A^{\downarrow}(m,a,w,b,s) & \\ \bot^{\uparrow}(m,a,w,b,s) & \to \emptyset & m \neq \Uparrow.m', \Downarrow.m' \text{ or } a \neq |.a' \\ \Gamma^{\uparrow}(m,a,w,b,s) & \to \Gamma_1^{\uparrow}(m,a,w,b,s) & \\ \Gamma_1^{\downarrow}(m,a,w,b,s) & \to \Gamma^{\downarrow}(m,a,w,b,s) & \end{array}$$

Fig. 4. $\bot$-transitions (exponential case).

Lemma 4. *If the type of the starting token is (p,q) with $q \geq 0$, at any step of the execution the type of the token is (p,q') with $q' \geq 0$.*

Lemma 5. $f_\pi(F,(m,a,w,b,s)) = (F',(m',a',w',b,s'))$, *and if* $\widetilde{b} = \widetilde{b_1}$ *then* $f_\pi(F,(m,a,w,b_1,s)) = (F',(m',a',w',b_1,s'))$.

Theorem 1.b (Soundness *continued*). *The Theorem 1 is still true in $\mathsf{LL}^\flat$ for a proof without any ? in its conclusions and a token of type (p,q) with $p \geq 0$ and $q \geq 0$.*

Proof. We keep the same notations as in the proof of Theorem 1. We will look for each exponential cut at the sequence $s_1, s_2, \ldots$ (resp. $s'_1, s'_2, \ldots$) of the states $F^{\uparrow}(m,a,w,b,s)$ in the conclusions of the sub-proofs π_1 and π_2 during the computation of M_π (resp. $M_{\pi'}$) on the state associated to j. With the notations given below, we can remark that s_{2i} (resp. s_{2i+1}) will always be in the conclusions of π_1 (resp. π_2) and also for s'_{2i} and s'_{2i+1}.

We only prove the digging case, the others are left to the reader.

Digging cut. In this case, we cannot prove $\forall i, s_i = s'_i$ but only the weaker result $\forall i, s_i = F^{\uparrow}(m,a,w,b,s) \iff s'_i = F^{\uparrow}(m,a,w,b',s)$ with $\widetilde{b} = \widetilde{b'}$. Lemma 5 proves that it doesn't really matter.

$$\dfrac{\dfrac{\dfrac{\pi_1}{\vdash A}}{\vdash !A}\,! \qquad \dfrac{\dfrac{\pi_2}{\vdash \Gamma_1, ??A^\bot}}{\vdash \Gamma_2, ?A^\bot}\,??}{\vdash \Gamma}\,cut \quad \to \quad \dfrac{\dfrac{\dfrac{\dfrac{\pi_1}{\vdash A}}{\vdash !A}\,!}{\vdash !!A}\,! \qquad \dfrac{\pi_2}{\vdash \Gamma_1, ??A^\bot}}{\vdash \Gamma}\,cut$$

If s_{2i+1} exists then:

- either s_{2i+2} doesn't exist because f_{π_2} is not defined on s_{2i+1} or because $f_{\pi_2}(s_{2i+1}) \in \Gamma_1$,
- or $f_{\pi_2}(s_{2i+1}) = (??A^\bot,(m,a,w,b,s))$ with $s = \sigma.\sigma'.s'$ (Lemma 4) and $s_{2i+2} = A^{\uparrow}(m,a,w,(\ulcorner\sigma'\urcorner.\sigma).b,s')$.

Remark that the stack b of s_{2i} is always of the shape $(\ulcorner\sigma'\urcorner.\sigma).b'$. If s_{2i} exists then:

$$!$$

$$\begin{array}{lll}
!A^{\uparrow}(m,a,w,b,\sigma.s) & \rightarrow A^{\uparrow}(m,a,w,\sigma.b,s) & \neg\mathrm{weak}(\sigma) \\
!A^{\uparrow}(m,a,w,b,\sigma.s) & \rightarrow !A^{\downarrow}(m,a,w,b,\sigma.s) & \mathrm{weak}(\sigma) \\
A^{\downarrow}(m,a,w,\sigma.b,s) & \rightarrow !A^{\downarrow}(m,a,w,b,\sigma.s) & \\
?\Gamma^{\uparrow}(m,a,w,b,\sigma.s) & \rightarrow \Gamma^{\uparrow}(m,a,w,\sigma.b,s) & \\
\Gamma^{\downarrow}(m,a,w,\sigma.b,s) & \rightarrow ?\Gamma^{\downarrow}(m,a,w,b,\sigma.s) &
\end{array}$$

$$?d$$

$$\begin{array}{lll}
?A^{\uparrow}(m,a,w,b,\Box.s) & \rightarrow A^{\uparrow}(m,a,w,b,s) & \\
?A^{\uparrow}(m,a,w,b,s) & \rightarrow \emptyset & s \neq \Box.s' \\
A^{\downarrow}(m,a,w,b,s) & \rightarrow ?A^{\downarrow}(m,a,w,b,\Box.s) & \\
\Gamma^{\uparrow}(m,a,w,b,s) & \rightarrow \Gamma_1^{\uparrow}(m,a,w,b,s) & \\
\Gamma_1^{\downarrow}(m,a,w,b,s) & \rightarrow \Gamma^{\downarrow}(m,a,w,b,s) &
\end{array}$$

$$??$$

$$\begin{array}{lll}
?A^{\uparrow}(m,a,w,b,(\ulcorner\sigma'\urcorner.\sigma).s) & \rightarrow ??A^{\uparrow}(m,a,w,b,\sigma.\sigma'.s) & \\
?A^{\uparrow}(m,a,w,b,s) & \rightarrow \emptyset & s \neq (\ulcorner\sigma\urcorner.\sigma').s' \\
??A^{\downarrow}(m,a,w,b,\sigma.\sigma'.s) & \rightarrow ?A^{\downarrow}(m,a,w,b,(\ulcorner\sigma'\urcorner.\sigma).s) & \\
\Gamma^{\uparrow}(m,a,w,b,s) & \rightarrow \Gamma_1^{\uparrow}(m,a,w,b,s) & \\
\Gamma_1^{\downarrow}(m,a,w,b,s) & \rightarrow \Gamma^{\downarrow}(m,a,w,b,s) &
\end{array}$$

$$?c$$

$$\begin{array}{lll}
?A^{\uparrow}(m,a,w,b,(g.\sigma).s) & \rightarrow ?A_1{}^{\uparrow}(m,a,w,b,\sigma.s) & \\
?A^{\uparrow}(m,a,w,b,(d.\sigma).s) & \rightarrow ?A_2{}^{\uparrow}(m,a,w,b,\sigma.s) & \\
?A^{\uparrow}(m,a,w,b,s) & \rightarrow \emptyset & s \neq (g.\sigma).s', (d.\sigma).s' \\
?A_1{}^{\downarrow}(m,a,w,b,\sigma.s) & \rightarrow ?A^{\downarrow}(m,a,w,b,(g.\sigma).s) & \\
?A_2{}^{\downarrow}(m,a,w,b,\sigma.s) & \rightarrow ?A^{\downarrow}(m,a,w,b,(d.\sigma).s) & \\
\Gamma^{\uparrow}(m,a,w,b,s) & \rightarrow \Gamma_1^{\uparrow}(m,a,w,b,s) & \\
\Gamma_1^{\downarrow}(m,a,w,b,s) & \rightarrow \Gamma^{\downarrow}(m,a,w,b,s) &
\end{array}$$

$$?w$$

$$\begin{array}{lll}
B^{\uparrow}(m,a,w,b,s) & \rightarrow ?A^{\downarrow}(|.m,|.a,w,b,[s]^{\Uparrow}.[\varepsilon]^{\Uparrow^{k-1}}) & \\
B_1^{\downarrow}(m,a,w,b,s) & \rightarrow ?A^{\downarrow}(|.m,|.a,w,b,[s]^{\Downarrow}.[\varepsilon]^{\Downarrow^{k-1}}) & \\
?A^{\uparrow}(|.m,|.a,w,b,[s]^{\Uparrow}.s') & \rightarrow B_1^{\uparrow}(m,a,w,b,s) & \\
?A^{\uparrow}(|.m,|.a,w,b,[s]^{\Downarrow}.s') & \rightarrow B^{\downarrow}(m,a,w,b,s) & \\
?A^{\uparrow}(m,a,w,b,s) & \rightarrow \emptyset & s \neq [s'].s'' \text{ or } m \neq |.m' \text{ or } a \neq |.a' \\
\Gamma^{\uparrow}(m,a,w,b,s) & \rightarrow \Gamma_1^{\uparrow}(m,a,w,b,s) & \\
\Gamma_1^{\downarrow}(m,a,w,b,s) & \rightarrow \Gamma^{\downarrow}(m,a,w,b,s) &
\end{array}$$

Fig. 5. Exponential group. (in the $?w$-transitions, k is the number of ? and ! in front of $?A$ and $[\varepsilon]^{\Uparrow^{k-1}}$ is used to preserve a correct type of the token)

- either s_{2i+1} doesn't exist because f_{π_1} is not defined on s_{2i},
- or $f_{\pi_1}(s_{2i}) = (A, (m, a, w, b, s))$ and, by Lemma 5, $b = (\ulcorner\sigma'\urcorner.\sigma).b'$ thus $s_{2i+1} = A^{\perp\uparrow}(m, a, w, b', \sigma.\sigma'.s)$.

We also have $s'_{2i+1} = s_{2i+1}$ and if $s_{2i} = A^{\uparrow}(m, a, w, (\ulcorner\sigma'\urcorner.\sigma).b, s)$ then $s'_{2i} = A^{\uparrow}(m, a, w, \sigma'.\sigma.b, s)$ by Lemma 5 with $\widetilde{(\ulcorner\sigma'\urcorner.\sigma)}.b = \widetilde{\sigma'.\sigma}.b$. □

To conclude, we have to note that the Theorem 3, about computation for booleans, is still true in the full case!

Acknowledgements. Thanks to Laurent Regnier for his support and to the referees for their comments about presentation.

References

[1] Samson Abramsky, Radha Jagadeesan, and Pasquale Malacaria. Full abstraction for PCF (extended abstract). In M. Hagiya and J. C. Mitchell, editors, *Theoretical Aspects of Computer Software*, volume 789 of *Lecture Notes in Computer Science*, pages 1–15. Springer, April 1994.

[2] Andrea Asperti, Cecilia Giovannetti, and Andrea Naletto. The bologna optimal higher-order machine. *Journal of Functional Programming*, 6(6):763–810, November 1996.

[3] Vincent Danos, Hugo Herbelin, and Laurent Regnier. Games semantics and abstract machines. In *Proceedings of Logic In Computer Science*, New Brunswick, 1996. IEEE Computer Society Press.

[4] Vincent Danos and Laurent Regnier. Reversible, irreversible and optimal λ-machines. In J.-Y. Girard, M. Okada, and A. Scedrov, editors, *Proceedings Linear Logic Tokyo Meeting*, volume 3 of *Electronic Notes in Theoretical Computer Science*. Elsevier, 1996.

[5] Jean-Yves Girard. Geometry of interaction I: an interpretation of system F. In Ferro, Bonotto, Valentini, and Zanardo, editors, *Logic Colloquium '88*. North-Holland, 1988.

[6] Jean-Yves Girard. Geometry of interaction III: accommodating the additives. In J.-Y. Girard, Y. Lafont, and L. Regnier, editors, *Advances in Linear Logic*, volume 222 of *London Mathematical Society Lecture Note Series*, pages 329–389. Cambridge University Press, 1995.

[7] Jean-Yves Girard. Proof-nets: the parallel syntax for proof-theory. In Ursini and Agliano, editors, *Logic and Algebra*, New York, 1996. Marcel Dekker.

[8] Georges Gonthier, Martin Abadi, and Jean-Jacques Lévy. The geometry of optimal lambda reduction. In *Proceedings of Principles of Programming Languages*, pages 15–26. ACM Press, 1992.

[9] Olivier Laurent. Polarized proof-nets and $\lambda\mu$-calculus. To appear in *Theoretical Computer Science*, 2001.

[10] Ian Mackie. The geometry of interaction machine. In *Proceedings of Principles of Programming Languages*, pages 198–208. ACM Press, January 1995.

[11] Marco Pedicini and Francesco Quaglia. A parallel implementation for optimal lambda-calculus reduction. In *Proceedings of Principles and Practice of Declarative Programming*, 2000.

Second-Order Pre-logical Relations and Representation Independence

Hans Leiß

Centrum für Informations- und Sprachverarbeitung
Universität München, Oettingenstr. 67, D-80538 München
leiss@cis.uni-muenchen.de

Abstract. We extend the notion of pre-logical relation between models of simply typed lambda-calculus, recently introduced by F. Honsell and D. Sannella, to models of second-order lambda calculus. With pre-logical relations, we obtain characterizations of the lambda-definable elements of and the observational equivalence between second-order models. These are are simpler than those using logical relations on extended models.

We also characterize representation independence for abstract data types and abstract data type constructors by the existence of a pre-logical relation between the representations, thereby varying and generalizing results of J.C. Mitchell to languages with higher-order constants.

1 Introduction

"Logical" relations between terms or elements of models of the simply typed λ-calculus allow one to prove, by induction on the structure of types, syntactical properties like normalization and Church-Rosser theorems ([Tai67], [Sta85]) and semantical properties like λ-definability of elements ([Plo80], [Sta85]) or observational equivalence of models ([Mit91]).

In first-order, i.e. simply typed, λ-calculus, a logical relation $\{R_\tau \mid \tau \in \mathit{Type}\} = \mathcal{R}$ is built by induction on types using

$$(f, g) \in R_{(\rho\to\sigma)} \iff \forall a, b : \rho\ ((a, b) \in R_\rho \to (f \cdot a,\ g \cdot b) \in R_\sigma), \qquad (1)$$

starting from relations R_τ on base types τ. In second-order λ-calculus, the domain of (semantic) types can no longer be constructed inductively; yet, central results of the first-order case can be transferred, such as the characterization of λ-definable elements or that of observational equivalence (cf. J. Mitchell [MM85],[Mit86]).

Even in the first-order case, logical relations have some disturbing properties, viz. they are not closed under relation composition and do not admit a characterization of observational equivalence for languages with higher-order constants. To overcome these difficulties, several authors (cf. [HS99], [PPST00], [PR00]) recently have proposed to use a more general class of relations, called *pre-* or *lax logical relations*. They arise by restricting direction $\Leftarrow$ in (1) to functions that are

S. Abramsky (Ed.): TLCA 2001, LNCS 2044, pp. 298–314, 2001.

λ-definable (using related parameters). F. Honsell and D. Sannella [HS99] gave several characterizations and applications of pre-logical relations that demonstrate the stability and usefulness of the notion.

We here follow this route and generalize pre-logical relations to the second-order λ-calculus by similarly weakening the conditions for $R_{\forall\alpha\tau}$. As in the first-order case, the λ-definable elements of a model can then be characterized as the intersection of all pre-logical predicates on the model. It follows that observational equivalence of second-order models can be expressed as a suitable pre-logical relation between models.

But our main interst lies on the equivalence of different implementations or representations of an abstract datatype α and the operations $c : \sigma(\alpha)$ coming with it. Two representations $(\tau_i, t_i : \sigma[\tau_i/\alpha])$ of $(\alpha, c : \sigma)$ in a model $\mathcal{A}$ are equivalent if appropriately defined expansions $\mathcal{A}(\tau_i, t_i)$ of $\mathcal{A}$ are observationally equivalent with respect to the language not containing α and c.

For languages without higher-order constants, J. C. Mitchell[Mit86,Mit91] has characterized this equivalence by the existence of a suitable logical relation between expansions of the models. However, programming languages like Standard ML [MTHM97] not only have abstract data types in the presence of higher-order constants, but they also allow to abstract from data type *constructors* (and, in some versions, higher-order functors). We handle abstract type constructors as indeterminates on the kind level and their implementations as definable expansions of a second-order model. Representation independence for abstract type constructors can then be characterized as the existence of a second-order pre-logical relation between the representations. This also holds for languages with higher-order constants.

In section 2 we consider declarations of abstract data types as expansions of first-order models by definable types and elements, and compare a characterization of equivalent representations based on pre-logical relations with a criterion by Mitchell based on logical relations. Section 3 gives syntax and semantics for second-order λ-calculus. In section 4 we define second-order pre-logical relations, present some basic properties in 4.1, and use them in 4.2 to characterize definable elements and observational equivalence for second order models. Section 4.3 treats abstract type constructors and polymorphic constants and characterizes the equivalence of two representations by definable type constructors.

2 Pre-logical Relations for Simply Typed λ-Calculus

Let T be the simple types $\sigma, \tau ::= \beta \mid (\sigma \to \tau)$ over base types β. We write $\Gamma \rhd t : \tau$ for a term typed in the context $\Gamma : Var \to T$ using the standard typing rules. $\lambda_C^{\to}$ stands for the set of typed terms over a set C of typed constants, $c : \sigma$.

Definition 2.1. *An* (extensional) model $\mathcal{A} = (A, \Phi^{\to}, \Psi^{\to}, C^{\mathcal{A}})$ *of the simply typed* λ*-calculus* $\lambda_C^{\to}$ *consists of a family* $A = \langle A_\sigma \rangle_{\sigma \in T}$ *of sets, an interpretation*

$C^{\mathcal{A}}(c) \in A_\sigma$ of the constants $c : \sigma \in C$, and families $\langle \Phi^{\rightarrow}_{\sigma,\tau} \rangle_{\sigma,\tau \in T}$ and $\langle \Psi^{\rightarrow}_{\sigma,\tau} \rangle_{\sigma,\tau \in T}$ of mappings

$$\Phi^{\rightarrow}_{\sigma,\tau} : A_{(\sigma \to \tau)} \to [A_\sigma \to A_\tau] \quad \text{and} \quad \Psi^{\rightarrow}_{\sigma,\tau} : [A_\sigma \to A_\tau] \to A_{(\sigma \to \tau)},$$

where $[A_\sigma \to A_\tau] \subseteq \{h \mid h : A_\sigma \to A_\tau\}$ contains all definable functions, such that $\Phi^{\rightarrow}_{\sigma,\tau} \circ \Psi^{\rightarrow}_{\sigma,\tau} = Id_{(A_\sigma \to A_\tau)}$ (and $\Psi^{\rightarrow}_{\sigma,\tau} \circ \Phi^{\rightarrow}_{\sigma,\tau} = Id_{A(\sigma \to \tau)}$) for all $\sigma, \tau \in T$.

An environment η over $\mathcal{A}$ *satisfies the context* Γ, in symbols: $\eta : \Gamma \to \mathcal{A}$, if $\eta(x) \in A_\sigma$ for all $x : \sigma$ in Γ. Terms are interpreted in $\mathcal{A}$ unter $\eta : \Gamma \to \mathcal{A}$ via

$$[\![\Gamma \rhd (r \cdot s)]\!]\eta := \Phi_{\sigma,\tau}([\![\Gamma \rhd r : (\sigma \to \tau)]\!]\eta)([\![\Gamma \rhd s : \sigma]\!]\eta),$$
$$[\![\Gamma \rhd \lambda x t : (\sigma \to \tau)]\!]\eta := \Psi_{\sigma,\tau}(\lambda a \in A_\sigma.[\![\Gamma, x : \sigma \rhd t : \tau]\!]\eta[a/x]).$$

We write $[\![t]\!]\eta$ rather than $[\![\Gamma \rhd t : \tau]\!]\eta$, if Γ and τ are clear from the context.

Definition 2.2. *A* predicate $\mathcal{R} \subseteq \mathcal{A}$ on $\mathcal{A}$ *is a family $\mathcal{R} = \{R_\sigma \mid \sigma \in T\}$ with $R_\sigma \subseteq A_\sigma$ for each $\sigma \in T$. An environment $\eta : \Gamma \to \mathcal{A}$ over $\mathcal{A}$* respects $\mathcal{R}$, *in symbols: $\eta : \Gamma \to \mathcal{R}$, if $\eta(x) \in R_\sigma$ for each $x : \sigma \in \Gamma$.*

A predicate $\mathcal{R} \subseteq \mathcal{A}$ is called pre-logical, *if $[\![\Gamma \rhd t : \tau]\!]\eta \in R_\tau$ for each term $\Gamma \rhd t : \tau$ and environment $\eta : \Gamma \to \mathcal{R}$. A* pre-logical relation *between $\mathcal{A}$ and $\mathcal{B}$ is a pre-logical predicate on $\mathcal{A} \times \mathcal{B}$.*

The definition is easily adapted to extensions of T by cartesian products $(\sigma \times \tau)$ and disjoint unions $(\sigma + \tau)$, if the extension of terms is done by adding constants.

The property used here as a definition is called the "Basic Lemma" for pre-logical relations by Honsell and Sannella. The Basic Lemma for logical relations says that every logical relation is a pre-logical relation.

Let $Def^{\mathcal{A}} = \{Def^{\mathcal{A}}_\tau \mid \tau \in T\}$ be the predicate of definable elements of $\mathcal{A}$, where

$$Def^{\mathcal{A}}_\tau := \{[\![t]\!]^{\mathcal{A}} \mid \rhd t : \tau,\ t \in \lambda^{\rightarrow}_C\}.$$

Normally, $Def^{\mathcal{A}}$ is not a logical predicate since $[Def_\sigma \to Def_\tau] \not\subseteq Def_{\sigma \to \tau}$. But:

Theorem 2.3 *([HS99],Proposition 5.7, Example 3.10) Let $\mathcal{A}$ be a model of $\lambda^{\rightarrow}_C$. (i) For each predicate $\mathcal{C} \subseteq \mathcal{A}$, there is a least pre-logical predicate $\mathcal{R} \subseteq \mathcal{A}$ with $\mathcal{C} \subseteq \mathcal{R}$. (ii) $Def^{\mathcal{A}}$ is the least pre-logical predicate on $\mathcal{A}$.*

We use the following version of observational equivalence (it differs slightly from indistinguishability by closed equations of observable types as used in [HS99]):

Definition 2.4. *Let T^+ be the simple types generated from a superset of the base types of T, and C^+ an extension of C by new constants with types in T^+. Let $\mathcal{A}^+, \mathcal{B}^+$ be two models of $\lambda^{\rightarrow}_{C+}$ with $A^+_\tau = B^+_\tau$ for all $\tau \in T$. $\mathcal{A}^+$ and $\mathcal{B}^+$ are* observationally equivalent with respect to T, *in symbols: $\mathcal{A}^+ \equiv_T \mathcal{B}^+$, if $[\![t]\!]^{\mathcal{A}^+} = [\![t]\!]^{\mathcal{B}^+}$ for all $\rhd t : \tau$ of $\lambda^{\rightarrow}_{C+}$ where $\tau \in T$.*

Theorem 2.5 *(cf. [HS99], Theorem 8.4)* $\mathcal{A}^+ \equiv_T \mathcal{B}^+$ *iff there is a pre-logical relation* $\mathcal{R}^+ \subseteq \mathcal{A}^+ \times \mathcal{B}^+$ *such that* $R^+_\tau \cap (Def^{\mathcal{A}^+}_\tau \times Def^{\mathcal{B}^+}_\tau) \subseteq Id_\tau$ *for all* $\tau \in T$.

For $\Rightarrow$ use 2.3 and $\mathcal{R}^+ = Def^{\mathcal{A}^+\times\mathcal{B}^+}$. With logical relations we have $\Rightarrow$ only if C does not contain higher-order constants. (cf. Exercise 8.5.6 in [Mit96].)

2.1 Abstract Data Types and Representation Independence

The declaration of an abstract data type,

$$(\textbf{abstype } (\alpha, x : \sigma) \textbf{ with } x : \sigma \rhd e_1 = e_2 : \rho \textbf{ is } (\tau, t : \sigma[\tau/\alpha]) \textbf{ in } s), \tag{2}$$

introduces a "new" type α and a "new" object $x : \sigma(\alpha)$ with "defining" property $e_1 = e_2 : \rho(\alpha)$, which are interpreted in scope s by the type τ and the term t, where α does not occur in the type of s. Roughly, evaluation of (2) in a model $\mathcal{A}$ of $\lambda^{\rightarrow}_C$ is done by first expanding $\mathcal{A}$ by interpreting α by A_τ and x by the value of $t : \sigma[\tau/\alpha]$ and then evaluating s in the expanded model, $\mathcal{A}(\tau, t)$ (see below).

The data type $(\alpha, x : \sigma)$ is *abstract* in s if the value of s is independent from the representation (τ, t) used; i.e. if for any two representations $(\tau_i, t_i : \sigma[\tau_i/\alpha])$, $i = 1, 2$, that "satisfy" the equation $x : \sigma \rhd e_1 = e_2 : \rho$, we have

$$[\![s]\!]^{\mathcal{A}(\tau_1,t_1)} = [\![s]\!]^{\mathcal{A}(\tau_2,t_2)}. \tag{3}$$

To "satisfy" the equation $e_1 = e_2$ in an expansion $\mathcal{A}(\tau_i, t_i)$, the equality on the new type α normally is not true equality on A_{τ_i}, but a suitable partial equivalence relation on A_{τ_i}:

$$[\![(\textbf{abstype } (\alpha, x_0 : \sigma_0) \textbf{ with } x_0 : \sigma_0 \rhd e_1 = e_2 : \rho \textbf{ is } (\tau_0, t : \sigma_0[\tau_0/\alpha]) \textbf{ in } s)]\!]^{\mathcal{A}}\eta$$
$$:= \begin{cases} (\textit{let } \mathcal{A}^+ = \mathcal{A}(\tau_0, [\![t]\!]^{\mathcal{A}}\eta) \textit{ in } [\![s]\!]^{\mathcal{A}^+}\eta), & \text{if there is a pre-logical} \\ & \text{partial equivalence relation } \mathcal{E} \subseteq \mathcal{A}^+ \times \mathcal{A}^+ \text{ such that} \\ & ([\![e_1]\!]\eta^{\mathcal{A}^+}, [\![e_2]\!]\eta^{\mathcal{A}^+}) \in E_\rho \text{ and } E_\tau = Id_{A_\tau} \text{ for } \tau \in T, \\ \textit{error}, & \text{else.} \end{cases}$$

Note that in the first case, there is a least such $\mathcal{E}$, and $\mathcal{A}$ is canonically embedded in $\mathcal{A}^+/\mathcal{E}$. Since $\mathcal{E}$ is pre-logical, the value of x_0 in $\mathcal{A}^+$ is self-related by E_{σ_0}, and the additional operations provided by t actually belong to $\mathcal{A}^+/\mathcal{E}$, which satisfies $e_1 = e_2$. But since the type of s does not involve α, the value of s in $\mathcal{A}^+$ and $\mathcal{A}^+/\mathcal{E}$ is the same, so there is no need to actually construct $\mathcal{A}^+/\mathcal{E}$.

Definition 2.6. *Let* T^+ *be the types constructed from the base types of* T *and a new base type* α. *Let* $\sigma_0 \in T^+$ *and* $C^+ = C, x_0 : \sigma_0$. *For* $\tau_0 \in T$, $a \in A_{\sigma_0[\tau_0/\alpha]}$ *let* $\mathcal{A}(\tau_0, a) = \mathcal{A}^+ := (A^+, \Phi^+, \Psi^+, (C^+)^{\mathcal{A}^+})$ *be given by:*

$$A^+_\sigma := A_{\sigma[\tau_0/\alpha]}, \qquad (\Phi^+)^{\rightarrow}_{\sigma,\tau} := \Phi^{\rightarrow}_{\sigma[\tau_0/\alpha],\tau[\tau_0/\alpha]},$$
$$(C^+)^{\mathcal{A}^+}(c) = \begin{cases} C^{\mathcal{A}}(c), & c : \tau \in C, \\ a, & c : \tau \equiv x_0 : \sigma_0, \end{cases} \qquad (\Psi^+)^{\rightarrow}_{\sigma,\tau} := \Psi^{\rightarrow}_{\sigma[\tau_0/\alpha],\tau[\tau_0/\alpha]}.$$

We call the model $\mathcal{A}^+$ *of* $\lambda^{\rightarrow}_{C^+}$ *the* expansion of $\mathcal{A}$ defined by (τ_0, a).

From now on we focus on the equivalence of representations and ignore the restriction to implementations satisfying a specification. Two expansions $\mathcal{A}(\tau_1, t_1)$ and $\mathcal{A}(\tau_2, t_2)$ of $\mathcal{A}$ are equivalent representations of an abstype $(\alpha, x : \sigma)$ if (3) holds for all terms s whose type does not contain α, i.e. if $\mathcal{A}(\tau_1, t_1) \equiv_T \mathcal{A}(\tau_2, t_2)$.

Remark 2.7. For a careful categorical treatment of abstract data types with equational specifications, see [PR00]. The simpler form of abstype declarations (**abstype** $(\alpha, x : \sigma)$ **is** (τ, t) **in** s) does occur in programming languages like SML and has been investigated in [MP85,Mit91]. It can be viewed as a case of SML's restriction *Str* :> *Sig* of the *structure* $(\alpha = \tau, x = t)$ to the *signature* $(\alpha, x : \sigma)$.

Example 2.8 *Consider a type of multisets with some higher-order operations. (Assume that* `'a` = `''b` *is a fixed type of* T*. Actually, bag* : $T \Rightarrow T$ *is a* type constructor, *since* `'a` *and* `''b` *are type variables. See section 4.2 for this.)*

```
signature BAG = sig
    type 'a bag
    val empty : 'a bag
    val member : ''a -> ''a bag -> int
    val insert : ''a * int -> ''a bag -> ''a bag
    val map : ('a -> ''b) -> 'a bag -> ''b bag
    val union : ('a -> ''b bag) -> 'a bag -> ''b bag
  end
```

An A*-bag* B *represents the multiset* $\{(a, n) \mid \texttt{member}\ a\ B = n > 0,\ a \in A\}$. *Each implementation of* `member` *gives a partial equivalence relation* $\mathcal{E}$ *such that* $B_1\ \mathcal{E}_{\texttt{BAG}}\ B_2$ *iff* B_1 *and* B_2 *represent the same multiset. One implementation* `Bag1` *of bags is by lists of elements paired with their multiplicity:*

```
structure Bag1 :> BAG = struct
    type 'a bag = ('a * int) list
    fun member a [] = 0
      | member a ((b,n)::B) = if a=b then n else (member a B)
    ...
  end
```

Another implementation `Bag2` *is by lists of elements:*

```
structure Bag2 :> BAG = struct
    type 'a bag = 'a list
    fun member a [] = 0
      | member a (b::B) = if a=b then 1+(member a B) else (member a B)
    ...
  end
```

The implementations `Bag1` *and* `Bag2` *are observationally equivalent if they assign the same value to the closed terms* `(member s t) : int`.

Observational equivalence can be characterized by pre-logical relations. To cover nested abstype-declarations, we characterize the observational equivalence of two expansions of models that are related by a pre-logical relation $\mathcal{R}$ instead of *Id*.

Theorem 2.9 *Let $\mathcal{R} \subseteq \mathcal{A} \times \mathcal{B}$ be a pre-logical relation. Let $\mathcal{A}^+$ and $\mathcal{B}^+$ be definable expansions of $\mathcal{A}$ and $\mathcal{B}$ to models of $\lambda^{\rightarrow}_{C+}$. The following are equivalent:*

(i) for each closed $\lambda^{\rightarrow}_{C+}$-term $\triangleright t : \tau$ with $\tau \in T$ we have $[\![t]\!]^{\mathcal{A}^+} R_\tau [\![t]\!]^{\mathcal{B}^+}$.
(ii) there is a pre-logical relation $\mathcal{R}^+$ with $\mathcal{A}^+ \mathcal{R}^+ \mathcal{B}^+$ and $R^+_\tau = R_\tau$ for all $\tau \in T$.

Proof. For simplicity of notation we consider the unary case. i.e. pre-logical predicates $\mathcal{R} \subseteq \mathcal{A}$ and $\mathcal{R}^+ \subseteq \mathcal{A}^+$. (ii) $\Rightarrow$ (i) is obvious. (i) $\Rightarrow$ (ii): for $\tau \in T^+$ let

$$Def^+_\tau = \{[\![t]\!]^{\mathcal{A}^+} \mid t \in \lambda^{\rightarrow}_{C+}, \vdash t : \tau\} \quad \text{and} \quad C_\tau := \begin{cases} R_\tau, & \tau \in T, \\ Def^+_\tau, & \text{else.} \end{cases}$$

By assumption we have $Def^+_\tau \subseteq R_\tau$ for $\tau \in T$. Each logical predicate $\mathcal{R}^+$ on $\mathcal{A}^+$ as in (ii) satisfies the following conditions in positively occurring unknowns X_τ:

$$X_\tau \supseteq C_\tau \cup \bigcup\{X_{(\rho\rightarrow\tau)} \cdot X_\rho \mid \rho \in T^+\}, \qquad \tau \in T^+. \tag{4}$$

Let $\mathcal{R}^+ = \{R^+_\tau \mid \tau \in T^+\} \subseteq \mathcal{A}^+$ be the least solution of (4). It is a pre-logical predicate: if $\eta : \Gamma \rightarrow \mathcal{R}^+$ and $\Gamma \triangleright t : \tau$ is a $\lambda^{\rightarrow}_{C+}$-term, with $\Gamma = \{x : \sigma\}$ say, then

$$[\![\Gamma \triangleright t : \tau]\!]\eta = [\![\triangleright \lambda x : \sigma.t : \sigma \rightarrow \tau]\!] \cdot \eta(x) \in Def^+_{\sigma\rightarrow\tau} \cdot R^+_\sigma \subseteq R^+_{\sigma\rightarrow\tau} \cdot R^+_\sigma \subseteq R^+_\tau.$$

It is not hard to see that for each $\tau \in T^+$

$$R^+_\tau = \bigcup\{Def^+_{\rho_1\rightarrow\ldots\rightarrow\rho_n\rightarrow\tau} \cdot R_{\rho_1} \cdots R_{\rho_n} \mid n \in I\!N, \rho_1, \ldots \rho_n \in T\}.$$

For $\boldsymbol{\rho}, \tau \in T$ we have $Def^+_{\boldsymbol{\rho}\rightarrow\tau} \subseteq R_{\boldsymbol{\rho}\rightarrow\tau}$ by (i), which gives $Def^+_{\boldsymbol{\rho}\rightarrow\tau} \cdot R_{\boldsymbol{\rho}} \subseteq R_\tau$ and so $R^+_\tau \subseteq R_\tau$. Therefore we also have $R^+_\tau = R_\tau$ for $\tau \in T$. □

In the corresponding theorem for logical relations (cf. [Mit91], Cor.1), direction (i) $\Rightarrow$ (ii) is restricted to languages where C has no higher-order constants. The logical relation $\mathcal{R}^+$ in (ii) can then be inductively generated from the predicates $\{R^+_\tau \mid \tau \in T^+ \text{ a base type}\}$ of the least solution of (4). This works even when all constants $c : \tau \in C^+$ are first-order *in* α, like `map` in 2.8. The observational equivalence of two representations can be shown by (ii) $\Rightarrow$ (i). When used with *logical* relations, premise (ii) can be reformulated to a 'local' criterion for equivalence of representations, which fails for *pre-logical* relations:

Lemma 2.10. *(cf. [Mit86], Lemma 4) Let $\mathcal{A}$ and $\mathcal{B}$ be logically related by $\mathcal{R}$, and let $\mathcal{A}^+$ and $\mathcal{B}^+$ be definable expansions of $\mathcal{A}$ and $\mathcal{B}$ to models of $\lambda^{\rightarrow}_{C+}$. Then (ii) and (i) are equivalent:*

(i) There is a relation $R^+_\alpha \subseteq A^+_\alpha \times B^+_\alpha$, such that $c^{\mathcal{A}^+} R^+_\tau c^{\mathcal{B}^+}$ for $c : \tau \in C^+ \setminus C$, where $R^+_\tau := R_\tau$ for $\tau \in T$ and $R^+_{(\rho\rightarrow\sigma)} = [R^+_\rho \rightarrow R^+_\sigma]$ for $(\rho \rightarrow \sigma) \notin T$.
(ii) There is a logical relation $\mathcal{R}^+ \subseteq \mathcal{A}^+ \times \mathcal{B}^+$ with $R^+_\tau = R_\tau$ for all $\tau \in T$.

Unfortunately, if $\mathcal{R} \subseteq \mathcal{A} \times \mathcal{B}$ is *pre*-logical, then condition (i) does *not* imply the existence of a pre-logical relation $\mathcal{R}^+ \subseteq \mathcal{A}^+ \times \mathcal{B}^+$ that agrees with $\mathcal{R}$ on types of T, because the necessary condition (cf. Theorem 2.9, (i)) $Def^+_{(\sigma \to \tau)} \subseteq R_{(\sigma \to \tau)}$ for $(\sigma \to \tau) \in T$ may fail: Take $f \in [R_\sigma \to R_\tau] \setminus R_{(\sigma \to \tau)}$ and let $f = c_2 \circ c_1$ with constants $c_1 : \sigma \to \rho, c_2 : \rho \to \tau$ for some $\rho \notin T$.

Hence, although pre-logical relations give a complete *characterization* of the equivalence of representations when higher-order constants are present, they seem harder to use in *establishing* the equivalence of representations: instead of checking the existence of R^+_α with property (i) of 2.10, one has to check the global property of $\lambda^{\to}_{C^+}$-observational equivalence with respect to all types of T.

3 Second Order Lambda Calculus, $\lambda^{\to,\forall}_C$

We now look at definability and observational equivalence for the second order λ-calculus $\lambda^{\to,\forall}_C$. The characterizations by pre-logical relations are similar to the first-order case, but the construction of pre-logical relations is more elaborate since these can no longer be defined by induction on type expressions.

3.1 Syntax and Semantics of $\lambda^{\to,\forall}_C$

In second order lambda calculus, in addition to terms denoting objects one has constructors denoting types or functions on types.

Definition 3.1. *The* kinds κ *of* $\lambda^{\to,\forall}_C$ *are given by* $\kappa := T \mid (\kappa \Rightarrow \kappa)$. *The classification C of constants is split in two components,* $C = (C_{Kind}, C_{Type})$. *By* C_{Kind}, *a set of assumptions of the form* $c : \kappa$, *each constructor constant* c *is assigned a unique kind* κ. *In particular,* C_{Kind} *contains the type constructors* $\to : T \Rightarrow (T \Rightarrow T)$ *and* $\forall : (T \Rightarrow T) \Rightarrow T$.

A constructor $\Delta \rhd \mu : \kappa$ of kind κ *is a sequent derivable with the following rules of* $\lambda^{\Rightarrow}_{C_{Kind}}$, *where* Δ *is a context of kind assumptions for constructor variables:*

$$(Mon)\ \frac{\Delta \rhd \mu : \kappa}{\Delta, \upsilon : \kappa' \rhd \mu : \kappa}, \quad \upsilon \notin dom(\Delta)$$

$$(Const)\ \Delta \rhd c : \kappa, \quad \text{for } c : \kappa \in C_{Kind} \qquad (Var)\ \Delta, \upsilon : \kappa \rhd \upsilon : \kappa$$

$$(\Rightarrow E)\ \frac{\Delta \rhd \mu : (\kappa_1 \Rightarrow \kappa_2), \quad \Delta \rhd \nu : \kappa_1}{\Delta \rhd (\mu \cdot \nu) : \kappa_2} \qquad (\Rightarrow I)\ \frac{\Delta, \upsilon : \kappa_1 \rhd \mu : \kappa_2}{\Delta \rhd \lambda\upsilon.\mu : (\kappa_1 \Rightarrow \kappa_2)}.$$

For $\lambda^{\Rightarrow}_{C_{Kind}}$ *we assume a reduction relation* $\rightsquigarrow$ *subsuming* α, β, η*-reduction and satisfying the subject reduction, i.e. if* $\Delta \rhd \mu : \kappa$ *and* $\mu \rightsquigarrow \nu$, *then* $\Delta \rhd \nu : \kappa$.

$$(=_1)\ \frac{\Delta \rhd \mu : \kappa \qquad \mu \rightsquigarrow \nu}{\Delta \rhd \mu = \nu : \kappa} \qquad (=_2)\ \frac{\Delta \rhd \mu : \kappa \qquad \mu \rightsquigarrow \nu}{\Delta \rhd \nu = \mu : \kappa}$$

A type $\Delta \rhd \mu : T$ *is constructor of kind* T. *We use* τ, σ *for types and write* $(\sigma \to \tau)$ *and* $\forall v \tau$ *etc. By* C_{Type}*, a set of assumptions of the form* $c : \mu$*, each individual constant* c *is assigned a unique closed constructor* μ *of kind* T.

A typed term $\Delta; \Gamma \rhd t : \tau$ *is a sequent derivable with the following rules, where* Δ *is a context of kind assumptions for constructor variables and* Γ *is a context of type assumptions for individual variables:*

$$(mon)\ \frac{\Delta; \Gamma \rhd t : \tau \qquad \Delta \rhd \sigma : T}{\Delta; \Gamma, x : \sigma \rhd t : \tau}, \quad x \notin dom(\Gamma)$$

$$(const)\ \frac{\rhd \tau : T}{\Delta; \Gamma \rhd c : \tau}, \quad for\ c : \tau \in C_{Type}$$

$$(var)\ \frac{\Delta \rhd \tau : T}{\Delta; \Gamma, x : \tau \rhd x : \tau}, \quad x \notin dom(\Gamma)$$

$$(\to I)\ \frac{\Delta \rhd \sigma : T \qquad \Delta; \Gamma, x : \sigma \rhd t : \tau}{\Delta; \Gamma \rhd (\lambda x : \sigma . t) : (\sigma \to \tau)}$$

$$(\to E)\ \frac{\Delta \rhd \tau : T \qquad \Delta; \Gamma \rhd t : (\sigma \to \tau) \qquad \Delta; \Gamma \rhd s : \sigma}{\Delta; \Gamma \rhd (t \cdot s) : \tau}$$

$$(\forall I)\ \frac{\Delta \rhd \mu : (T \Rightarrow T) \qquad \Delta, \alpha : T; \Gamma \rhd t : (\mu \cdot \alpha)}{\Delta; \Gamma \rhd \lambda \alpha t : \forall \mu}, \quad if\ \alpha\ is\ not\ in\ \Gamma$$

$$(\forall E)\ \frac{\Delta \rhd \mu : (T \Rightarrow T) \qquad \Delta \rhd \tau : T \qquad \Delta; \Gamma \rhd t : \forall \mu}{\Delta; \Gamma \rhd (t \cdot \tau) : (\mu \cdot \tau)}$$

$$(Typ =)\ \frac{\Delta; \Gamma \rhd t : \tau \qquad \Delta \rhd \tau = \sigma : T}{\Delta; \Gamma \rhd t : \sigma}$$

By induction on the derivation, using the subject reduction property of $\lambda^{\Rightarrow}_{C_{Kind}}$ in the case of (Typ =), we obtain:

Proposition 3.2. *If* $\Delta; \Gamma \rhd t : \sigma$ *is a term, then* $\Delta \rhd \sigma : T$ *is a type.*

Definition 3.3. *A* model frame $\mathcal{M} = (\mathcal{U}, \mathcal{D}, \Phi, \Psi)$ *for* $\lambda^{\to,\forall}_{C}$ *consists of*

1. *an extensional model* $\mathcal{U} = (U, \Phi^{\Rightarrow}, \Psi^{\Rightarrow}, C^{\mathcal{U}}_{Kind}, [\![\cdot]\!]^{\mathcal{U}}\cdot)$ *of* $\lambda^{\Rightarrow}_{C_{Kind}}$ *to interprete the constructors,*
2. *a family of individual domains for the types,* $\mathcal{D} = (\langle D_A \rangle_{A \in U_T}, C^{\mathcal{D}}_{Type})$*, with individuals* $C^{\mathcal{D}}_{Type}(c) \in D_{[\![\tau]\!]^{\mathcal{U}}}$ *for* $c : \tau \in C_{Type}$*,*
3. *families* $\Phi = (\Phi^{\to}, \Phi^{\forall})$*,* $\Psi = (\Psi^{\to}, \Psi^{\forall})$ *with* $\Phi^{\to} = \langle \Phi^{\to}_{A,B} \rangle_{A,B \in U_T}$*,* $\Phi^{\forall} = \langle \Phi^{\forall}_{f} \rangle_{f \in U_{(T \Rightarrow T)}}$*,* $\Psi^{\to} = \langle \Psi^{\to}_{A,B} \rangle_{A,B \in U_T}$*,* $\Psi^{\forall} = \langle \Psi^{\forall}_{f} \rangle_{f \in U_{(T \Rightarrow T)}}$ *of mappings*

$$\Phi^{\to}_{A,B} : D_{(A \to B)} \longrightarrow [D_A \to D_B] \qquad \Psi^{\to}_{A,B} : [D_A \to D_B] \longrightarrow D_{(A \to B)}$$

$$\Phi^{\forall}_{f} : D_{\forall f} \longrightarrow [\Pi A \in U_T . D_{f \cdot A}] \qquad \Psi^{\forall}_{f} : [\Pi A \in U_T . D_{f \cdot A}] \longrightarrow D_{\forall f},$$

for subsets $[D_A \to D_B] \subseteq \{h \mid h : D_A \to D_B\}$ *and* $[\Pi A \in U_T.D_{f\cdot A}] \subseteq \Pi A \in U_T.D_{f\cdot A}$*, such that for all* $A, B \in U_T$ *and* $f \in U_{(T\Rightarrow T)}$

$$\Phi^{\to}_{A,B} \circ \Psi^{\to}_{A,B} = Id_{[D_A \to D_A]} \quad \text{and} \quad \Phi^{\forall}_{f} \circ \Psi^{\forall}_{f} = Id_{[\Pi A \in U_T.D_{f\cdot A}]}.$$

A (environment) model *of* $\lambda^{\to,\forall}_C$ *consist of a model frame* $\mathcal{M} = (\mathcal{U}, \mathcal{D}, \Phi, \Psi)$ *and an evaluation* $[\![\cdot]\!]^{\mathcal{D}}\cdot$ *of terms, such that every typed term gets a value using*

$$[\![\Delta; \Gamma \rhd x : \tau]\!]\eta = \eta(x), \qquad [\![\Delta; \Gamma \rhd c : \tau]\!]\eta = C^{\mathcal{D}}_{Type}(c), \quad \text{for } c : \tau \in C_{Type}$$

$$[\![\Delta; \Gamma \rhd (t\cdot s) : \tau]\!]\eta = \Phi^{\to}_{[\![\Delta\, \rhd\, \sigma]\!]\eta, [\![\Delta\, \rhd\, \tau]\!]\eta}([\![\Delta; \Gamma \rhd t : (\sigma \to \tau)]\!]\eta)([\![\Delta; \Gamma \rhd s : \sigma]\!]\eta)$$

$$[\![\Delta; \Gamma \rhd \lambda x : \sigma.t : (\sigma \to \tau)]\!]\eta =$$

$$\Psi^{\to}_{[\![\Delta\, \rhd\, \sigma]\!]\eta, [\![\Delta\, \rhd\, \tau]\!]\eta}(\lambda a \in D_{[\![\Delta\, \rhd\, \sigma:T]\!]\eta}.[\![\Delta; \Gamma \rhd t : \tau]\!]\eta[a/x])$$

$$[\![\Delta; \Gamma \rhd (t\cdot\tau) : (\mu \cdot \tau)]\!]\eta = \Psi^{\forall}_{[\![\Delta\, \rhd\, \mu:(T\Rightarrow T)]\!]\eta}([\![\Delta; \Gamma \rhd t : \forall\mu]\!]\eta)([\![\Delta \rhd \tau : T]\!]\eta)$$

$$[\![\Delta; \Gamma \rhd \lambda\alpha t : \forall\mu]\!]\eta = \Psi^{\forall}_{[\![\Delta\, \rhd\, \mu(T\Rightarrow T)]\!]\eta}(\lambda A \in U_T.[\![\Delta, \alpha : T; \Gamma \rhd t : (\mu \cdot \alpha)]\!]\eta[A/\alpha])$$

In particular, $[\![\Delta; \Gamma \rhd t : \tau]\!]\eta \in D_{[\![\Delta\, \rhd\, \tau:T]\!]\eta}$. If it is clear which context and type is meant, we write $[\![t]\!]\eta$ instead of $[\![\Delta; \Gamma \rhd t : \tau]\!]\eta$. We write $c^{\mathcal{D}}$ for $C^{\mathcal{D}}_{Type}(c)$.

We refer to Bruce e.a.[BMM90] for an overview of various models of $\lambda^{\to,\forall}_C$.

4 Pre-logical Relations for Second Order λ-Calculus

Definition 4.1. *Let* $\mathcal{A} = (\mathcal{U}, \mathcal{D}, \Phi, \Psi, [\![\cdot]\!]^{\mathcal{D}}\cdot)$ *be an environment model of* $\lambda^{\to,\forall}_C$*. A* predicate $\mathcal{R} = (\mathcal{R}_{Kind}, \mathcal{R}_{Type})$ on $\mathcal{A}$ *consists of a predicate* $\mathcal{R}_{Kind} = \{R_\kappa \mid \kappa \in Kind\}$ *on* $\mathcal{U}$ *where* $R_\kappa \subseteq U_\kappa$ *for each* $\kappa \in Kind$ *and a predicate* $\mathcal{R}_{Type} = \{R_A \mid A \in R_T\}$ *on* $\mathcal{D}$ *where* $R_A \subseteq D_A$ *for each* $A \in R_T$.

An environment $\eta : \Delta; \Gamma \to \mathcal{A}$ respects *the predicate* $\mathcal{R}$*, in symbols:* $\eta : \Delta; \Gamma \to \mathcal{R}$*, if* $\eta(v) \in R_\kappa$ *for each* $v : \kappa \in \Delta$ *and* $\eta(x) \in R_{[\![\Delta\, \rhd\, \tau:T]\!]\eta}$ *for each* $x : \tau \in \Gamma$.

The predicate $\mathcal{R} \subseteq \mathcal{A}$ *is* algebraic, *if* $\mathcal{R}_{Kind}$ *is algebraic (Def. 3.2 of [HS99]) and*

a) $c^{\mathcal{D}} \in R_A$ *for each* $c : \tau \in C_{Type}$ *where* $A = [\![\ \rhd \tau : T]\!]^{\mathcal{U}}$,
b) $R_{(A\to B)} \subseteq \{h \in D_{(A\to B)} \mid h \cdot R_A \subseteq R_B\}$ *for all* $A, B \in R_T$,
c) $R_{\forall f} \subseteq \{h \in D_{\forall f} \mid \forall A \in R_T\ h \cdot A \in R_{f\cdot A}\}$ *for all* $f \in R_{(T\Rightarrow T)}$.

An algebraic predicate $\mathcal{R}$ *is* logical, *if in b) and c) we also have* $\supseteq$.

Definition 4.2. *A predicate* $\mathcal{R} = (\mathcal{R}_{Kind}, \mathcal{R}_{Type})$ *on* $\mathcal{A}$ *is* pre-logical, *if*

(i) $\mathcal{R}_{Kind}$ *is a pre-logical predicate on* $\mathcal{U}$*, and*
(ii) $\mathcal{R}_{Type}$ *is a 'pre-logical predicate on* $\mathcal{D}$*', i.e.* $[\![\Delta; \Gamma \rhd t : \tau]\!]\eta \in R_{[\![\Delta\, \rhd\, \tau:T]\!]\eta}$ *for each term* $\Delta; \Gamma \rhd t : \tau$ *and every environment* $\eta : \Delta; \Gamma \to \mathcal{R}$.

An algebraic (logical, pre-logical) relation $\mathcal{R}$ *between* $\mathcal{A}_1, \ldots, \mathcal{A}_n$ *is an algebraic (logical, pre-logical) predicate* $\mathcal{R} \subseteq \mathcal{A}_1 \times \ldots \times \mathcal{A}_n$.

4.1 Basic Properties of Second-Order Pre-logical Relations

The Basic Lemma for second order logical relations just says they are pre-logical:

Theorem 4.3 *([MM85], Theorem 2) Let $\mathcal{R} \subseteq \mathcal{A} \times \mathcal{B}$ be a logical relation between environment models $\mathcal{A}$ and $\mathcal{B}$ of $\lambda_C^{\rightarrow,\forall}$. For each term $\Delta\,;\,\Gamma \rhd t : \tau$ and every environment $\eta : \Delta; \Gamma \to \mathcal{R}$, $[\![\Delta \rhd \tau : T]\!]\eta \in R_T$ and $[\![\Delta\,;\,\Gamma \rhd t : \tau]\!]\eta \in R_{[\![\Delta\, \rhd\, \tau:T]\!]\eta}$.*

As in the first-order case one can view definition 4.2 as a Basic Lemma for relations defined using the following properties:

Lemma 4.4. *A predicate $\mathcal{R} \subseteq \mathcal{A}$ is pre-logical iff:*

1. *$\mathcal{R}_{Kind}$ is pre-logical and $\mathcal{R}$ is algebraic,*
2. *for all terms $\Delta\,;\,\Gamma, x : \sigma \rhd t : \tau$ and environments $\eta : \Delta; \Gamma \to \mathcal{R}$ such that*
$$\forall a \in R_{[\![\sigma]\!]\eta}\ [\![\Delta\,;\,\Gamma, x : \sigma \rhd t : \tau]\!]\eta[a/x] \in R_{[\![\tau]\!]\eta},$$
we have $[\![\Delta\,;\,\Gamma \rhd \lambda x : \sigma\, t : (\sigma \to \tau)]\!]\eta \in R_{[\![\sigma \to \tau]\!]\eta}$,
3. *for all terms $\Delta, \alpha : T\,;\,\Gamma \rhd t : \tau$ and environments $\eta : \Delta; \Gamma \to \mathcal{R}$ such that*
$$\forall A \in R_T\ [\![\Delta, \alpha : T\,;\,\Gamma \rhd t : \tau]\!]\eta[A/\alpha] \in R_{[\![\Delta,\alpha:T\, \rhd\, \tau:T]\!]\eta[A/\alpha]},$$
we have $[\![\Delta\,;\,\Gamma \rhd \lambda\alpha\, t : \forall\alpha\,\tau]\!]\eta \in R_{[\![\forall\alpha\tau]\!]\eta}$.

Proof. $\Rightarrow$: Let $\mathcal{R}$ be pre-logical. Then $\mathcal{R}_{Kind}$ is pre-logical and algebraic. To show that $\mathcal{R}_{Type}$ is algebraic, i.e. that a), b), c) of definition 4.1 hold, we focus on c): If $f \in R_{(T\Rightarrow T)}$, $h \in R_{\forall f}$ and $A \in R_T$, and say $\Delta = \{v : (T \Rightarrow T), \alpha : T\}$, $\Gamma = \{x : \forall v\}$, then since $\Delta\,;\,\Gamma \rhd x \cdot \alpha : v \cdot \alpha$ one has
$$h \cdot A = [\![\Delta\,;\,\Gamma \rhd x \cdot \alpha : v \cdot \alpha]\!]\eta \in R_{[\![\Delta\, \rhd\, v\cdot\alpha:T]\!]\eta} = R_{fA}$$
for all environments η with $\eta(v) = f$, $\eta(\alpha) = A$, and $\eta(x) = h$. That $\mathcal{R}$ satisfies 2. and 3. on definable functions is seen for 2. via $\Delta\,;\,\Gamma \rhd \lambda x : \sigma t : (\sigma \to \tau)$ and for 3. via $\Delta\,;\,\Gamma \rhd \lambda\alpha t : \forall\alpha\tau$.

$\Leftarrow$: Condition (i) of definition 4.2 is clear, and for (ii) the assumptions are just what is needed to prove the claim $[\![\Delta\,;\,\Gamma \rhd t : \tau]\!]\eta \in R_{[\![\tau]\!]\eta}$ by induction on the derivation of $\Delta\,;\,\Gamma \rhd t : \tau$. □

The class of first-order binary pre-logical relations is closed under composition. For second-order relations $\mathcal{R} \subseteq \mathcal{A} \times \mathcal{B}$ and $\mathcal{B} \subseteq \mathcal{B} \times \mathcal{C}$, define $\mathcal{R} \circ \mathcal{S}$ by taking

$$(\mathcal{R} \circ \mathcal{S})_{Kind} = \{(R \circ S)_\kappa \mid \kappa \in Kind\}, \quad \text{where } (R \circ S)_\kappa := R_\kappa \circ S_\kappa,$$
$$(\mathcal{R} \circ \mathcal{S})_{Type} = \{(R \circ S)_{(A,C)} \mid (A, C) \in (R \circ S)_T\}, \quad \text{where}$$
$$(R \circ S)_{(A,C)} = \bigcup \{R_{(A,B)} \circ S_{(B,C)} \mid B \in U_T^{\mathcal{B}}, (A, B) \in R_T, (B, C) \in S_T\}.$$

It is then routine to check the following

Proposition 4.5. *Let $\mathcal{R} \subseteq \mathcal{A} \times \mathcal{B}$ and $\mathcal{S} \subseteq \mathcal{B} \times \mathcal{C}$ be pre-logical relations between models of $\lambda_C^{\to,\forall}$. Let $\eta = (\eta^{\mathcal{A}}, \eta^{\mathcal{C}}) : \Delta; \Gamma \to \mathcal{R} \circ \mathcal{S}$ and suppose there is some $\eta^{\mathcal{B}} : \Delta; \Gamma \to \mathcal{B}$ such that $(\eta^{\mathcal{A}}, \eta^{\mathcal{B}}) : \Delta; \Gamma \to \mathcal{R}$ and $(\eta^{\mathcal{B}}, \eta^{\mathcal{C}}) : \Delta; \Gamma \to \mathcal{S}$. Then*

$$[\![\Delta; \Gamma \triangleright t : \tau]\!]\eta \in (R \circ S)_{[\![\Delta \,\triangleright\, \tau:T]\!]\eta}$$

for each term $\Delta; \Gamma \triangleright t : \tau$.

This does not quite mean that $\mathcal{R} \circ \mathcal{S}$ is a pre-logical relation: of course, for each $\eta : \Delta; \Gamma \to \mathcal{R} \circ \mathcal{S}$ there is some $\eta^{\mathcal{B}}_{Kind} : \Delta \to \mathcal{B}$ such that $(\eta^{\mathcal{A}}, \eta^{\mathcal{B}}_{Kind}) : \Delta \to \mathcal{R}_{Kind}$ and $(\eta^{\mathcal{B}}_{Kind}, \eta^{\mathcal{C}}) : \Delta \to \mathcal{S}_{Kind}$. Yet, there may be no extension $\eta^{\mathcal{B}} = \eta^{\mathcal{B}}_{Kind} \cup \eta^{\mathcal{B}}_{Type}$ as needed in the proposition: if $\eta(x : \sigma) = (a, c) \in (R \circ S)_{(A,C)}$ for $(A, C) = [\![\Delta \triangleright \sigma : T]\!]\eta$, there is *some* B such that $(A, B) \in R_T, (B, C) \in S_T$, and $(a, b) \in R_{(A,B)}, (b, c) \in S_{(B,C)}$, but B may depend on (a, c), not just (A, C), and may be different from $[\![\sigma]\!]\eta^{\mathcal{B}}_{Kind}$.

If for each $(A, C) \in (R \circ S)_T$, the relations $R_{(A,B)} \circ S_{(B,C)}$ for all B with $(A, B) \in R_T$ and $(B, C) \in S_T$ coincide, we can extend $\eta^{\mathcal{B}}_{Kind}$ by a suitable $\eta^{\mathcal{B}}_{Type}$. So we still have the following special case, which may be sufficient for applications of second-order pre-logical relations to step-wise data refinement (cf. [HLST00]):

Corollary 4.6. *If $\mathcal{R} \subseteq \mathcal{A} \times \mathcal{B}$ and $\mathcal{S} \subseteq \mathcal{B} \times \mathcal{A}$ are pre-logical and R_T (or the inverse of S_T) is functional, then $\mathcal{R} \circ \mathcal{S}$ is pre-logical.*

For the same reason, the projection of a pre-logical $\mathcal{R} \subseteq \mathcal{A} \times \mathcal{B}$ to the first component is a pre-logical predicate on $\mathcal{A}$ if $\mathcal{R}$ is functional, but not in general.

Proposition 4.7. *Let $\{\mathcal{R}_i \mid i \in I\}$ be a family of pre-logical predicates on $\mathcal{A}$. Then $\bigcap\{\mathcal{R}_i \mid i \in I\}$ is a pre-logical predicate.*

Remark 4.8. By Proposition 7.1 of [HS99], every first-order pre-logical relation $\mathcal{R} \subseteq \mathcal{A} \times \mathcal{B}$ is the composition of three logical relations, $embed_A \subseteq \mathcal{A} \times \mathcal{A}[X]$, $\mathcal{R}[X] \subseteq \mathcal{A}[X] \times \mathcal{B}[X]$, and $embed_B^{-1} \subseteq \mathcal{B}[X] \times \mathcal{B}$, where X is a set of indeterminates. Since the embedding relations are functional, we expect that this result extends to the second-order case, using Corollary 4.6.

4.2 Definability and Observational Equivalence

Definition 4.9. *Let $\mathcal{A} = (\mathcal{U}, \mathcal{D}, \Phi, \Psi, [\![\cdot]\!]\cdot)$ and $A \in U_T$. Element $a \in D_A$ is* definable of type A, *if there is a closed term $; \triangleright t : \tau$ of $\lambda_C^{\to,\forall}$ with $A = [\![\ \triangleright \tau : T]\!]^{\mathcal{A}}$ and $a = [\![; \triangleright t : \tau]\!]^{\mathcal{A}}$. We denote the set of definable elements of type A by*

$$Def_A^{\mathcal{A}} := \{[\![; \triangleright t : \tau]\!]^{\mathcal{A}} \mid\ ; \triangleright t : \tau \text{ a closed term, } A = [\![\ \triangleright \tau : T]\!]^{\mathcal{A}}\}.$$

Mitchell and Meyer ([MM85], Theorem 4) have characterized the set of definable elements of a model $\mathcal{A}$ of $\lambda_C^{\to,\forall}$ as the intersection of all logical relations on an extension $\mathcal{A}^*$ of $\mathcal{A}$ by infinitely many unknowns of each type. Using pre-logical relations, we get a simpler characterization:

Theorem 4.10 *An element a of $\mathcal{A}$ is definable of type $A \in U_T$ iff for each pre-logical predicate $\mathcal{R} \subseteq \mathcal{A}$ we have $A \in R_T$ and $a \in R_A$.*

Proof. $\Rightarrow$: There is a term $; \rhd t : \tau$ with $A = [\![\rhd \tau : T]\!]$ and $a = [\![; \rhd t : \tau]\!] \in D_A$. For each pre-logical predicate $\mathcal{R}$ on $\mathcal{A}$, $[\![; \rhd t : \tau]\!] \in R_{[\![\rhd \tau : T]\!]}$ by definition, and $A = [\![\rhd \tau : T]\!] \in R_T$ follows from $\rhd \tau : T$, by 2.3.

$\Leftarrow$: We show that $Def = (Def_{Kind}, Def_{Type})$ is a pre-logical predicate on $\mathcal{A}$. Then $A \in Def_T$, so $A = [\![\rhd \sigma : T]\!]$ for some type $\rhd \sigma : T$, and for $a \in Def_A$ there is a term $; \rhd t : \tau$ with $a = [\![; \rhd t : \tau]\!]$ and $[\![\rhd \tau : T]\!] = A$. So a is definable of type A.

By theorem 2.3, $Def_{Kind} = \{Def_\kappa \mid \kappa \in Kind\}$ where $Def_\kappa = \{[\![\mu]\!] \mid \rhd \mu : \kappa\}$, is pre-logical, since $\mathcal{U}$ is a model of $\lambda^{\Rightarrow}_{C_{Kind}}$.

For Def_{Type}, suppose $\Delta ; \Gamma \rhd t : \tau$ and $\eta : \Delta; \Gamma \to Def$. For each $v_i : \kappa_i \in \Delta$ and $x_j : \tau_j \in \Gamma$ there is $\rhd \mu_i : \kappa$ and $; \rhd t_j : \tau_j$ such that $\eta(v_i) = [\![\mu_i]\!] \in Def_\kappa$ and $\eta(x_j) = [\![; \rhd t_j : \tau_j]\!] \in Def_{[\![\rhd \tau_j : T]\!]}$. With the substitution lemma (cf. [BMM90]),

$$\begin{aligned} [\![\Delta ; \Gamma \rhd t : \tau]\!]\eta &= [\![\Delta ; \Gamma \rhd t : \tau]\!][[\![\rhd \mu_1 : \kappa_1]\!]/v_1, \ldots, [\![; \rhd t_1 : \tau_i]\!]/x_1, \ldots] \\ &= [\![; \rhd (t : \tau)[\mu_1/v_1, \ldots, t_1/x_1, \ldots]]\!] \\ &\in Def_{[\![\rhd \tau[\mu_1/v_1,\ldots,t_1/x_1,\ldots]:T]\!]} \;=\; Def_{[\![\Delta \rhd \tau:T]\!]\eta}. \end{aligned}$$

Hence Def is the least pre-logical predicate on $\mathcal{A}$. □

Definition 4.11. *Let OBS be a set of closed types and $\mathcal{A}, \mathcal{B}$ be models of $\lambda^{\to,\forall}_{C}$ such that $D^{\mathcal{A}}_{[\![\tau]\!]} = D^{\mathcal{B}}_{[\![\tau]\!]}$ for each $\tau \in OBS$. $\mathcal{A}$ and $\mathcal{B}$ are* observationally equivalent, *in symbols: $\mathcal{A} \equiv_{OBS} \mathcal{B}$, if $[\![; \rhd t : \tau]\!]^{\mathcal{A}} = [\![; \rhd t : \tau]\!]^{\mathcal{B}}$ for all closed terms $; \rhd t : \tau$ where $\tau \in OBS$.*

Theorem 4.12 *Suppose $\vdash \sigma = \tau$ whenever $[\![\rhd \sigma : T]\!]^{\mathcal{A}\times\mathcal{B}} = [\![\rhd \tau : T]\!]^{\mathcal{A}\times\mathcal{B}}$. Then $\mathcal{A} \equiv_{OBS} \mathcal{B}$ iff there is a pre-logical relation $\mathcal{R} \subseteq \mathcal{A} \times \mathcal{B}$ with*

$$R_{(A,B)} \cap (Def^{\mathcal{A}}_A \times Def^{\mathcal{B}}_B) \;\subseteq\; Id_{A,B} \subseteq D^{\mathcal{A}}_A \times D^{\mathcal{B}}_B$$

for each observable type $\rhd \tau : T \in OBS$, where $(A, B) = [\![\rhd \tau : T]\!]^{\mathcal{A}\times\mathcal{B}}$.

Proof. $\Leftarrow$ Let $; \rhd t : \tau$ be a term. Then $\rhd \tau : T$ and since $\mathcal{R}$ is pre-logical, we have $(A, B) \in R_T$ for $(A, B) = [\![\rhd \tau : T]\!]^{\mathcal{A}\times\mathcal{B}}$, and $(a, b) \in R_{(A,B)}$ for $(a, b) = [\![\rhd t : \tau]\!]^{\mathcal{A}\times\mathcal{B}}$. If $\tau : T \in OBS$ then $a = b$ by the assumption.

$\Rightarrow$: Take $\mathcal{R} := Def^{\mathcal{A}\times\mathcal{B}}$, which is pre-logical by the previous proof. Suppose $\rhd \tau : T \in OBS$ and $(A, B) = [\![\rhd \tau : T]\!]^{\mathcal{A}\times\mathcal{B}}$. For $(a, b) \in R_{(A,B)}$, there is a term $; \rhd s : \sigma$ such that

$$(a, b) = [\![; \rhd s : \sigma]\!]^{\mathcal{A}\times\mathcal{B}} \quad \text{and} \quad (A, B) = [\![\rhd \sigma : T]\!]^{\mathcal{A}\times\mathcal{B}}.$$

By assumption, $\vdash \sigma = \tau$, hence $; \rhd s : \tau$ by $(Typ =)$, and $(a, b) = [\![; \rhd s : \tau]\!]^{\mathcal{A}\times\mathcal{B}}$. Since $\mathcal{A} \equiv_{OBS} \mathcal{B}$ and $\tau \in OBS$, we get $a = b$. □

4.3 Abstract Type Constructors and Representation Independence

To simplify the setting, we had assumed that the declaration in example 2.8 introduces an abstract *type*, α *bag*, and constants $x : \sigma$ of simple type. But in fact we would like to model a declaration like

$$(\textbf{abstype}\ (bag : T \Rightarrow T, x : \sigma)\ \textbf{is}\ (\lambda\alpha : T.\tau, t : \sigma[\lambda\alpha : T.\tau/bag])\ \textbf{in}\ s),$$

introducing a *type constructor* and a constant of *polymorphic* type σ. In $\lambda_C^{\rightarrow,\forall}$ we can do this as follows: we extend $\mathcal{A} = (\mathcal{U}, \Phi, \Psi, C^{\mathcal{A}}, [\![\cdot]\!]\cdot)$ to a model $\mathcal{A}^+$, by first adjoining (cf. definition 6.1) an indeterminate, *bag*, to the kind structure $\mathcal{U}$:

$$\mathcal{U}^+ = \mathcal{U}[bag : T \Rightarrow T] = \{U^+_\kappa \mid \kappa \in Kind\},\ \text{where}\ U^+_\kappa := U_{(T\Rightarrow T)\Rightarrow\kappa} \cdot bag.$$

This provides us with new types $(A\ bag), ((A\ bag)\ bag) \in U^+_T$, for each $A \in U_T$, and we can introduce new constants of polymorphic type, like

$$empty : \forall\alpha.(\alpha\ bag) \qquad \text{or} \qquad member : \forall\alpha(\alpha \to ((\alpha\ bag) \to int)).$$

Next, when assigning domains to the new types, we interprete *bag* by a definable constructor, for example the list constructor $^* : T \Rightarrow T$, and use the domains of the resulting types of U_T as domains of the new types, i.e.

$$D^+_{A\ bag} := D_{A^*} = (D_A)^*,$$

and as interpretation of the new constants the elements of these domains that are the values of the defining terms, for example

$$[\![empty : \forall\alpha.(\alpha\ bag)]\!]^+ := [] \in D_{[\![\forall\alpha.\alpha^*]\!]} = D^+_{[\![\forall\alpha(\alpha\ bag)]\!]}.$$

Notice that a predicate $\mathcal{R}$ on $\mathcal{A}^+$ will have different predicates $R_{A\ bag}, R_{A^*} \subseteq D_{A\ bag} = D_{A^*}$, since the types $A\ bag$ and A^* are not the same.

Definition 4.13. *Let* $v_0 : \kappa_0 \rhd \sigma_0 : T$ *and* $C^+ := C_{Kind}, v_0 : \kappa_0\,; C_{Type}, x_0 : \sigma_0$ *be an extension of* C *by new 'constants'* v_0, x_0. *Let* $\mathcal{A} = (\mathcal{U}, \mathcal{D}, \Phi, \Psi, [\![\cdot]\!])$ *be an environment model of* $\lambda_C^{\rightarrow,\forall}$, $\rhd\ \mu_0 : \kappa_0$ *a closed constructor of* $\lambda^{\Rightarrow}_{C_{Kind}}$ *with value* $k_0 = [\![\ \rhd \mu_0 : \kappa_0]\!]$, *and* $A = [\![\sigma_0[\mu_0/v_0]]\!] \in U_T$ *and* $a \in D_A$.

Define the expansion $\mathcal{A}(\mu_0, a) := \mathcal{A}^+ = (\mathcal{U}^+, \mathcal{D}^+, \Phi^+, \Psi^+, [\![\cdot]\!]^+)$ *as follows:* $\mathcal{U}^+ = \mathcal{U}[v_0 : \kappa_0]$ *is the extension of* $\mathcal{U}$ *by an indeterminate* $v_0 : \kappa_0$ *(cf. definition 6.1). Each* $k \in U^+_\kappa$ *can be seen as an element of* $U_{(\kappa_0 \Rightarrow \kappa)}$ *and hence determines an element* $k(k_0) \in U_\kappa$*; therefore,* $\mathcal{D}^+, \Phi^+, \Psi^+, ([\![\cdot]\!]\cdot)^+$ *are given by*

$$\begin{array}{llll}
D^+_A & := D_{A(k_0)}, & & \\
x_0^{\mathcal{D}^+} & := a, & c^{\mathcal{D}^+} & := c^{\mathcal{D}}\ \text{for}\ c : \sigma \in C_{Type}, \\
(\Phi^+)^{\rightarrow}_{A,B} & := \Phi^{\rightarrow}_{A(k_0),B(k_0)}, & (\Psi^+)^{\rightarrow}_{A,B} & := \Psi^{\rightarrow}_{A(k_0),B(k_0)}, \\
(\Phi^+)^{\forall}_f & := \Phi^{\forall}_{f(k_0)}, & (\Psi^+)^{\forall}_f & := \Psi^{\forall}_{f(k_0)},
\end{array}$$

and $[\![\Delta\,; \Gamma \rhd t : \tau]\!]^{\mathcal{D}^+}\eta := [\![\Delta\,; \Gamma \rhd t : \tau]\!]^{\mathcal{D}}\eta_{k_0}$, *where* $\eta_{k_0}(v) := \eta(v)(k_0) \in U_\kappa$ *for* $v : \kappa \in \Delta$ *and* $\eta_{k_0}(x) := \eta(x) \in D_{A(k_0)}$ *for* $x : \sigma \in \Gamma$ *with* $A = [\![\Delta \rhd \sigma : T]\!] \in U^+_T$.

We can now give a generalization of the representation independence theorem 2.9 to $\lambda_C^{\to,\forall}$ that covers abstract type constructors. (This is also the reason why we did not use $\lambda_C^{\to,\forall,\exists}$: adding $\exists : (T \Rightarrow T) \Rightarrow T$ to $\lambda_C^{\to,\forall}$ to handle abstract *types* (cf. [MP85]) would not be sufficient; we'd need existential quantifiers of other kinds as well.) For simplicity of notation, we only state the unary version.

Theorem 4.14 *Let $\mathcal{R}$ be a pre-logical predicate on a model $\mathcal{A}$ of $\lambda_C^{\to,\forall}$, such that $\mathcal{R}_{Kind}$ is logical and its elements are definable using parameters of R_T. For each definable expansion $\mathcal{A}^+ = \mathcal{A}(\mu_0, a)$ of $\mathcal{A}$ to a model of $\lambda_{C+}^{\to,\forall}$, the following are equivalent:*

(i) $Def_A^+ \subseteq R_A$ for each $A \in R_T$.
(ii) There is a pre-logical predicate $\mathcal{R}^+ \subseteq \mathcal{A}^+$ such that $R_T \subseteq R_T^+$ and $R_A^+ = R_A$ for all $A \in R_T$.

Proof. (ii) $\Rightarrow$ (i): for $A \in R_T \subseteq R_T^+$ we have $Def_A^+ \subseteq R_A^+$ since $\mathcal{R}^+$ is pre-logical, so $Def_A^+ \subseteq R_A$ by $R_A^+ = R_A$.

(i) $\Rightarrow$ (ii): By construction, elements of U_κ^+ are equivalence classes $[v_0 : \kappa_0 \rhd \mu : \kappa]$ of constructors $\mu \in \lambda_{C_{Kind},\mathcal{U}}^{\Rightarrow}$ with parameters for elements of $\mathcal{U}$; each one can be written as $[k \cdot v_0]$ with a unique $k \in U_{(\kappa_0 \Rightarrow \kappa)}$. We define $\mathcal{R}_{Kind}^+ \subseteq \mathcal{U}^+$ by

$$\mathcal{R}_{Kind}^+ := \{R_\kappa^+ \mid \kappa \in Kind\}, \qquad \text{where} \quad R_\kappa^+ := R_{(\kappa_0 \Rightarrow \kappa)} \cdot v_0 \text{ for } \kappa \in Kind.$$

For $\mathcal{R}_{Type}^+ = \{R_A^+ \mid A \in R_T^+\}$, let R_A^+ be the set of elements of D_A^+ that can be defined in $\lambda_{C+}^{\to,\forall}$ with *type* parameters from R_T and individual parameters from $\mathcal{R}_{Type}$, i.e. for $A \in R_T^+$ (and $\eta = (\eta_{Kind}; \eta_{Type})$) we put

$$\begin{aligned} R_A^+ := \{\, [\![\Delta; \Gamma \rhd t : \tau]\!]\eta \mid{} & \Delta; \Gamma \rhd t : \tau \text{ is a term of } \lambda_{C+}^{\to,\forall}, \\ & \eta_{Kind} : \Delta \to R_T,\ A = [\![\Delta \rhd \tau : T]\!]\eta_{Kind}, \\ & \text{for all } x : \sigma \in \Gamma \text{ is } \sigma \text{ a type of } \lambda_C^{\to,\forall}, \\ & \eta_{Type} : \Gamma \to \mathcal{R}_{Type}\ \}. \end{aligned} \tag{5}$$

Claim 1 $\mathcal{R}^+ := (\mathcal{R}_{Kind}^+, \mathcal{R}_{Type}^+)$ is a pre-logical predicate on $\mathcal{A}^+$.

Proof: (Sketch) Since $\mathcal{R}_{Kind}$ is logical, not just pre-logical, by lemma 6.2, $\mathcal{R}_{Kind}^+ \subseteq \mathcal{U}^+$ is a logical, hence pre-locgial predicate. For $\mathcal{R}_{Type}$ we use lemma 4.4. To see that $\mathcal{R}^+$ is algebraic, use the fact that types $A \in R_T^+$ can be represented as $\mu \cdot v_0$ for some $\mu \in R_{(\kappa_0 \Rightarrow T)}$, and by the assumption on $\mathcal{R}$, μ is definable by a term of $\lambda_{C_{Kind}}^{\Rightarrow}$ with parameters from R_T.

To show 2. and 3. of Lemma 4.4, suppose f is a function λ-definable with parameters from $\mathcal{R}^+$. Use the substitution lemma to replace in f's defining term all parameters from $\mathcal{R}^+$ by their defining $\lambda_{C+}^{\to,\forall}$-terms with parameters from $\mathcal{R}_{Type}$.

Claim 2 For each $A \in R_T$ we have $A \in R_T^+$ and $R_A^+ = R_A$.

Proof: Let $A \in R_T$. Then $A = [\![\alpha : T; \rhd (\lambda v\, \alpha) \cdot v_0 : T]\!][A/\alpha] \in R_{(\kappa \Rightarrow T)} \cdot v_0 = R_T^+$. To show $R_A \subseteq R_A^+$, let η be an environment with $\eta(\alpha : T) = A$ and $\eta(x : \alpha) = a \in R_A$. Then $a = [\![\alpha : T; x : \alpha \rhd x : \alpha]\!]\eta \in R_A^+$ by definition of R_A^+.

To show $R_A^+ \subseteq R_A$, let $a = [\![\Delta; \Gamma \rhd t : \tau]\!]\eta \in R_A^+$ where $\Delta; \Gamma \rhd t : \tau$ and η have the properties given in (5). Then Δ has the form $\alpha_1 : T, \ldots, \alpha_n : T$, and with $\Gamma = x_1 : \sigma_1, \ldots, x_m : \sigma_m$ the abstraction

$$\bar{t} : \bar{\tau} := \lambda\alpha_1 \ldots \lambda\alpha_n \lambda x_1 : \sigma_1 \ldots \lambda x_m : \sigma_m.\, t : \forall\alpha_1 \ldots \forall\alpha_n(\sigma_1 \to \ldots \to \sigma_m \to \tau)$$

is a closed term of $\lambda_{C+}^{\to;\forall}$ with type $\bar{A} := [\![\ \rhd \bar{\tau} : T]\!] \in R_T$. By (i),

$$[\![; \rhd \bar{t} : \bar{\tau}]\!] \in Def_{\bar{A}}^+ \subseteq R_{\bar{A}}.$$

Let $A_i = \eta(\alpha_i)$, $b_j = \eta(x_j)$ and $B_j = [\![\Delta \rhd \sigma_j : T]\!]\eta$. By assumption on η and since $\mathcal{R}$ is pre-logical we have $A_i, B_j \in R_T$ and $b_j \in R_{B_j}$. This gives

$$\begin{aligned} a = [\![; \rhd \bar{t} : \bar{\tau}]\!] \cdot A_1 \cdots A_n \cdot b_1 \cdots b_m &\in R_{\bar{A}} \cdot A_1 \cdots A_n \cdot b_1 \cdots b_m \\ &\subseteq R_{B_1 \to \ldots \to B_m \to A} \cdot R_{B_1} \cdots R_{B_m} \subseteq R_A. \end{aligned}$$

□

Remark 4.15. For $\lambda_C^{\to,\forall,\exists}$, Mitchell ([Mit86], Theorem 7) states (without proof) a criterion for the equivalence of two representations of an abstract datatype where $\mathcal{R}^+$ in (ii) is a logical predicate. I am unable to construct such a logical predicate from his version of (i). It seems unlikely that we can modify the pre-logical predicate from the proof of 4.14 to obtain a logical predicate satisfying (ii) (when $\exists$ is dropped), in particular since C may have higher-order constants.

5 Directions for Future Work

First, we conjecture that the characterizations of pre-logical relations by logical relations given in [HS99] for first-order models can be transferred to the class of second-order models considered here, but details remain to be checked (cf. remark 4.8). Next, a categorical generalization of second-order pre-logical relations, extending the work on lax logical relations [PPST00], would be useful for applications to categorical models and probably to imperative languages. Third, representation independence theorems for languages with $\exists$-types or dependent types and for a general form of abstraction like

(**abstype** *context* **with** *specification* **is** *representation* **in** *scope*)

or SML's restriction construct `structure :> signature` would be very useful, especially for SML with higher-order functors (cf. [Ler95], section 4). Finally, observational equivalence as the logical relation at $\exists$-types (cf. [Mit86]) may have a flexible variant with pre-logical relations, in particular in connection with specification refinement as studied in [Han99].

Acknowledgement. Thanks to Furio Honsell for directing me to [HS99], and to the referees for helpful comments and criticism.

6 Appendix: Adjunction of Indeterminates

Definition 6.1. *Let $\mathcal{A} = (A, \Phi, \Psi, C^{\mathcal{A}})$ be an extensional model of $\lambda_C^{\vec{\to}}$, $\tau \in T$. Extend $\lambda_C^{\vec{\to}}$ to $\lambda_{C,A}^{\vec{\to}}$-terms by adding a constant $\underline{a}$ for each $a \in A_\sigma$. Consider the equivalece of $\lambda_{C,A}^{\vec{\to}}$-terms of type σ in the free variable $x : \tau$ given by*

$$s(x) =_{\mathcal{A}} t(x) :\iff \forall a \in A_\tau \ [\![s]\!][a/x] = [\![t]\!][a/x] \in A_\sigma.$$

Each equivalence class $[s : \sigma]$ of $=_{\mathcal{A}}$ can be represented in the form $[\underline{f} \cdot x]$ where $f \in A_{\tau\to\sigma}$. Since $\mathcal{A}$ is extensional, $f = [\![\lambda x\, s]\!]$ is unique and $a \mapsto [\underline{a}]$ is injective.

Let $\mathcal{A}[x : \tau] := (A', \Phi', \Psi', C')$, the extension of $\mathcal{A}$ by the indeterminate $x : \tau$, *be*

$$\begin{aligned} A'_\sigma = A[x : \tau]_\sigma &:= \{[s] \mid s \in \lambda_{C,A}^{\vec{\to}},\ x : \tau \rhd s : \sigma\} \\ \Phi'_{\rho,\sigma}([\underline{f} \cdot x])([\underline{g} \cdot x]) &:= [\underline{\Phi_{\tau\to\rho,\tau\to\sigma}(S_{\tau,\rho,\sigma} \cdot f)(g)} \cdot x] \\ \Psi'_{\rho,\sigma}(\lambda[\underline{g} \cdot x].[\underline{h(g)} \cdot x]) &:= [\underline{\Psi_{\tau\to\rho,\tau\to\sigma}(h)} \cdot x] \\ C'(c) &:= [\underline{C^{\mathcal{A}}(c)}] \quad \text{for } c : \sigma \in C, \end{aligned}$$

where $S_{\tau,\rho,\sigma} := \lambda f \lambda g \lambda x (fx(gx)) : (\tau \to \rho \to \sigma) \to (\tau \to \rho) \to (\tau \to \sigma)$. We write $f \cdot x$ instead of $[\underline{f} \cdot x]$ and, correspondingly, $A[x : \tau]_\sigma = A_{\tau\to\sigma} \cdot x$.

Lemma 6.2. *Let $\mathcal{A}$ be an extensional model of $\lambda_C^{\vec{\to}}$ and $\mathcal{R} \subseteq \mathcal{A}$ a logical predicate. There is a logical predicate $\mathcal{S}$ on $\mathcal{A}[x : \tau]$ with $[x] \in S_\tau$ and $R_\sigma \subseteq S_\sigma$ for all types σ. If $R_\tau \neq \emptyset$, then $S_\sigma \cap A_\sigma = R_\sigma$.*

Proof. Putting $S_\sigma := R_{\tau\to\sigma} \cdot x$ for all types σ, the claim is easily verified.

Note that C may contain higher-order constants. This is used for $\forall$ when applying 6.2 in the proof of theorem 4.14. Unfortunately, the lemma apparently is *wrong* with pre-logical instead of logical relations.

References

[BMM90] Kim B. Bruce, Albert R. Meyer, and John C. Mitchell, *The semantics of second-order lambda calculus*, Information and Computation **85** (1990), 76–134.

[Han99] Jo Erskine Hannay, *Specification refinement with system F*, In Proc. CSL'99, LNCS 1683, Springer Verlag, 1999, pp. 530–545.

[HLST00] Furio Honsell, John Longley, Donald Sannella, and Andrzej Tarlecki, *Constructive data refinement in typed lambda calculus*, 3rd Intl. Conf. on Foundations of Software Science and Computation Structures. European Joint Conferences on Theory and Practice of Software (ETAPS'2000), LNCS 1784, Springer Verlag, 2000, pp. 149–164.

[HS99] Furio Honsell and Donald Sannella, *Pre-logical relations*, Proc. Computer Science Logic, CSL'99, LNCS 1683, Springer Verlag, 1999, pp. 546–561.

[Ler95] Xavier Leroy, *Applicative functors and fully transparent higher-order modules*, Proc. of the 22nd Annual ACM Symposium on Principles of Programming Languages, ACM, 1995, pp. 142–153.

[Mit86] John C. Mitchell, *Representation independence and data abstraction*, Proceedings of the 13th ACM Symposium on Principles of Programming Languages, January 1986, pp. 263–276.

[Mit91] John C. Mitchell, *On the equivalence of data representations*, Artificial Intelligence and Mathematical Theory of Computation: Papers in Honour of John C. McCarthy (V. Lifschitz, ed.), Academic Press, 1991, pp. 305–330.

[Mit96] John C. Mitchell, *Foundations for programming languages*, The MIT Press, Cambridge, Mass., 1996.

[MM85] John C. Mitchell and Albert Meyer, *Second-order logical relations*, Proc. Logics of Programs, LNCS 193, Springer Verlag, 1985, pp. 225–236.

[MP85] John C. Mitchell and Gordon D. Plotkin, *Abstract types have existential type*, 12-th ACM Symposium on Principles of Programming Languages, 1985, pp. 37–51.

[MTHM97] Robin Milner, Mads Tofte, Robert Harper, and David MacQueen, *The definition of Standard ML (revised)*, The MIT Press, Cambridge, MA, 1997.

[Plo80] Gordon D. Plotkin, *Lambda definability in the full type hierarchy*, To H.B. Curry: Essays on Combinatory Logic, Lambda Calculus and Formalism, Academic Press, 1980, pp. 363–373.

[PPST00] Gordon Plotkin, John Power, Don Sannella, and Robert Tennent, *Lax logical relations*, ICALP 2000, Springer LNCS 1853, 2000, pp. 85–102.

[PR00] John Power and Edmund Robinson, *Logical relations and data abstraction*, Proc. Computer Science Logic, CSL 2000, LNCS 1862, Springer Verlag, 2000, pp. 497–511.

[Sta85] R. Statman, *Logical relations and the typed lambda calculus*, Information and Control **65** (1985), 85–97.

[Tai67] W.W. Tait, *Intensional interpretation of functionals of finite type*, Journal of Symbolic Logic **32** (1967), 198–212.

Characterizing Convergent Terms in Object Calculi via Intersection Types

Ugo de'Liguoro

Dipartimento di Informatica, Università di Torino,
Corso Svizzera 185, 10149 Torino, Italy
deligu@di.unito.it
http://www.di.unito.it/~deligu

Abstract. We give a simple characterization of convergent terms in Abadi and Cardelli untyped Object Calculus (ς-calculus) via intersection types. We consider a λ-calculus with records and its intersection type assignment system. We prove that convergent λ-terms are characterized by their types. The characterization is then inherited by the object calculus via self-application interpretation.

1 Introduction

Concerning type systems for object oriented languages, theoretical research over the last decades has focused on subtyping, having as correctness criterion that typed programs will never rice "message not understood" exception at run time. Undoubtedly these are central issues in the field; nevertheless there are questions for which a different understanding of typing could be useful.

We move from the remark that, at least in case of major theoretical models, like the Objects Calculus of [1], or the Lambda Calculus of Objects of [18], typed terms do not normalize, in general. This is not surprising, since objects are essentially recursive, and these calculi are Turing complete; but this has the unpleasant consequence that types are completely insensitive with respect to termination. For example, the ς-term $[l = \varsigma(x)x.l].l$, which diverges, has type $[]$ in the basic first order type system OB_1; but an inspection of the derivation shows that it has any type A, since the object term $[l = \varsigma(x)x.l]$ may be typed by $[l{:}A]$, for any A (more precisely, for any A there exists a typed version $[l = \varsigma(x{:}[l{:}A])x.l].l$ of the diverging term, which has type A). Exactly the same remark applies to the Lambda Calculus of Objects, where a coding of the paradoxical combinator (following notation of [18]) $\mathbf{Y} \stackrel{\Delta}{=} \lambda f.\langle rec = \lambda x.f(x \Leftarrow rec)\rangle \Leftarrow rec$ is typable by $(\sigma \to \sigma) \to \sigma$, for any σ: then $\mathbf{Y}(\lambda x.x) : \sigma$ for all σ.

Should we care about termination in OO systems? If one's focus is on event driven systems, modularization, encapsulation or software reusability, as much as these features are supported by OO languages, probably not. But, after all, object orientation is a programming paradigm: if a program enters some infinite loop, without exhibiting any communication capability, it is true that no system inconsistency is responsible for this fact; it might also be clear that, due to some

S. Abramsky (Ed.): TLCA 2001, LNCS 2044, pp. 315–328, 2001.

clever typing, one knows in advance that no object will receive some unexpected message; nevertheless such a program is hardly useful.

In the present paper we show that a simple characterization of converging terms in the ς-calculus is achievable via a combination of (untyped) interpretations of object calculi and type assignment systems. We consider untyped λ-calculus with records as target language in which object terms from ς-calculus are translated, according to interpretations which have been proposed in the literature. We restrict attention to self-application interpretation (see [1], chapter 18), since it is easily proved that convergency in the ς-calculus (see [1], chapter 6) and in the λ-calculus with records are equivalent under such interpretation. We provide a characterization of convergency in the λ-calculus with records via an intersection type assignment system; the characterization is then inherited by the (untyped) ς-calculus.

Intersection types are better understood as "functional characters" (namely computational properties), than as sets of values: this is in accordance with the fact that in such systems any term has a (possibly trivial) type, and, moreover, that each term is typable by infinitely many types. On the other hand the set of types that can be given to a term describes its functional behaviour, that is its meaning (see e.g. [12,6,22,14]). That convergency is characterized by typability within the system by types of some specific shape is basic with respect to the construction of denotational models using types (see [6,3,4]).

As a matter of fact, we consider the study of reduction properties via type assignment systems a preliminary step toward a theory of equivalence and of models for object calculi, based on domain logic and type assignment, which is left for further research.

1.1 Related Work

The use of intersection types as a tool for the study of computational properties of the untyped λ-calculus begins with [25] and [12], and it is an established theory: see [22] for an exposition. These technique has been recently applied to the study of lazy λ-calculus [4], of parallel extension of λ-calculus [7,16], and of call-by-value λ-calculus [17,20]. Further studies of reduction properties via intersection types are reported in [15].

Intersection types have been also used by Reynolds in the design of his FORSYTHE language (see among many others [26,24]). Although intersection semantics is the same as that in the literature quoted above, the meaning of types is close to the standard interpretation of polymorphism, and does not provide a tool for characterizing properties of reduction. However, the typing of records is strikingly similar to that one we have used here.

The source of the λ-calculus of records is [1], chapter 8, where it is contrasted to the ς-calculus. Interpretations have a long story, both as formalizations of the informal notion of object, and as translation of formal calculi: in particular the self-application interpretation originates from Kamin work [21]. Sources of further information on the subject of interpretations are [8,2,9], as well as [1], chapter 18. More recently Crary [10] advocated a use of intersection types for object encoding, together with existential and recursive types. The paper discusses encoding into typed calculi, and is in the line of Reynolds and Pierce understanding of intersection types.

2 A λ-Calculus with Records

The syntax of terms is obtained from that of untyped λ-terms by adding records, equipped with selection and update operators:

$$M ::= x \mid \lambda x.M \mid MN \mid \langle l_i = M_i \;^{i\in I}\rangle \mid M\cdot l \mid M\cdot l := N,$$

where I varies over finite sets of indexes, and $l_i \neq l_j$ if $i \neq j$. Note that $M\cdot l$ is not a subterm of $M\cdot l := N$ (much as $a.l$ is not a subterm of $a.l \Leftarrow \varsigma(x)b$ in the ς-calculus). We call Λ_R the resulting set of terms. The set of closed terms in Λ_R is denoted by Λ^0_R.

To give semantics to the calculus we first define a notion of reduction, and then choose a suitable subset of normal forms to be considered as values.

Definition 1. *The* weak reduction *relation,* $\longrightarrow_w$*, over* Λ_R *is defined by the following axioms and rules:*

(β) $(\lambda x.M)N \longrightarrow_w M[N/x]$,
(ν) $M \longrightarrow_w M' \;\Rightarrow\; MN \longrightarrow_w M'N$,
$(R1)$ $\langle l_i = M_i \;^{i\in I}\rangle\cdot l_j \longrightarrow_w M_j$, *if* $j \in I$,
$(R2)$ $M \longrightarrow_w M' \;\Rightarrow\; M\cdot l \longrightarrow_w M'\cdot l$,
$(R3)$ $\langle l_i = M_i \;^{i\in I}\rangle\cdot l_j := N \longrightarrow_w \langle l_i = M_i \;^{i\in I\setminus\{j\}}, l_j = N\rangle$, *if* $j \in I$,
$(R4)$ $M \longrightarrow_w M' \;\Rightarrow\; M\cdot l := N \longrightarrow_w M'\cdot l := N$.

This is lazy reduction of the λ-calculus plus extra reduction rules for record terms.

Definition 2. *The set of* values $\mathcal{V}$ *is the union of the set* $\mathcal{R} = \{\langle l_i = M_i \;^{i\in I}\rangle \mid \forall i \in I_i \;^{M_i \in \Lambda^0_R}\}$ *and the set* $\mathcal{F}$ *of closed abstractions. Then we define a convergency predicate w.r.t. weak-reduction by:*

i) $M \Downarrow V \triangleq V \in \mathcal{V} \wedge M \stackrel{*}{\longrightarrow}_w V$,
ii) $M \Downarrow \triangleq \exists V.\; M \Downarrow V$.

Definition 2 rules out closed normal forms like $(\lambda x.x)\cdot l := (\lambda x.x)$. Any value is a normal form w.r.t. $\longrightarrow_w$, but not viceversa: in particular any term representing a selection or an update over a label which is not defined in its operand in normal form, results in a blocked term which is not a value.

Terms:

$$a ::= x \mid [l_i = \varsigma(x_i)b_i \ ^{i \in I}] \mid a.l \mid a.l \Leftarrow \varsigma(x)b$$

Values:

$$v ::= [l_i : \varsigma(x_i)b_i \ ^{i \in I}]$$

Evaluation Rules:

$$\frac{}{v \downarrow v} \qquad \frac{a \downarrow [l_i : \varsigma(x_i)b_i \ ^{i \in I}] \quad b_j\{[l_i : \varsigma(x_i)b_i \ ^{i \in I}]/x_j\} \downarrow v \quad j \in I}{a.l_j \downarrow v}$$

$$\frac{a \downarrow [l_i : \varsigma(x_i)b_i \ ^{i \in I}] \quad j \in I}{a.l_j \Leftarrow \varsigma(y)b \downarrow [l_i : \varsigma(x_i)b_i \ ^{i \in I \setminus j}, l_j = \varsigma(y)b]}$$

Fig. 1. The untyped ς-calculus and its operational semantics.

3 Self-Application Interpretation and Convergency

Syntax and operational semantics of the untyped ς-calculus are from [1], chapter 6; they are reported in figure 1. Note that substitution is written $a\{b/x\}$, instead of using square braces, to avoid confusion with notation of object terms.

We then introduce the self-application interpretation $[\![\_]\!]^S$, which is a mapping sending ς-terms into Λ_R. We take this definition from [1], chapter 18.

Definition 3. *Under self-application interpretation, ς-terms are translated according to the following rules:*

$$\begin{array}{ll}
[\![x]\!]^S & \stackrel{\Delta}{=} x \\
[\![[l_i = \varsigma(x_i)b_i \ ^{i \in I}]]\!]^S & \stackrel{\Delta}{=} \langle l_i = \lambda x_i.[\![b_i]\!]^S \ ^{i \in I} \rangle \\
[\![a.l]\!]^S & \stackrel{\Delta}{=} ([\![a]\!]^S \cdot l)[\![a]\!]^S \\
[\![a.l \Leftarrow \varsigma(y)b]\!]^S & \stackrel{\Delta}{=} [\![a]\!]^S \cdot l := \lambda y.[\![b]\!]^S
\end{array}$$

For the sake of relating convergency predicates via self-application interpretation, we give a one-step operational semantics of the ς-calculus which is equivalent to the big-step one.

Definition 4. *The reduction relation $\longrightarrow_\varsigma$ over ς-terms is defined by :*

i) $[l_i = \varsigma(x_i)b_i \ ^{i \in I}].l_j \longrightarrow_\varsigma b_j\{[l_i = \varsigma(x_i)b_i \ ^{i \in I}]/x_j\}$, *if* $j \in I$,
ii) $[l_i = \varsigma(x_i)b_i \ ^{i \in I}].l_j \Leftarrow \varsigma(y)b \longrightarrow_\varsigma [l_i = \varsigma(x_i)b_i \ ^{i \in I \setminus \{j\}}, l_j = \varsigma(y)b]$, *if* $j \in I$,
iii) $a \longrightarrow_\varsigma b \Rightarrow a.l \longrightarrow_\varsigma b.l$,
iv) $a \longrightarrow_\varsigma b \Rightarrow a.l \Leftarrow \varsigma(y)c \longrightarrow_\varsigma b.l \Leftarrow \varsigma(y)c$.

This reduction relation is weaker than the one-step reduction defined in [1] 6.2-1; on the other hand the subsequent lemma does not hold for that relation.

Lemma 1. *For any ς-term a and value v:*

$$a \downarrow v \Leftrightarrow a \stackrel{*}{\longrightarrow}_\varsigma v.$$

Proof. By induction over the definition of $a \xrightarrow{*}_{\varsigma} v$ (if part), and over the definition of $a \downarrow v$ (only if part).

□

The next theorem states that operational semantics is faithfully mirrored under self-application interpretation. We write $a \downarrow$ if there exists some v such that $a \downarrow v$.

Theorem 1. *For any ς-term a, $a \downarrow$ if and only if $[\![a]\!]^S \Downarrow$, that is the self-application interpretation preserves and respects the convergency predicate.*

Proof. We observe that:

a) if $a \longrightarrow_{\varsigma} b$, then $[\![a]\!]^S \xrightarrow{+}_w [\![b]\!]^S$ (proof by induction over $\longrightarrow_{\varsigma}$);
b) if $[\![a]\!]^S \longrightarrow_w M$, then, for some N and b,

$$M \xrightarrow{*}_w N \equiv [\![b]\!]^S \quad \text{and} \quad a \longrightarrow_{\varsigma} b,$$

(proof by induction over a);
c) v is a value in the ς-calculus if and only if $[\![v]\!]^S$ is such in Λ_R (immediate from the shape of $[\![v]\!]^S$);
d) if $V \in \mathcal{V}$ and $V \equiv [\![a]\!]^S$ for some a, then a is a value (by inspection of the definition of $[\![\cdot]\!]^S$).

If $a \downarrow$ then $a \xrightarrow{*}_{\varsigma} v$ for some value v, by lemma 1; hence $[\![a]\!]^S \xrightarrow{*}_w [\![v]\!]^S$, by (a), and $[\![v]\!]^S \in \mathcal{V}$ by (c); therefore $[\![a]\!]^S \Downarrow$.

Viceversa, if $[\![a]\!]^S \Downarrow$, then $[\![a]\!]^S \xrightarrow{*}_w V$ for some $V \in \mathcal{V}$; since V is a normal form, by (b) we know that $V \equiv [\![b]\!]^S$, with $a \xrightarrow{*}_{\varsigma} b$; it follows that b is a value, by (d), so that $a \downarrow$ by lemma 1.

□

4 A Type Assignment System

In this section we introduce the basic tool we use for analyzing the computational behaviour of λ-terms. It is a type assignment system, which is an extension of system CDV_ω (see [12,5,14]), also called $\mathcal{D}\Omega$ in [22]. To arrow and intersection type constructors we add a constructor for record types, which is from [1].

Types are defined according to the grammar:

$$\sigma ::= \alpha \mid \omega \mid \sigma \rightarrow \tau \mid \sigma \wedge \tau \mid \langle l_i : \sigma_i \ ^{i \in I} \rangle,$$

where α ranges over a denumerable set of type variables, ω is a constant, σ and τ are types, I ranges over finite sets of indexes.

Definition 5. *The type assignment system CDV_ω^R is defined in figure 2. Judgments take the usual form $\Gamma \vdash M : \tau$, with $M \in \Lambda_R$. The context Γ is a finite set of assumptions $x : \sigma$, where each variable occurs at most once; the writing $\Gamma, x : \sigma$ is an abbreviation of $\Gamma \cup \{x : \sigma\}$, for $x \notin \Gamma$. We write $\Gamma \vdash_{CDV_\omega^R} M : \tau$ to abbreviate "$\Gamma \vdash M : \tau$ is derivable in system CDV_ω^R".*

$$\frac{}{\Gamma, x:\sigma \vdash x:\sigma}\ (Var) \qquad \frac{\Gamma, x:\sigma \vdash M:\tau}{\Gamma \vdash \lambda x.M : \sigma \to \tau}\ (\to I) \qquad \frac{\Gamma \vdash M : \sigma \to \tau \quad \Gamma \vdash N:\sigma}{\Gamma \vdash MN : \tau}\ (\to E)$$

$$\frac{}{\Gamma \vdash M:\omega}\ (\omega) \qquad \frac{\Gamma \vdash M:\sigma \quad \Gamma \vdash M:\tau}{\Gamma \vdash M : \sigma \wedge \tau}\ (\wedge I) \qquad \frac{\Gamma \vdash M : \sigma_1 \wedge \sigma_2 \quad i = 1,2}{\Gamma \vdash M : \sigma_i}\ (\wedge E)$$

$$\frac{\Gamma \vdash M_i : \sigma_i \quad \forall i \in I \quad J \subseteq I}{\Gamma \vdash \langle l_i = M_i\ ^{i\in I}\rangle : \langle l_j : \sigma_j\ ^{j\in J}\rangle}\ (\langle\rangle I) \qquad \frac{\Gamma \vdash M : \langle l_i : \sigma_i\ ^{i\in I}\rangle \quad j \in I}{\Gamma \vdash M\cdot l_j : \sigma_j}\ (\langle\rangle E)$$

$$\frac{\Gamma \vdash M : \langle l_i : \sigma_i\ ^{i\in I}\rangle \quad \Gamma \vdash N : \tau \quad j \in I}{\Gamma \vdash M\cdot l_j := N : \langle l_i : \sigma_i\ ^{i\in I\setminus j}, l_j : \tau\rangle}\ (\langle\rangle U)$$

Fig. 2. CDV_ω^R, intersection type assignment system for Λ_R.

Adding rules (ω), $(\wedge I)$ and $(\wedge E)$ to Curry rules (Var), $(\to I)$ and $(\to E)$ yields system CDV_ω. Rules $(\langle\rangle E)$ and $(\langle\rangle U)$ are from [1]; rule $(\langle\rangle I)$ is slightly more liberal than usual record type introduction rule: it is however a good example of the distinctive feature of intersection types. In fact, in ordinary typed systems, records are elements of some cartesian product; with respect to such interpretation rule $(\langle\rangle I)$ is unsound. But the meaning of the record type $\langle l_i : \sigma_i\ ^{i\in I}\rangle$ in our system (as it is formally defined below) is the property to reduce to some record such that, if some l_i is selected, for $i \in I$, then something having property σ_i is returned. Hence the extension of the property $\langle l_1 : \sigma_1, l_2 : \sigma_2\rangle$ is included in the extension of $\langle l_i : \sigma_i\rangle$, for any $i = 1, 2$. Finally, rule $(\langle\rangle U)$ is sound both w.r.t. the standard interpretation of record types and w.r.t. our interpretation: in fact, if the component types have to express properties of record components, and some of these is changed by an update, then it is reasonable that its type changes too. It would be unsound, instead, in system $OB_{1<:}$ (see figure 4), as this rule immediately conflicts with the self type.

The essential reason for using intersection types and ω is type invariance under subject reduction and expansion, as stated in the next theorem; its proof follows a standard pattern and it is omitted.

Theorem 2 (Subject reduction and expansion). *Let $M, N \in \Lambda_R$ and σ be any type:*

i) if $\Gamma \vdash_{CDV_\omega^R} M : \sigma$ and $M \longrightarrow_w N$ then $\Gamma \vdash_{CDV_\omega^R} N : \sigma$;
ii) if $\Gamma \vdash_{CDV_\omega^R} N : \sigma$ and $M \longrightarrow_w N$ then $\Gamma \vdash_{CDV_\omega^R} M : \sigma$.

□

We observe that a restrictive rule for update, closer to rule (*Val Update*) of system $OB_{1<:}$, such as

$$\frac{\Gamma \vdash M : \langle l_i : \sigma_i\ ^{i\in I}\rangle \quad \Gamma \vdash N : \sigma_j \quad j \in I}{\Gamma \vdash M\cdot l_j := N : \langle l_i : \sigma_i\ ^{i\in I}\rangle}\ (\langle\rangle U')$$

would break (ii) of theorem 2. Indeed, if $\mathbf{\Omega} \triangleq (\lambda x.xx)(\lambda x.xx)$ and $\mathbf{I} \triangleq \lambda x.x$, then $\langle l = \mathbf{\Omega}\rangle \cdot l := \mathbf{I} \longrightarrow_w \langle l = \mathbf{I}\rangle$; now $\langle l = \mathbf{I}\rangle$ has type $\langle l : \sigma \to \sigma\rangle$, for any σ, but, with $(\langle\rangle U')$ in place of $(\langle\rangle U)$, $\langle l = \mathbf{\Omega}\rangle \cdot l := \mathbf{I}$ has type $\langle l : \omega\rangle$ at best.

That types are "functional characters" is formalized by the following type interpretation, which associates to each type its extension. Note that it is a closed interpretation, namely extensions of types are subsets of Λ^0_R.

For $X, Y, X_i \subseteq \Lambda^0_R$ let us set (by overloading $\to$ and $\langle\rangle$):

$$\begin{array}{ll} X \to Y & = \{M \in \Lambda^0_R \mid \exists F \in \mathcal{F}.\ M \Downarrow F \wedge \ \forall N \in X.\ FN \in Y\}, \\ \langle l_i : X_i\rangle^{i \in I} & = \{M \in \Lambda^0_R \mid \exists R \in \mathcal{R}.\ M \Downarrow R \wedge \ \forall i \in I.\ R \cdot l_i \in X_i\}. \end{array}$$

We are now in place to define type interpretation, by associating to each type a subset of Λ^0_R, given an interpretation of type variables.

Definition 6 (Type interpretation). *A* closed interpretation $\mathcal{I}$ *is any map from type variables to subsets of* Λ^0_R. *It extends to a* closed type interpretation *(type interpretation for short)* $[\![\sigma]\!]_{\mathcal{I}} \subseteq \Lambda^0_R$ *as follows:*

$$\begin{array}{ll} [\![\alpha]\!]_{\mathcal{I}} & = \mathcal{I}(\alpha), \\ [\![\omega]\!]_{\mathcal{I}} & = \Lambda^0_R, \\ [\![\sigma \to \tau]\!]_{\mathcal{I}} & = [\![\sigma]\!]_{\mathcal{I}} \to [\![\tau]\!]_{\mathcal{I}}, \\ [\![\langle l_i : \sigma_i \ {}^{i \in I}\rangle]\!]_{\mathcal{I}} & = \langle l_i : [\![\sigma_i]\!]_{\mathcal{I}}^{i \in I}\rangle, \\ [\![\sigma \wedge \tau]\!]_{\mathcal{I}} & = [\![\sigma]\!]_{\mathcal{I}} \cap [\![\tau]\!]_{\mathcal{I}}. \end{array}$$

With this definition of type interpretation there may be empty types, and even types which are empty under any interpretation $\mathcal{I}$, e.g. $(\omega \to \omega) \wedge \langle l : \omega\rangle$, which also shows that the problem would not be solved neither by some clever definition of interpretation of type variables, nor by eliminating type variables at all.

On the other hand one could weaken the definition of arrow and record type interpretations, by asking only that $M \Downarrow V$, for some V, and observing that both $(\lambda x.M) \cdot l$ and $\langle l = M\rangle N$ are in the interpretation of ω, for any closed M, N. But all this results into unnecessary weakening of the theorem below, and contradicts the philosophy of types as functional characters.

As a final remark about type interpretation let us define:

$$\sigma \leq \tau \triangleq \forall \mathcal{I}.\ [\![\sigma]\!]_{\mathcal{I}} \subseteq [\![\tau]\!]_{\mathcal{I}}, \quad \text{and} \quad \sigma = \tau \triangleq \sigma \leq \tau \leq \sigma.$$

Then we have some expected inequations: among them note (i) and (ii), which are subtyping in width and in depth respectively.

Proposition 1. *The following (in)equations hold:*

i) $I \supseteq J \Rightarrow \langle l_i : \sigma_i \ {}^{i \in I}\rangle \leq \langle l_j : \sigma_j \ {}^{j \in J}\rangle$;
ii) $\forall i \in I.\ \sigma_i \leq \tau_i \Rightarrow \langle l_i : \sigma_i \ {}^{i \in I}\rangle \leq \langle l_i : \tau_i \ {}^{i \in I}\rangle$;
iii) $\langle l_i : \sigma_i \ {}^{i \in I}\rangle = \bigwedge_{i \in I}\langle l_i : \sigma_i\rangle$;
iv) $\langle l : \sigma\rangle \wedge \langle l : \tau\rangle = \langle l : \sigma \wedge \tau\rangle$.

□

Types:

$$A ::= \alpha \mid \texttt{Top} \mid [l_i : A_i \ ^{i \in I}]$$

Subtyping rules:

$$\frac{}{E \vdash A <: \texttt{Top}}\ (\text{Sub Top}) \qquad \frac{I \supseteq J}{E \vdash [l_i : B_i \ ^{i \in I}] <: [l_i : B_i \ ^{i \in J}]}\ (\text{Sub Object})$$

$$\frac{}{E \vdash A <: A}\ (\text{Sub Refl}) \qquad \frac{E \vdash A <: B \quad E \vdash B <: C}{E \vdash A <: C}\ (\text{Sub Trans})$$

Fig. 3. Subtyping rules for system $\mathrm{OB}_{1<:}$.

5 Typing Interpretations of ς-Calculus

We now turn to untyped interpretations of ς-terms into Λ_R. Since we are also interested in appreciating the closeness or the distance between our typing of the interpretations and what can be deduced for typed versions of the same ς-terms, we consider a type assignment version of Abadi and Cardelli system $\mathrm{OB}_{1<:}$, which still we call $\mathrm{OB}_{1<:}$.

To make reading more comfortable, we report the definition of system $\mathrm{OB}_{1<:}$ in figures 3 and 4 (we omit both rules for kinds and premises concerning the assumption that types and contexts are well formed, being first order types easily defined by a grammar and supposing that in a context each term variable occurs at most once). In the examples below we add term constants to make reading easier: they are typed according to some obvious rules, which we collectively name (*Const*) in both systems ($\mathrm{OB}_{1<:}$ and ours).

Self-application interpretation has been introduced in definition 3 of section 3. In [1] the criticism to this interpretation is that it is unsuitable w.r.t. subtyping, because the abstractions in front of method bodies makes the type of the interpretation of an object term contravariant in the self type.

Let us consider the ς-term:

$$a_2 \triangleq [l_1 = \varsigma(x)3, l_2 = \varsigma(x)x.l_1 \Leftarrow \varsigma(y)x.l_1 + 1].$$

In system $\mathrm{OB}_{1<:}$ it can be typed by $\sigma \triangleq [l_1 : int, l_2 : []]$:

$$\dfrac{\dfrac{}{x : \sigma \vdash 3 : int}\ (\text{Const}) \qquad \dfrac{\dfrac{\dfrac{}{x : \sigma \vdash x : \sigma}\ (\text{Var}) \qquad \dfrac{\dfrac{\dfrac{}{x : \sigma, y : \sigma \vdash x : \sigma}\ (\text{Var})}{x : \sigma, y : \sigma \vdash x.l_1 : int}\ (\text{Val Select})}{x : \sigma, y : \sigma \vdash x.l_1 + 1 : int}\ (\text{Const})}{x : \sigma \vdash x.l_1 \Leftarrow \varsigma(y)x.l_1 + 1 : \sigma}\ (\text{Val Update})}{x : \sigma \vdash x.l_1 \Leftarrow \varsigma(y)x.l_1 + 1 : []}\ (\text{Val Sub})}{\vdash a_2 : \sigma}\ (\text{Type Object})$$

Its interpretation is

$$[\![a_2]\!]^S \triangleq \langle l_1 = \lambda x.3, l_2 = \lambda x.x{\cdot}\, l_1 := \lambda y.(x{\cdot}\, l_1)x + 1\rangle$$

$$\frac{}{E, x : A \vdash x : A}\ \text{(Var)} \qquad \frac{E, x_i : [l_i : B_i\ ^{i \in I}] \vdash b_i : B_i \quad \forall i \in I}{E \vdash [l_i = \varsigma(x_i)b_i\ ^{i \in I}] : [l_i : B_i\ ^{i \in I}]}\ \text{(Type Object)}$$

$$\frac{E \vdash a : [l_i : B_i\ ^{i \in I}] \quad j \in I}{E \vdash a.l_j : B_j}\ \text{(Val Select)} \qquad \frac{\begin{array}{c} A \equiv [l_i : B_i\ ^{i \in I}] \\ E \vdash a : A \quad E, y : A \vdash b : B_j \quad j \in I \end{array}}{E \vdash a.l_j \Leftarrow \varsigma(y)b : A}\ \text{(Val Update)}$$

$$\frac{E \vdash a : A \quad E \vdash A <: B}{E \vdash a : B}\ \text{(Val Sub)}$$

Fig. 4. Typing rules of type assignment system $\mathrm{OB}_{1<:}$.

In CDV_ω^R we may assign to $[\![a_2]\!]^S$ the type $\sigma_1 \triangleq \langle l_1 : \omega \to int, l_2 : \omega \rangle$, which is close to the original type of a_2 in $\mathrm{OB}_{1<:}$; but we can also deduce

$$\sigma_2 \triangleq \langle l_1 : \sigma_1 \to int, l_2 : \sigma_1 \to \sigma_1 \rangle.$$

In fact:

$$\dfrac{\dfrac{\dfrac{}{x : \sigma_1 \vdash 3 : int}\ \text{(Const)}}{\vdash \lambda x.3 : \sigma_1 \to int}\ (\to\mathrm{I}) \qquad \dfrac{\dfrac{x : \sigma_1 \vdash \lambda y.(x \cdot l_1)x + 1 : \omega \to int}{x : \sigma_1 \vdash x \cdot l_1 := \lambda y.(x \cdot l_1)x + 1 : \sigma_1}\ (\langle\rangle\mathrm{U})}{\vdash \lambda x.x \cdot l_1 := \lambda y.(x \cdot l_1)x + 1 : \sigma_1 \to \sigma_1}\ (\to\mathrm{I})}{\vdash [\![a_2]\!]^S : \sigma_2}\ (\langle\rangle\mathrm{I})$$

since

$$\dfrac{\dfrac{\dfrac{\dfrac{\dfrac{}{x : \sigma_1, y : \omega \vdash x : \sigma}\ \text{(Var)}}{x : \sigma_1, y : \omega \vdash x \cdot l_1 : \omega \to int}\ (\langle\rangle\mathrm{E}) \qquad \dfrac{}{x : \sigma_1, y : \omega \vdash x : \omega}\ (\omega)}{x : \sigma_1, y : \omega \vdash (x \cdot l_1)x : int}\ (\to\mathrm{E})}{x : \sigma_1, y : \omega \vdash (x \cdot l_1)x + 1 : int}\ \text{(Const)}}{x : \sigma_1 \vdash \lambda y.(x \cdot l_1)x + 1 : \omega \to int}\ (\to\mathrm{I})$$

These types are not that different from those which are derivable for a_2 in $\mathrm{OB}_{1<:}$; moreover the occurrence of ω seems to be connected to their recursive nature. But $[\![a_2]\!]^S$ is a normal form (and a value): by analogy with untyped λ-calculus and the characterization of strongly normalizing terms in system CDV (see [14,22]), we expect that it should be typable without any occurrence of ω, both in the conclusion and in the derivation. This is actually the case. Let τ, ρ be any types (possibly type variables) without occurrences of ω; define

$$\sigma_3 \triangleq \langle l_1 : \tau \to int, l_2 : \rho \rangle.$$

Then:

$$
\dfrac{
\dfrac{\dfrac{\dfrac{}{x:\sigma_3\wedge\tau, y:\tau \vdash x:\sigma_3\wedge\tau}\,(\text{Var})}{x:\sigma_3\wedge\tau, y:\tau \vdash x:\sigma_3}\,(\wedge\text{E})}{x:\sigma_3\wedge\tau, y:\tau \vdash x\cdot l_1 : \tau\to \mathit{int}}\,(\langle\rangle\text{E})
\qquad
\dfrac{\dfrac{}{x:\sigma_3\wedge\tau, y:\tau \vdash x:\sigma_3\wedge\tau}\,(\text{Var})}{x:\sigma_3\wedge\tau, y:\tau \vdash x:\tau}\,(\wedge\text{E})
}{
\dfrac{\dfrac{x:\sigma_3\wedge\tau, y:\tau \vdash (x\cdot l_1)x : \mathit{int}}{x:\sigma_3\wedge\tau, y:\tau \vdash (x\cdot l_1)x+1 : \mathit{int}}\,(\text{Const})}{x:\sigma_3\wedge\tau \vdash \lambda y.(x\cdot l_1)x+1 : \tau\to \mathit{int}}\,(\to\text{I})
}\,(\to\text{E})
$$

Therefore, writing $N \triangleq (x\cdot l_1)x+1$, we have

$$
\dfrac{
\dfrac{\dfrac{}{x:\tau\vdash 3:\mathit{int}}\,(\text{Const})}{\vdash \lambda x.3:\tau\to\mathit{int}}\,(\to\text{I})
\qquad
\dfrac{\dfrac{\dfrac{\dfrac{}{x:\sigma_3\wedge\tau\vdash x:\sigma_3\wedge\tau}\,(\text{Var})}{x:\sigma_3\wedge\tau\vdash x:\sigma_3}\,(\wedge\text{E}) \qquad x:\sigma_3\wedge\tau\vdash \lambda y.N:\tau\to\mathit{int}}{x:\sigma_3\wedge\tau\vdash x\cdot l_1 := \lambda y.N : \sigma_3}\,(\langle\rangle\text{U})}{\vdash \lambda x.x\cdot l_1 := \lambda y.N : \sigma_3\wedge\tau\to\sigma_3}\,(\to\text{I})
}{
\vdash [\![a_2]\!]^S : \langle l_1:\tau\to\mathit{int}, l_2:\sigma_3\wedge\tau\to\sigma_3\rangle
}\,(\langle\rangle\text{I})
$$

As a matter of fact we conjecture the stronger statement: if CDV^R is obtained from CDV^R_ω by deleting ω from the type definition, and eliminating rule (ω) from the system, then $M \in \Lambda_R$ is typable in CDV^R if and only if it is strongly normalizing. If the full reduction of ς-calculus is considered, we also conjecture that any term $[\![a]\!]^S$, such that a is typable in $\mathrm{OB}_{1<:}$, is typable in system CDV^R if and only if a is strongly normalizable in the ς-calculus.

As the last example shows, the interpretation of the self-variable can be typed in a non uniform way. Indeed $[\![[l_i = \varsigma(x_i)b_i\ ^{i\in I}]]\!]^S$ is typed (among many other possibilities) by $\langle l_i : \sigma_i \to \tau_i\ ^{i\in I}\rangle$, for some σ_i, τ_i, where the σ_i are not necessarily equal.

This, which surely sounds odd to those familiar with typings of object calculi, is sound in our perspective: in fact in the derivation of the type of object interpretations the judgment $x_i : \sigma_i$ does not mean "the type of this object is σ_i", being the type we derive just a predicate of records. It is indeed clear that the notion of self is not immediately translated into the interpretation of object terms, rather it is implicit in the translation of method invocation.

We only observe that it is possible to collect all the assumptions made about the self-variable into a uniform typing: indeed any derivation of $[\![[l_i = \varsigma(x_i)b_i\ ^{i\in I}]]\!]^S : \langle l_i : \sigma_i \to \tau_i\ ^{i\in I}\rangle$ can be transformed into a derivation of $[\![[l_i = \varsigma(x_i)b_i\ ^{i\in I}]]\!]^S : \langle l_i : \bigwedge_{j\in I}\sigma_j \to \tau_i\ ^{i\in I}\rangle$.

6 The Characterization Theorem

In this section we provide a characterization of convergent λ-terms with records using the type assignment system of section 4. Combining this with theorem 1, we obtain a characterization of convergent ς-terms, which is the main result of the paper. Henceforth by types we mean intersection types for Λ_R. This characterization has a strict analogy with the characterization of those terms from the (classical) λ-calculus which are reducible to some head normal form (see e.g. [22]).

A type is *trivial* if its interpretation is Λ_R^0 for any $\mathcal{I}$: then a trivial type is either ω or an intersection of trivial types. A subset $X \subseteq \Lambda_R^0$ is *saturated* if it is closed under closed expansions (M is a closed expansion of N if $M \longrightarrow_w N$ and $M \in \Lambda_R^0$); we also say that $\mathcal{I}$ is a *saturated interpretation* if $\mathcal{I}(\alpha)$ is a saturated set, for all type variable α. A straightforward induction shows that, if $\mathcal{I}$ is saturated then $[\![\sigma]\!]_\mathcal{I}$ is saturated, for all σ.

A *closed substitution* is some mapping $\vartheta : \mathit{TermVar} \rightarrow \Lambda_R^0$; $M\vartheta$ denotes the result of substituting all free occurrences of x in M by $\vartheta(x)$. We say that ϑ *respects* $\Gamma, \mathcal{I}$ if for all $x : \sigma \in \Gamma$ it is the case that $\vartheta(x) \in [\![\sigma]\!]_\mathcal{I}$. Observe that, if ϑ respects $\Gamma, \mathcal{I}$ then $[\![\sigma]\!]_\mathcal{I} \neq \emptyset$, for all σ occurring in Γ.

Lemma 2 (Soundness of Type Interpretation). *Let $\Gamma \vdash_{CDV_\omega^R} M : \tau$ and suppose that $\mathcal{I}$ is a saturated interpretation. If ϑ is some closed substitution respecting $\Gamma, \mathcal{I}$, then $M\vartheta \in [\![\tau]\!]_\mathcal{I}$.*

Proof. By induction over the derivation of $\Gamma \vdash M : \tau$. Cases *(Var)* and (ω) are immediate by the hypothesis. Cases $(\wedge I), (\wedge E)$ and $(\rightarrow E)$ follow by induction hypothesis. The fact that the interpretation of some types may be empty is relevant just in case the derivation ends with an application of rule $(\rightarrow I)$.

Case $(\rightarrow I)$: the derivation ends by

$$\frac{\Gamma, x : \sigma \vdash M : \tau}{\Gamma \vdash \lambda x.M : \sigma \rightarrow \tau}\ (\rightarrow I)$$

Clearly $(\lambda x.M)\vartheta \Downarrow (\lambda x.M)\vartheta$. If $[\![\sigma]\!]_\mathcal{I} = \emptyset$, then $(\lambda x.M)\vartheta \in [\![\sigma \rightarrow \tau]\!]_\mathcal{I}$ vacuously. Else, for any $N \in [\![\sigma]\!]_\mathcal{I}$ let ϑ' be such that $\vartheta'(x) = N$, $\vartheta'(y) = \vartheta(y)$, if $y \not\equiv x$. Since ϑ' respects $\Gamma, x : \sigma, \mathcal{I}$, by induction we have $M\vartheta' \equiv (M[N/x])\vartheta \in [\![\tau]\!]_\mathcal{I}$; now $((\lambda x.M)N)\vartheta \longrightarrow_w (M[N/x])\vartheta$, and the thesis follows being $[\![\tau]\!]_\mathcal{I}$ a saturated set.

Case $(\langle\rangle I)$: the derivation ends by

$$\frac{\Gamma \vdash M_i : \sigma_i \quad \forall i \in I \quad J \subseteq I}{\Gamma \vdash \langle l_i = M_i\ ^{i \in I}\rangle : \langle l_j : \sigma_j\ ^{j \in J}\rangle}\ (\langle\rangle I)$$

Since $\langle l_i = M_i\ ^{i \in I}\rangle\vartheta \equiv \langle l_i = M_i\vartheta\rangle$, we have, for all $j \in J \subseteq I$,

$$\langle l_i = M_i\vartheta\ ^{i \in I}\rangle{\cdot}\, l_j \longrightarrow_w M_j\vartheta \in [\![\sigma_j]\!]_\mathcal{I}$$

by induction; the thesis now follows since $[\![\sigma_j]\!]_\mathcal{I}$ is saturated.

Case $(\langle\rangle E)$: the derivation ends by

$$\frac{\Gamma \vdash M : \langle l_i : \sigma_i\ ^{i \in I}\rangle \quad j \in I}{\Gamma \vdash M{\cdot}\, l_j : \sigma_j}\ (\langle\rangle E)$$

By induction $M\vartheta \in [\![\langle l_i : \sigma_i\ ^{i\in I}\rangle]\!]_{\mathcal{I}}$, hence for some $R \in \mathcal{R}$, $M\vartheta \Downarrow R$ and $R\cdot l_i \in [\![\sigma_i]\!]_{\mathcal{I}}$, for all $i \in I$. $M\vartheta \xrightarrow{*}_w R$ implies $M\vartheta\cdot l_i \xrightarrow{*}_w R\cdot l_i$, and therefore $M\vartheta\cdot l_i \in [\![\sigma_i]\!]_{\mathcal{I}}$, being $[\![\sigma_i]\!]_{\mathcal{I}}$ saturated.

Case $(\langle\rangle U)$: the derivation ends by

$$\frac{\Gamma \vdash M : \langle l_i : \sigma_i\ ^{i\in I}\rangle \quad \Gamma \vdash L : \sigma \quad j \in I}{\Gamma \vdash M\cdot l_j := L : \langle l_i : \sigma_i\ ^{i\in I\setminus\{j\}}, l_j : \sigma\rangle}\ (\langle\rangle U)$$

By induction there exists some $R \in \mathcal{R}$ such that $M\vartheta \xrightarrow{*}_w R$ and $R\cdot l_i \in [\![\sigma_i]\!]_{\mathcal{I}}$, for all $i \in I$. By definition R has the shape $\langle l_i : M_i\ ^{i\in I}\rangle$; therefore

$$\begin{aligned}(M\cdot l_j := L)\vartheta &\equiv M\vartheta\cdot l_j := L\vartheta \\ &\xrightarrow{*}_w \langle l_i : M_i\ ^{i\in I}\rangle\cdot l_j := L\vartheta \\ &\longrightarrow_w \langle l_i : M_i\ ^{i\in I\setminus\{j\}}, l_j = L\vartheta\rangle\end{aligned}$$

and $\langle l_i : M_i\ ^{i\in I\setminus\{j\}}, l_j = L\vartheta\rangle \in [\![\langle l_i : \sigma_i\ ^{i\in I\setminus\{j\}}, l_j : \sigma\rangle]\!]_{\mathcal{I}}$ by the above and the induction hypothesis. The thesis follows since $[\![\langle l_i : \sigma_i\ ^{i\in I\setminus\{j\}}, l_j : \sigma\rangle]\!]_{\mathcal{I}}$ is saturated.

□

Given the soundness of type interpretation we use it, together with type invariance under reduction and expansion, to characterize convergent λ-terms with records:

Theorem 3. *For any closed term M, $M \Downarrow$ if and only if it is typable by some non trivial type in CDV_ω^R; moreover $M \Downarrow F$ for some $F \in \mathcal{F}$ if and only if M is typable by $\omega \to \omega$, and $M \Downarrow R$, for some $R \in \mathcal{R}$, if and only if M is typable by $\langle l_i : \omega\ ^{i\in I}\rangle$, for some I.*

Proof. The only if part follows by theorem 2 (ii) and the fact that $\vdash_{CDV_\omega^R} \lambda x.M : \omega \to \omega$ and $\vdash_{CDV_\omega^R} \langle l_i = M_i\ ^{i\in i}\rangle : \langle l_i : \omega\ ^{i\in I}\rangle$, for all $\lambda x.M \in \mathcal{F}$ and $\langle l_i = M_i\ ^{i\in i}\rangle \in \mathcal{R}$. The if part is consequence of lemma 2.

□

A further consequence of this theorem is that terms reducing to a selection $M\cdot l$ or an update $M\cdot l := N$ over some label l which is undefined in M have only trivial types; in particular ill-formed terms like $\mathbf{I}\cdot l$ are only typable by conjunctions of ω.

We are eventually in place to state the main result of the paper.

Corollary 1. *For all pure (i.e. constant free) untyped ς-term a, $a \downarrow$ if and only if $[\![a]\!]^S \Downarrow$ if and only if, for some Γ and l, $\Gamma \vdash_{CDV_\omega^R} [\![a]\!]^S : \langle l : \omega\rangle$.*

Proof. By theorems 1 and 3.

□

7 Conclusion and Further Work

We have shown that a piece of theory of type assignment nicely yields a characterization of convergent ς-terms, up to the modest overhead of self-application interpretation. But it seems that we have just scratched the surface of a subject which deserves further investigation.

First, a suitable extension of the notion of saturated sets should give the tool to settle the conjecture in section 5 that exactly the interpretations of strongly normalizing objects (w.r.t. the full reduction relation) are typable in system CDV^R. In the same vein one may also consider the problem of characterizing other properties of reduction in object calculi that have been studied for the λ-calculus [15].

A further step is to build filter models of object calculi using Λ_R and its typings as an auxiliary tool. This opens the question of the structure of the model, namely its theory; conversely one may investigate whether, given a theory such as bisimulation theory of objects [19], a filter model can be devised such that the theory is complete w.r.t. that model.

An obvious task is investigation of subtyping: if we consider the containment induced by type interpretation in section 4, this is subtyping in depth and width; but a simple and direct correspondence with subtyping in object calculi is unlikely. If instead of containment semantics one consider the coercion semantics of subtyping (see e.g. [23], chapter 10), however, our framework looks more promising: it is also tempting to consider the retraction as types proposal by Scott [27], and see what happens.

Finally, looking for some practical application, it should not be difficult to find out a type reconstruction method based on the notion of principal types, even if, of course, the typability of normalizing objects is undecidable in our system. Also it is worthy to see whether certain abstract interpretation and static analysis techniques based on type systems (see e.g. [24,13,11]) carry over to object calculi using our approach.

Acknowledgment. We wish to thank Mariangiola Dezani for reading a preliminary version of this paper, and Viviana Bono for representing the point of view of object oriented people. The final version of the paper profits of the careful reading and detailed comments of two anonymous referees.

References

1. M. Abadi, L. Cardelli, *A Theory of Objects*, Springer 1996.
2. M. Abadi, L. Cardelli, R. Viswanathan, "An interpretation of objects and object types" *Proc. of of POPL'96* 1996, 396-409.
3. S. Abramsky, "Domain Theory in Logical Form", *APAL* 51, 1991, 1-77.
4. S. Abramsky, C.-H. L. Ong "Full abstraction in the lazy lambda calculus", *Inf. Comput.* 105(2), 1993, 159-267.
5. S. van Bakel, *Intersection Type Disciplines in Lambda Calculus and Applicative Term Rewriting*, Ph.D. Thesis, Mathematisch Centrum, Amsterdam 1993.

6. H.P. Barendregt, M. Coppo, M. Dezani, "A Filter Lambda Model and the Completeness of Type Assignment", *JSL* 48, 1983, 931-940.
7. G. Boudol, "A Lambda Calculus for (Strict) Parallel Functions", *Info. Comp.* 108, 1994, 51-127.
8. K.B. Bruce, "A paradigmatic Object-Oriented design, static typing and semantics", *J. of Fun. Prog.* 1(4), 1994, 127-206.
9. K.B. Bruce, L. Cardelli, B.C. Pierce, "Comparing object encodings", *Proc. of TACS'97, LNCS* 1281, 1997, 415-438.
10. K. Crary, "Simple, Efficient Object Encoding using Intersection Types", CMU Technical Report CMU-CS-99-100.
11. M. Coppo, F. Damiani, P. Giannini, "Inference based analyses of functional programs: dead-code and strictness", in [28], 143-176.
12. M. Coppo, M. Dezani, B. Venneri, "Functional characters of solvable terms", *Grund. der Math.*, 27, 1981, 45-58.
13. M. Coppo, A. Ferrari, "Type inference, abstract interpretation and strictness analysis", *TCS* 121, 1993, 113-144.
14. M. Dezani, E. Giovannetti, U. de' Liguoro, "Intersection types, λ-models and Böhm trees", in [28], 45-97.
15. M. Dezani, F. Honsell, Y. Motohama, "Compositional Characterizations of λ-terms using Intersection Types", *Proc. of MFCS'00, LNCS* 1893, 2000, 304-313.
16. M. Dezani, U. de' Liguoro, A. Piperno. "A filter model for concurrent lambda-calculus", *Siam J. Comput.* 27(5), 1998,1376-1419.
17. L. Egidi, F. Honsell, S. Ronchi della Rocca, "Operational, denotational and logical descriptions: a case study", *Fund. Inf.* 16, 1992, 149-169.
18. K. Fisher, F. Honsell, J.C. Mitchell, "A lambda calculus of objects and method specialization", *Nordic J. Comput.* 1, 1994, 3-37.
19. A. Gordon, G. Rees, "Bisimilarity for first-order calculus of objects with subtyping", *Proc. of POPL'96*, 1996, 386-395.
20. H. Ishihara, T. Kurata, "Completeness of intersection and union type assignment systems for call-by-value λ-models", to appear in *TCS*.
21. S. Kamin, "Inheritance in Smalltalk-80: a denotational definition", *Proc. of POPL'88*, 1988, 80-87.
22. J.L. Krivine, *Lambda-calcul, types et modèles*, Masson 1990.
23. J.C. Mitchell, *Foundations for Programming Languages*, MIT Press, 1996.
24. B. Pierce, *Programming with Intersection Types and Bounded Polymorphism*, Ph.D. Thesis, 1991.
25. G. Pottinger, "A Type Assignment System for Strongly Normalizable λ-terms", in R. Hindley, J. Seldin eds., *To H.B. Curry, Essays on Combinatory Logic, Lambda Calculus and Formalisms*, Academic Press, 1980, 561-527.
26. J. Reynolds, "The Coherence of Languages with Intersection Types", *LNCS* 526, 1991, 675-700.
27. D. Scott, "Data types as lattices", *SIAM J. Comput.* 5, n. 3, 1976, 522-587.
28. M. Takahashi, M. Okada, M. Dezani eds., *Theories of Types and Proofs*, Mathematical Society of Japan, vol. 2, 1998.

Parigot's Second Order $\lambda\mu$-Calculus and Inductive Types

Ralph Matthes*

Institut für Informatik der Ludwig-Maximilians-Universität München
Oettingenstraße 67, D-80538 München, Germany
matthes@informatik.uni-muenchen.de

Abstract. A new proof of strong normalization of Parigot's (second order) $\lambda\mu$-calculus is given by a reduction-preserving embedding into system • (second order polymorphic λ-calculus). The main idea is to use the least stable supertype for any type. These non-strictly positive inductive types and their associated iteration principle are available in system F, and allow to give a translation vaguely related to CPS translations (corresponding to the Kolmogorov embedding of classical logic into intuitionistic logic). However, they simulate Parigot's μ-reductions whereas CPS translations hide them.

As a major advantage, this embedding does not use the idea of reducing stability ($\neg\neg\phi \to \phi$) to that for atomic formulae. Therefore, it even extends to non-interleaving positive fixed-point types. As a non-trivial application, strong normalization of $\lambda\mu$-calculus, extended by primitive recursion on monotone inductive types, is established.

1 Introduction

$\lambda\mu$-calculus [12] essentially is the extension of natural deduction by "reductio ad absurdum" (RAA), i. e., by a term formation rule corresponding to stability ($\neg\neg\rho \to \rho$, also called "duplex negatio affirmat") and by rewrite rules for the simplification of the application of elimination rules to RAA (in the case of $\to$-elimination this corresponds to the fact that the stability of $\rho \to \sigma$ is derivable from that of σ).

In [14]—besides a direct proof of strong normalization via saturated sets—we find a reduction of the proof of strong normalization of $\lambda\mu$-calculus to the well-known strong normalization of system F. The proof is quite intricate since the considered CPS translation maps the RAA redexes and their contracta to the same term, and hence needs additional arguments on RAA reductions alone to guarantee strong normalization of the whole calculus.

Since these μ-reductions have the flavour of iteration (the application occurs in the contractum in a controlled fashion), it has been tempting to explain μ-reductions via inductive types. This is achieved by studying the "stabilization"

* I am grateful for an invitation to present a preliminary version of the present results at the "Séminaire Preuves, Programmes et Systèmes" at Paris VII in October 2000.

S. Abramsky (Ed.): TLCA 2001, LNCS 2044, pp. 329–343, 2001.

$\sharp\rho$ of any type ρ. It is the least stable type (i.e., there is a constant of type $\neg\neg\sharp\rho \to \sharp\rho$) such that ρ is included in $\sharp\rho$. From the minimality, we get an iteration principle which in fact simulates μ-reductions and therefore illustrates nicely what can be achieved with non-strictly positive inductive types.

There is an even easier embedding into system F [6] which exploits the fact that the elimination rules deconstruct the type to be eliminated. Our approach is not based on this observation, and it even turns out that also type concepts can be treated where this observation cannot be made any longer, i.e., where the elimination does not deconstruct the type to be eliminated. This is exemplified with the addition of non-interleaving positive fixed-point types. It is very unlikely that they can be embedded into system F [17]. Therefore, we cannot expect to get an embedding into system F but we do get an embedding into system F augmented with those fixed-point types. μ-reductions for fixed-point types still have a connection between the type eliminated and the result type. This is not true of primitive recursion: The inductive type and the type of results of the functional defined by primitive recursion may be completely unrelated. Nevertheless, we can also treat this extension. Firstly, the most general formulation of primitive recursion is introduced: primitive recursion on monotone inductive types. Secondly, an embedding into non-interleaving positive fixed-point types is established, which moreover interacts nicely with μ-reductions. Therefore, we can even prove that $\lambda\mu$-calculus with primitive recursion on monotone inductive types (and μ-reductions for every type construct) is strongly normalizing.

The next section recalls system F, in section 3 we review $\lambda\mu$-calculus and discuss variations on the definition. Section 4 presents the stabilization $\sharp\rho$ and the associated iteration principle and explains it via an impredicative encoding, whereas in section 5 this system becomes the target of an embedding of $\lambda\mu$-calculus. Substitution lemmas are proven in great detail in order to give a feeling for the embedding. Fixed-point types are introduced in section 6, and the embedding of the previous section is extended to cover fixed-point types (in any of the systems studied before). This extension is astonishingly straightforward. In the final section 7 monotone inductive types with primitive recursion are described. A new embedding of monotone inductive types with primitive recursion into non-interleaving positive fixed-point types gives the final normalization theorem.

2 System F

We consider system F (see e.g. [5]) only with function types and universal types, i.e., we have infinitely many type variables (denoted by α, β, ...) and with types ρ and σ we also have the function type $\rho \to \sigma$. Moreover, given a variable α and a type ρ we form the universal type $\forall\alpha\rho$. The $\forall$ binds α in ρ. The renaming convention for bound variables is adopted. Let $\mathrm{FV}(\rho)$ be the set of type variables occurring free in ρ.

The terms of F are presented as in [5], i.e., without contexts and with fixed types (see [1, p. 159] for comments on this original typing à la Church). We have infinitely many term variables with types (denoted e. g. by x^ρ), lambda abstraction $(\lambda x^\rho r^\sigma)^{\rho\to\sigma}$ for terms binding x^ρ in r^σ, term application $(r^{\rho\to\sigma}s^\rho)^\sigma$, lambda abstraction $(\Lambda\alpha r^\rho)^{\forall\alpha\rho}$ for types (under the usual proviso that $\alpha \notin \mathrm{FV}(\sigma)$ for any x^σ free in r^ρ) and type application $(r^{\forall\alpha\rho}\sigma)^{\rho[\alpha:=\sigma]}$. We freely use the renaming convention for bound term and type variables of terms, and moreover omit type superscripts of the terms. We do this even in case of typed variables which are in fact pairs of variable names and types, but only if the type can be reconstructed from a lambda abstraction binding a variable with the same name. (The interested reader may consult the discussion of these issues in [7, sections 2.1.2 and 2.2.6.].) Instead of r^ρ we often write that r has type ρ or even $r : \rho$. We let application associate to the left and write $\lambda x^\rho.r$ to avoid parenthesizing of r.

Beta reduction $\rhd$ for system F is as usual given by

$$\begin{array}{ll}(\beta_\to) & (\lambda x^\rho r)s \rhd r[x^\rho := s] \\ (\beta_\forall) & (\Lambda\alpha r)\sigma \rhd r[\alpha := \sigma] \ .\end{array}$$

Here, we used the result $r[x^\rho := s]$ of the capture-free substitution of the typed variable x^ρ by the term s of type ρ in r and the result $r[\alpha := \sigma]$ of the capture-free substitution of the type variable α by the type σ in r. The reduction relation $\to$ is defined as the term closure of $\rhd$. The main theorem on F (due to Girard) is that $\to$ is strongly normalizing, i. e., there are no infinite reduction sequences $r_1 \to r_2 \to r_3 \to \cdots$ (In [7] all the definitions and lemmas are given quite carefully so as to make sure that the main theorem can also be proved in this presentation without contexts.)

3 Second Order $\lambda\mu$-Calculus

We present $\lambda\mu$-calculus [12] not with sequents but in the same style as system F. Although $\lambda\mu$-calculus extends system F by the classical law of *reductio ad absurdum*, there is neither falsity nor negation in the type system. Hence, we keep the type system of F. However, there is a second kind of variables (called μ-variables in [12]), denoted by $a, b, c, \ldots$ They are also paired with types, e. g. a^ρ, whose interpretation is the assumption of the *negation* of ρ.

The term system is essentially that of F, extended by the rule of *reductio ad absurdum*: $\mu a^\sigma.[b^\rho]r^\rho$ is a term of type σ, binding a^σ in $[b^\rho]r^\rho$. The latter is no term but will be called a named term as in [12].

The named term $[b^\rho]r^\rho$ is to be understood as the application of the μ-variable b^ρ—assuming the negation of ρ—to the term r proving ρ. Hence, b^ρ is free in $[b^\rho]r^\rho$. Moreover, a named term would prove falsity (if it were a legal term and falsity were included into the type system), hence $[b^\rho]r^\rho$ gives falsity under the assumption of the negation of σ, expressed by the bound variable a^σ. *Reductio ad absurdum* yields σ, hence justifying the type assignment for $\mu a^\sigma.[b^\rho]r^\rho$.

As a word of caution, a^ρ is *not* a term and not legal for lambda abstraction. (Although we allow renaming of bound variables without explicit mention, we have to keep to the same sort: either normal variable—called λ-variable in [12]—or μ-variable.)

The beta reduction relation $\triangleright$ of system F is extended by μ-reductions as follows:

$$
\begin{array}{ll}
(\mu_{\to}) & (\mu a^{\rho\to\sigma}.r)s \triangleright \mu b^{\sigma}.r\left[x^{\rho\to\sigma}.[a^{\rho\to\sigma}]x := [b](xs)\right] \\
(\mu_{\forall}) & (\mu a^{\forall\alpha\rho}.r)\sigma \triangleright \mu b^{\rho[\alpha:=\sigma]}.r\left[x^{\forall\alpha\rho}.[a^{\forall\alpha\rho}]x := [b](x\sigma)\right] .
\end{array}
$$

Here, r is always a named term, and $r\left[x^{\rho}.[a^{\rho}]x := [b^{\sigma}]t^{\sigma}\right]$ denotes the result of replacing inductively (bottom-up) in r each subexpression of the form $[a^{\rho}]u^{\rho}$ by $[b^{\sigma}](t[x^{\rho} := u])$ for any term u of type ρ ("x^{ρ}." binds x^{ρ} in this substitution notation, especially in t). In fact, this is only another notation for the same substitution concept in [12].

Remark 1. For our present purposes it seems that this unusual notion of substitution cannot be avoided. If we had type $\bot$, we could formulate *reductio ad absurdum* without reference to μ-variables: Write $\neg\rho$ for $\rho \to \bot$ and extend F only by terms $\mu x^{\neg\rho}.r^{\bot}$ of type ρ which bind $x^{\neg\rho}$ in r. (We do not need the extra concept of named terms since we simply have $x^{\neg\rho}r^{\rho}$ of type $\bot$ instead of some named term $[a^{\rho}]r^{\rho}$.) Now, we can extend $\triangleright$ of system F by

$$
\begin{array}{ll}
(\mu_{\to})' & (\mu x^{\neg(\rho\to\sigma)}.r)s \triangleright \mu y^{\neg\sigma}.r[x^{\neg(\rho\to\sigma)} := \lambda z^{\rho\to\sigma}.y(zs)] \\
(\mu_{\forall})' & (\mu x^{\neg\forall\alpha\rho}.r)\sigma \triangleright \mu y^{\neg\rho[\alpha:=\sigma]}.r[x^{\neg\forall\alpha\rho} := \lambda z^{\forall\alpha\rho}.y(z\sigma)] .
\end{array}
$$

This formulation[1] only needs standard substitution and is obviously slightly more general than λμ (λμ imposes a stronger discipline on term formation). By the method of saturated sets, it is possible to show strong normalization of the term closure $\to$ of $\triangleright$. Of course, this requires an appropriate formulation of the notion of saturation. Moreover, we would reprove strong normalization of F instead of using this fact via an embedding into system F. Unfortunately, our embedding shown below does not extend to this reformulation which in some sense abuses the function types (via negation) for the explanation of *reductio ad absurdum*. Therefore, we abandon this reformulation.

Another approach to avoid the peculiar substitution is taken in [18, pp.17–20] and [2]: Instead of named terms $[a]r$, we have responses/commands $[C]r$ (resp. $\langle r \mid C\rangle$) where C is a stack of terms with a μ-variable a at the bottom. This more general view of continuations/contexts allows to define the equality relation associated with μ-reduction by help of ordinary substitution. While this is sufficient for the study of equality in [18], we are interested in strong normalization. Hence, also the call-by-value and call-by-name formulations in [2] which are possible with ordinary substitution do not suffice.

Second order λμ-calculus is strongly normalizing [13].

[1] It amounts to a λ-calculus notation for classical natural deduction in the style of Prawitz [16].

4 Extension of F by Iteration on Stabilization

If $\neg\neg\rho \to \rho$ is provable, then ρ is called stable. We consider an extension $\mathsf{F}^\sharp$ of system F by a least stable supertype $\sharp\rho$ for any type ρ. This is expressed as follows: We add a type constant $\bot$ for falsity[2] (and set $\neg\rho := \rho \to \bot$) and for every type ρ we assume the type $\sharp\rho$ ($\sharp$ is a unary type former) called the stabilization of ρ.

The term system reflects that $\sharp\rho$ is stable, that ρ embeds into $\sharp\rho$ and that $\sharp\rho$ is the least type with these properties.[3] We assume constants $I_{\sharp\rho}$ of type $\rho \to \sharp\rho$ and $S_{\sharp\rho}$ of type $\neg\neg\sharp\rho \to \sharp\rho$ and add a term formation rule: If r has type $\sharp\rho$, τ is a type and $s_1 : \rho \to \tau$ and $s_2 : \neg\neg\tau \to \tau$ then $rE_\tau s_1 s_2$ is a term of type τ.

$$\frac{}{I_{\sharp\rho} : \rho \to \sharp\rho} \qquad \frac{}{S_{\sharp\rho} : \neg\neg\sharp\rho \to \sharp\rho} \qquad \frac{r : \sharp\rho \qquad s_1 : \rho \to \tau \qquad s_2 : \neg\neg\tau \to \tau}{rE_\tau s_1 s_2 : \tau}$$

Extend $\rhd$ of system F by

$$(\sharp I) \qquad (I_{\sharp\rho} r)E_\tau s_1 s_2 \rhd s_1 r$$
$$(\sharp S) \qquad (S_{\sharp\rho} r)E_\tau s_1 s_2 \rhd s_2\Big(\lambda y^{\neg\tau}.r(\lambda z^{\sharp\rho}.y(zE_\tau s_1 s_2))\Big) .$$

In the second rule, we assume that $y^{\neg\tau}$ and $z^{\sharp\rho}$ do not occur free in s_1 or s_2. Let $\to$ be the term closure of $\rhd$. Clearly, it enjoys subject reduction, i. e., if $r^\rho \to s^\sigma$ then $\rho = \sigma$.

The presently defined system $\mathsf{F}^\sharp$ will be called F with iteration on stabilization. Let $\mathsf{F}^\bot$ be F extended only by the type constant $\bot$. Trivially, $\mathsf{F}^\bot$ inherits strong normalization from F since we consider neither a term formation nor a rewrite rule for $\bot$.

Definition 1. *A type-respecting reduction-preserving embedding (embedding for short) of a typed term rewrite system $\mathcal{S}$ into a typed term rewrite system $\mathcal{S}'$ is a function $-'$ (the $-$ sign represents the indefinite argument of the function $'$) which assigns to every type ρ of $\mathcal{S}$ a type ρ' of $\mathcal{S}'$ and to every term r of type ρ of $\mathcal{S}$ a term r' of the (image) type ρ' of $\mathcal{S}'$ such that the following implication holds: If $r \to s$ in $\mathcal{S}$, then $r' \to^+ s'$ in $\mathcal{S}'$. ($\to^+$ denotes the transitive closure of $\to$.)*

Obviously, if there is an embedding of $\mathcal{S}$ into $\mathcal{S}'$, then strong normalization of $\mathcal{S}'$ is inherited by $\mathcal{S}$.

[2] $\bot$ is just some constant: We do not assume an elimination rule expressing *ex falsum quodlibet*.

[3] The author has recently been informed that already [15, p. 110] describes a similar idea: An inductively defined predicate may be "made classical" by adding stability as another clause to the definition. This turns the definition into a non-strictly positive one and enforces stability. Note, however, that non-strictly positive inductive definitions lead to inconsistencies in higher-order predicate logic, see the example reported in [15, p. 108]. In the framework of system •, we are in the fortunate situation that arbitrary types may be stabilized without harm to consistency.

Lemma 1. *There is an embedding of* $\mathsf{F}^\sharp$ *into* $\mathsf{F}^\perp$.

Proof. By the very description, $\sharp\rho$ is nothing but the non-strictly positive inductive type $\mu\alpha.\rho+\neg\neg\alpha$ with $\alpha \notin \mathrm{FV}(\rho)$, formulated without the sum type (do not mix up the notation $\mu\alpha\rho$ for inductive types with μar for *reductio ad absurdum*). Its canonical polymorphic encoding would be $\forall\alpha.((\rho+\neg\neg\alpha) \to \alpha) \to \alpha$ which can be simplified to $\forall\alpha.(\rho \to \alpha) \to (\neg\neg\alpha \to \alpha) \to \alpha$.

By iteration on the type ρ of $\mathsf{F}^\sharp$ define the type ρ' of $\mathsf{F}^\perp$. This shall be done homomorphically in all cases except:

$$(\sharp\rho)' := \forall\alpha.(\rho' \to \alpha) \to (\neg\neg\alpha \to \alpha) \to \alpha$$

with $\alpha \notin \mathrm{FV}(\rho)$. Clearly, $\mathrm{FV}(\rho') = \mathrm{FV}(\rho)$ and $(\rho[\alpha := \sigma])' = \rho'[\alpha := \sigma']$.

By iteration on the term $r : \rho$ of $\mathsf{F}^\sharp$ define the term $r' : \rho'$ of $\mathsf{F}^\perp$. (Simultaneously, one has to show that the free variables of r' are the $x^{\rho'}$ with x^ρ free in r.) We only consider the non-homomorphic cases.

$$\begin{aligned}(I_{\sharp\rho})' &:= \lambda x^{\rho'} \Lambda\alpha\lambda x_1^{\rho'\to\alpha}\lambda x_2^{\neg\neg\alpha\to\alpha}.x_1 x\\ (S_{\sharp\rho})' &:= \lambda x^{\neg\neg(\sharp\rho)'} \Lambda\alpha\lambda x_1^{\rho'\to\alpha}\lambda x_2^{\neg\neg\alpha\to\alpha}.x_2\Big(\lambda y^{\neg\alpha}.x(\lambda z^{(\sharp\rho)'}.y(z\alpha x_1 x_2))\Big)\\ (rE_\tau s_1 s_2)' &:= r'\tau' s_1' s_2' \ .\end{aligned}$$

Since $(r[x^\rho := s])' = r'[x^{\rho'} := s']$ and $(r[\alpha := \rho])' = r'[\alpha := \rho']$, this translation respects $(\beta_\to)$ and $(\beta_\forall)$. It is a trivial calculation to prove that

$$((I_{\sharp\rho} r)E_\tau s_1 s_2)' \to^4 (s_1 r)' \qquad \text{and}$$

$$\Big((S_{\sharp\rho} r)E_\tau s_1 s_2\Big)' \to^4 \Big(s_2\Big(\lambda y^{\neg\tau}.r(\lambda z^{\sharp\rho}.y(zE_\tau s_1 s_2))\Big)\Big)'.$$

(We use $\to^n$ to express n steps of $\to$.) Therefore, $-'$ is indeed an embedding. □

5 Embedding Second Order $\lambda\mu$-Calculus into F with Iteration on Stabilization

In contrast to the proof of strong normalization of second order $\lambda\mu$-calculus in [14] by a CPS translation which maps the terms to the left and the right side of $(\mu_\to)$ and $(\mu_\forall)$ to the same term, respectively, and therefore needs an additional argument for the strong normalization of $(\mu_\to)$ and $(\mu_\forall)$ alone (without $(\beta_\to)$ and $(\beta_\forall)$), our translation given below also simulates those μ-reductions and hence is an *embedding*. Note, however, that [6] presents an embedding which is even easier—the only non-homomorphic rule is $(\forall\alpha\rho)' := \forall\alpha.(\neg\neg\alpha \to \alpha) \to \rho'$—but which heavily uses the fact that stability may be proved for those translated types from stability of their free type variables. This will rule out the extension to fixed-point types to be studied in the next section.

Define the type ρ^* of $\mathsf{F}^\sharp$ by iteration on the type ρ of second order $\lambda\mu$-calculus (in the sequel denoted by $\lambda\mu$) as follows:

$$\begin{aligned}\alpha^* &:= \alpha\\ (\rho \to \sigma)^* &:= \sharp\rho^* \to \sharp\sigma^*\\ (\forall\alpha\rho)^* &:= \forall\alpha\,\sharp\rho^*\end{aligned}$$

By induction on ρ one verifies that $(\rho[\alpha := \sigma])^* = \rho^*[\alpha := \sigma^*]$ and $\mathrm{FV}(\rho^*) = \mathrm{FV}(\rho)$.
The type translation of our embedding is given by $\rho' := \sharp\rho^*$. Therefore,

$$\begin{aligned}\alpha' &= \sharp\alpha\\ (\rho \to \sigma)' &= \sharp(\rho' \to \sigma')\\ (\forall\alpha\rho)' &= \sharp(\forall\alpha\rho')\end{aligned}$$

and $(\rho[\alpha := \sigma])' = \rho'[\alpha := \sigma^*]$, and also $\mathrm{FV}(\rho') = \mathrm{FV}(\rho)$.

Of course, ρ' could have been defined directly by iteration on ρ without reference to ρ^*. However, the substitution property would not have looked so natural. Note that if we had used $\neg\neg\rho$ instead of $\sharp\rho$ everywhere in this definition, we would have arrived at Kolmogorov's negative translation used in [3,14] (the corresponding term translation would have been a translation in continuation-passing style).

By iteration on the term $r : \rho$ of $\lambda\mu$ define the term $r' : \rho'$ of $\mathsf{F}^\sharp$. (Simultaneously, one has to show that the free variables of r' are the $x^{\rho'}$ with x^ρ a free normal variable in r and the $a^{\neg\sigma'}$ with a^σ a free μ-variable in r. Hence, we assume that the names of the μ-variables are also names of variables of our $\mathsf{F}^\sharp$.)

$$\begin{aligned}(x^\rho)' &:= x^{\rho'}\\ (\lambda x^\rho r^\sigma)' &:= I_{(\rho\to\sigma)'}(\lambda x^{\rho'} r')\\ (r^{\rho\to\sigma} s^\rho)' &:= r' E_{\sigma'}(\lambda z^{\rho'\to\sigma'}.zs')S_{\sigma'}\\ (\Lambda\alpha r^\rho)' &:= I_{(\forall\alpha\rho)'}(\Lambda\alpha r')\\ (r^{\forall\alpha\rho}\sigma)' &:= r' E_{(\rho[\alpha:=\sigma])'}(\lambda z^{\forall\alpha\rho'}.z\sigma^*)S_{(\rho[\alpha:=\sigma])'}\\ (\mu a^\sigma.[b^\rho]r^\rho)' &:= S_{\sigma'}(\lambda a^{\neg\sigma'}.b^{\neg\rho'}r')\end{aligned}$$

In the third clause, we assume that $z^{\rho\to\sigma}$ is not free in s. Note that the fourth clause is legal since the proviso on the formation of $\Lambda\alpha r'$ is fulfilled by our statement which is proved simultaneously with the definition. The substitution property $(\rho[\alpha := \sigma])' = \rho'[\alpha := \sigma^*]$ is heavily used in the fifth clause. Finally observe that no definition is given for a' since a is no term.

It will now be useful to treat named terms as if they were terms. Therefore, the sixth rule decomposes into:

$$\begin{aligned}([b^\rho]r^\rho)' &:= b^{\neg\rho'}r' : \bot\\ (\mu a^\sigma.r)' &:= S_{\sigma'}(\lambda a^{\neg\sigma'}.r')\end{aligned}$$

Lemma 2. *$(r[x^\rho := s])' = r'[x^{\rho'} := s']$ and $(r[\alpha := \rho])' = r'[\alpha := \rho^*]$.*

Proof. Induction on r. □

Corollary 1. *$((\lambda x^\rho r)s)' \to^3 (r[x^\rho := s])'$ and $((\Lambda\alpha r)\sigma)' \to^3 (r[\alpha := \sigma])'$.*

Let $\to^*$ denote the reflexive transitive closure of $\to$.

Lemma 3. *$r'[a^{\neg\rho'} := \lambda x^{\rho'}.b^{\neg\sigma'}t'] \to^* \big(r[x^\rho.[a^\rho]x := [b^\sigma]t]\big)'$ for t of type σ.*

Proof. By induction on named terms and terms r. ("Special" substitution can readily be extended to terms r.) We show the only non-trivial case where r has the form $[a^\rho]s^\rho$, hence with the same μ-variable a^ρ. The left-hand side becomes

$$(\lambda x^{\rho'}.b^{\neg\sigma'}t')s'[a^{\neg\rho'} := \lambda x^{\rho'}.b^{\neg\sigma'}t'].$$

One beta reduction step yields

$$b^{\neg\sigma'}t'\Big[x^{\rho'} := s'[a^{\neg\rho'} := \lambda x^{\rho'}.b^{\neg\sigma'}t']\Big].$$

By induction hypothesis,

$$s'[a^{\neg\rho'} := \lambda x^{\rho'}.b^{\neg\sigma'}t'] \to^* \big(s[x^\rho.[a^\rho]x := [b^\sigma]t]\big)'.$$

Hence, $\to^*$ leads to

$$b^{\neg\sigma'}t'\Big[x^{\rho'} := \big(s[x^\rho.[a^\rho]x := [b^\sigma]t]\big)'\Big].$$

By the previous lemma,

$$t'\Big[x^{\rho'} := \big(s[x^\rho.[a^\rho]x := [b^\sigma]t]\big)'\Big] = \Big(t\Big[x^\rho := s[x^\rho.[a^\rho]x := [b^\sigma]t]\Big]\Big)'.$$

To sum up, the left-hand side is in relation $\to^*$ to

$$\Big([b^\sigma]t\Big[x^\rho := s[x^\rho.[a^\rho]x := [b^\sigma]t]\Big]\Big)'$$

which is the right-hand side by the definition of "special" substitution. □

Theorem 1. *$-'$ is an embedding of $\lambda\mu$ into $\mathsf{F}^\sharp$.*

Proof. We only check that μ-reduction steps give rise to at least one rewrite step of F with stabilization. Since we already convinced ourselves that there are no problems with types, we will neglect the types altogether. As an additional benefit, we may treat both μ-reductions uniformly (and will later profit from this uniformity in the extension by fixed-point types). Write R for a term or a type. Define the term or type $\hat{R}$ as follows: $\hat{r} := r'$ and $\hat{\rho} := \rho^*$. The μ-reduction rules now both become:

$$(\mu a.[c]r)R \rhd \mu b.([c]r)[x.[a]x := [b](xR)].$$

Moreover, we uniformly have $(rR)' = r'E(\lambda z.z\hat{R})S$. Therefore,

$$((\mu a.[c]r)R)' = (S(\lambda a.cr'))E(\lambda z.z\hat{R})S$$

and one application of $(\sharp S)$ leads to

$$S\Big(\lambda b.(\lambda a.cr')\Big(\lambda x.b(xE(\lambda z.z\hat{R})S)\Big)\Big) = S\Big(\lambda b.(\lambda a.([c]r)')(\lambda x.b(xR)')\Big).$$

One beta reduction step yields

$$S\Big(\lambda b.([c]r)'[a := \lambda x.b(xR)']\Big) \to^* S\Big(\lambda b.\Big(([c]r)[x.[a]x := [b](xR)]\Big)'\Big)$$

by the previous lemma. This is $\Big(\mu b.([c]r)[x.[a]x := [b](xR)]\Big)'$. □

Hence, second order $\lambda\mu$-calculus has been proven to be strongly normalizing once more since also $\mathsf{F}^\sharp$ has been embedded into the strongly normalizing system $\mathsf{F}^\perp$.

6 Extension to Fixed-Point Types

We extend each system by non-interleaving positive fixed-point types: system F, system $\mathsf{F}^\sharp$ (with iteration on stabilization) and second order $\lambda\mu$-calculus. Our aim is to extend the embedding of $\lambda\mu$ via $\mathsf{F}^\sharp$ into $\mathsf{F}^\perp$ to the variants $\lambda\mu^f$, $\mathsf{F}^{\sharp f}$ and $\mathsf{F}^{\perp f}$ with those fixed-point types.

System F with non-interleaving positive fixed-point types (in the sequel called F^f) essentially has been studied in [4] under the name F_{ret}, and its strong normalization shown by an embedding into Mendler's system [10]. A direct proof of strong normalization by saturated sets has been given in [9] under the name NPF. There is strong evidence that no embedding into system F exists [17].

We now extend system F by types $f\alpha\rho$ which are supposed to describe arbitrary fixed-points of $\lambda\alpha\rho$, i.e., of the operation $\sigma \mapsto \rho[\alpha := \sigma]$. We confine ourselves to (non-strictly) positive dependencies which moreover have to be non-interleaved, i.e., $f\alpha\rho$ may only be formed when every occurrence of α in ρ is "to the left of an even number of $\to$" and not free in some subexpression $f\beta\sigma$ of $f\alpha\rho$. The last clause may be rephrased as follows: If fixed-point types $f\beta\sigma$ are formed with a free parameter α then the formation of a fixed-point type $f\alpha\rho$—hence w.r.t. that parameter α—is forbidden.[4] More formally:

Definition 2. *We inductively define the set $\mathcal{T}_{npf}$ of non-interleaving positive fixed-point types and simultaneously for every $\rho \in \mathcal{T}_{npf}$ the sets $N_+(\rho)$ and $N_-(\rho)$ of type variables which occur only positively or occur only negatively, respectively, and moreover do not occur in the scope of a fixed-point type formation (the set $FV(\rho)$ of free type variables is defined as before with the additional $FV(f\alpha\rho) := FV(\rho)\setminus\{\alpha\}$). Let always range p over the set $\{+,-\}$ of polarities and set $-+ := -$ and $-- := +$. Let TV be the set of type variables.*

$\alpha \in \mathcal{T}_{npf}$. $N_+(\alpha) := TV$. $N_-(\alpha) := TV\setminus\{\alpha\}$.
If $\rho, \sigma \in \mathcal{T}_{npf}$ then $\rho \to \sigma \in \mathcal{T}_{npf}$ and $N_p(\rho \to \sigma) := N_{-p}(\rho) \cap N_p(\sigma)$.
If $\rho \in \mathcal{T}_{npf}$ then $\forall\alpha\rho \in \mathcal{T}_{npf}$ and $N_p(\forall\alpha\rho) := N_p(\rho) \cup \{\alpha\}$.
If $\rho \in \mathcal{T}_{npf}$ and $\alpha \in N_+(\rho)$ (the only place where the $N_p(\rho)$ enter the conditions) then $f\alpha\rho \in \mathcal{T}_{npf}$ and $N_p(f\alpha\rho) := TV\setminus FV(f\alpha\rho)$.

[4] Note that otherwise there would be a very high degree of freedom in the interpretation of $f\alpha\rho$ since $f\beta\sigma$ is intended only to model an arbitrary fixed-point.

Note the change of the polarity in the second rule which substantiates the slogan that α's occurrences may only be to the left of an even number of →. In the last rule we achieve non-interleavedness by removing any free variable of $f\alpha\rho$.

It is somewhat awkward to prove that $\mathcal{T}_{npf}$ is closed under substitution [9, p. 303]. Since it nevertheless holds, we may extend system F to $\mathcal{T}_{npf}$ and moreover add the following term formation rules: If $t : \rho[\alpha := f\alpha\rho]$ then $C_{f\alpha\rho}t : f\alpha\rho$. If $r : f\alpha\rho$ then $rE_f : \rho[\alpha := f\alpha\rho]$.

$$\frac{t : \rho[\alpha := f\alpha\rho]}{C_{f\alpha\rho}t : f\alpha\rho} \qquad \frac{r : f\alpha\rho}{rE_f : \rho[\alpha := f\alpha\rho]}$$

Beta reduction for fixed-point types will extend ⊳ of F by

$$(\beta_f) \qquad (C_{f\alpha\rho}t)E_f \rhd t\ .$$

This constitutes F^f. It has been mentioned above that the term closure of ⊳ is strongly normalizing, i. e., F^f is strongly normalizing. Let $\mathsf{F}^{\bot f}$ be its extension by the type constant ⊥. As before, strong normalization is inherited.

Likewise extend system $\mathsf{F}^\sharp$ by non-interleaving positive fixed-point types to yield system $\mathsf{F}^{\sharp f}$: We also write $\mathcal{T}_{npf}$ for its set of types which additionally has the rules:

$\bot \in \mathcal{T}_{npf}$. $N_p(\bot) := \mathrm{TV}$.
If $\rho \in \mathcal{T}_{npf}$ then $\sharp\rho \in \mathcal{T}_{npf}$ and $N_p(\sharp\rho) := N_p(\rho)$.

We may now add the same term formation rules and (β_f) as above. It is clear that the embedding of $\mathsf{F}^\sharp$ into $\mathsf{F}^\bot$ immediately extends to an embedding of $\mathsf{F}^{\sharp f}$ into $\mathsf{F}^{\bot f}$: one only has to add homomorphic clauses for the new type former and the new term formation rules. (Note that $(\rho[\alpha := f\alpha\rho])' = \rho'[\alpha := (f\alpha\rho)'] = \rho'[\alpha := f\alpha\rho']$ indicates that the new type former does not pose any problem with the embedding.)

$\lambda\mu$ may as well be extended to the types $\mathcal{T}_{npf}$ (the original definition). We again add the two term formation rules and (β_f) but also a μ-rule pertaining to the fixed-point types:

$$(\mu_f) \qquad (\mu a^{f\alpha\rho}.r)E_f \rhd \mu b^{\rho[\alpha:=f\alpha\rho]}.r\Big[x^{f\alpha\rho}.[a^{f\alpha\rho}]x := [b](xE_f)\Big]$$

with r a named term. It strictly follows the pattern given in the proof of Theorem 1 if we also consider E_f as a possible value of R. Hence, the μ-reduction rules are still uniformly described by (the untyped pattern)

$$(\mu a.[c]r)R \rhd \mu b.([c]r)[x.[a]x := [b](xR)].$$

The resulting system shall be denoted by $\lambda\mu^f$. Let us extend the embedding of section 5. Define the type ρ^* of $\mathsf{F}^{\sharp f}$ by iteration on the type ρ of $\lambda\mu^f$ as follows (and simultaneously prove that $N_p(\rho^*) = N_p(\rho)$ and $\mathrm{FV}(\rho^*) = \mathrm{FV}(\rho)$):

$$\begin{aligned} \alpha^* &:= \alpha \\ (\rho \to \sigma)^* &:= \sharp\rho^* \to \sharp\sigma^* \\ (\forall\alpha\rho)^* &:= \forall\alpha\,\sharp\rho^* \\ (f\alpha\rho)^* &:= f\alpha\,\sharp\rho^* \end{aligned}$$

The last clause is legal since $\alpha \in N_+(\rho) = N_+(\rho^*) = N_+(\sharp\rho^*)$ by the simultaneously proved statement.

By induction on ρ one again verifies that $(\rho[\alpha := \sigma])^* = \rho^*[\alpha := \sigma^*]$.

Again set $\rho' := \sharp\rho^*$ which implies that $(f\alpha\rho)' = \sharp(f\alpha\rho')$. Also we still have $(\rho[\alpha := \sigma])' = \rho'[\alpha := \sigma^*]$ and $\mathrm{FV}(\rho') = \mathrm{FV}(\rho)$.

The previous translation $-'$ of the terms is extended by clauses for the new term formation rules. If we set $\hat{E}_f := E_f$, then the crucial rule follows the usual (untyped) pattern:

$$(rR)' := r'E(\lambda z.z\hat{R})S$$

The new rules are

$$(C_{f\alpha\rho}t^{\rho[\alpha:=f\alpha\rho]})' := I_{(f\alpha\rho)'}(C_{f\alpha\rho'}t')$$
$$(r^{f\alpha\rho}E_f)' := r'E_{(\rho[\alpha:=f\alpha\rho])'}(\lambda z^{f\alpha\rho'}.zE_f)S_{(\rho[\alpha:=f\alpha\rho])'}$$

Notice that the first term is well-typed since $t' : (\rho[\alpha := f\alpha\rho])'$ and

$$(\rho[\alpha := f\alpha\rho])' = \rho'[\alpha := (f\alpha\rho)^*] = \rho'[\alpha := f\alpha\rho'].$$

The crucial equation $(\rho[\alpha := f\alpha\rho])' = \rho'[\alpha := f\alpha\rho']$ also justifies the second definition.

Theorem 2. *$-'$ is an embedding of $\lambda\mu^f$ into $\mathsf{F}^{\sharp f}$.*

Proof. Lemma 2 and Lemma 3 clearly still hold, $((C_{f\alpha\rho}t)E_f)' \to^2 (C_{f\alpha\rho'}t')E_f \to t'$, and the treatment of (μ_f) is already captured by the uniform proof of Theorem 1. □

Corollary 2. *The system $\lambda\mu^f$ of second order $\lambda\mu$-calculus with non-interleaving positive fixed-point types is strongly normalizing.*

Remark 2. There seems to be a widespread belief that in some sense μ-reductions are nothing but an exploitation of the fact that stability of a type may be proved from the stability of its atoms. This immediately works for first-order $\lambda\mu$-calculus (only function types) since the stability of $\rho \to \sigma$ is derivable from stability of σ. Since also the stability of $\forall\alpha\rho$ is derivable from that of ρ, one may expect that the universal quantifier also is well-behaved. But notice that the set of atoms (type variables) varies with quantification. Nevertheless, it is possible to give an embedding on grounds of this view [6]. As remarked in the introduction to section 5, the crucial clause is $(\forall\alpha\rho)' := \forall\alpha.(\neg\neg\alpha \to \alpha) \to \rho'$, hence a relativization which neatly solves the problem.

What is the problem with fixed-points? It is again possible to derive the stability of $f\alpha\rho$ from that of $\rho[\alpha := f\alpha\rho]$:

$$\lambda u^{\neg\neg\rho[\alpha:=f\alpha\rho]\to\rho[\alpha:=f\alpha\rho]}\lambda x^{\neg\neg f\alpha\rho}.C_{f\alpha\rho}(u(\lambda y^{\neg\rho[\alpha:=f\alpha\rho]}.x(\lambda z^{f\alpha\rho}.y(zE_f)))).$$

But the latter type is usually more complex than the former! Therefore, I do not see a way to extend the idea of reducing the proof of stability to that of the free type variables. How could one define stability proofs for ρ' by iteration on ρ with some easy translation $-'$? Our embedding shown above does not at all care about such a definition since every ρ' is stable by construction.

Remark 3. One could ask for other type constructions where stability is not inherited from the constituent types. It is well-known that sum types (disjunction) provide an example of this phenomenon. Unfortunately, the embedding into $\mathsf{F}^\sharp$ does not extend to sum types with permutative conversions but only without them. A solution could be to introduce permutative conversions for the stabilization types. At present, we could as well take the impredicative encoding of sums and also get the respective μ-reduction for free (inside $\lambda\mu$). This idea will be demonstrated in the next section for the more interesting case of monotone inductive types with primitive recursion.

7 Second Order $\lambda\mu$-Calculus with Primitive Recursion on Monotone Inductive Types Is Strongly Normalizing

Inductive types are a syntactic representation of least pre-fixed-points of operations $\sigma \mapsto \rho[\alpha := \sigma]$. Typically, they are studied as long as α only occurs positively in ρ. (Often, even interleaving is ruled out.) Nevertheless, it turned out [15, sect. 6.3 in chap. 2] that the only needed ingredient for a useful notion of inductive type is a monotonicity witness, i. e., a term of type $\forall\alpha\forall\beta.(\alpha \to \beta) \to \rho \to \rho[\alpha := \beta]$ (in [15] monotone specifications are considered instead). If those terms are not given beforehand but incorporated into the term system they even do not need to be closed in order to guarantee strong normalization of the rewrite rules associated with them [7]. There is a choice whether the monotonicity witnesses are attached to the introduction rule or to the elimination rule which is primitive recursion. In both cases one has strong normalization [7], in the second case this even has been shown by an embedding into system F^f [8]. However, for practical purposes the first variant seems more adequate. Fortunately, it also embeds into system F^f which will be the key to this section's result.

Recall that the product type $\rho \times \sigma$ may be impredicatively encoded in system F by $\rho \times \sigma := \forall\alpha.(\rho \to \sigma \to \alpha) \to \alpha$ for $\alpha \notin \mathrm{FV}(\rho) \cup \mathrm{FV}(\sigma)$. If $r : \rho$ and $s : \sigma$ then $\langle r, s\rangle := \Lambda\alpha\lambda z^{\rho\to\sigma\to\alpha}.zrs : \rho \times \sigma$.

Second order $\lambda\mu$-calculus with primitive recursion on monotone inductive types (in the sequel denoted by $\lambda\mu^\mu$) is defined as an extension of second order $\lambda\mu$-calculus by arbitrary types $\mu\alpha\rho$ (μ binds α in ρ) and by the following term formation rules:
If $m : \forall\alpha\forall\beta.(\alpha \to \beta) \to \rho \to \rho[\alpha := \beta]$ and $t : \rho[\alpha := \mu\alpha\rho]$ then $C_{\mu\alpha\rho}mt : \mu\alpha\rho$.
If $r : \mu\alpha\rho$ and $s : \rho[\alpha := \mu\alpha\rho \times \sigma] \to \sigma$ then $rE_\mu\sigma s : \sigma$.

$$\frac{m : \forall\alpha\forall\beta.(\alpha \to \beta) \to \rho \to \rho[\alpha := \beta] \qquad t : \rho[\alpha := \mu\alpha\rho]}{C_{\mu\alpha\rho}mt : \mu\alpha\rho}$$

$$\frac{r : \mu\alpha\rho \qquad s : \rho[\alpha := \mu\alpha\rho \times \sigma] \to \sigma}{rE_\mu\sigma s : \sigma}$$

The associated beta reduction rule of primitive recursion is

$$(\beta_\mu) \qquad (C_{\mu\alpha\rho}mt)E_\mu\sigma s \rhd s\Big(m(\mu\alpha\rho)(\mu\alpha\rho \times \sigma)\Big(\lambda x^{\mu\alpha\rho}.\langle x, (\lambda x^{\mu\alpha\rho}.xE_\mu\sigma s)x\rangle\Big)t\Big)$$

The μ-reduction rule (μ in the sense of Parigot) follows the standard pattern if we now even allow R to be $E_\mu \sigma s$:

$$(\mu_\mu) \qquad (\mu a^{\mu\alpha\rho}.r)E_\mu \sigma s \rhd \mu b^\sigma.r\Big[x^{\mu\alpha\rho}.[a^{\mu\alpha\rho}]x := [b](xE_\mu\sigma s)\Big]$$

Again, r denotes a named term in this rule. Note that σ is not at all related to $\mu\alpha\rho$.

Theorem 3. *There is an embedding of $\lambda\mu^\mu$ into $\lambda\mu^f$.*

Proof. $\rho' \in \mathcal{T}_{npf}$ is defined by iteration on ρ. The only non-homomorphic clause is that for $\mu\alpha\rho$:

$$(\mu\alpha\rho)' := f\beta\forall\gamma.\Big(\Big(\forall\alpha.(\beta \times \gamma \to \alpha) \to \rho'\Big) \to \gamma\Big) \to \gamma\ .$$

We assume that $\beta, \gamma \notin \{\alpha\} \cup \mathrm{FV}(\rho)$. In fact, the only occurrence of β is 6 times to the left of $\to$ (do not forget that the coding of $\beta \times \gamma$ provides 2 of them). It is easy to see that $(\rho[\alpha := \sigma])' = \rho'[\alpha := \sigma']$ and $\mathrm{FV}(\rho') = \mathrm{FV}(\rho)$.

The translation of the terms is also defined homomorphically, with the exception of the two clauses for monotone inductive types:

$$(C_{\mu\alpha\rho}mt)' := C_{(\mu\alpha\rho)'}\Big(\Lambda\gamma\lambda z^{(\forall\alpha.((\mu\alpha\rho)'\times\gamma\to\alpha)\to\rho')\to\gamma}.z\Big(\Lambda\alpha\lambda u^{(\mu\alpha\rho)'\times\gamma\to\alpha}.$$
$$m'(\mu\alpha\rho)'\alpha\Big(\lambda x^{(\mu\alpha\rho)'}.u\langle x, (\lambda x^{(\mu\alpha\rho)'}.xE_f\gamma z)x\rangle\Big)t'\Big)\Big)$$

and

$$(rE_\mu\sigma s)' := r'E_f\sigma'\Big(\lambda z^{\forall\alpha.((\mu\alpha\rho)'\times\sigma'\to\alpha)\to\rho'}.s'\Big(z((\mu\alpha\rho)' \times \sigma')(\lambda x^{(\mu\alpha\rho)'\times\sigma'}x)\Big)\Big)$$

Since $(r[x^\rho := s])' = r'[x^{\rho'} := s']$ and $(r[\alpha := \rho])' = r'[\alpha := \rho']$, also this translation respects $(\beta_\to)$ and $(\beta_\forall)$. It is an interesting exercise to prove that also (β_μ) is simulated.

Since the other term formation rules including *reductio ad absurdum* are translated homomorphically, we may set $([b^\rho]r)' := [b^{\rho'}]r'$ and hence have a translation also for named terms. Consequently, $(\mu a^\sigma.r)' = \mu a^{\sigma'}.r'$ with r a named term. In order to treat the μ-reduction rules we need $(r[x.[a]x := [b]t])' = r'[x.[a]x := [b]t']$ for named terms r, but this is proved for named terms and for ordinary terms r simultaneously by induction on the size of r.

By induction on natural numbers n, one easily gets (only with μ-reductions) that

$$(\mu a.r)R_1 \dots R_n \to^n \mu b.r[x.[a]x := [b](xR_1 \dots R_n)]$$

(where $R_1, \dots, R_n$ are terms or types or objects of the form $E_\mu\sigma s$). Of course, this is compatible with the typing requirements.

Therefore, also (μ_μ) is simulated:
$((\mu a.r)E_\mu\sigma s))' = (\mu a.r')E_f\sigma'(\lambda z^{\cdots}.s' \dots) \to^3$
$\mu b.r'[x.[a]x := [b](xE_f\sigma'(\lambda z^{\cdots}.s' \dots))] = \mu b.r'[x.[a]x := [b](xE_\mu\sigma s)'] =$
$\mu b.(r[x.[a]x := [b](xE_\mu\sigma s)])' = (\mu b.r[x.[a]x := [b](xE_\mu\sigma s)])'$.
The other μ-reductions are treated slightly easier. □

Remark 4. There is no hope for an embedding in the style of the previous section with fixed-point types replaced by inductive types, i. e., for a direct embedding of $\lambda\mu^{\mu}$ into $\mathsf{F}^{\sharp}$, extended by primitive recursion on monotone inductive types. Firstly, the function spaces are overly used for the purpose of defining the typing for $rE_{\mu}\sigma s$. Secondly, if we replaced it by

$$\frac{r : \mu\alpha\rho \qquad s_0 : \sigma}{rE_{\mu}\sigma(x^{\rho[\alpha:=\mu\alpha\rho\times\sigma]}.s_0) : \sigma}$$

where $x^{\rho[\alpha:=\mu\alpha\rho\times\sigma]}$ is bound in s_0, we cannot use $(\rho[\alpha := \mu\alpha\rho \times \sigma])'$ since $\times$ is encoded. Assume it were explicitly included into the system. Then we would get $(\rho[\alpha := \mu\alpha\rho \times \sigma])' = \rho'[\alpha := (\mu\alpha\rho \times \sigma)^*] = \rho'[\alpha := (\mu\alpha\rho)' \times \sigma']$. But we would need $\mu\alpha\rho'$ instead of $(\mu\alpha\rho)'$. And this would have to be lifted with a monotonicity witness for ρ' w. r. t. α. But we do not even have a monotonicity witness for $\mu\alpha\rho$ at hand since those only come with the introduction rule for $\mu\alpha\rho$. Consequently, we first have to get rid of the inductive types (in favour of fixed-point types) before we can attack *reductio ad absurdum.*

Corollary 3. *Second order $\lambda\mu$-calculus with primitive recursion on monotone inductive types is strongly normalizing.*

8 Conclusions and Future Work

An alternative to the Kolmogorov translation of classical logic into minimal logic has been presented which simplifies proofs of normalization for a classical version of λ-calculus (Parigot's $\lambda\mu$). The translation using stabilization types properly simulates Parigot's μ-reductions and carries over to extensions of system F. The logical reading of the main theorem gives consistency for classical second-order propositional logic with "extended induction" on monotone inductive propositions (where extended induction is given by reading the typing rule for E_{μ} as an inference rule of natural deduction).

On the computational side, the result gives an application of iteration on non-strictly positive inductive types. However, the exact nature of the computation involved in this translation should be further studied. Does it exemplify a programming style comparable to continuation-passing style? Moreover, does the method help in understanding other λ-calculi for classical logic such as symmetric λ-calculus?

References

1. Henk P. Barendregt. *The Lambda Calculus: Its Syntax and Semantics.* North–Holland, Amsterdam, second revised edition, 1984.
2. Pierre-Louis Curien and Hugo Herbelin. The duality of computation. In *Proceedings of the fifth ACM SIGPLAN International Conference on Functional Programming (ICFP '00), Montréal,* pages 233–243. ACM Press, 2000.

3. Philippe de Groote. A CPS-translation of the $\lambda\mu$-calculus. In Sophie Tison, editor, *Trees in Algebra and Programming - CAAP'94, 19th International Colloquium*, volume 787 of *Lecture Notes in Computer Science*, pages 85–99, Edinburgh, 1994. Springer Verlag.
4. Herman Geuvers. Inductive and coinductive types with iteration and recursion. In Bengt Nordström, Kent Pettersson, and Gordon Plotkin, editors, *Proceedings of the 1992 Workshop on Types for Proofs and Programs, Båstad, Sweden, June 1992*, pages 193–217, 1992. Only published electronically: `ftp://ftp.cs.chalmers.se/pub/cs-reports/baastad.92/proc.dvi.Z`
5. Jean-Yves Girard, Yves Lafont, and Paul Taylor. *Proofs and Types*, volume 7 of *Cambridge Tracts in Theoretical Computer Science*. Cambridge University Press, 1989.
6. Thierry Joly. Un plongement de la logique classique du 2nd ordre dans AF_2. Unpublished manuscript. In French, 5 pp., January 1996.
7. Ralph Matthes. *Extensions of System F by Iteration and Primitive Recursion on Monotone Inductive Types*. Doktorarbeit (PhD thesis), University of Munich, 1998. Available via `http://www.tcs.informatik.uni-muenchen.de/~matthes/`.
8. Ralph Matthes. Monotone (co)inductive types and positive fixed-point types. *Theoretical Informatics and Applications*, 33(4/5):309–328, 1999.
9. Ralph Matthes. Monotone fixed-point types and strong normalization. In Georg Gottlob, Etienne Grandjean, and Katrin Seyr, editors, *Computer Science Logic, 12th International Workshop, Brno, Czech Republic, August 24–28, 1998, Proceedings*, volume 1584 of *Lecture Notes in Computer Science*, pages 298–312. Springer Verlag, 1999.
10. Nax P. Mendler. Recursive types and type constraints in second-order lambda calculus. In *Proceedings of the Second Annual IEEE Symposium on Logic in Computer Science, Ithaca, N.Y.*, pages 30–36. IEEE Computer Society Press, 1987. Forms a part of [11].
11. Paul F. Mendler. Inductive definition in type theory. Technical Report 87-870, Cornell University, Ithaca, N.Y., September 1987. Ph.D. Thesis (Paul F. Mendler = Nax P. Mendler).
12. Michel Parigot. $\lambda\mu$-calculus: an algorithmic interpretation of classical natural deduction. In Andrei Voronkov, editor, *Logic Programming and Automated Reasoning, International Conference LPAR'92, St. Petersburg, Russia*, volume 624 of *Lecture Notes in Computer Science*, pages 190–201. Springer Verlag, 1992.
13. Michel Parigot. Strong normalization for second order classical natural deduction. In *Proceedings, Eighth Annual IEEE Symposium on Logic in Computer Science*, pages 39–46, Montreal, Canada, 1993. IEEE Computer Society Press.
14. Michel Parigot. Proofs of strong normalisation for second order classical natural deduction. *The Journal of Symbolic Logic*, 62(4):1461–1479, 1997.
15. Christine Paulin-Mohring. *Définitions Inductives en Théorie des Types d'Ordre Supérieur*. Habilitation à diriger les recherches, ENS Lyon, 1996.
16. Dag Prawitz. *Natural Deduction. A Proof-Theoretical Study*. Almquist and Wiksell, 1965.
17. Zdzisław Spławski and Paweł Urzyczyn. Type Fixpoints: Iteration vs. Recursion. *SIGPLAN Notices*, 34(9):102–113, 1999. Proceedings of the 1999 International Conference on Functional Programming (ICFP), Paris, France.
18. Thomas Streicher and Bernhard Reus. Classical logic, continuation semantics and abstract machines. *Journal of Functional Programming*, 8(6):543–572, 1998.

The Implicit Calculus of Constructions

Extending Pure Type Systems with an Intersection Type Binder and Subtyping

Alexandre Miquel

INRIA Rocquencourt – Projet LogiCal
BP 105, 78 153 Le Chesnay cedex, France
Alexandre.Miquel@inria.fr

Abstract. In this paper, we introduce a new type system, the *Implicit Calculus of Constructions*, which is a Curry-style variant of the Calculus of Constructions that we extend by adding an intersection type binder—called the *implicit dependent product*. Unlike the usual approach of Type Assignment Systems, the implicit product can be used at every place in the universe hierarchy. We study syntactical properties of this calculus such as the $\beta\eta$-subject reduction property, and we show that the implicit product induces a rich subtyping relation over the type system in a natural way. We also illustrate the specificities of this calculus by revisiting the impredicative encodings of the Calculus of Constructions, and we show that their translation into the implicit calculus helps to reflect the computational meaning of the underlying terms in a more accurate way.

1 Introduction

In the last two decades, the proofs-as-programs paradigm—the Curry-Howard isomorphism—has been used successfully both for understanding the computational meaning of intuitionistic proofs and for implementing proof-assistant tools based on Type Theory. Since work of Martin-Löf in the 70's, a large scale of rich formalisms have been proposed to enhance expressiveness of Type Theory. Among those formalisms, the theory of Pure Type Systems (PTS) [2][1] plays an important role since it attempts to give a unifying framework to what seems to be a 'jungle of formalisms' for the one who enters for the first time into the field of Type Theory. Most modern proof assistants based on the Curry-Howard isomorphism such as Alf [11], Coq [3], LEGO [10] or Nuprl [7] implement a formalism which belongs to this family.[2]

Despite of this, PTS-based formalisms have some practical and theoretical drawbacks, due to the inherent 'verbosity' of their terms, which tends to over-use

[1] Formerly called *Generalized Type Systems*.

[2] In fact, this is only true for the core language of those proof-assistants, since they also implement features that go beyond the strict framework of PTS, such as sigma-types, primitive inductive data-type declarations and recursive function definitions.

S. Abramsky (Ed.): TLCA 2001, LNCS 2044, pp. 344–359, 2001.

abstraction and application, particularly for type arguments. This is especially true when compared with ML-style languages.

From a practical point of view, writing polymorphic functional programs may become difficult since the programmer has to explicitly instanciate each polymorphic function with the appropriate type arguments before applying its 'real' arguments. However, there are good reasons to write those extra annotations in a PTS. The first reason is that there is in general no syntactic distinction between types and terms: type abstraction (type application) is only a particular case of λ-abstraction (term application). Another reason is that without such type annotations, decidability of type-checking may be lost when the considered PTS is expressive enough. This is the case of system F for example [17].

From a more theoretical point of view, the verbosity of PTS-terms also tends to hide the real computational contents of proof-terms behind a lot of 'noise' induced by all those type abstractions and applications. A simple example is given by the Leibniz equality which can be defined impredicatively in the Calculus of Constructions[3] by

$$\begin{aligned} \mathsf{eq} &= \lambda A\!:\!\mathsf{Set}\,.\,\lambda x,y\!:\!A\,.\,\Pi P\!:\!A\to\mathsf{Prop}\,.\,P\;x\to P\;x \\ &:\ \Pi A\!:\!\mathsf{Set}\,.\,A\to A\to\mathsf{Prop} \end{aligned}$$

Using that definition, we can prove reflexivity of equality by the following term:

$$\lambda A\!:\!\mathsf{Set}\,.\,\lambda x\!:\!A\,.\,\lambda P\!:\!A\to\mathsf{Prop}\,.\,\lambda p\!:\!P\;x\,.\,p \quad : \quad \Pi A\!:\!\mathsf{Set}\,.\,\Pi x\!:\!A\,.\,\mathsf{eq}\;A\;x\;x.$$

What is the computational meaning of this proof ? It is simply the identity function $\lambda p\,.\,p$. To understand that point, let us remove type annotations in all λ-abstractions (since they play no role in the process of computation) to obtain :

$$\lambda A\,.\,\lambda x\,.\,\lambda P\,.\,\lambda p\,.\,p \quad : \quad \Pi A\!:\!\mathsf{Set}\,.\,\Pi x\!:\!A\,.\,\mathsf{eq}\;A\;x\;x.$$

The term above shows that the first three arguments are only used for type-checking purposes, and that only the fourth one is really involved in the computation process.

Many solutions have been proposed to that problem, both on the theoretical and practical sides. Most proof assistants (Coq [3,15], LEGO [14]) implement some kind of 'implicit arguments' to avoid the user the nuisance of writing redundant applications that the system can automatically infer.

A Common Practical Approach. Generally, implementations dealing with implicit arguments are based on a distinction between two kinds of products, abstractions and applications, which may be either 'explicit' or 'implicit'. Although explicit and implicit constructions do not semantically differ, the proof-checking system distinguishes them by allowing the user to omit arguments of implicit applications—the 'implicit arguments'—provided the system is able to infer them. Such arguments are reconstructed during the type-checking process

[3] For an explanation about the distinction • ••• /• ••, see paragraph 2.1.

and then silently kept into the internal representation of terms, since they might be needed later by the conversion test.

The major advantage of this method is to keep the semantics of the original calculus—*modulo* the coloring of the syntax—since implicit arguments are only implicit for the user, but not for the system. Nevertheless, the user may sometimes be confused by the fact that the system keeps implicit arguments behind its back, especially when two (dependent) types are printed identically although they are not internally identical, due to hidden implicit arguments.

A Calculus with 'Really Implicit' Arguments. In [6], M. Hagyia and Y. Toda have studied the possibility of dropping implicit arguments out of the internal representation of the terms of the bicolored Calculus of Constructions—that is, the Calculus of Constructions with explicit and implicit constructors. Their work is based on the following idea: if we ensure uniqueness of the reconstruction of implicit arguments (up to β-conversion), then we can drop implicit arguments out of the internal representation of terms, since the β-conversion test on implicit terms (*i.e* terms where implicit arguments have been erased) will give the same result as if performed on the corresponding reconstructed explicit terms.

To achieve this goal, they propose a restriction of the syntax of implicit terms in order to ensure decidability and uniqueness (up to β-conversion) of the reconstruction of implicit arguments. But their restriction actually seems to be too drastic, since it forbids the use of the implicit abstraction in order to avoid dynamic type-checking during β-reduction [6].

The Theoretical Approach of Type Assignment Systems. On the theoretical side, many Curry-style formalisms have been proposed as 'implicit' counterparts of usual Pure Type Systems, such as the Curry-style system F [8]. In [5], P. Giannini et al. proposed an uniform description of Curry-style variants of the systems of the cube, which they call the *Type Assignment Systems* (TAS)—as opposed to (Pure) Type Systems. This work follows the idea that from a purely computational point of view, polymorphic terms of the systems of the cube do not depend on their type arguments (this is called 'structural polymorphism'). As a consequence, the authors define an erasing function from Barendregt's cube to the cube of TAS, which precisely erases all the type dependencies in proof terms, thus mapping PTS-style proof-terms to ordinary pure λ-terms.

The major difference between this work and the approaches described above is that the implicit use of the dependent product is not determined by some coloring of the syntax, but by the stratification of terms. In other words, a dependent product of TAS is 'implicit' if and only if it is formed by the rule of polymorphism and, in all other cases, it is an 'explicit' product. Also notice that in the TAS framework, the erasing function does not only erase polymorphic applications, but it also erases polymorphic abstractions and type annotations in proof-term abstractions.

It is interesting to mention that the (theoretical) approach of TAS raises the same problem as the (practical) approach of M. Hagiya and Y. Toda: if the erasing function erases too much information, then it will identify terms which were not originally convertible. The isomorphism between 'explicit' and 'implicit' formalisms is then irremediably lost. In the framework of TAS, this problem arises in the systems of the cube involving dependent types [16].

Towards Implicit Pure Type Systems. The main limitation of the approach of TAS is that it restricts the 'implicit' use of the dependent product to polymorphism. If we want to generalize this approach to all PTS—which are not necessarily impredicative—it seems natural to equip them with an implicit product binder (written $\forall x : T . U$). Such a syntactic distinction naturally disconnects the kind of dependent product (explicit or implicit) from the stratification. Nevertheless, this approach raises two important issues:

The first one is that the presence of an implicit product binder (which can be used at any level of the hierarchy) induces a deep change of the underlying semantics. In particular, the isomorphism between explicit and implicit formalisms is definitively lost. This is not necessarily a negative aspect: it simply means that in our approach, 'implicit arguments' are now really implicit, in the sense that they can no more be interpreted by some invisible applications or abstractions. (In particular, the domain-theoretical model described in [13] really interprets implicit products as intersections.)

The other point raised by the introduction of an implicit product binder is that the arguments which may become really implicit (without jeopardizing the consistency of the system) have little to do with the arguments that today's algorithms are able to infer (this will be illustrated by our examples in Sect. 5). For that reason, our approach has mostly a theoretical significance, especially to understand the computational meaning of proofs, but the formalism seems to be a bad candidate for being used practically in a real proof-checking environment.

In the following, we will concentrate our study to the case of the Implicit Calculus of Constructions. However, our approach is general enough to be extended to all the other PTS. In particular, most syntactic results of Sect. 3 can be generalized to what we could call *Implicit Pure Type Systems*.

2 The Implicit Calculus of Constructions

2.1 Syntax

The Implicit Calculus of Constructions (ICC)—or, shortly, the *implicit calculus*—is a Curry-style variant of the Calculus of Constructions with universes—a.k.a. ECC [9]—in which we make a distinction between two forms of dependent products: the *explicit product*, denoted by $\Pi x : T . U$, and the *implicit product*, denoted by $\forall x : T . U$. The syntax of *sorts*, *terms* and *contexts* is given in Fig. 1. We follow here the convention of the Calculus of Inductive Constructions [18] by making a distinction between two impredicative sorts : a sort Prop for propositional types, and a sort Set for impredicative data types. However, both impredicative sorts are isomorphic for the typing rules.

Sorts	s	$::=$	$\bullet\bullet\bullet \mid \bullet\bullet\bullet\bullet \mid \bullet\bullet\bullet\bullet_i \quad (i>0)$
Terms	M,N,T,U	$::=$	$x \mid s$
		$\mid$	$\Pi x:T.U \mid \forall x:T.U$
		$\mid$	$\lambda x.M \mid M\ N$
Contexts	Γ,Δ	$::=$	$[] \mid \Gamma;[x:T]$

Fig. 1. Syntax of the Implicit Calculus of Constructions

Terms will be considered up to α-conversion. The set of free variables of a term M is written $FV(M)$, and $M\{x:=N\}$ denotes the (external) substitution operation. Notice that the product binders $\Pi x:T.U$ and $\forall x:T.U$ bind all the free occurrences of the variable x in U, but none of the occurrences of x in T.

The *non-dependent* explicit product $\Pi x:T.U$ (where $x \notin FV(U)$) is written $T \to U$.[4] We will also follow the usual writing conventions of the λ-calculus by associating type arrows to the right, multiple applications to the left, and by factorizing consecutive λ-abstractions.

A *declaration* is an ordered pair denoted by $(x:T)$, where x is a variable and T a term. A *typing context*—or shortly, a *context*—is simply a finite ordered list of declarations denoted by $\Gamma = [x_1:T_1;\ldots;x_n:T_n]$. Concatenation of contexts Γ and Δ is denoted by $\Gamma;\Delta$. A declaration $(x:T)$ *belongs* to a context Γ if $\Gamma = \Gamma_1;[x:T];\Gamma_2$ for some contexts Γ_1 and Γ_2, that we write $(x:T)\in\Gamma$. Contexts are ordered by

- the *prefix* ordering, denoted by $\Gamma \sqsubset \Gamma'$, which means that $\Gamma' = \Gamma;\Delta$ for some context Δ;
- the *inclusion* ordering, denoted by $\Gamma \subset \Gamma'$, which means that any declaration belonging to Γ also belongs to Γ'.

If $\Gamma = [x_1:T_1;\ldots;x_n:T_n]$ is a context, the set of *declared variables* of Γ is the set defined by $DV(\Gamma) = \{x_1;\ldots;x_n\}$. We also extend the notations $FV(M)$ and $M\{x:=N\}$ to contexts by setting

$$\begin{aligned} FV(\Gamma) &= FV(T_1)\cup\cdots\cup FV(T_n) \\ \text{and}\quad \Gamma\{x:=N\} &= [x_1:T_1\{x:=N\};\ldots;x_n:T_n\{x:=N\}], \end{aligned}$$

the latter notation making sense only if $x \notin DV(\Gamma)$. Finally, we will write $\forall\Delta.U = \forall x_1:T_1.\ldots.\forall x_n:T_n.U$ for any context $\Delta = [x_1:T_1;\ldots;x_n:T_n]$ and for any term U.

[4] There is no equivalent notation for the non-dependent implicit product, whose meaning will be discussed in paragraph 2.3.

2.2 Reduction Rules

As for the untyped λ-calculus, we will use the notions of β and η-reduction. (The need of the η-reduction rule, which is not assumed in the theory of Pure Type Systems, will be explained in paragraphs 2.3 and 3.2.) For each reduction rule $R \in \{\beta;\ \eta;\ \beta\eta\}$, we define

- the *one-step R-reduction*, denoted $\rightarrow_R$, as the contextual closure of $\triangleright_R$;
- the *R-reduction*, denoted $\twoheadrightarrow_R$, as the reflexive and transitive closure of $\rightarrow_R$;
- the *R-convertibility* equivalence, denoted $\cong_R$, as the reflexive, symmetric and transitive closure of $\rightarrow_R$.

Proposition 1 (Church-Rosser). *The β-, η- and $\beta\eta$-reduction are Church-Rosser.*

In the strict framework of Pure Type Systems, the $\beta\eta$-reduction does not satisfy the Church-Rosser property [4], due to the presence of a type annotation in the λ-abstraction. However, such a problem does not arise in the implicit calculus, since we use a Curry-style λ-abstraction.

As for the untyped λ-calculus, any sequence of $\beta\eta$-reductions can be decomposed as a sequence of β-reductions followed by a sequence of η-reductions. This is a consequence of the following lemma, which will be useful for proving the $\beta\eta$-subject reduction property :

Lemma 1 (η-reduction delaying). — *For any terms M_0, M_1 and M_2 such that $M_0 \twoheadrightarrow_\eta M_1$ and $M_1 \twoheadrightarrow_\beta M_2$, there exists a term M_1' such that $M_0 \twoheadrightarrow_\beta M_1'$ and $M_1' \twoheadrightarrow_\eta M_2$.*

2.3 Typing Rules

The typing rules of the implicit calculus are parametrized by a set **Axiom** $\subset \mathcal{S}^2$ for typing sorts, a set **Rule** $\subset \mathcal{S}^3$ for typing both explicit and implicit products, and a cumulative ordering $s_1 \leq s_2$ between sorts, which are summarized in Fig. 2. Typing rules of the implicit calculus involve two judgments:

- $\Gamma \vdash$, which means: "the context Γ is well-formed";
- $\Gamma \vdash M : T$, which means: "under the context Γ, the term M has type T".

Validity of those judgments is defined by mutual induction using rules of Fig. 2.

The rules (Var), (Sort), (ExpProd), (ImpProd), (Lam), (App), (Conv) and (Cum) are the usual rules of ECC, except that we have an extra rule for the implicit product—which shares the same premises as the rule for the explicit product. Moreover, the convertibility rule (Conv) now identifies types up to $\beta\eta$-convertibility.

The rules (Gen) and (Inst) are the introduction and elimination rules for implicit product types. In contrast to the rules (Lam) and (App), the rules (Gen) and (Inst) have no associated constructors. Remark that the rule (Gen) involves

Axioms, product formation rules and cumulative ordering

$$\textbf{Axiom} = \{(\bullet\bullet\bullet\bullet, \bullet\bullet\bullet\bullet_1);\ (\bullet\bullet\bullet, \bullet\bullet\bullet\bullet_1);\ (\bullet\bullet\bullet\bullet_i, \bullet\bullet\bullet\bullet_{i+1});\quad i > 0)\}$$

$$\textbf{Rule} = \{(s, \bullet\bullet\bullet\bullet, \bullet\bullet\bullet\bullet);\ (s, \bullet\bullet\bullet, \bullet\bullet\bullet);\ (\bullet\bullet\bullet\bullet_i, \bullet\bullet\bullet\bullet_i, \bullet\bullet\bullet\bullet_i);\quad s \in \mathcal{S},\ i > 0\}$$

$$\bullet\bullet\bullet\bullet \leq \bullet\bullet\bullet\bullet;\quad \bullet\bullet\bullet \leq \bullet\bullet\bullet;\quad \bullet\bullet\bullet\bullet \leq \bullet\bullet\bullet\bullet_i;\quad \bullet\bullet\bullet \leq \bullet\bullet\bullet\bullet_i;\quad \bullet\bullet\bullet\bullet_i \leq \bullet\bullet\bullet\bullet_j \text{ if } i \leq j$$

Rules for well-formed contexts

$$\frac{}{[\,] \vdash}\ (\text{WF-E}) \qquad \frac{\Gamma \vdash T : s \quad x \notin DV(\Gamma)}{\Gamma; [x : T] \vdash}\ (\text{WF-S})$$

Rules for well-typed terms

$$\frac{\Gamma \vdash \quad (x : T) \in \Gamma}{\Gamma \vdash x : T}\ (\textsc{Var}) \qquad \frac{\Gamma \vdash \quad (s_1, s_1) \in \textbf{Axiom}}{\Gamma \vdash s_1 : s_2}\ (\textsc{Sort})$$

$$\frac{\Gamma \vdash T : s_1 \quad \Gamma; [x : T] \vdash U : s_2 \quad (s_1, s_2, s_3) \in \textbf{Rule}}{\Gamma \vdash \Pi x{:}T\,.\,U : s_3}\ (\textsc{ExpProd})$$

$$\frac{\Gamma \vdash T : s_1 \quad \Gamma; [x : T] \vdash U : s_2 \quad (s_1, s_2, s_3) \in \textbf{Rule}}{\Gamma \vdash \forall x{:}T\,.\,U : s_3}\ (\textsc{ImpProd})$$

$$\frac{\Gamma; [x : T] \vdash M : U \quad \Gamma \vdash \Pi x{:}T\,.\,U : s}{\Gamma \vdash \lambda x\,.\,M : \Pi x{:}T\,.\,U}\ (\textsc{Lam}) \qquad \frac{\Gamma \vdash M : \Pi x{:}T\,.\,U \quad \Gamma \vdash N : T}{\Gamma \vdash M\ N : U\{x := N\}}\ (\textsc{App})$$

$$\frac{\Gamma; [x : T] \vdash M : U \quad \Gamma \vdash \forall x{:}T\,.\,U : s \quad x \notin FV(M)}{\Gamma \vdash M : \forall x{:}T\,.\,U}\ (\textsc{Gen})$$

$$\frac{\Gamma \vdash M : \forall x{:}T\,.\,U \quad \Gamma \vdash N : T}{\Gamma \vdash M : U\{x := N\}}\ (\textsc{Inst})$$

$$\frac{\Gamma \vdash M : T \quad \Gamma \vdash T' : s \quad T \cong_{\beta\eta} T'}{\Gamma \vdash M : T'}\ (\textsc{Conv}) \qquad \frac{\Gamma \vdash T : s_1 \quad s_1 \leq s_2}{\Gamma \vdash T : s_2}\ (\textsc{Cum})$$

$$\frac{\Gamma \vdash \lambda x\,.\,(M\ x) : T \quad x \notin FV(M)}{\Gamma \vdash M : T}\ (\textsc{Ext})$$

$$\frac{\Gamma; [x : T] \vdash M : U \quad x \notin FV(M) \cup FV(U)}{\Gamma \vdash M : U}\ (\textsc{Str})$$

Fig. 2. Typing rules of the Implicit Calculus of Constructions

a side-condition ensuring that the variable x whose type has to be generalized does not appear free in the term M.

The purpose of the next rule, called (Ext) for 'extensionality', is to enforce the η-subject reduction property in the implicit calculus. Such a rule cannot be derived from the other rules, for the same reasons that it cannot be derived in Curry-style system F, which is included in ICC. This rule is desirable here, since it gives smoother properties to the subtyping relation, such as the contravariant/covariant subtyping rules in products.[5]

The Meaning of the Non-dependent Implicit Product. The presence of the last rule—called (Str) for "strengthening"—may be surprising, since the corresponding rule is admissible in the (Extended) Calculus of Constructions, and more generally in all functional PTS [4]. In the implicit calculus, this is not the case, due to the presence of non-dependent implicit products. The main consequence of rule (Str)—an the reason for introducing it—is the following:

Lemma 2 (Non-dependent implicit product). — *Let Γ be a context, and let T and U be terms such that $x \notin FV(U)$ and $\forall x\!:\!T\,.\,U$ is a well-formed type in Γ. Then, for any term M we have the equivalence:*

$$\Gamma \vdash M : \forall x\!:\!T\,.\,U \quad \Leftrightarrow \quad \Gamma \vdash M : U$$

In other words, a non-dependent implicit product $\forall x\!:\!T\,.\,U$ has the very same inhabitants as the type U, obtained by removing the 'dummy' quantification $\forall x\!:\!T$. Without the rule (Str), this result would hold only if the type T is not empty in the context Γ.

3 Typing Properties

3.1 Subject Reduction

The $\beta\eta$-subject reduction of the implicit calculus is surprisingly hard to prove due to the presence of the rule (Ext) whose premise involves a term structurally larger than the term in the conclusion. For that, we have to use a trick based on lemma 1 in order to isolate the rule (Ext).

Step 1: Preliminary Results. We first prove the following three lemmas by an immediate induction on the structure of derivations :

Lemma 3 (Well-formed contexts). — *Let Γ be a context.*

1. *If Γ is well-formed, then each prefix of Γ is also well-formed.*
2. *If $\Gamma \vdash M : T$, then Γ is well-formed.*

[5] See lemma 14 in paragraph 3.2.

Lemma 4 (Weakening). — *Let Γ and Γ' be two contexts such that $\Gamma \subset \Gamma'$. If $\Gamma \vdash M : T$ and Γ' is well-formed, then $\Gamma' \vdash M : T$.*

Lemma 5 (Substitutivity). — *If $\Gamma_1 \vdash M_0 : T_0$ and $\Gamma_1; [x_0 : T_0]; \Gamma_2 \vdash M : T$, then*

$$\Gamma_1; (\Gamma_2\{x_0 := M_0\}) \vdash M\{x_0 := M_0\} : T\{x_0 := M_0\}.$$

Step 2: The η-Subject Reduction Property. We now need to show that rule (Ext) can only be used at some places in a derivation. For that, we have to introduce the notion of *stable form*. A term M is said to be

1. a *sort form* if $M \cong_{\beta\eta} \forall\Delta\,.\,s$ for some context Δ and some sort s.
2. a *product form* if $M \cong_{\beta\eta} \forall\Delta\,.\,\Pi x{:}T\,.\,U$ for some context Δ and some terms T and U;
3. a *stable form* if M is either a sort form or a product form.

The terminology of 'stable form' comes from the fact that stable forms are preserved at the right-hand side of judgments by subtyping rules such as (Inst), (Gen), (Ext), (Str), (Conv) or (Cum).

Lemma 6 (Stable forms). — *If $\Gamma \vdash M : T$, then*

1. *if M is a sort, an explicit or an implicit product, then T is a sort form;*
2. *if M is a λ-abstraction, then T is a product form*

Using this lemma, we prove the inversion lemma for explicit and implicit products, which is necessary to establish the η-subject reduction property.

Lemma 7 (Inversion of products). — *If $\Gamma \vdash Bx{:}T\,.\,U : R$ (where B is one of Π or $\forall$), then there exists a context Δ and four sorts s_1, s_2, s_3, s such that*

1. *$R \cong_{\beta\eta} \forall\Delta\,.\,s$;*
2. *$\Gamma; \Delta \vdash T : s_1$;*
3. *$\Gamma; \Delta; [x : T] \vdash U : s_2$;*
4. *$(s_1, s_2, s_3) \in$ **Rule**;*
5. *$s_3 \leq s$.*

Lemma 8 (Type of types). — *If $\Gamma \vdash M : T$, there there exists a sort s such that $\Gamma \vdash T : s$.*

Lemma 9 (Context conversion). — *Let Γ and Γ' be contexts such that $\Gamma \cong_{\beta\eta} \Gamma'$. If $\Gamma \vdash M : T$ and if Γ' is well-formed, then $\Gamma' \vdash M : T$.*

Proposition 2 (η-subject reduction). — *If $\Gamma \vdash M : T$ and $M \to_\eta M'$, then $\Gamma \vdash M' : T$.*

Step 3: η-direct Derivations. We now need to isolate rule (Ext). For that, we say that a derivation of $\Gamma \vdash M : T$ is *η-direct* if one of the following conditions is satisfied :

- the last rule is (Var) or (Sort);
- the last rule is (ExpProd), (ImpProd) or (App), and the derivation of both premises are η-direct;
- the last rule is (Lam) and the derivation of the first premise is η-direct;
- the last rule is (Gen), (Inst), (Conv), (Cum) or (Str) and the derivation of the first premise is η-direct.

Intuitively, an η-direct derivation of a judgement $\Gamma \vdash M : T$ is a derivation in which the rule (Ext) can not appear in the parts of the derivation corresponding to the destructuration of the term M.

In the following, we will write $\Gamma \vdash_d M : T$ when a judgment $\Gamma \vdash M : T$ has an η-direct derivation. This notion has good closure properties: lemmas 4, 5, 6, 7, 8 and 9 still hold even if we replace $\Gamma \vdash M : T$ by $\Gamma \vdash_d M : T$ everywhere. (However, the η-subject reduction property does not hold when considering η-direct derivations only.)

Lemma 10 (η-direct inversion of abstraction). — *If $\Gamma \vdash_d \lambda x \,.\, M : R$, then there exists a context Δ and two terms T, U such that :*

1. *$R \cong_{\beta\eta} \forall \Delta \,.\, \Pi x : T \,.\, U$;*
2. *$\Gamma; \Delta; [x : T] \vdash_d M : U$.*

Lemma 11 (η-direct β-subject reduction). — *If $\Gamma \vdash_d M : T$ and $M \to_\beta M'$, then $\Gamma \vdash_d M' : T$.*

Step 4: β-Subject Reduction. Before concluding, we need to show that any derivation of $\Gamma \vdash M : T$ can be transformed into an η-direct derivation provided we make some η-expansions in the term M.

Lemma 12 (η-direct expansion). — *If $\Gamma \vdash M : T$, then there exists a term M_0 such that $M_0 \twoheadrightarrow_\eta M$ and $\Gamma \vdash_d M_0 : T$.*

The β-subject reduction property is then an immediate consequence of lemmas 1, 11 and 12.

Proposition 3 (β-subject reduction). — *If $\Gamma \vdash M : T$ and $M \twoheadrightarrow_\beta M'$, then $\Gamma \vdash M' : T$.*

3.2 Subtyping

One of the most interesting aspects of the Implicit Calculus of Constructions is the rich subtyping relation induced by the implicit product. This subtyping relation, which is denoted by $\Gamma \vdash T \leqslant T'$, can be defined directly from the typing judgment as the following 'macro':

$$\Gamma \vdash T \leqslant T' \quad \equiv \quad \Gamma; x : T \vdash x : T' \qquad (x \text{ a fresh variable})$$

Using that definition, we can prove that in a given context, subtyping is a pre-ordering on well-formed types which satisfies the expected (Sub) rule:

Lemma 13 (Subtyping preordering). — *The following rules are admissible:*

$$\frac{\Gamma \vdash T : s}{\Gamma \vdash T \leqslant T} \qquad \frac{\Gamma \vdash T_1 \leqslant T_2 \quad \Gamma \vdash T_2 \leqslant T_3}{\Gamma \vdash T_1 \leqslant T_3} \qquad \frac{\Gamma \vdash M : T \quad \Gamma \vdash T \leqslant T'}{\Gamma \vdash M : T'}\ (\text{Sub})$$

Moreover, product formation acts in a contravariant way for the domain part, and in a covariant way for the codomain part:

Lemma 14 (Subtyping in products). — *The following rules are admissible:*

$$\frac{\Gamma \vdash T' \leqslant T \quad \Gamma; [x : T'] \vdash U \leqslant U'}{\Gamma \vdash \Pi x\!:\!T \,.\, U \leqslant \Pi x\!:\!T' \,.\, U'} \qquad \frac{\Gamma \vdash T' \leqslant T \quad \Gamma; [x : T'] \vdash U \leqslant U'}{\Gamma \vdash \forall x\!:\!T \,.\, U \leqslant \forall x\!:\!T' \,.\, U'}$$

The subtyping rule for explicit products would not hold without the rule (Ext). This is the main motivation for introducing the rule (Ext), which has been proven equivalent to the subtyping rule for explicit products in [12].

Besides the notion of subtyping, we can also define a notion of *typing equivalence*, denoted by $\Gamma \vdash T \sim T'$, which is simply the symmetric closure of the subtyping judgment $\Gamma \vdash T \leqslant T'$. We can prove the following equivalences:

Lemma 15 (Product commutations). — *The following rules are admissible:*

$$\frac{\Gamma \vdash \forall x_1\!:\!T_1 \,.\, \forall x_2\!:\!T_2 \,.\, U : s \qquad \Gamma \vdash \forall x_2\!:\!T_2 \,.\, \forall x_1\!:\!T_1 \,.\, U : s'}{\Gamma \ \vdash \ \forall x_1\!:\!T_1 \,.\, \forall x_2\!:\!T_2 \,.\, U \ \sim \ \forall x_2\!:\!T_2 \,.\, \forall x_1\!:\!T_1 \,.\, U}$$

$$\frac{\Gamma \vdash \forall x_1\!:\!T_1 \,.\, \forall x_2\!:\!T_2 \,.\, U : s \qquad \Gamma \vdash \forall x_2\!:\!T_2 \,.\, \forall x_1\!:\!T_1 \,.\, U : s'}{\Gamma \ \vdash \ \Pi x_1\!:\!T_1 \,.\, \forall x_2\!:\!T_2 \,.\, U \ \sim \ \forall x_2\!:\!T_2 \,.\, \Pi x_1\!:\!T_1 \,.\, U}$$

(Notice that the premises imply that there is no mutual dependency in the quantifications of conclusions, *i.e.* $x_1 \notin FV(T_2)$ and $x_2 \notin FV(T_1)$.)

3.3 Consistency Results

In the implicit calculus, there are two propositions for representing the falsity: the *explicit falsity* $\Pi A\!:\!\mathsf{Prop} \,.\, A$ and the *implicit falsity* $\forall A\!:\!\mathsf{Prop} \,.\, A$. However, both falsities are provably equivalent:

$$\begin{aligned} \lambda f \,.\, f\ (\forall A\!:\!\mathsf{Prop} \,.\, A) \quad &: \quad (\Pi A\!:\!\mathsf{Prop} \,.\, A) \to (\forall A\!:\!\mathsf{Prop} \,.\, A) \\ \lambda p, A \,.\, p \quad &: \quad (\forall A\!:\!\mathsf{Prop} \,.\, A) \to (\Pi A\!:\!\mathsf{Prop} \,.\, A) \end{aligned}$$

The last proof is quite general, since we have

$$\lambda p, x \,.\, p \quad : \quad (\forall x : T \,.\, U) \to (\Pi x : T \,.\, U),$$

which means that an explicit product has at least as much inhabitants as the corresponding implicit product.

The main consistency result of the Implicit Calculus of Constructions is a consequence of the following lemma :

Lemma 16 (Stable forms). — *In the empty context, the type of a term which has a weak head normal form is a stable form.*

Since the implicit falsity is not a stable form, we have :

Proposition 4. — *If the Implicit Calculus of Constructions is strongly normalizing, then it is logically consistent.*

4 Semantics and Strong Normalization

Building a model of the Implicit Calculus of Constructions is a fascinating challenge, especially because its rich subtyping relation. The main difficulty is caused by the interpretation of the Curry-style λ-abstraction which imply the traditional typing ambiguity, but also a *stratification ambiguity*. For instance, the identity $\lambda x \,.\, x$ has several types such as $\forall A : \mathsf{Prop} \,.\, A \to A$ or $\forall A : \mathsf{Type}_i \,.\, A \to A$ $(i > 0)$ which are not defined at the same level of the universe hierarchy.

In [13], we have proposed a domain-theoretical model of the restricted implicit calculus—that is the implicit calculus without the rule (STR). This model is based on a untyped interpretation of terms in a large coherence space. The corresponding interpretation has nice properties: it allows to interpret all the terms—even the ill-typed ones—independently of their possible types.

More recently, we have transformed this model into a strong normalization model, using the ideas of [1] by incorporating reducibility information into the denotation of types. This normalization model now interprets the full calculus—including the strengthening rule—thus proving the following result :[6]

Theorem 1 (Strong normalization). — *Every well-typed term of the Implicit Calculus of Constructions is strongly normalizing.*

Corollary 1. — *The Implicit Calculus of Constructions is logically consistent.*

[6] The manuscript of the strong normalization proof is available on the author's web page at `http://pauillac.inria.fr/~miquel`.

5 Impredicative Encodings

In this section we shall illustrate the expressiveness of the Implicit Calculus of Constructions by comparing impredicative encodings of lists and dependent lists (vectors), and by studying their relationships with respect to subtyping.

In the implicit calculus, lists are encoded as follows:

$$\begin{array}{rcl} \mathsf{list} & : & \mathsf{Set} \to \mathsf{Set} \quad := \quad \lambda A\,.\,\forall X : \mathsf{Set}\,.\,X \to (A \to X \to X) \to X \\ \mathsf{nil} & : & \forall A : \mathsf{Set}\,.\,\mathsf{list}\ A \quad := \quad \lambda x f\,.\,x \\ \mathsf{cons} & : & \forall A : \mathsf{Set}\,.\,A \to \mathsf{list}\ A \to \mathsf{list}\ A \quad := \quad \lambda a l x f\,.\,f\ a\ (l\ x\ f) \end{array}$$

Notice that here, the polymorphic constructors nil and cons are exactly the usual constructors of (untyped) lists in the pure λ-calculus. In fact, this result is not specific to the implicit calculus: this example could have been encoded the same way in the Curry-style equivalent of system $F\omega$ in the cube of TAS, since the implicit quantification was precisely used for impredicative products.

In such a framework, it is not necessary to give an extra argument at each 'cons' operation to build a list:

$$\mathsf{cons\ true\ (cons\ false\ (cons\ true\ nil))} \quad : \quad \mathsf{list\ bool}.$$

Using the traditional encoding of lists in the Calculus of Constructions, the same list would have been written

$$\mathsf{cons\ bool\ true\ (cons\ bool\ false\ (cons\ bool\ true\ (nil\ bool)))} \quad : \quad \mathsf{list\ bool}$$

by explicitly instanciating the type of constructors at each construction step.

In the implicit calculus anyway, the constructor of lists has the good covariance property with respect to the subtyping relation:

Proposition 5 (Covariance of the type of lists). — *For all context Γ and for all terms A and B of type* Set *in Γ we have:*

$$\Gamma \vdash A \leqslant B \quad \Rightarrow \quad \Gamma \vdash \mathsf{list}\ A \leqslant \mathsf{list}\ B.$$

In fact, the situation becomes far more interesting if we consider the type of dependent lists—that we call *vectors*. The type of vectors is like the type of lists, except that it also depends on the size of the list. In the implicit calculus, the type of vectors can be encoded as follows

$$\begin{array}{rl} \mathsf{vect} & : \quad \mathsf{Set} \to \mathsf{nat} \to \mathsf{Set} \\ & := \quad \lambda A n\,.\,\forall P : \mathsf{nat} \to \mathsf{Set}\,.\,P\ 0 \to (\forall p : \mathsf{nat}\,.\,A \to P\ p \to P\ (\mathsf{S}\ p)) \to P\ n, \end{array}$$

where nat, 0 and S are defined according to the usual encoding of Church integers in Curry-style system F. The interesting point is that we do not need to define a new nil and a new cons for vectors. Indeed, it is straightforward to check that the nil and cons that we defined for building lists have also the following types:

$$\begin{array}{lcl}\mathsf{nil} & : & \forall A\!:\!\mathsf{Set}\,.\,\mathsf{vect}\;A\;0\\ \mathsf{cons} & : & \forall A\!:\!\mathsf{Set}\,.\,\forall n\!:\!\mathsf{nat}\,.\,A\to\mathsf{vect}\;A\;n\to\mathsf{vect}\;A\;(\mathsf{S}\;n)\end{array}$$

In other words, lists and (fixed-length) vectors share the very same constructors, so we can take back the list of booleans above and assign to it the following more accurate type:

$$\mathsf{cons\ true\ (cons\ false\ (cons\ true\ nil))} \quad : \quad \mathsf{vect\ bool\ (S\ (S\ (S\ 0)))}.$$

In the Calculus of Constructions, such a sharing of constructors is not possible between lists and dependent lists, so we have to define a new pair of constructors nil′ and cons′ to write the term

$$\begin{array}{l}\mathsf{cons'\ bool\ (S\ (S\ 0))\ true}\\ \quad\mathsf{(cons'\ bool\ (S\ 0)\ false}\\ \qquad\mathsf{(cons'\ bool\ 0\ true\ (nil'\ bool)))} \quad : \quad \mathsf{vect\ bool\ (S\ (S\ (S\ 0)))}.\end{array}$$

whose real computational contents is completely hidden by the type and size arguments given to the constructors nil′ and cons′.

In the implicit calculus, we can even derive that the type of vectors (of a given size) is a subtype of the type of lists:

Proposition 6. — *For all context Γ and for all terms A and n such that $\Gamma \vdash A : \mathsf{Set}$ and $\Gamma \vdash n : \mathsf{nat}$, one can derive the subtyping judgment:*

$$\Gamma \vdash \mathsf{vect}\;A\;n \leqslant \mathsf{list}\;A.$$

To give another illustration of the expressive power of the Implicit Calculus of Constructions, let us study the case of Leibniz equality. In the implicit calculus, the natural impredicative encoding of equality is the following:

$$\mathsf{eq} \;:\; \Pi A\!:\!\mathsf{Set}\,.\,A\to A\to\mathsf{Prop} \quad := \quad \lambda A,x,y\,.\,\forall P\!:\!A\to\mathsf{Prop}\,.\,P\;x\to P\;y.$$

The reflexivity of equality is simply proven by the identity function

$$\lambda p\,.\,p \quad : \quad \forall A\!:\!\mathsf{Set}\,.\,\forall x\!:\!A\,.\,\mathsf{eq}\;A\;x\;x$$

whereas the proof of transitivity is given by the composition operator

$$\lambda fgp\,.\,g\;(f\;p) \quad : \quad \forall A\!:\!\mathsf{Set}\,.\,\forall x,y,z\!:\!A\,.\,\mathsf{eq}\;A\;x\;y\to\mathsf{eq}\;A\;y\;z\to\mathsf{eq}\;A\;x\;z.$$

A Remark about Implicit Positions. In the example above, the type parameter A is an implicit argument of the reflexivity and transitivity proofs, but it is an explicit argument of the equality predicate, although it can be easily infered in that context. On the contrary, the implicit calculus allows the use of implicit elimination predicates (see for instance the encoding of vectors), although the inference of such predicates require complex techniques based on higher-order unification in practice. Those examples show that the arguments that can be automatically infered and the arguments that can be dropped out of the syntax without harm for the consistency are generally not the same.

6 Future Work

Undecidability of Type-Checking. Decidability of type-checking in the implicit calculus is still an open problem. However, we strongly conjecture that type-checking is undecidable, at least because it contains the Curry-style system F. In fact, the inclusion of Curry-style system F into the implicit calculus seems to be only a minor point, since the implicit product allows to hide far more typing information than in the TAS. For that reason, the implicit calculus is not suitable for being used in a proof assistant system. Nevertheless, it could be fruitful to study *ad hoc* restrictions of the implicit calculus, in which decidability of type-checking is preserved.

Extending this Approach to All PTS. The approach described here can be easily extended to all Pure Type Systems. Within the more general framework of *Implicit Pure Type Systems*, it is possible to have different formation rules for explicit and implicit products (by introducing two sets $\mathbf{Rule}^{\Pi}, \mathbf{Rule}^{\forall} \subset \mathcal{S}^3$ instead of the single set **Rule** of the Implicit Calculus of Constructions). In that framework, most of the results exposed in Sect. 3 still hold (including the $\beta\eta$-subject reduction property), since their proofs do not rely on the assumption that explicit and implicit products share the same formation rules.

References

1. T. Altenkirch. *Constructions, Inductive types and Strong Normalization.* PhD thesis, University of Edinburgh, 1993.
2. Henk Barendregt. Introduction to generalized type systems. Technical Report 90-8, University of Nijmegen, Department of Informatics, May 1990.
3. B. Barras, S. Boutin, C. Cornes, J. Courant, J.C. Filliâtre, E. Giménez, H. Herbelin, G. Huet, C. Muñoz, C. Murthy, C. Parent, C. Paulin, A. Saïbi, and B. Werner. The Coq Proof Assistant Reference Manual – Version V6.1. Technical Report 0203, INRIA, August 1997.
4. J. H. Geuvers and M. J. Nederhof. A modular proof of strong normalization for the calculus of constructions. In *Journal of Functional Programming*, volume 1,2(1991), pages 155–189, 1991.
5. P. Giannini, F. Honsell, and S. Ronchi della Rocca. Type inference: some results, some problems. In *Fundamenta Informaticæ*, volume 19(1,2), pages 87–126, 1993.
6. M. Hagiya and Y. Toda. On implicit arguments. Technical Report 95-1, Department of Information Science, Faculty of Science, University of Tokyo, 1995.
7. Paul B. Jackson. The Nuprl proof development system, version 4.1 reference manual and user's guide. Technical report, Cornell University, 1994.
8. D. Leivant. Polymorphic type inference. In *Proceedings of the 10th ACM Symposium on Principles of Programming Languages*, pages 88–98, 1983.
9. Z. Luo. *Computation and Reasoning: A Type Theory for Computer Science.* Oxford University Press, 1994.
10. Zhaohui Luo and Randy Pollack. Lego proof development system: User's manual. Technical Report 92-228, LFCS, 1992.

11. Lena Magnusson. Introduction to ALF — an interactive proof editor. In Uffe H. Engberg, Kim G. Larsen, and Peter D. Mosses, editors, *Proceedings of the 6th Nordic Workshop on Programming Theory* (Aarhus, Denmark, 17–19 October, 1994), number NS-94-6 in Notes Series, page 269, Department of Computer Science, University of Aarhus, December 1994. BRICS. vi+483.
12. Alexandre Miquel. Arguments implicites dans le calcul des constructions: étude d'un formalisme à la Curry. Master's thesis, Université Denis-Diderot Paris 7, octobre 1998.
13. Alexandre Miquel. A model for impredicative type systems with universes, intersection types and subtyping. In *Proceedings of the 15 th Annual IEEE Symposium on Logic in Computer Science (LICS'00)*, 2000.
14. R. Pollack. Implicit syntax. In Gérard Huet and Gordon Plotkin, editors, *Proceedings of the First Workshop on Logical Frameworks (Antibes)*, may 1990.
15. A. Saïbi. *Algèbre Constructive en Théorie des Types, Outils génériques pour la modélisation et la démonstration, Application à la théorie des Catégories.* PhD thesis, Université Paris VI, 1998.
16. S. van Bakel, L. Liquori, R. Ronchi della Rocca, and P. Urzyczyn. Comparing Cubes. In A. Nerode and Yu. V. Matiyasevich, editors, *Proceedings of LFCS '94. Third International Symposium on Logical Foundations of Computer Science,* St. Petersburg, Russia, volume 813 of *Lecture Notes in Computer Science*, pages 353–365. Springer-Verlag, 1994.
17. J. B. Wells. Typability and type checking in system F are equivalent and undecidable. In *Annals of Pure and Applied Logic*, volume 98(1-3), pages 111–156, 1999.
18. B. Werner. *Une théorie des Constructions Inductives.* PhD thesis, Université Paris VII, 1994.

Evolving Games and Essential Nets for Affine Polymorphism

Andrzej S. Murawski* and C.-H. Luke Ong**

Oxford University Computing Laboratory
Wolfson Building, Parks Rd, Oxford OX1 3QD, UK
{Andrzej.Murawski,Luke.Ong}@comlab.ox.ac.uk

Abstract. This paper presents a game model of Second-order Intuitionistic Multiplicative Affine Logic (IMAL2). We extend Lamarche's essential nets to the second-order affine setting and use them to show that the model is fully and faithfully complete.

Keywords: Full Completeness, Game Semantics, Linear Logic, Polymorphism.

1 Introduction

This paper is about a second-order extension of AJM games [2,1], which we call *evolving games*. A play begins with O making an opening move, and the two players alternate thereafter. An evolving game has two kinds of tokens: ground and second-order. Ground tokens are standard; they are playable at once (if reachable). *Second-order tokens* are (descriptions of) game evolutions, which cannot be played right away on their own. If a second-order token θ is reachable at a given position, a player may import a game A as an argument for the evolution operator θ, thus causing the current game to grow locally, with the game $\theta(A)$ grafted at where the token θ was. Transported now to the new and expanded game, the same player may play a ground token if one is reachable and so complete the second-order move; or he may continue the evolution process (if a second-order token is reachable) by importing another game, and so on. However after finitely many such evolution steps, the player is required to play a ground token, thus finally completing the second-order move. A version of this approach first appeared in a LICS'97 paper [8] by Hughes. He shows how a fully complete model for System F can be constructed. A more abstract presentation has been considered in [15].

Our goal is to construct a simple evolving game model for Second-order Intuitionistic Multiplicative Affine Logic (IMAL2) (see Figure 1 for the rules of the IMAL2 Sequent Calculus; note that we do not consider unit) in the style of [15]. In Section 2 we introduce evolving games and define the playable moves and

* On leave from Nicholas Copernicus University, Toruń, Poland.

** http://www.comlab.ox.ac.uk/oucl/work/luke.ong.html
Tel:

S. Abramsky (Ed.): TLCA 2001, LNCS 2044, pp. 360–375, 2001.

(atom)	$a \vdash a$	**(var)**	$X \vdash X$
(exch)	$\dfrac{\Gamma, A, B, \Delta \vdash C}{\Gamma, B, A, \Delta \vdash C}$	**(wk)**	$\dfrac{\Gamma \vdash B}{\Gamma, A \vdash B}$
($\otimes$-l)	$\dfrac{A, B, \Gamma \vdash C}{A \otimes B, \Gamma \vdash C}$	**($\otimes$-r)**	$\dfrac{\Gamma \vdash A \quad \Delta \vdash B}{\Gamma, \Delta \vdash A \otimes B}$
($\multimap$-l)	$\dfrac{\Gamma \vdash A \quad B, \Delta \vdash C}{A \multimap B, \Gamma, \Delta \vdash C}$	**($\multimap$-r)**	$\dfrac{\Gamma, A \vdash B}{\Gamma \vdash A \multimap B}$
($\forall$-l)	$\dfrac{\Gamma, A[B/X] \vdash C}{\Gamma, \forall X.A \vdash C}$	**($\forall$-r)**	$\dfrac{\Gamma \vdash A}{\Gamma \vdash \forall X.A}$

where the side condition of ($\forall$-r) is: X does not occur free in Γ.

Fig. 1. The rules defining valid IMAL2 sequents

positions of an evolving game. As our aim is to construct a fully complete model for IMAL2, our treatment here is restricted to a version of free such games, which are in one-one correspondence with closed IMAL2 types. IMAL2 proofs are modelled by strategies which we present in two stages. We consider first the simple scenario in which the games imported by O (as arguments for his second-order tokens) are guaranteed to be singleton games consisting of a ground token. P-strategies for playing such evolving games are called *symbolic*. Strategies that denote proofs are total, (ground) token-reflecting and finitely presentable; in addition, the evolution arguments P imports are determined schematically by those which O has imported thus far in the play. We call symbolic strategies that possess these properties *regular*, and they are introduced in Section 3. Regular strategies cannot be composed; however strategies that are generated from regular strategies by a process of copycat expansion do compose, as we show in Section 4. (It is worth noting that our proof of compositionality is direct and syntax-independent, or rather, independent of the formal system IMAL2.) We call such strategies *good*, and they give rise to a model of IMAL2.

In Section 5 we turn our attention to essential nets [11], which are a kind of oriented proof nets, for IMAL2. We give a *correctness criterion* for such nets, and prove that all correct nets are sequentializable. The main result of the paper is Theorem 2:

Evolving games and good strategies are fully and faithfully complete for IMAL2

which is proved in Section 6. The key step is a correspondence result (Lemma 4 and Proposition 1) which shows that each regular strategy determines a correct essential net for the associated end-sequent. Faithful completeness is proved with respect to a notion of equivalence of such nets.

The only game model for IMAL2 in the literature is given in [1]. Recently Abramsky and Lenisa [3] have constructed a linear combinatory algebra of partial involutions on the natural numbers, arising from Geometry of Interaction constructions; they show that a fully and faithfully complete model for ML polymorphic types of system F can be obtained in this way. To the best of our knowledge, our game model is the first fully (and faithfully) complete model for IMAL2; indeed all results in this paper are new.

2 Evolving Games

We assume the notions of games (and singleton games) as defined in [1,12]; we shall call them ***IMAL games***. Recall that new games can be constructed from old using standard game constructions. We write $s \upharpoonright A$ to mean the subsequence of s consisting only of moves from A, and define $\overline{P} = O$ and $\overline{O} = P$. For a game G, we write $M_G^{\circledast}$ to mean the set of finite alternating sequences of moves from M_G. The first two, *tensor games* $A \otimes B$ and *linear function space games* $A \multimap B$, are standard. For $\circledcirc = \otimes$ and $\multimap$, we have

$$M_{A \circledcirc B} = M_A + M_B$$
$$P_{A \circledcirc B} = \{\, s \in M_{A \circledcirc B}^{\circledast} \mid s \upharpoonright A \in P_A,\ s \upharpoonright B \in P_B \,\}$$

where $\lambda_{A \otimes B}$ is defined to be the canonical map $[\lambda_A, \lambda_B] : M_A + M_B \longrightarrow \{P, O\}$, and $\lambda_{A \multimap B} = [\overline{\lambda_A}, \lambda_B]$. Note that it is a consequence of the definition that every $s \in P_{A \otimes B}$ satisfies the *O-Switching Condition*: for each pair of consecutive moves mm' in s, if m and m' are from different components (i.e. one is from A the other from B), then m' is an O-move. Similarly it follows that every $s \in P_{A \multimap B}$ satisfies the *P-Switching Condition* i.e. only P can switch component.

Tokens: ground and second-order. Fix an infinite set $\mathcal{T}_g$ of *ground tokens* which are ranged over by a, b, c, etc. The set $\mathcal{G}$ of ***free games*** is generated from the ground tokens by the constructors of IMAL2 as follows:

- every ground token $a \in \mathcal{T}_g$, considered as a singleton game, is in $\mathcal{G}$
- if G_1 and G_2 are in $\mathcal{G}$ then $G_1 \otimes G_2$ and $G_1 \multimap G_2$ are in $\mathcal{G}$
- if $G \in \mathcal{G}$ then for each $e \in \mathcal{T}_g$, the singleton game $[\forall e.G]$ is in $\mathcal{G}$.

(By abuse of notation, we confuse a singleton game with its sole token.) Tokens of the form $[\forall e.G]$ are called ***second-order tokens***. Thus $\mathcal{G}$ contains games generated from both ground and second-order tokens using the constructors $\otimes$ and $\multimap$.

Remark 1. As we shall see shortly, a second-order token $[\forall a.G]$ is a description of the game operation: $A \mapsto G[A/a]$. As usual "α-equivalent" second-order tokens are considered identical. Clearly free games are in one-one correspondence with closed IMAL2 types.

Given any IMAL2 formula F and a map ρ from its free variables to $\mathcal{G}$, there is an obvious denotation $[\![F]\!]_\rho$ in $\mathcal{G}$. E.g. take $F = X \otimes \forall Y.(Y \otimes (X \multimap \forall Z.Z))$ and $\rho : X \mapsto [\forall b.(e \multimap b)] \otimes e$; we have

$$[\![F]\!]_\rho = ([\forall b.(e \multimap b)] \otimes e) \otimes [\forall Y.(Y \otimes (([\forall b.(e \multimap b)] \otimes e) \multimap [\forall Z.Z]))].$$

Playable moves of an evolving game. When considered as an IMAL game, a move of a free game G is either a ground or a second-order token, which can be named by its occurrence in the syntax tree of G. For this purpose the branches of $\otimes$ are labelled l and r, and those of $\multimap$ are labelled L and R. For instance, the tokens of the game $a \otimes (b \multimap [\forall c.c])$ (from left to right) are l, rL and rR.

In the setting of evolving games, starting from G (say), the interim game (which we can think of as the "current game board") grows as the play unfolds, so that over the life time of a play, many more moves than are specified in the initial game G become available. We call these generalized moves *playable moves* of the evolving game G and define it by recursion as follows.

Definition 1. A string $s \in (\{l, r, L, R, \} \cup \mathcal{G})^*$ is a ***playable move*** of an evolving game G if:

- either $s \in \{l, r, L, R\}^*$ names a ground token in G, called the *playable token* of s – then we say that G does not evolve (as a consequence of playing s)
- or there exist $A \in \mathcal{G}$ (called ***evolution argument***), $s_1 \in \{l, r, L, R\}^*$, $s_2 \in (\{l, r, L, R\} \cup \mathcal{G})^*$ such that $s = s_1 A s_2$, s_1 names a second-order token $[\forall e.T]$ in G, and s_2 is a playable move of the game $T[A/e]$. The *playable token* of s is defined to be the same as that of s_2 (although playability is defined with respect to a different game). If s_2 causes $T[A/e]$ to evolve into G', then we say that s causes G to evolve into $G[G'/[\forall e.T]]$.

Note that in the second case above it may take several evolution arguments to make a playable move. We shall write the evolution arguments as superscripts (e.g. $s_1^A \cdots$ above). From the moment such an A is played we say that the ***slot*** s_1 is defined. It may turn out that s_2 begins with a contiguous segment of evolution arguments in which case all of them form part of the definition of slot s_1. If $s_2 = A_1 \cdots A_n s_3$ for $A_i \in \mathcal{G}$ and s_3 begins with one of l, r, L or R, we write $\lceil s_1 \rceil = A A_1 \cdots A_n$. If s is a playable move for G, we call $s \upharpoonright \{l, r, L, R\}$ the ***location*** of s, and we write it as $\lfloor s \rfloor$.

Example 1. (i) $(a \otimes [\forall d.d])r(b \otimes [\forall c.c])l$ is a playable move for the singleton game $[\forall b.b]$, which consequently evolves into $a \otimes (b \otimes [\forall c.c])$. The playable token is b and the slots defined by that move are $\lceil \epsilon \rceil = a \otimes [\forall d.d]$ and $\lceil r \rceil = b \otimes [\forall c.c]$.

(ii) The playable move $\underbrace{[\forall e.e] \cdots [\forall e.e]}_{k} a$ causes the singleton game $[\forall b.b]$ to evolve into a. There are k intermediate steps - all equal to $[\forall b.b]$, and a is the playable token. The move defines $\lceil \epsilon \rceil$ to be $\underbrace{[\forall e.e] \cdots [\forall e.e]}_{k} a$.

Observe that in general $\lceil - \rceil : \{l, r, L, R\}^* \rightharpoonup \mathcal{G}^+$, where X^+ means the set of non-empty words over the alphabet-set X. (A playable move m can be seen as defining a set $\lceil s_1^m \rceil, \cdots, \lceil s_{n^m}^m \rceil$ of slot definitions for $s_j^m \in \{l, r, L, R\}^*$.)

Plays of an evolving game. We define *positions* of the evolving game G as follows.

Definition 2 (Positions). The empty sequence is a position. If the finite sequence of playable moves, s, is a position and G' is the evolved game at that point, then sm is a position provided

(i) m is a playable move of G', and
(ii) the sequence consisting of the respective locations of the elements in sm is a position of G'' (here in the standard sense of IMAL games [12]), where G'' is the game that has evolved from G' after m is played.

Example 2. We examine two maximal positions for the game

$$[\forall X.((X \multimap X) \otimes [\forall Y.(Y \multimap Y)])] \multimap [\forall X.((X \multimap X) \otimes [\forall Y.(Y \multimap Y)])]$$

(i) We display the position as follows:

$$\begin{array}{lcccccccc} & (R^e l R) & (L^e l R) & (L l L) & (R l L) & (R r^c R) & (L r^c R) & (L r L) & (R r L) \\ \text{evolution} & G_1 & G_2 & G_2 & G_2 & G_3 & G_4 & G_4 & G_4 \end{array}$$

where:

$$\begin{array}{ll} G_1 & [\forall X.((X \multimap X) \otimes [\forall Y.(Y \multimap Y)])] \multimap ((e \multimap e) \otimes [\forall Y.(Y \multimap Y)]) \\ G_2 & (e \multimap e) \otimes [\forall Y.(Y \multimap Y)] \multimap (e \multimap e) \otimes [\forall Y.(Y \multimap Y)] \\ G_3 & (e \multimap e) \otimes [\forall Y.(Y \multimap Y)] \multimap (e \multimap e) \otimes (c \multimap c) \\ G_4 & (e \multimap e) \otimes (c \multimap c) \multimap (e \multimap e) \otimes (c \multimap c) \end{array}$$

After the first move we have $\lceil R \rceil = e$, after the second $\lceil L \rceil = e$. The fifth move defines $\lceil Rr \rceil$ to be c and the next makes $\lceil Lr \rceil$ equal c.

(ii) After the first move we have $\lceil R \rceil = e$ and $\lceil Rr \rceil = c$. The second move sets $\lceil L \rceil = e$ and $\lceil Lr \rceil = c$. The position is displayed below:

$$\begin{array}{lcccccccc} & (R^e r^c R) & (L^e r^c R) & (L r L) & (R r L) & (R l R) & (L l R) & (L l L) & (R l L) \\ \text{evolution} & H_1 & H_2 & H_2 & H_2 & H_2 & H_2 & H_2 & H_2 \end{array}$$

where:

$$\begin{array}{ll} H_1 & [\forall X.((X \multimap X) \otimes [\forall Y.(Y \multimap Y)])] \multimap (e \multimap e) \otimes (c \multimap c) \\ H_2 & (e \multimap e) \otimes (c \multimap c) \multimap (e \multimap e) \otimes (c \multimap c) \end{array}$$

Call a string $s \in \{l, r, L, R\}^*$ an ***O-location*** if s contains an even number of L's. Otherwise it is a ***P-location***. The sets containing them are called L_O and L_P respectively. Intuitively an O-location is the occurrence of an O-move in a game. Thus if $\lceil s \rceil$ is defined in a position and s is an O-location (respectively P-location), then it was first defined by O (respectively P). In other words, a player cannot define his opponent's slots.

3 Symbolic Strategies

In order to prove a definability result, it is convenient to consider first the simple scenario in which the evolution arguments provided by O are guaranteed to be (sequences of) ground tokens (regarded as singleton games) – we say that O plays *symbolically* in this case. This assumption makes it easy to see that denotable P-strategies have a finite description.

Formally we say that O plays ***symbolically*** in a position if for each $s \in L_O$ defined therein we have $\lceil s \rceil \in \mathcal{T}_g^+$. For the rest of the section we assume that O plays symbolically. We say that a P-strategy σ is ***symbolic*** just in case for every even-length $s \in \sigma$, if sm is a position then $sm \in \sigma$ if and only if the O-move m is symbolic.

The purpose of the next definition is to introduce strategies that are parametric in the sense that the evolution arguments given by P are determined by those given by O in a schematic fashion. First we set $\mathcal{G}(L_O)$ to be the collection of formal objects defined by the following rules:

- every ground token $a \in \mathcal{T}_g$ and every O-location are in $\mathcal{G}(L_O)$
- if G_1 and G_2 are in $\mathcal{G}(L_O)$ then $G_1 \otimes G_2$ and $G_1 \multimap G_2$ are in $\mathcal{G}(L_O)$
- if $G \in \mathcal{G}(L_O)$ then for each $e \in \mathcal{T}_g$, $[\forall e.G]$ is in $\mathcal{G}(L_O)$.

For example we have $c \multimap Rr \otimes [\forall a.(LLr \multimap a \otimes b)] \in \mathcal{G}(L_O)$. (Informally $\mathcal{G}(L_O)$ is the collection of free games in which some ground tokens are replaced by O-locations.)

Definition 3. The pair $f : L_O \rightharpoonup L_P$ and $F : L_P \rightharpoonup (\mathcal{G}(L_O))^+$ define a symbolic strategy σ if for all even-length $s m_O m_P \in \sigma$ (m_O necessarily symbolic):

- $\lfloor m_P \rfloor = f(\lfloor m_O \rfloor)$
- for all $u, v \in (\{l, r, L, R\} \cup \mathcal{G})^*$ and $\overline{A} \in \mathcal{G}^+$ such that $m_P = u^{\overline{A}} v$ where $\lfloor u \rfloor$ is a P-location and v begins with one of l, r, L or R, we have $\overline{A}$ is determined by $F(\lfloor u \rfloor)$ as follows: Suppose the O-locations that appear in $F(\lfloor u \rfloor)$ are $w_1, \cdots, w_k$, then $\overline{A}$ is obtained from $F(\lfloor u \rfloor)$ by replacing each w_i by $\lceil w_i \rceil$, each $\lceil w_i \rceil$ must already be defined in $s m_O$.

We write $\sigma = \sigma_{f,F}$ if f, F are the least such functions. □

By definition, it is easy to see:

1. f and F are related: each P-location in $dom(F)$ is a prefix of some element in $cod(f)$.

2. By leastness of f and F, elements of $dom(f) \cup cod(f)$ are incomparable.

Note that a symbolic strategy is *location-wise* history-free, in the sense that the location of P's response at any position depends only on the preceding O-move.

To model proofs, f and F should be finite (it is enough to require that of f). The finiteness is needed to eliminate infinite strategies like the one for the game $[\forall X.X] \multimap [\forall Y.Y]$ in which P replies by importing $([\forall X.X] \multimap [\forall Y.Y]) \multimap a$ where a is the token of the last O-move, and then playing the ground token a. This puts O in the same position as at the beginning of the game, and the play could continue indefinitely, if f and F were allowed to be infinite.

Definition 4. A ***regular strategy*** is a total, ground-token-reflecting[1] symbolic strategy $\sigma_{f,F}$ given by some finite f and F, where f is an injection.

It follows that if $s m_O m_P \in \sigma$, $m_P = u^{\overline{A}} v$ for a P-location $\lfloor u \rfloor$, v beginning with one of l, r, L and R, and

$$F(\lfloor u \rfloor) \;=\; G_1(w_1, \cdots, w_k) \cdots G_j(w_1, \cdots, w_k) \;\in\; (\mathcal{G}(L_O))^+$$

(where each $G_i(w_1, \cdots, w_k)$ is in $\mathcal{G}(L_O)$ such that the O-locations that appear in it are among $w_1, \cdots, w_k$) then in the *P-view* [9] of $s m_O$ there must be O-moves which define $\lceil w_i \rceil$ for each i (i.e. the evolution argument which O has supplied at each w_i).

Example 3. The strategy considered in Example 2 given by:

$$f = \{(RlR, LlR), (LlL, RlL), (RrR, LrR), (LrL, RrL)\}$$
$$F = \{(L, R), (Lr, Rr)\}$$

is regular. Note that this is the identity strategy.

Remark 2. In general, it is too weak to require the dependence of evolution arguments on P-views. Consider the following sequent:

$$\vdash \forall X.X \otimes (\forall X.X \multimap b \otimes \forall Y.Y) \multimap b \otimes \forall Y.Y.$$

P's response at the two positions

$$(Rl)(LrRl)(LrL^a) \text{ and } (Rr^a)(LrRr^a)(LrL^a)$$

should be the same, say, $(Ll^{a \otimes a} r)$, but if the way P imports evolution arguments is to depend on P-views, we are free to play $Ll^{a \otimes a} r$ in the first instance and $Ll^{[\forall V.V] \otimes a} r$ in the second. For regular strategies, this is not possible. Since $\lceil Ll \rceil$ must be defined in terms of the evolution arguments O has imported (at certain O-locations) defined in the history of the play, it must depend solely on $\lceil LrL \rceil$. This forces P to import the same evolution argument in both instances.

It is easy to see that regular strategies are finite, in the sense that it is a finite set of positions, and so, every play terminates. In the following we shall see how they can be used to generate possibly infinite plays.

[1] I.e. for every even-length $s m^- m \in \sigma$, if the O-move m^- is a *ground* token a, then the token of m is also a (see [12]).

4 Good Strategies for Composition

Unfortunately regular strategies cannot be composed. This section is concerned with a notion of strategies that are good for composition. *Good strategies* are generated from regular ones by a process of *copycat expansion*, which is essentially a semantic form of η-expansion [8,10]. We shall see that good strategies compose by the standard mechanism of "parallel composition with hiding" [2].

Definition 5 (Good strategies). Take a regular strategy $\sigma = \sigma_{f,F}$. The *good strategy generated from* σ, written $\overline{\sigma}$, is defined by the following algorithm:

Suppose the odd-length position $sm \in \overline{\sigma}$ such that $\lfloor m \rfloor = t$. Find $u \leq t$ such that $u \in cod(f) \cup dom(f)$. (Note that a unique such u exists by induction on the length of positions.) This decomposes the O-move m into $m_u m_v$ such that $t = uv$, m_u ends in one of l, r, L and R, and $\lfloor m_u \rfloor = u$. There are two cases:

(i) If $u \in dom(f)$, play $\overline{f(u)}m_v$
(ii) If $u \in cod(f)$, play $f^{-1}(u)m_v$

In case (i) if u is encountered in the play for the first time, $\overline{f(u)}$ is obtained from $f(u)$ by inserting appropriate evolution arguments using F, based on the evolution arguments which O has already given in sm; otherwise $\overline{f(u)} = f(u)$

Remark 3. In a play, the evolution argument for each *slot* (whether O or P) is given explicitly only once. Thus in case (i), $\overline{f(u)} = f(u)$ if the u in question has already been met in the history of play. In case (ii), all slots should have already been defined, since the corresponding instance of case (i) must have been encountered first. This explains the apparent asymmetry in the two cases.

Lemma 1 (Zigzag). *In any position of a good strategy generated by f and F, for any $u \in dom(f)$, f is applied in case (i) first if at all; thereafter each time before (i) is applied again for the same u, an instance of case (ii) must have already occurred for that u.*

Proof. Otherwise, in the subgame rooted at t, we would have two consecutive O-moves. □

It is straightforward to see that good strategies are (location-wise) history-free and total. Good strategies can be infinite, as the following example illustrates.

Example 4. Take the good strategy extending the symbolic identity $\{\epsilon, R^a, R^a L^a\}$ for the game $[\forall X.X] \multimap [\forall Y.Y]$. Suppose O plays $R^{[\forall X.X] \multimap a}R$. P plays copycat and so responds by $L^{[\forall X.X] \multimap a}R$. From this point onwards, both players behave analogously, thus engaging in an infinite exchange.

Now composition (of good strategies) can be defined by the conventional definition. In fact

Theorem 1. *Good strategies compose.* □

We prove the Theorem directly, without recourse to the formal system IMAL2, as a simple consequence of the following lemma.

Lemma 2. *(i) The composite strategy of two good strategies is total i.e. there is no infinite chattering when the two strategies interact.*
(ii) The composite of two good strategies is generated from a regular strategy $\sigma_{f,F}$ for some finite f (and so F is also finite). □

Remark 4. The composition algorithm that works is the standard one [2] but appropriate adjustments should be made to respect the convention that the locations of B-moves in $A \multimap B$ begin with R, whereas those in $B \multimap C$ begin with L. Therefore if σ (for $A \multimap B$) tells P to play (the move whose location is) Ru in B, he should ask τ (for $B \multimap C$) what to do with (the move whose location is) Lu (rather than Ru).

We devote the next sections to the introduction of essential nets for IMAL2 with the aim of proving the main result of the paper:

Theorem 2 (Full Completeness). *Evolving games and good strategies are fully and faithfully complete for IMAL2.* □

5 Essential Nets for IMAL2 and IMLL2

We extend Lamarche's essential nets [11,14] for (quantifier-free) IMLL to the second-order affine fragment. Our correctness criterion is motivated by the evolving game model and corresponds to the rule that the game imported by P is determined "uniformly" by those imported by O earlier. To our knowledge, the results in this section are also new. Proof nets for first-order MLL which extend the DR-criterion [5] for MLL have already been proposed in [6], from which it is an easy step to derive a correctness criterion for MLL2 (i.e. second-order *classical* MLL). Although IMLL2 is a sublogic of MLL2, essential nets are quite a different kind of structure from proof nets; knowing the corresponding DR-criterion is not much of a help in finding a correctness criterion for essential nets.

Essential nets are a graphical representation of derivations of IMAL2 sequents expressed in ***polarized form*** $\Vdash N_1, \ldots, N_k, P$, where the N_is are negative formulas and P is a positive formula; note that the polarized sequents are one-sided. In the following we shall assume the correspondence between the two-sided sequents $\vdash$ and one-sided polarized sequents $\Vdash$, and use them interchangeably.

Definition 6. An ***essential net*** for an IMAL2 sequent $\Vdash N_1^-, \cdots, N_k^-, P^+$ in polarized form is a directed graph which is constructed from the following components:

- axiom link, which is an oriented edge from a positive copy to a negative copy of a propositional atom or second-order variable:

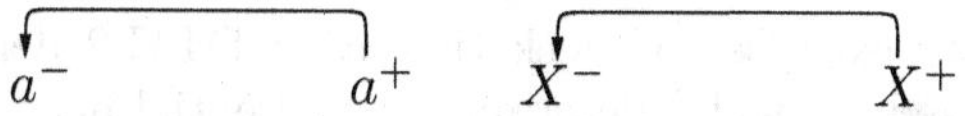

– polarized ⊗-nodes and ⅋-nodes:

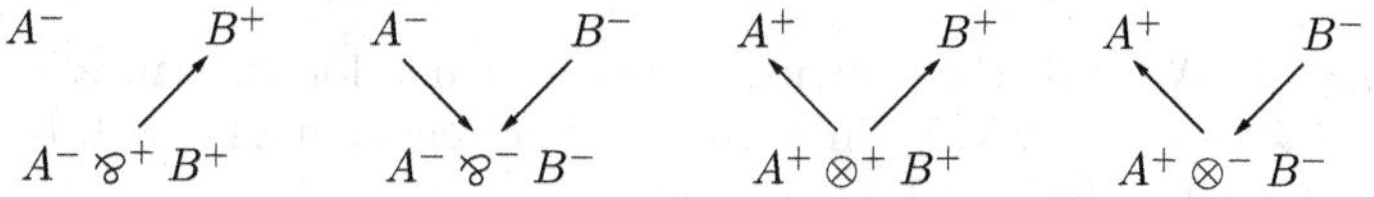

In each of the four components above, the two nodes situated above are called ***premises*** of the node below (similarly for the $\forall^+$-node and $\forall^-$-node in the following). We call the left premise of a $⅋^+$-node its ***sink***; note that there is no edge linking a $⅋^+$-node to its sink.

– $\forall^+$-nodes:

$$\begin{array}{c} A^+ \\ \uparrow \\ \forall^+ X.A^+ \end{array}$$

where X is called the ***eigenvariable*** of the node $\forall^+ X.A^+$.

– $\forall^-$-nodes:

$$\begin{array}{c} A^-[T/X] \\ \downarrow T \\ \forall^- X.A^- \end{array}$$

Note the *labelled* edge. If X occurs in A, the label T can be retrieved from the net. We call T the ***eigentype*** of the $\forall^-$-node.

– and ***weakening nodes*** (note the bar over the formula)

$$\overline{A^-}$$

such that the ***conclusions*** of the net (nodes that are not premises of any node) are exactly $N_1^-, \cdots, N_k^-, P^+$, polarized as indicated. The positive conclusion is called the ***root*** of the net.

Important Convention. In addition we require that:

1. the eigenvariables be distinct
2. every variable that occurs in the net be the eigenvariable of some $\forall^+$-node, otherwise we replace all its occurrences throughout the net by a fresh propositional atom
3. conclusions are closed formulas.

An essential net for an IMLL2 sequent is defined in exactly the same way by construction from the above components but *less* the weakening nodes. Note that an IMLL2 net is also an IMAL2 net. □

It is an easy exercise to check that each IMAL2 derivation (expressed in terms of polarized sequents) determines an essential net for the end-sequent. Of course there are essential nets that do not arise in this way, as the next example shows.

Example 5. We give the "obvious" essential net for the (invalid) sequent $\vdash \forall Z.(a \otimes Z) \multimap (a \otimes \forall X.X)$ in Figure 2. The reader might wish to construct an essential net for the (invalid) sequent $\vdash \forall XV.(X \otimes V) \multimap (\forall Y.Y \otimes \forall Z.Z)$.

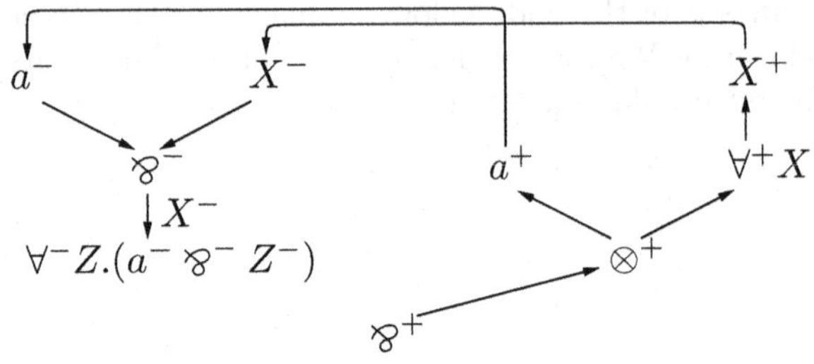

Fig. 2. An essential net for $\vdash \forall^- Z.(a^- \wp^- Z^-) \wp^+ (a^+ \otimes^+ \forall^+ X.X^+)$.

[**A note on figures:** Owing to typographical constraints, we have been economical with the node-labels in our drawing of essential nets, often displaying only the outermost connective rather than the whole formula.]

The main challenge in finding a *correctness criterion* (a characterization of essential nets that arise from derivations) is to capture the side condition of the rule ($\forall$-r). It turns out that the notion of essential nets alone (and in particular the Convention) already rules out several invalid sequents.

Example 6. (i) There is no essential net for $\vdash X^-, \forall^+ X.X^+$ because of the convention that conclusions must be closed formulas – note that X^- may not be replaced by a fresh atom because it is an eigenvariable.
(ii) There is no essential net for

$$\vdash \forall^- Y.(Y^- \wp^- Y^-), \forall^+ X.X^+ \otimes^+ \forall^+ Z.Z^+$$

because X^+ and Z^+ are different and it is impossible to axiom-link each of them with Y^-.

Correctness Criterion for IMAL2 and IMLL2 Essential Nets

It is straightforward to see that if an IMAL2 essential net is acyclic then every node x is reachable from either the root or from a weakening node (or from both). In the case of IMLL2 nets, acyclicity implies that every node is reachable from the root.

Definition 7. An IMLL2 essential net is said to be ***correct*** just in case:

(1) the digraph is acyclic
(2) for any $\wp^+$-node p, every path from the root to p's sink passes through p
(3) for any $\forall^-$-node q whose eigentype is T (say), for any free variable X that occurs in T, every path from the root to q passes through the $\forall^+$-node whose eigenvariable is X. □

Example 7. (i) In the essential net in Figure 2, note that the $\forall^-$-node does not satisfy condition (3): there is a path from the root to the $\forall^-$-node without passing through the $\forall^+$-node.
(ii) The sequent $\vdash \forall Y.Y \otimes \forall Z.Z \multimap \forall X.(X \otimes X)$ has (infinitely) many correct essential nets.

Definition 8. An *explicit IMAL2 net* N' is a directed graph that is obtained from an IMAL2 net N by joining each weakening node to some positive node by a directed edge (pointing from the positive node to the negative weakening node); we say that N *extends* to N'. Clearly N' has the same node-set as N; by abuse of terminology, we say that a node in N' is a $\wp^+$-node (similarly for $\forall^+$-node, $\otimes^-$-node etc.) just in case it is in N.

An IMAL2 essential net N is said to be ***correct*** if it is possible to extend N to an explicit IMAL2 net N' which satisfies conditions (1), (2) and (3) of Definition 7. □

Now for some examples of essential nets that have weakening nodes:

Example 8. (i) Consider the essential net given by the following derivation

$$\dfrac{\dfrac{\dfrac{\vdash Y^- \wp^+ Y^+}{\vdash Y^-, \; Y^- \wp^+ Y^+}}{\vdash \forall^- X.X^-, \; Y^- \wp^+ Y^+}}{\vdash \forall^- X.X^-, \; \forall^+ Y.Y^- \wp^+ Y^+}$$

which has two connected components. The net can be strengthened to a net that satisfies the IMLL2 correctness criterion by connecting Y^+ to the weakening node Y^-.
(ii) The sequent $(a \multimap b) \otimes (b \multimap a) \multimap c, d \vdash d$ has no correct essential net in which c is a weakening node.
(iii) The sequent $a \multimap a, b \vdash b$ has no correct essential net which contains no weakening node. Of course it has a correct essential net in which $a \multimap a$ is a single weakening node; in which case the net is suitably strengthened by linking b^+ with $a \multimap a$.

Theorem 3 (Sequentialization). *IMAL2 (and hence also IMLL2) essential nets $\mathcal{E}$ satisfying the correctness criterion are sequentializable i.e. there is a derivation of the end-sequent that gives rise to $\mathcal{E}$.* □

Canonical Essential Nets

Although essential nets respect the commuting conversions of IMLL, the addition of weakening brings out new ones which are not handled so well. For example the proofs below should be deemed equivalent, but their essential nets differ:

$$\dfrac{z \vdash z \qquad \dfrac{x \vdash x}{y, x \vdash x}}{z, z \multimap y, x \vdash x} \qquad \dfrac{\dfrac{x \vdash x}{z \multimap y, x \vdash x}}{z, z \multimap y, x \vdash x}$$

Another problem is the commutation of weakening with the $\otimes^-$-rule. The two proofs

$$\dfrac{\dfrac{\dfrac{x \vdash x}{y, x \vdash x}}{y, z, x \vdash x}}{y \otimes z, x \vdash x} \qquad \dfrac{x \vdash x}{y \otimes z, x \vdash x}$$

are equivalent, but the corresponding nets are different. This motivates the following notion of equivalence:

Definition 9. Two essential nets for the same end-sequent are said to be ***equivalent*** if they are identical when restricted to the parts that are reachable from the root (the unique positive conclusion).

Good strategies respect this notion of equivalence. For the purpose of proving a strong definability result, it is convenient to reason in terms of appropriate representatives of equivalence classes of essential nets. We call these representatives ***canonical*** essential nets, and they satisfy an additional condition: If a node is *not* reachable from the root, it is a weakening node. (We know that the converse holds, because weakening nodes have no premises and are of negative polarity, so no edge can lead to them.)

The new condition has a number of consequences:

1. a weakening node cannot be the negative premise of a $\otimes^-$-node, because it would render that $\otimes^-$-node unreachable
2. at most one premise of a $\otimes^-$-node may be a weakening node, for otherwise the $\otimes^-$-node would be unreachable
3. the premise of a $\forall^-$-node cannot be introduced by weakening
4. all axiom links are reachable.

Canonical essential nets arise "naturally" from sequent calculus derivations in which the use of weakening is delayed for "as long as possible". A derivation may be normalized to a canonical form by first deleting each unreachable connected component and then adding the component back as a *single* weakening node. A nice consequence is that all $\forall^-$-nodes are reachable (from the root), which promises an exact correspondence with strategies. Another is a simplified version of the correctness criterion:

Lemma 3. *A canonical IMAL2 net is correct if and only if it satisfies all conditions of Definition 7.* □

6 Proof of Full and Faithful Completeness

Our final task is to prove a strong definability theorem:

Theorem 4 (Strong Definability). *Correct canonical IMAL2 essential nets and shortsighted regular strategies are in 1-1 correspondence.* □

A strategy σ is said to be ***shortsighted*** just in case for every even-length $s \in \sigma$, if sm is a position then $sm \in \sigma$ if and only if the O-move m is *enabled* by the last move of s. (See [12] for a definition of the enabling ordering.)

Lemma 4. *Fix an IMAL2 sequent* $\vdash \Gamma$. *(There is no harm in assuming that* Γ *is a formula.) Every shortsighted regular strategy* $\sigma = \sigma_{f,F}$ *for the associated game determines a canonical essential net* $\mathcal{E}_\sigma$ *for* $\vdash \Gamma$. *Further:*

(i) f *and* F *respectively define the axiom links and the eigentypes of* $\mathcal{E}_\sigma$.
(ii) There is a one-one correspondence between positions in σ *and paths (starting from the root) in* $\mathcal{E}_\sigma$ *which end in a leaf.* □

The Lemma can be proved by first establishing a lemma similar to [12, Lemma 19], which we omit. Instead we set out in detail the correspondence in (i) of the Lemma:

- For each $(u, v) \in f$, u is the occurrence (in $\mathcal{E}_\sigma$) of the positive end of a unique axiom link, and v the occurrence of the negative end.
- Each t in the domain of F is the occurrence of a unique $\forall^-$-node. Suppose its eigentype has free variables $Y_1, \cdots, Y_n$ (say), then $F(t)$ is expressed in terms of O-locations which are the occurrences of the $\forall^+$-nodes whose eigenvariables are the Y_is.

and illustrate it by the following example.

Example 9. Consider the IMAL2 sequent $\vdash \forall Z.Z \multimap \forall X.(\forall Y.Y \multimap X)$ and a regular strategy $\sigma_{f,F}$ generated by

$$f = \{(RR, LR), (LLR, RLR), (RLL, LLLl)\}$$
$$F = \{(L, ([\forall b.b] \otimes R \multimap R) \multimap R), (RL, R \multimap R), (LLLl, R)\}$$

The strategy has the following maximal position

$$(R^a R)\,(L^{([\forall b.b] \otimes a \multimap a) \multimap a} R)\,(LLR)\,(RL^{a \multimap a} R)\,(RLL)\,(LLLl^a)$$

The reader may wish to check that f and F respectively specify the axiom links and the eigentypes of the essential net which is determined by $\sigma_{f,F}$ as shown in Figure 3.

Following on from Lemma 4 we can show that:

Proposition 1. $\mathcal{E}_\sigma$ *is correct.*

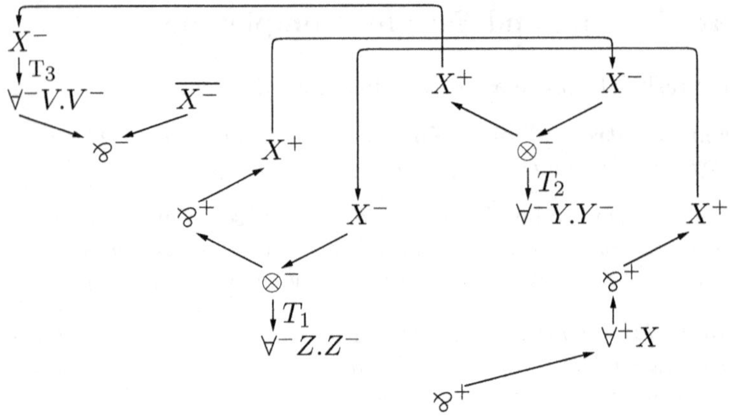

where $T_1 = ((\forall^- V.V^- \parr^- X^-) \parr^+ X^+) \otimes^- X^-$, $T_2 = X^+ \otimes^- X^-$, $T_3 = X^-$.

Fig. 3. A correct canonical essential net for $\vdash \forall^- Z.Z^- \parr^+ (\forall^+ X.(\forall^- Y.Y^- \parr^+ X^+))$.

Proof. We take advantage of Lemma 3 and prove that $\mathcal{E}_\sigma$ satisfies condition (3) of Definition 7 for illustration; the other two conditions can be shown to hold by arguments similar to those in the proof of [12, Theorem 22]. Suppose there is a node $\forall^- Z.A$ – call it p – whose eigentype has a free variable Y (say). Since $\mathcal{E}_\sigma$ is canonical, p is reachable from the root. For a contradiction, suppose there is a path from the root to p, whose occurrence (a P-location) in $\mathcal{E}_\sigma$ is t (say), without passing through a $\forall^+ Y.G$-node, whose occurrence (an O-location) is t' (say). By Lemma 4(ii), there is a position (pick the shortest) in σ, $s m_O m_P$ say, that ends in a move m_P of the form $u^{\overline{B}} v$ where the location of u is t; further no move from the position has a location that has t' as a prefix. By definition of symbolic strategy and by Lemma 4(i), we have $F(t)$ is expressed in terms of $\lceil t' \rceil$ (among possibly other O-locations), which should already be defined in the position. But this contradicts our assumption. □

This proves Theorem 4 and hence Theorem 2.

Further directions. For us, this work is a necessary step towards the construction of a fully complete model for Second-order Intuitionistic Multiplicative *Light* Affine Logic (IMLAL2) ([7,4]), and so, a game characterization of the PTIME functions. We believe that an appropriate synthesis of this work with our recent discreet strategies model for (quantifier-free) IMLAL (see [13]) should produce the result. It would also be interesting to investigate full completeness for IMLL2, and for IMAL2 with unit.

Acknowledgements. This work was partially funded by EU TMR projects AppSem and LINEAR. Ong is grateful to the National University of Singapore for funding a visit during which part of the work was carried out. Murawski is supported by a University of Oxford Scatcherd European Scholarship and a British Overseas Research Students Award 98032135.

References

1. Abramsky, S.: Semantics of Interaction: an introduction to Game Semantics. *Semantics and Logics of Computation.* Cambridge Univ. Press, 1–32
2. Abramsky, S., Jagadeesan, R.: Games and full completeness for multiplicative linear logic. *Journal of Symbolic Logic* **59** (1994) 543–574
3. Abramsky, S., Lenisa, M.: A Fully-complete PER Model for ML Polymorphic Types. In *Proceedings of CSL2000*, Annual Conference of the European Association of Computer Science Logic, August 2000, Fischbachau, Germany. LNCS Vol. **1862**, Springer-Verlag, 2000
4. Asperti, A.: Light affine logic. In *Proc. LICS'98.* IEEE Computer Society (1998)
5. Danos, V., Regnier, L.: The Structures of Multiplicatives. *Archive for Mathematical Logic*, Vol. **28**, 1989, 181–203
6. Girard, J.-Y.: Quantifiers in Linear Logic II. Preprint, Équipe de Logique de Paris VII, 1991
7. Girard, J.-Y.: Light linear logic. *Information & Computation* **143** (1998) 175–204
8. Hughes, D. H. D.: Games and definability for System F. In *Proceedings of 12th IEEE Symposium on Logic in Computer Science*, 1997, IEEE Computer Science Society
9. Hyland, J. M. E., Ong, C.-H. L.: On Full Abstraction for PCF: I, II & III. *Information & Computation* (2000) 130 pp, in press
10. Ker, A. D., Nickau, H., Ong, C.-H. L.: A universal innocent game model for the Böhm tree lambda theory. In *Proceedings of the 8th Annual Conference on the EACSL*, Madrid, Spain, September 1999, LNCS, Vol. **1683** Springer-Verlag,1999, 405–419
11. Lamarche, F.: Proof Nets for Intuitionistic Linear Logic 1: Essential Nets. Preprint ftp-able from *Hypatia*, 1994
12. Murawski, A. S., Ong, C.-H. L.: Exhausting Strategies, Joker Games and Full Completeness for IMLL with Unit. In *Proc. 8th Conf. CTCS'99.* ENTCS **29** (1999)
13. Murawski, A. S., Ong, C.-H. L.: Discreet Games, Light Affine Logic and PTIME Computation. In *Proceedings of CSL2000*, Annual Conference of the European Association of Computer Science Logic, Fischbachau, Germany. LNCS Vol. **1862**, Springer-Verlag, 2000, 427–441
14. Murawski, A. S., Ong, C.-H. L.: Dominator Trees and Fast Verification of Proof Nets. In *Proceedings of 15th IEEE Symposium on Logic in Computer Science*, IEEE Computer Society Press, 2000, 181–191
15. Nickau, H., Ong, C.-H. L.: Games for System F. Working paper, 1999

Retracts in Simple Types

Vincent Padovani

Equipe Preuves, Programmes et Systèmes
CNRS - Université Paris 7
padovani@pps.jussieu.fr

Abstract. In this paper we prove the decidability of the existence of a definable retraction between two given simple types. Instead of defining some extension of a former type system from which these retractions could be inferred, we obtain this result as a corollary of the decidability of the minimal model of simply typed λ-calculus.

Although definably isomorphic simple types are fully characterized in [BL85] as a sequel to a classical result by Dezani for untyped $\lambda\beta\eta$-calculus [Dez76], the more general problem of definable retractions has been left open: given two simple types A and B, this problem is to decide whether there exists $u : A \to B$ and $v : B \to A$ such that $(v \circ u) =_{\beta\eta} I$.

In [BL85], Bruce and Longo exhibit a type system from which one can infer all retractions that are definable without the η-rule. In the same paper, they prove the soundess and the incompleteness, with respect to the general case, of a proper extension of this system. In [LPS92], de'Liguoro, Piperno and Statman show how to extend the latter system to derive all retractions that are definable by linear λ-terms.

This paper attacks the retraction problem from a much more syntactical angle. Instead of trying to give a complete extension of a former system, we use the decision algorithm for the minimal model of simply-typed λ-calculus (see [Pad95]) to build an new algorithm that decides whether a simple type is a retract of another.

For the sake of simplicity, we will focus on the special case where the calculus contains a single ground type. However, this restriction is not crucial: as shown in [Pad95], a minimal model built with more than one ground type is still decidable, and the result can be easily extended to the general case. Also, we will add to the calculus a constant of ground type - again, this addition is harmless, and is only intended to simplify the proof. Both restrictions are discussed in the conclusion.

The paper is divided in two main parts. In section 1.2, we introduce two sets of terms $\mathcal{C}$ and $\mathcal{D}$, the sets of *coders* and *decoders*. In section 1.3, we show that the definability of a retraction by some pair in $\mathcal{C} \times \mathcal{D}$, is a decidable property. In section 2, we prove that if a retraction is definable, then it is definable by a pair in $\mathcal{C} \times \mathcal{D}$. For that purpose we define in sections 2.1 and 2.2 an algorithm which transforms any retraction pair into a pair in $\mathcal{C} \times \mathcal{D}$. The faithfulness of the conversion is proved at section 2.3.

S. Abramsky (Ed.): TLCA 2001, LNCS 2044, pp. 376–384, 2001.

1 Coders and Decoders

We consider the simply-typed λ-calculus with a single ground type $\circ$, a single constant $\perp$ of ground type, with a typing *à la Church.* In the sequel, by terms and types, we understand simply-typed terms and simple types.

The notation $M : A$ indicates that M is a term of type A. Types will be frequently ommited whenever they are irrelevant, or implicit from the context. We will often make use of the following notations:

- $A_1 \dots A_n \to \circ$ denotes the type $(A_1 \to (A_2 \to \dots (A_n \to \circ) \dots))$.
- $\lambda x_1 \dots x_n.(MN_1 \dots N_m)$ denotes $\lambda x_1 \dots \lambda x_n(\dots(MN_1)\dots N_m)$. When there is no amibugity, this term will be also denoted by $\lambda\overline{x}.(M\overline{N})$.
- $\lambda\overline{d}.M$ denotes a term of the form $\lambda x_1 \dots x_n.M$ with no x_i free in M.

1.1 Retracts and Products

We say that the *product of* $A^1, \dots, A^n$ *is a retract of* B if and only if the following property holds:

> There exists a term $M : B$, distinct variables $f^1 : A^1, \dots, f^n : A^n$ free in M, and closed terms $N^1 : B \to A^1, \dots, N^n : B \to A^n$, such that for all $i \in \{1, \dots, n\}$ we have $(N^i\, M) =_{\beta\eta} f^i$.

We denote this property by $(A^1 \times \dots \times A^n) \lhd B$, and say that the $n+1$-uplet $(M, N^1, \dots, N^n)$ is a *witness for* $(A^1 \times \dots \times A^n) \lhd B$. In the special case where $n = 1$ with $A^1 = A$, we simply write $A \lhd B$, and say that A is a retract of B.

Remark 1. In this definition, we do not require $f^1, \dots, f^n$ to be the only free variables of M. Thus, for all subset $\{i_1, \dots, i_p\}$ of $\{1, \dots, n\}$, $(M, N^{i_1}, \dots, N^{i_p})$ is also a witness for $(A^{i_1} \times \dots \times A^{i_p}) \lhd B$. In particular, if 0 denotes the empty product, every term of type B is a witness for $0 \lhd B$.

Our first aim will be to build two sets of terms, $\mathcal{C}$ and $\mathcal{D}$, and to show that the existence of a witness in $\mathcal{C} \times \mathcal{D}^n$ for a relation of the form $(A^1 \times \dots \times A^n) \lhd B$, is decidable.

1.2 Projections, Coders, and Decoders

- we let $(\circ^0 \to \circ) = \circ$, and $(\circ^{n+1} \to \circ) = (\circ \to (\circ^n \to \circ))$.
- we call *projection* every term of the form
  $\Pi^n_i = \lambda x_1 \dots x_n.x_i : (\circ^n \to \circ)\ (i \in \{1, \dots, n\})$.
- we call *selector* every closed term whose type is of the form
  $B_1 \dots B_m \to (\circ^n \to \circ)$.
- we let $\mathcal{C}$, $\mathcal{D}$ be the least sets of terms satisfying the following:
  - $f : \circ \in \mathcal{C}$, for all variable f of ground type.
  - $\lambda x.x : \circ \to \circ \in \mathcal{D}$.
  - Suppose the types $A^1, \dots, A^n, B$ are of the following form:

$$A^i = A^i_1 \dots A^i_{p_i} \to \circ \quad (i \in \{1, \dots, n\}),$$
$$B = B_1 \dots B_m \to \circ.$$

Let $f^1 : A^1, \dots, f^n : A^n$, $z_1 : B_1, \dots, z_m : B_m$ be distinct variables.
Let Σ be a selector of type $B_1 \dots B_m \to (\circ^n \to \circ)$.
For each $i \in \{1, \dots, n\}$:
Let ϕ_i be an arbitrary function from $\{1, \dots, p_i\}$ to $\{1, \dots, m\}$.
For each $j \in \{1, \dots, p_i\}$, let $D^i_j : B_{\phi_i(j)} \to A^i_j \in \mathcal{D}$.

Then, the following term $C : B$ belongs to $\mathcal{C}$:

$$C = \lambda z_1 \dots z_m.(\Sigma\, z_1 \dots z_m)\ (f^1\ (D^1_1\ z_{\phi_1(1)}) \dots (D^1_{p_1}\ z_{\phi_1(p_1)}))$$
$$\vdots$$
$$(f^n\ (D^n_1\ z_{\phi_n(1)}) \dots (D^n_{p_n}\ z_{\phi_n(p_n)}))$$

The term Σ will be called the *main selector* of C.

- Let $y_1 : A_1, \dots, y_p : A_p$ be distinct variables. Let $C_1 : B_1, \dots, C_m : B_m$ be terms in $\mathcal{C}$ whose free variables are all amongst $y_1, \dots, y_p$. Let g be a fresh variable of type $B = B_1 \dots B_m \to \circ$. Then:

$$D = \lambda g\, y_1 \dots y_p.(g\, C_1 \dots C_m) : B \to (A_1 \dots A_p \to \circ) \in \mathcal{D}$$

We call *coder* every element of $\mathcal{C}$, and *decoder* every element of $\mathcal{D}$.

1.3 Observational Equivalence

We define the equivalence relation $\equiv$ on the set of selectors with:

let Σ, Σ' be selectors of same type $B_1 \dots B_m \to (\circ^n \to \circ)$. Then $\Sigma \equiv \Sigma'$ if and only if for all projection $\Pi : (\circ^n \to \circ)$ and for all term F, $(F\,\Sigma) =_\beta \Pi \Leftrightarrow (F\,\Sigma') =_\beta \Pi$.

Proposition 1. *The number of selector classes of any given type is finite. Moreover, there exists a computable function which, given an arbitrary selector type, returns a set of selectors of this type which is complete, up to* $\equiv$.

Proof. This follows immediately from the fact that, on one hand, the relation $\equiv$ on selectors coincide with observational equivalence in the minimal model of simply typed λ-calculus, on the other hand, this model is decidable. See [Pad95] for details, or [Sch98] for an alternate proof of decidability, or [Loa97] for a brief presentation of Schmidt-Schauß' algorithm.

We extend $\equiv$ to coders and decoders with the following definition:

- let C, C' be coders of same type, same free variables $f^1 : A^1, \dots, f^n : A^n$, of the following forms:
  - $C = \lambda \overline{z}.(\Sigma\, \overline{z})(f^1\, \overline{u}^1) \dots (f^n\, \overline{u}^n)$,

  - $C' = \lambda\overline{z}.(\Sigma'\,\overline{z})(f^1\,\overline{v}^1)\ldots(f^n\,\overline{v}^n)$.

  Then $C \equiv C'$ if and only if:
  - $\Sigma \equiv \Sigma'$, *i.e.* the main selectors of C, C' are equivalent,
  - for all i, j, if $u^i_j = (D\,z)$ and $v^i_j = (D'\,z')$, then $z = z'$ and $D \equiv D'$.
- let D, D' be decoders of same type, of the following forms:
  - $D = \lambda g\,\overline{y}.(g\,C_1 \ldots C_m)$,
  - $D' = \lambda g\,\overline{y}.(g\,C'_1 \ldots C'_m)$.

  Then $D \equiv D'$ if and only if for all $k \in \{1, \ldots, m\}$ we have $C_k \equiv C'_k$.

Lemma 1. *Let C, C' be coders. Suppose $C \equiv C'$. Then for all term F and all projection Π, we have $F\,C =_\beta \Pi$ if and only if $F\,C' =_\beta \Pi$.*

Proof. Assuming $C \equiv C'$, the terms C, C' are of the following forms:

- $C = \lambda\overline{z}.(\Sigma\,\overline{z})(f^1\,\overline{u}^1)\ldots(f^n\,\overline{u}^n)$,
- $C' = \lambda\overline{z}.(\Sigma'\,\overline{z})(f^1\,\overline{v}^1)\ldots(f^n\,\overline{v}^n)$,

with $\Sigma \equiv \Sigma'$. Suppose for instance $(F\,C) =_\beta \Pi$. Then $f_1, \ldots, f_n$ are not free in the normal form of $(F\,C)$, therefore:

$$\begin{aligned}\Pi &=_\beta (F\,C)\\ &=_\beta (F\,C[\lambda\overline{d}.\bot/f_1, \ldots, \lambda\overline{d}.\bot/f_n])\\ &=_\beta F\,(\lambda\overline{z}.(\Sigma\,\overline{z}\,\overline{\bot}))\end{aligned}$$

where $\overline{\bot}$ denotes a sequence of $\bot$ of length n. Now, the main selectors Σ, Σ' of C, C' are equivalent, therefore:

$$\begin{aligned}\Pi &=_\beta F\,(\lambda\overline{z}.(\Sigma'\,\overline{z}\,\overline{\bot}))\\ &=_\beta (F\,C'[\lambda\overline{d}.\bot/f_1, \ldots, \lambda\overline{d}.\bot/f_n])\\ &=_\beta (F\,C')\end{aligned}$$

Lemma 2. *Let C be a coder, D a decoder. If (C, D) is a witness for $A \lhd B$, and if $C \equiv C'$ and $D \equiv D'$, then (C', D') is a witness for $A \lhd B$.*

Proof. Let f be the variable such that $(D\,C) =_{\beta\eta} f$. We use induction on the type B of C and C' to show that $(D'\,C') =_{\beta\eta} f$. According to the definition of equivalence, the considered terms are of the following forms:

- $C = \lambda z_1 \ldots z_m.(\Sigma\,\overline{z})(f^1\,\overline{u}^1)\ldots(f^n\,\overline{u}^n)$,
- $C' = \lambda z_1 \ldots z_m.(\Sigma'\,\overline{z})(f^1\,\overline{v}^1)\ldots(f^n\,\overline{v}^n)$,
- $D = \lambda g\,y_1 \ldots y_p.(g\,C_1 \ldots C_m) = \lambda g\,\overline{y}.(g\,\overline{C})$,
- $D' = \lambda g\,y_1 \ldots y_p.(g\,C'_1 \ldots C'_m) = \lambda g\,\overline{y}.(g\,\overline{C}')$,

with $\Sigma \equiv \Sigma'$ and $C_k \equiv C'_k$ for all $k \in \{1, \ldots, m\}$. Suppose $f = f^i$. In that case, the sequences $\overline{u}^i$, $\overline{v}^i$ are of the forms $(u^i_1, \ldots, u^i_p)$, $(v^i_1, \ldots, v^i_p)$, and:

- $(\Sigma\,\overline{C}) =_\beta \Pi^n_i$,
- $(D\,C) =_\beta \lambda\overline{y}.(f^i\,\overline{u}^i)[\overline{C}/\overline{z}] =_{\beta\eta} \lambda\overline{y}.(f\,\overline{y})$.

Since $\Sigma \equiv \Sigma'$, we have $(\Sigma \overline{C}') =_\beta \Pi_i^n$. Since $C_k \equiv C'_k$ for all $k \in \{1, \ldots, m\}$, by lemma 1 we have $(\Sigma' \overline{C}') =_\beta \Pi_i^n$. Therefore $(D' C') =_\beta \lambda\overline{y}.(f\,\overline{v}^i)[\overline{C}'/\overline{z}]$. Now, for all j, for k and d such that $u_j^i = (d\, z_k)$, we have:

- v_j^i is of the form $(d'\, z_k)$ with $d \equiv d'$,
- $u_j^i[\overline{C}/\overline{z}] = (d\, C_k) =_{\beta\eta} y_j$,
- by induction hypothesis, $v_j^i[\overline{C}'/\overline{z}] = (d'\, C'_k) =_{\beta\eta} y_j$.

We conclude that $(D' C') =_\beta \lambda\overline{y}.(f^i\,\overline{v}^i)[\overline{C}'/\overline{z}] =_{\beta\eta} \lambda\overline{y}.(f^i\,\overline{y}) =_{\beta\eta} f^i$.

Theorem 1. *Let $f^1 : A^1, \ldots, f^n : A^n$ be typed variables. Let B be an arbitrary type. Then:*

- *the number of classes of coders of type B and free variables $f^1, \ldots, f^n$, is finite,*
- *for all i, the number of classes of decoders of type $B \to A^i$ is finite.*

Moreover, there exists a computable function which takes as an input the typed variables $f^1 : A^1, \ldots, f^n : A^n$ and the type B, and returns a representative of each class.

Proof. The proof of finiteness and the simultaneous construction of complete sets of representatives is straightforward, using induction on B, proposition 1 and lemmas 1 and 2.

Corollary 1. *Given the types $A^1, \ldots, A^n$ and the type B, the existence of a witness in $\mathcal{C} \times \mathcal{D}^n$ for $(A^1 \times \ldots \times A^n) \lhd B$ is decidable.*

Our next aim will be to prove that if $(A^1 \times \ldots \times A^n) \lhd B$, then there is indeed a witness in $\mathcal{C} \times \mathcal{D}^n$ for this relation. As an immediate consequence, we will get the decidability of $\lhd$.

2 Witness Conversion

This section presents an algorithm, *Convert*, which takes as an input an arbitrary witness $(M, N^1, \ldots, N^n)$ for $(A^1 \times \ldots \times A^n) \lhd B$, and returns a witness $(C, D^1, \ldots, D^n)$ in $\mathcal{C} \times \mathcal{D}^n$ for this relation.

2.1 Linearization

Note that if (M, N) is a witness for $A \lhd B$, then the β-normal, η-long form of N is necessarily of the form $\lambda g \lambda\overline{y}.(g \ldots)$ with g of type B. The following function is intended to transform N into a term with a single (head) occurrence of g.

Function Linearize $(M : B, f : A, N : B \to A)$

assuming N closed and $(N\, M) =_{\beta\eta} f$,

0. Redefine N as the β-normal, η-long form of N.
   Let $M_\perp = M[\lambda\overline{d}.\perp/f^1, \ldots, \lambda\overline{d}.\perp/f^n]$, where $f^1, \ldots, f^n$ are all free variables of M.
1. Let $\lambda g \lambda\overline{y}.(g\, u_1 \ldots u_m) = N$.
2. If $\lambda\overline{y}.(M_\perp\, u_1[M/g] \ldots u_m[M/g]) =_{\beta\eta} f$,
   redefine N as the normal form of $\lambda g \lambda\overline{y}.(M_\perp\, u_1 \ldots u_m)$, and goto 1.
3. If $\lambda\overline{y}.(M\, u_1[M_\perp/g] \ldots u_m[M_\perp/g]) =_{\beta\eta} f$,
   redefine N as the normal form of $\lambda\overline{y}.(g\, u_1[M_\perp/g] \ldots u_m[M_\perp/g])$, and return N.

Remark 2. At step 1, $\lambda\overline{y}.(M\, u_1[M/g] \ldots u_m[M/g]) =_{\beta\eta} f$, where f is free in M and not free in $u_1, \ldots, u_m$. Therefore, one and only one of the conditions at steps 2 or 3 is satisfied.

Note also that all reductions of $(N\, M_\perp)$ are of finite length, therefore this algorithm terminates. It is not hard to check that $N' = \mathit{Linearize}(M, f, N)$ is closed, and that we still have $(N'\, M) =_{\beta\eta} f$.

2.2 Conversion

Function Convert $(M : B, f^1 : A^1, \ldots, f^n : A^n, N^1 : B \to A^1, \ldots, N^n : B \to A^n)$

assuming $B = B_1 \ldots B_m \to o$, $f^1, \ldots, f^n$ *distinct, and for all* $i \in \{1, \ldots, n\}$*:* $A^i = A^i_1 \ldots A^i_{p_i} \to o$, N^i *closed,* $(N^i\, M) =_{\beta\eta} f^i$*,*

0. If $n = 0$, return $\lambda\overline{d}.\perp : B$ and exit.
1. For each $f \notin \{f^1, \ldots, f^n\}$ free in M, redefine M as $M[\lambda\overline{d}.\perp/f]$.
   For each $i \in \{1, \ldots, n\}$:
   let $\lambda g y^i_1 \ldots y^i_{p_i}.(g\, u^i_1 \ldots u^i_m) = \mathit{Linearize}(M, f^i, N^i)$.
2. Define $\Sigma : B_1 \ldots B_m \to (o^n \to o)$ as the normal form of
   $\lambda\overline{z}\lambda x_1 \ldots x_n.(M[\lambda\overline{d}.x_1/f^1, \ldots, \lambda\overline{d}.x_n/f^n]\ \overline{z})$.
3. For each $i \in \{1, \ldots, n\}$, for each $j \in \{1, \ldots, p_i\}$,
   - let σ be the least substitution satisfying:
     - $\sigma(f^i) = \lambda\overline{y}^i.(y^i_j\, \overline{X}^i_j) : A^i$ where the $\overline{X}^i_j$ are fresh variables,
     - $\sigma(f^k) = \lambda\overline{d}.\perp : A^k$ for all $k \in \{1, \ldots, n\}$, $k \neq i$,
   - let $v^i_j : A^i_j$ be the normal form of $\lambda\overline{X}^i_j.(\sigma(M)\ \overline{z})$.
4. for each $i \in \{1, \ldots, n\}$,
   - for each $k \in \{1, \ldots, m\}$, let $b^i_k = u^i_k[\lambda\overline{d}.\perp/y^i_1, \ldots, \lambda\overline{d}.\perp/y^i_{p_i}]$.
   - for each $j \in \{1, \ldots, p_i\}$, define $\phi_i(j)$ as the unique $k \in \{1, \ldots, m\}$ such that the term
     $w^i_j = v^i_j[b^i_1/z_1, \ldots, b^i_{k-1}/z_{k-1}, b^i_{k+1}/z_{k+1}, \ldots, b^i_m/z_m]$
     satisfies $w^i_j[u^i_k/z_k] =_{\beta\eta} y^i_j$.
5. For each $i \in \{1, \ldots, n\}$, for each $k \in \{1, \ldots, m\}$,

- let $\{j_1, \dots, j_q\} = \phi_i^{-1}(k)$,
- let $(C_k^i : B_k, D_{j_1}^i : B_k \to A_{j_1}^i, \dots, D_{j_q}^i : B_k \to A_{j_q}^i)$
  $= Convert(u_k^i, y_{j_1}^i, \dots, y_{j_q}^i, \lambda z_k\, w_{j_1}^i, \dots, \lambda z_k\, w_{j_q}^i)$.

6. Define $C : B$ as $\lambda \overline{z}.(\Sigma\, \overline{z})\ (f^1\ (D_1^1\ z_{\phi_1(1)}) \dots (D_{p_1}^1\ z_{\phi_1(p_1)}))$

$$\vdots$$

$$(f^n\ (D_1^n\ z_{\phi_n(1)}) \dots (D_{p_n}^n\ z_{\phi_n(p_n)}))$$

   For each $i \in \{1, \dots, n\}$, define $D^i : B \to A^i$ as
   $\lambda g\, y_1^i \dots y_{p_i}^i.(g\, C_1^i \dots C_m^i)$.
7. Return $(C, D^1, \dots, D^n)$.

2.3 Faithfullness

Lemma 3. *Suppose $(C, \dots) = Convert(M, \dots, \dots)$. Then, for all term F and all projection Π, we have $(F\, M) =_\beta \Pi$ iff $(F\, C) =_\beta \Pi$.*

Proof. Suppose $(F\, M) =_\beta \Pi$ or $(F\, C) =_\beta \Pi$. Then $f_1, \dots, f_m$ are not free in the normal form of $(F\, M)$, or not free in the normal form of $(F\, C)$. According to steps 2 and 6, we have:

$$\begin{aligned} & (F\, M) \\ =_\beta\ & (F\, M[\lambda \overline{d}.\bot/f^1, \dots, \lambda \overline{d}.\bot/f^n]) \\ =_\beta\ & F\, (\lambda \overline{z}.(\Sigma\, \overline{z}\, \overline{\bot})) \\ =_\beta\ & (F\, C[\lambda \overline{d}.\bot/f^1, \dots, \lambda \overline{d}.\bot/f^n]) \\ =_\beta\ & (F\, C) \end{aligned}$$

where $\overline{\bot}$ denotes a sequence of $\bot$ of length n.

Lemma 4. *Let $(M, N^1, \dots, N^n)$ be a witness for $(A_1 \times \dots \times A^n) \lhd B$, such that $(N^i\, M) =_{\beta\eta} f^i$ for all i. Then $(C, D^1, \dots, D^n) =$ Convert$(M, \overline{f}, \overline{N})$ belongs to $\mathcal{C} \times \mathcal{D}^n$, and is still a witness for this relation.*

Proof. Assuming $n > 0$, we use induction on the type B of M and C. Let us trace the construction of $C, D^1, \dots D^n$ in $Convert$.

At step 2, $(N^i\, M) =_\beta \lambda \overline{y}^i.(M\, \overline{u}^i) =_{\beta\eta} f^i$ implies $(\Sigma\, \overline{u}^i) =_\beta \Pi_i^n$.

At step 3, $(M\, \overline{u}^i) =_{\beta\eta} (f^i\, \overline{y}^i)$ implies $v_j^i[\overline{u}^i/\overline{z}] =_{\beta\eta} \lambda \overline{X}_j^i.(y_j^i\, \overline{X}_j^i) =_{\beta\eta} y_j^i$. Note that all free variables of v_j^i belong to $\{z_1, \dots, z_m\}$.

At step 4, $v_j^i[\overline{u}^i/\overline{z}] =_{\beta\eta} y_j^i$ implies the existence of a unique $k \in \{1, \dots, m\}$ such that the free occurence of y_j^i in the $\beta\eta$-normal form of $v_j^i[\overline{u}^i/\overline{z}]$ is a residual of a free occurence of y_j^i in u_k^i. Thus, $\phi_i(j)$ is well defined. Note that $z_{\phi_i(j)}$ is the only free variable of w_j^i.

At step 5, for all $i \in \{1, \ldots, n\}$ and all $j \in \{1, \ldots, p_i\}$, we have $((\lambda z_{\phi_i(j)} w^i_j)\, u^i_{\phi_i(j)}) =_{\beta\eta} y^i_j$. Therefore, for all i and all $k \in \{1, \ldots, m\}$, $(u^i_k, \lambda z_k w^i_{j_1}, \ldots, \lambda z_k w^i_{j_q})$ is a witness for $(A^i_{j_1} \times \ldots \times A^i_{j_q}) \lhd B_k$. By induction hypothesis, C^i_k is a coder, $D^i_{j_1}, \ldots, D^i_{j_q}$ are decoders, and $(C^i_k, D^i_{j_1}, \ldots, D^i_{j_q})$ is a witness for $(A^i_{j_1} \times \ldots \times A^i_{j_q}) \lhd B_k$. In other words, for all i, j we have $(D^i_j\, C^i_{\phi_i(j)}) =_{\beta\eta} y^i_j$, therefore we have $((\lambda \overline{z}.f^i\, (D^i_1\, z_{\phi_i(1)}) \ldots (D^i_{p_i}\, z_{\phi_i(p_i)}))\, \overline{C}^i) =_{\beta\eta} (f^i\, y^i_1 \ldots y^i_{p_i})$.

At step 6, obviously C is a coder and the $D^1, \ldots, D^n$ are decoders. According to the analysis of step 5, it remains to check that for all i we have $(\Sigma\, \overline{C}^i) =_\beta \Pi^n_i$. We already know that $(\Sigma\, \overline{u}^i) =_\beta \Pi^n_i$, and that each C^i_k was computed by feeding *Convert* with u^i_k as first argument. By lemma 3, for all F, $(F\, u^i_k) =_\beta \Pi^n_k$ implies $(F\, C^i_k) =_\beta \Pi^n_i$, so we are done.

The faithfullness of the conversion immediately implies:

Lemma 5. *If* $(A^1 \times \ldots \times A^n) \lhd B$ *then there is a witness in* $\mathcal{C} \times \mathcal{D}^n$ *for this relation.*

At last, we obtain our expected main result:

Theorem 2. *The relation* $\lhd$ *is decidable.*

3 Conclusion

So far, we have proved that the existence of a definable retraction between two simple types is decidable, in the special case of one ground type $\circ$, with a constant $\bot$ of type $\circ$. The generalization to many ground types, with one constant of each ground type, requires only minor modifications of our proof, thanks to the following lemma:

Lemma 6. *If* $A^1 \times \ldots \times A^n \lhd B$ *then* $A^1, \ldots, A^n, B$ *are of same rightmost ground type.*

Proof. Suppose $A \lhd B$, with $A = A_1 \ldots A_p \to \circ$ and $B = B_1 \ldots B_m \to \circ'$. Take $M : B$ and a closed $N = \lambda g\, y_1 \ldots y_p.(g\, \overline{u}) : B \to A$ such that $(NM) =_{\beta\eta} f$. Then the $\beta\eta$-normal form of $(M\, \overline{u}[M/g]) : \circ'$ is equal to $(f\, \overline{y}) : \circ$, hence $\circ = \circ'$.

As a consequence, if we consider many ground types, the linearization and conversion algorithms still work as they are,[1] and all we need to do is to abstract $\circ$ at the right places in the definitions of selectors, coders, decoders. The decidability still holds, due to the fact that a minimal model built with finitely

[1] The assumptions made in *Convert* ensure that $A^1, \ldots, A^n, B$ are of same rightmost ground type $\circ$, so the selector built at step 2 is indeed of type $B_1 \ldots B_m \to (\circ^n \to \circ)$. At step 1, $\bot$ becomes implicitely a constant of adequate ground type. *Linearize* can be leaved unchanged, because it is always called with $f_1, \ldots, f_n$ of same rightmost ground type: a single constant $\bot$ is required to compute $M_\bot$.

many ground types is decidable, and that we do not need any other ground types than the ones appearing in $A^1, \ldots A^n, B$ in order to build in $\mathcal{C} \times \mathcal{D}^n$ a witness for $A^1 \times \ldots \times A^n \lhd B$. Furthermore, we can easily get rid of all constants by using instead fresh variables, distinct from every other variable mentioned in our proofs and definitions, and by modifying accordingly the notion of "closed" term.

In conclusion, let us remark that this proof leaves open the existence of a type system from which all definable retractions could be inferred, whithout appealing to some syntactical criterion - this question probably requires a further understanding of the minimal model itself. Still, it is possible to extract from the proof of lemma 4 an alternate proof of the following theorem, which is an immediate corollary of the Witness Theorem in [LPS92], and which provides an incomplete characterization of definable retractions:

Theorem 3. *Let* $A = A_1 \ldots A_p \to \circ$, $B = B_1 \ldots B_m \to \circ$ *be simple types. If a definable retraction exists between* A *and* B *then, for all* j*, there exists a* k *such that a definable retraction exists between* A_j *and* B_k.

Indeed, in lemma 4, as we reconstruct a witness for $(A^1 \times \ldots \times A^n) \lhd B$ we prove, for each i and for each $j \in \{1, \ldots, p_i\}$, the existence of an $k = \phi_i(j)$ such that $A^i_j \lhd B_k$.

Acknowledgements. Thanks to Laurent Regnier and Pawel Urzyczyn, for submitting the right problem at the right time.Special thanks to Paul-André Mellies, Vincent Danos and François Maurel, for their patience and wisdom.

References

[BL85] Bruce, K., Longo, G. (1985) Provable isomorphims and domain equations in models of typed languages. *A.C.M. Symposium on Theory of Computing (STOC 85).*

[Dez76] Dezani-Ciancaglini, M. (1976) Characterization of normal forms possessing inverse in the $\lambda\beta\eta$-calculus. *TCS* **2**.

[LPS92] de'Liguoro, U., Piperno, A., Statman, R. (1992) Retracts in simply typed $\lambda\beta\eta$-calculus. *Proceeding of the 7th annual IEEE symposium on Logic in Computer Science (LICS92)*, IEEE Computer Society Press.

[Loa97] Loader, R. (1997) An Algorithm for the Minimal Model. *Manuscript.* Available at `http://www.dcs.ed.ac.uk/home/loader/`.

[Pad95] Padovani, V. (1995) Decidability of all minimal models. *Proceedings of the annual meeting Types for Proof and Programs - Torino 1995, Lecture Notes in Computer Science* 1158, Springer-Verlag.

[Sch98] Schmidt-Schauß, M. (1998) Decidability of Behavioral Equivalence in Unary PCF. *Theorical Computer Science* 216.

Parallel Implementation Models for the λ-Calculus Using the Geometry of Interaction (Extended Abstract)

Jorge Sousa Pinto

Departamento de Informática
Universidade do Minho
Campus de Gualtar, 4710-057 Braga, Portugal
jsp@di.uminho.pt

Abstract. An examination of Girard's execution formula suggests implementations of the Geometry of Interaction at the syntactic level. In this paper we limit our scope to ground-type terms and study the *parallel* aspects of such implementations, by introducing a family of abstract machines which can be directly implemented. These machines address all the important implementation issues such as the choice of an inter-thread communication model, and allow to incorporate specific strategies for dividing the computation of the execution path into smaller tasks.

1 Introduction

This paper proposes novel parallel implementation techniques for the λ-calculus based on the geometry of interaction (GoI) [6,5,7]. GoI-based implementation is quite different from other techniques: it uses a graph representation of each term, from which its value is derived by performing *path* computations, which can be done locally and asynchronously. This encompasses both β-reduction and the variable substitution mechanism.

Informally, for every ground-type term there is a path which leaves from the root of the respective graph, and traverses the term, finishing back at root. This path will survive reduction; in particular in the normal form of the term, it will simply go from the root to the constant which is the value of the term. The GoI treats this path algebraically, by assigning a weight to every edge in the initial graph. This allows, on one hand, to identify the unique path which survives reduction, and on the other hand, to calculate algebraically its weight, which is invariant throughout reduction, and equal, in fact, to the value of the term.

The geometry of interaction has been developed as a semantics for linear logic proof-nets [4]. Combined with a standard translation of the λ-calculus into these nets, the results may then be lifted to the scope of functional programs. The nodes in the graph of each term are logical symbols with premises and conclusions, and each orientated edge links a conclusion of a node to a premise of another node. Paths are sequences of (direct $\cdot \longrightarrow \cdot$ or reverse $\cdot \longleftarrow \cdot$) edges. *Straight* paths are those that do not bounce (i.e., no edge is followed by the same

S. Abramsky (Ed.): TLCA 2001, LNCS 2044, pp. 385–399, 2001.

edge in the opposite direction) and do not twist (a path arriving at a premise of a node is not followed by an edge leaving from the other premise).

Persistent paths are those that remain invariant with respect to reduction, and the geometry of interaction is a tool for calculating them: Girard's *execution formula* gives the interpretation of a term as a set of straight paths (called *regular* paths) which are proved [3] to be exactly the persistent paths.

Regular paths are calculated algebraically: edges in the graph are labelled with a *weight*, a term in the GoI *dynamic algebra* $\mathcal{L}^*$. The weight $w(\cdot)$ of a path is defined inductively: $w(\varepsilon) = 1$ for the empty path ε, and $w(t\gamma) = w(\gamma) \cdot w(t)$ where γ is a path and t an edge, and $\cdot$ denotes composition in $\mathcal{L}^*$. For the case of ground-type terms a single persistent path exists which starts and ends at the root of the graph, and the term can be evaluated by calculating its weight.

Implementation via GoI. The first work which proposed to use the GoI as an implementation mechanism [2] defined *virtual reduction* (VR), a local and confluent reduction on graphs, which already suggested the use of parallelism. Virtual reduction allows to add to a graph new edges representing composed paths. Since it preserves the execution of terms, VR provides a way of calculating regular paths. In order to avoid compositions corresponding to bouncing paths, as well as the repeated compositions of pairs of edges, virtual reduction *filters* the weights of the composed edges, for which an extension of the algebraic structure is required. This makes the calculations rather complex.

Directed Virtual Reduction [1], applied with the *combustion* strategy, eliminates this complexity and achieves strong local confluence, but at a cost: the introduction of many bureaucratic reduction steps. To the best of our knowledge, only the directed version has been implemented [11], with the introduction of the *half-combustion* strategy, which allows for a higher degree of parallelism.

The other way in which GoI has been used for implementation was by turning graphs into bideterministic automata, attaching an action to each edge. Actions act on *contexts* (which play the role of words), as given by the context semantics of [8]. The first GoI implementation [10] was in fact obtained in this way, for the **PCF** language: the Geometry of Interaction Machine compiles terms into assembly code of a generic register machine, which runs an automaton.

Our Approach. In this paper we apply the execution formula directly. Our approach resembles VR in that it is syntactic, however it uses $\mathcal{L}^*$ rather than the more complicated structure of VR, thus algebraic manipulation is kept simple. This simplicity is a result of the representation of terms using matrices of weights, following Girard's presentation of the GoI.

From a data-structures point of view, matrices are a convenient representation for graphs. We have studied elsewhere [12] sequential algorithms for calculating execution paths, derived from the execution formula and using the same matrix representation that will be used here.

We achieve concurrent execution by calculating segments of the path (assigned to different threads of computation) starting from different nodes of its graph, and allowing the threads to communicate so that the weight of a finished segment can be used to calculate the weight of another (longer) segment.

Each implementation in this paper will be presented as an abstract machine, an abstract rewriting system working on *machine configurations*. While strongly based on the theory, each machine addresses all the major implementation questions, including the choice of a model for inter-thread communication (shared-memory vs. message-passing) and synchronization issues. The machines are parameterized, allowing to impose different path reduction strategies.

Plan. In Sect. 2 we briefly review basic concepts of the geometry of interaction. Section 3 defines the computational tasks that will be distributed for parallel execution, and in Sect. 4 a shared-memory abstract machine is defined. This is optimized in Sect. 5 to eliminate the need for synchronization mechanisms. In Sect. 6 we study redundant path computations and propose an abstract machine that gets rid of redundancy. Section 7 defines a distributed-memory machine, and in Sect. 8 we conclude with some comments about implementing these ideas.

The long version of this paper contains proofs and examples of execution.

2 Background

Our treatment of the theory here is necessarily superficial; for a more thorough introduction to GoI (including VR and DVR), see [6,5,2,3].

The Language. We will use a typed λ-calculus with a single base type. The syntax of our terms (ranged over by t, u, v) will be (with x, y, z variables, n an integer constant and S the successor function):

$$t ::= \mathsf{n} \mid \mathsf{S} \mid x \mid uv \mid \lambda x.u$$

The typing rules are the standard ones: if with $x : \sigma$ we have $M : \tau$ then $\lambda x.M : \sigma \to \tau$; if $M : \sigma \to \tau$ and $N : \sigma$ then $MN : \tau$. The constants $\mathsf{S} : \mathbf{nat} \to \mathbf{nat}$ and $\mathsf{n} : \mathbf{nat}$ have the expected types. The reduction rules we wish to implement are β-reduction and a δ-rule for the constants:

$$\begin{aligned} (\lambda x.t)u &\longrightarrow t[u/x] \\ \mathsf{Sn} &\longrightarrow \mathsf{n}+1 \end{aligned}$$

All the results in the paper can be extended to include conditionals and recursion; we choose to keep the language as simple as possible for the sake of clarity.

The Geometry of Interaction Dynamic Algebra $\mathcal{L}^*$. We define a single-sorted signature with constants $0, 1, p, q, r, s, t, d$; two unary operators $(\cdot)^*$ and $!(\cdot)$, and an infix (denoted by .) binary composition. The equational theory $\mathcal{L}^*$ is defined over this signature as follows (where variables x, y stand for arbitrary terms):

- The structure is monoidal, with identity 1 and composition as multiplicative operation, and 0 is an absorbing element for composition. Associativity allows to write $u.v$ as uv, and both $u.(v.w)$ and $(u.v).w$ as uvw.

- The *inversion operator* $(\cdot)^*$ is an involutive antimorphism for 0, 1, and composition:

$$0^* = 0 \qquad 1^* = 1$$
$$(x^*)^* = x \qquad (xy)^* = y^*x^*$$

- The *exponential operator* ! is a morphism for 0, 1, inversion, and composition:

$$!(0) = 0 \qquad !(1) = 1$$
$$!(x)^* = !(x^*) \qquad !(x)!(y) = !(xy)$$

- The constants verify the *annihilation* equations:

$$c^*c = 1 \qquad \text{for } c = q, p, r, s, t, d$$
$$q^*p = p^*q = 0$$
$$r^*s = s^*r = 0$$

- The following *commutation equations* are verified:

$$!(x)r = r!(x) \qquad !(x)s = s!(x)$$
$$!(x)t = t!!(x) \qquad !(x)d = dx$$

- To accomodate our language, we extend this theory following [9] with constants n (for each natural number) and S, with equations:

$$\mathsf{n}\,\mathsf{S} = \mathsf{S}\,(\mathsf{n}+1) \qquad \mathsf{S}^*\mathsf{S} = 1$$

Each commutation equation has a dual form as a consequence of $(xy)^* = y^*x^*$, for instance, $d^*!(x) = xd^*$. A binary sum operator may also be included in this theory, which is commutative and associative, has 0 as identity, and composition distributes over it. We call $\mathcal{L}^*_+$ the theory $\mathcal{L}^*$ extended with this operator.

Execution Paths. The standard presentation consists in first defining the weight of paths in proof-nets. *Execution* paths are *regular* (i.e. their weight does not equal 0 in $\mathcal{L}^*$) and have as source and goal conclusions of the net. A translation of λ-terms into nets allows to lift the interpretation to the λ-calculus.

Decidability of Regularity. The term-rewriting system $\mathcal{R}_{\mathcal{L}^*}$ is obtained by orientating from left to right all the equations in $\mathcal{L}^*$ (including the dual commutation equations). $\mathcal{R}_{\mathcal{L}^*}$ is confluent [13]. A *stable form* is 1 or any term $a!(m)b^*$ in $\mathcal{L}^*$ where a and b are positive flat terms, i.e., they contain no applications of $(\cdot)^*$ or $!(\cdot)$, and m is stable. Stable forms are normal with respect to $\mathcal{R}_{\mathcal{L}^*}$. Every stable form is equal to some term AB^* with A and B positive but not necessarily flat.

Proposition 1 (AB^* property [13,3]). *If γ is a straight path in some net then its weight $w(\gamma)$ can be rewritten either to a stable form or to 0.*

Let γ be a path with $w(\gamma)$ rewritable to a stable form ab^*. Since for a positive monomial x, $\mathcal{L}^* \vdash x^*x = 1$, then $\mathcal{L}^* \vdash a^*ab^*b = 1$, and $\mathcal{L}^* \vdash a^*w(\gamma)b = 1$. Thus $\mathcal{L}^* \not\vdash w(\gamma) = 0$ (otherwise $\mathcal{L}^* \vdash 0 = 1$, contradicted by the existence of non-trivial models for $\mathcal{L}^*$). This gives a decidable process for checking regularity.

Matrix Presentation. The interpretation or *execution* of a net is the set of all tuples (s, d, φ) such that there is an execution path with source s, goal d, and weight φ, with s, d (the indexes of) two conclusions of the net. One way to represent this is as a matrix of weights indexed by the conclusions of the net.

Following Girard [6] we associate to a proof a pair of matrices $(\Pi^\bullet, \sigma)$, indexed by the *terminal ports* of the corresponding proof-net. These are either conclusions of symbols which are connected to cut links, or conclusions of the net. An *up-down* path is a path that starts upwards at a terminal port and ends downwards at a terminal port. An *elementary* path is an up-down path which doesn't cross any cut link. The matrix $\Pi^\bullet$ associated to a net contains all the (sums of the weights of) elementary paths in it, and σ contains information relative to the cuts in the net (it corresponds to an involutive permutation on the non-conclusion terminal ports). In the long version of the paper we show how to build these matrices directly from a λ-term.

Definition 1 (Execution Formula). *Let t be a term and $(\Pi^\bullet, \sigma)$ the matrices associated to it. The* execution *of t is defined as follows, where C is called the* central part *of the formula:*

$$\mathcal{E}x(t) = (1 - \sigma^2)C(1 - \sigma^2) \quad \textit{where} \quad C = \Pi^\bullet \sum_{k=0}^{\infty} (\sigma\Pi^\bullet)^k$$

In this paper we only interpret ground-type terms. In these conditions the execution formula expresses an invariant on computation: if $t \longrightarrow t'$, then $\mathcal{E}x(t) = \mathcal{E}x(t')$. The unique conclusion (or *root*) of the corresponding proof-net will by convention have the highest index in the matrices. $\mathcal{E}x(t)$ is a square matrix of dimension N containing 0 everywhere except $\mathcal{E}x(t)_{N,N}$, which is the weight of the execution path of t. The following result from [10] is the last ingredient required for using the execution formula as an evaluation device:

Proposition 2. *If t has ground type and reduces to a constant c, and ϕ is the weight of its execution path, then $\mathcal{L}^* \vdash \phi = c$.*

3 Basic Computation Tasks

As far as the design of an abstract machine is concerned, the first step is to choose a notion of basic task of computation. Then the operation of the machine simply manages the concurrent execution of these tasks by the available threads.

Let φ be a path ending at a terminal port connected to a cut link C. A basic task is the action of composing φ with all the paths consisting of the cut link C followed by (i) an elementary path or (ii) any other up-down path. An element $(\sigma\Pi^\bullet)_{i,j}$ is the weight of a path starting downwards at terminal port i, crossing a cut, and traversing an elementary path. If φ ends at port i, multiplying its weight by the row vector $(\sigma\Pi^\bullet)_i$ captures the first case above. By adding the weights of other up-down paths to a copy of $\sigma\Pi^\bullet$, the second case is also captured.

Auxiliary Functions. Let N be the number of terminal ports in a net $\mathcal{N}$. We consider defined (in the context of a configuration) a matrix of weights B of dimension N (initially containing a copy of $\sigma\Pi^\bullet$), and a predicate $\mathsf{storePred}$ on paths, represented as tuples (s, d, φ), with s and d the source and goal ports, and φ their weight. $\mathsf{storePred}$ identifies paths which will cease to be grown. Instead, their weight will be stored to be reused later. The path starting at root should never stop being grown, thus one imposes $\mathsf{storePred}(N, d, \varphi) = \mathsf{false}$.

The function $\mathcal{I}_{cs}$ takes as argument a path (s, d, φ) and returns a pair (l_1, l_2) of lists of paths. This pair is obtained by composing the weight φ with all the weights in the row indexed by d in matrix B, and including each of the resulting weights (s, m, τ) in l_1 or l_2 according to whether $\mathsf{storePred}(s, m, \tau)$ holds or not.

A related function $\mathcal{I}'_{cs}$ also takes a path (s, d, φ) and returns a pair (l_1, l_2) of lists, now obtained by composing every weight stored in the column indexed by s in matrix X (part of the current configuration) with φ, and splitting the resulting paths using $\mathsf{storePred}$. The use of $\mathcal{I}'_{cs}$ will become clear in section 6.

4 Shared-Everything Abstract Machines

In this section we define a first abstract machine (i.e., a notion of configurations and a reduction relation) corresponding to a shared-memory implementation.

Definition 2. *An* SE-configuration *is a tuple* $\langle B \mid S \mid C \mid [t_1, \ldots t_m]\rangle$ *where*

- B *is a matrix of weights of paths of dimension* N*, representing a net.*
- $S, C \in (\mathbb{N} \times \mathbb{N} \times \mathcal{L}^*)^*$ *are the* storage *and* composition task lists, *respectively.*
- *Each* thread t_k *is a* State, *a term built from the following signature:*

$$\bullet\bullet\bullet\bullet\bullet : \mathbb{N} \times \mathbb{N} \times \mathcal{L}^* \to \text{State} \qquad \bullet\bullet\bullet\ \bullet\bullet\bullet\bullet : \mathbb{N} \times \mathbb{N} \times \mathcal{L}^* \to \text{State}$$
$$\bullet\bullet\bullet\bullet\bullet : \text{State} \qquad \bullet\bullet\bullet\bullet : \mathcal{L}^* \to \text{State}$$
$$\bullet\bullet\bullet\bullet\bullet : (\mathbb{N} \times \mathbb{N} \times \mathcal{L}^*)^* \times (\mathbb{N} \times \mathbb{N} \times \mathcal{L}^*)^* \to \text{State}$$

We will use the following definitions in the context of a configuration: $N_D = \{n \in \mathbb{N} \mid n \leq N\}$ and $N_T = \{n \in \mathbb{N} \mid n \leq m\}$.

The abstract machine rules are given in Table 1, where we omit some unchanged components. Standard notation is used for lists. The auxiliary function $add(B, i, j, \alpha)$ gives the matrix obtained by adding the weight α to $B_{i,j}$. The rules define a reduction relation $\longrightarrow$ on single-threaded configurations.

Definition 3 (SE Reduction). *The* $\stackrel{se}{\longrightarrow}$ *reduction relation on multi-threaded configurations is the smallest relation verifying:*

$$\frac{\langle B \mid S \mid C \mid [t_i]\ \rangle \longrightarrow \langle \widehat{B} \mid \widehat{S} \mid \widehat{C} \mid [\widehat{t_i}]\ \rangle}{\langle B \mid S \mid C \mid [t_1, \ldots t_i, \ldots t_m]\ \rangle \stackrel{se}{\longrightarrow} \langle \widehat{B} \mid \widehat{S} \mid \widehat{C} \mid [t_1, \ldots \widehat{t_i}, \ldots t_m]\ \rangle}$$

Definition 4. *A tuple* (s, d, ϕ)*, where* $\phi \in \mathcal{L}^*_+$*, is said to* belong to the execution *of a term iff* ϕ *can be written as a sum* $\phi = \sum_n \phi_n$*, such that for each* ϕ_n *one has* $\phi_n + \alpha_n = (\Pi^\bullet(\sigma\Pi^\bullet)^{i_n})_{s,d}$ *for some* i_n *and some term* $\alpha_n \in \mathcal{L}^*_+$*.*

Table 1. Shared-everything (SE) abstract machine

0 $s = d = N$	Net Thread	B ••• ••••(s, d, φ)	B ••••(φ)
I	Net Thread	B •••••(s, d, φ)	$add(B, \sigma(s), d, \varphi)$ •••••
II $s \neq N$ or $d \neq N$	Net Thread	B ••• ••••(s, d, φ)	B •• •••$(\mathcal{I}_{cs}(s, d, \varphi))$
III	STasks Thread	S •• •••$((s, d, \varphi) : T_s, T_c)$	$(s, d, \varphi) : S$ •• •••(T_s, T_c)
IV	CTasks Thread	C •• •••$(\varepsilon, (s, d, \varphi) : T_c)$	$(s, d, \varphi) : C$ •• •••(ε, T_c)
V	Thread	•• •••$(\varepsilon, \varepsilon)$	•• •••
VI	STasks Thread	$(s, d, \varphi) : S$ •• •••	S •••••(s, d, φ)
VII	CTasks STasks Thread	$(s, d, \varphi) : C$ ε •• •••	C ε ••• ••••(s, d, φ)

The following propositions establish sufficient conditions for the correctness of the machine and for the execution path to be computed.

Proposition 3. *Let t be a term represented as $(\Pi^\bullet, \sigma)$, and $\Sigma_0 = \langle \sigma\Pi^\bullet \mid \varepsilon \mid C_0 \mid [\mathsf{delist} \ldots \mathsf{delist}] \rangle$ a configuration where C_0 contains only paths $(i, j, \Pi^\bullet_{i,j})$ taken from $\Pi^\bullet$, excluding repetitions. Let $\Sigma_0 \xrightarrow{se}_* \Sigma = \langle B \mid S \mid C \mid [t_1 \ldots t_m] \rangle$.*

1. *If t_k is some thread in Σ and $t_k = \mathsf{compose}(s, d, \varphi)$ or $t_k = \mathsf{store}(s, d, \varphi)$, then (s, d, φ) belongs to the execution of t.*
2. *If $t_k = \mathsf{enlist}(T_s, T_c)$ is some thread in Σ and $(s, d, \varphi) \in T_s$ or $(s, d, \varphi) \in T_c$, then (s, d, φ) belongs to the execution of t.*
3. *If $(s, d, \varphi) \in S$ or $(s, d, \varphi) \in C$, then (s, d, φ) belongs to the execution of t.*
4. *For all s, d, the tuple $(\sigma(s), d, B_{s,d})$ belongs to the execution of t.*

Proposition 4. *Consider an initial configuration in the conditions of Prop. 3, where additionally C_0 contains the paths $(N, j, \Pi^\bullet_{N,j})$ such that $\Pi^\bullet_{N,j} \neq 0$. Then the machine stops with a final configuration $\overline{\Sigma}$ containing a thread in the state $\mathsf{stop}(\varphi)$, where φ is the weight of the unique execution path of the term.*

Deterministic Execution. For initial configurations Σ_0 in the conditions of Prop. 4 with C_0 containing *only* the weight of the path in row N of $\Pi^\bullet$, execution of the abstract machine is deterministic and equivalent to the Geometry of Interaction Machine [10]. There is always a single path inside the machine, that results from growing the unique elementary path with source the root of the term, and given the ground-type of the term, the result of each step (and each invocation of $\mathcal{I}_{cs}$) must be a unique path, for which the storePred predicate always returns false, thus a new compose task will be generated with the new path as argument.

Concurrent Execution. The present machine is a formalization of a variant of the producer-consumers model for (shared-memory) parallel programming, where the consumer threads are also producers, running the following cycle: first dequeue a task from the shared queue, then process it (possibly enqueueing new tasks); restart. In the abstract machine there are two different task lists, with different priorities (store tasks have higher priority than compose tasks).

The concurrent behaviour of $\xrightarrow{se}_*$ comes from sequentiality with non-determinism. A single thread executes a machine rule at each step of the $\xrightarrow{se}$ reduction, allowing to capture synchronization when accessing shared data-structures. For instance, if two threads may execute at the same time rule I, there will be a (2 step) $\xrightarrow{se}_*$ reduction with the correct result (B is changed by the two threads).

If parallel computation is desired, one must add more paths to C_0 than those in row N of $\Pi^\bullet$. These paths will be concurrently grown into longer regular paths. When a path φ computed by thread t_1 reaches a port from which another path φ' has been grown by t_2, φ' will be used for extending φ. At this point, t_1 and t_2 communicate using matrix B: if s and d are the source and goal ports of φ', then the weight φ' will be added to the current weight in $B_{\sigma(s),d}$ and composition of φ with φ' will happen naturally since $\sigma(s)$ is the goal port of φ.

One could devise a strategy for virtual reduction mimicking the abstract machine – given a set of nodes, grow all the paths leaving from those nodes. In VR each composition results in a new edge immediately incorporated in the net, whereas the machine will only perform store operations (which will add paths to the net represented by matrix B) with selected paths (which then cease to be grown). To illustrate this point, consider the net in Fig. 1, where a path is to be grown starting from the source of the edge φ_0. Virtual reduction produces the net on the left, where φ_0 has been composed with γ_1 to give φ_1, which has then been composed with γ_2 and so on. The abstract machine grows the path φ_0 until storePred is verified, and only then does it store the result path φ.

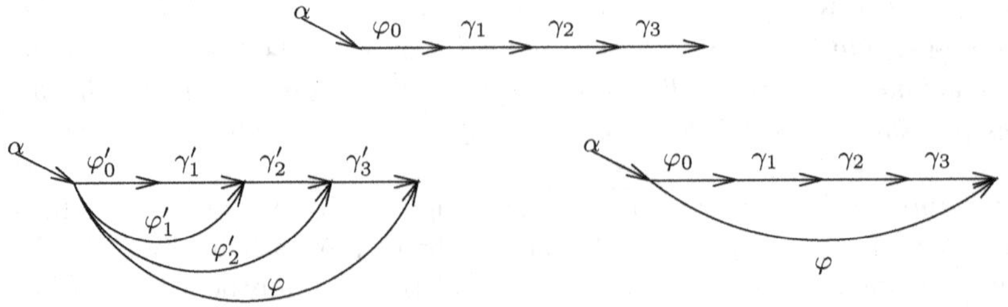

Fig. 1. Example path reductions

The machine is parameterized on which paths to start growing (included in C_0), and on the predicate storePred. One possible criterion is given in Sect. 6.

Implementation. The abstract machine may be implemented in any shared-memory architecture; threads in a configuration will be mapped into machine threads independently running the machine rules; threads will run in true parallelism, with synchronization introduced for accessing the shared data-structures. This ensures that the behaviour of the implementation corresponds to the interleaving reduction of the abstract machine. Any shared-memory library contains the appropriate synchronization devices, which we will here call *locks*.

Synchronization is required for accessing the shared task lists S and C, which may be read and written by any thread. This is done by associating a lock to each list: the C-lock must be acquired before and released after execution of rules IV and VII, and the same happens for the S-lock with rules III and VI.

Since the elements in a matrix are stored in independent memory positions, they can be protected individually, rather than treating B as a monolithic structure with a single lock. Synchronization is needed when two threads execute simultaneously rule I with the same s, d arguments: the individual lock associated to $B_{\sigma(s),d}$ must be acquired by a thread $\mathsf{store}(s, d, \varphi)$ executing rule I.

5 Distributed-Task-Lists Abstract Machines

Much synchronization is required for the parallel implementation of the SE-machine. We will eliminate this by including in threads private task lists.

Definition 5. *A* DTL-configuration *is a tuple* $\langle B \mid [t_1, \dots t_m] \rangle$ *where*

- B *is a matrix of weights of paths of dimension* N*, representing a net.*
- *Each* t_k *is a* thread, $t_k = \langle S_k \mid C_k \mid st_k \rangle$*, where* $S_k, C_k \in (\mathbb{N} \times \mathbb{N} \times \mathcal{L}^*)^*$ *are the* storage *and* composition task lists *of the thread, respectively, and* st_k *its* state, *built from the same signature as before.*

Table 2 defines a reduction relation $\longrightarrow$ on single-threaded configurations (rule 0 will henceforth be considered implicit). We then define the following:

Definition 6 (DTL reduction). $\xrightarrow{dtl}$ *is the smallest relation verifying*

$$\frac{\langle B \mid [t_i] \rangle \longrightarrow \langle \widehat{B} \mid [\widehat{t_i}] \rangle}{\langle B \mid [t_1, \dots t_i, \dots t_m] \rangle \xrightarrow{dtl} \langle \widehat{B} \mid [t_1, \dots \widehat{t_i}, \dots t_m] \rangle}$$

Properties. Proposition 3 holds slightly modified, with initial configurations $\Sigma_0 = \langle \sigma\Pi^\bullet \mid [t_1^0, \dots t_m^0] \rangle$ where each thread $t_k^0 = \langle \varepsilon \mid C_k^0 \mid \mathsf{delist} \rangle$, each C_k^0 contains only paths $(s, d, \Pi^\bullet_{s,d})$ from $\Pi^\bullet$, and the same path does not occur repeatedly in any two or in the same C_k^0. Also the third condition in the proposition is changed to "if t_k is some thread and $(s, d, \varphi) \in S_k$ or $(s, d, \varphi) \in C_k$, then (s, d, φ) belongs to the execution of t". Proposition 4 holds as well with small modifications: in the initial configurations, the paths $(N, j, \Pi^\bullet_{N,j})$ must be contained in some C_0^k. Finally another interesting property holds:

Table 2. Distributed-task-lists (DTL) abstract machine

I		Net	B	$add(B, \sigma(s), d, \varphi)$
	t_k	State	••••••(s, d, φ)	••••••
II		Net	B	B
$s \neq N$ or $d \neq N$	t_k	State	••• ••••(s, d, φ)	••••••$(\mathcal{I}_{cs}(s, d, \varphi))$
III	t_k	STasks	S	$(s, d, \varphi) : S$
		State	••••••$((s, d, \varphi) : T_s, T_c)$	••••••(T_s, T_c)
IV	t_k	CTasks	C	$(s, d, \varphi) : C$
		State	••••••$(\varepsilon, (s, d, \varphi) : T_c)$	••••••(ε, T_c)
V	t_k	State	••••••$(\varepsilon, \varepsilon)$	••••••
VI	t_k	STasks	$(s, d, \varphi) : S$	S
		State	••••••	••••••(s, d, φ)
VII		CTasks	$(s, d, \varphi) : C$	C
	t_k	STasks	ε	ε
		State	••••••	••• ••••(s, d, φ)

Proposition 5. *Let $i : N_D \to N_T$ be any map. Consider a configuration $\Sigma_0 = \langle B^0 \mid [t_1^0, \ldots t_m^0] \rangle$ with $t_k^0 = \langle \varepsilon \mid C_k^0 \mid \mathsf{delist} \rangle$, and C_k^0 containing only paths $(s, d, \Pi^{\bullet}_{s,d})$ such that $i(s) = k$. If $\Sigma_0 \overset{dtl}{\longrightarrow} \Sigma$ and t_k is any thread in Σ with state $\mathsf{compose}(s, d, \varphi)$ or $\mathsf{store}(s, d, \varphi)$, then $i(s) = k$.*

Remarks. Suppose i is the identity function (and there are enough threads in the configuration). Then this proposition means that (with an appropriate initial configuration) thread t_k will handle all paths with source k, and only those. An immediate consequence is that when executing rule I, each thread writes to positions located in a unique row (indexed by $\sigma(k)$) in matrix B.

If not enough threads are available for all terminal ports, the function i will map terminal ports to threads. In this case each thread t_k will process paths with source ports from a distinct set, and will thus write to positions in different rows of B. t_k will however read from positions in any row of B.

Implementation. Each thread reads from and writes to its own task list only (so no synchronization is needed for accessing those lists). As to matrix B, no protection is needed, because of the previous remark. Thus this abstract machine can be implemented as a wait-free shared-memory program (no locks used).

6 Eliminating Redundancy

Consider again Fig. 1, and a path α ending where φ_0 starts. Two paths (φ_0 and φ) are available for composition with α. If α is composed with φ_0, the resulting path may continue being extended by composing with γ_1 and so on. These are redundant computations, since φ has already been computed. In terms of the abstract machine, after the path $\varphi : s \to d$ has its weight stored in $B_{\sigma(s),d}$, the path α with goal the port $\sigma(s)$ may be composed not only with φ, but also with the elementary φ_0 from which φ was grown, whose weight still stands in the row indexed by $\sigma(s)$ in B. The current machine either follows *both* paths, performing many redundant computations, or, if some thread extends α before φ has been stored in B, it does not follow φ at all, which sequentializes execution.

To eliminate these redundancies, it is sufficient to remove from the net the path φ_0. In the abstract machine it can be removed from matrix B:

Definition 7. *A redundancy-free initial DTLW-configuration is any* $\Sigma_0 = \langle B^0 \mid [0]_N \mid [t_1^0, \ldots t_m^0] \rangle$, *all* $t_k^0 = \langle \varepsilon \mid C_k^0 \mid \mathsf{delist} \rangle$, *and*

$$B^0_{\sigma(s),d} = \begin{cases} 0 & \textit{if } \Pi^\bullet_{s,d} \in C^0_k \textit{ for some } k, \\ (\sigma\Pi^\bullet)_{\sigma(s),d} & \textit{otherwise} \end{cases}$$

There is a problem with this if φ has not yet been computed: whereas before the path α could continue being extended by composing with φ_0, now it will die, preventing computation of the execution path. This is solved by keeping account of all the paths candidates for composition with paths in B, and performing the corresponding compositions when new weights are stored in B. These "waiting paths" (such as α in the example) will be kept in a matrix X in configurations (DTL-configurations with this X component are called DTLW), and the function $\mathcal{I}'_{cs}$ will be used to perform the necessary compositions.

Proposition 5 cannot hold, since thread t_k will handle paths with arbitrary sources, generated by composing elements of X (of arbitrary source) with a path (s, d, φ) to be stored. Thus this machine cannot be implemented without synchronization. A change of perspective will allow to recover Prop. 5, at the expense of allowing threads to write to each other's task lists.

Table 3 contains the rules for the new abstract machine. Rules III and IV involve two threads. We will consider that the list of threads may be accessed as an *array*, a partial map from indexes to threads, $L : NT \hookrightarrow \text{State}$. When $i \notin dom(L)$, $L[i \mapsto t_i]$ denotes the union of L with the singleton $\{(i, t_i)\}$. We will still use list notation if convenient, and t_i will abbreviate $L(i)$.

$$\frac{\langle B \mid X \mid [t_i] \rangle \longrightarrow \langle \widehat{B} \mid \widehat{X} \mid [\widehat{t_i}] \rangle \qquad i \notin dom(L)}{\langle B \mid X \mid L[i \mapsto t_i] \rangle \stackrel{dtlmr}{\longrightarrow} \langle \widehat{B} \mid \widehat{X} \mid L[i \mapsto \widehat{t_i}] \rangle}$$

$$\frac{\langle B \mid X \mid [t_a, t_b] \rangle \longrightarrow_{III,IV} \langle \widehat{B} \mid \widehat{X} \mid [\widehat{t_a}, \widehat{t_b}] \rangle \qquad a, b \notin dom(L)}{\langle B \mid X \mid L[a \mapsto t_a, b \mapsto t_b] \rangle \stackrel{dtlmr}{\longrightarrow} \langle \widehat{B} \mid \widehat{X} \mid L[a \mapsto \widehat{t_a}, b \mapsto \widehat{t_b}] \rangle}$$

$\longrightarrow_{III,IV}$ denotes reduction using one of rules III or IV. A possible particular case for these two rules is that $i(s) = k$. For this reason the definition of $\stackrel{dtlmr}{\longrightarrow}$ includes the possibility of a single-thread reduction using rule III or IV.

Properties. Propositions 3 and 5 hold, with Σ_0 in the conditions of Def. 7 and $\stackrel{dtlmr}{\longrightarrow}$ replacing $\stackrel{dtl}{\longrightarrow}$. A consequence of Prop. 5 is that

Table 3. DTLW abstract machine with mutual writing

I		Net	B	$add(B, \sigma(s), d, \varphi)$
		WPaths	X	X
	t_k	State	••••• (s, d, φ)	•• ••• $(\mathcal{I}'_{cs}(\sigma(s), d, \varphi))$
II		Net	B	B
$s \neq N$ or $d \neq N$		WPaths	X	$add(X, s, d, \varphi)$
	t_k	State	••• •••• (s, d, φ)	•• ••• $(\mathcal{I}_{cs}(s, d, \varphi))$
III	$t_{i(s)}$	STasks	S	$(s, d, \varphi) : S$
	t_k	State	•• ••• $((s, d, \varphi) : T_s, T_c)$	•• ••• (T_s, T_c)
IV	$t_{i(s)}$	CTasks	C	$(s, d, \varphi) : C$
	t_k	State	•• ••• $(\varepsilon, (s, d, \varphi) : T_c)$	•• ••• (ε, T_c)
V	t_k	State	•• ••• $(\varepsilon, \varepsilon)$	•• •••
VI	t_k	STasks	$(s, d, \varphi) : S$	S
		State	•• •••	••••• (s, d, φ)
VII		CTasks	$(s, d, \varphi) : C$	C
	t_k	STasks	ε	ε
		State	•• •••	••• •••• (s, d, φ)

- the element indexed by (s, d) in matrix B is only written by thread $t_{i(\sigma(s))}$ but can be read by any thread;
- the element indexed by (s, d) in matrix X is only written by thread $t_{i(s)}$ and only read by thread $t_{i(\sigma(d))}$ (when applying function $\mathcal{I}'_{cs}$).

Proposition 4 no longer holds for free: a judicious choice of initial configurations and definition of storePred are now necessary, guaranteeing that storePred is verified at some point for all the paths calculated concurrently. Notably, this means these paths should not overlap. Let α and β be two subpaths of the execution path such that β starts inside α. If when the port where β starts is reached, β has already been stored, then the part of α that has been computed will be composed with β, and the storePred predicate will never be applied to α.

We propose as an example the following criterion: consider a set P of terminal ports, and include in C_0 all the paths $\Pi^{\bullet}_{i,j}$ such that $i \in P$, and let

$$\mathsf{storePred}(s, d, \varphi) = (\sigma(d) \in P \textbf{ or } d = N) \textbf{ and } s \neq N$$

where the condition $d = N$ is necessary to store the last subpath. This guarantees that paths do not overlap since each path ends where another one starts.

Implementation. The access to matrices B and X is naturally protected – no two threads can write to the same position in B or X. Locks are required, for the individual lists of all threads, to be used as follows:

- for executing rule III, thread t_k must own the S-lock of thread $t_{i(s)}$;
- for executing rule IV, thread t_k must own the C-lock of thread $t_{i(s)}$;
- for executing rule VI, thread t_k must own its own S-lock;
- for executing rule VII, thread t_k must own its own C-lock.

Table 4. Distributed-everything abstract machine

I		Net	B	$add(B, \sigma(s), d, \varphi)$
	t_k	WPaths	X	X
		State	•••••(s, d, φ)	•• •••$(\mathcal{I}'_{cs}(\sigma(s), d, \varphi))$
II		Net	B	B
$s \neq N$ or $d \neq N$	t_k	WPaths	X	$add(X, s, d, \varphi)$
		State	••• ••••(s, d, φ)	•• •••$(\mathcal{I}_{cs}(s, d, \varphi))$
III	$t_{i(\sigma(s))}$	STasks	S	$(s, d, \varphi) : S$
	t_k	State	•• •••$((s, d, \varphi) : T_s, T_c)$	•• •••(T_s, T_c)
IV	$t_{i(d)}$	CTasks	C	$(s, d, \varphi) : C$
	t_k	State	•• •••$(\varepsilon, (s, d, \varphi) : T_c)$	•• •••(ε, T_c)
V	t_k	State	•• •••$(\varepsilon, \varepsilon)$	•• •••
VI	t_k	STasks	$(s, d, \varphi) : S$	S
		State	•• •••	•••••(s, d, φ)
VII		CTasks	$(s, d, \varphi) : C$	C
	t_k	STasks	ε	ε
		State	•• •••	••• ••••(s, d, φ)

7 Distributed-Everything Abstract Machine

In our final machine, threads keep individual copies of the B and X matrices.

Definition 8. *A* DE-configuration *is a list* $[t_1, \ldots t_m]$ *where each* $t_k = \langle B_k \mid X_k \mid S_k \mid C_k \mid st_k \rangle$ *is a* thread, *with* B_k *and* X_k *matrices of weights of dimension* N*;* S_k *and* C_k *the* storage *and* composition task lists *of* t_k*, and* st_k *its* state.

Table 4 defines a reduction $\longrightarrow$ on one- and two-thread configurations. Then:

Definition 9 (DE reduction). $\stackrel{de}{\longrightarrow}$ *is the smallest relation verifying:*

$$\frac{[t_i] \longrightarrow [\widehat{t_i}] \qquad i \notin dom(L)}{L[i \mapsto t_i] \stackrel{de}{\longrightarrow} L[i \mapsto \widehat{t_i}]} \qquad \frac{[t_a, t_b] \longrightarrow_{III,IV} [\widehat{t_a}, \widehat{t_b}] \qquad a, b \notin dom(L)}{L[a \mapsto t_a, b \mapsto t_b] \stackrel{de}{\longrightarrow} L[a \mapsto \widehat{t_a}, b \mapsto \widehat{t_b}]}$$

Each thread now writes composition tasks to the task list of the thread corresponding to the *goal* port of the respective path.

Proposition 6. *Let* $i : N_D \rightarrow N_T$ *be a map, and consider a configuration* $\Sigma_0 = [t_1^0, \ldots t_m^0]$*, each* $t_k^0 = \langle B_k^0 \mid [0]_N \mid \varepsilon \mid C_k^0 \mid \mathsf{delist} \rangle$*, and* C_k^0 *containing only paths* $(s, d, \Pi^\bullet_{s,d})$ *with* $i(d) = k$*. If* $\Sigma_0 \stackrel{de}{\longrightarrow} \Sigma$ *and* t_k *is any thread in* Σ *with state* $\mathsf{compose}(s, d, \varphi)$*, then* $i(d) = k$*; if* t_k *has state* $\mathsf{store}(s, d, \varphi)$*, then* $i(\sigma(s)) = k$*.*

Corollary 1. *With* Σ_0 *in the conditions of Prop. 6,* $B_{s,d}$ *is only read and written by thread* $t_{i(s)}$*, and* $X_{s,d}$ *is only read and written by thread* $t_{i(d)}$*.*

Each thread only needs to read from exactly the same positions of B and X that it writes to, thus the local copies of B and X do not need to be kept consistent.

Implementation. In an implementation of this machine, synchronization is only needed for accessing the task lists of individual threads, used for communication.

In practice it is not necessary that threads keep copies of the entire matrices: rows of B and columns of X can be distributed so that thread t_k keeps only the rows of B and columns of X indexed by d such that $i(d) = k$.

A Message-passing Machine. We now propose a change of perspective: consider that the task lists are *communication buffers*, where messages sent to a thread are kept before they are received by the thread. Then the enlist operation is a synchronous buffered *send* operation, which puts a task into the destination thread's buffer. delist is a *receive* operation, by which a thread removes a message from one of its buffers. Two types of messages (compose and store) may be sent to a thread, which will be stored in different buffers.

The message-passing mechanisms provided by any parallel-programming library ensure that messages are naturally ordered on arrival and placed sequentially in the corresponding buffer (thus replacing synchronization).

8 Conclusions and Further Work

The fact that the abstract machines allow to identify the necessary synchronization mechanisms is of great importance: an important product of this is the wait-free abstract machine of Sect. 5. Wait-free implementations are typically difficult to obtain (to understand the need for synchronization in VR the reader should think of a situation like $\cdot \longrightarrow \cdot \longrightarrow \cdot \longrightarrow \cdot$ where a critical pair exists).

The abstract machines are parameterized on the initial paths to be extended, as well as on the criterion to stop extending paths. This allows to implement different strategies for path computations (unlike virtual reduction, which, being a local reduction relation, has no built-in strategy). An instance of the abstract machines given here always incorporates a precise strategy, and this allows notably to eliminate synchronization as well as useless computations.

The parameterization we have given in Sect. 6 guarantees the correctness of the machines in sections 6 and 7, but does not allow for a subpath ϕ of the execution path to be used to extend another subpath ϕ': only the execution path can be extended using already computed subpaths. This has the advantage of simplicity, but it remains to study other efficient criteria.

The appropriate technologies exist for implementing the given machines in widely available architectures, both for shared-memory (for instance POSIX threads on SMP architectures) and distributed-memory (message-passing libraries such as MPI or PVM). It is worth mentioning that we have implemented the DE-machine using MPI, and started testing it over a local-area network. With respect to shared-memory implementations, it will be important to compare the wait-free (Sect. 5) and the redundancy-free (Sect. 6) machines.

References

1. Vincent Danos, Marco Pedicini, and Laurent Regnier. Directed virtual reductions. In M. Bezem and D. van Dalen, editors, *Computer Science Logic, 10th International Workshop, CSL '96*, volume 1258 of *Lecture Notes in Computer Science*. Springer Verlag, 1997.
2. Vincent Danos and Laurent Regnier. Local and asynchronous beta-reduction (an analysis of Girard's execution formula). In *Proceedings of the 8th Annual IEEE Symposium on Logic in Computer Science (LICS'93)*, pages 296–306. IEEE Computer Society Press, 1993.
3. Vincent Danos and Laurent Regnier. Proof-nets and the Hilbert space. In Jean-Yves Girard, Yves Lafont, and Laurent Regnier, editors, *Advances in Linear Logic*, number 222 in London Mathematical Society Lecture Note Series, pages 307–328. 1995.
4. Jean-Yves Girard. Linear Logic. *Theoretical Computer Science*, 50(1):1–102, 1987.
5. Jean-Yves Girard. Geometry of interaction 2: Deadlock-free algorithms. In Per Martin-Löf and G. Mints, editors, *International Conference on Computer Logic, COLOG 88*, pages 76–93. Springer-Verlag, 1988. Lecture Notes in Computer Science 417.
6. Jean-Yves Girard. Geometry of interaction 1: Interpretation of System F. In R. Ferro, C. Bonotto, S. Valentini, and A. Zanardo, editors, *Logic Colloquium 88*, volume 127 of *Studies in Logic and the Foundations of Mathematics*, pages 221–260. North Holland Publishing Company, Amsterdam, 1989.
7. Jean-Yves Girard. Geometry of interaction III : accommodating the additives. In J.-Y. Girard, Y. Lafont, and L. Regnier, editors, *Advances in Linear Logic*, number 222 in London Mathematical Society Lecture Note Series, pages 329–389. 1995.
8. Georges Gonthier, Martín Abadi, and Jean-Jacques Lévy. Linear logic without boxes. In *Proceedings of the 7th IEEE Symposium on Logic in Computer Science (LICS'92)*, pages 223–234. IEEE Press, 1992.
9. Ian Mackie. *The Geometry of Implementation.* PhD thesis, Department of Computing, Imperial College of Science, Technology and Medicine, September 1994.
10. Ian Mackie. The geometry of interaction machine. In *Proceedings of the 22nd ACM Symposium on Principles of Programming Languages (POPL'95)*, pages 198–208. ACM Press, January 1995.
11. M. Pedicini and F. Quaglia. A parallel implementation for optimal lambda-calculus reduction. In *Proceedings of the 2nd International Conference on Principles and Practice of Declarative Programming (PPDP 2000)*. ACM press, 2000.
12. Jorge Sousa Pinto. *Parallel Implementation with Linear Logic (Applications of Interaction Nets and of the Geometry of Interaction).* PhD thesis, École Polytechnique, 2001.
13. Laurent Regnier. *Lambda-Calcul et Réseaux.* PhD thesis, Université Paris VII, January 1992.

The complexity of β-reduction in low orders

Aleksy Schubert*

Institute of Informatics
Warsaw University
alx@mimuw.edu.pl

Abstract. This paper presents the complexity of β-reduction for redexes of order $1, 2$ and 3. It concludes with the following results — evaluation of Boolean expressions can be reduced to β-reduction of order 1 and β-reduction of order 1 is in $O(n \log n)$, β-reduction of order 2 is complete for PTIME, and β-reduction of order 3 is complete for PSPACE.

1 Introduction

The mechanism of evaluation in functional languages is based on β-reduction. Thus, it is interesting to study the complexity of the decision problem to answer if a given value (a lambda term) is a result of some program (another lambda term). As most functional programs do not use functions of a very high order, we restrict the research to low orders. This paper concerns reductions in 1st, 2nd and 3rd orders of simply typed lambda calculus.

Another good reason to study these problems is application of the results and techniques to the study of the problem of higher-order matching. Known higher-order matching algorithms are usually based on check if a term obtained by some calculation actually reduces to particular normal form. This exactly corresponds to the situation in our problems. Additionally, the obtained proofs shed more light on the nature of β-reduction which is essential for the final solution of the higher-order matching problem.

Related research There is a similar problem of β-equivalence. It was studied in [Sta79] and non-elementary bound on the complexity of the problem was found. The problem was also discussed in [Mai92] where an alternative proof of the result was described. Another similar problem of finding the length of a β-reduction sequence for a term was studied in [Sch91]. The first attempt to analyse the complexity of β-reduction was presented in [HK96] where a whole hierarchy of orders and complexities was discussed but for a slightly different problem in which restricted syntax is considered and some δ-rules are allowed.

The content of the paper A reduction of evaluation of Boolean expressions to 1st-order β-reduction is proved in Section 3 together with a $O(n \log n)$ algorithm for the reduction, PTIME-completeness for 2nd-order β-reduction is proved in Section 4 and PSPACE-completeness for 3rd-order β-reduction is proved in Section 5.

* This work was partly supported by KBN grant no 8 T11C 035 14.

S. Abramsky (Ed.): TLCA 2001, LNCS 2044, pp. 400–414, 2001.

2 Basic notions

We deal with the simply typed λ-calculus denoted by $\lambda_{\rightarrow}$ as in [Bar92]. The results obtained match both Curry and Church-style version of the calculus. We study the β-reduction relation here. One step reduction is denoted by $\rightarrow_\beta$. The transitive-reflexive closure of the relation is denoted by $\rightarrow_\beta^*$. The β-normal form of a term M is denoted by $\mathrm{NF}(M)$. The relation of α-equivalence is denoted by $\equiv_\alpha$. We also use the notion of a context which is usually denoted by $C[\cdot]$ and it is a term with a single hole that may be filled in by a term of a suitable type. The operation of 'filling in' does not perform any variable renaming. The context in which its hole is filled in with the term M is denoted by $C[M]$.

The notion of order is defined as: $\mathrm{ord}(\alpha) = 0$ for α atomic and $\mathrm{ord}(\sigma_1 \rightarrow \sigma_2) = \max(\mathrm{ord}(\sigma_1) + 1, \mathrm{ord}(\sigma_2))$.

In the Church-style calculus, the order of the redex $(\lambda x.M)N$ in the term $P = C[(\lambda x.M)N]$ is the order of the type of $\lambda x.M$ assigned in the derivation of the type of P. In the Curry-style calculus, the order of such a redex is the minimum of orders assigned to types of $\lambda x.M$ in type derivations for P.

As far as the Curry-style definition is concerned then there occurs a question whether there is a uniform derivation of a type for P in which all redexes have minimal orders. The answer is 'there is'. The derivation for principal type of P has this property.

The general formulation of the problem we deal with will follow

Problem 1. Input: A $\lambda_{\rightarrow}$ term M_1 with redexes of order at most n and a normal form $\lambda_{\rightarrow}$ term M_2. *Question:* Does M_1 β-reduce to M_2?

We consider the problem for $n = 2, 3, 4$. Note, that we assume that the input is already a term in $\lambda_{\rightarrow}$ and has redexes of suitable order. We do not make any checks that the input values are correct in presented algorithms. These checks require at least essentially polynomial time algorithm which majorises bounds on the resources needed in some constructions presented in the paper. In fact, all presented reductions and algorithms may be performed for both Curry and Church terms.

3 The order 1

3.1 First-order β-reduction is in $O(n \log n)$

The first-order reduction can be performed in $O(n \log n)$ time. Our algorithm uses the notion of graph reduction. We assume here that the reader is familiar with this notion. The recommended readings about graph reduction include: [Lam90,AL93] and [AG98]. Due to limited room, we do not present definitions pertinent to optimal reductions algorithms. We use the presentation included in the latter paper. For the sake of clarity we use the version of graph reduction in which fan-nodes have more than 2 auxiliary ports. This approach can easily be translated into the one with 2-port fan-nodes without affecting the complexity.

Definition 1 (`algorithm for 1st order` β`-reduction`)
Let M_1 and M_2 be the input for the algorithm (we assume here w.l.o.g. that these terms are closed). The algorithm `reduce_1st` is described as follows. We need an additional stack S and a counter i. Some nodes of the graph will be marked during the reduction. We proceed as follows:

1. Translate M_1 into its graph of reduction, initiate S to the empty stack.
2. Walk through the starting λ-nodes without any change.
3. Initiate i to 0.
4. Go through @-nodes incrementing i at each one and taking their left branch until you meet a fan-node, an auxiliary port of a λ-node, a marked node or a principal port of a λ-node.
   (a) if it is a fan-node, an auxiliary port of a λ-node or a marked node then check if S is empty if it is go to the point (5) if it is not, pop the value of i from S, then pop a node A from the stack, and perform the β-redex above the node A marking the topmost node of the argument of the redex; finally, go to the point 4;
   (b) if it is a principal port of a λ-node and $i > 0$ then decrement i, push the λ-node on S, push i, step to the right branch of the last @-node and begin the whole procedure from the point (3);
   (c) if it is a principal port of a λ-node and $i = 0$ then go through the λ-node without any change and step to the point 4.
5. Perform the read-back of the graph; the resulting term is M_3.
6. Check the α-equivalence of M_3 and M_2.

Theorem 1. *Let M_1 have redexes of order at most 1 and M_2 be in normal form. The algorithm* `reduce_1st` *results in success on these terms iff $M_1 \to_\beta^* M_2$.*
Moreover, `reduce_1st` *needs only $O(n \log n)$ time to run.*

Proof. The algorithm is correct as it is only a strategy in an optimal reduction algorithm.

Let us analyse the complexity of the algorithm. Let n be the size of the input for `reduce_1st`.

The translation of the term to the graph can be performed in $O(n)$ time using usual syntax analysis methods. The rest of the algorithm visits each node at most 2 times and the number of steps performed for each node is bounded by a constant except for the time needed to store i and a node on the stack. The last operation takes $O(\log n)$ time because of the length of the counter and the pointer to the node. This altogether gives $O(n \log n)$ time.

3.2 Boolean expressions reduce to first-order β-reduction

Boolean expression is an expression that is built of the connectives $\wedge, \vee$ and values `true` and `false`. An example is (`true` $\wedge$ `false`) $\vee$ `true`. We can associate with each expression of this kind its value which is generated according to the truth tables of logical connectives $\wedge$ and $\vee$. The problem of evaluation of Boolean expressions is:

Definition 2 (evaluation of Boolean expressions)
Input: A Boolean expression E.
Question: Is `true` the value of E?

The problem is in ALOGTIME (see [Bus87]). In order to relate the 1st order situation to 2nd and 3rd orders we present a first-order (i.e. in first-order logic) reduction of the problem to the first-order β-reduction problem. This presentation is only for the sake of completeness with the rest of the paper where some variations of Boolean formulas are dealt with. In fact, we only relate β-reduction to the evaluation of expressions which itself has no proof of ALOGTIME-hardness. A helpful definition of logical values is

Definition 3 (Boolean values)
We define terms corresponding to Boolean values as TRUE $= \lambda x_1 x_2.x_1$ and FALSE $= \lambda x_1 x_2.x_2$.

The translation from Boolean expressions is:

Definition 4 (translation from Boolean expressions to $\lambda_{\rightarrow}$)
The translation from Boolean expressions to $\lambda_{\rightarrow}$ has as an input a Boolean expression E and results in two terms M_1 and M_2. We put $M_1 = \mathtt{E2L}(E)$ and M_2 = TRUE. The function E2L is defined by induction on the form of the Boolean expression (we may assume w.l.o.g. that the expressions do not contain negation):

$$\mathtt{E2L}((E_1 \wedge E_2)) = \lambda xy.(\mathtt{E2L}(E_1))((\mathtt{E2L}(E_2))xy)y; \qquad \mathtt{E2L}(\mathtt{true}) = \mathrm{TRUE};$$
$$\mathtt{E2L}((E_1 \vee E_2)) = \lambda xy.(\mathtt{E2L}(E_1))x((\mathtt{E2L}(E_2))xy); \qquad \mathtt{E2L}(\mathtt{false}) = \mathrm{FALSE}.$$

Theorem 2. *Let E be a Boolean expression. E has the result* `true` *iff the term* $\mathtt{E2L}(E)$ *reduces to* TRUE.

Moreover, the term $\mathtt{E2L}(E)$ *has redexes of order at most 1.*

Proof. The main claim is obtained by a routine induction on the expression E.

The only redexes in the term occur during the translation in cases for $\wedge$ and $\vee$. By induction on E, we can show that $\mathtt{E2L}(E)$ is of the type $\alpha \rightarrow \alpha \rightarrow \alpha$ so these redexes are of order 1.

Theorem 3. *The term* $\mathtt{E2L}(E)$ *may be represented by a first-order formula over the signature of Boolean expressions.*

Proof. The formula that constitutes the universe has 5 variables $x_1, \ldots, x_5$. The first one is used to determine which operator is encoded the rest is used to encode the Boolean representation of nodes needed to represent a Boolean connective. The first lambda node is encoded as 0000, then x as 0001, the second λ node as 0010 and so on. The edge relation (in a λ) term is defined so that the first coordinate is constant and the other coordinates represent suitable bits as in the above-mentioned encoding. Details are left for the reader.

4 The order 2

4.1 Second-order β-reduction is in PTIME

The second-order reduction can be performed in polynomial time. Our algorithm uses again the notion of graph reduction.

Let us see what is the graph reduction look like in this case. The starting point for this reduction is shown in Figure 1(a). The figure presents a β-redex located in a term (the omission of a part of the context of the redex is denoted by the dotted line). The star marked by G_0 symbolizes the body of the λ-abstraction that takes part in β-reduction. The circle marked by G_1 symbolizes the body of the argument that takes part in β-reduction. For the sake of clarity we denote a set of fan-nodes by a single fan with many entry ports.

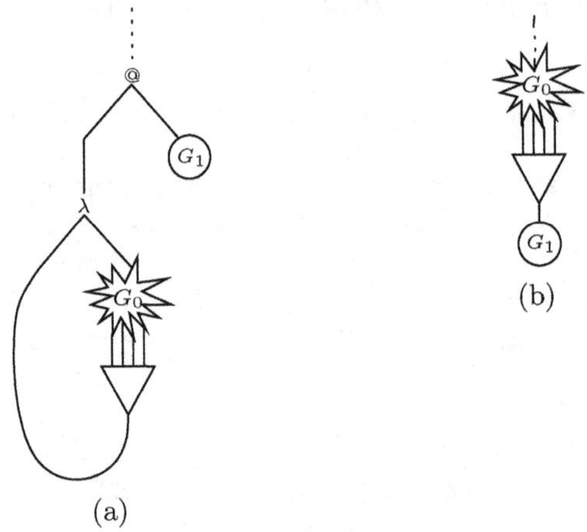

Fig. 1. (a) The starting point for 2nd-order β-reduction. (b) The result of the first phase of β-reduction

The result of the first β-reduction step is shown in Figure 1(b). As we see the argument G_1 goes into several places of the subterm G_0. Since we use a fan-node the argument is not copied. This kind of reduction is performed during the first phase of our algorithm. Note that the performing of some β-redexes may introduce other ones. There are two ways in which this redex may occur: the one as in the term $(\lambda x_1.(\lambda x_2.M))N_1N_2$ or the other as in the term $(\lambda x_1.C[x_1M])(\lambda x_2.N)$. We conduct our reduction in such a way that redexes of the first kind are contracted in this phase whereas the redexes of the second kind are not. We achieve this behaviour later in definitions by marking the edge outgoing from G_1 (see the point 2 in Definition 5). Note that this does not force us to reduce some redexes but these redexes are certainly of order 1. We repeat this kind of reduction until there are no redexes. The result of the process is a term that has no 2nd-order redexes.

Although there are no explicit redexes (except for the marked ones), we have some redexes hidden behind fan-nodes. We can extract these redexes as in Figure 2(a) and then contract them to the form presented in Figure 2(b).

This process should be repeated until there are no λ-nodes behind fan-nodes (in other words, until there are no paths which enter a fan-node and then after some number of brackets and croissants immediately enter a λ-node).

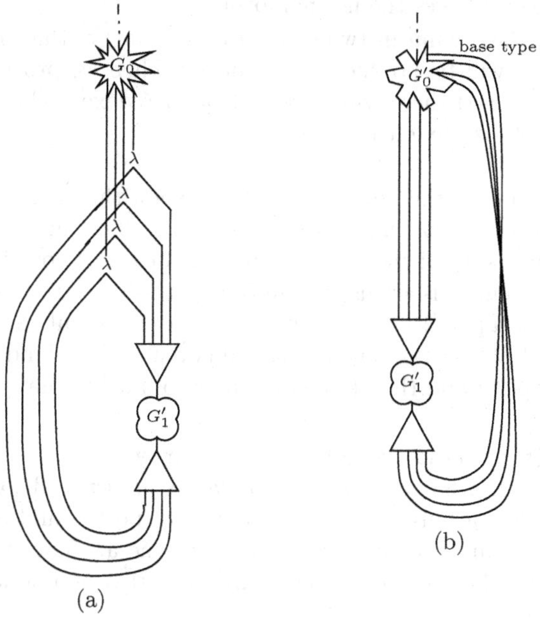

Fig. 2. (a) The λ-nodes are extracted from the fan-nodes. (b) After the first-order reduction

Definition 5 `(the algorithm for 2nd order)`
Let M_1 and M_2 be the input data for the algorithm. The algorithm `reduce_2nd` proceeds performing the following steps:

1. Translate the term M_1 into the corresponding graph.
2. Perform one by one all existing β-reductions (after performing a reduction step mark the edge that goes from the argument; in future reductions in this phase, omit redexes with such edges going out of λ-nodes).
3. Push all λ-nodes through fans.
4. Perform one by one all existing β-reductions and if necessary go to Point 3.
5. Reduce all matching fan-nodes so that they disappear.
6. Perform the read-back of the resulting graph; if the result is larger than the term M_2: fail. Let M_3 be the result of the read-back.
7. If $M_2 \equiv_\alpha M_3$ then success otherwise failure.

For the sake of clarity, we omit reductions for brackets and croissants in the description assuming that they are performed implicitly, resulting in the disappearance of nodes that occur at principal ports of fan-nodes, λ-nodes or @-nodes.

Theorem 4. *If* `reduce_2nd` *stops with success then* $\mathrm{NF}(M_1) \equiv_\alpha M_2$. *If the algorithm* `reduce_2nd` *fails then* $\mathrm{NF}(M_1) \not\equiv_\alpha M_2$.

Proof. The correctness of `reduce_2nd` is implied by the correctness of the graph reduction. The only thing to be proved is that before entering Point (6) in Definition 5, we obtain a graph that has no β-redexes in any reduction sequence so that the read-back gives the normal form.

We prove the last claim in two steps: First, we prove that after performing the step (2) there will be no 2nd-order redexes. Then we prove that after performing the steps (3-4) there will be no 1st-order redexes. These are proved by contradiction. Details are left for the reader.

In order to analyse the complexity of the algorithm `reduce_2nd`, we have to introduce the notion of *mixed bracket property.* This notion formalises and generalises the property of the old-fashioned arithmetical notation in which different kinds of parenthesis are used as in the expression: $[(2+3)\cdot 5+6]\cdot 11$. This property says that a parenthesis of A kind may be closed only if each parenthesis of any other kind B that is opened after the parenthesis of kind A is closed. For example, if we open [and then (then in order to put] we have to put) first.

Definition 6 (`the mixed parenthesis property`)
We say that a path ρ has the *mixed parenthesis property* iff for any fan-nodes A and B if ρ enters a fan-node A at an auxiliary port α and afterword a fan-node B at an auxiliary port β then it must exit the auxiliary port β of a node corresponding to B before it exits the auxiliary port α of a node corresponding to A.

Fact 1 *During the reductions performed in* `reduce_2nd` *all paths have mixed parenthesis property.*

Proof. The proof is by induction on the number of steps of reduction during the algorithm `reduce_2nd`. Details are left for the reader.

Theorem 5. *The procedure* `reduce_2nd` *runs in* $O(n^2)$ *time.*

Proof. Let $n = |M_1| + |M_2|$. Most of the points have easily verified linear time complexity. Point (6) needs $O(n^2)$ steps since the mixed parenthesis property (Fact 1) ensures that there are no two fan-nodes that meet with principal ports. If the matching fan-nodes exist then they are reduced in Point (5), if there are two non-matching fan-nodes then they break the mixed parenthesis property. As there are no fan-nodes that meet with principal ports, each path that exits a principal port of a fan-node and then after some, possibly non-zero, number of other (non-bracket and non-croissant) nodes enters a principal port of a fan-node must go through either @-node or λ-node. This ensures that such a node is visited once at least after visiting all fan-nodes. At last (7) can be performed in $O(n)$ time since M_2 is a part of input and α conversion can be performed in $O(m)$ where m is the size of terms to be checked. This altogether gives the time $O(n^2)$.

4.2 Second-order β-reduction is PTIME-hard

The problem of the evaluation of Boolean circuits is reduced to the problem of β-reduction in second-order in this section. The reduction is in LOGSPACE. This implies that second-order β-reduction is PTIME-hard.

Definition 7 (`Boolean circuit`)
A Boolean circuit is a directed acyclic graph such that:

- its nodes are labeled with $\vee, \wedge, \neg,$ `true`, or `false` and a single node labeled with `result`;
- nodes labeled with $\vee$ and $\wedge$ have two outgoing edges;
- nodes labeled with $\neg$ and `result` have a single outgoing edge;
- nodes labeled with `true` and `false` have no outgoing edges.

The result of a Boolean circuit can be defined recursively in an obvious way e.g. the value of a $\vee$-node is $v_1 \vee v_2$ where v_1 is the value of the node at the end of the first outgoing edge and v_2 is the value of the node at the end of the second outgoing edge, the value of `result` is v where v is the value of the node at the end of the outgoing edge.

Definition 8 (`the problem of evaluation of a Boolean circuit`)
The problem of the evaluation of a Boolean circuit is:
Input: A Boolean circuit $\mathcal{C}$
Question: Does the circuit have the result `true`?

The above-mentioned problem is PTIME-hard (see Theorem 8.1 in [Pap95]).

We define *level* of a node in a Boolean circuit. This notion helps use define the reduction.

Definition 9 (`level of a node`)
In a Boolean circuit $\mathcal{C}$, the node `result` has the level 0. A node n has the level l if $l = \max\{l_1, \ldots, l_k\} + 1$ where $\{l_1, \ldots, l_k\}$ is the set of levels for nodes n' such that (n', n) is an edge in $\mathcal{C}$.

We denote by $\mathcal{C}_n$ the set of nodes of the level n.

As Boolean circuits use logical connectives $\vee, \wedge$ and $\neg$, we should define their counterparts in λ-calculus. We also define logical values and quantifiers which are needed later.

Definition 10 (`connectives for translations`)

$$
\begin{array}{ll}
\mathrm{TRUE} = \lambda x_1 x_2.x_1 & \forall = \lambda \phi x_1 x_2.\mathrm{AND}(\phi \mathrm{TRUE} x_1 x_2)(\phi \mathrm{FALSE} x_1 x_2) \\
\mathrm{FALSE} = \lambda x_1 x_2.x_2 & \exists = \lambda \phi x_1 x_2.\mathrm{OR}(\phi \mathrm{TRUE} x_1 x_2)(\phi \mathrm{FALSE} x_1 x_2) \\
\mathrm{AND} = \lambda b_1 b_2 x_1 x_2.b_1(b_2 x_1 x_2)x_2 & \mathrm{NOT} = \lambda b_1 x_1 x_2.b_1 x_2 x_1 \\
\mathrm{OR} = \lambda b_1 b_2 x_1 x_2.b_1 x_1(b_2 x_1 x_2) &
\end{array}
$$

Definition 11 (`reduction from Boolean circuits`)
This reduction is recursively defined on the level of nodes. We introduce variables $\{x_i^j \mid 1 \leq i \leq k_j j\}$ where k_j is the number of nodes on the level j.

– The term LEVEL_{-1} is defined as $x^0_{\texttt{result}}$.
– The term LEVEL_{n+1} is defined on the basis of the term LEVEL_n as

$$(\lambda x_1^{n+1} \ldots x_{k_{n+1}}^{n+1}.\text{LEVEL}_n)B_1 \ldots B_{k_{n+1}}$$

where

- $B_i = \text{AND} x^l_k x^{l'}_{k'}$ if the i-th node on the level $n+1$ is $\wedge$ and one of its outgoing edges leads to k-th node on the l-th level and the other to k'-th node on the l'-th level;
- $B_i = \text{OR} x^l_k x^{l'}_{k'}$ if the i-th node on the level $n+1$ is $\vee$ and one of its outgoing edges leads to k-th node on the l-th level and the other to k'-th node on the l'-th level;
- $B_i = \text{NOT} x^l_k$ if the i-th node on the level $n+1$ is $\neg$ and its outgoing edge leads to k-th node on the l-th level;
- $B_i = \text{TRUE}$ if the i-th node on the level $n+1$ is `true`;
- $B_i = \text{FALSE}$ if the i-th node on the level $n+1$ is `false`.
- $B_i = \text{FALSE}$ if the i-th node on the level $n+1$ is `false`.

(Note that $\text{LEVEL}_0 = (\lambda x^0_{\texttt{result}}.x^0_{\texttt{result}})A$, where A is either TRUE or FALSE.)

Theorem 6. *Let G be a Boolean circuit and n its maximum level of nodes. G has the result* `true` *iff the term* LEVEL_n *reduces to* TRUE.

Moreover, the term LEVEL_n *has redexes of order at most 2.*

Proof. The proof is by induction on the maximal level of the graph G. The induction step consists in suitable reduction of the highest level so that it disappears and the number of levels is decreased.

Theorem 7. *The term* LEVEL_n *may be generated with use of additional space of size* $O(\log|G|)$.

Proof. W.l.o.g. we may assume that Boolean circuits have assigned to each node its level. This allows us to use a counter that says on which level we are. This is enough to identify where should be placed appropriate variables and terms AND, OR, NOT, TRUE and FALSE. Such a counter needs $O(\log n)$ space. Another counter is needed for names of variables, but $O(\log n)$ is sufficient here too. Details are left for the reader.

5 The order 3

5.1 Third-order β-reduction is in PSPACE

The third-order reduction can be performed in polynomial space. Our algorithm, similarly to the second-order case, uses the notion of graph reduction.

Let us see how does the process of graph reduction look like in this case. The starting point of such a reduction may look like in Figure 3(a). The figure

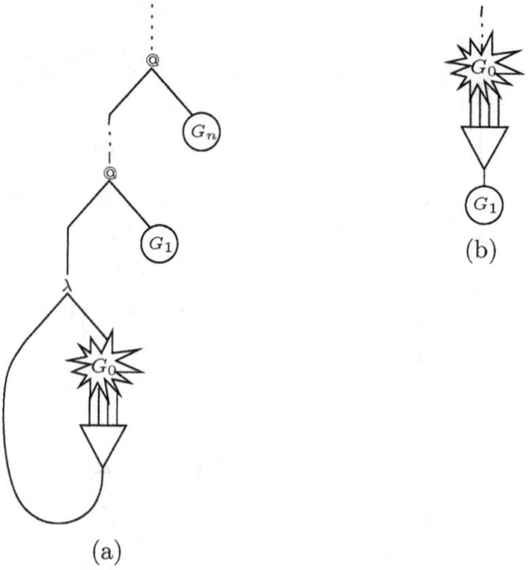

Fig. 3. (a) The starting point for 3rd-order β-reduction. (b) The result of the first phase of β-reduction

presents a β-redex in a λ-term. The star marked by G_0 denotes the body of the λ-abstraction that takes part in β-reduction. The circle marked by G_1 denotes the body of the argument that takes part in β-reduction. The dotted lines represent parts of the term that are missing in the picture.

The result of the first β-reduction step is shown in Figure 3(b). As we see, the argument G_1 goes into several places of the subterm G_0 as in the 2nd-order case. This kind of reduction is performed during the first phase of our algorithm. Again, the process of reduction may introduce other ones. Again, we perform only some of the new redexes similarly to the 2nd-order case. We repeat this kind of reduction until there are no redexes. The result of the process is a term that has no 3rd-order redexes.

Although there are no explicit redexes (except for the marked ones) we have some redexes hidden behind fan-nodes. We can extract these redexes as in Figure 4(a) and then contract them with @-nodes that come from G_0 as depicted in Figure 4(b). This process should be repeated until there are no λ-nodes behind fan-nodes (in other words, until there are no paths which enter a fan-node and then after some number of brackets and croissants immediately enter a λ-node).

The result of such reduction is depicted in Figure 5(a). We have two fan-nodes surrounding G'_1 — the upper one because the term occurs in several places and the lower one because different terms are substituted for a variable depending on which place is taken into account. This ends the second phase of the reduction (the reduction of 2nd-order redexes).

The last phase of the reduction begins — the reduction of 1st-order redexes. These redexes occur as in Figure 5(b) and begin to interact with the graph G'_1. As the 1st-order variable that took part in the 2nd-order reduction (the lambdas of which were multiplied in Figure 4(a)) can occur in several places inside G'_1, several @-nodes will take part in the reduction of 1st-order redexes. We can see

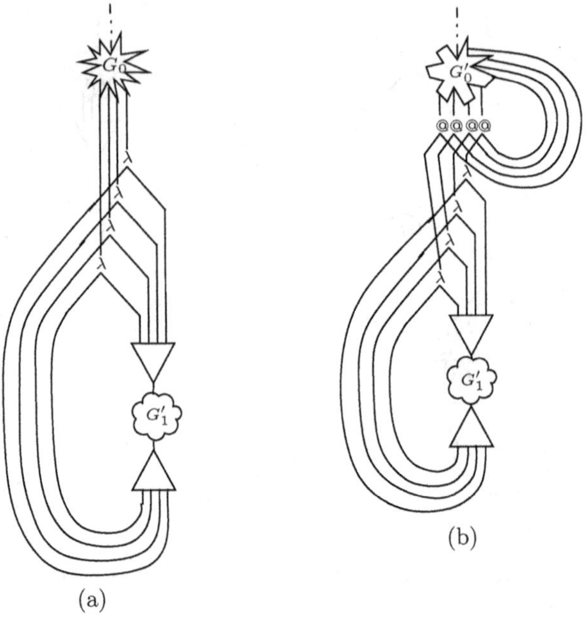

Fig. 4. (a) The λ-nodes are extracted from fan-nodes. (b) The λ-nodes meet suitable @-nodes

these @-nodes in Figure 6(a). As this multiplication concerns only one variable, we have a fan-node that performs this operation — also visible in Figure 6(a).

The fan-nodes that meet begin to interact. The result of the interaction is depicted in Figure 6(b) where it is denoted by the letter F. When we zoom the area denoted by F we will see a complicated web of links which is shown in Figure 7. The next step to perform is to push @-nodes through fan-nodes. The result of performing this step is partially shown in Figure 8(a). Each upper fan-node gets multiplied as it must go into two edges outgoing from each @-node. The next phase is to push λ-nodes through fan-nodes and perform β-redexes. The result of these operations is depicted in Figure 8(b). The left, big fan-node indicates that the body of the applied function goes into the place where application was situated previously. The right, big fan-node indicates that arguments of the application are placed in variables.

Definition 12 (`the algorithm for 3rd order`)
Let M_1 and M_2 be the input data for the algorithm. The algorithm `reduce_3rd` proceeds as follows:

1. Translate the term M_1 into the corresponding graph.
2. Perform one by one all existing β-reductions (after performing a reduction step mark the edge that goes from the argument; in future reductions within this phase, omit redexes with such an edge going out of a λ-node).
3. Clear all markings.
4. Push all fans through λ-nodes.
5. Perform one by one all existing β-reductions (again with marking).

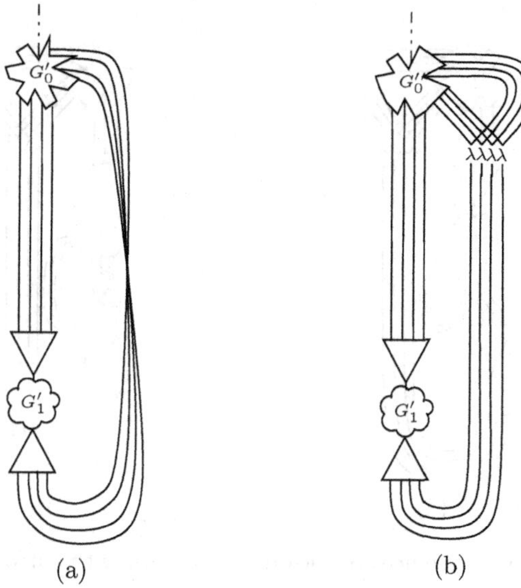

Fig. 5. (a) The result of the second phase of reduction. (b) First-order λ-nodes begin to reduce

6. Perform all interactions between fans and afterword push all fans through λ- and @-nodes.
7. Perform one by one all existing β-reductions.
8. Perform the read-back of the resulting graph; if the result is larger than the term M_2 to be equated: fail. Let M_3 be the result of the read-back.
9. If $M_2 \equiv_\alpha M_3$ then success otherwise failure.

In order to precisely describe the complexity we need a special notion called the level of a redex.

Definition 13 (level of a redex)
Let us define a special kind of reduction in which $(\lambda x.M)N \to_{\beta'} M[x := N^*]$ where N^* is the term N with a special marking (the marking should be understood as a new kind of language symbol similar to the application or abstraction, i.e. the marking is applied locally not throughout the whole term N and thus is not visible in redexes inside N). Note that we forbid the reduction $(\lambda x.M)^*N \to_\beta M[x := N^*]$. Of course, all reductions performed in this framework may be performed as the usual β-reduction. Thus paths of β'-reduction may be treated as paths of β-reduction. On the other hand, each path of β-reduction $M_1, \ldots, M_n$ may be presented as $M_1, \ldots, M_{i_1}, M_{i_1+1}, \ldots, M_{i_2}, \ldots, M_{i_{k-1}+1}, \ldots, M_{i_k}$ where redexes between terms $M_{i_j+1}, \ldots, M_{i_{j+1}+1}$ can be performed using β'-reduction and $M_{i_{j+1}+1}$ is a β'-normal form. The β-redexes in j-th such section are called *redexes of the level* j.

It is easily verified that each reduction of a term with redexes with order at most n has redexes of order at most $n-2$. If n is the highest order of the redex in a term then redexes of the order n are reduced during the 0-level section, the redexes of the level $n-1$ are reduced during the 1-level section and so on.

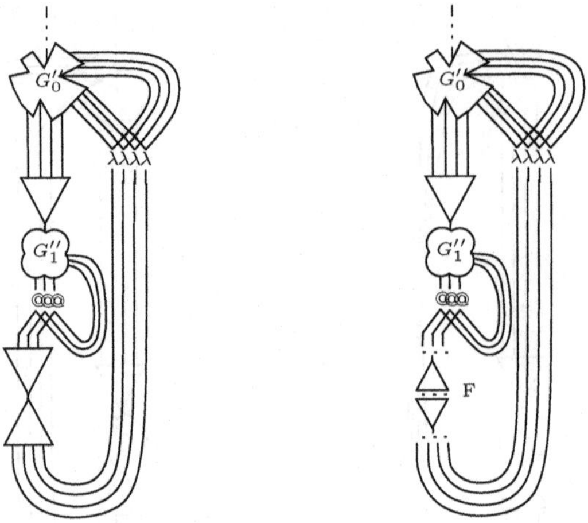

(a) (b)

Fig. 6. (a) Multiple occurrences of 1st-order variables with surrounding @-nodes. (b) Fan-nodes interact

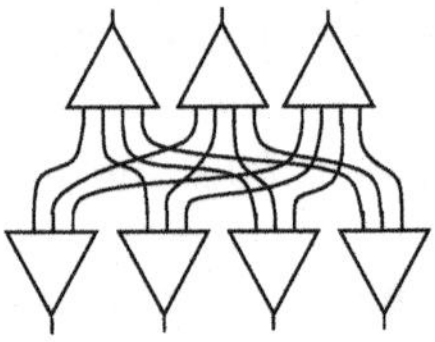

Fig. 7. The interaction of fan-nodes

Also the notion of the level of a redex straightforwardly translates to graph reduction. The algorithm for 3rd-order reduction needs redexes of order at most 1. The redexes of the level 0 are reduced in the step (2) of the algorithm, the redexes of the level 1 are reduced in the step (5) of the algorithm and at last redexes of the level 2 are reduced in the step (7) of the algorithm.

Theorem 8. *Let M_1 have redexes of order at most 3 and M_2 be in normal form. The algorithm* `reduce_3rd` *results in success on these terms iff $M_1 \to_\beta^* M_2$.*

Moreover, `reduce_3rd` *needs only $O(n^3)$ space to run.*

Proof. The algorithm is correct as it is only a strategy in an optimal reduction algorithm.

The analysis of the complexity of the algorithm is quite routine. Here are the most difficult cases:

Point (4) requires the multiplication of λ-nodes and fan-nodes. This multiplication is performed as in Figure 4(a) and so the number of new λ-nodes is bounded by $k_1 \cdot k_2$ where k_1 is the number of variables that take part in the step (2) of the algorithm and k_2 is the number of variables that are in the arguments of the former variables in the input. This gives the $O(n^3)$ space. The fan-nodes are replicated only $O(n)$ times as the number of variables that take part in the step (2) majorises the number of replications. The last number is of $O(n^2)$ magnitude.

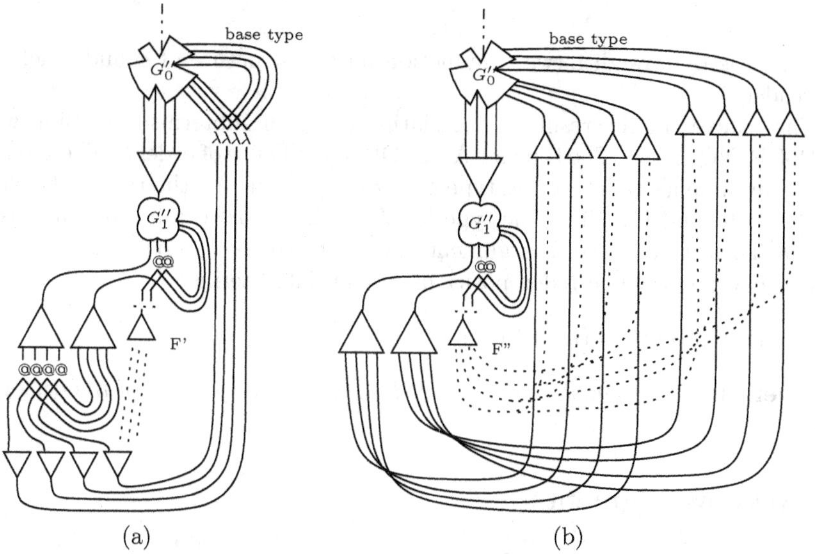

Fig. 8. (a) One application goes between fan-nodes. (b) After pushing λ-nodes through fans, applications are reduced

Point (7) is a usual walk through the graph in hand. As the size of the graph is $O(n^3)$, the time and thus the space is $O(n^3)$.

This altogether gives $O(n^3)$ space.

5.2 Third-order β-reduction is PSPACE-hard

We relay here on *quantified Boolean formulae* problem (QBF). Which consists in deciding whether a given formula with quantified Boolean variables is true. This problem in known to be PSPACE-complete. We present a PTIME reduction of the QBF problem to the 3rd-order reduction problem.

The translation is defined as follows

Definition 14 (`translation from QBF to` $\lambda_{\rightarrow}$)
The translation from QBF to $\lambda_{\rightarrow}$ has as an input a QBF sentence ϕ and as a result two terms M_1 and M_2. We put $M_1 = \texttt{Q2L}(\phi)$ and $M_2 = \text{TRUE}$. The function Q2L is defined by induction on the form of the QBF formula:

$\texttt{Q2L}(\texttt{true}) = \text{TRUE}$; $\texttt{Q2L}(\texttt{false}) = \text{FALSE}$;
$\texttt{Q2L}(x) = x$ where x is a variable; $\texttt{Q2L}(\neg\phi) = \text{NOT}(\texttt{Q2L}(\phi))$;
$\texttt{Q2L}(\phi_1 \wedge \phi_2) = \text{AND}(\texttt{Q2L}(\phi_1))(\texttt{Q2L}(\phi_2))$; $\texttt{Q2L}(\phi_1 \vee \phi_2) = \text{OR}(\texttt{Q2L}(\phi_1))(\texttt{Q2L}(\phi_2))$;
$\texttt{Q2L}(\forall x.\phi) = \forall(\lambda x.\texttt{Q2L}(\phi))$ $\texttt{Q2L}(\exists x.\phi) = \exists(\lambda x.\texttt{Q2L}(\phi))$

Theorem 9. *A QBF sentence ϕ is true iff the term* Q2L(ϕ) *reduces to* TRUE. *Moreover, the term* Q2L(ϕ) *has redexes of order at most 3.*

Proof. We need a little bit extended version of the claim:

Let ϕ be a QBF formula with free variables in $A = \{x_1, \ldots, x_n\}$. The formula ϕ is true under the valuation $v : A \rightarrow \{\texttt{true}, \texttt{false}\}$ iff the term $\texttt{Q2L}(\phi)[x_1 := \texttt{Q2L}(v(x_1)), \ldots, x_n := \texttt{Q2L}(v(x_n))]$ reduces to TRUE.

The proof is by straightforward induction on the structure of ϕ and is left for the reader.

The redexes in the result of translation occur in subterms beginning with AND, OR, NOT, $\forall$, $\exists$. The type for AND, OR and NOT is of order 2. These terms take as arguments values of the type $\alpha \to \alpha \to \alpha$ (which is the type of Boolean terms TRUE and FALSE). The type for $\forall$ and $\exists$ is more complicated and is of order 3. These terms take an argument of the type $(\alpha \to \alpha \to \alpha) \to \alpha \to \alpha \to \alpha$. No other terms occur in redex positions in translated terms.

We have also by routine analysis:

Theorem 10. *The translation from QBF to $\lambda_\to$ can be performed in $O(n \log n)$ time.*

6 Acknowledgments

I would like to thank Damian Niwiński for his hints concerning good problems to reduce from. Thanks also go to anonymous referees who helped to improve the paper. I also thank dear Joanna for her help in improving my English.

References

[AG98] Andrea Asperti and Stefano Guerrini, *The optimal implementation of functional programming languages*, Cambridge University Press, 1998.

[AL93] Andrea Asperti and Cosimo Laneve, *Interaction Systems II: the practice of optimal reductions*, Tech. Report UBLCS-93-12, Laboratory for Computer Science, Universitá di Bologna, 1993.

[Bar92] H. P. Barendregt, *Lambda calculi with types*, Handbook of Logic in Computer Science (S. Abramsky, D. M. Gabbay, and T. S. E. Mainbaum, eds.), vol. 2, Oxford University Press, 1992, pp. 117–309.

[Bus87] S.R. Buss, *The boolean formula value problem is in ALOGTIME*, Proceedings of the 19th Annual ACM Symposium on Theory of Computing, ACM Press, 1987, pp. 123–131.

[HK96] G. Hillebrand and P. Kanellakis, *On the expressive power of simply typed and let-polymorphic lambda calculi*, Proceedings of the 11th IEEE Conference on Logic in Computer Science, 1996, pp. 253–263.

[Lam90] John Lamping, *An algorithm for optimal lambda calculus reductions*, Proceedings of 17th ACM Symposium on Principles of Programming Languages, 1990, pp. 16–30.

[Mai92] H. Mairson, *A simple proof of a theorem of statman*, Theoretical Computer Science (1992), no. 103, 213–226.

[Pap95] Ch. H. Papadimitriou, *Computational complexity*, Addison–Wesley, 1995.

[Sch91] H. Schwichtenberg, *An upper bound for reduction sequences in the typed λ-calculus*, Archive for Mathematical Logic (1991), no. 30, 405–408.

[Sta79] R. Statman, *The typed λ-calculus is not elementary recursive*, Theoretical Computer Science (1979), no. 9, 73–81.

Strong Normalisation for a Gentzen-like Cut-Elimination Procedure

Christian Urban*

Department of Pure Mathematics and Mathematical Statistics,
University of Cambridge.
cu200@dpmms.cam.ac.uk

Abstract. In this paper we introduce a cut-elimination procedure for classical logic, which is both strongly normalising and consisting of local proof transformations. Traditional cut-elimination procedures, including the one by Gentzen, are formulated so that they only rewrite neighbouring inference rules; that is they use local proof transformations. Unfortunately, such local proof transformation, if defined naïvely, break the strong normalisation property. Inspired by work of Bloo and Geuvers concerning the $\lambda\mathtt{x}$-calculus, we shall show that a simple trick allows us to preserve this property in our cut-elimination procedure. We shall establish this property using the recursive path ordering by Dershowitz.

Keywords. Cut-Elimination, Classical Logic, Explicit Substitution, Recursive Path Ordering.

1 Introduction

Gentzen showed in his seminal paper [6] that all cuts can be eliminated from sequent proofs in LK and LJ. He not only proved that cuts can be eliminated, but also gave a simple procedure for doing so. This procedure consists of proof transformations, or cut-reductions, that do not eliminate all cuts from a proof immediately, but rather replace every instance of a cut with simpler cuts, and by iteration one eventually ends up with a cut-free proof. We refer to a proof transformation as being local, or *Gentzen-like*, if it only rewrites neighbouring inference rules, possibly by duplicating a subderivation. Most of the traditional cut-elimination procedures, including Gentzen's original procedure, consist of such local proof transformations.

In [11] and [12] three criteria for a cut-elimination procedure were introduced:

1. the cut-elimination procedure should *not* restrict the collection of normal forms reachable from a given proof in such a way that "essential" normal forms are no longer reachable,

* I should like to thank Roy Dyckhoff for many helpful discussions. I am currently funded with a research fellowship from Corpus Christi College, Cambridge.

S. Abramsky (Ed.): TLCA 2001, LNCS 2044, pp. 415–429, 2001.

2. the cut-elimination procedure should be *strongly normalising*, i.e., all possible reduction strategies should terminate, and
3. the cut-elimination procedure should allow cuts to pass over other cuts.

Owing to space restrictions we cannot defend these criteria here and refer the reader to [11,12] where it is explained why they play an important rôle in investigating the computational content of classical logic.

The main purpose of this paper is to present a cut-elimination procedure that satisfies the three criteria *and* that consists of only Gentzen-like cut-reductions. At the time of writing, we are not aware of any other cut-elimination procedure fulfilling both demands. The problem with Gentzen-like cut-reductions is that they, if defined naïvely, break the strong normalisation property, as illustrated in the following example given in [2,5]. Consider the following LK-proof.

$$\dfrac{\dfrac{\dfrac{\dfrac{}{A\vdash A}\quad\dfrac{}{A\vdash A}}{A\lor A\vdash A,A}\lor_L}{A\lor A\vdash A}\mathbf{Contr}_R\qquad\dfrac{\dfrac{\dfrac{}{A\vdash A}\quad\dfrac{}{A\vdash A}}{A,A\vdash A\land A}\land_R}{A\vdash A\land A}\mathbf{Contr}_L}{A\lor A\vdash A\land A}\mathbf{Cut}\tag{1}$$

Using Gentzen-like proof transformations there are two possibilities for permuting the cut upwards: it can either be permuted upwards in the left proof branch or in the right proof branch. In both cases a subproof has to be duplicated. Taking the former possibility, we obtain the proof

$$\dfrac{\dfrac{\dfrac{\dfrac{\dfrac{}{A\vdash A}\quad\dfrac{}{A\vdash A}}{A\lor A\vdash A,A}\lor_L\qquad\dfrac{\dfrac{\dfrac{}{A\vdash A}\quad\dfrac{}{A\vdash A}}{A,A\vdash A\land A}\land_R}{A\vdash A\land A}\mathrm{Contr}_L}{A\lor A\vdash A,A\land A}\mathrm{Cut}\qquad\dfrac{\dfrac{\dfrac{}{A\vdash A}\quad\dfrac{}{A\vdash A}}{A,A\vdash A\land A}\land_R}{A\vdash A\land A}\mathrm{Contr}_L}{A\lor A\vdash A\land A,A\land A}\mathrm{Cut}}{A\lor A\vdash A\land A}\mathrm{Contr}_R$$

where two copies of the right subproof are created. Now permute the upper cut to the right, which gives the following proof.

$$\dfrac{\dfrac{\dfrac{\dfrac{\dfrac{\dfrac{\dfrac{}{A\vdash A}\quad\dfrac{}{A\vdash A}}{A\lor A\vdash A,A}\lor_L\qquad\dfrac{\dfrac{\dfrac{}{A\vdash A}\quad\dfrac{}{A\vdash A}}{A\lor A\vdash A,A}\lor_L\qquad\dfrac{\dfrac{}{A\vdash A}\quad\dfrac{}{A\vdash A}}{A,A\vdash A\land A}\land_R}{A\lor A,A\vdash A,A\land A}\mathrm{Cut}}{A\lor A,A\lor A\vdash A,A,A\land A}\mathrm{Cut}}{A\lor A\vdash A,A,A\land A}\mathrm{Contr}_L}{A\lor A\vdash A,A\land A}\mathbf{Contr}_R\qquad\dfrac{\dfrac{\dfrac{}{A\vdash A}\quad\dfrac{}{A\vdash A}}{A,A\vdash A\land A}\land_R}{A\vdash A\land A}\mathbf{Contr}_L}{A\lor A\vdash A\land A,A\land A}\mathbf{Cut}}{A\lor A\vdash A\land A}\mathrm{Contr}_R$$

This proof contains an instance of the reduction applied in the first step (compare the rule names in bold face). Even worse, it is bigger than the proof with which we started, and so in effect we can construct infinite reduction sequences.

Another problem with Gentzen-like cut-reductions arises from the third criterion. If one introduces the following reduction rule, which allows a cut (Suffix 2) to pass over another cut (Suffix 1)

$$\dfrac{\dfrac{\ldots\vdash\ldots \quad \ldots\vdash\ldots}{\ldots\vdash\ldots}\,\text{Cut}_1 \quad \ldots\vdash\ldots}{\ldots\vdash\ldots}\,\text{Cut}_2 \longrightarrow \dfrac{\dfrac{\ldots\vdash\ldots \quad \ldots\vdash\ldots}{\ldots\vdash\ldots}\,\text{Cut}_2 \quad \ldots\vdash\ldots}{\ldots\vdash\ldots}\,\text{Cut}_1 \qquad (2)$$

then clearly one loses the strong normalisation property—the reduct is again an instance of this rule, and one can loop by constantly applying this reduction. Thus a common restriction is to not allow a cut to pass over another cut in any circumstances. Unfortunately, this has several serious drawbacks, as noted in [4,7]; it limits, for example, in the intuitionistic case the correspondence between cut-elimination and beta-reduction. In particular, strong normalisation of beta-reduction cannot be inferred from strong normalisation of cut-elimination. Therefore we shall introduce cut-reductions that avoid the infinite reduction sequence illustrated in (1), but allow cuts to pass over other cuts without breaking the strong normalisation property.

Because of the conflicting demands of being very liberal (e.g. allowing cuts to pass over other cuts) and at the same time preserving the strong normalisation property, such a cut-elimination procedure seems difficult to obtain. So rather surprisingly we found that if one adds to the usual cut-rule

$$\dfrac{\Gamma_1 \vdash \Delta_1, C \quad C, \Gamma_2 \vdash \Delta_2}{\Gamma_1, \Gamma_2 \vdash \Delta_1, \Delta_2}\,\text{Cut}$$

the following two, referred to as *labelled* cut-rules,

$$\dfrac{\Gamma_1 \vdash \Delta_1, C \quad C, \Gamma_2 \vdash \Delta_2}{\Gamma_1, \Gamma_2 \vdash \Delta_1, \Delta_2}\,\text{C}\overleftarrow{\text{u}}\text{t} \qquad \dfrac{\Gamma_1 \vdash \Delta_1, C \quad C, \Gamma_2 \vdash \Delta_2}{\Gamma_1, \Gamma_2 \vdash \Delta_1, \Delta_2}\,\text{C}\overrightarrow{\text{u}}\text{t}$$

then one can define a cut-elimination procedure that satisfies the three criteria *and* that only consists of Gentzen-like cut-reductions.

Reconsider the proof given in (1). There the infinite reduction sequence could be constructed by permuting cuts into alternating directions. Clearly, this reduction sequence can be avoided if commuting cuts have to be permuted into only one direction. (A cut is said to be a *logical* cut when both cut-formulae are introduced by axioms or logical inference rules; otherwise the cut is said to be a *commuting* cut.) Furthermore, instead of the cut-reduction shown in (2), we can introduce the following cut-reduction

$$\dfrac{\dfrac{\ldots\vdash\ldots \quad \ldots\vdash\ldots}{\ldots\vdash\ldots}\,\text{Cut} \quad \ldots\vdash\ldots}{\ldots\vdash\ldots}\,\text{C}\overleftarrow{\text{u}}\text{t} \longrightarrow \dfrac{\dfrac{\ldots\vdash\ldots \quad \ldots\vdash\ldots}{\ldots\vdash\ldots}\,\text{C}\overleftarrow{\text{u}}\text{t} \quad \ldots\vdash\ldots}{\ldots\vdash\ldots}\,\text{Cut}$$

which allows cuts to pass over other cuts, but which does not break the strong normalisation property—the reduction cannot be applied to the reduct.

Although the "trick" with the labelled cuts seems to be trivial, the corresponding strong normalisation proof is rather non-trivial (mainly because we allow cuts to pass over other cuts). To prove this property, we shall make use of a technique developed in [1]. This technique appeals to the recursive path ordering theorem by Dershowitz. Our proof is more difficult than the one given in [4], which also appeals to the recursive path ordering theorem, because we allow, as mentioned above, cuts to pass over other cuts. To be able to present our proof in a manageable form, we found it extremely useful to annotate sequent proofs with terms. In consequence, our contexts are sets of (label,formula) pairs, as in type theory, and *not* multisets, as in LK or LJ.

The paper is organised as follows. In Section 2 our sequent calculus and the corresponding term annotations will be given. To save space, we shall restrict our attention in this paper to the $\wedge$-fragment of classical logic, but it should be emphasised that the strong normalisation result for cut-elimination may be obtained for all connectives by a simple adaptation of the proof we shall give. The cut-elimination procedure will be defined in Section 3. A comparison with the λ**x**-calculus will be given in Section 4. In Section 5 we shall describe the proof of strong normalisation, and conclude and give suggestions for further work in Section 6.

2 Sequent Calculus, Terms, and Typing Judgements

In this section we shall introduce our sequent calculus for classical logic. As mentioned earlier, we shall restrict our attention to the $\wedge$-fragment. Thus the formulae are given by the grammar

$$B ::= A \mid B{\wedge}B$$

in which A ranges over propositional symbols.

Our sequents contain two contexts—an *antecedent* and a *succedent*—both of which are sets of (label,formula) pairs. As we shall see, the use of sets allows us to define the sequent calculus so that the structural rules, i.e., weakening and contraction, are completely implicit in the form of the logical inference rules. Since there are two sorts of contexts, it will be convenient to separate the labels into *names* and *co-names*; in what follows a, b, c, ... will stand for co-names and similarly ..., x, y, z for names. Thus, antecedents are built up by (name,formula) pairs and succedents by (co-name,formula) pairs. We shall employ some shorthand notation for contexts: rather than writing, for example, $\{(x,B),(y,C),(z,D)\}$, we shall simply write $x:B, y:C, z:D$.

Whereas in LK the sequents consists of an antecedent and succedent only, in our sequent calculus the sequents have another component: a *term*. Terms encode the structure of sequent proofs and thus allow us to define a complete cut-elimination procedure as a term rewriting system. The set of raw terms, $\mathfrak{R}_\wedge$, is defined by the grammar

$$\begin{array}{rlll}
M,N ::= & \mathsf{Ax}(x,a) & \text{Axiom} & \\
\mid & \mathsf{And}_R(\langle a{:}B\rangle M, \langle b{:}C\rangle N, c) & \text{And-R} & \\
\mid & \mathsf{And}^i_L((x{:}B)M, y) & \text{And-L}_i & (i=1,2)\\
\mid & \mathsf{Cut}(\langle a{:}B\rangle M, (x{:}B)N) & \text{Cut} & \\
\mid & \overleftarrow{\mathsf{Cut}}(\langle a{:}B\rangle M, (x{:}B)N) & \text{Cut with label `}\leftarrow\text{'} & \\
\mid & \overrightarrow{\mathsf{Cut}}(\langle a{:}B\rangle M, (x{:}B)N) & \text{Cut with label `}\rightarrow\text{'} &
\end{array}$$

where x, y are taken from a set of names and a, b, c from a set of co-names; B and C are types (formulae). In a term we use round brackets to signify that a name becomes bound and angle brackets that a co-name becomes bound. In what follows we shall often omit the types on the bindings for brevity, regard terms as equal up to alpha-conversions and adopt a Barendregt-style convention for the names and co-names. These conventions are standard in term rewriting. Notice however that names and co-names are not the same notions as a variable in the lambda-calculus: whilst a term can be substituted for a variable, a name or a co-name can only be "renamed". Rewriting a name x to y in a term M is written as $M[x \mapsto y]$, and similarly rewriting a co-name a to b is written as $M[a \mapsto b]$. The routine formalisation of these rewriting operations is omitted. In our proof it will be useful to have the following notions: a term is said to be *labelled* provided its top-most term constructor is either $\overleftarrow{\mathsf{Cut}}$ or $\overrightarrow{\mathsf{Cut}}$; otherwise the term is said to be *unlabelled*; a term is said to be *completely unlabelled* provided all subterms are unlabelled. Other useful notions are as follows.

- A term, M, *introduces* the name z or co-name c iff M is of the form

  for z: $\mathsf{Ax}(z,c)$, $\mathsf{And}_L^i((x)S, z)$ for c: $\mathsf{Ax}(z,c)$, $\mathsf{And}_R(\langle a\rangle S, \langle b\rangle T, c)$

- A term, M, *freshly introduces* a name iff M introduces this name, but none of its proper subterms. In other words, the name must not be free in a proper subterm, just in the top-most term constructor. Similarly for co-names.

We can now formally introduce sequents, or *typing judgements*. They are of the form $\Gamma \triangleright M \triangleright \Delta$ with Γ being an antecedent, M a term and Δ a succedent. Henceforth we shall be interested in only *well-typed* terms; this means those for which there are two contexts, Γ and Δ, such that $\Gamma \triangleright M \triangleright \Delta$ holds given the inference rules in Figure 1. We shall write $\mathfrak{T}_\wedge$ for the set of well-typed terms.

Whilst the structural rules are *implicit* in our sequent calculus, i.e., the calculus has fewer inference rules, there are a number of subtleties concerning contexts. First, we assume the convention that a context is ill-formed, if it contains more than one occurrence of a name or co-name. For example the antecedent $x{:}B, x{:}C$ is not allowed. Hereafter, this will be referred to as the context convention, and it will be assumed that all inference rules respect this convention.

Second, we have the following conventions for the commas in Figure 1: a comma in a conclusion stands for set union and a comma in a premise stands for *disjoint* set union. Consider for example the $\wedge_{L_i}$-rule. This rule introduces the (name,formula) pair $y{:}B_1{\wedge}B_2$ in the conclusion, and consequently, y is a free name in $\mathsf{And}_L^i((x)M, y)$. However, y can already be free in the subterm M, in which case $y{:}B_1{\wedge}B_2$ belongs to Γ. We refer to this as an *implicit contraction.* Hence the antecedent of the conclusion of $\wedge_{L_i}$ is of the form $y{:}B_1{\wedge}B_2 \oplus \Gamma$ where $\oplus$ denotes set union. Clearly, if the term $\mathsf{And}_L^i((x)M, y)$ freshly introduces y, then this antecedent is of the form $y{:}B_1{\wedge}B_2 \otimes \Gamma$ where $\otimes$ denotes *disjoint* set union. Note that $x{:}B_i$ cannot be part of the conclusion: x becomes bound in the term. Thus the antecedent of the premise must be of the form $x{:}B_i \otimes \Gamma$.

There is one point worth mentioning in the cut-rules, because they are the only inference rules in our sequent calculus that do not share the contexts, but require that two contexts are joined on each side of the conclusion. Thus we take the cut-rule labelled with '$\leftarrow$', for example, to be of the form

$$\frac{\Gamma_1 \triangleright M \triangleright \Delta_1 \otimes a{:}B \qquad x{:}B \otimes \Gamma_2 \triangleright N \triangleright \Delta_2}{\Gamma_1 \oplus \Gamma_2 \triangleright \overleftarrow{\mathsf{Cut}}(\langle a\rangle M, (x)N) \triangleright \Delta_1 \oplus \Delta_2}\ \overleftarrow{\mathrm{Cut}}\,.$$

In effect, this rule is only applicable, if it does not break the context convention, which can always be achieved by renaming some labels appropriately. Notice that we do not require that cut-rules have to be "fully" multiplicative: the Γ_i's (respectively the Δ_j's) can share some formulae.

3 Cut-Reductions

We are now ready to define our Gentzen-like cut-elimination procedure. For this we shall introduce four sorts of cut-reduction, each of which is assumed to be

$$\frac{}{x{:}B, \Gamma \triangleright \mathsf{Ax}(x,a) \triangleright \Delta, a{:}B}\;\mathrm{Ax}$$

$$\frac{x{:}B_i, \Gamma \triangleright M \triangleright \Delta}{y{:}B_1 \wedge B_2, \Gamma \triangleright \mathsf{And}_L^i((x)M, y) \triangleright \Delta}\;\wedge_{L_i} \qquad \frac{\Gamma \triangleright M \triangleright \Delta, a{:}B \qquad \Gamma \triangleright N \triangleright \Delta, b{:}C}{\Gamma \triangleright \mathsf{And}_R(\langle a\rangle M, \langle b\rangle N, c) \triangleright \Delta, c : B \wedge C}\;\wedge_R$$

$$\frac{\Gamma_1 \triangleright M \triangleright \Delta_1, a{:}B \quad x{:}B, \Gamma_2 \triangleright N \triangleright \Delta_2}{\Gamma_1, \Gamma_2 \triangleright \overleftarrow{\mathsf{Cut}}(\langle a\rangle M, (x)N) \triangleright \Delta_1, \Delta_2}\;\overleftarrow{\mathrm{Cut}} \qquad \frac{\Gamma_1 \triangleright M \triangleright \Delta_1, a{:}B \quad x{:}B, \Gamma_2 \triangleright N \triangleright \Delta_2}{\Gamma_1, \Gamma_2 \triangleright \overrightarrow{\mathsf{Cut}}(\langle a\rangle M, (x)N) \triangleright \Delta_1, \Delta_2}\;\overrightarrow{\mathrm{Cut}}$$

$$\frac{\Gamma_1 \triangleright M \triangleright \Delta_1, a{:}B \quad x{:}B, \Gamma_2 \triangleright N \triangleright \Delta_2}{\Gamma_1, \Gamma_2 \triangleright \mathsf{Cut}(\langle a\rangle M, (x)N) \triangleright \Delta_1, \Delta_2}\;\mathrm{Cut}$$

Fig. 1. Term assignment for sequent proofs in the $\wedge$-fragment of classical logic.

closed under context formation. This is a standard convention in term rewriting. The first sort of cut-reduction, written $\stackrel{l}{\longrightarrow}$, deals with logical cuts.

Logical Reductions:

1. $\mathsf{Cut}(\langle b\rangle \mathsf{And}_R(\langle a_1\rangle M_1, \langle a_2\rangle M_2, b), (y)\mathsf{And}_L^i((x)N, y)) \stackrel{l}{\longrightarrow} \mathsf{Cut}(\langle a_i\rangle M_i, (x)N)$
   if $\mathsf{And}_R(\langle a_1\rangle M_1, \langle a_2\rangle M_2, b)$ and $\mathsf{And}_L^i((x)N, y)$ freshly introduce b and y

2. $\mathsf{Cut}(\langle a\rangle M, (x)\mathsf{Ax}(x, b)) \stackrel{l}{\longrightarrow} M[a \mapsto b]$
   if M freshly introduces a

3. $\mathsf{Cut}(\langle a\rangle \mathsf{Ax}(y, a), (x)M) \stackrel{l}{\longrightarrow} M[x \mapsto y]$
   if M freshly introduces x

As can be seen, these cut-reductions are restricted so that they are applicable only if the immediate subterms of the cuts freshly introduce the names and co-names corresponding to the cut-formulae. Without this restriction bound names or bound co-names might become free during cut-elimination, as demonstrated in [11,12]. Note that in Reduction 2 (resp. 3) it is permitted that b (resp. y) is free in M.

The next sort of cut-reduction applies to commuting cuts, that means to those where at least one immediate subterm of the cut does not freshly introduce the name or co-name of the cut-formula.

Commuting Reductions:

5. $\mathsf{Cut}(\langle a\rangle M, (x)N) \stackrel{c}{\longrightarrow} \overleftarrow{\mathsf{Cut}}(\langle a\rangle M, (x)N)$
   if M does not freshly introduce a and is unlabelled, *or*
6. $\mathsf{Cut}(\langle a\rangle M, (x)N) \stackrel{c}{\longrightarrow} \overrightarrow{\mathsf{Cut}}(\langle a\rangle M, (x)N)$
   if N does not freshly introduce x and is unlabelled.

A point to note is that Reductions 5 and 6 may be applicable at the same time. Take for example the term $\mathsf{Cut}(\langle a\rangle \mathsf{Ax}(x, b), (y)\mathsf{Ax}(z, c))$, which can reduce to either $\overleftarrow{\mathsf{Cut}}(\langle a\rangle \mathsf{Ax}(x, b), (y)\mathsf{Ax}(z, c))$ or $\overrightarrow{\mathsf{Cut}}(\langle a\rangle \mathsf{Ax}(x, b), (y)\mathsf{Ax}(z, c))$—the choice to which term it reduces is not specified. Therefore, our cut-elimination procedure is non-deterministic.

Once a cut is "labelled" by Reduction 5 or 6, then cut-reductions written as $\xrightarrow{x}$ apply (see Figure 2). Each of them pushes labelled cuts inside the subterms until they reach a place where the cut-formula is introduced. However care needs to be taken when applying an $\xrightarrow{x}$-reduction to ensure that no name or co-name clash occurs. This can always be achieved by appropriate alpha-conversions, and we shall assume that these conversions are done implicitly.

It is worthwhile to comment on the reductions $\xrightarrow{c}$ and $\xrightarrow{x}$. We required in Reduction 5 (similarly in 6) that the term M is unlabelled, i.e., the top-most term constructor is not $\mathsf{C\overleftarrow{u}t}$ or $\mathsf{C\overrightarrow{u}t}$. This restriction is to avoid certain reduction sequences. Suppose M and N are cut-free, and assume the term $\mathsf{Cut}(\langle a\rangle M, (x)N)$ is a logical cut. Furthermore assume c is not free in this term. Then consider the reduction sequence

$$\begin{aligned}\mathsf{C\overleftarrow{u}t}(\langle c\rangle \mathsf{Cut}(\langle a\rangle M, (x)N), (y)P) &\xrightarrow{x} \mathsf{Cut}(\langle a\rangle \mathsf{C\overleftarrow{u}t}(\langle c\rangle M, (y)P), (x)\mathsf{C\overleftarrow{u}t}(\langle c\rangle N, (y)P))\\ &\xrightarrow{c} \mathsf{C\overleftarrow{u}t}(\langle a\rangle \mathsf{C\overleftarrow{u}t}(\langle c\rangle M, (y)P), (x)\mathsf{C\overleftarrow{u}t}(\langle c\rangle N, (y)P))\\ &\xrightarrow{x}{}^{+} \mathsf{C\overleftarrow{u}t}(\langle a\rangle M, (x)N)\end{aligned}$$

where the logical cut has become labelled ($\xrightarrow{c}$-reduction), because another cut passed over it (first $\xrightarrow{x}$-reduction). While this reduction is harmless with respect to strong normalisation (this cut becomes a logical cut again), it causes the strong normalisation proof to be much harder. To save space, we thus exclude reduction sequences in which a logical cut becomes labelled, and the side-conditions in Reduction 5 and 6 are doing just that.

Another point worth mentioning is that the first and second rule in Figure 2 (similarly the fourth and fifth) can be replaced with the reduction

$$\mathsf{C\overleftarrow{u}t}(\langle c\rangle \mathsf{Ax}(x, c), (y)P) \longrightarrow P[y \mapsto x] \tag{3}$$

which is equally effective, in that all cut-rules are eliminable from a proof. However, this reduction has subtle defect, as explained in [11,12]. Consider a term N in which x is not free and a term P in which b is not free. We would expect that from $\mathsf{C\overrightarrow{u}t}(\langle a\rangle N, (x)\mathsf{C\overleftarrow{u}t}(\langle b\rangle M, (y)P))$ and $\mathsf{C\overleftarrow{u}t}(\langle b\rangle \mathsf{C\overrightarrow{u}t}(\langle a\rangle N, (x)M), (y)P)$ the same collection of normal forms can be reached (the order of "independent" labelled cuts should not matter). Unfortunately, using the rule in (3) this does *not* hold. Therefore we have formulated the $\xrightarrow{x}$-reductions so that the the order of labelled cuts—as long as they are "independent"—is irrelevant with respect to which normal forms are reachable. This is an important property for analysing the computational content of classical proofs [8].

The last sort of cut-reduction, named *garbage reduction*, deals with labelled cuts whose name or co-name of the cut-formula is not free in the corresponding subterm. In LK this corresponds to a cut on a weakened formula.

Garbage Reductions:

7. $\mathsf{C\overleftarrow{u}t}(\langle a\rangle M, (x)N) \xrightarrow{gc} M$ if a is not a free co-name in M
8. $\mathsf{C\overrightarrow{u}t}(\langle a\rangle M, (x)N) \xrightarrow{gc} N$ if x is not a free name in N

$$
\begin{aligned}
\overleftarrow{\mathsf{Cut}}(\langle c\rangle\mathsf{Ax}(x,c),(y)P) &\overset{x}{\longrightarrow} \mathsf{Cut}(\langle c\rangle\mathsf{Ax}(x,c),(y)P)\\
\overleftarrow{\mathsf{Cut}}(\langle b\rangle\mathsf{Cut}(\langle a\rangle M,(x)\mathsf{Ax}(x,b)),(y)P) &\overset{x}{\longrightarrow} \mathsf{Cut}(\langle a\rangle\overleftarrow{\mathsf{Cut}}(\langle b\rangle M,(y)P),(y)P)\\
\overleftarrow{\mathsf{Cut}}(\langle c\rangle\mathsf{And}_R(\langle a\rangle M,\langle b\rangle N,c),(y)P) &\overset{x}{\longrightarrow}\\
&\mathsf{Cut}(\langle c\rangle\mathsf{And}_R(\langle a\rangle\overleftarrow{\mathsf{Cut}}(\langle c\rangle M,(y)P),\langle b\rangle\overleftarrow{\mathsf{Cut}}(\langle c\rangle N,(y)P),c),(y)P)\\[1ex]
\overrightarrow{\mathsf{Cut}}(\langle c\rangle P,(y)\mathsf{Ax}(y,a)) &\overset{x}{\longrightarrow} \mathsf{Cut}(\langle c\rangle P,(y)\mathsf{Ax}(y,a))\\
\overrightarrow{\mathsf{Cut}}(\langle b\rangle P,(x)\mathsf{Cut}(\langle a\rangle\mathsf{Ax}(x,a),(y)M)) &\overset{x}{\longrightarrow} \mathsf{Cut}(\langle b\rangle P,(y)\overrightarrow{\mathsf{Cut}}(\langle b\rangle P,(x)M))\\
\overrightarrow{\mathsf{Cut}}(\langle c\rangle P,(y)\mathsf{And}_L^i((x)M,y)) &\overset{x}{\longrightarrow} \mathsf{Cut}(\langle c\rangle P,(y)\mathsf{And}_L^i((x)\overrightarrow{\mathsf{Cut}}(\langle c\rangle P,(y)M),y))\\[1ex]
\overleftarrow{\mathsf{Cut}}(\langle b\rangle\mathsf{Ax}(x,a),(y)P) &\overset{x}{\longrightarrow} \mathsf{Ax}(x,a)\\
\overleftarrow{\mathsf{Cut}}(\langle b\rangle\mathsf{Cut}(\langle a\rangle M,(x)N),(y)P) &\overset{x}{\longrightarrow} \mathsf{Cut}(\langle a\rangle\overleftarrow{\mathsf{Cut}}(\langle b\rangle M,(y)P),(x)\overleftarrow{\mathsf{Cut}}(\langle b\rangle N,(y)P))\\
\overleftarrow{\mathsf{Cut}}(\langle a\rangle\mathsf{And}_L^i((x)M,y),(z)P) &\overset{x}{\longrightarrow} \mathsf{And}_L^i((x)\overleftarrow{\mathsf{Cut}}(\langle a\rangle M,(z)P),y)\\
\overleftarrow{\mathsf{Cut}}(\langle d\rangle\mathsf{And}_R(\langle a\rangle M,\langle b\rangle N,c),(y)P) &\overset{x}{\longrightarrow}\\
&\mathsf{And}_R(\langle a\rangle\overleftarrow{\mathsf{Cut}}(\langle d\rangle M,(y)P),\langle b\rangle\overleftarrow{\mathsf{Cut}}(\langle d\rangle N,(y)P),c)\\[1ex]
\overrightarrow{\mathsf{Cut}}(\langle b\rangle P,(y)\mathsf{Ax}(x,a)) &\overset{x}{\longrightarrow} \mathsf{Ax}(x,a)\\
\overrightarrow{\mathsf{Cut}}(\langle b\rangle P,(y)\mathsf{Cut}(\langle a\rangle M,(x)N)) &\overset{x}{\longrightarrow} \mathsf{Cut}(\langle a\rangle\overrightarrow{\mathsf{Cut}}(\langle b\rangle P,(y)M),(x)\overrightarrow{\mathsf{Cut}}(\langle b\rangle P,(y)N))\\
\overrightarrow{\mathsf{Cut}}(\langle a\rangle P,(z)\mathsf{And}_L^i((x)M,y)) &\overset{x}{\longrightarrow} \mathsf{And}_L^i((x)\overrightarrow{\mathsf{Cut}}(\langle a\rangle P,(z)M),y)\\
\overrightarrow{\mathsf{Cut}}(\langle d\rangle P,(y)\mathsf{And}_R(\langle a\rangle M,\langle b\rangle N,c)) &\overset{x}{\longrightarrow}\\
&\mathsf{And}_R(\langle a\rangle\overrightarrow{\mathsf{Cut}}(\langle d\rangle P,(y)M),\langle b\rangle\overrightarrow{\mathsf{Cut}}(\langle d\rangle P,(y)N),c)
\end{aligned}
$$

Fig. 2. Cut-reductions for labelled cuts.

We are now ready to define our Gentzen-like cut-elimination procedure. Since we annotated terms to our sequent proofs, we can define it as a term rewriting system.

Definition 1 (Gentzen-like Cut-Elimination Procedure). The Gentzen-like cut-elimination procedure is the term rewriting system $(\mathcal{T}_\wedge, \overset{loc}{\longrightarrow})$ where:

- $\mathcal{T}_\wedge$ is the set of terms well-typed by the rules shown in Figure 1, and
- $\overset{loc}{\longrightarrow}$ consists of the reduction rules for logical, commuting and labelled cuts as well as the garbage reductions; that is

$$\overset{loc}{\longrightarrow} \overset{\text{def}}{=} \overset{l}{\longrightarrow} \cup \overset{c}{\longrightarrow} \cup \overset{x}{\longrightarrow} \cup \overset{gc}{\longrightarrow} .$$

Notice that by assumption all reductions are closed under context formation. The completeness of $\overset{loc}{\longrightarrow}$ is simply the fact that every term beginning with a cut matches at least one left-hand side of the reduction rules. So each irreducible term is cut-free. We shall however omit a proof of this fact. The theorem for which we are going to give a proof for is as follows, but we delay the proof until Section 5.

Theorem 1. For all terms in $\mathcal{T}_\wedge$ the reduction $\overset{loc}{\longrightarrow}$ is strongly normalising.

As said earlier, this theorem can be generalised to include all connectives, and our proof can be easily adapted to the more general case.

4 Comparison with Explicit Substitution Calculi

There is a close correspondence between our cut-elimination procedure and explicit substitution calculi, as we shall illustrate in this section.

Explicit substitution calculi have been developed to internalise the substitution operation—a meta-level operation on lambda-terms—arising from beta-reductions. For example in $\lambda\mathbf{x}$ [10], the beta-reduction $(\lambda x.M)N \xrightarrow{\beta} M[x := N]$ is replaced by the reduction $(\lambda x.M)N \xrightarrow{b} M\langle x := N\rangle$ where the reduct contains a new syntactic constructor. The following reduction rules apply to this constructor.

$$\begin{aligned} y\langle x := P\rangle &\xrightarrow{x} P \text{ if } x \equiv y \text{ otherwise } y \\ (\lambda y.M)\langle x := P\rangle &\xrightarrow{x} \lambda y.\, M\langle x := P\rangle \\ (MN)\langle x := P\rangle &\xrightarrow{x} M\langle x := P\rangle\, N\langle x := P\rangle \end{aligned}$$

Similarly, our labelled cuts internalise a proof substitution introduced in [11, 12]. This substitution operation is written as $M\{a := (x)N\}$ and $N\{x := \langle a\rangle M\}$ where M and N belong to $\mathcal{TU}_\wedge$ that is defined as the set of terms well-typed by the typing rules given in Figure 1 excluding the rules for labelled cuts. Thus $\mathcal{TU}_\wedge$ consists of well-typed but completely unlabelled terms, and clearly, we have $\mathcal{TU}_\wedge \subset \mathcal{T}_\wedge$. In terms of the reductions given above the proof substitution can be defined as the juxtaposition of a $\xrightarrow{c}$-reduction and a series of $\xrightarrow{x}$-reductions, which need to be applied until no further $\xrightarrow{x}$-reduction is applicable (later we shall refer to such a term as x-normal form). Here we omit an inductive definition of the proof substitution, which can be found in [11,12]. Using this proof substitution we can reformulate the reduction for commuting cuts as follows.

5'. $\mathsf{Cut}(\langle a\rangle M, (x)N) \xrightarrow{c'} M\{a := (x)N\}$ if M does not freshly introduce a, *or*
6'. $\mathsf{Cut}(\langle a\rangle M, (x)N) \xrightarrow{c'} N\{x := \langle a\rangle M\}$ if N does not freshly introduce x.

This leads to the following cut-elimination procedure, which satisfies the three criteria given in the introduction, but which is *not* Gentzen-like (the proof substitution is a "global" operation).

Definition 2 (Global Cut-Elimination Procedure). The cut-elimination procedure $(\mathcal{TU}_\wedge, \xrightarrow{gbl})$ is the term rewriting system where:

- $\mathcal{TU}_\wedge$ is the set of well-typed but completely unlabelled terms, and
- $\xrightarrow{gbl}$ consists of the reduction rules for logical and commuting cuts; that is

$$\xrightarrow{gbl} \stackrel{\text{def}}{=} \xrightarrow{l} \cup \xrightarrow{c'} .$$

A proof of strong normalisation for $(\mathcal{TU}_\wedge, \xrightarrow{gbl})$ is given in [11,12]. There is no known technique that would give a strong normalisation result for $(\mathcal{T}_\wedge, \xrightarrow{loc})$ via a simple translation from $\mathcal{T}_\wedge$ to $\mathcal{TU}_\wedge$. This is similar to the situation with the lambda-calculus and $\lambda\mathbf{x}$: strong normalisation for the explicit substitution calculus does not follow directly from strong normalisation of the lambda-calculus. Indeed as shown in [9] explicit substitution calculi, if defined naïvely, may break the strong normalisation property. So the proof we shall present next is rather involved.

5 Proof of Strong Normalisation

In this section we shall give a proof for Theorem 1. In this proof we shall make use of the recursive path ordering by Dershowitz [3].

Definition 3 (Recursive Path Ordering). Let $s \equiv f(s_1, \ldots, s_m)$ and $t \equiv g(t_1, \ldots, t_n)$ be terms, then $s >^{rpo} t$ iff

	(i) $s_i \geq^{rpo} t$ for some $i = 1, \ldots, m$	(subterm)
or	(ii) $f \gg g$ and $s >^{rpo} t_j$ for all $j = 1, \ldots, n$	(decreasing heads)
or	(iii) $f = g$ and $\{\!\{s_1, \ldots, s_m\}\!\} >^{rpo}_{mult} \{\!\{t_1, \ldots, t_n\}\!\}$	(equal heads)

where $\gg$ is a precedence defined over term constructors, $>^{rpo}_{mult}$ is the extension of $>^{rpo}$ to finite multisets and $\{\!\{\ldots\}\!\}$ stands for a multiset of terms; $\geq^{rpo}$ means $>^{rpo}$ or equivalent up to permutation of subterms.

The recursive path ordering theorem says that $>^{rpo}$ is well-founded iff the precedence of the term constructors, $\gg$, is well-founded. Unfortunately, two problems preclude a direct application of this theorem.

- First, the theorem requires a well-founded precedence for our term constructors. However, our reduction rules include the two reductions (written schematically)

$$\mathsf{Cut}(\_,\_) \xrightarrow{c} \overleftarrow{\mathsf{Cut}}(\_,\_)$$
$$\overleftarrow{\mathsf{Cut}}(\mathsf{Cut}(\_,\_),\_) \xrightarrow{x} \mathsf{Cut}(\overleftarrow{\mathsf{Cut}}(\_,\_), \overleftarrow{\mathsf{Cut}}(\_,\_))$$

  and consequently we have a cycle between Cut and $\overleftarrow{\mathsf{Cut}}$. In [1] a clever solution for an analogous problem in λx was presented. We shall adapt this solution for our rewrite system. The essence of this solution is that we take into account (in a non-trivial way) that $(\mathcal{TU}_\wedge, \xrightarrow{gbl})$ is strongly normalising.
- The second problem arises from the fact that the recursive path ordering theorem applies only to first-order rewrite systems, i.e., no binding operations are allowed. In our term calculus however we have two binding operations: one for names and one for co-names. We solve this problem by introducing another term calculus, denoted by $\mathcal{H}$, for which we can apply this theorem, and then prove strong normalisation for $(\mathcal{T}_\wedge, \xrightarrow{loc})$ by translation.

The first important fact in our proof is that $\xrightarrow{x}$ is confluent, in contrast to $\xrightarrow{loc}$, which is clearly not.

Lemma 1. The reduction $\xrightarrow{x}$ is strongly normalising and confluent.

Proof. We can show the first part of the lemma by a simple calculation using the measure, $[\![\_]\!]$, that is 1 for axioms and that is the sum of the measures of the subterms increased by 1 for $\mathsf{And}_R(\_,\_)$; similarly for $\mathsf{And}^i_L(\_)$ and $\mathsf{Cut}(\_,\_)$. For the labelled cuts we have:

$$[\![\overrightarrow{\mathsf{Cut}}(\langle a\rangle M, (x)N)]\!] \stackrel{\text{def}}{=} ([\![M]\!] + 1) * (4[\![N]\!] + 1)$$
$$[\![\overleftarrow{\mathsf{Cut}}(\langle a\rangle M, (x)N)]\!] \stackrel{\text{def}}{=} (4[\![M]\!] + 1) * ([\![N]\!] + 1)$$

This gives $[M] > [N]$ whenever $M \xrightarrow{x} N$. Confluence of $\xrightarrow{x}$ follows from local confluence, which can be easily established, and strong normalisation. □

As a result, we can define the unique x-normal form of a term belonging to $\mathcal{T}_\wedge$.

Definition 4. The unique x-normal form of a term $M \in \mathcal{T}_\wedge$ is denoted by $|M|_x$.

By a careful case analysis we can show that for all $M \in \mathcal{T}_\wedge$ the x-normal form $|M|_x$ is an element $\mathcal{TU}_\wedge$, i.e., is well-typed and completely unlabelled. The details are omitted.

Next we shall prove that $\xrightarrow{x}$ correctly simulates the proof substitution operation of $\xrightarrow{gbl}$.

Lemma 2. For all $M, N \in \mathcal{T}_\wedge$ we have

$$\begin{array}{ll}\text{(i)} & |\overleftarrow{\mathsf{Cut}}(\langle a\rangle M, (y)N)|_x \equiv |M|_x\{a := (y)\,|N|_x\} \\ \text{(ii)} & |\overrightarrow{\mathsf{Cut}}(\langle a\rangle N, (y)M)|_x \equiv |M|_x\{y := \langle a\rangle\,|N|_x\}\end{array}$$

Proof. We can show the lemma by induction on M in case M is completely unlabelled. We can then prove the lemma for all terms by a simple calculation, as illustrated next for (i): by uniqueness of the x-normal form we have that $|\overleftarrow{\mathsf{Cut}}(\langle a\rangle M, (y)N)|_x \equiv |\overleftarrow{\mathsf{Cut}}(\langle a\rangle\,|M|_x, (y)\,|N|_x)|_x$ and, because $|M|_x$ is completely unlabelled, this is $|M|_x\{a := (y)\,|N|_x\}$. □

Now we are in a position to show another important fact in our proof, namely that the $\xrightarrow{loc}$-reductions project onto $\xrightarrow{gbl}$-reductions.

Lemma 3. For all terms $M, N \in \mathcal{T}_\wedge$ if $M \xrightarrow{loc} N$ then $|M|_x \xrightarrow{gbl} |N|_x$.

Proof. By induction on the definition of $\xrightarrow{loc}$. □

As mentioned earlier, this lemma is not strong enough to prove strong normalisation of $\xrightarrow{loc}$. To prove this property we shall use a translation that maps every $\xrightarrow{loc}$-reduction onto a pair of terms belonging to the set $\mathcal{H}$, defined as follows.

Definition 5. Let $\mathcal{H}$ be the set of all terms generated by the grammar

$$M, N ::= \star \mid M \cdot_n N \mid M\langle N\rangle_n \mid \langle M\rangle_n N \mid (\!|M, N|\!) \mid (\!|M|\!)$$

where n is a natural number. The well-founded precedence $\gg$ is given by

$$\_\cdot_{n+1}\_ \;\gg\; \langle\_\rangle_n\_\,,\; \_\langle\_\rangle_n \;\gg\; \_\cdot_n\_ \;\gg\; \star,\, (\!|\_|\!),\, (\!|\_,\_|\!)\,.$$

To define the translation we shall use, as it turns out later, an alternative definition of the set $\mathcal{T}_\wedge$. This alternative definition is required in order to strengthen an induction hypothesis.

Definition 6. The set of *bounded* terms, $\mathcal{B}_\wedge$, consists of well-typed terms M wherby for every subterm N of the M the corresponding x-normal form, $|N|_x$, must be strongly normalising with respect to $\xrightarrow{gbl}$.

Clearly, we now have to show the fact that

Lemma 4. The set of bounded terms is closed under $\xrightarrow{loc}$-reductions.

Proof. By induction on the definition of $\xrightarrow{loc}$ using Lemma 3.

Next we define the translation from bounded terms to terms of $\mathcal{H}$.

Definition 7. The translation $\underline{} : \mathcal{B} \rightarrow \mathcal{H}$ is inductively defined by the clauses

$$\underline{\mathsf{Ax}(x,a)} \stackrel{\text{def}}{=} \star \qquad \underline{\mathsf{And}_R(\langle a\rangle S, \langle b\rangle T, c)} \stackrel{\text{def}}{=} (\!|\underline{S}, \underline{T}|\!) \qquad \underline{\mathsf{And}^i_L(\langle x\rangle S, y)} \stackrel{\text{def}}{=} (\!|\underline{S}|\!)$$

$$\begin{array}{llll}
\underline{\mathsf{Cut}(\langle a\rangle S, \langle x\rangle T)} \stackrel{\text{def}}{=} \underline{S} \cdot_l \underline{T} & & l \stackrel{\text{def}}{=} \text{MAXRED}_{gbl}(|\mathsf{Cut}(\langle a\rangle S, \langle x\rangle T)|_x) \\
\underline{\overleftarrow{\mathsf{Cut}}(\langle a\rangle S, \langle x\rangle T)} \stackrel{\text{def}}{=} \underline{S}\langle \underline{T}\rangle_m & & m \stackrel{\text{def}}{=} \text{MAXRED}_{gbl}(|\overleftarrow{\mathsf{Cut}}(\langle a\rangle S, \langle x\rangle T)|_x) \\
\underline{\overrightarrow{\mathsf{Cut}}(\langle a\rangle S, \langle x\rangle T)} \stackrel{\text{def}}{=} \langle \underline{S}\rangle_n \underline{T} & & n \stackrel{\text{def}}{=} \text{MAXRED}_{gbl}(|\overrightarrow{\mathsf{Cut}}(\langle a\rangle S, \langle x\rangle T)|_x)
\end{array}$$

where $\text{MAXRED}_{gbl}(|M|_x)$ denotes the number of steps of the longest $\xrightarrow{gbl}$-reduction sequence starting from the x-normal form of M. Clearly, this translation is well-defined since it is restricted to bounded terms. The next lemma will be applied when we need to compare labels of terms in $\mathcal{H}$.

Lemma 5. For all terms $M, N \in \mathcal{B}$ we have

(i) $\text{MAXRED}_{gbl}(|M|_x) \geq \text{MAXRED}_{gbl}(|N|_x)$, provided $M \xrightarrow{loc} N$.
(ii) $\text{MAXRED}_{gbl}(|M|_x) \geq \text{MAXRED}_{gbl}(|N|_x)$, provided N is an immediate subterm of M and M is unlabelled.
(iii) $\text{MAXRED}_{gbl}(|M|_x) > \text{MAXRED}_{gbl}(|N|_x)$, provided $M \xrightarrow{l} N$ or $M \xrightarrow{c} N$ on the outermost level.

Proof. (i) follows from Lemma 3; for (ii) note that all reductions which $|N|_x$ can perform can be performed by $|M|_x$; (iii) is by a simple calculation and the fact that the side conditions put on $\xrightarrow{c}$ ensures that $|M|_x \xrightarrow{gbl} |N|_x$. □

We shall now prove the (main) lemma, which relates a $\xrightarrow{loc}$-reduction to a pair of terms belonging to $\mathcal{H}$ and ordered decreasingly according to $>^{rpo}$.

Lemma 6. For all terms $M, N \in \mathcal{B}$ if $M \xrightarrow{loc} N$, then $\underline{M} >^{rpo} \underline{N}$.

Proof. By induction on the definition of $\xrightarrow{loc}$. As there are many possible reductions, we shall present only a few representative cases. First we give one case where an inner reduction occurs (we shall write rpo for Definition 3).

- $M \equiv \mathsf{Cut}(\langle a\rangle S, \langle x\rangle T) \xrightarrow{loc} \mathsf{Cut}(\langle a\rangle S', \langle x\rangle T) \equiv N$

(1) $S \xrightarrow{loc} S'$ and $\underline{S} >^{rpo} \underline{S'}$	by assumption and induction
(2) $\underline{M} = \underline{S} \cdot_m \underline{T}$ and $\underline{N} = \underline{S'} \cdot_n \underline{T}$	by Definition 7
(3) $m \geq n$	by Lemma 5(i)
(4) $\underline{S} \cdot_m \underline{T} >^{rpo} \underline{S'}$, $\underline{S} \cdot_m \underline{T} >^{rpo} \underline{T}$, $\{\!\{\underline{S}, \underline{T}\}\!\} >^{rpo}_{mult} \{\!\{\underline{S'}, \underline{T}\}\!\}$	by (1) and rpo(i)
(5) $\underline{M} >^{rpo} \underline{N}$	by (4) and rpo(ii,iii)

We now show two typical cases where an $\xrightarrow{x}$-reduction is performed

- $M \equiv \overleftarrow{\mathsf{Cut}}(\langle c\rangle \mathsf{And}_R(\langle a\rangle S, \langle b\rangle T, c), (x)U)$
  $\xrightarrow{x} \mathsf{Cut}(\langle c\rangle \mathsf{And}_R(\langle a\rangle \overleftarrow{\mathsf{Cut}}(\langle c\rangle S, (x)U), \langle b\rangle \overleftarrow{\mathsf{Cut}}(\langle c\rangle T, (x)U), c), (x)U) \equiv N$
  (1) $\underline{M} = (\!|\underline{S}, \underline{T}|\!)\,\langle \underline{U}\rangle_m$ and $\underline{N} = (\!|\underline{S}\langle \underline{U}\rangle_r, \underline{T}\langle \underline{U}\rangle_s|\!) \cdot_t \underline{U}$ by Definition 7
  (2) $m \geq t, r, s$ and $\_\langle\_\rangle_m \gg \_ \cdot_t \_$ by Lemma 5(i,ii) and Definition 5
  (3) $(\!|\underline{S}, \underline{T}|\!)\langle \underline{U}\rangle_m >^{rpo} \underline{S}$, $(\!|\underline{S}, \underline{T}|\!)\langle \underline{U}\rangle_m >^{rpo} \underline{U}$, $\{\!|(\!|\underline{S}, \underline{T}|\!), \underline{U}|\!\} >^{rpo}_{mult} \{\!|\underline{S}, \underline{U}|\!\}$ by rpo(i)
  (4) $(\!|\underline{S}, \underline{T}|\!)\langle \underline{U}\rangle_m >^{rpo} \underline{S}\langle \underline{U}\rangle_r$ by (3) and rpo(ii,iii)
  (5) $(\!|\underline{S}, \underline{T}|\!)\langle \underline{U}\rangle_m >^{rpo} \underline{T}\langle \underline{U}\rangle_s$ analogous to (3,4)
  (6) $(\!|\underline{S}, \underline{T}|\!)\langle \underline{U}\rangle_m >^{rpo} (\!|\underline{S}\langle \underline{U}\rangle_r, \underline{T}\langle \underline{U}\rangle_s|\!)$ by (4,5) and rpo(ii)
  (7) $(\!|\underline{S}, \underline{T}|\!)\langle \underline{U}\rangle_m >^{rpo} \underline{U}$ by rpo(i)
  (8) $\underline{M} >^{rpo} \underline{N}$ by (2,6,7) and rpo(ii)
- $M \equiv \overleftarrow{\mathsf{Cut}}(\langle d\rangle \mathsf{And}_R(\langle a\rangle S, \langle b\rangle T, c), (x)U)$
  $\xrightarrow{x} \mathsf{And}_R(\langle a\rangle \overleftarrow{\mathsf{Cut}}(\langle d\rangle S, (x)U), \langle b\rangle \overleftarrow{\mathsf{Cut}}(\langle d\rangle T, (x)U), c) \equiv N$
  (1) $\underline{M} = (\!|\underline{S}, \underline{T}|\!)\,\langle \underline{U}\rangle_m$ and $\underline{N} = (\!|\underline{S}\langle \underline{U}\rangle_r, \underline{T}\langle \underline{U}\rangle_s|\!)$ by Definition 7
  (2) $m \geq r, s$ by Lemma 5(i,ii)
  (3) $(\!|\underline{S}, \underline{T}|\!)\langle \underline{U}\rangle_m >^{rpo} \underline{S}$, $(\!|\underline{S}, \underline{T}|\!)\langle \underline{U}\rangle_m >^{rpo} \underline{U}$, $\{\!|(\!|\underline{S}, \underline{T}|\!), \underline{U}|\!\} >^{rpo}_{mult} \{\!|\underline{S}, \underline{U}|\!\}$ by rpo(i)
  (4) $(\!|\underline{S}, \underline{T}|\!)\langle \underline{U}\rangle_m >^{rpo} \underline{S}\langle \underline{U}\rangle_r$ by (3) and rpo(ii,iii)
  (5) $(\!|\underline{S}, \underline{T}|\!)\langle \underline{U}\rangle_m >^{rpo} \underline{T}\langle \underline{U}\rangle_s$ analogous to (3,4)
  (6) $\underline{M} >^{rpo} \underline{N}$ by (4,5) and rpo(ii)

Last we tackle two cases, one where a commuting reduction and one where a logical reduction occurs.

- $M \equiv \mathsf{Cut}(\langle a\rangle S, (x)T) \xrightarrow{c} \overleftarrow{\mathsf{Cut}}(\langle a\rangle S, (x)T) \equiv N$
  (1) $\underline{M} = \underline{S} \cdot_m \underline{T}$ and $\underline{N} = \underline{S}\langle \underline{T}\rangle_n$ by Definition 7
  (2) $m > n$ and $\_ \cdot_m \_ \gg \_\langle\_\rangle_n$ by Lemma 5(iii) and Definition 5
  (3) $\underline{S} \cdot_m \underline{T} >^{rpo} \underline{S}$ and $\underline{S} \cdot_m \underline{T} >^{rpo} \underline{T}$ by rpo(i)
  (4) $\underline{M} >^{rpo} \underline{N}$ by (2,3) and rpo(ii)
- $M \equiv \mathsf{Cut}(\langle c\rangle \mathsf{And}_R(\langle a\rangle S, \langle b\rangle T, c), (y)\mathsf{And}^1_L((x)U, y)) \xrightarrow{l} \mathsf{Cut}(\langle a\rangle S, (x)U) \equiv N$
  (1) $\underline{M} = (\!|\underline{S}, \underline{T}|\!) \cdot_m (\!|\underline{U}|\!)$ and $\underline{N} = \underline{S} \cdot_n \underline{U}$ by Definition 7
  (2) $m > n$ by Lemma 5(iii)
  (3) $(\!|\underline{S}, \underline{T}|\!) \cdot_m (\!|\underline{U}|\!) >^{rpo} \underline{S}$, $(\!|\underline{S}, \underline{T}|\!) \cdot_m (\!|\underline{U}|\!) >^{rpo} \underline{U}$ by rpo(i)
  (4) $\underline{M} >^{rpo} \underline{N}$ by (2,3) and rpo(ii)

Using this lemma we can show that every $\xrightarrow{loc}$-reduction sequence starting from a term belonging to $\mathcal{TU}_\wedge$ is terminating.

Lemma 7. Every $\xrightarrow{loc}$-reduction sequence starting with a term that belongs to $\mathcal{TU}_\wedge$ is terminating.

Proof. Suppose for the sake of deriving a contradiction that from M the infinite reduction sequence $M \equiv M_1 \xrightarrow{loc} M_2 \xrightarrow{loc} M_3 \xrightarrow{loc} M_4 \xrightarrow{loc} \ldots$ starts. Because M is completely unlabelled we have for all subterms N of M that $|N|_x \equiv N$, and because M is well-typed we know that each of them is strongly normalising under $\xrightarrow{gbl}$. Consequently, every $\textsc{maxred}_{glb}(|N|_x)$ is finite, and thus M is bounded. By Lemmas 4 and 7 we have that the infinite reduction sequence starting from M can be mapped onto the decreasing chain $\underline{M_1} >^{rpo} \underline{M_2} >^{rpo} \underline{M_3} >^{rpo} \underline{M_4} >^{rpo} \ldots$ which however contradicts the well-foundedness of $>^{rpo}$. Thus all $\xrightarrow{loc}$-reduction sequences starting with a term that is an element in $\mathcal{TU}_\wedge$ must terminate. □

Next, we extend this lemma to all terms of $\mathcal{T}_\wedge$. To do so, we shall first show that for every $M \in \mathcal{T}_\wedge$ there is a term $N \in \mathcal{TU}_\wedge$, such that $N \xrightarrow{loc}{}^{*} M$. Because N is an element in $\mathcal{TU}_\wedge$, we have that N is strongly normalising by the lemma just given, and so M, too, must be strongly normalising.

Lemma 8. For every term $M \in \mathcal{T}_\wedge$ with the typing judgement $\Gamma \triangleright M \triangleright \Delta$, there is a term $N \in \mathcal{TU}_\wedge$ with the typing judgement $\Gamma', \Gamma \triangleright N \triangleright \Delta, \Delta'$ such that $N \xrightarrow{loc}{}^{*} M$.

Proof. We construct N by inductively replacing in M all occurrences of $\overleftarrow{\mathsf{Cut}}$ and $\overrightarrow{\mathsf{Cut}}$ by some instances of Cut. We analyse the case where $\overleftarrow{\mathsf{Cut}}(\langle a\rangle S, (x)T)$ is a subterm of M.

- If the subterm S does not freshly introduce a, then we replace $\overleftarrow{\mathsf{Cut}}(\langle a\rangle S, (x)T)$ simply by $\mathsf{Cut}(\langle a\rangle S, (x)T)$ (both terms have the same typing judgement). In this case we have $\mathsf{Cut}(\langle a\rangle S, (x)T) \xrightarrow{c} \overleftarrow{\mathsf{Cut}}(\langle a\rangle S, (x)T)$.
- The more interesting case is where S freshly introduces a. Here we cannot simply replace $\overleftarrow{\mathsf{Cut}}$ with Cut, because there is no reduction with $N \xrightarrow{loc}{}^{*} M$. Therefore we replace $\overleftarrow{\mathsf{Cut}}(\langle a\rangle S, (x)T)$ by $\mathsf{Cut}(\langle a\rangle \mathsf{Cut}(\langle b\rangle S, (y)\mathsf{Ax}(y, c)), (x)T)$ in which b and c are fresh co-names that do not occur anywhere else (this ensures that the new cut-instances are well-typed). Now we show how the new term can reduce. Because $\mathsf{Cut}(\langle b\rangle S, (y)\mathsf{Ax}(y, c))$ does not freshly introduce a, we can first perform two commuting reductions and subsequently we can remove the labelled cut by a $\xrightarrow{gc}$-reduction, *viz.*

$$\begin{aligned}\mathsf{Cut}(\langle a\rangle \mathsf{Cut}(\langle b\rangle S, (y)\mathsf{Ax}(y, c)), (x)T) &\xrightarrow{c} \overleftarrow{\mathsf{Cut}}(\langle a\rangle \mathsf{Cut}(\langle b\rangle S, (y)\mathsf{Ax}(y, c)), (x)T)\\ &\xrightarrow{c} \overleftarrow{\mathsf{Cut}}(\langle a\rangle \overleftarrow{\mathsf{Cut}}(\langle b\rangle S, (y)\mathsf{Ax}(y, c)), (x)T)\\ &\xrightarrow{gc} \overleftarrow{\mathsf{Cut}}(\langle a\rangle S, (x)T)\end{aligned}$$ □

Now the proof of Theorem 1 is by a simple contradiction argument.

Proof of Theorem 1. Suppose $M \in \mathcal{T}_\wedge$ is not strongly normalising. Then by the lemma just given there is a term $N \in \mathcal{TU}_\wedge$ such that $N \xrightarrow{loc}{}^{*} M$. Clearly, if M is not strongly normalising, then so is N, which however contradicts Lemma 7. Consequently, M must be strongly normalising. □

6 Conclusion

In this paper we considered the problem of defining a strongly normalising cut-elimination procedure for classical logic that satisfies the three criteria given in the introduction and that is Gentzen-like. While Gentzen-like cut-elimination procedures tend to break strong normalisation, in this paper we have shown that this property can be retained by introducing labelled cuts. For reasons of space we have given our system for only the $\wedge$-fragment. However, our techniques apply to the other connectives and to the first-order quantifiers. This should provide us with a bridge between our earlier calculus [11,12] and an implementation.

There are many directions for further work. For example what is the precise correspondence in the intuitionistic case between normalisation in the lambda-calculus (with explicit substitutions) and our strongly normalising cut-elimination procedure? This is of interest since the Gentzen-like cut-elimination procedure presented in this paper is rather helpful in proving strong normalisation of other reduction systems by simple translations (e.g. the lambda-calculus, λx and Parigot's $\lambda\mu$). Some of these issues are addressed in [11].

References

1. R. Bloo and H. Geuvers. Explicit Substitution: On the Edge of Strong Normalisation. *Theoretical Computer Science*, 211(1–2):375–395, 1999.
2. V. Danos, J.-B. Joinet, and H. Schellinx. A New Deconstructive Logic: Linear Logic. *Journal of Symbolic Logic*, 62(3):755–807, 1997.
3. N. Dershowitz. Orderings for Term Rewriting Systems. *Theoretical Computer Science*, 17:279–301, 1982.
4. R. Dyckhoff and L. Pinto. Cut-Elimination and a Permutation-Free Sequent Calculus for Intuitionistic Logic. *Studia Logica*, 60(1):107–118, 1998.
5. J. Gallier. Constructive Logics. Part I: A Tutorial on Proof Systems and Typed λ-calculi. *Theoretical Computer Science*, 110(2):249–239, 1993.
6. G. Gentzen. Untersuchungen über das logische Schließen I and II. *Mathematische Zeitschrift*, 39:176–210, 405–431, 1935.
7. H. Herbelin. A λ-calculus Structure Isomorphic to Sequent Calculus Structure. In *Computer Science Logic*, volume 933 of *LNCS*, pages 67–75. Springer Verlag, 1994.
8. J. M. E. Hyland. Proof Theory in the Abstract. *Annals of Pure and Applied Logic*, 2000. To appear.
9. P. A. Melliès. Typed Lambda Calculi with Explicit Substitutions May Not Terminate. In *Typed Lambda Calculi and Applications*, volume 902 of *LNCS*, pages 328–334. Springer Verlag, 1995.
10. K. H. Rose. Explicit Substitution: Tutorial & Survey. Technical report, BRICS, Department of Computer Science, University of Aarhus, 1996.
11. C. Urban. *Classical Logic and Computation.* PhD thesis, Cambridge University, October 2000.
12. C. Urban and G. M. Bierman. Strong Normalisation of Cut-Elimination in Classical Logic. *Fundamenta Informaticae*, 45(1–2):123–155, 2001.

Author Index

Lecture Notes in Computer Science

For information about Vols. 1–1953
please contact your bookseller or Springer-Verlag

Vol. 1954: W.A. Hunt, Jr., S.D. Johnson (Eds.), Formal Methods in Computer-Aided Design. Proceedings, 2000. XI, 539 pages. 2000.

Vol. 1955: M. Parigot, A. Voronkov (Eds.), Logic for Programming and Automated Reasoning. Proceedings, 2000. XIII, 487 pages. 2000. (Subseries LNAI).

Vol. 1956: T. Coquand, P. Dybjer, B. Nordström, J. Smith (Eds.), Types for Proofs and Programs. Proceedings, 1999. VII, 195 pages. 2000.

Vol. 1957: P. Ciancarini, M. Wooldridge (Eds.), Agent-Oriented Software Engineering. Proceedings, 2000. X, 323 pages. 2001.

Vol. 1959: M. Bouzeghoub, Z. Kedad, E. Métais (Eds.), Natural Language Processing and Information Systems. Proceedings, 2000. X, 372 pages. 2001.

Vol. 1960: A. Ambler, S.B. Calo, G. Kar (Eds.), Services Management in Intelligent Networks. Proceedings, 2000. X, 259 pages. 2000.

Vol. 1961: J. He, M. Sato (Eds.), Advances in Computing Science – ASIAN 2000. Proceedings, 2000. X, 299 pages. 2000.

Vol. 1963: V. Hlaváč, K.G. Jeffery, J. Wiedermann (Eds.), SOFSEM 2000: Theory and Practice of Informatics. Proceedings, 2000. XI, 460 pages. 2000.

Vol. 1964: J. Malenfant, S. Moisan, A. Moreira (Eds.), Object-Oriented Technology. Proceedings, 2000. XI, 309 pages. 2000.

Vol. 1965: Ç. K. Koç, C. Paar (Eds.), Cryptographic Hardware and Embedded Systems – CHES 2000. Proceedings, 2000. XI, 355 pages. 2000.

Vol. 1966: S. Bhalla (Ed.), Databases in Networked Information Systems. Proceedings, 2000. VIII, 247 pages. 2000.

Vol. 1967: S. Arikawa, S. Morishita (Eds.), Discovery Science. Proceedings, 2000. XII, 332 pages. 2000. (Subseries LNAI).

Vol. 1968: H. Arimura, S. Jain, A. Sharma (Eds.), Algorithmic Learning Theory. Proceedings, 2000. XI, 335 pages. 2000. (Subseries LNAI).

Vol. 1969: D.T. Lee, S.-H. Teng (Eds.), Algorithms and Computation. Proceedings, 2000. XIV, 578 pages. 2000.

Vol. 1970: M. Valero, V.K. Prasanna, S. Vajapeyam (Eds.), High Performance Computing – HiPC 2000. Proceedings, 2000. XVIII, 568 pages. 2000.

Vol. 1971: R. Buyya, M. Baker (Eds.), Grid Computing – GRID 2000. Proceedings, 2000. XIV, 229 pages. 2000.

Vol. 1972: A. Omicini, R. Tolksdorf, F. Zambonelli (Eds.), Engineering Societies in the Agents World. Proceedings, 2000. IX, 143 pages. 2000. (Subseries LNAI).

Vol. 1973: J. Van den Bussche, V. Vianu (Eds.), Database Theory – ICDT 2001. Proceedings, 2001. X, 451 pages. 2001.

Vol. 1974: S. Kapoor, S. Prasad (Eds.), FST TCS 2000: Foundations of Software Technology and Theoretical Computer Science. Proceedings, 2000. XIII, 532 pages. 2000.

Vol. 1975: J. Pieprzyk, E. Okamoto, J. Seberry (Eds.), Information Security. Proceedings, 2000. X, 323 pages. 2000.

Vol. 1976: T. Okamoto (Ed.), Advances in Cryptology – ASIACRYPT 2000. Proceedings, 2000. XII, 630 pages. 2000.

Vol. 1977: B. Roy, E. Okamoto (Eds.), Progress in Cryptology – INDOCRYPT 2000. Proceedings, 2000. X, 295 pages. 2000.

Vol. 1978: B. Schneier (Ed.), Fast Software Encryption. Proceedings, 2000. VIII, 315 pages. 2001.

Vol. 1979: S. Moss, P. Davidsson (Eds.), Multi-Agent-Based Simulation. Proceedings, 2000. VIII, 267 pages. 2001. (Subseries LNAI).

Vol. 1808: S. Pande, D.P. Agrawal (Eds.), Compiler Optimizations for Scalable Parallel Systems. XXVII, 779 pages. 2001.

Vol. 1980: M. Agosti, F. Crestani, G. Pasi (Eds.), Lectures on Information Retrieval. Proceedings, 2000. XI, 311 pages. 2001.

Vol. 1981: J.M.L.M. Palma, J. Dongarra, V. Hernández (Eds.), Vector and Parallel Processing – VECPAR 2000. Proceedings, 2000. XVI, 580 pages. 2001.

Vol. 1983: K.S. Leung, L.-W. Chan, H. Meng (Eds.), Intelligent Data Engineering and Automated Learning – IDEAL 2000. Proceedings, 2000. XVI, 573 pages. 2000.

Vol. 1984: J. Marks (Ed.), Graph Drawing. Proceedings, 2001. XII, 419 pages. 2001.

Vol. 1985: J. Davidson, S.L. Min (Eds.), Languages, Compilers, and Tools for Embedded Systems. Proceedings, 2000. VIII, 221 pages. 2001.

Vol. 1987: K.-L. Tan, M.J. Franklin, J. C.-S. Lui (Eds.), Mobile Data Management. Proceedings, 2001. XIII, 289 pages. 2001.

Vol. 1988: L. Vulkov, J. Wasniewski, P. Yalamov (Eds.), Numerical Analysis and Its Applications. Proceedings, 2000. XIII, 782 pages. 2001.

Vol. 1989: M. Ajmone Marsan, A. Bianco (Eds.), Quality of Service in Multiservice IP Networks. Proceedings, 2001. XII, 440 pages. 2001.

Vol. 1990: I.V. Ramakrishnan (Ed.), Practical Aspects of Declarative Languages. Proceedings, 2001. VIII, 353 pages. 2001.

Vol. 1991: F. Dignum, C. Sierra (Eds.), Agent Mediated Electronic Commerce. VIII, 241 pages. 2001. (Subseries LNAI).

Vol. 1992: K. Kim (Ed.), Public Key Cryptography. Proceedings, 2001. XI, 423 pages. 2001.

Vol. 1993: E. Zitzler, K. Deb, L. Thiele, C.A.Coello Coello, D. Corne (Eds.), Evolutionary Multi-Criterion Optimization. Proceedings, 2001. XIII, 712 pages. 2001.

Vol. 1995: M. Sloman, J. Lobo, E.C. Lupu (Eds.), Policies for Distributed Systems and Networks. Proceedings, 2001. X, 263 pages. 2001.

Vol. 1997: D. Suciu, G. Vossen (Eds.), The World Wide Web and Databases. Proceedings, 2000. XII, 275 pages. 2001.

Vol. 1998: R. Klette, S. Peleg, G. Sommer (Eds.), Robot Vision. Proceedings, 2001. IX, 285 pages. 2001.

Vol. 1999: W. Emmerich, S. Tai (Eds.), Engineering Distributed Objects. Proceedings, 2000. VIII, 271 pages. 2001.

Vol. 2000: R. Wilhelm (Ed.), Informatics: 10 Years Back, 10 Years Ahead. IX, 369 pages. 2001.

Vol. 2001: G.A. Agha, F. De Cindio, G. Rozenberg (Eds.), Concurrent Object-Oriented Programming and Petri Nets. VIII, 539 pages. 2001.

Vol. 2002: H. Comon, C. Marché, R. Treinen (Eds.), Constraints in Computational Logics. Proceedings, 1999. XII, 309 pages. 2001.

Vol. 2003: F. Dignum, U. Cortés (Eds.), Agent Mediated Electronic Commerce III. XII, 193 pages. 2001. (Subseries LNAI).

Vol. 2004: A. Gelbukh (Ed.), Computational Linguistics and Intelligent Text Processing. Proceedings, 2001. XII, 528 pages. 2001.

Vol. 2006: R. Dunke, A. Abran (Eds.), New Approaches in Software Measurement. Proceedings, 2000. VIII, 245 pages. 2001.

Vol. 2007: J.F. Roddick, K. Hornsby (Eds.), Temporal, Spatial, and Spatio-Temporal Data Mining. Proceedings, 2000. VII, 165 pages. 2001. (Subseries LNAI).

Vol. 2009: H. Federrath (Ed.), Designing Privacy Enhancing Technologies. Proceedings, 2000. X, 231 pages. 2001.

Vol. 2010: A. Ferreira, H. Reichel (Eds.), STACS 2001. Proceedings, 2001. XV, 576 pages. 2001.

Vol. 2011: M. Mohnen, P. Koopman (Eds.), Implementation of Functional Languages. Proceedings, 2000. VIII, 267 pages. 2001.

Vol. 2012: D.R. Stinson, S. Tavares (Eds.), Selected Areas in Cryptography. Proceedings, 2000. IX, 339 pages. 2001.

Vol. 2013: S. Singh, N. Murshed, W. Kropatsch (Eds.), Advances in Pattern Recognition – ICAPR 2001. Proceedings, 2001. XIV, 476 pages. 2001.

Vol. 2015: D. Won (Ed.), Information Security and Cryptology – ICISC 2000. Proceedings, 2000. X, 261 pages. 2001.

Vol. 2018: M. Pollefeys, L. Van Gool, A. Zisserman, A. Fitzgibbon (Eds.), 3D Structure from Images – SMILE 2000. Proceedings, 2000. X, 243 pages. 2001.

Vol. 2020: D. Naccache (Ed.), Topics in Cryptology – CT-RSA 2001. Proceedings, 2001. XII, 473 pages. 2001

Vol. 2021: J. N. Oliveira, P. Zave (Eds.), FME 2001: Formal Methods for Increasing Software Productivity. Proceedings, 2001. XIII, 629 pages. 2001.

Vol. 2022: A. Romanovsky, C. Dony, J. Lindskov Knudsen, A. Tripathi (Eds.), Advances in Exception Handling Techniques. XII, 289 pages. 2001

Vol. 2024: H. Kuchen, K. Ueda (Eds.), Functional and Logic Programming. Proceedings, 2001. X, 391 pages. 2001.

Vol. 2025: M. Kaufmann, D. Wagner (Eds.), Drawing Graphs. XIV, 312 pages. 2001.

Vol. 2026: F. Müller (Ed.), High-Level Parallel Programming Models and Supportive Environments. Proceedings, 2001. IX, 137 pages. 2001.

Vol. 2027: R. Wilhelm (Ed.), Compiler Construction. Proceedings, 2001. XI, 371 pages. 2001.

Vol. 2028: D. Sands (Ed.), Programming Languages and Systems. Proceedings, 2001. XIII, 433 pages. 2001.

Vol. 2029: H. Hussmann (Ed.), Fundamental Approaches to Software Engineering. Proceedings, 2001. XIII, 349 pages. 2001.

Vol. 2030: F. Honsell, M. Miculan (Eds.), Foundations of Software Science and Computation Structures. Proceedings, 2001. XII, 413 pages. 2001.

Vol. 2031: T. Margaria, W. Yi (Eds.), Tools and Algorithms for the Construction and Analysis of Systems. Proceedings, 2001. XIV, 588 pages. 2001.

Vol. 2033: J. Liu, Y. Ye (Eds.), E-Commerce Agents. VI, 347 pages. 2001. (Subseries LNAI).

Vol. 2034: M.D. Di Benedetto, A. Sangiovanni-Vincentelli (Eds.), Hybrid Systems: Computation and Control. Proceedings, 2001. XIV, 516 pages. 2001.

Vol. 2035: D. Cheung, G.J. Williams, Q. Li (Eds.), Advances in Knowledge Discovery and Data Mining – PAKDD 2001. Proceedings, 2001. XVIII, 596 pages. 2001. (Subseries LNAI).

Vol. 2037: E.J.W. Boers et al. (Eds.), Applications of Evolutionary Computing. Proceedings, 2001. XIII, 516 pages. 2001.

Vol. 2038: J. Miller, M. Tomassini, P.L. Lanzi, C. Ryan, A.G.B. Tettamanzi, W.B. Langdon (Eds.), Genetic Programming. Proceedings, 2001. XI, 384 pages. 2001.

Vol. 2039: M. Schumacher, Objective Coordination in Multi-Agent System Engineering. XIV, 149 pages. 2001. (Subseries LNAI).

Vol. 2040: W. Kou, Y. Yesha, C.J. Tan (Eds.), Electronic Commerce Technologies. Proceedings, 2001. X, 187 pages. 2001.

Vol. 2044: S. Abramsky (Ed.), Typed Lambda Calculi and Applications. Proceedings, 2001. XI, 431 pages. 2001.

Vol. 2045: B. Pfitzmann (Ed.), Advances in Cryptology – EUROCRYPT 2001. Proceedings, 2001. XII, 545 pages. 2001.

Vol. 2053: O. Danvy, A. Filinski (Eds.), Programs as Data Objects. Proceedings, 2001. VIII, 279 pages. 2001.

Vol. 2054: A. Condon, G. Rozenberg (Eds.), DNA Computing. Proceedings, 2000. X, 271 pages. 2001.